Handbook of Human Factors and Ergonomics Methods

Handbook of Human Factors and Ergonomics Methods

Neville Stanton
Alan Hedge
Karel Brookhuis
Eduardo Salas
Hal Hendrick

CRC Press
Taylor & Francis Group
Boca Raton London New York

CRC Press is an imprint of the
Taylor & Francis Group, an **informa** business

CRC Press
Taylor & Francis Group
6000 Broken Sound Parkway NW, Suite 300
Boca Raton, FL 33487-2742

First issued in paperback 2019

ISBN-13: 978-0-415-28700-5 (hbk)
ISBN-13: 978-0-367-86452-1 (pbk)

Library of Congress Card Number 2003012359

Library of Congress Cataloging-in-Publication Data
The handbook of human factors and ergonomics methods / edited by Neville Stanton ... [et al.].
p. cm.
Includes bibliographical references and index.
ISBN 0-415-28700-6 (alk. paper)
1. Human engineering—Handbooks, manuals, etc. I. Stanton, Neville, 1960–
TA166.H275 2004
620.8′2—dc21   2003012359
CIP

Preface

I must confess to a love of human factors and ergonomics methods. This is a love bordering on obsession. Ever since I was taught how to use hierarchical task analysis (HTA) almost 20 years ago, I have been hooked. Since that time, I have learned how to use dozens of methods. Each time, it is a mini-adventure. I sometimes wonder if I will understand a new method properly, but when it clicks, I feel euphoric. I have also spent a good deal of time training others in the use of methods. This is an extremely rewarding experience, particularly when a trainee presents an analysis of his/her own that shows a clear grasp of how the method works. I have also enjoyed developing some new methods. For example, in collaboration with Chris Baber at the University of Birmingham, I have developed an error-prediction methodology called "task analysis for error identification" (TAFEI). As with HTA, we have sought to underpin TAFEI with a theory of human performance. We are still discovering new aspects of the TAFEI analysis, and it gives us both a thrill to see other people reporting their studies using TAFEI.

The inspiration for this handbook came after I wrote *A Guide to Methodology in Ergonomics* with Mark Young, which was also published by Taylor & Francis. It was clear to me that, although the human factors and ergonomics literature is full of references to methods, there are few consistent standards for how these methods are described and reported. This handbook began in 2000 with a proposal to Taylor & Francis. Fortunately, Tony Moore smiled on this book. With his go-ahead, I contacted experts in each of the various domains of ergonomics methods and asked them to edit different sections of the book. I feel very fortunate that I managed to recruit such an eminent team. To be fair, they did not take much persuasion, as they also agreed that this project was a worthwhile undertaking. The next step was to ask experts in the various ergonomics methodologies to summarize their methods in a standardized format. It was a pleasant surprise to see how willingly the contributors responded.

Now, some 4 years after the initial conception, all of the contributions have been gathered and edited. On behalf of the editorial team, I hope that you, the reader, will find this to be a useful handbook. We hope that this book will encourage developers of methods to structure the reporting of their methods in a consistent manner. Equally important, we hope that this handbook will encourage users of the methods to be more adventurous.

Neville A. Stanton
August 2004

Acknowledgments

On behalf of the editorial team, I would like to thank all of the contributors to this handbook for their professionalism and diligence. I would also like to thank the book commissioning and production team at Taylor & Francis and CRC Press, especially Tony Moore, Sarah Kramer, Matt Gibbons, Jessica Vakili, Cindy Carelli, and Naomi Lynch.

Editors

Neville A. Stanton is a professor of human-centered design at Brunel University in the U.K. He has a bachelor's degree in psychology from the University of Hull as well as master and doctoral degrees in human factors from Aston University. Professor Stanton has published over 70 peer-reviewed journal papers and 7 books on human-centered design. He was a visiting fellow of the Department of Design at Cornell University in 1998. He was awarded the Institution of Electrical Engineers Divisional Premium Award for a paper on engineering psychology and system safety in 1998. The Ergonomics Society awarded him the Otto Edholm Medal in 2001 for his contribution to basic and applied ergonomics research. Professor Stanton is on the editorial boards of *Ergonomics, Theoretical Issues in Ergonomics Science*, and the *International Journal of Human Computer Interaction*. Professor Stanton is a chartered psychologist and a fellow of the British Psychological Society, a fellow of the Ergonomics Society, and a fellow of the Royal Society for the Arts.

Eduardo Salas is a professor of psychology at the University of Central Florida, where he also holds an appointment as program director for the Human Systems Integration Research Department at the Institute for Simulation and Training. He is also the director of UCF's Ph.D. Applied Experimental & Human Factors Program. Previously, he served as a senior research psychologist and head of the Training Technology Development Branch of the Naval Air Warfare Center Training Systems Division for 15 years. During this period, Dr. Salas served as a principal investigator for numerous R&D programs focusing on teamwork, team training, decision making under stress, and performance assessment.

Dr. Salas has coauthored over 200 journal articles and book chapters and has coedited 11 books. He has served on the editorial boards of the *Journal of Applied Psychology, Personnel Psychology, Military Psychology, Interamerican Journal of Psychology, Applied Psychology: an International Journal, International Journal of Aviation Psychology, Group Dynamics*, and the *Journal of Organizational Behavior*.

His expertise includes helping organizations to foster teamwork, to design and implement team training strategies, to facilitate training effectiveness, to manage decision making under stress, to develop performance measurement tools, and to design learning environments. He is currently working on designing tools and techniques to minimize human errors in aviation, law enforcement, and medical environments. He has served as a consultant in a variety of manufacturing settings, pharmaceutical laboratories, and industrial and governmental organizations. Dr. Salas is a fellow of the American Psychological Association (SIOP and Division 21) and the Human Factors and Ergonomics Society, and he is a recipient of the Meritorious Civil Service Award from the Department of the Navy. He received his Ph.D. degree (1984) in industrial and organizational psychology from Old Dominion University.

Hal W. Hendrick, Ph.D., CPE, DABFE, is emeritus professor of human factors and ergonomics at the University of Southern California and principal of Hendrick and Associates, an ergonomics and industrial and organizational psychology consulting firm. He is a certified professional ergonomist, diplomate of the American Board of Forensic Examiners, and holds a Ph.D. in industrial psychology and an M.S. in human factors from Purdue University, with a minor in industrial engineering. He is a past chair of USC's Human Factors Department, former executive director of the university's Institute of Safety and Systems Management, and a former dean at the University of Denver. He earlier was an associate professor at the U.S. Air Force Academy, where he helped develop the psychology major and developed the Cooperative MS Program in Human Factors with Purdue University. Hal is a past president of the Human Factors and Ergonomics Society (HFES), the International Ergonomics Association, and the Board of Certification in Professional Ergonomics. He is a fellow of the International Ergonomics Association (IEA), HFES,

American Psychological Association, and American Psychological Society. He is a recipient of the USC outstanding teaching award and both the HFES Jack A. Kraft Innovator Award and Alexander C. Williams, Jr., Design Award. He is the author or coauthor of over 180 professional publications, including 3 books, and editor or coeditor of 11 books. Hal conceptualized and initiated the subdiscipline of macroergonomics.

Alan Hedge is a professor in the Department of Design and Environmental Analysis at Cornell University. His work focuses on the effects of workplace design on the health, comfort, and performance of people. Recent projects have investigated alternative input device design, ergonomic chairs, and other furniture workstation elements that can reduce musculoskeletal disorder risk factors. He also researches indoor environmental design issues, especially air quality, ventilation, and the sick-building syndrome as well as office lighting and computer-vision syndrome. He has coauthored a book, *Keeping Buildings Healthy*, 25 chapters, and over 150 professional publications. He is active in several professional societies.

Karel Brookhuis studied psychology at Rijksuniversiteit Groningen, specializing in experimental psychology, in 1980. He then became a research fellow (Ph.D. student) at the Institute for Experimental Psychology, with a specialization in psychophysiology. In 1983 he became a senior researcher at the Traffic Research Centre, which later merged into the Centre for Environmental and Traffic Psychology, at the University of Groningen. In 1986 he became head of the Department of Biopsychological Aspects of Driving Behaviour, later renamed the Department of Task Performance and Cognition. In 1994 he was appointed as a research manager, responsible for the centre's research planning and quality control. After the centre was closed on January 1, 2000, he became associate professor (UHD) in the Department of Experimental and Work Psychology. Since 2001, Brookhuis has served as a part-time full professor at the Section of Transport Policy and Logistics of the Technical University of Delft.

Contributors

Torbjörn Åkerstedt
National Institute for Psychosocial
  Factors and Health
Stockholm, Sweden

W.G. Allread
Ohio State University
Institute for Ergonomics
Columbus, OH

Dee H. Andrews
U.S. Air Force Research Laboratory
Warfighter Training Research
  Division
Mesa, AZ

John Annett
University of Warwick
Department of Psychology
Coventry, U.K.

Amelia A. Armstrong
Klein Associates Inc.
Fairborn, OH

Christopher Baber
University of Birmingham
Computing Engineering
Birmingham, U.K.

David P. Baker
American Institutes for Research
Washington, D.C.

Natale Battevi
EPM-CEMOC
Milan, Italy

J. Matthew Beaubien
American Institutes for Research
Washington, D.C.

Artem Belopolsky
University of Illinois
Department of Psychology
Champaign, IL

Jennifer Blume
National Space Biomedical
Research Institute
Houston, TX

Gunnar Borg
Stockholm University
Department of Psychology
Stockholm, Sweden

Wolfram Boucsein
University of Wuppertal
Physiological Psychology
Wuppertal, Germany

Clint A. Bowers
University of Central Florida
Department of Psychology
Orlando, FL

Peter R. Boyce
Rensselaer Polytechnic Institute
Lighting Research Center
Troy, NY

Karel A. Brookhuis
University of Groningen
Experimental & Work Psychology
Groningen, the Netherlands

Ogden Brown, Jr.
University of Denver
Denver, CO

Peter Buckle
University of Surrey
Robens Center for Health
  Ergonomics
Guildford, U.K.

C. Shawn Burke
University of Central Florida
Institute for Simulation & Training
Orlando, FL

Pascale Carayon
University of Wisconsin
Center for Quality & Productivity
  Improvement
Madison, WI

Daniela Colombini
EPM-CEMOC
Milan, Italy

Nancy J. Cooke
Arizona State University East
Applied Psychology Program
Mesa, AZ

Lee Cooper
University of Birmingham
Computing Engineering
Birmingham, U.K.

Nigel Corlett
University of Nottingham
Institute for Occupational
  Ergonomics
Nottingham, U.K.

Dana M. Costar
American Institutes for Research
Washington, D.C.

Pamela Dalton
Monell Chemical Senses Center
Philadelphia, PA

Renée E. DeRouin
University of Central Florida
Institute for Simulation & Training
Orlando, FL

Dick de Waard
University of Groningen
Experimental & Work Psychology
Groningen, the Netherlands

David F. Dinges
University of Pennsylvania
School of Medicine
Philadelphia, PA

James E. Driskell
Florida Maxima Corporation
Winter Park, FL

Robin Dunkin-Chadwick
NIOSH
Division of Applied Research
  & Technology
Cincinnati, OH

J.R. Easter
Aegis Research Corporation
Pittsburgh, PA

W.C. Elm
Aegis Research Corporation
Pittsburgh, PA

Eileen B. Entin
Aptima, Inc.
Wodburn, MA

Elliot E. Entin
Aptima, Inc.
Wodburn, MA

Gary W. Evans
Cornell University
Department of Design &
   Environmental Analysis
Ithaca, NY

Stephen M. Fiore
University of Central Florida
Institute for Simulation & Training
Orlando, FL

M.M. Fleischer
University of Southern California
Los Angeles, CA

Jennifer E. Fowlkes
Chi Systems, Inc.
Orlando, FL

Philippe Geslin
Institut National de la Recherche
   Agronomique (INRA)
Toulouse, France
and
Université de Neuchâtel Institut
   d'ethnologie
Neuchâtel, Switzerland

Matthias Göbel
Berlin University of Technology
Department of Human Factors
   Engineering and Product
   Ergonomics
Berlin, Germany

Thad Godish
Ball State University
Department of Natural Resources
Muncie, IN

Gerald F. Goodwin
U.S. Army Research Institute
Alexandria, VA

Paul Grossman
Freiburg Institute for Mindfulness
   Research
Freiburg, Germany

J.W. Gualtieri
Aegis Research Corporation
Pittsburgh, PA

Bianka B. Hahn
Klein Associates Inc.
Fairborn, OH

Thomas R. Hales
NIOSH
Division of Applied Research
   & Technology
Cincinnati, OH

George Havenith
Loughborough University
Department of Human Sciences
Loughborough, U.K.

Alan Hedge
Cornell University
Department of Design &
   Environmental Analysis
Ithaca, NY

Hal W. Hendrick
Hendrick and Associates
Greenwood Village, CO

Sue Hignett
Loughborough University
Department of Human Sciences
Loughborough, U.K.

Vincent H. Hildebrandt
TNO Work & Employment
Hoofddorp, the Netherlands
and
Body@Work Research Center on
   Physical Activity, Work and
   Health TNO Vumc
Amsterdam, the Netherlands

Hermann Hinrichs
University of Magdeburg
Clinic for Neurology
Magdeburg, Germany

Peter Hoonakker
University of Wisconsin
Center for Quality & Productivity
   Improvement
Madison, WI

Karen Jacobs
Boston University Programs
 in Occupational Therapy
Boston, MA

Florian Jentsch
University of Central Florida
Department of Psychology
Orlando, FL

R.F. Soames Job
University of Sydney
School of Psychology
Sydney, Australia

Debra G. Jones
SA Technologies, Inc.
Marietta, GA

David B. Kaber
North Carolina State University
Department of Industrial
   Engineering
Raleigh, NC

Jussi Kantola
University of Louisville
Center for Industrial Ergonomics
Louisville, KY

Waldemar Karwowski
University of Louisville
Center for Industrial Ergonomics
Louisville, KY

Kristina Kemmlert
National Institute for Working Life
Solna, Sweden

Mark Kirby
University of Huddersfield
School of Computing and
   Engineering
Huddersfield, U.K.

Gary Klein
Klein Associates Inc.
Fairborn, OH

Brian M. Kleiner
Virginia Polytechnical Institute
   and State University
Grado Department of Industrial
   and Systems Engineering
Blacksburg, VA

David W. Klinger
Klein Associates Inc.
Fairborn, OH

Arthur F. Kramer
University of Illinois
Department of Psychology
Champaign, IL

Guangyan Li
Human Engineering Limited
Bristol, U.K.

Jean MacMillan
Aptima, Inc.
Wodburn, MA

Ann Majchrzak
University of Southern California
Marshall School of Business
Los Angeles, CA

Melissa M. Mallis
NASA Ames Research Center
Fatigue Countermeasures Group
Moffett Field, CA

W.S. Marras
Ohio State University
Institute for Ergonomics
Columbus, OH

Philip Marsden
University of Huddersfield
School of Computing and
    Engineering
Huddersfield, U.K.

Laura Martin-Milham
University of Central Florida
Institute for Simulation & Training
Orlando, FL

Lorraine E. Maxwell
Cornell University
Design & Environmental Analysis
Ithaca, NY

Lynn McAtamney
COPE Occupational Health and
    Ergonomics Services Ltd.
Nottingham, U.K.

Olga Menoni
EPM-CEMOC
Milan, Italy

J. Mokray
University of Southern California
Los Angeles, CA

J. Steven Moore
Texas A&M University
School of Rural Public Health
Bryan, TX

**Lambertus (Ben) J.M.
Mulder**
University of Groningen
Experimental & Work Psychology
Groningen, the Netherlands

Brian Mullen
Syracuse University
Syracuse, NY

Mitsuo Nagamachi
Hiroshima International University
Hiroshima, Japan

Leah Newman
Pennsylvania State University
The Harold & Inge Marcus
    Department of Industrial &
    Manufacturing Engineering
University Park, PA

Enrico Occhipinti
EPM-CEMOC
Milan, Italy

Michael J. Paley
Aptima, Inc.
Wodburn, MA

Daniela Panciera
EPM-CEMOC
Milan, Italy

Brian Peacock
National Space Biomedical
    Research Institute
Houston, TX

S.S. Potter
Aegis Research Corporation
Pittsburgh, PA

Heather A. Priest
University of Central Florida
Institute for Simulation & Training
Orlando, FL

Renate Rau
University of Technology
Occupational Health Psychology
Dresden, Germany

Mark S. Rea
Rensselaer Polytechnic Institute
Lighting Research Center
Troy, NY

Maria Grazia Ricci
EPM-CEMOC
Milan, Italy

Hannu Rintamäki
Oulu Regional Institute of
    Occupational Health
Oulu, Finland

Michelle M. Robertson
Liberty Mutual Research Institute
    for Safety
Hopkinton, MA

Suzanne H. Rodgers
Consultant in Ergonomics
Rochester, NY

D. Roitman
University of Southern California
Los Angeles, CA

E.M. Roth
Roth Cognitive Engineering
Brookline, MA

Eduardo Salas
University of Central Florida
Department of Psychology
Orlando, FL

Steven L. Sauter
NIOSH
Division of Applied Research
    & Technology
Cincinnati, OH

Steven M. Shope
US Positioning Group, LLC
Mesa, AZ

Monique Smeets
Utrecht University
Department of Social Sciences
Utrecht, the Netherlands

Tonya L. Smith-Jackson
Virginia Polytechnic Institute and
    State University
Grado Department of Industrial
    and Systems Engineering
Blacksburg, VA

Kimberly A. Smith-Jentsch
University of Central Florida
Department of Psychology
Orlando, FL

Stover H. Snook
Harvard School of Public Health
Boston, MA

Neville A. Stanton
Brunel University
School of Engineering
London, U.K.

Naomi G. Swanson
NIOSH
Division of Applied Research
  && Technology
Cincinnati, OH

Jørn Toftum
Technical University of Denmark
International Centre for Indoor
  Environment & Energy
Lyngby, Denmark

Rendell R. Torres
Rensselaer Polytechnic Institute
School of Architecture
Troy, NY

Susan Vallance
Johnson Engineering
Houston, TX

Gordon A. Vos
Texas A&M University
School of Rural Public Health
Bryan, TX

Guy Walker
Brunel University
School of Engineering
London, U.K.

Donald E. Wasserman
University of Tennessee
Institute for the Study of Human
  Vibration
Knoxville, TN

Jack F. Wasserman
University of Tennessee
Institute for the Study of Human
  Vibration
Knoxville, TN

Thomas R. Waters
NIOSH
Division of Applied Research
  && Technology
Cincinnati, OH

Christopher D. Wickens
University of Illinois at Urbana-
  Champaign
Institute of Aviation
Aviation Human Factors Division
Savoy, IL

Cornelis J.E. Wientjes
NATO Research & Technology
  Agency
Brussels, Belgium

David Wilder
University of Tennessee
Institute for the Study of Human
  Vibration
Knoxville, TN

Mark S. Young
University of New South Wales
Department of Aviation
Sydney, Australia

Contents

Psychophysiological Methods

Behavioral and Cognitive Methods

Team Methods

Environmental Methods

Macroergonomic Methods

1

Human Factors and Ergonomics Methods

Neville A. Stanton
Brunel University

1.1 Aims of the Handbook

The main aim of this handbook is to provide a comprehensive, authoritative, and practical account of human factors and ergonomics methods. It is intended to encourage people to make full use of human factors and ergonomics methods in system design. Research has suggested that even professional ergonomists tend to restrict themselves to two or three of their favorite methods, despite variations in the problems that they address (Baber and Mirza, 1988; Stanton and Young, 1998). If this book leads people to explore human factors and ergonomics methods that are new to them, then it will have achieved its goal.

The page constraints of this handbook meant that coverage of the main areas of ergonomics had to be limited to some 83 methods. The scope of coverage, outlined in Table 1.1, was determined by what ergonomists do.

From these definitions, it can be gleaned that the domain of human factors and ergonomics includes:

- Human capabilities and limitations
- Human–machine interaction
- Teamwork
- Tools, machines, and material design
- Environmental factors
- Work and organizational design

These definitions also put an emphasis (sometimes implicit) on analysis of human performance, safety, and satisfaction. It is no wonder, then, that human factors and ergonomics is a discipline with a strong tradition in the development and application of methods.

Hancock and Diaz (2002) argue that, as a scientific discipline, ergonomics holds the moral high ground, with the aim of bettering the human condition. They suggest that this may be at conflict with other aims of improving system effectiveness and efficiency. No one would argue with the aims of improved comfort, satisfaction, and well-being, but the drawing of boundaries between the improvements for individuals and improvements for the whole system might cause some heated debate. Wilson (1995) suggests that the twin interdependent aims of ergonomics might not be easy to resolve, but ergonomists have a duty to both individual jobholders and the employing organization. Ethical concerns about the issue of divided

TABLE 1.1   Definitions of Human Factors and Ergonomics

Author	Definition of Human Factors and Ergonomics
Murrell, 1965	…the scientific study of the relationship between man and his working environment. In this sense, the term environment is taken to cover not only the ambient environment in which he may work but also his tools and materials, his methods of work and the organization of the work, either as an individual or within a working group. All these are related to the nature of man himself; to his abilities, capacities and limitations.
Grandjean, 1980	…is a study of man's behavior in relation to his work. The object of this research is man at work in relation to his spatial environment…the most important principle of ergonomics: Fitting the task to the man. Ergonomics is interdisciplinarian: it bases its theories on physiology, psychology, anthropometry, and various aspects of engineering.
Meister, 1989	…is the study of how humans accomplish work-related tasks in the context of human–machine system operation and how behavioral and nonbehavioral variables affect that accomplishment.
Sanders and McCormick, 1993	…discovers and applies information about human behavior, abilities, limitations, and other characteristics to the design of tools, machines, tasks, jobs, and environments for productive, safe, comfortable, and effective human use.
Hancock, 1997	…is that branch of science which seeks to turn human–machine antagonism into human–machine synergy.

Source: Dempsey, P.G., Wolgalter, M.S., and Hancock, P.A. (2000), *Theor. Issues Ergonomics Sci.*, 1, 3–10. With permission.

responsibilities might only be dealt with satisfactorily by making it clear to all concerned where one's loyalties lie.

The *International Encyclopedia of Human Factors and Ergonomics* (Karwowski, 2001) has an entire section devoted to methods and techniques. Many of the other sections of the encyclopedia also provide references to, if not actual examples of, ergonomics methods. In short, the importance of human factors and ergonomics methods cannot be overstated. These methods offer the ergonomist a structured approach to the analysis and evaluation of design problems. The ergonomist's approach can be described using the scientist-practitioner model. As a scientist, the ergonomist is:

- Extending the work of others
- Testing theories of human–machine performance
- Developing hypotheses
- Questioning everything
- Using rigorous data-collection and data-analysis techniques
- Ensuring repeatability of results
- Disseminating the finding of studies

As a practitioner, the ergonomist is:

- Addressing real-world problems
- Seeking the best compromise under difficult circumstances
- Looking to offer the most cost-effective solution
- Developing demonstrators and prototype solutions
- Analyzing and evaluating the effects of change
- Developing benchmarks for best practice
- Communicating findings to interested parties

Most ergonomists will work somewhere between the poles of scientist and practitioner, varying the emphasis of their approach depending upon the problems that they face. Human factors and ergonomist methods are useful in the scientist-practitioner model because of the structure, and the potential for repeatability, that they offer. There is an implicit guarantee in the use of methods that, provided they are used properly, they will produce certain types of useful products. It has been suggested that human factors and ergonomist methods are a route to making the discipline accessible to all (Diaper, 1989; Wilson, 1995). Despite the rigor offered by methods, however, there is still plenty of scope for the role of experience. Stanton and Annett (2000) summarized the most frequently asked questions raised by users of ergonomics methods as follows:

- How deep should the analysis be?
- Which methods of data collection should be used?
- How should the analysis be presented?
- Where is the use of the method appropriate?
- How much time and effort does each method require?
- How much and what type of expertise is needed to use the method?
- What tools are there to support the use of the method?
- How reliable and valid is the method?

It is hoped that the contributions to this book will help answer some of those questions.

1.2   Layout of the Handbook

The handbook is divided into six sections, each section representing a specialized field of ergonomics with a representative selection of associated methods. The sequence of the sections and a brief description of their contents are presented in Table 1.2. The six sections are intended to represent all facets of human factors and ergonomics in systems analysis, design, and evaluation. Three of the methods sections (Sections I through III) are concerned with the individual person and his or her interaction with the world (i.e., physical methods, psychophysiological methods, and behavioral–cognitive methods). One of the methods sections (Section IV) is concerned with the social groupings and their interaction with the world (i.e., team methods). Another of the methods sections (Section V) is concerned with the effect

TABLE 1.2   Description of the Contents of the Six Methods Sections of the Handbook

Methods Sections in Handbook	Brief Description of Contents
Section I: Physical Methods	This section deals with the analysis and evaluation of musculoskeletal factors
	The topics include: measurement of discomfort, observation of posture, analysis of workplace risks, measurement of work effort and fatigue, assessing lower back disorder, and predicting upper-extremity injury risks
Section II: Psychophysiological Methods	This section deals with the analysis and evaluation of human psychophysiology
	The topics include: heart rate and heart rate variability, event-related potentials, galvanic skin response, blood pressure, respiration rate, eyelid movements, and muscle activity
Section III: Behavioral–Cognitive Methods	This section deals with the analysis and evaluation of people, events, artifacts, and tasks
	The topics include: observation and interviews, cognitive task analysis methods, human error prediction, workload analysis and prediction, and situational awareness
Section IV: Team Methods	This section deals with the analysis and evaluation of teams
	The topics include: team training and assessment requirements, team building, team assessment, team communication, team cognition, team decision making, and team task analysis
Section V: Environmental Methods	This section deals with the analysis and evaluation of environmental factors
	The topics include: thermal conditions, indoor air quality, indoor lighting, noise and acoustic measures, vibration exposure, and habitability
Section VI: Macroergonomics Methods	This section deals with the analysis and evaluation of work systems
	The topics include: organizational and behavioral research methods, manufacturing work systems, anthropotechnology, evaluations of work system intervention, and analysis of the structure and processes of work systems

that the environment has on people (i.e., environmental methods). Finally, the last of the methods sections (Section VI) is concerned with the overview of work systems (i.e., macroergonomics methods). These sets of methods are framed by the classic onion-layer analysis model, working from the individual, to the team, to the environment, to the work system. In theoretical system terms, the level of analysis can be set at all four levels, or it may focus at only one or two levels. The system boundaries will depend upon the purpose of the analysis or evaluation.

Each section of the handbook begins with an introduction written by the editor of that section. The introduction provides a brief overview of the field along with a description of the methods covered in the sequence that they appear. The editor responsible for that section determined the contents of each section. Their brief was to provide a representative set of contemporary methods that they felt were useful for ergonomic analyses and evaluation. Given the restrictions on page length for the handbook, this was a tall order. Nonetheless, the final set of chapters does present a good overview of contemporary developments in ergonomics methods and serves as a useful handbook. Some of the methods in Section V, Environmental Methods, do not follow the template approach, especially in lighting and thermal methods. This is because there is no single method that is favored or complete. Therefore, it would be very misleading to select any single method.

Wilson (1995) divides ergonomics methods into five basic types of design data:

1. Methods for collecting data about people (e.g., collection of data on physical, physiological, and psychological capacities)
2. Methods used in system development (e.g., collection of data on current and proposed system design)
3. Methods to evaluate human–machine system performance (e.g., collection of data on quantitative and qualitative measures)
4. Methods to assess the demands and effects on people (e.g., collection of data on short-term and longer-term effects on the well-being of the person performing the tasks being analyzed)
5. Methods used in the development of an ergonomics management program (e.g., strategies for supporting, managing, and evaluating sustainable ergonomics interventions).

These five basic types of design data have been put into a table to help in assessing their relationship with the six methods section in this book, as shown in Table 1.3.

As Table 1.3 shows, the methods in this handbook cover all of the five basic types of design data. The darker shading represents a primary source of design data, and the lighter shading represents a secondary, or contributory, source of design data.

TABLE 1.3   Mapping Wilson's Five Basic Types of Design Data onto the Method Sections in the Handbook

	Data about People	Systems Development	Human–Machine Performance	Demand and Effects on People	Ergonomics Management Programs
Physical	■	▨	■	■	
Psychophysiological	■	▨		■	
Behavioral–Cognitive	■	■	■	■	
Team	■	▨		■	
Environmental		▨	■	■	
Macroergonomics		■	■	■	■

1.3 Layout of Each Entry

The layout of each chapter is standardized to assist the reader in using the handbook. This approach was taken so that the reader would easily be able to locate the relevant information about the method. All of the information is given in a fairly brief form, and the reader is encouraged to consult other texts and papers for more background research on the methods and more case examples of application of the methods. The standard layout is described in Table 1.4.

The standardized approach should support quick reference to any particular method and encourage the readers to browse through potential methods before tackling the particular problem that they face. It is certainly the intention of this text to encourage the use of ergonomics methods, provided that suitable support and mentoring is in place to ensure that the methods are used properly.

1.4 Other Methods Books

The number of methods books continues to grow, making it impossible to keep up with every text and to choose or recommend a single method book for all purposes. The best advice is to select two or three that meet most of your needs, unless you can afford to stock a comprehensive library. There tend to be four types of methods books. The first type is the specialized and single authored, such as *Hierarchical Task Analysis* (Shepherd, 2001). The second type of book is specialized and edited, such as *Task Analysis* (Annett and Stanton, 2000). The third type of book is generalized and edited, such as *Evaluation of Human Work* (Wilson and Corlett, 1995). The fourth kind of book is generalized and authored, such as *A Guide to Methodology in Ergonomics* (Stanton and Young, 1999). This classification in presented in Table 1.5.

TABLE 1.4 Layout of the Chapters in the Handbook

Section Chapter	Description of Contents
Name and acronym	Name of the method and its associated acronym
Author name and affiliation	Names and affiliations of the authors
Background and applications	Introduces the method, its origins and development, and applications
Procedure and advice	Describes the procedure for applying the method and general points of expert advice
Advantages	A list or description of the advantages associated with using the method
Disadvantages	A list or description of the disadvantages associated with using the method
Example	Provides one or more examples of the application to show the output of the method
Related methods	Lists any closely related methods, particularly if the input comes from another method or the method's output feeds into another method
Standards and regulations	Lists any national or international standards or regulations that have implications for the use of the method
Approximate training and application times	Provides estimates of the training and application times to give the reader an idea of the commitment
Reliability and validity	Cites any evidence on the reliability or validity of the method
Tools needed	A description of the tools, devices, and software needed to carry out the method
References	A bibliographic list of recommended further reading on the method and the surrounding topic area

TABLE 1.5 Methods Books Taxonomy

	Specialized	Generalized
Authored	*Hierarchical Task Analysis* by Andrew Shepherd	*A Guide to Methodology in Ergonomics* by Neville Stanton and Mark Young
Edited	*Task Analysis* by John Annett and Neville Stanton	*Evaluation of Human Work* by John Wilson and Nigel Corlett

TABLE 1.6   Overview of Other Methods Books

Author(s)	Title	Edited/Authored	Date (ed.)	Pages	Coverage [a]
Annett and Stanton	*Task Analysis*	Edited	2000 (1st)	242	B/C, T
Corlett and Clarke	*The Ergonomics of Workspace and Machines*	Edited	1995 (2nd)	128	P, B/C
Diaper and Stanton	*Task Analysis in Human–Computer Interaction*	Edited	2004 (1st)	760	B/C, T
Helender et al.	*Handbook of Human–Computer Interaction*	Edited	1997 (2nd)	1582	P, B/C, T, M
Jacko and Sears	*The Human–Computer Interaction Handbook*	Edited	2003 (1st)	1277	P, B/C, T, M
Jordan et al.	*Usability Evaluation in Industry*	Edited	1996 (1st)	252	P, B/C
Karwowski and Marras	*The Occupational Ergonomics Handbook*	Edited	1999 (1st)	2065	P, PP, B/C, T, E, M
Kirwan	*A Guide to Practical Human Reliability Assessment*	Authored	1994 (1st)	592	B/C
Kirwan and Ainsworth	*A Guide to Task Analysis*	Edited	1992 (1st)	417	B/C
Salvendy	*Handbook of Human Factors and Ergonomics*	Edited	1997 (2nd)	2137	P, PP, B/C, T, E, M
Schraagen et al.	*Cognitive Task Analysis*	Edited	1999 (1st)		B/C
Seamster et al.	*Applied Cognitive Task Analysis*	Authored	1997 (1st)	338	B/C
Shepherd	*Hierarchical Task Analysis*	Authored	2001 (1st)	270	B/C
Stanton	*Human Factors in Consumer Products*	Edited	1998 (1st)	287	P, B/C
Stanton and Young	*A Guide to Methodology in Ergonomics*	Authored	1999 (1st)	150	B/C
Wilson and Corlett	*Evaluation of Human Work*	Edited	1995 (2nd)	1134	P, PP, B/C, T, E, M

[a] Key to coverage: physical methods (P), psychophysiological methods (PP), behavioral and cognitive methods (B/C), team methods (T), environmental methods (E), macroergonomic methods (M).

An analysis of 15 other methods books published over the past decade shows the range of edited and authored texts in this field, the length of the books, and their coverage. Any of these books could complement this handbook. Where they differ is in their scope (e.g., either being focused on human–computer interaction or more generalized) and their coverage (e.g., either covering one or two areas of ergonomics or having more general coverage). A summary of the texts is presented in Table 1.6.

As Table 1.6 indicates, there is certainly no shortage of ergonomics methods texts. Selection of the appropriate text will depend on the intended scope and coverage of the ergonomics intervention required.

1.5   Challenges for Human Factors and Ergonomics Methods

Ergonomics science abounds with methods and models for analyzing tasks, designing work, predicting performance, collecting data on human performance and interaction with artifacts and the environment in which this interaction takes place. Despite the plethora of methods, there are several significant challenges faced by the developers and users of ergonomics methods. These challenges include:

- Developing methods that integrate with other methods
- Linking methods with ergonomics theory
- Making methods easy to use

- Providing evidence of reliability and validity
- Showing that the methods lead to cost-effective interventions
- Encouraging ethical application of methods

Annett (2002) questions the relative merits for construct and criterion-referenced validity in the development of ergonomics theory. He distinguishes between construct validity (how acceptable the underlying theory is), predictive validity (the usefulness and efficiency of the approach in predicting the behavior of an existing or future system), and reliability (the repeatability of the results). Investigating the matter further, Annett identifies a dichotomy of ergonomics methods: analytical methods and evaluative methods. Annett argues that analytical methods (i.e., those methods that help the analyst gain an understanding of the mechanisms underlying the interaction between human and machines) require construct validity, whereas evaluative methods (i.e., those methods that estimate parameters of selected interactions between human and machines) require predictive validity. This distinction is made in Table 1.7.

This presents an interesting debate for ergonomics: Are the methods really this mutually exclusive? Presumably, methods that have dual roles (i.e., both analytical and evaluative, such as task analysis for error identification) must satisfy both criteria. It is possible for a method to satisfy three types of validity: construct (i.e., theoretical validity), content (i.e., face validity), and predictive (i.e., criterion-referenced empirical validity). The three types of validity represent three different stages in the design, development, and application of the methodology, as illustrated in Figure 1.1. There is also the question of reliability, and a method should be demonstrably stable over time and between people. Any differences in analyses should be due entirely to differences in the aspect of the world being assessed rather than differences in the assessors.

Theoretical and criterion-referenced empirical validation should be an essential part of the method development and reporting process. This in turn should inform the method selection process. Stanton and Young (1999) have recommended a structured approach for selecting methods for ergonomic analysis, design, and evaluations. This has been adapted for more generic method selection and is presented in Figure 1.2.

As shown in Figure 1.1, method selection is a closed-loop process with three feedback loops. The first feedback loop validates the selection of the methods against the selection criteria. The second feedback loop validates the methods against the adequacy of the ergonomic intervention. The third feedback loop validates the initial criteria against the adequacy of the intervention. There could be errors in the development of the initial criteria, the selection of the methods, and the appropriateness of the intervention. Each should be checked. The main stages in the process are identified as: determine criteria (where the criteria for assessment are identified), compare methods against criteria (where the pool of methods are compared for their suitability), application of methods (where the methods are applied), implementation of ergonomics intervention (where an ergonomics program is chosen and applied), and evaluation of the effectiveness of the intervention (where the assessment of change brought about by the intervention is assessed).

TABLE 1.7 Annett's Dichotomy of Ergonomics Methods

	Analytic	Evaluative
Primary purpose	Understand a system	Measure a parameter
Examples	Task analysis, training needs analysis, etc.	Measures of workload, usability, comfort, fatigue, etc.
Construct validity	Based on an acceptable model of the system and how it performs	Construct is consistent with theory and other measures of parameter
Predictive validity	Provides answers to questions, e.g., structure of tasks	Predicts performance
Reliability	Data collection conforms to an underlying model	Results from independent samples agree

Source: Adapted from Annett, J. (2002), *Theor. Issues Ergonomics Sci.*, 3, 229–232. With permission.

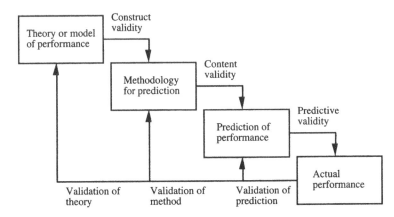

FIGURE 1.1 Validation of methods. (Adapted from Diaper, D. and Stanton, N.A. [2004], *The Handbook of Task Analysis for Human-Computer Interaction*, Lawrence Erlbaum Associates, Mahwah, NJ. With permission.)

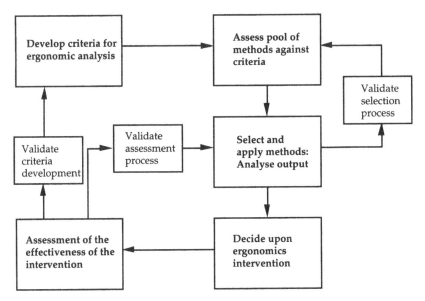

FIGURE 1.2 Validating the methods selection ergonomics intervention process. (Adapted from Stanton, N.A. and Young, M.S. [1999], *A Guide to Methodology in Ergonomics*, Taylor & Francis, London. With permission.)

The ultimate criteria determining the usefulness of ergonomics methods will be whether or not they help in analyzing tasks, designing work, predicting performance, collecting data on human performance and interaction with artifacts and the environment in which this interaction takes place. This requires that the twin issues of theoretical validity and predictive validity be addressed when developing and testing old and new methods. The approach taken in this handbook provides a benchmark on reporting on human factors and ergonomics methods. The information provided here is what all developers should ask of their own methods and, at the very least, all users of methods should demand of the developers.

References

Annett, J. (2002), A note on the validity and reliability of ergonomics methods, *Theor. Issues Ergonomics Sci.*, 3, 229–232.

Annett, J. and Stanton, N.A. (2000), *Task Analysis*, Taylor & Francis, London.

Baber, C. and Mirza, M.G. (1988), Ergonomics and the evaluation of consumer products: surveys of evaluation practices, in *Human Factors in Consumer Product Design*, Stanton, N.A., Ed., Taylor & Francis, London.

Corlett, E.N. and Clarke, T.S. (1995), *The Ergonomics of Workspaces and Machines*, 2nd ed., Taylor & Francis, London.

Dempsey, P.G., Wolgalter, M.S., and Hancock, P.A. (2000), What's in a name? Using terms from definitions to examine the fundamental foundation of human factors and ergonomics science, *Theor. Issues Ergonomics Sci.*, 1, 3–10.

Diaper, D. (1989), *Task Analysis in Human Computer Interaction*, Ellis Horwood, Chichester, U.K.

Diaper, D. and Stanton, N.A. (2004), *The Handbook of Task Analysis for Human-Computer Interaction*, Lawrence Erlbaum Associates, Mahwah, NJ.

Diaper, D. and Stanton, N.A. (2004), Wishing on a star: the future of task analysis, in *The Handbook of Task Analysis for Human-Computer Interaction*, Diaper, D. and Stanton, N.A., Eds., Lawrence Erlbaum Associates, Mahwah, NJ, pp. 603–619.

Grandjean, E. (1980), *Fitting the Task to the Man*, Taylor & Francis, London.

Hancock, P.A. (1997), *Essays on the Future of Human-Machine Systems*, Banta, Minneapolis, MN.

Hancock, P.A. and Diaz, D.D. (2002), Ergonomics as a foundation for a science of purpose, *Theor. Issues Ergonomics Sci.*, 3 (2), 115–123.

Helender, M.G., Landauer, T.K., and Prabhu, P.V. (1997), *Handbook of Human-Computer Interaction*, 2nd ed., Elsevier, Amsterdam.

Jacko, J.A. and Sears, A. (2003), *The Human-Computer Interaction Handbook*, Lawrence Erlbaum Associates, Mahwah, NJ.

Jordan, P.W., Thomas, B., Weerdmeester, B.A., and McClelland, I.L. (1996), *Usability Evaluation in Industry*, Taylor & Francis, London.

Karwowski, W. and Marras, W.S. (1998), *The Occupational Ergonomics Handbook*, CRC Press, Boca Raton, FL.

Karwowski, W. (2001), *International Encyclopedia of Ergonomics and Human Factors*, Vols. I–III, Taylor & Francis, London.

Kirwan, B. (1994), *A Guide to Practical Human Reliability Assessment*, Taylor & Francis, London.

Kirwan, B. and Ainsworth, L. (1992), *A Guide to Task Analysis*, Taylor & Francis, London.

Meister, D. (1989), *Conceptual Aspects of Human Factors*, Johns Hopkins University Press, Baltimore, MD.

Murrell, K.F.H. (1965), *Human Performance in Industry*, Reinhold Publishing, New York.

Salvendy, G. (1997), *Handbook of Human Factors and Ergonomics*, 2nd ed., Wiley, New York.

Sanders, M.S. and McCormick, E.J. (1993), *Human Factors Engineering and Design*, McGraw-Hill, New York.

Schraagen, J.M., Chipman, S., and Shalin, V. (1999), *Cognitive Task Analysis*, Lawrence Erlbaum Associates, Mahwah, NJ.

Seamster, T.L., Redding, R.E., and Kaempf, G.L. (1997), *Applied Cognitive Task Analysis in Aviation*, Avebury, Aldershot, U.K.

Shepherd, A. (2001), *Hierarchical Task Analysis*, Taylor & Francis, London.

Stanton, N.A. (1998), *Human Factors in Consumer Product Design*, Taylor & Francis, London.

Stanton, N.A. and Young, M. (1998), Is utility in the mind of the beholder? A review of ergonomics methods, *Appl. Ergonomics*, 29, 41–54.

Stanton, N.A. and Annett, J. (2000), Future directions for task analysis, in *Task Analysis*, Annett, J. and Stanton, N.A., Eds., Taylor & Francis, London, pp. 229–234.

Stanton, N.A. and Young, M.S. (1999), *A Guide to Methodology in Ergonomics*, Taylor & Francis, London.

Wilson, J.R. (1995), A framework and context for ergonomics methodology, in *Evaluation of Human Work*, 2nd ed., Wilson, J.R. and Corlett, E.N., Eds., Taylor & Francis, London, pp. 1–39.

Wilson, J.R. and Corlett, E.N. (1995), *Evaluation of Human Work*, 2nd ed., Taylor & Francis, London.

Physical Methods

2

Physical Methods

Alan Hedge
Cornell University

The use of physical methods to assess how work is being performed is crucial to the work of many ergonomists. The physical methods included in this section can be used to obtain essential surveillance data for the management of injury risks in the workforce. It is generally accepted that many musculoskeletal injuries begin with the worker experiencing discomfort. If ignored, the risk factors responsible for the discomfort eventually will lead to an increase in the severity of symptoms, and what began as mild discomfort will gradually become more intense and will be experienced as aches and pains. If left unchecked, the aches and pains that signal some cumulative trauma eventually may result in an actual musculoskeletal injury, such as tendonitis, tenosynovitis, or serious nerve-compression injury like carpal tunnel syndrome. Sensations of discomfort are the body's early warning signs that some attribute of the worker's job should be changed. Discomfort will also adversely affect work performance, either by decreasing the quantity of work, decreasing the quality of work through increased error rates, or both. Reducing the levels of discomfort actually decreases the risk of an injury occurring. Consequently, changes in levels of discomfort can also be used to gauge the success of the design of an ergonomic product or the implementation of an ergonomic program intervention.

Three methods are presented (Chapters 3 through 5) that can be used to assess levels of musculoskeletal discomfort among workers. These methods all use self-report surveys to quantify discomfort, because discomfort cannot be directly observed or objectively measured. The methods in this section are representative of the range of methods available to the ergonomist. The section does not present a comprehensive set of all available methods for assessing discomfort. Other methods are available, and several of these are referenced in the chapters included in this section. The three chosen methods presented here are PLIBEL, the U.S. National Institute of Occupational Safety and Health (NIOSH) discomfort surveys, and the Dutch Musculoskeletal Survey.

The PLIBEL method is one of the earliest methods developed to gauge a worker's degree of musculoskeletal discomfort. It comprises a checklist of items derived from a comprehensive review of the ergonomics literature. It allows workers to systematically assess workplace ergonomic hazards associated with five body regions by completing a simple checklist. An assessment can be made for a task or several tasks or for a complete job. PLIBEL results can serve as the basis for discussions on improvements to job design. PLIBEL is available in several languages.

The NIOSH discomfort questionnaires have been extensively used in U.S. studies of ergonomic hazards. This self-report method allows the ergonomist to easily assess measures of musculoskeletal discomfort in numerous body regions, such as the intensity, frequency, and duration of discomfort. This chapter also gives a comprehensive list of NIOSH research reports.

The Dutch Musculoskeletal Survey represents one of the most comprehensive and thoroughly validated survey measures of musculoskeletal discomfort. It exists in short and long forms, depending on the intent of its use. It comprises a collection of scales that deal with a broad range of workplace ergonomic hazards, and thus the ergonomist can selectively choose the relevant scales. Analytical software is also available for this survey, though only in Dutch at present.

Other survey questionnaires are also available to researchers, such as the Cornell Musculoskeletal Discomfort Survey (Hedge et al., 1999); the Standardized Nordic Questionnaire (SNQ), which focuses on general body, low back, and neck/shoulder complaints (Kuorinka et al., 1987); and a more recent revision of this (Dickinson et al., 1992) called the Nordic Musculoskeletal Questionnaire (NMQ). These instruments can be self-administered or interview administered.

Although self-reports of discomfort provide valuable information to the ergonomist, they are intrusive and they do require some effort on the part of the worker to answer the various questions, and this may be disrupting to work activities. There is considerable value in using unobtrusive methods to gauge injury risks. Consequently, several methods have been developed to systematically assess a worker's posture while performing work. Posture is an observable reflection of musculoskeletal activity, and these methods all allow the ergonomist to assess risks by systematic observation alone. This means that ergonomic analyses can be performed on visual recordings of workplaces, such as videotapes or photographs. It is assumed that every body segment moves through a range of motion, termed the "neutral zone," within which the anatomical stresses and strains are insufficient to initiate an injury process. However, the further the worker makes excursions away from this neutral zone, the greater the injury risk, especially when such excursions are frequently repeated and/or sustained for extended periods. These postural observation methods also offer the advantage that they allow high-risk postures to be readily identified for corrective action, often even before the worker has been exposed for a sufficient time to develop significant musculoskeletal discomfort. Thus, when correctly used, posture targeting methods provide even earlier risk detection capabilities than do discomfort surveys.

Four methods (Chapters 6 through 9) are presented that provide the ergonomist with an excellent arsenal of postural evaluation tools. The Quick Exposure Checklist has a high level of usability and sensitivity, and it allows for quick assessment of the exposure to risks for work-related musculoskeletal disorders. This method has the advantage that it can be used to analyze interactions between various workplace risks, even by relatively inexperienced raters. The RULA and REBA posture-targeting methods are probably the most well-known methods for rapid assessment of risks. The RULA method is well suited to analyzing sedentary work, such as computer work. The REBA method is ideal for rapid assessment of standing work. Both of these methods have been extensively used in ergonomic research studies and also in evaluating the impact of workplace design changes on body posture. The Strain Index is a more comprehensive method that specifically focuses on the risks of developing distal upper extremity musculoskeletal disorders, i.e., injuries of the elbow, forearm, wrist, and hand. All of these methods take little time to administer and can be used in a wide variety of work situations. The methods can be used to assess overall postural risks and/or those to specific body segments.

Other similar posture-targeting methods, such as the Ovako Working Posture Analysis System (OWPAS) (Karhu et al., 1977) and the Portable Ergonomics Observation (PEO) method (Fransson-Hall et al., 1995), have not been included but can also be used. The OWPAS method involves direct observation and sampling of tasks using a whole-body posture-coding system to estimate injury risks. The PEO method records hand, neck, trunk, and knee postures and also evaluates manual handling activities, such as lifting. Real-time observations are directly entered into a computer. Ergonomists can use posture-targeting methods to measure the success of any ergonomic design changes to equipment or to the layout of a workplace, and the ability to quantify changes in likely injury risk can be a valuable aid to management decision making. With the advent of handheld personal digital assistants (PDAs), the ergonomist can easily carry an extensive ergonomics toolkit into any workplace and generate almost instant analyses and reports, as is shown in Chapter 10, which discusses the use of PDAs.

The measurement of work effort and fatigue was one of the earliest challenges that ergonomists faced, and this challenge remains today. Although the performance of work in more-deviated postures invariably requires more muscular effort, which in turn may accelerate muscular fatigue, none of the methods used to assess discomfort or posture actually yields information on the degree of work effort or on the level of accumulated fatigue that could amplify an injury risk. Two methods are included that quantify effort and fatigue. The Borg Ratings of Perceived Exertion scale (Chapter 11) provides a physiologically validated method for quantifying how much effort is involved in performing physical work. The Muscle Fatigue Assessment method (Chapter 12) characterizes discomfort and identifies the ways that workers change their behavior in an attempt to cope with accumulated fatigue. Both methods are invaluable to the successful design of physical jobs so that neither the quantity nor quality of work performance will suffer over the course of a work shift, and so that the worker will not experience undue physical demands or fatigue that could increase the risks of an injury or accident.

Evaluations of discomfort, work posture, and effort provide valuable insights into possible injury risks in the workplace. However, such approaches do not necessarily help to predict the risks of potentially acute injuries, such as back injuries, and they do not set safe limits on work or predict how changes in a job will impact the level of safety. Back injuries account for up to 50% of musculoskeletal injuries in the U.S., costing the U.S. economy up to $60 billion each year (NAS, 2002), and ergonomic research has been undertaken to set safe limits for lifting work. Two methods for assessing back injury risks are presented here. Using a psychophysical approach to assessing strength, Snook tables (Chapter 13) set safe weight limits for men and women who perform lifting, pushing, and pulling tasks at work. These give the ergonomist a quick method for assessing the injury potential of a specific work task that involves these actions. Mital tables are a similar tabular method for determining lifting limits (Mital, 1984), and revised tables were introduced in 1989 to also deal with asymmetrical lifting tasks and confined lifting situations (Mital, 1989; Mital, 1992). However, Snook tables are presented here because they also set the maximum acceptable weights for lifting and lowering, and set the maximum forces for pushing, pulling, and carrying tasks.

A predictive method for determining back injury risks was pioneered by the NIOSH in 1981 and has undergone substantial revision and enhancement since then, with a revised equation being introduced in 1991. The NIOSH lifting equation has not been included because it is widely used and is a well-known method (see http://www.cdc.gov/niosh/94-110.html). Also, the lifting equation method does not account for motion at the time of lifting, and it does not use any measurement of actual spinal loading. The more recent Lumbar Motion Monitor method described in Chapter 14 was developed to overcome these limitations of the NIOSH lifting equation by providing a more direct assessment of the dynamic components of low back disorder risks at work.

The final two methods that are described, the OCRA and MAPO methods (Chapters 15 and 16), are the most comprehensive, yet they are also somewhat complex and the most time consuming. The OCRA index is a detailed analytical and reliable method that can be predictive of upper-extremity injury risks in exposed worker populations. The OCRA index can also be used as the basis for identifying opportunities for task and/or workstation redesign, and as a means of evaluating the success of any interventions. The MAPO method has been specifically developed to analyze health-care workplaces, especially those places where workers are involved in the care and handling of disabled patients, paralyzed patients, and wheelchair-bound patients. Nursing work that involves patient handling poses the greatest risks for developing a lower-back injury, and without ergonomic attention, this situation may worsen as the nursing workforce ages and as the patient population gets heavier. Unlike the majority of the other physical methods, the MAPO method also incorporates an assessment of the environment in which the work is being performed.

The range and scope of the methods described in this section provide the reader with the tools to undertake a range of studies, including epidemiological ergonomic research, evaluations of ergonomic programs and design interventions, surveillance of workplace ergonomic hazards, and the detection and quantification of exposures to adverse workplace physical ergonomic stressors. Armed with this battery of tools, the ergonomist will be well positioned to systematically tackle a wide range of workplace issues and to implement effective solutions to the problems that are uncovered.

References

Dickinson, C.E., Campion, K., Foster, A.F., Newman, S.J., O'Rourke, A.M.T., and Thomas, P.G. (1992), Questionnaire development: an examination of the Nordic Musculoskeletal Questionnaire, *Appl. Ergonomics*, 23, 197–201.

Fransson-Hall, C., Gloria, R., Kilbom, A., Winkel, J., Larlqvist, L., and Wiktorin, C. (1995), A portable ergonomic observation method (PEO) for computerized online recording of postures and manual handling, *Appl. Ergonomics*, 26, 93–100.

Hedge, A., Morimoto, S., and McCrobie, D. (1999), Effects of keyboard tray geometry on upper body posture and comfort, *Ergonomics*, 42, 1333–1349; see also http://ergo.human.cornell.edu/ahm-squest.html.

Karhu, O., Kansi, P., and Kuorinka, I. (1977), Correcting working postures in industry: a practical method for analysis, *Appl. Ergonomics*, 8, 199–201.

Kuorinka, I., Jonsson, B., Kilbom, A., Vinterberg, H., Biering-Sorensen, F., Andersson, G., and Jorgensen, K. (1987), Standardised Nordic questionnaires for the analysis of musculoskeletal symptoms, *Appl. Ergonomics*, 18, 907–916.

Mital, A. (1984), Comprehensive maximum acceptable weight of lift database for regular 8-hr workshifts, *Ergonomics*, 27, 1127–1138.

Mital, A. (1989), Consideration of load asymmetry, placement restrictions, and type of lifting in a design database for industrial workers, *J. Safety Res.*, 20, 93–101.

Mital, A. (1992), Psychophysical capacity of industrial workers for lifting symmetrical and asymmetrical loads symmetrically and asymmetrically for 8 h work shifts, *Ergonomics*, 35, 745–754.

National Academy of Sciences (NAS) (2001), Musculoskeletal disorders and the workplace: low back and upper extremities, National Academy Press, Washington, D.C.

3

PLIBEL — The Method Assigned for Identification of Ergonomic Hazards

Kristina Kemmlert
National Institute for Working Life

3.1 Background and Applications

The Swedish Work Environment Act stipulates that the employer shall investigate occupational injuries, draw up action plans, and organize and evaluate job modifications. Hence it is also of interest for the government's Labour Inspectorate to study conditions and improvements in the workplace.

The "method for the identification of musculoskeletal stress factors which may have injurious effects" (PLIBEL) was designed to meet such needs (Figure 3.1). PLIBEL has been used in several studies, in practical on-site ergonomic work, and as an educational tool. It has been presented in various parts of the world and translated into several languages (Kemmlert, 1995, 1996a, 1996b, 1997).

PLIBEL is a simple checklist screening tool intended to highlight musculoskeletal risks in connection with workplace investigations. Time aspects as well as environmental and organizational considerations also have to be considered as modifying factors.

The checklist was designed so that items ordinarily checked in a workplace assessment of ergonomic hazards would be listed and linked to five body regions (Figure 3.1). Only specific work characteristics, defined and documented as ergonomic hazards in scientific papers or textbooks, are listed (Figure 3.2 and Figure 3.3). Whenever a question is irrelevant to a certain body region, and/or if documentation has not been found in the literature, it is represented by a gray field in the checklist and need not be answered.

The list was made in 1986, and new references have since then been read continuously and the list updated. Mostly, these only add knowledge to the primary list, which accordingly has not been changed.

Kemmlert, K. and Kilbom, A. (1986) National Board of Occupational Safety and Health, Research Department, Work Physiology Unit, 17184 Solna, Sweden

Body region	neck/shoulders, upper part of back	elbows, forearms hands	feet	knees and hips	low back
1. Is the walking surface uneven, sloping, slippery, or nonresilient?			1.	1.	1.
2. Is the space too limited for work movements or work materials?	2.	2.	2.	2.	2.
3. Are tools and equipment unsuitably designed for the worker or the task?	3.	3.	3.	3.	3.
4. Is the working height incorrectly adjusted?	4.				4.
5. Is the working chair poorly designed or incorrectly adjusted?	5.				5.
6. (If the work is performed while standing) Is there no possibility to sit and rest?				6.	6.
7. Is fatiguing foot-pedal work performed?				7.	7.
8. Is fatiguing leg work performed, e.g.: a) repeated stepping up on stool, step, etc.? b) repeated jumps, prolonged squatting, or kneeling? c) one leg being used more often in supporting the body?				8. a b c	8. a b c
9. Is repeated or sustained work performed when the back is: a) mildly flexed forward? b) severely flexed forward? c) bent sideways or mildly twisted? d) severely twisted?	9. a b c d				9. a b c d
10. Is repeated or sustained work performed when the neck is: a) flexed forward? b) bent sideways or mildly twisted? c) severely twisted? d) extended backward?	10. a b c d				
11. Are loads lifted manually? Notice factors of importance as: a) periods of repetitive lifting e) handling beyond forearm length b) weight of load f) handling below knee height c) awkward grasping of load g) handling above shoulder height d) awkward location of load at onset or end of lifting	11. a—e b—f c—g d				11. a—e b—f c—g d
12. Is repeated, sustained, or uncomfortable carrying, pushing, or pulling of loads performed?	12.	12.			12.
13. Is sustained work performed when one arm reaches forward or to the side without support?	13.				
14. Is there repetition of: a) similar work movements? b) similar work movements beyond comfortable reaching distance?	14. a b	14. a b			
15. Is repeated or sustained manual work performed? Notice factors of importance as: a) weight of working materials or tools b) awkward grasping of working materials or tools	15. a b	15. a b			
16. Are there high demands on visual capacity?	16.				
17. Is repeated work, with forearm and hand, performed with: a) twisting movements? c) uncomfortable hand positions? b) forceful movements? d) switches or keyboards?	17. a—c b—d	17. a b			

Method of application.

- Find the injured body region.
- Follow white fields to the right.
- Do the work tasks contain any of the factors described?
- If so, tick where appropriate.

Also take these factors into consideration:

a) the possibility to take breaks and pauses
b) the possibility to choose order and type of work tasks or pace of work
c) if the job is performed under time demands or psychological stress
d) if the work can have unusual or unexpected situations
e) presence of cold, heat, draught, noise, or troublesome visual conditions
f) presence of jerks, shakes, or vibrations

FIGURE 3.1 The PLIBEL form.

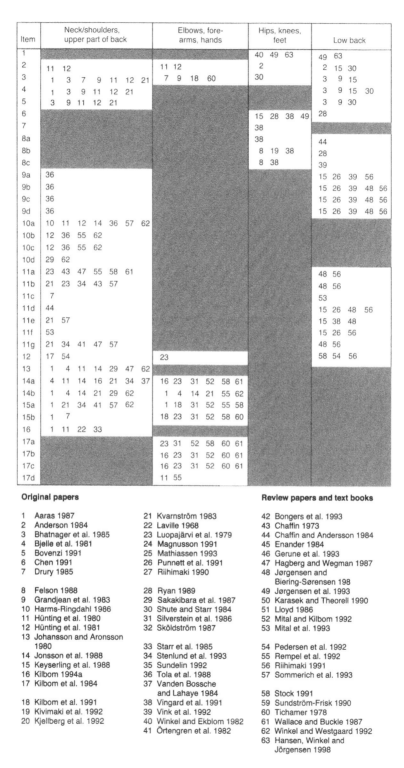

Item	Neck/shoulders, upper part of back	Elbows, fore-arms, hands	Hips, knees, feet	Low back
1			40 49 63	49 63
2	11 12	11 12	2	2 15 30
3	1 3 7 9 11 12 21	7 9 18 60	30	3 9 15
4	1 3 9 11 12 21			3 9 15 30
5	3 9 11 12 21			3 9 30
6			15 28 38 49	28
7			38	
8a			38	44
8b			8 19 38	28
8c			8 38	39
9a	36			15 26 39 56
9b	36			15 26 39 48 56
9c	36			15 26 39 48 56
9d	36			15 26 39 48 56
10a	10 11 12 14 36 57 62			
10b	12 36 55 62			
10c	12 36 55 62			
10d	29 62			
11a	23 43 47 55 58 61			48 56
11b	21 23 34 43 57			48 56
11c	7			53
11d	44			15 26 48 56
11e	21 57			15 38 48
11f	53			15 26 56
11g	21 34 41 47 57			48 56
12	17 54	23		58 54 56
13	1 4 11 14 29 47 62			
14a	4 11 14 16 21 34 37	16 23 31 52 58 61		
14b	1 4 14 21 29 62	1 4 14 21 55 62		
15a	1 21 34 41 57 62	1 18 31 52 55 58		
15b	1 7	18 23 31 52 58 60		
16	1 11 22 33			
17a		23 31 52 58 60 61		
17b		16 23 31 52 60 61		
17c		16 23 31 52 60 61		
17d		11 55		

Original papers

1 Aaras 1987
2 Anderson 1984
3 Bhatnager et al. 1985
4 Bjelle et al. 1981
5 Bovenzi 1991
6 Chen 1991
7 Drury 1985

8 Felson 1988
9 Grandjean et al. 1983
10 Harms-Ringdahl 1986
11 Hünting et al. 1980
12 Hünting et al. 1981
13 Johansson and Aronsson 1980
14 Jonsson et al. 1988
15 Keyserling et al. 1988
16 Kilbom 1994a
17 Kilbom et al. 1984

18 Kilbom et al. 1991
19 Kivimaki et al. 1992
20 Kjellberg et al. 1992

21 Kvarnström 1983
22 Laville 1968
23 Luopajärvi et al. 1979
24 Magnusson 1991
25 Mathiassen 1993
26 Punnett et al. 1991
27 Riihimaki 1990

28 Ryan 1989
29 Sakakibara et al. 1987
30 Shute and Starr 1984
31 Silverstein et al. 1986
32 Sköldström 1987

33 Starr et al. 1985
34 Stenlund et al. 1993
35 Sundelin 1992
36 Tola et al. 1988
37 Vanden Bossche and Lahaye 1984
38 Vingard et al. 1991
39 Vink et al. 1992
40 Winkel and Ekblom 1982
41 Örtengren et al. 1982

Review papers and text books

42 Bongers et al. 1993
43 Chaffin 1973
44 Chaffin and Andersson 1984
45 Enander 1984
46 Gerune et al. 1993
47 Hagberg and Wegman 1987
48 Jørgensen and Biering-Sørensen 198
49 Jørgensen et al. 1993
50 Karasek and Theorell 1990
51 Lloyd 1986
52 Mital and Kilbom 1992
53 Mital et al. 1993

54 Pedersen et al. 1992
55 Rempel et al. 1992
56 Riihimaki 1991
57 Sommerich et al. 1993

58 Stock 1991
59 Sundström-Frisk 1990
60 Tichamer 1978
61 Wallace and Buckle 1987
62 Winkel and Westgaard 1992
63 Hansen, Winkel and Jörgensen 1998

FIGURE 3.2 Documented background for PLIBEL. References, as numbered in the footnote, are given for each risk factor in relation to body regions, as in the PLIBEL form. Note, however, that in this presentation the distribution is by four body regions. Hips, knees, and feet are combined in the table.

Also take these factors into consideration

The possibility to take breaks and pauses	21	25	35	50	57	59	61		
The possibility to choose order and type or work tasks or pace of work	13	21	35	50	57				
If the job is performed under time demands or psychological stress	13	21	35	42	50	59	61		
If the work can have unusual or unexpected situations	34	38	50	56					
Presence of:									
cold	6	45	51						
heat	32	53							
draught	35	36							
noise	20								
troublesome visual conditions	1	11	22	33	61				
jerks, shakes or vibration	5	24	34	37	46	48	52	54	56

FIGURE 3.3 Documented background for modifying factors (for references, see footnote to Figure 3.2).

Only one — concerning hips, knees, feet, and the lower spinal region — has the kind of new information searched for and has therefore been added to the documented background (Figure 3.2).

3.2 Procedure

A workplace assessment using PLIBEL starts with an introductory interview with the employee and with a preliminary observation. The assessments focus on representative parts of the job, the tasks that are conducted for most of the working hours, and tasks that the worker and/or the observer look upon as particularly stressful to the musculoskeletal system. Thus several PLIBEL forms may have to be filled in for each employee. The assessments should be related to the capacity of the individual observed. Unusual or personal ways of doing a task are also recorded.

When an ergonomic hazard is observed, the numbered area on the form is checked or a short note is made. In the concluding report, where the crude dichotomous answers are arranged in order of importance, quotations from the list of ergonomic hazards can be used. Modifying factors — duration and quantities of environmental or organizational factors — are then taken into consideration (Figure 3.1).

Usually PLIBEL is used to identify musculoskeletal injury risk factors for a specific body region, and only questions relevant to that body region need be answered. In cases where a more general application is desired, the whole list is used, and the result can be referred to one or more body regions.

To use PLIBEL, first locate the injured body region, then follow the white fields to the right and check any observed risk factor(s) for the work task. The continued assessment is more difficult, as it requires consideration of questions a through f (Figure 3.1). These can either upgrade or minimize the problem. Additional evident risks, not mentioned in the checklist, are noted and addressed.

For example, there are no duration criteria for a PLIBEL record, and so cumbersome but short-lasting and/or rare events can also be recorded. In fact, the purpose of the interview with the worker that precedes the observation is to make such aspects of the task manifest.

A participatory approach of this kind has also been suggested by other authors (e.g., Drury, 1990), who recommend that observers talk to operators to get a feel for what is important. If only "normal" subjects and work periods are chosen for assessments, many of the unusual conditions that may constitute main hazards can be missed.

A handbook (unpublished material) has been compiled to provide the scientific background for each item and help identify the cutoff point for "yes" or "no" answers. This facilitates the assessments, which are to be performed by knowledgeable and experienced observers. To make the checklist easy to handle and applicable in many different situations, the questions are basic.

The analysis of possible ergonomic hazards is done at the workplace, and only relevant risk information from the assessment is considered. The issues identified as risks are arranged in order of importance. The concluding report gives an interpretation of the ergonomic working conditions, starting with the most tiresome movements and postures.

3.3   Advantages

The PLIBEL method is a general assessment method and is not intended for any specific occupations or tasks. It observes a part, or the whole, of the body and summarizes the actual identification of ergonomic hazards in a few sentences.

It is simple and is designed for primary checking. For labor inspectors and others observing many tasks every day, it is certainly enough to be equipped and well acquainted with the checklist.

PLIBEL is an initial investigative method for the workplace observer to identify ergonomic hazards, and it can be supplemented by other measurements, for instance weight and time, or quotations/observations from other studies.

Although it is tempting to add items to the checklist, to obtain a simple and quantitative measure of ergonomic conditions after a workplace assessment, PLIBEL should not be modified or used in this way. Different ergonomic hazards do not have an equal influence on worker injury, and certain problems can appear with more than one hazardous factor in the checklist.

3.4   Disadvantages

The PLIBEL method is a general assessment method and is not intended for any specific occupations or tasks. Many other methods are intended for a specific occupation or body region and can record more-detailed answers. If necessary, these more specific methods can easily be used to supplement the PLIBEL questions.

3.5   Example

PLIBEL analysis of the following task (Figure 3.4) reveals that the task entails a risk of musculoskeletal stress to the lower region of the back due to the nonresilient walking surface, the unsuitably designed tools and equipment, and the lack of any possibility to sit and rest. Repetitive and sustained work is performed with the back flexed slightly forward, bent sideways, and slightly twisted. Loads are repeatedly lifted manually and often above shoulder height. Note that the text order has been expressed by giving the most exposed body region and the environmental and instrumental conditions first. The second sentence above gives "the answers" from the body, followed by a description of the tiresome, and perhaps individual, way of performing the task.

3.6   Related methods

To provide a reference instrument for PLIBEL, an inventory of available scientific literature on occupational risk factors for musculoskeletal disorders was performed. Original papers, review papers, and

FIGURE 3.4   Example of a task posing ergonomic hazards that was analyzed using PLIBEL.

textbooks were studied. After a thorough review of the literature, the German ergonomics job analysis procedure AET (Arbeitswissenshaftliche Erhebungsverfahren zur Tätigkeitsanalyse) was chosen as the referent instrument for field testing (Rohmert and Landau, 1983).

Like PLIBEL, AET is applicable to all sorts of occupational tasks. It covers workplaces, tools and objects, degree of repetitiveness, work organization, cognitive demands, and environmental factors such as visual conditions, noise, and vibration. But while AET analyzes all components in a man-at-work system, PLIBEL focuses on one extreme phenomenon, i.e., the occurrence of an ergonomic hazard.

Two researchers, who each had been practicing AET and PLIBEL and clearing very many workplace assessments, identified 18 matching items in the two methods. For PLIBEL, only dichotomous answers are used, whereas multilevel codes in steps 0 through 5, are applied in AET. For each of the items, the corresponding level between the two methods was identified. The two observation methods were then used simultaneously for observations on a total of 25 workers, men and women in different tasks.

When the results of PLIBEL and AET were compared, the agreement between matching items was considerable. However, the modifications of AET scores for a dichotomous coding could not completely eliminate the differences between the methods. In concordance with its purposes, PLIBEL was more sensitive to ergonomic hazards.

3.7   Standards and Regulations

PLIBEL was designed to meet the needs of a standardized and practical method for the identification of ergonomic hazards and for a preliminary assessment of risk factors. An ergonomics screening tool, for the assessment of ergonomic conditions at workplaces, has been suggested as a feasible instrument by other researchers.

Moreover, it is valuable to have a systematic assessment method when doing follow-ups and when analyzing how intervention after the onset of occupational musculoskeletal injuries could be made more effective.

PLIBEL follows the standards and regulations of the day, and although it is a self-explanatory, subjective assessment method, registering only at a dichotomous level, it requires a solid understanding of ergonomics. To use the method skillfully, a certain degree of practice is firmly recommended.

3.8   Approximate Training and Application Time

Identifying an awkward situation is not difficult, nor is it difficult to find such a situation with the aid of the checklist. PLIBEL is quick to use and easy to understand, and users will become familiar with this tool within hours. However, although PLIBEL is a self-explanatory subjective assessment method that makes dichotomous judgments about risks, it requires a solid ergonomics understanding, and skillful use of the method requires practice.

3.9   Reliability and Validity

A reliability and validity study of the method has been performed according to Carmines and Zeller (1979). It was tested (Kemmlert, 1995) for:

Construct validity
Criterion validity
Reliability
Applicability

The agreement between matching items was considerable, and the interobserver reliability yielded kappa values expressing a fair to moderate agreement on the following questions:

Is the content of PLIBEL and the set of items consistent with theoretically derived expectations?

Can occurrence of the criterion (ergonomic hazard) be validated by comparison with another method?
Are the results from different users of the PLIBEL method consistent when observing the same working situation?
How has the method been used? What are the experiences?

PLIBEL has been translated into several languages, including English, Dutch, French, Spanish (Serratos-Pérez and Kemmlert, 1998), and Greek (Serratos-Pérez and Kemmlert, 1998).

Observational findings have provided a basis for recommended improvements, for discussion of ergonomics problems, and for work-site education. Moreover, PLIBEL has been used for ergonomics education both in industry and in the Swedish school system.

3.10   Tools Needed

Paper, pencil, a folding rule, and a camera are sufficient for ordinary workplace observations and for initial identification of ergonomic hazards.

References

Carmines, E.G. and Zeller, R.A. (1979), *Reliability and Validity Assessment*, Sage Publications, London, pp. 1–71.

Drury, C.G. (1990), *Evaluation of Human Work*, Wilson, J.R. and Corlett, E.N., Eds., Taylor & Francis, London, pp. 35–57.

Kemmlert, K. (1995), A method assigned for the identification of ergonomic hazards — PLIBEL, *Appl. Ergonomics*, 126, 199–211.

Kemmlert, K. (1996a), Prevention of occupational musculo-skeletal injuries, *Scand. J. Rehabil. Med.*, Suppl. 35, 1–34.

Kemmlert, K. (1996b), New Analytic Methods for the Prevention of Work-Related Musculoskeletal Injuries: Fifteen Years of Occupational-Accident Research in Sweden, Swedish Council for Work Life Research, Stockholm, pp. 176–185.

Kemmlert, K. (1997), On the Identification and Prevention of Ergonomic Risk Factors, Ph.D. thesis, National Institute of Occupational Health, Luleå University of Technology, Sweden.

Rohmert, W. and Landau, K. (1983), *A New Technique for Job Analysis*, Taylor & Francis, London, pp. 1–95.

Serratos-Pérez, N. and Kemmlert, K. (1998), Assessing ergonomic conditions in industrial operations: a field for global cooperation, *Asian Pac. Newsl. Occup. Health Saf.*, 5, 67–69.

4
Musculoskeletal Discomfort Surveys Used at NIOSH

Steven L. Sauter
NIOSH

Naomi G. Swanson
NIOSH

Thomas R. Waters
NIOSH

Thomas R. Hales
NIOSH

Robin Dunkin-Chadwick
NIOSH

4.1 Background

Self-report measures of musculoskeletal discomfort are widely used and generally accepted as a proxy or risk factor for musculoskeletal disorders in epidemiologic research and workplace health surveillance. Discomfort measures are also commonly used to evaluate ergonomic interventions, or as a screening tool in the context of hazard surveillance to detect exposures to workplace physical stressors.

As popularized by the classic work of Corlett and Bishop (1976), the most familiar forms of musculoskeletal discomfort surveys employ body maps together with rating scales for assessing attributes of discomfort in multiple regions of the body. Cameron (1996) and Straker (1999) provide excellent reviews of the literature on measurement of body part discomfort, including the wide range of methods employed.

Of the many methods for surveying musculoskeletal discomfort, few have been used repeatedly in a standardized fashion. Exceptions include the Standardized Nordic Questionnaire (SNQ) (Kourinka et al., 1987) and the University of Michigan Upper Extremity Questionnaire (UMUEQ) (Franzblau et al., 1997). These two instruments are similar in many respects to the discomfort surveys used by the National Institute for Occupational Safety and Health (NIOSH).

Table 4.1 describes characteristics of major NIOSH studies of musculoskeletal discomfort that have been conducted in the last decade, including laboratory and epidemiologic investigations and workplace health hazard evaluations. Most of these studies (all but 8, 10, 20) were conducted in actual workplaces and involved work with video display terminals, meat processing, and the handling of grocery products. Nine studies were prospective in nature (8, 10, 13, 16, 17, 19, 20, 22, 23) and featured various interventions with follow-up, primarily in field settings. Physical demands and task design were examined in all of the studies. Broader psychosocial aspects of the job (e.g., participation in decision making, social support, job satisfaction) were also investigated in ten studies (3, 7, 9, 11, 12, 14, 15, 18, 19, 21), and in three of these studies (3, 7, 11), multiple regression models were able to discern unique effects of psychosocial

TABLE 4.1   Representative NIOSH Studies Employing Body-Part Discomfort Measures

Studies [a]	Type of Work Investigated	Factors Studied [b]	Study Design [c]	Body Regions Targeted [d]	Severity Measures [e]
1. NIOSH (1989a)	poultry processing	pt	c	ue	d, f
2. NIOSH (1989b)	beef processing	pt	c	ue	d, f
3. NIOSH (1990a)	newspaper/VDT	pt, ps	c	ue, b	d, f
4. NIOSH (1990b)	poultry processing	pt	c	ue	d, f
5. Sauter, Schleifer, and Knutson (1991)	data processing/VDT	pt	c	ue, le	d, f
6. NIOSH (1991); Baron and Habes (1992)	grocery scanning	pt	c	ue, b	f
7. NIOSH (1992); Hales, Sauter, Petersen, Fine, Putz-Anderson, and Schleifer (1994)	telecommunications/VDT	pt, ps	c	ue	d, f , i
8. Sauter and Swanson (1992)	data processing/VDT (lab)	pt	pi	ue, le, b	i
9. NIOSH (1993a)	grocery warehouse	pt, ps	c	ue, le, b	i
10. Swanson and Sauter (1993)	data processing/VDT (lab)	pt	pi	ue, le, b	i
11. NIOSH (1993b); Bernard, Sauter, Fine, Petersen, and Hales (1994)	newspaper/VDT	pt, ps	c	ue, b	d, f , i
12. NIOSH (1994); Hoekstra, Hurrell, Swanson, and Tepper (1996)	customer service/VDT	pt, ps	c	ue, b	d, f, i
13. Becker, Swanson, Sauter, and Galinsky (1995)	data processing/VDT	pt	pi	ue, le, b	i
14. NIOSH (1995)	grocery warehouse	pt, ps	c	ue, le, b	i
15. NIOSH (1996a)	medical laboratory	pt, ps	c	ue, b	d, f, i
16. NIOSH (1996b)	beverage distribution	pt	pi	ue, le, b	i
17. Galinsky, Swanson, Sauter, Hurrell, and Dunkin (1997)	data processing/VDT	pt	pi	ue, le, b	i
18. NIOSH (1997)	small appliance manufacturing	pt, ps	c	ue, le, b	d, f, i
19. Sauter, Swanson, Conway, Lim, and Galinsky (1997)	data processing/VDT	pt, ps	pi	ue, le, b	d, f, i
20. Swanson, Galinsky, Cole, Pan, and Sauter (1997)	data processing/VDT (lab)	pt	pi	ue, le, b	i
21. NIOSH (1998)	textile manufacturing	pt, ps	c	ue, le, b	i
22. Galinsky, Swanson, Sauter, Hurrell, and Schleifer (2000)	data processing/VDT	pt	pi	ue, le, b	i
23. Lowe, Moore, Swanson, Perez, and Alderson (2001)	clerical/VDT	pt	pi	ue, b	d, f, i

[a] Second entry in rows denotes journal publication of the study.
[b] pt = physical demands and task design; ps = psychosocial factors.
[c] c = cross sectional; pi = prospective with intervention.
[d] ue = upper extremity; le = lower extremity; b = back.
[e] d = duration; f = frequency; i = intensity.

factors on discomfort outcomes. Ten of the studies (1–4, 6, 7, 11, 12, 15, 23) focused on the upper extremities alone or together with back discomfort. All of the remaining studies examined discomfort in both the upper and lower extremities. In total, musculoskeletal discomfort data have been collected from nearly 6,000 subjects in these studies.

4.2   Discomfort Survey Methods at NIOSH

Nearly all the NIOSH studies listed in Table 4.1 employed surveys that combined body maps and rating scales to assess discomfort in multiple regions of the body. The discomfort survey employed in the NIOSH (1993b) study of newspaper workers illustrates the body maps and rating scales used in many of the NIOSH studies (this report and the study survey can be viewed and printed from http://www.cdc.gov/niosh/hhe/reports/pdfs/1990-0013-2277.pdf). Although most NIOSH studies have shared common survey elements, there have been some variations in the way body regions were mapped and in the measures used for discomfort ratings.

The body maps used in many NIOSH studies are very close to standardized diagrams used to distinguish various upper- and lower-extremity body regions in the SNQ (neck, shoulders, elbow, wrists-hands, upper and lower back, hips/thighs, knees, ankles/feet), in contrast to the UMUEQ, which employs verbal descriptors to distinguish body regions (a diagram is used only to localize discomfort in the hand). However, discomfort in different body regions is characterized in NIOSH surveys using procedures more similar to the UMUEQ, which captures richer information on discomfort attributes (e.g., intensity and temporal aspects) than the SNQ.

Except for four investigations (8, 10, 16, 20), where surveys were self-administered by computer, in all of the studies paper surveys were administered individually or in small groups by a research team.

4.2.1   Defining the Location of Discomfort

Musculoskeletal discomfort surveys collect information on the location of discomfort by reference to specific body regions or by use of partial- or whole-body diagrams that designate specific regions to be assessed. Less commonly, body maps are shaded by respondents to identify regions of discomfort. The number of regions targeted varies in relation to the interests of the study. A general purpose survey proposed by Cameron (1996) targeted over 100 regions, involving permutations of the left, right, front, and back sides of the body.

The ten NIOSH studies that focused mainly on upper-extremity discomfort targeted the same upper-extremity sites as the SNQ (neck, shoulders, elbows, wrists-hands) but, unlike the SNQ and UMUEQ, discomfort assessment did not differentiate the left and right side of the body in nine of these studies. The 13 remaining NIOSH studies that targeted both the upper and lower extremities evaluated discomfort on the left and right sides of the body separately. Except for some small clusters of studies that used identical body maps, these remaining studies exhibited considerable variation in body regions targeted. One NIOSH study (1996b) separately mapped the front and back and left and right sides of the body, similar to Cameron (1996). Differences in regions targeted in NIOSH studies were dictated to some extent by the physical stressors under investigation. For example, the nine studies targeting upper-extremity regions only were focused on food processing and other tasks involving exertions of mainly the arms and hands.

Two different display formats have been used for identifying body parts in the NIOSH studies. For nearly one half of the studies, including all of the ten upper-extremity studies, partial-body diagrams provided multiple views of designated regions of interest, as illustrated in Figure 4.1 (top). Each of these targeted regions was accompanied in the survey with a series of questions and rating scales (described below) for assessing multiple facets of discomfort at that location. In most of the remaining studies (which surveyed both upper- and lower-extremity discomfort), only a single attribute of discomfort (usually intensity) was rated. Thus it was possible to target all regions of interest in a single integrated diagram, with a space for recording ratings contiguous to each designated region, as illustrated in Figure 4.1 (bottom). Similar to the SNQ, rear-view perspectives of the body are presented in most of these

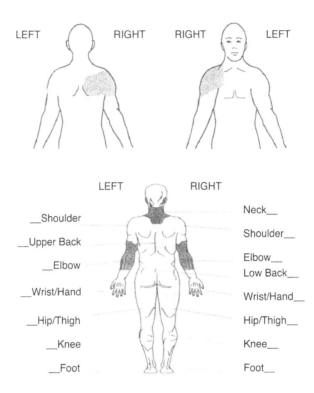

FIGURE 4.1 Example of partial- and whole-body diagrams used to target discomfort assessments in NIOSH studies.

whole-body diagrams. Two NIOSH studies (Galinsky et al. 1997, 2000) used only verbal descriptions of body sites to target discomfort assessments, similar to the UMUEQ.

4.2.2 Assessing the Nature of Discomfort

Most discomfort surveys, including the SNQ, use descriptors such as "pain," "bother," "problems," and "discomfort" without further definition of these conditions to screen for the presence of an untoward state within a specified time period (usually one year), and then rate these conditions using various severity indicators. The UMUEQ similarly asks about the presence and severity of a "problem" in a specific location, but also asks the respondent to qualify the problem in terms of the types of symptoms experienced (burning, stiffness, tingling, etc.). Discomfort surveys used in most of the NIOSH studies in Table 4.1 take a slightly different approach. Rather than beginning by asking about the presence of discomfort or a similar condition in general terms, the survey begins with a single question that screens for the presence of one or more of six symptoms (pain, aching, stiffness, burning, numbness, or tingling) in each body region. An affirmative response is then followed by a rating of this "problem" using as many as three severity measures (duration, frequency, and intensity).

As shown in Table 4.1, all but six of the NIOSH studies (Baron and Habes, 1992; NIOSH, 1989a, 1989b, 1990a, 1990b; Sauter et al., 1991) rated the intensity of discomfort, over one half of the studies used two or more severity measures, and nearly one third of the studies used all three measures. Table 4.2 describes the actual discomfort rating scales most commonly used in the NIOSH studies.

The discomfort duration scales used in NIOSH studies were adapted from work at the University of Michigan (Silverstein and Fine, 1984; Silverstein, 1985), and are similar to the discomfort duration scale in the UMUEQ. Eleven of the 12 NIOSH studies (all but study 1) that rated discomfort duration used the scale shown in Table 4.2 or a close variation of this scale. Nine of the 13 studies (all but 1, 2, 5, 18) that rated discomfort frequency used the Table 4.2 scale or a slightly altered version of this scale. Less consistency is seen in the rating of discomfort intensity among NIOSH studies. Only six studies (7, 11,

TABLE 4.2   Discomfort Rating Scales Commonly Used in NIOSH Studies

Discomfort Duration	Discomfort Frequency	Discomfort Intensity
Less than 1 hour	Almost never (every 6 months)	No pain
1 to 24 hours	Rarely (every 2 to 3 months)	Mild
25 hours to 1 week	Sometimes (once a month)	Moderate
More than 1 to 2 weeks	Frequently (once a week)	Severe
More than 2 weeks to 1 month	Almost always (daily)	Worst pain ever in life
More than 1 to 2 months		
More than 3 months		

12, 15, 18, 23) used the intensity scale shown in Table 4.2. Eight other studies employed verbal-numeric scales with four to six intensity levels ranging from "comfortable" or "no pain/discomfort" at one extreme to "very/extreme/severe discomfort" or "worst pain ever imaginable" at the other extreme. Similar to the UMUEQ, two studies (16, 20) used ten-point numeric scales for rating discomfort intensity. But with these two exceptions, neither the frequency nor intensity scales used by NIOSH have close parallels in either the SNQ or UMUEQ.

Several NIOSH surveys also incorporated questions asking about medical follow-up of discomfort and effects of discomfort on performance, and a series of questions to ascertain the work-relatedness of discomfort (e.g., onset in relation to current employment; prior traumatic injury and underlying medical conditions). Similar items appear in the UMUEQ, whereas the SNQ asks only about medical follow-up and activity restrictions.

Nearly one half of the NIOSH studies (1–4, 6, 7, 11, 12, 15, 18) used discomfort ratings together with information from questions on the work-relatedness of discomfort to define cases of work-related musculoskeletal "disorders" and to examine the prevalence of disorders and their relationships to various job factors. The case definition most commonly employed by NIOSH required satisfaction of all of the following criteria:

- Discomfort within the past year
- Discomfort began after employment in the current job
- No prior accident or sudden injury (affecting focal area of discomfort)
- Discomfort episodes occur monthly or the duration exceeds a week

Seven studies (3, 6, 7, 11, 12, 15, 18) used this definition or a close variant. As an additional step, in seven of the NIOSH studies (1, 2, 4, 7, 9, 11, 23), physical examinations were conducted. Positive findings were used together with symptom information to derive cases for calculation of incidence rates of musculoskeletal disorders and for statistical analysis of these disorders in relation to the work situation.

4.3   Quality of NIOSH Discomfort Survey Methods

Whether for epidemiologic research or surveillance purposes, discomfort surveys need to be practical to use (i.e., quick and easy to administer in a variety of populations and workplaces, readily analyzed, etc.). They also need to demonstrate acceptable psychometric properties (reliability and validity). However, with just a few exceptions (e.g., recent evaluations of the reliability and validity of the UMUEQ), these aspects of discomfort surveys have received surprisingly little study.

In a recent examination of musculoskeletal symptom surveys, NIOSH researchers (Baron et al., 1996) cite the widespread use of the SNQ and NIOSH surveys as evidence of their practical quality. A cursory literature search reveals over three dozen studies in many countries since 1990 that relied on the SNQ or a variation of this instrument. Additionally, Baron et al. (1996) note that the typical NIOSH survey requires an average of just 30 minutes to administer and has been used among thousands of workers in occupations with widely varying literacy requirements.

4.3.1   Reliability

Test-retest reliability data on NIOSH discomfort surveys have not been previously reported. However, recent analysis of data from repeat administration (within 48 hours) of the survey employed by Lowe et al. (2001) produced encouraging findings in a sample of 89 office workers. Identical responses across survey administrations were examined for items denoting discomfort (yes/no) during the last seven days in eight upper-extremity regions. Agreement rates across survey administrations for these items ranged from 91 to 99% (mode = 93%). Kappa values ranged from 0.75 to 0.89 for seven items (0.95 for the eighth item). Strong inflation of these agreement rates and kappa values relating to memory effects might be suspected owing to the short test-retest interval. However, these agreement rates and kappa values compare very favorably with findings reported in two reliability studies of the UMUEQ (Franzblau et al., 1997; Salerno et al., 2001) where the test-retest interval was three weeks for items asking about the presence of upper-extremity discomfort within the past year. For example, agreement rates for the left- and right-hand discomfort questions were 92% and 96% (kappa = 0.82 and 0.87), respectively, in the retest of the NIOSH survey, and the agreement rates in the retest of the hand discomfort item (both hands together) in the UMUEQ studies were 93% in each study (kappa = 0.84 and 0.86 in the two studies).

In addition to studies of the UMUEQ, test-retest reliability studies of the SNQ are described by Kourinka et al. (1987) and Dickinson et al. (1992). One-week, 15-day, and 3-week retests in these studies resulted in identical response rates across questions ranging from 70 to 100%. Additionally, van der Grinten (1991) reports good retest reliability of a body-part discomfort survey over a fortnight. However, the absence of statistical analysis in these studies limits interpretation.

4.3.2   Validity

Like "fatigue" or "effort," discomfort is a psychological construct. As such, the validity of a discomfort measure can be assessed by judging the appropriateness of the items comprising the measure (content validation) and by examining its association with other measures that should in theory be related to discomfort (construct validation).

Many of the discomfort surveys employed in NIOSH studies sample a wide domain of discomfort attributes, including multiple qualities of discomfort and multiple indicators of severity, thereby suggesting strong content validity of these surveys. Baron et al. (1996) reported significant correlations (0.27 to 0.38) among duration, frequency, and intensity measures of hand discomfort in NIOSH surveys, indicating a common underlying construct. Analyses of data subsequently collected in the context of two other NIOSH studies (Lowe et al., 2001; Sauter et al., 1997) show even stronger correlations among these variables for hand discomfort (0.39 to 0.64 and 0.31 to 0.59, respectively, for these two studies).

With regard to construct validity of NIOSH surveys, two types of evidence are relevant. Perhaps most compelling is that discomfort measures proved sensitive to ergonomic factors investigated in an overwhelming majority of NIOSH studies in the last decade. These factors varied widely, from heavy lifting in grocery warehouses to brief rest pauses in data processing.

Additionally, associations between NIOSH discomfort survey measures and other health-relevant outcomes are found in Baron et al. (1996) and in analyses of data from the Lowe et al. (2001) sample. Baron et al. (1996) reported significantly increased odds of seeking medical care among workers with elevated duration, frequency, and intensity of hand discomfort (odds ratio [OR] = 2.1). Similar results were found for analyses of discomfort severity measures and medical-care seeking in a sample of nearly 200 office workers in the Lowe et al. (2001) study (OR = 2.7 for hands and 1.6 to 4.0 for other regions). Using a survey similar to NIOSH surveys, Marley and Kumar (1994) were able to predict medical-care seeking with 82% sensitivity from an algorithm based on musculoskeletal discomfort frequency and duration (although specificity was just 56%).

Baron et al. (1996) also reported reasonable sensitivity (71%) and specificity (72%) of elevated hand discomfort in detecting cases of hand disorders as defined by physical exam. Additionally, Lowe et al. (2001) found that NIOSH measures of upper-limb discomfort were significant predictors (more so than electromyography [EMG] measures of upper-limb loading) of upper-limb disorders defined by physical exam.

Related studies have also been conducted by Scandinavian and University of Michigan researchers. Using the SNQ, Ohlsson et al. (1994) found similar levels of sensitivity and specificity as Baron et al. (1996) for detecting upper-extremity diagnoses, although the sensitivity for hand disorders was lower (67%). Michigan researchers (Homan et al., 1999) found poor agreement between carpal tunnel cases as defined by the UMUEQ, physical examinations, and electrodiagnostic criteria. However, based on analyses of these data, the investigators concluded that discomfort surveys were the only procedures that could be used alone for surveillance of carpal tunnel syndrome in the workplace.

4.4 Summary and Implications

Surveys of musculoskeletal discomfort vary widely along many dimensions: the time frame for assessment (e.g., last year, last 30 days, last 7 days, current discomfort); assessment of qualitative aspects (pain or problem vs. specific symptoms); assessment of quantitative aspects (intensity and temporal characteristics) and scaling methods employed (from binary yes/no choices to Borg scales); and the derivation of summary indices of discomfort (ranging from region-specific cases of musculoskeletal disorders to continuous measures of whole-body discomfort). Many of these variations can be seen when comparing versions of the NIOSH survey, the UMUEQ, and variations of the SNQ.

It is of interest that, irrespective of these design differences, musculoskeletal discomfort surveys have proved remarkably effective in ergonomics applications. Research has shown the NIOSH surveys, UMUEQ, and SNQ to be sensitive to a wide range of physical stressors across many occupations, and to have prognostic value for more objective measures of musculoskeletal disorders. Psychosocial factors believed to influence musculoskeletal disorders are also associated with discomfort in NIOSH studies (Bernard et al., 1994; Hales et al., 1994). These findings provide strong convergent evidence of the validity and robustness of discomfort surveys.

The diversity of discomfort surveys, however, raises the question about the best survey measures of discomfort. This question cannot be answered without specifying the criterion measure (i.e., the standard for musculoskeletal disorders against which the measure will be judged). While no gold standard exists, physical signs, electrodiagnostic findings, and disability represent common outcomes of interest. Surprisingly few studies, however, have examined the relationship between the design of discomfort surveys and their predictive power for these different outcomes. In one such study, Homan et al. (1999) evaluated combinations of hand discomfort measures obtained with the Michigan questionnaire (recurrent symptoms of the hand-wrist-fingers, current symptoms, nocturnal symptoms, symptom intensity, and hand diagrams scores) for their relationship to electrodiagnostic evidence of carpal tunnel syndrome in a working population. Of interest, recurrent symptoms alone proved the best predictor of electrical abnormalities. Further investigations of this nature may lead to improvements in the design of musculoskeletal discomfort surveys by enabling researchers to optimize survey content and economy in relation to predictive capacity (sensitivity, specificity, positive and negative predictive value) for different outcomes.

References

Baron, S., Hales, T., and Hurrell, J. (1996), Evaluation of symptom surveys for occupational musculoskeletal disorders, *Am. J. Ind. Med.*, 29, 609–617.

Baron, S. and Habes, D. (1992), Occupational musculoskeletal disorders among supermarket cashiers, *Scand. J. Work Environ. Health*, 18, 127–129.

Becker, A., Swanson, N., Sauter, S., and Galinsky, T. (1995), Compatibility of Job Rotation Subtasks in Data Entry Work, poster session presented at the 39th Annual Meeting of the Human Factors and Ergonomics Society, Oct. 1995, San Diego, CA.

Bernard, B., Sauter, S., Fine, L., Petersen, M., and Hales, T. (1994), Job task and psychosocial risk factors for work-related musculoskeletal disorders among newspaper employees, *Scand. J. Rehabil. Med.*, 15, 417–426.

Cameron, J.A. (1996), Assessing work-related body-part discomfort: current strategies and a behaviorally oriented assessment tool, *Int. J. Ind. Ergonomics*, 18, 389–398.

Corlett, E. and Bishop, R. (1976), A technique for assessing postural discomfort, *Ergonomics*, 19, 175–182.

Dickinson, C., Campion, K., Foster, A., Newman, S., O'Rourke, A., and Thomas, P. (1992), Questionnaire development: an examination of the Nordic Musculoskeletal Questionnaire, *Appl. Ergonomics*, 23, 197–201.

Franzblau, A., Salerno, D., Armstrong, T., and Werner, R. (1997), Test–retest reliability of an upper-extremity discomfort questionnaire in an industrial population, *Scand. J. Work Environ. Health*, 23, 299–307.

Galinsky, T., Swanson, N., Sauter, S., Hurrell, J., and Dunkin, R. (1997), Discomfort and Performance Effects of Supplemental Restbreaks for Data Entry Operators: Empirical Evidence from Two Worksites, paper presented at HCI International '97, San Francisco, August 1997.

Galinsky, T., Swanson, N., Sauter, S., Hurrell, J., and Schleifer, L. (2000), A field study of supplementary rest breaks for data-entry operators, *Ergonomics*, 43, 622–638.

Hales, T., Sauter, S., Petersen, M., Fine, L., Putz-Anderson, V., and Schleifer, L. (1994), Musculoskeletal disorders among visual display terminal users in a telecommunications company, *Ergonomics*, 37, 1603–1621.

Hoekstra, E., Hurrell, J., Swanson, N., and Tepper, A. (1996), Ergonomic, job task, and psychosocial risk factors for work-related musculoskeletal disorders among teleservice center representatives, *Int. J. Hum.–Comput. Interact.*, 8, 421–431.

Homan, M., Franzblau, A., Werner, R., Albers, J., Armstrong, T., and Bromberg, M. (1999), Agreement between symptom surveys, physical examination procedures and electrodiagnostic findings for carpal tunnel syndrome, *Scand. J. Work Environ. Health*, 25, 115–124.

Kourinka, I., Jonsson, B., Kilbom, A., Vinterberg, H., Biering-Sorensen, F., Andersson, G., and Jorgensen, K. (1987), Standardized Nordic questionnaire for the analysis of musculoskeletal symptoms, *Appl. Ergonomics*, 18, 233–237.

Lowe, B., Moore, J.S., Swanson, N., Perez, L., and Alderson, M. (2001), Relationships between upper limb loading, physical findings, and discomfort associated with keyboard use, in *Proceedings of the Human Factors and Ergonomics Society 45th Annual Meeting*, Santa Monica, CA, pp. 1087–1091.

Marley, R. and Kumar, N. (1994), An improved musculoskeletal discomfort assessment tool, in *Advances in Industrial Ergonomics and Safety VI*, Aghazadeh, F., Ed., Taylor & Francis, London, pp. 45–52.

NIOSH (1989a), Health evaluation and technical assistance report: Cargill Poultry Division, Buena Vista, GA, Report HETA 89-251-1997, U.S. Department of Human Services, Public Health Service, Centers for Disease Control and Prevention, National Institute for Occupational Safety and Health, Cincinnati.

NIOSH (1989b), Health evaluation and technical assistance report: John Morrell & Co., Sioux Falls, SD, Report HETA 88-180-1958, U.S. Department of Human Services, Public Health Service, Centers for Disease Control and Prevention, National Institute for Occupational Safety and Health, Cincinnati.

NIOSH (1990a), Health evaluation and technical assistance report: Newsday, Inc., Melville, NY, Report HETA 89-250-2046, U.S. Department of Human Services, Public Health Service, Centers for Disease Control and Prevention, National Institute for Occupational Safety and Health, Cincinnati.

NIOSH (1990b), Health evaluation and technical assistance report: Perdue Farms, Inc., Lewington, NC, Robersonville, NC, Report HETA 89-307-2009, U.S. Department of Human Services, Public Health Service, Centers for Disease Control and Prevention, National Institute for Occupational Safety and Health, Cincinnati.

NIOSH (1991), Health evaluation and technical assistance report: Shoprite Supermarkets, New Jersey–New York, Report 88-344-2092, U.S. Department of Human Services, Public Health Service, Centers for Disease Control and Prevention, National Institute for Occupational Safety and Health, Cincinnati.

NIOSH (1992), Health evaluation and technical assistance report: US West Communications, Phoenix, AR, Minneapolis, MN, Denver, CO, Report HETA 89-299-2230, U.S. Department of Human Services, Public Health Service, Centers for Disease Control and Prevention, National Institute for Occupational Safety and Health, Cincinnati.

NIOSH (1993a), Health evaluation and technical assistance report: Big Bear grocery warehouse, Report HETA 91-405-2340, U.S. Department of Human Services, Public Health Service, Centers for Disease Control and Prevention, National Institute for Occupational Safety and Health, Cincinnati.

NIOSH (1993b), Health evaluation and technical assistance report: Los Angeles Times, Los Angeles, CA, Report HETA 90-013-2277, U.S. Department of Human Services, Public Health Service, Centers for Disease Control and Prevention, National Institute for Occupational Safety and Health, Cincinnati.

NIOSH (1994), Health evaluation and technical assistance report: Social Security Administration teleservice centers, Boston, MA, Fort Lauderdale, FL, Report HETA 92-0382-2450, U.S. Department of Human Services, Public Health Service, Centers for Disease Control and Prevention, National Institute for Occupational Safety and Health, Cincinnati.

NIOSH (1995), Health evaluation and technical assistance report: Kroger grocery warehouse, Nashville, TN, Report HETA 93-0920-2548, U.S. Department of Human Services, Public Health Service, Centers for Disease Control and Prevention, National Institute for Occupational Safety and Health, Cincinnati.

NIOSH (1996a), Health evaluation and technical assistance report: Scientific Application International Corp., Frederick, MD, Report HETA 95-0294-2594, U.S. Department of Human Services, Public Health Service, Centers for Disease Control and Prevention, National Institute for Occupational Safety and Health, Cincinnati.

NIOSH (1996b), Ergonomics Interventions for the Soft Drink Beverage Delivery Industry, Publication 96-109, U.S. Department of Human Services, Public Health Service, Centers for Disease Control and Prevention, National Institute for Occupational Safety and Health, DHHS, Cincinnati.

NIOSH (1997), Health evaluation and technical assistance report: Frigidare laundry products, Webster City, IA, Report HETA 95-0154, U.S. Department of Human Services, Public Health Service, Centers for Disease Control and Prevention, National Institute for Occupational Safety and Health, Cincinnati.

NIOSH (1998), Health evaluation and technical assistance report: Foss Manufacturing Company, Inc, Hampton, NH, Report HETA 96-0258-2673, U.S. Department of Human Services, Public Health Service, Centers for Disease Control and Prevention, National Institute for Occupational Safety and Health, Cincinnati.

Ohlsson, K., Attewell, R., Johnsson, B., Ahlm, A., and Skerfving, S. (1994), An assessment of neck and upper extremity disorders by questionnaire and clinical examination, *Ergonomics*, 37, 891–897.

Salerno, D., Franzblau, A., Armstrong, T., Werner, R., and Becker, M. (2001), Test-retest reliability of the upper extremity questionnaire among keyboard operators, *Am. J. Ind. Med.*, 40, 655–666.

Sauter, S., Schleifer, L., and Knutson, S. (1991), Work posture, workstation design, and musculoskeletal discomfort in a VDT data entry task, *Hum. Factors*, 33, 151–167.

Sauter, S., Swanson, N., Conway, F., Lim, S., and Galinsky T. (1997), Prospective Study of Restbreak Interventions in Keyboard Intensive Work, paper presented at the Marconi '97 Research Conference, April 1997, Marshall, CA.

Sauter, S. and Swanson, N. (1992), The Effects of Frequent Rest Breaks on Performance and Musculoskeletal Comfort in Repetitive VDT Work, paper presented at the Work with Display Units '92 International Conference, September 1992, Berlin.

Silverstein, B. and Fine, L. (1984), Evaluation of Upper Extremity and Low Back Cumulative Trauma Disorders, School of Public Health, University of Michigan, Ann Arbor.

Silverstein, B. (1985), The Prevalence of Upper Extremity Cumulative Trauma Disorders in Industry, Ph.D. dissertation, University of Michigan, Ann Arbor.

Straker, L.M. (1999), Body discomfort assessment tools, in *The Occupational Ergonomics Handbook*, Karwowski, W. and Marras, W.S., Eds., CRC Press, Boca Raton, FL, pp. 1239–1252.

Swanson, N., Galinsky, L., Cole, L., Pan, C., and Sauter, S. (1997), The impact of keyboard design on comfort and productivity in a text-entry task, *Appl. Ergonomics*, 28, 9–16.

Swanson, N.G. and Sauter, S.L. (1993), The effects of exercise on the health and performance of data entry operators, in *Work with Display Units 92*, Luczak, H., Cakir, A., and Cakir, G., Eds., North Holland, Amsterdam, pp. 288–291.

van der Grinten, M.P. (1991), Test-retest reliability of a practical method for measuring body part discomfort, in *Designing for Everyone*, Proceedings of the 11th Congress of the International Ergonomics Association, Quennec, Y. and Daniellou, R., Eds., Taylor & Francis, Paris, pp. 54–56.

5

The Dutch Musculoskeletal Questionnaire (DMQ)

Vincent H. Hildebrandt
TNO Work and Employment
Body@Work Research Center

5.1   Background and Application

The Dutch musculoskeletal questionnaire (DMQ) allows ergonomists and occupational health professionals to measure work-related musculoskeletal risk factors and symptoms in worker populations in a quick yet standardized way. The standard version of the questionnaire consists of 9 pages with around 25 questions per page, to be filled in by the workers themselves. Completion time is around 30 minutes. A short version (4 pages) and an extended version (14 pages) are also available. The questionnaire includes the following sections:

Background variables: age, gender, education, duration of employment, work history, shift work.
Tasks: prevalence rates and perceived heaviness of task demands.
Musculoskeletal workload: postures, forces, movements.
Work pace and psychosocial working conditions: demands, control and autonomy, work organization
        and social support, work satisfaction; such factors may play an important role for workers with
        musculoskeletal disorders (Bongers et al., 1993).
Health: in particular musculoskeletal symptoms; the phrasing of questions on prevalence is comparable
        with the Nordic questionnaire on musculoskeletal disorders (Kuorinka et al., 1987), including the
        definition of areas of the body pictorially; in addition, the extended version contains more-detailed
        questions on the nature and severity of these symptoms.
Lifestyle: e.g., sports, smoking (in the extended version of the questionnaire only).
Perceived bottlenecks and ideas for improvements: suggested by the workers themselves (optional).

The questionnaire seeks to obtain a simple representation of the relationships between work tasks and musculoskeletal symptoms (Dul et al., 1992; Paul, 1993). Work-related musculoskeletal symptoms are seen as the result of high internal physical loads, caused by the poor postures, movements, and forceful exertions needed to perform the work tasks. Other factors, such as other working conditions, individual factors (gender, age), psychosocial aspects, or lifestyle can also influence these relationships.

To ensure an optimal content validity of the questionnaire, the choice of variables to be measured was based on reviews of the epidemiological literature (Hildebrandt, 1987; Riihimäki, 1991; Stock, 1991; Walsh et al., 1989) that identified a large number of potentially harmful postures, movements, force exertions, and other potentially hazardous working conditions (Ariëns, 2000; Bernard, 1997; Hoogendoorn, 1999).

Musculoskeletal workload (postures, forces, movements) is addressed in 63 questions. These questions are categorized into seven indices and four separate questions, as seen in Table 5.1.

The questions are formulated to indicate the presence or absence of exposure and not the amount of discomfort caused by the exposure, which is addressed in a separate part of the questionnaire. The precise formulation is based on several field studies using preliminary versions of the questionnaire. For brevity, the exposures addressed in the questions are not defined, explained, or illustrated. No training is necessary to complete the questionnaire. Most questions require dichotomous answers (yes/no). This qualitative approach, does not quantify frequency and duration of variables. The validity of quantitative MSD (musculoskeletal disorders) surveys has been seriously questioned (Baty et al., 1987; Kilbom, 1994; Kumar, 1993; Viikari-Juntura et al., 1996; Wiktorin et al., 1993; Winkel and Mathiassen, 1994).

Because participation of the workers is required, the DMQ fits well into a participatory ergonomics approach to problems (Vink et al., 1992). A full copy of the DMQ can be downloaded from http://www.workandhealth.org; search for "DMQ."

TABLE 5.1   Name, Content, and Cronbach's Alpha of Seven Indices and Four Separate Questions

	Name	Content	n [a]	Alpha [b]
1	Force exertion	Lifting, pushing and pulling, carrying, forceful movements with arms, high physical exertion, lifting in unfavorable postures, lifting with the load away from the body, lifting with twisted trunk, lifting with loads above the chest, lifting with bad grip, lifting with very heavy loads, short-force exertions, exerting great force on hands	13	0.90
2	Dynamic loads	Trunk movements (bending and/or twisting); movements of neck, shoulders or wrists; reaching; making sudden and/or unexpected movements; pinching; working under, at, or above shoulder level	12	0.83
3	Static loads	Lightly bent, twisted trunk posture; heavily bent, twisted trunk posture; postures of neck or wrists	11	0.87
4	Repetitive loads	Working in the same postures; making the same movements with trunk, arms, hands, wrists or legs; making small movements with hands at a high pace	6	0.85
5	Ergonomic environment	Available working space, no support, slipping and falling, trouble with reaching things with tools, not enough room above to perform work without bending	6	0.78
6	Vibration	Whole-body vibration, vibrating tools, driving	3	0.57
7	Climate	Cold, draught, changes in temperature, moisture	4	0.84
	Uncomfortable postures	Having often to deal with uncomfortable postures at work	1	—
	Sitting	Sitting often at work	1	—
	Standing	Standing often at work	1	—
	Walking	Walking often at work	1	—
	Overall index	Indices 1–7	55	0.95

[a] Number of questions. The maximum score of the index equals the number of questions in the index and corresponds to a positive answer to all questions in the index. The higher the mean score, the higher the self-reported exposure.
[b] Cronbach's alpha, a measure of reliability indicating the homogeneity of the index.
Source: Hildebrandt et al., 2001.

5.2 Procedure

5.2.1 Preparation

5.2.1.1 Defining the Population at Risk

Groups of workers should be selected who are performing more or less identical task(s), e.g., occupational groups, departments, companies, or branches. This enables the identification of an association between symptoms found and specific working situations. The selection depends on the questions to be answered. Defining these questions is crucial in this phase.

A reference group known to be exposed to a lesser physical workload should also be used to enable better comparative interpretation of results. A minimum of 20 workers per group is recommended to enable valid conclusions.

5.2.1.2 Introduction of the Project in the Worker Groups Involved

Management support and commitment to implement any recommendations based upon the results is important to ensure high worker response rates. The reasons for the survey as well as its goals and content should be communicated clearly to all involved workers, along with how the results will be communicated and subsequent projects or interventions will be implemented.

5.2.1.3 Analysis of Prevalent Tasks in the Groups

Using existing documents as well as discussions with management and workers, an inventory of the number and type of the most prevalent tasks should be compiled. If tasks are very heterogeneous, groups should be subdivided into more homogeneous units according to their physical workload. Up to nine tasks can be accommodated in the group-specific section of the DMQ.

5.2.1.4 Defining the Way the DMQ Is Administered to the Workers

Three possible options are:

- Postal survey (inexpensive and easy, but risk of low response)
- Distribution in the workplace with a request to complete the questionnaire during or after working hours
- Group sessions during working hours in which each worker is invited to complete his/her questionnaire (recommended)

5.2.2 The Actual Survey

In this phase, the DMQ has to be printed and distributed. Response rates have to be monitored to be able to send reminders to those who do not initially respond. It is extremely important that a large proportion of the workers selected actually participate in the survey, since the results should be representative for all workers with the same tasks.

5.2.3 Data Entry, Data Analysis, and Report

The use of a statistical program to calculate results is recommended. However, a spreadsheet may be sufficient for computing the most important indices. Reporting the results of the survey can be done on the basis of a few tables summarizing the main findings:

- Response and general characteristics of the workers involved (e.g., age, gender, education)
- Tasks: prevalence and perceived exertion
- Physical workload
- Psychosocial workload
- 12-month prevalence of musculoskeletal symptoms per body area

Data should be presented for all worker groups selected and the reference population. If the number of respondents is greater than 20, differences between groups should be statistically tested for significance.

To enable experts to easily work with this questionnaire, a software package (LOQUEST) has been developed for data entry, data analysis, and autoreporting of the main results, but this program (MS-DOS) is only available in Dutch (http://www.arbeid.tno.nl/kennisgebieden/bewegen_bewegingsapparaat/files/loquest.zip).

5.2.4 Implementation of Results

On the basis of the screening results, those worker groups or workplaces requiring a more thorough ergonomic analysis, using more sophisticated methods, can be identified and prioritized. Those worker groups with relatively high symptom rates and/or relatively high workloads will require further ergonomic analysis to determine appropriate ergonomic interventions. Following implementation, a second survey using the DMQ can be conducted to quantify improvements in workload and health among the workers involved.

5.2.5 Advantages

* Standardized method
* Relatively inexpensive and easy
* Broad comprehensive overview of possible risk factors and morbidity (data on both exposure and effect)
* No technical equipment necessary
* Input from workers themselves
* Can be used to evaluate the effects of solutions implemented

5.2.6 Disadvantages

* Self-reported data (detailed exposure measurement not possible)
* Reference group recommended but may not always be available
* Less suitable for smaller groups of workers
* No quantification of risks
* Cooperation of management and workers is crucial
* Data entry can become laborious when groups are large
* More-detailed data analysis requires statistical knowledge

5.2.7 Example Output

Figure 5.1 shows the results of a DMQ survey in five departments of a steel company (Hildebrandt et al., 1996). High risks are easily identified and presented to the management and workers in graphical form. An example of a DMQ survey in agriculture can be found in Hildebrandt (1995).

5.2.8 Related Methods

Ergonomists, occupational physicians, nurses, and hygienists need simple and quick methods to obtain relevant information on work-related factors that contribute to the musculoskeletal workload and related disorders. Based on such screening, priorities can be set for worker groups or workplaces requiring a more thorough ergonomic analysis. A detailed measurement of musculoskeletal workload (postures, movements, and force exertions) by direct methods, like observations or inclinometry, is complicated and time-consuming when large worker groups are involved and skilled analysts are needed for reliable measurements (Buckle, 1987; Hagberg, 1992; Kilbom, 1994; Winkel and Mathiassen, 1994). Simpler screening instruments for identifying groups of workers at risk (jobs, departments, tasks, etc.) include

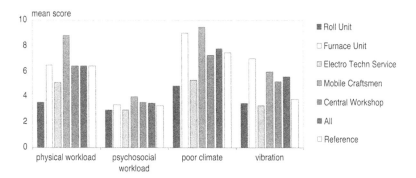

FIGURE 5.1 Mean scores of four indices of workload and working conditions reported by maintenance workers in five departments of a steel company (n = 436) and a reference group of nonsedentary workers (n = 396).

checklists (Keyserling et al., 1992), rating of physical job requirements (Buchholz et al., 1996), surveys (Bishu, 1989), or periodic surveillance (Weel et al., 2000). Although the quantification of the absolute exposure levels has its limitations using these methods, the information gathered can be sufficient to rank groups according to their levels of exposures (Burdorf, 1999). Subsequently, a more laborious detailed ergonomic analysis can be restricted to those high-risk workers and workplaces.

5.2.9   Standard and Regulations

The DMQ can be used to comply with the need for employers to carry out a risk assessment of physical working conditions. However, it should be stressed that a DMQ survey is only a tool to set priorities for further analyses and development of solutions.

5.2.10   Approximate Training and Application Times

No special training is required to conduct a DMQ survey. Some background on epidemiology is recommended to conduct a more detailed data analysis. The time to complete the DMQ (standard edition) is between 20 and 60 min, depending in particular on the educational background of the worker. For workers with less education, the short edition of the DMQ is most appropriate.

5.2.11   Costs

Costs are restricted to the printing (and sometimes the mailing) of the questionnaire. Other costs include the ergonomist's time for preparing, conducting, analyzing, and reporting the survey. This may be up to 120 hours, depending on the amount of effort necessary to motivate the company and workers and the number of workers actually responding to the survey.

5.2.12   Reliability and Validity

Several validity aspects have been addressed by analyses of a database containing data for 1575 workers in various occupations who completed the questionnaire. From factor analysis, questions on musculoskeletal workload and associated potentially hazardous working conditions were grouped into seven indices: force, dynamic load, static load, repetitive load, climatic factors, vibration, and ergonomic environmental factors. Together with four separate questions on standing, sitting, walking, and uncomfortable postures, the indices constitute a brief overview of the main findings on musculoskeletal workload and associated potentially hazardous working conditions. Homogeneity of the indices, assessed by computing Cronbach's alphas, was found to be satisfactory, as was the divergent validity of the indices assessed by computing intercorrelations with an index of psychosocial working conditions. Discriminative power was good: worker groups with contrasting musculoskeletal loads could be differentiated on the basis of the indices. Significant associations of most indices with musculoskeletal symptoms demonstrated concurrent validity (Hildebrandt et al., 2001).

To study the validity of the questions on physical workload, four homogeneous worker groups (VDU [visual display unit] workers, office workers, dispatch workers, and assembly workers) completed the questionnaire and were observed (through video) performing their main tasks. Self-reported exposure to postures was computed for each group, as well as the mean frequency and duration of postures in different categories of trunk flexion and rotation angle. Both methods identified the same groups with the highest exposures to unfavorable postures. Simple qualitative questions seemed adequate (Hildebrandt, 2001). To study the validity of the questions on musculoskeletal symptoms, four other homogeneous worker groups (VDU workers, office workers, vehicle drivers, and printers) were scrutinized; they completed the questionnaire as well as a standardized physical examination. The questionnaire appeared to identify the same worker groups with high prevalence rates of low back pain as the physical examination. Seven-day prevalence rates resulted in the highest specificity, while lifetime prevalence rates resulted in the best sensitivity. Overall, the one-year prevalence rates turned out to be a reasonable intermediate choice (Hildebrandt, 2001).

5.2.13 Tools Needed

The DMQ can be completed using a pen and paper. Data entry requires specialized software. For large groups, data entry using OCR (optical character recognition) software should be considered; for small groups, a spreadsheet program could be sufficient. Statistical software is recommended to analyze data of larger worker groups.

References

Ariëns, G.A.M., Van Mechelen, W., Bongers, P.M., Bouter, L.M., and Van der Wal, G. (2000), Physical risk factors for neck pain, *Scand. J. Work Environ. Health*, 26, 7–19.

Baty, D., Buckle, P.W., and Stubbs, D.A. (1987), Posture recording by direct observation, questionnaire assessment and instrumentation: a comparison based on a recent field study, in *The Ergonomics of Working Postures*, Corlett, E.N., Manenica, I., and Wilson, ! R., Eds., Taylor & Francis, London, pp. 283–292.

Bernard, B.P., Ed. (1997), Musculoskeletal Disorders and Workplace Factors: A Critical Review of Epidemiologic Evidence for Work Related Musculoskeletal Disorders of the Neck, Upper Extremities and Low Back, National Institute for Occupational Safety and Health, U.S. Department of Health and Human Services, Cincinnati.

Bishu, R.R. (1989), Risks of back pain: can a survey help? A discriminant analytic approach, *J. Occup. Accid.*, 11, 51–68.

Bongers, P.M., De Winter, C.R., Kompier, M.A.J., and Hildebrandt, V.H. (1993), Psychosocial factors at work and musculoskeletal disease, *Scand. J. Work Environ. Health*, 19, 297–312.

Buchholz, B., Paquet, V., Punnett, L., Lee, D., and Moir, S. (1996), PATH: a work sampling-based approach to ergonomic job analysis for construction and other non-repetitive work, *Appl. Ergonomics*, 27, 177–187.

Buckle, P. (1987), Musculoskeletal disorders of the upper extremities: the use of epidemiological approaches in industrial settings, *J. Hand Surg.*, 12A, 885–889.

Burdorf, A. (1999), In musculoskeletal epidemiology are we asking the unanswerable in questionnaires on physical load? *Scand. J. Work Environ. Health*, 25, 81–83.

Dul, J., Delleman, N.J., and Hildebrandt, V.H. (1992), Posture and movement analysis in ergonomics: principles and research, in *Biolocomotion: A Century of Research Using Moving Pictures*, Cappozzo, A., Marchetti, M., and Tosi, V., Eds., proceedings of symposium held at Formia, Italy, April 1989, ISB Series Vol. 1, Promograph, Rome.

Hagberg, M. (1992), Exposure variables in ergonomic epidemiology, *Am. J. Ind. Med.*, 21, 91–100.

Hildebrandt, V.H. (1987), A review of epidemiological research on risk factors of low back pain, in *Musculoskeletal Disorders at Work*, Buckle, P., Ed., proceedings of conference held at the University of Surrey, Guildford, U.K., April 1987, Taylor & Francis, London, pp. 9–16.

Hildebrandt, V.H. (1995), Musculoskeletal symptoms and workload in 12 branches of Dutch agriculture, *Ergonomics*, 38, 2576–2587.

Hildebrandt, V.H. (2001), Prevention of Work Related Musculoskeletal Disorders: Setting Priorities Using the Standardized Dutch Musculoskeletal Questionnaire, Ph.D. thesis, TNO Work and Employment, Hoofddorp, Netherlands.

Hildebrandt, V.H., Bongers, P.M., Dul, J., Dijk, F.J.H., and Van Kemper, H.C.G. (1996), Identification of high-risk groups among maintenance workers in a steel company with respect to musculoskeletal symptoms and workload, *Ergonomics*, 39, 232–242.

Hildebrandt, V.H., Bongers, P.M., van Dijk, F.J., Kemper, H.C., and Dul, J. (2001), Dutch musculoskeletal questionnaire: description and basic qualities, *Ergonomics*, 44, 1038–1055.

Hoogendoorn, W.E., Van Poppel, M.N.M., Bongers, P.M., Koes, B.W., and Bouter, L.M. (1999), Physical load during work and leisure time as risk factors for back pain, *Scand. J. Work Environ. Health*, 25, 387–403.

Keyserling, W.M., Brouwer, M., and Silverstein, B.A. (1992), A checklist for evaluating ergonomic risk factors resulting from awkward postures of the legs, trunk and neck, *Int. J. Ind. Ergonomics*, 9, 283–301.

Kilbom, A. (1994), Assessment of physical exposure in relation to work related musculoskeletal disorders: what information can be obtained from systematic observations? *Scand. J. Work Environ. Health*, 20(special issue), 30–45.

Kumar, S. (1993), The accuracy of trunk posture perception among young males subjects, in *Advances in Industrial Ergonomics and Safety*, Nielsen, R. and Jorgensen, V., Eds., Taylor & Francis, London, pp. 225–229.

Kuorinka, I., Jonsson, B., Kilbom, Å., et al. (1987), Standardised Nordic questionnaires for the analysis of musculoskeletal symptoms, *Appl. Ergonomics*, 18, 233–237.

Paul, J.A. (1993), Pregnancy and the Standing Working Posture, Ph.D. thesis, University of Amsterdam.

Riihimäki, H. (1991), Low back pain, its origin and risk indicators, *Scand. J. Work Environ. Health*, 17, 81–90.

Stock, S.R. (1991), Workplace ergonomic factors and the development of musculoskeletal disorders of the neck and upper limbs: a meta-analysis, *Am. J. Ind. Med.*, 19, 87–107.

Viikari-Juntura, E., Rauas, S., Martikainen, R., Kuosma, E., Riihimäki, H., Takala, E.P., and Saarenmaa K. (1996), Validity of self-reported physical work load in epidemiologic studies on musculoskeletal disorders, *Scand. J. Work Environ. Health*, 22, 251–259.

Vink, P., Lourijsen, E., Wortel, E., and Dul, J. (1992), Experiences in participatory ergonomics: results of a roundtable session during the 11th IEA Congress, Paris, July 1991, *Ergonomics*, 35, 123–127.

Walsh, K., Varnes, N., Osmond, C., Styles, R., and Coggon, D. (1989), Occupational causes of low back pain, *Scand. J. Work Environ. Health*, 15, 54–59.

Weel, A.N.H., Broersenn, J.P.J., and Van Dijk, F.J.H. (2000), Questionnaire surveys on health and working conditions: development of an instrument for risk assessment in companies, *Int. Arch. Occup. Environ. Health*, 73, 47–55.

Wiktorin, C., Karlqvist, L., and Winkel, J. (1993), Validity of self-reported exposures to work postures and manual materials handling, *Scand. J. Work Environ. Health*, 19, 208–214.

Winkel, J. and Mathiassen, S.E. (1994), Assessment of physical work load in epidemiologic studies: concepts, issues and operational considerations, *Ergonomics*, 37, 979–988.

6

Quick Exposure Checklist (QEC) for the Assessment of Workplace Risks for Work-Related Musculoskeletal Disorders (WMSDs)

Guangyan Li
Human Engineering Limited

Peter Buckle
University of Surrey

6.1 Background and Applications

The quick exposure checklist (QEC) quickly assesses the exposure to risks for work-related musculoskeletal disorders (WMSDs) (Li and Buckle, 1999a). QEC is based on the practitioners' needs and research on major WMSD risk factors (e.g., Bernard, 1997). About 150 practitioners have tested QEC and modified and validated it using both simulated and real tasks. QEC has a high level of sensitivity and usability and largely acceptable inter- and intraobserver reliability. Field studies confirm that QEC is applicable for a wide range of tasks. With a short training period and some practice, assessment can normally be completed quickly for each task.

QEC gives an evaluation of a workplace and of equipment design, which facilitates redesign. QEC helps to prevent many kinds of WMSDs from developing and educates users about WMSD risks in their workplaces.

6.2   Procedure

QEC uses five steps:

6.2.1   Step 1: Self-Training

First-time QEC users must read the "QEC User Guide" (Appendix 6.1) to understand the terminology and assessment categories that are used in the checklist. Experienced users can skip step 1.

6.2.2   Step 2: Observer's Assessment Checklist

The QEC user (the observer) uses the "Observer's Assessment" checklist (Appendix 6.2) to conduct a risk assessment for a particular task. Most checklist assessment items are self-explanatory. New users can consult the "QEC User Guide" (Appendix 6.1). At least one complete work cycle is observed before making the assessment. If a job consists of a variety of tasks, each task can be assessed separately. Where a job cannot easily be broken down into tasks, the "worst" event within that job when a particular body part in question is most heavily loaded should be observed. The assessment can be carried out by direct observation or by using video footage (if the information about the "Worker's Assessment" can be obtained at another time; see step 3).

6.2.3   Step 3: Worker's Assessment Checklist

The worker being observed must complete the "Worker's Assessment" checklist (Appendix 6.3).

6.2.4   Step 4: Calculation of Exposure Scores

Use the "Table of Exposure Scores" (Appendix 6.4) to calculate the exposure scores for each task assessed as follows:

1. Circle all the letters corresponding to the answers from the "Observer's Assessment" and the "Worker's Assessment."
2. Mark the numbers at the crossing point of every pair of circled letters. For example, for the exposure to the back, number 8 should be selected as score 1, corresponding to the assessment items A2 and A3.
3. Calculate a total score for each body part.

Exposure score calculations can be done with the help of software (http://www.geocities.com/qecuk).

6.2.5   Step 5: Consideration of Actions

QEC quickly identifies the exposure levels for the back, shoulder/arm, wrist/hand, and the neck, and the method evaluates whether an ergonomic intervention can effectively reduce these exposure levels. Preliminary action levels for the QEC, based on QEC and RULA (McAtamney and Corlett, 1993) assessments of a variety of tasks, have been suggested (Brown and Li, 2003) as seen in Table 6.1.

The exposure level E in Table 6.1 is calculated as a percentage rate between the actual total exposure score X and the maximum possible total X_{max}. For manual handling tasks, $X_{maxMH} = 176$; for other tasks, $X_{max} = 162$.

$$E \, (\%) = X/X_{max} \times 100\%$$

TABLE 6.1   Preliminary Action Levels for the QEC

QEC Score (E) (percentage total)	Action	Equivalent RULA Score
≤40%	acceptable	1–2
41–50%	investigate further	3–4
51–70%	investigate further and change soon	5–6
>70%	investigate and change immediately	7+

6.3   Advantages

- Covers some major physical risk factors for WMSDs.
- Considers user needs and can be used by inexperienced users.
- Considers combination and interaction of multiple workplace risk factors.
- Provides good level of sensitivity and usability.
- Provides encouraging level of inter- and intraobserver reliability.
- Is easy to learn and quick to use.

6.4   Disadvantages

- Method focuses on physical workplace factors only.
- Hypothetical exposure scores with the suggested "action levels" need validating.
- Additional training and practice may be needed for novice users to improve assessment reliability.

6.5   Example of QEC Output

The following is an example of an observer assessing a manual handling task.

The task involves unloading boxes from a trolley onto a shelf. The operator's back is nearly straight, with infrequent movement during work. The boxes are sometimes placed at a height above shoulder level, with frequent adjustment for repositioning of the boxes, often using one hand. The hands/wrists were seen to bend and move between 11 and 20 times per minute, and the neck is occasionally seen to rotate to either side. The boxes weigh around 4 kg each, and the work lasts up to 6 hours per day. The visual demand for the task is considered low. The assessment results are shown in Figure 6.1.

The overall exposure is: $E = (106/176) \times 100\% = 60.2\%$. According to Table 6.1, the score indicates a need to "investigate further and change soon." The major concerns are the exposure to the shoulder/arm and the wrist regions because the operator has to handle the load at or above shoulder level. Possible solutions include providing the worker with a footstool to avoid raising the arms too high, using machinery (a forklift) for high-level loading, or introducing more frequent breaks to reduce the frequency of the repeated tasks.

After the workplace/task intervention, reassess the task using the same QEC approach as described in steps 2 to 5, and compare the pre- and postintervention results to see whether the exposures have been effectively reduced, preferably below an "acceptable" level.

6.6   Related Methods

The suggested "action levels" of the QEC system were based on equivalent RULA scores (McAtamney and Corlett, 1993). The tool was developed with a critical review and analysis of existing methods available at the time (Li and Buckle, 1999b), by adopting "user participation" approaches (e.g., using

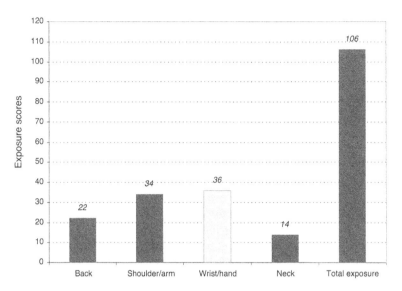

FIGURE 6.1   Assessment results of the example task.

questionnaires and focus groups and by asking the potential users — health and safety practitioners — to design an exposure tool for themselves [Li and Buckle, 1999a]), as well as by using a "think aloud" approach to understand the methods that health and safety practitioners adopt when undertaking a risk assessment in the workplace (Bainbridge and Sanderson, 1995).

6.7   Standards and Regulations

U.K. Management of Health and Safety at Work Regulations 1992:

Regulation 3 requires that "every employer shall make a suitable and sufficient assessment of the risks to the health and safety of his employees to which they are exposed whilst they are at work…. The purpose of the risk assessment is to help the employer or self-employed person to determine what measures should be taken to comply with the employer's or self-employed person's duties under the 'relevant statutory provisions.'"

Regulation 12 mentions that "employees have a duty under Section 7 of the Health and Safety at Work etc. Act 1974 to co-operate with their employer to enable the employer to comply with statutory duties for health and safety."

U.K. Manual Handling Operations Regulations 1992, Regulation 4(1)(b) requires that "each employer shall, where it is not reasonably practicable to avoid the need for his employees to undertake any manual handling operations at work which involve a risk of their being injured, make a suitable and sufficient assessment of all such manual handling operations to be undertaken by them…. The views of staff can be particularly valuable in identifying manual handling problems and practical solutions to them."

6.8   Approximate Training and Application Times

The initial training (self-learning) time of the QEC for a new user is approximately 15 to 20 min, but some practice is suggested for novice users, with exercise assessments either on real tasks or on video-recorded tasks. It normally takes about 10 min to complete an assessment for each task.

6.8.1   Reliability and Validity

The construction validity of the QEC is reported in Li and Buckle (1999a). The tool is found to have a high sensitivity (the ability to identify a change in exposure before and after an ergonomic intervention), a good intraobserver reliability, and a practically acceptable interobserver reliability (Li and Buckle, 1999a).

6.8.2 Tools Needed

QEC is a pen-and-paper-based exposure assessment tool. Calculation of exposure scores can be done with a computer program available at www.geocities.com/qecuk.

References

Bainbridge, L. and Sanderson, P. (1995), Verbal protocol analysis, in *Evaluation of Human Work*, 2nd ed., Wilson, J.R. and Corlett, E.N., Eds., Taylor & Francis, London, pp. 169–201.

Bernard, B.P., Ed. (1997), Musculoskeletal Disorders and Workplace Factors: A Critical Review of Epidemiologic Evidence for Work-Related Musculoskeletal Disorders of the Neck, Upper Extremity, and Low Back, 2nd ed., National Institute for Occupational Safety and Health (NIOSH), Cincinnati.

Brown, R. and Li, G. (2003), The development of action levels for the 'Quick Exposure Check' (QEC) system, in *Contemporary Ergonomics 2003*, McCabe, P.T., Ed., Taylor & Francis, London, pp. 41–46.

Li, G. and Buckle, P. (1999a), Evaluating Change in Exposure to Risk for Musculoskeletal Disorders: A Practical Tool, HSE contract report 251/1999, HSE Books, Suffolk.

Li, G. and Buckle, P. (1999b), Current techniques for assessing physical exposure to work-related musculoskeletal risks, with emphasis on posture-based methods, *Ergonomics*, 42, 674–695.

McAtamney, L. and Corlett, E.N. (1993), RULA: a survey method for the investigation of work-related upper limb disorders, *Appl. Ergonomics*, 24, 91–99.

Appendix 6.1 QEC User Guide

This exposure tool has been designed to assess the change in exposure to musculoskeletal risks before and after an ergonomic intervention. Before making the risk assessment, a preliminary observation of the job should be made for at least one work cycle. Record all information as listed at the top of the checklist in Appendix 6.2.

Exposure Assessment for the Back

Back Posture (A1 to A3)

The assessment for the back posture should be made at the moment when the back is most heavily loaded. For example, when lifting a box, the back may be considered under highest loading at the point when the person leans or reaches forward to pick up the load.

- The back can be regarded as "**almost neutral**" (Level A1) if the person is seen to work with his/her back flexion/extension, twisting, or side bending less than 20°, as shown in Figure A1.

- The back can be regarded as "**moderately flexed or twisted**" (Level A2) if the person is seen to work with his/her back flexion/extension, twisting, or side bending more than 20° but less than 60°, as shown in Figure A2.

FIGURE A1 The back is "almost neutral."

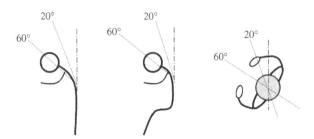

FIGURE A2 The back is "moderately flexed or twisted."

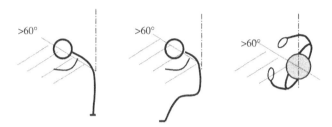

FIGURE A3 The back is "excessively flexed or twisted."

- The back can be regarded as "**excessively flexed or twisted**" (Level A3) if the person is seen to work with his/her back flexion or twisting more than 60º (or close to 90º), as shown in Figure A3.

Back Movement (B1 to B5)
- For manual material handling tasks, assess B1 to B3. This refers to how often the person needs to bend, rotate his/her back when performing the task. Several back movements may happen within one task cycle.
- For tasks other then manual handling, such as sedentary work or repetitive tasks performed in standing or seated position, ignore B1 to B3 and assess B4 and B5 only.

Exposure Assessment for the Shoulder/Arm

Shoulder/Arm Posture (C1 to C3)
Assessment should be made when the shoulder/arm is most heavily loaded during work, but not necessarily at the same time as the back is assessed. For example, the load on the shoulder may not be at the highest level when the operator bends down to pick up a box from the floor, but may become greater subsequently when the box is placed at a higher level.

Shoulder/Arm Movement (D1 to D3)
The movement of the shoulder/arm is regarded as

- "**Infrequent**" if there is no regular motion pattern
- "**Frequent**" if there is a regular motion pattern with some short pauses
- "**Very frequent**" if there is a regular continuous motion pattern during work

Exposure Assessment for the Wrist/Hand

Wrist/Hand Posture (E1 to E2)
This is assessed during the performance of the task at the point when the most awkward wrist posture is adopted, including wrist flexion/extension, side bending (ulnar/radial deviation), and rotation of the

wrist around the axis of the forearm. The wrist is regarded as **"almost straight"** (Level E1) if its movement is limited within a small angular range (e.g., <15°) of the neutral wrist posture (Figure E1). Otherwise, if an obvious wrist angle can be observed during the performance of the task, the wrist is considered to be **"deviated or bent"** (Figure E2).

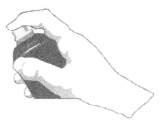

FIGURE E1 The wrist is "almost straight."

FIGURE E2 The wrist is "deviated or bent."

Wrist/Hand Movement (F1 to F3)

This refers to the movement of the wrist/hand and forearm, excluding the movement of the fingers. One motion is counted every time when the same or similar motion pattern is repeated over a set period of time (e.g., 1 min).

Exposure Assessment for the Neck

The neck can be considered to be **"excessively bent or twisted"** if it is bent or twisted at an obvious angle (or more than 20°) relative to the torso.

Worker's Assessment of the Same Task

After the observer's assessment is made, ask the worker to answer the questions in the checklist in Appendix 6.3. Explain the meaning of the terms to him/her when necessary.

Calculation of the Total Exposure Scores

The total exposure scores can be obtained by combining the assessments from the observer (QEC checklist in Appendix 6.2, items A to G) and the worker (QEC checklist in Appendix 6.3, items a through g). Ensure that the correct combined scores have been determined before adding them into the total.

Additional Points

- For group work, ensure that a sufficiently representative number of individual workers are assessed.
- Workers whose daily pattern of work and job demands are variable should be observed more than once.

Appendix 6.2 Quick Exposure Checklist (QEC) for Work-Related Musculoskeletal Risks

Observer's Assessment

Job title: Task: Assessment conducted by: Worker's name: Date: Time:

Back

• When performing the task, is the back

A1. Almost neutral?
A2. Moderately flexed or twisted or side bent?
A3. Excessively flexed or twisted or side bent?

• For manual handling tasks only: Is the movement of the back-

B1: Infrequent?   (Around 3 times per minute or less)
B2: Frequent?   (Around 8 times per minute)
B3: Very frequent?   (Around 12 times per minute or more)

• Other Tasks: Is the task performed in static postures most of the time?
                   (either seated or standing)

B4: No.   B5: Yes

Shoulder/arm

• Is the task performed
C1: At or below waist height?
C2: At about chest height?
C3: At or above shoulder height?

• Is the arm movement repeated

D1: Infrequently?   (Some intermittent arm movement)
D2: Frequently?   (Regular arm movement with some pauses)
D3: Very frequently?   (almost continuous arm movement)

Wrist/Hand

• Is the task performed
E1: With almost a straight wrist?
E2: With a deviated or bent wrist position?

• Is the task performed with similar repeated motion patterns
F1: 10 times per minute or less?
F2: 11 to 20 times per minute?
F3: More than 20 times per minute?

Neck

• When performing the task, is the head/neck bent or twisted excessively?
G1: No
G2: Yes, occasionally
G3: Yes, continuously

Appendix 6.3 Worker's Assessment

Name:	Job title:	Date:

• What is the maximum weight handled in this task?
A1: Light   (5 kg or less)
A2: Moderate   (6 to 10 kg)
A3: Heavy   (11 to 20 kg)

• How much time on average do you spend per day doing this task?
B1: Less than 2 hours
B2: 2 to 4 hours
B3: More than 4 hours

• When performing this task (single or double handed), what is the
  maximum force level exerted by one hand?
C1: Low   (e.g., Less than 1 kg)
C2: Medium   (e.g., 1 to 4 kg)
C3: High   (e.g., More than 4 kg)

• Do you experience any vibration during work?
D1: Low (or no)
D2: Medium
D3: High

• Is the visual demand of this task-
E1: Low? (There is almost no need to view fine details)
E2: High? (There is a need to view some fine details)

• Do you have difficulty keeping up with this work?
F1: Never
F2: Sometimes
F3: Often

• How stressful do you find this work?
G1: Not at all
G2: Low
G3: Medium
G4: High

Appendix 6.4 Table of Exposure Scores

Exposure to the Back:

	A1	A2	A3	Score 1	B1	B2	B3	Score 2	b1	b2	b3	Score 3
A1	2	4	6		2	4	6		2	4	6	
A2	4	6	8		4	6	8		4	6	8	
A3	6	8	10		6	8	10		6	8	10	
A4	8	10	12		8	10	12		8	10	12	

				Score 4				B4	B5	Score 5	Total score for the back = Sum of scores 1 to 5
B1	2	4	6		2	4	6	2	4		
B2	4	6	8		4	6	8	4	6		
B3	6	8	10		6	8	10	6	8		

Exposure to the Shoulder/Arm:

	C1	C2	C3	Score 1	D1	D2	D3	Score 2	b1	b2	b3	Score 3
A1	2	4	6		2	4	6		2	4	6	
A2	4	6	8		4	6	8		4	6	8	
A3	6	8	10		6	8	10		6	8	10	
A4	8	10	12		8	10	12		8	10	12	

				Score 4				Score 5	Total score for shoulder/arm = Sum of scores 1 to 5
B1	2	4	6		2	4	6		
B2	4	6	8		4	6	8		
B3	6	8	10		6	8	10		

Exposure to the Wrist/Hand:

	F1	F2	F3	Score 1	E1	E2	Score 2	b1	b2	b3	Score 3
C1	2	4	6		2	4		2	4	6	
C2	4	6	8		4	6		4	6	8	
C3	6	8	10		6	8		6	8	10	

				Score 4			Score 5	Total score for the wrist/hand = Sum of scores 1 to 5
B1	2	4	6		2	4		
B2	4	6	8		4	6		
B3	6	8	10		6	8		

Exposure to the Neck:

	G1	G2	G3	Score 1	e1	e2	Score 2	Total score for the neck = Scores 1+ 2
b1	2	4	6		2	4		
b2	4	6	8		4	6		
b3	6	8	10		6	8		

Exposure scores: Back:_____ Shoulder/arm:_____ Wrist/hand:_____ Neck:_____

7

Rapid Upper Limb Assessment (RULA)

Lynn McAtamney
*COPE Occupational Health and
Ergonomics Services Ltd.*

Nigel Corlett
University of Nottingham

7.1 Background

Rapid upper-limb assessment (RULA) (McAtamney and Corlett, 1993) provides an easily calculated rating of musculoskeletal loads in tasks where people have a risk of neck and upper-limb loading. The tool provides a single score as a "snapshot" of the task, which is a rating of the posture, force, and movement required. The risk is calculated into a score of 1 (low) to 7 (high). These scores are grouped into four action levels that provide an indication of the time frame in which it is reasonable to expect risk control to be initiated.

7.2 Applications

RULA is used to assess the posture, force, and movement associated with sedentary tasks. Such tasks include screen-based or computer tasks, manufacturing, or retail tasks where the worker is seated or standing without moving about.

The four main applications of RULA are to:

1. Measure musculoskeletal risk, usually as part of a broader ergonomic investigation
2. Compare the musculoskeletal loading of current and modified workstation designs

3. Evaluate outcomes such as productivity or suitability of equipment
4. Educate workers about musculoskeletal risk created by different working postures

In all applications, it is strongly recommended that users receive training in RULA prior to use, although no previous ergonomic assessment skills are required.

7.2.1 Measuring Musculoskeletal Risk

RULA assesses a working posture and the associated level of risk in a short time frame and with no need for equipment beyond pen and paper. RULA was not designed to provide detailed postural information, such as the finger position, which might be relevant to the overall risk to the worker. It may be necessary for RULA to be used with other assessment tools as part of a broader or more detailed ergonomic investigation. When using RULA, the assessor can benefit from establishing the following information when making recommendations for change (McAtamney and Corlett, 1992): a knowledge of the products, processes, tasks, previous musculoskeletal injuries, training, workplace layout and dimensions, and relevant environmental risks or constraints.

RULA can be used to assess a particular task or posture for a single user or group of users (Herbert et al., 1996). It may be necessary to assess a number of different postures during a work cycle to establish a profile of the musculoskeletal loading. In such cases, it is useful to videotape or photograph workers from both sides and from the back while they are performing the tasks.

7.2.2 Compare Current and Modified Workstation Designs

RULA is useful in comparing existing and proposed workstation designs as part of a justification or proposal for ergonomic changes. The RULA scores provide any nonergonomist or stakeholder with evidence that proposed modifications can reduce musculoskeletal loading, which can facilitate funding approval.

For example, RULA was used in the following studies:

- Gutierrez (1998), who evaluated assembly workers in an electronics factory and compared postures when improvements were introduced.
- Hedge et al. (1995), who assessed different computer equipment.
- Cook and Kothiyal (1998), who assessed the influence of mouse position on muscular activity using RULA and EMG (electromyography).
- Leuder (1996), who modified RULA (http://www.humanics-es.com/rulacite.htm) to include broader risks associated with office-based tasks, for example glare on the computer screen (see http://www.humanics-es.com/files/rula.pdf). While the modified tool has not been validated, it provides useful information on workstation risks.

7.2.3 Evaluate Outcomes

As part of a detailed ergonomic investigation, RULA scores can be compared against other outcome factors. In a manufacturing plant, Axellson (1997) found a correlation between high RULA scores and a higher proportion of products that were discarded as defective when processed at a particular workstation. As part of a macroergonomic management program, the company subsequently improved the identified high-risk workstation, leading to a 39% drop in quality deficiencies and an annual cost savings of $25,000.

7.2.4 Educate Workers

Many adults have developed habitual postural movement patterns that they find very difficult to change. The use of photographs of trainees at work along with the RULA score anecdotally helps motivate trainees to make the effort to change techniques.

7.3 Procedure

The procedure for using RULA is explained in three steps:

1. The posture or postures for assessment are selected.
2. The postures are scored using the scoring sheet, body-part diagrams, and tables.
3. These scores are converted to one of the four action level.

7.3.1 Observing and Selecting the Postures to Assess

A RULA assessment represents a moment in the work cycle, and it is important to observe the postures adopted during the full work cycle or a significant working period prior to selecting the postures for assessment. Depending upon the type of study, the selection could be the longest held posture or what appears to be the worst postures adopted. It also can be useful to estimate the proportion of time spent in the various postures being evaluated (McAtamney and Corlett, 1992).

7.3.2 Scoring and Recording the Posture

Decide whether the left, right, or both upper arms are at risk and need to be assessed. Then score the posture of each body part using the free software found on-line at http://www.ergonomics.co.uk/Rula/Ergo/index.html or the paper version (McAtamney and Corlett, 1993), which is on the Web at http://ergo.human.cornell.edu/ahRULA.html.

Use the RULA assessment diagrams (Figure 7.1 shows the software version) to score the posture for each body part, along with the forces/loads and the muscle use required for that particular posture. Follow the score sheet to calculate the posture scores for Groups A and B if using the paper version (the software version does this for you). Use the calculation button on the software or use Table C to calculate the grand score.

7.3.3 Action Level

The grand score can be compared with the list of action levels. In most cases, to ensure that this guide is used as an aid in efficient and effective control of any risks identified, the actions lead to a more detailed investigation. The action levels are listed in Table 7.1.

7.4 Example

Ergonomics takes high priority in the design and development of pallets for Jaguar Cars Ltd., U.K., and RULA has become one of the key criteria used by them and the suppliers.

7.4.1 RULA Used in Design and Development Process

Jaguar uses RULA in its standards documentation for pallet/box manufacture to enable the company and its suppliers to assess and improve the ergonomics of their pallets before production commences, thereby avoiding disruption and risk of injury during steady-state production. As part of the development process, all suppliers complete a full RULA risk assessment on the use of each pallet prototype, and they make any necessary changes to eliminate any risk to the operators. The engineers and physiotherapist at Jaguar have found that this process minimizes bending, stretching, or twisting.

When materials for the X200 series were supplied, RULA was used to assess the unpacking task. The initial unpacking task forced the operator to reach into the box, as seen in Figure 7.2. The musculoskeletal risk increased as the box was emptied. The RULA scoring of the posture depicted in Figure 7.2 is presented in Table 7.2 along with relevant comments on the scoring. It is useful to follow the score sheet and body part diagrams (Figure 7.1) while reading Table 7.2.

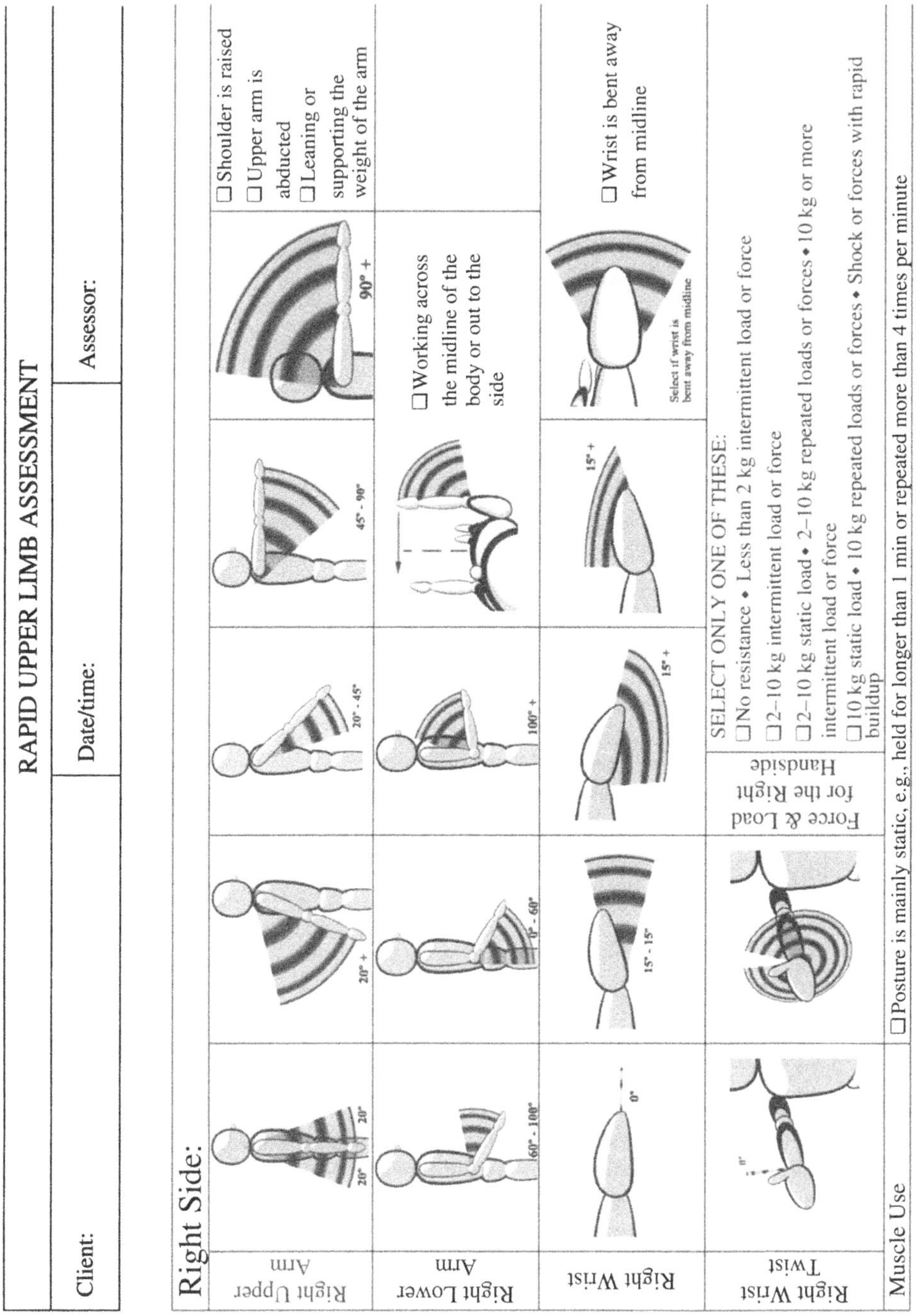

RAPID UPPER LIMB ASSESSMENT		
Client:	Date/time:	Assessor:

Right Side:

Right Upper Arm	20° | 20°	20° +	20° - 45°	45° - 90°	90° +	☐ Shoulder is raised ☐ Upper arm is abducted ☐ Leaning or supporting the weight of the arm
Right Lower Arm	60° - 100°	0° - 60°	100° +		☐ Working across the midline of the body or out to the side	
Right Wrist	0°	15° - 15°	15° +	15° +	Select if wrist is bent away from midline	☐ Wrist is bent away from midline
Right Wrist Twist			Force & Load for the Right Handside			

SELECT ONLY ONE OF THESE:
☐ No resistance ◆ Less than 2 kg intermittent load or force
☐ 2–10 kg intermittent load or force
☐ 2–10 kg static load ◆ 2–10 kg repeated loads or forces ◆ 10 kg or more intermittent load or force
☐ 10 kg static load ◆ 10 kg repeated loads or forces ◆ Shock or forces with rapid buildup

Muscle Use ☐ Posture is mainly static, e.g., held for longer than 1 min or repeated more than 4 times per minute

FIGURE 7.1A RULA software assessment form.

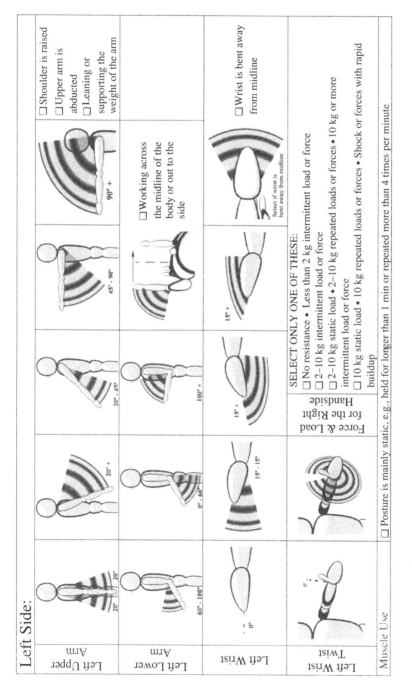

FIGURE 7.1B   RULA software assessment form.

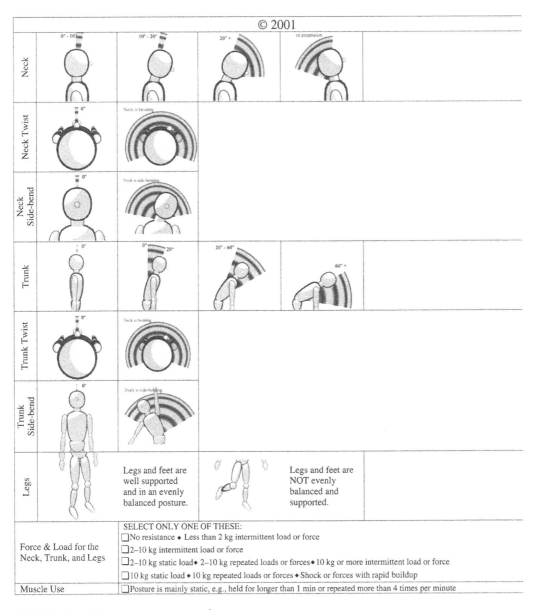

FIGURE 7.1C   RULA software assessment form.

TABLE 7.1   RULA Action Levels

Action level 1	Score of 1 or 2 indicates that the posture is acceptable if it is not maintained or repeated for long periods
Action level 2	Score of 3 or 4 indicates that further investigation is needed, and changes may be required
Action level 3	Score of 5 or 6 indicates that investigation and changes are required soon
Action level 4	Score of 7 indicates that investigation and changes are required immediately

FIGURE 7.2 Unpacking task (before intervention).

It is clear from the RULA assessment of Figure 7.2 that ergonomic risks were present that required modification. The body parts at risk were the back and neck postures, with overreaching causing risk to the upper arms. Figure 7.3 illustrates the solution, a hydraulic tilter controlled by the operator to allow the height and angle of the pallet to be adjusted. This device provided a RULA score of 1 to 4, depending on the position of the item in the box.

7.4.2 RULA Used in Early Intervention and Risk Assessment Related to Musculoskeletal Strains and Sprains

As part of production operations, Jaguar has introduced RULA assessments as one of its procedures when:

1. There has been a work-related strain/sprain reported
2. The risk assessment identifies that an assessment is required
3. There has been a change to a process
4. There is an operator concern

Figure 7.4 provides an outline of the processes in place to find practical solutions to any of the above criteria. If a change cannot be made in the short term, then containment measures are introduced, such as increasing recovery time or increasing job rotation. RULA assessments are made accessible to the personnel carrying out the task by keeping them in the immediate vicinity.

The process outlined in Figure 7.4 offers a participatory systematic approach to problem solving using skills and knowledge from relevant personnel in the company. In this process, RULA provides an objective measure around which changes can be suggested and investigated, with the ultimate goal of implementing the best-practice solution.

7.5 Approximate Training and Application Times

RULA was developed to require minimal training. Dismukes (1996) reported that people untrained in ergonomics could accurately assess upper-limb disorders using RULA. However, it is strongly recom-

TABLE 7.2 Scoring of Posture in Figure 7.2

Body Part		Score	Comments
Group A	upper arms	3 + 1 for shoulders raised	This score is taken in relation to the trunk; shoulders are raised because of the excessive reaching
	lower arms	2	Lower arms are straight
	wrist	1	Wrist posture is often obscured, so particular note needs to be taken of wrist position when on site
	wrist twist	1	A score of 1 is given if the hand is in a handshake position; otherwise, the score is 2
	Using Table A, these produce a posture score A = 4		
Group B	neck	4	Neck posture is taken in relation to the trunk; in this position, the neck can actually be extended to provide vision when the trunk is flexed forward
	trunk	3	There is no twisting or side flexion
	legs	2	If the operator is reaching by standing on the toes, then the risk of slipping increases; 1 is given only if the weight is evenly distributed onto both feet and they are in a good position
	Using Table B, these scores produce posture score B = 7		
Group A	force	3	Handling at end of range increases the biomechanical risk, which creates a jolt and rapid buildup of load on the shoulders
	muscle use	0	This task is repeated, but not as frequently as four times every minute
Posture score A (4) + 3 = 7			
Group B	force	3	The handling at end of range creates a jolt on the spine
	muscle use	0	Walking and change of posture occur before and after the task
Posture score B (7) + 3= 10			
	Using Table C, the grand score for this posture is 7		

mended that users have training so that they use the tool correctly. A common fault is unskilled users trying to add posture scores when Tables A and B must be used.

It is suggested that new users practice using photographs and videotape of postures prior to using the tool in an assessment. One difficulty with any observation tool is deciding the angle of joint range, particularly if the angle of vision is not in line with the side and back of the body. Where the user is unable to decide on the posture score, it is recommended that the higher of the two scores be taken. For example, if it is difficult to establish whether the upper arm is in range 2 or 3, then 3 should be selected. This approach ensures that all risks are included rather than excluded.

The user of the RULA software (http://www.ergonomics.co.uk/Rula/Ergo/index.html) need not be concerned with using the tables. Users of the paper-based system follow the guide on the scoring sheet to calculate the RULA grand score. Familiarization with the tables and method requires 1 to 2 hours of time.

7.6 Reliability and Validity

The reliability and validity studies undertaken during the development of RULA are detailed in McAtamney and Corlett (1993). The validity was assessed using a laboratory-based DSE workstation, where

FIGURE 7.3 Unpacking task (after intervention).

the RULA scores and body-part discomfort were analyzed. Further validity and reliability studies were conducted in both industrial and office-based settings by ergonomists and physiotherapists as part of their postgraduate training.

7.7 Costs and Tools Needed

RULA is freely available on the Web. While the scoring form can be downloaded, the user must do the scoring on-line. The paper-based system (McAtamney and Corlett, 1993) requires photocopying and a pencil.

In using RULA, it is suggested that a camera may assist the user in recording the posture for later scoring. Photographs need to be taken directly from the side and back to avoid parallax error. Likewise, video recordings should be taken from back, side, and front if possible.

7.8 Related Methods

RULA is one of a number of observational posture-assessment tools that are useful in task analysis. RULA is useful as an initial tool in ergonomic investigations, although additional task-specific investigation may be required following a RULA assessment. REBA (rapid entire body assessment, see Chapter 8) should be used instead of RULA when there are tasks involving manual handling, whole body movement, or risk to the back and legs as well as the upper limbs and neck.

Acknowledgments

Our thanks to Mike Huthnance, Anita McDonald, and Janet King from Jaguar Cars for their assistance and support in the preparation of this chapter. Thanks also to The Osmond Group for reproduction of the Web-based RULA software.

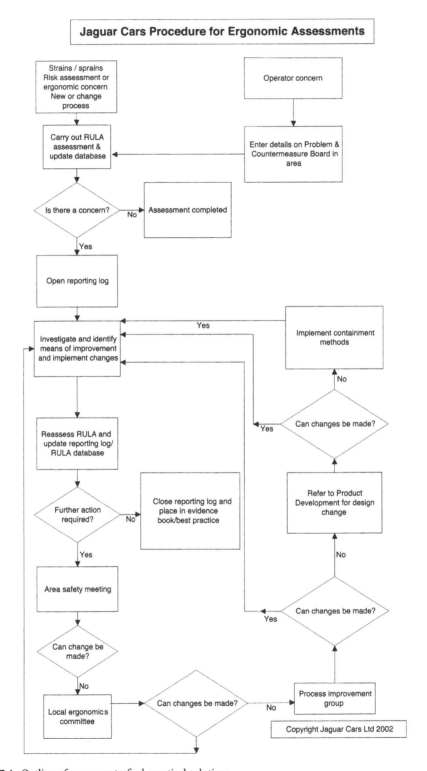

FIGURE 7.4 Outline of processes to find practical solutions.

References

Axelsson, J.R.C. (1997), RULA in action: enhancing participation and continuous improvements, from experience to innovation, in *Proceedings of the 13th Triennial Congress of the International Ergonomics Association*, Seppala, P., Luopajarvi, T., Nygard, C.H., and Mattila, M., Eds., Finnish Institute of Occupational Health, Helsinki, Vol. 4, pp. 251–253.

Cook, C.J. and Kothiyal, K. (1998), Influence of mouse position on muscular activity in the neck, shoulder and arm in computer users, *Appl. Ergonomics*, 29, 439–443.

Dismukes, S. (1996), An Ergonomic Assessment Method for Non-Ergonomists, in *Proceedings of the Silicon Valley Ergonomics Conference and Exposition: ErgoCon'96*, Silicon Valley Ergonomics Institute, San Jose State University, San Jose, CA, pp. 106–117.

Gutierrez, A.M.J.A. (1998), A workstation design for a Philippine semiconductor, in *Proceedings of the First World Congress on Ergonomics for Global Quality and Productivity*, Bishu, R.R, Karwowski, W., and Goonetilleke, R.S., Eds., Hong Kong University of Science and Technology, Clear Water Bay, Hong Kong, pp. 133–136.

Hedge, A., McRobie, D., Land, B., Morimoto, S., and Rodriguez, S. (1995), Healthy keyboarding: effects of wrist rests, keyboard trays, and a preset tiltdown system on wrist posture, seated posture, and musculoskeletal discomfort for the global village, in *Proceedings of the Human Factors and Ergonomics Society 39th Annual Meeting*, Human Factors and Ergonomics Society, Santa Monica, CA, Vol. 1, pp. 630–634.

Herbert, R., Dropkin, J., Sivin, D., Doucette, J., Kellogg, L., Bardin, J., Janeway, K., and Zoloth, S. (1996), Impact of adjustable chairs on upper extremity musculoskeletal symptoms among garment workers, in *Advances in Occupational Ergonomics and Safety I*, Mital, A., Krueger, H., Kumar, S., Menozzi, M., and Fernandez, J., Eds., International Society for Occupational Ergonomics and Safety, Cincinnati, Vol. 2, pp. 832–837.

Leuder, R. (1996), A Proposed RULA for Computer Users, paper presented at Ergonomics Summer Workshop, University of California, Berkeley, August 1996.

McAtamney, L. and Corlett, E.N. (1993), RULA: a survey method for the investigation of work-related upper limb disorders, *Appl. Ergonomics*, 24, 91–99.

McAtamney, L. and Corlett, E.N. (1992), Reducing the Risks of Work Related Upper Limb Disorders: A Guide and Methods, Institute for Occupational Ergonomics, University of Nottingham, U.K.

Mechan, J.E. and Porter, M.L. (1997), Stereophotogrammetry: a three-dimensional posture measuring tool, in *Contemporary Ergonomics*, Robertson, S.A., Ed., Taylor & Francis, London, pp. 456–460.

Smyth, G. and Haslam, R. (1995), Identifying risk factors for the development of work related upper limb disorders, in *Contemporary Ergonomics*, Robertson, S.A., Ed., Taylor & Francis, London, pp. 440–445.

8

Rapid Entire Body Assessment

Lynn McAtamney
*COPE Occupational Health and
Ergonomics Services Ltd.*

Sue Hignett
Loughboroough University

8.1   Background

Rapid entire body assessment (REBA) (Hignett and McAtamney, 2000) was developed to assess the type of unpredictable working postures found in health-care and other service industries. Data are collected about the body posture, forces used, type of movement or action, repetition, and coupling. A final REBA score is generated to give an indication of the level of risk and urgency with which action should be taken.

In the spectrum of postural analysis tools, REBA lies between the detailed event-driven systems and time-driven tools. Examples of detailed event-driven tools include a three-dimensional observation system (Hsiao and Keyserling, 1990) or the NIOSH (National Institute for Occupational Safety and Health) equation (Waters et al., 1993), which requires information about specific parameters to give high sensitivity. Time-driven field tools such as OWAS (Ovako working posture analysis system) (Karhu et al., 1977) provide high generality but low sensitivity (Fransson-Hall et al., 1995). REBA was designed to be used as an event-driven tool due the complexity of data collection. However it has recently been computerized by Janik et al. (2002) for field use on a Palm PC, and it can now be used as a time-driven tool.

The initial development was based on the ranges of limb positions using concepts from RULA (rapid upper limb assessment) (McAtamney and Corlett, 1993), OWAS (Karhu et al., 1977), and NIOSH (Waters et al., 1993). The baseline posture is the functional anatomically neutral posture (American Academy of Orthopedic Surgeons, 1965). As the posture moves away from the neutral position, the risk score increases. Tables are available to transform the 144 posture combinations into a single score that represents the level of musculoskeletal risk. These scores are then banded into five action levels that advise on the urgency of avoiding or reducing the risk of the assessed posture.

8.2   Application

REBA can be used when an ergonomic workplace assessment identifies that further postural analysis is required and:

- The whole body is being used.
- Posture is static, dynamic, rapidly changing, or unstable.
- Animate or inanimate loads are being handled either frequently or infrequently.
- Modifications to the workplace, equipment, training, or risk-taking behavior of the worker are being monitored pre/post changes.

8.3   Procedure

REBA has six steps:

1. Observe the task.
2. Select the postures for assessment.
3. Score the postures.
4. Process the scores.
5. Establish the REBA score.
6. Confirm the action level with respect to the urgency for control measures.

8.3.1   Observe the Task

Observe the task to formulate a general ergonomic workplace assessment, including the impact of the work layout and environment, use of equipment, and behavior of the worker with respect to risk taking. If possible, record data using photographs or a video camera. However, as with any observational tool, multiple views are recommended to control for parallax errors.

8.3.2   Select Postures for Assessment

Decide which postures to analyze from the observations in step one. The following criteria can be used:

- Most frequently repeated posture
- Longest maintained posture
- Posture requiring the most muscular activity or the greatest forces
- Posture known to cause discomfort
- Extreme, unstable, or awkward posture, especially where a force is exerted
- Posture most likely to be improved by interventions, control measures, or other changes

The decision can be based on one or more of the above criteria. The criteria for deciding which postures to analyze should be reported with the results/recommendations.

8.3.3   Score the Postures

Use the scoring sheet (Figure 8.1) and body-part scores (Table 8.4 and Table 8.5) to score the posture. The initial scoring is by group:

- Group A: trunk, neck, legs (Figure 8.2)
- Group B: upper arms, lower arms, wrists (Figure 8.3)

Group B postures are scored separately for the left and right sides, as indicated on the scoring sheet (Figure 8.1). Note that additional points can be added or subtracted, depending on the position. For example, in Group B, the upper arm can be supported in its position, and so 1 point is deducted from

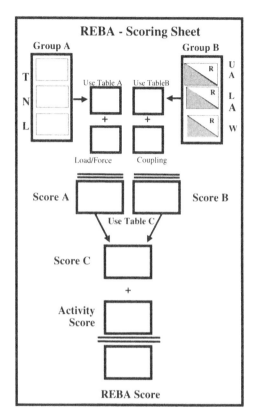

FIGURE 8.1 REBA scoring sheet.

its score. The load/force score (Table 8.1), the coupling score (Table 8.2), and the activity score (Table 8.3) are allocated at this stage. This process can be repeated for each side of the body and for other postures.

8.3.4 Process the Scores

Use Table A (Table 8.4) to generate a single score from the trunk, neck, and legs scores. This is recorded in the box on the scoring sheet (Figure 8.1) and added to the load/force score (Table 8.1) to provide score A. Similarly the upper arms, lower arms, and wrist scores are used to generate the single score using Table B (Table 8.5). This is repeated if the musculoskeletal risk (and therefore the scores for the left and right arms) is different. The score is then added to the coupling score (Table 8.2) to produce score B. Scores A and B are entered into Table C (Table 8.6), and a single score is read off. This is score C.

8.3.5 Calculate REBA Score

The type of muscle activity being performed is then represented by an activity score (Table 8.3), which is added to give the final REBA score.

8.3.6 Confirm the Action Level

The REBA score is then checked against the action levels (Table 8.7). These are bands of scores corresponding to increasing urgency for the need to make changes.

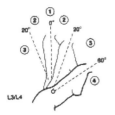

Trunk			
	Movement	**Score**	
	Upright	1	
	0°-20° flexion 0°-20° extension	2	Change score: +1 if twisting or side flexed
	20°-60° flexion > 20° extension	3	
	> 60° flexion	4	

Neck

	Movement	**Score**	
	0°-20° flexion	1	Change score: +1 if twisting or side flexed
	> 20° flexion or extension	2	

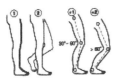

Legs			
	Position	**Score**	Change score:
	Bilateral weight bearing, walking or sitting	1	+1 if knee(s) between 30° and 60° flexion
	Unilateral weight-bearing, feather weight-bearing, or an unstable posture	2	+2 if knee(s) >60° flexion (N.B. not for sitting)

FIGURE 8.2 Group A scoring.

Upper arms

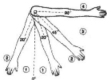

	Position	Score	Change score:
	20° extenion to 20° flexion	1	+1 if arm is:
	> 20° extension 20° - 45° flexion	2	• abducted • rotated
	45°-90° flexion	3	+1 if shoulder is raised
	> 90° flexion	4	-1 of leaning, supporting weight of arm or if posture is gravity assisted

Lower arms

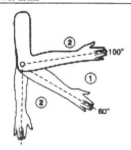

	Movement	Score
	60°-100° flexion	1
	<60° flexion > 100° flexion	2

Wrists

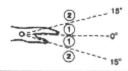

	Movement	Score	Change score:
	0°-15° flexion/extension	1	+1 if wrist is deviated or twisted
	>15° flexion/extension	2	

FIGURE 8.3 Group B scoring.

TABLE 8.1 Load/Force Score

0	1	2	+1
<5 kg	5–10 kg	>10 kg	Shock or rapid buildup of force

TABLE 8.2   Load Coupling Score

0 (Good)	1 (Fair)	2 (Poor)	3 (Unacceptable)
Well-fitting handle and a midrange power grip	Handhold acceptable but not ideal **or** Coupling is acceptable via another part of the body	Handhold not acceptable, although possible	Awkward, unsafe grip; no handles **or** Coupling is unacceptable using other parts of the body

TABLE 8.3   Activity Score

Score	Description
+1	If one or more body parts are static, e.g., held for longer than 1 min
+1	If repeated small-range actions occur, e.g., repeated more than 4 times per minute (not including walking)
+1	If the action causes rapid large-range changes in postures or an unstable base

TABLE 8.4   Table A: Scoring for Body Parts A (Trunk, Neck, Legs)

		Neck											
		1				2				3			
	Legs	1	2	3	4	1	2	3	4	1	2	3	4
Trunk													
1		1	2	3	4	1	2	3	4	3	3	5	6
2		2	3	4	5	3	4	5	6	4	5	6	7
3		2	4	5	6	4	5	6	7	5	6	7	8
4		3	5	6	7	5	6	7	8	6	7	8	9
5		4	6	7	8	6	7	8	9	7	8	9	9

TABLE 8.5   Table B: Scoring for Body Parts B (Upper Arms, Lower Arms, Wrists)

		Lower Arm					
		1			2		
	Wrist	1	2	3	1	2	3
Upper Arm							
1		1	2	2	1	2	3
2		1	2	3	2	3	4
3		3	4	5	4	5	5
4		4	5	5	5	6	7
5		6	7	8	7	8	8
6		7	8	8	8	9	9

TABLE 8.6 Table C: Grand Score

		1	2	3	4	5	6	7	8	9	10	11	12
						Group B Score							
G	1	1	1	1	2	3	3	4	5	6	7	7	7
R	2	1	2	2	3	4	4	5	6	6	7	7	8
O	3	2	3	3	3	4	5	6	7	7	8	8	8
U	4	3	4	4	4	5	6	7	8	8	9	9	9
P	5	4	4	4	5	6	7	8	8	9	9	9	9
A	6	6	6	6	7	8	8	9	9	10	10	10	10
	7	7	7	7	8	9	9	9	9	10	10	11	11	11
S	8	8	8	8	9	10	10	10	10	10	11	11	11
C	9	9	9	9	10	10	10	11	11	11	12	12	12
O	10	10	10	10	11	11	11	11	12	12	12	12	12
R	11	11	11	11	11	12	12	12	12	12	12	12	12
E	12	12	12	12	12	12	12	12	12	12	12	12	12

TABLE 8.7 REBA Action Levels

REBA Score	Risk Level	Action Level	Action (including further assessment)
1	negligible	0	none necessary
2–3	low	1	may be necessary
4–7	medium	2	necessary
8–10	high	3	necessary soon
11–15	very high	4	necessary now

8.3.7 Subsequent Reassessment

If or when the task changes due to interventions or control measures, the process can be repeated, and the new REBA score can be compared with the previous one to monitor the effectiveness of the changes.

8.4 Example

REBA has proved to be useful in educating care workers as part of the risk-assessment process for patient handling. In the example of using slide sheets to roll a client, the correct posture is shown in Figure 8.4. The incorrect, although commonly adopted, posture is shown in Figure 8.5. Scoring each of the postures with REBA enables the care worker to see the different risk levels to which workers are exposed. This can focus the attention of care workers on the task and control measures (e.g., raising the bed height).

To make the REBA scoring easier, lines of reference have been added to the photographs in Figures 8.4 and 8.5.

8.4.1 Scoring Figure 8.4 — Correct Posture

Use the scoring sheet (Figure 8.1) and group A (Figure 8.2) and group B (Figure 8.3) to score the respective body parts. In the photograph in Figure 8.4, the trunk angle (T) is between 20 and 60°, giving a score of 3. The neck posture (N) is neutral, with a score of 1. The leg score (L) is in two parts: weight is taken on both feet, giving a score of 1; the knee is bent between 30 and 60°, giving + 1. The load/force score (Table 8.1) is between 5 and 10 kg, giving a score of 1.

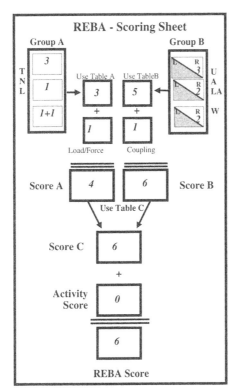

FIGURE 8.4 The correct posture.

Using Table A (Table 8.4) for group A (trunk, neck, and legs), the three posture scores are entered to produce a score of 3. This is added to the load/force score of 1 to produce a score A equal to 4.

Only the right arm is visible in Figure 8.4, so this is the limb that has been scored. It is likely that the left arm was in a similar posture. If so, this could have been recorded using multiangle photography. The upper arm (UA) is in a posture between 45 and 90°, giving a score of 3, while the lower arm (LA) is between 0 and 60°, giving a score of 2. The wrist is obscured in the photograph, but the position was recorded when the photograph was taken. The wrist (W) was extended with the fingers gripping the sheet, giving a score of 2. The coupling is fair, giving a score of 1.

Use Table B (Table 8.5) to find the single posture score from the upper arms, lower arms, and wrist posture scores. This gives a score of 5, which is added to the coupling score (1) to produce a score B of 6.

Score A (4) and score B (6) are entered into Table C (Table 8.6) to produce score C (6). The activity score (0) (no repeated, static, or sudden large range changes in posture) is added to score C. The final REBA score is 6. The action level (Table 8.7) is confirmed as a medium risk level.

8.4.2 Scoring Figure 8.5 — Poor Posture

The posture in Figure 8.5 is different, mostly due the lower bed height. The same process for scoring is used as described for Figure 8.1, and the final REBA score is 11. This is categorized as a very high risk level, with immediate action needed to control the risks. One immediate control measure is to raise the bed height, and an electric bed could facilitate this action (Dhoot and Georgieva, 1996; Hampton, 1998).

Other factors in the risk assessment might include the worker's cooperation, the worker's size, the surrounding environment, the frequency of other manual handling operations, the available resources and equipment (e.g., hoist), and the worker's ability to recognize when a task is beyond his/her capacity and request help.

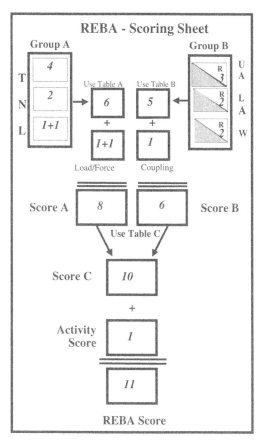

FIGURE 8.5 Incorrect posture.

8.5 Related Methods

There are a number of observational postural analysis tools, with each being developed to meet slightly different aims. To date, REBA is the only postural analysis tool developed to assess animate load handling. Other related tools include:

- OWAS (Karhu et al., 1977)
- RULA (McAtamney and Corlett, 1993)
- HARBO (Wiktorin et al., 1995)
- PEO (portable ergonomic observation) method (Fransson-Hall et al., 1995)
- QEC (quick exposure checklist) (Li and Buckle, 1999)

8.6 Standards and Regulations

REBA was not specifically designed to comply with a particular standard. However, it has been used in the U.K. for assessments related to the Manual Handling Operations regulations (HSE, 1998). It has also been widely used internationally and was included in the draft U.S. Ergonomic Program Standard (OSHA, 2000).

8.7   Approximate Training and Application Times

The training time for REBA is approximately 3 hours, although experience in the use of OWAS and RULA will reduce the time considerably.

It takes less than 2 min to score a posture using pen and paper, and less than 30 sec using a Palm PC.

8.8   Reliability and Validity

Reliability of REBA was established in two stages. The first stage involved three ergonomists/physiotherapists independently coding the 144 posture combinations. They discussed and resolved any conflicts in the scores and then incorporated the additional risk scores for load, coupling, and activity to generate the final REBA score on a range of 1 to 15. The second stage involved two workshops with 14 health professionals using REBA to code over 600 examples of work postures from the health-care, manufacturing, and electricity industries. This established good face validity, and REBA has continued to be widely used, particularly in the health-care industry. Note, however, that small changes were made to the upper-arm code (introduction of the gravity-assisted concept) during the validation process, so additional work is planned to undertake more-detailed reliability and validity testing.

8.9   Tools Needed

REBA is available in the public domain and requires only a photocopy of the tool and the scoring sheets (Figure 8.1) along with a pen. A video recorder or camera may also be useful, but neither is essential.

References

Ahonen, M., Launuis, M., and Kuorinka, I., Eds. (1989), *Ergonomic Workplace Analysis*, Finnish Institute of Occupational Health.

American Academy of Orthopaedic Surgeons (1965), *Joint Motion: Method of Measuring and Recording*, Churchill Livingstone, Edinburgh.

Borg, G. (1970), Perceived exertion as an indicator of somatic stress, *Scand. J. Med. Rehabil.*, 92–93.

Dhoot, R. and Georgieva, C. (1996), The Evolution Bed in the NHS Hospital Environment, unpublished report, The Management School, Lancaster University, Lancaster, U.K.

Fransson-Hall, C., Gloria, R., Kilbom, A., and Winkel, J. (1995), A portable ergonomic observation method (PEO) for computerized on-line recording of postures and manual handling, *Appl. Ergonomics*, 26, 93–100.

Hampton, S. (1998), Can electric beds aid pressure sore prevention in hospitals? *Br. J. Nursing*, 7, 1010–1017.

Hignett, S. and McAtamney, L. (2000), Rapid entire body assessment (REBA), *Appl. Ergonomics*, 31, 201–205.

Hignett, S. (1998), Ergonomics, in *Rehabilitation of Movement: Theoretical Basis of Clinical Practice*, Pitt-Brooke, Ed., W.B. Saunders, London, pp. 480–486.

HSE (1998), Manual Handling Operations Regulations, 1992, 1998, Guidance of Regulations, L23, HSE Books, London.

Hsiao, H. and Keyserling, W.M. (1990), A three-dimensional ultrasonic system for posture measurement, *Ergonomics*, 33, 1089–1114.

Janik, H., Münzbergen, E., and Schultz, K. (2002), REBA-verfahren (rapid entire body assessment) auf einem Pocket Computer, in *Proceedings of 42, Jahrestagung der Deutschen Gesellschaf für Arbeitsmedizin un Umweltmedizin e. V. (DGAUM)*, April 2002, München, p. V25.

Karhu, O., Kansi, P., and Kuorinka, I. (1977), Correcting working postures in industry: a practical method for analysis, *Appl. Ergonomics*, 8, 199–201.

Kuorinka, I., Jonsson, B., Kilbom, A., Vinterberg, H., Biering-Sorenson, F., Andersson, G., and Jorgenson, K. (1987), Standardised Nordic questionnaires for the analysis of musculoskeletal symptoms, *Appl. Ergonomics*, 18, 233–237.

Li, G. and Buckle, P. (1999), Evaluating Change in Exposure to Risk for Musculoskeletal Disorders: A Practical Tool, contract research report 251/1999 (HSE), Her Majesty's Stationery Office, Norwich, U.K.

McAtamney, L. and Corlett, E.N. (1993), RULA: a survey method for the investigation of work-related upper limb disorders, *Appl. Ergonomics*, 24, 91–99.

OSHA (2000), Final Ergonomics Program Standard, regulatory text (draft), OSHA, U.S. Department of Labor, Washington, D.C.

Waters, T.R., Putz-Anderson, V., Garg, A., and Fine, L.J. (1993), Revised NIOSH equation for the design and evaluation of manual lifting tasks, *Ergonomics*, 36, 749–776.

Wiktorin, C., Mortimer, M., Ekenvall, L., Kilbom, Å., and Wigaeus Hjelm, E. (1995), HARBO, a simple computer-aided observation method for recording work postures, *Scand. J. Work Environ. Health*, 21, 440–449.

9

The Strain Index

J. Steven Moore
Texas A&M University

Gordon A. Vos
Texas A&M University

9.1 Background and Application

The strain index (SI) is a method of evaluating jobs to determine if they expose workers to increased risk of developing musculoskeletal disorders of the distal upper extremity (DUE) (Moore and Garg, 1995). The DUE is defined as the elbow, forearm, wrist, and hand. Musculoskeletal disorders of the DUE include specific diagnoses (e.g., epicondylitis, peritendinitis, tendon entrapment at the wrist or finger, and carpal tunnel syndrome) and less specific symptomatic conditions related to the muscle-tendon units of the DUE.

The strain index was derived from physiological, biomechanical, and epidemiological principles. (See Figure 9.1.) According to work physiology, intensity of exertion (as a percentage of task-specific maximal effort), duration of exertion, and duration of recovery time between exertions are the critical parameters for predicting the onset and magnitude of localized muscle fatigue. According to biomechanics, the tensile load of a muscle–tendon unit is the sum of contractile force from the muscle component and elastic force related to elongation (stretching). In addition, when loaded tendons cross joints and change direction, there are localized compressive forces that are proportional to the tensile load and the degree of deviation (joint posture) at that location. Epidemiological studies demonstrate that the magnitude, duration, and frequency of forces related to hand activity are associated with DUE morbidity.

Using these principles, one can propose an index based on two measurements derived from a single cycle of work — total exertion time divided by recovery time. Total exertion time is the sum of the durations of the individual exertions applied by a hand within one job cycle (there may be one or many). Recovery time is cycle time minus exertion time. As exertion time increases (by increasing duration or frequency), recovery time decreases and the index value increases. As exertion time decreases, recovery time increases and the index value decreases. Since the physical stress on the body also depends on the magnitudes of these exertions, more forceful exertions represent greater stress than less forceful exertions. Therefore, the exertion durations in the numerator are "weighted" by their respective intensities.

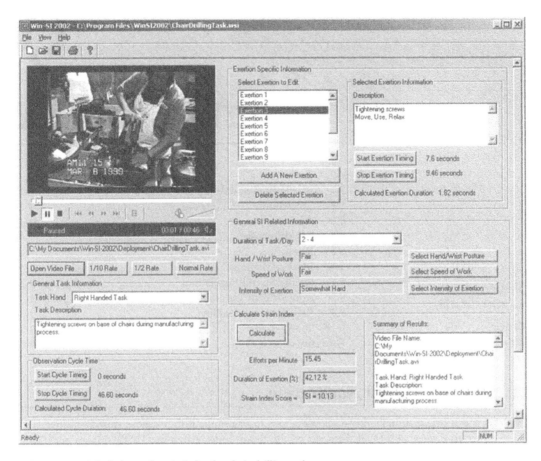

FIGURE 9.1   Calculations of strain index for chair-drilling task.

When watching someone work, it is usually easy to see workers apply forces with their hands, such as grasping, pinching, or pressing, and it may be possible to measure these forces directly. But these external applied forces arise from internal forces in the muscle-tendon units of the DUE. These internal forces are primarily tensile and, depending on joint posture, may also be compressive. Higher applied forces (forceful gripping) imply higher tensile forces in the relevant muscle–tendon units (finger flexor and wrist extensor muscles). If the wrist is straight, the tendons of these muscle–tendon units will not be deflected, so there will be minimal compressive forces. As wrist deviation increases, however, compressive forces will increase. Models of pathogenesis for DUE disorders emphasize the roles of these tensile and compressive forces.

The strain index uses six task variables to describe hand exertions: intensity of exertion, duration of exertion, exertions per minute, hand/wrist posture, speed of work (how fast), and duration per day. Intensity, duration, and posture were discussed above. Exertions per minute accounts for effects related to frequency. Speed of work accounts for reduced recovery efficiency when exertions are highly dynamic. Duration of task per day integrates these stresses across varying durations of task performance. The strain index involves the direct measurement of duration of exertion, efforts per minute, and duration per day and the estimation or direct measurement of intensity of exertion, hand/wrist posture, and speed of work. The values of these task variables represent descriptions of *exposure* (external physical stress). Translation of this information into *dose* and *dosage* (internal physical strain) is done by a set of linking functions that specify multiplier values for the values of the task variables. The strain index score is the product of these six multipliers.

9.2 Procedure

To analyze a job with the strain index, it is important to observe or videotape a representative sample of the job (Moore and Garg, 1995). It is easier to perform the analysis from a videotape, and there is free software available to facilitate the analysis from digitized video files (AVI format). Strain index calculators have been developed by several individuals and organizations. The right and left sides are analyzed separately. The higher score should be used to characterize the job as a whole.

In terms of procedure, there are five steps:

1. Collect data on the six task variables.
2. Assign ordinal ratings using the ratings table (Table 9.1).
3. Determine multiplier values using the multiplier table (Table 9.2).
4. Calculate the SI score (the product of the six multiplier values).
5. Interpret the result.

The simplest analysis, described here, occurs when the job involves a single task and the intensities and postures for each hand exertion are approximately equal. Data collection usually begins with establishing a timeline of the exertions of one hand during a representative job cycle, similar to time and motion study. Durations of individual exertions and the total cycle time can be measured manually with a stopwatch or facilitated by computer software that allows marking of the beginnings and ends of the exertions and job cycle. The duration-of-exertion task variable represents the percent exertion time per job cycle and is calculated by dividing the total exertion time by the cycle time and multiplying by 100. Counts of exertions can be made manually or via software, and the efforts-per-minute task variable is calculated by dividing the number of exertions per job cycle by the total cycle time (minutes). Data on duration of task per day can be measured, but it is usually ascertained by interviewing workers and supervisors. Ratings corresponding to these data are assigned using the ratings table (Table 9.1).

TABLE 9.1   Ratings Table for Finding the Rating Values for Each Task Variable

Rating Values	Intensity of Exertion	Duration of Exertion (%)	Efforts/ Minute	Hand/Wrist Posture	Speed of Work	Duration per Day (hours)
1	light	<10	<4	very good	very slow	≤1
2	somewhat hard	10–29	4–8	good	slow	1–2
3	hard	30–49	9–14	fair	fair	2–4
4	very hard	50–79	15–19	bad	fast	4–8
5	near maximal	≥80	≥20	very bad	very fast	≥8

Note: The rating value for each task variable is found by selecting the appropriate cell from the corresponding column based on the measured or estimated value for that task variable. The corresponding rating value is at the far left on the same row.

TABLE 9.2   Multiplier Table for Finding the Multipliers for Each Task Variable

Rating Value	Intensity of Exertion	Duration of Exertion	Efforts/ Minute	Hand/Wrist Posture	Speed of Work	Duration per Day
1	1	0.5	0.5	1.0	1.0	0.25
2	3	1.0	1.0	1.0	1.0	0.5
3	6	1.5	1.5	1.5	1.0	0.75
4	9	2.0	2.0	2.0	1.5	1.0
5	13	3.0	3.0	3.0	2.0	1.5

Note: The multiplier for each task variable is found by identifying the cell within the task variable's column that corresponds to its rating value at the far left.

Data collection for intensity of exertion, hand/wrist posture, and speed of work is usually done qualitatively using the ratings table directly. Multiplier values for each task variable are determined using the multiplier table (Table 9.2). The strain index score is the product of the six multipliers

Interpretation of the strain index score is the last step. To date, studies of the strain index's predictive validity suggest that a strain index score of 5.0 best discriminates between jobs with and jobs without a history of workers developing distal upper extremity disorders (Knox and Moore, 2001; Moore and Garg, 1995, 1997; Moore et al., 2001; Rucker and Moore, 2002). A job or task with a strain index score less than 5.0 would be considered "safe." A job or task with a strain index score greater than 5.0 would be considered "hazardous." When a hazard is predicted, examination of the multiplier values may reveal intervention strategies that would make the job or task safer.

9.3 Advantages

- It is based on principles relevant to the assessment of exposure for and the pathogenesis of DUE disorders (Moore and Garg, 1995; Moore, 2002).
- It accounts for (1) adverse effects related to the magnitude, duration, and frequency of tensile and compressive forces in the DUE and (2) the beneficial effects of recovery time and limited task duration.
- It is a semiquantitative method using procedures related to time and motion study.
- The outcome allows for dichotomous classification of a job or task that is familiar and practical and allows for simulation of potential interventions.
- Its predictive validity has been demonstrated and statistically modeled in several settings.

9.4 Disadvantages

- It is not a quick pencil-and-paper screening method.
- It is best used by individuals with experience and training.
- It does not account for DUE hazards related to localized compression or hand–arm vibration.
- Methods to analyze jobs characterized by multiple simple tasks performed per day (job rotation) or multiple tasks performed within a job cycle (complex tasks) are under development, but these tend to be complicated and are not validated.

9.5 Example

Examples of the strain index are in the published literature (Knox and Moore, 2001; Moore and Garg, 1995, 1997; Moore et al., 2001; Rucker and Moore, 2002). The output below is from the "Win-SI" program. This program is a freely available and downloadable software tool for implementation of the strain index. Using a digital video clip of a job task, the program obtains all timing-related information directly from the video clip, and automatically calculates the SI score based upon user inputs. The program is intended to improve the training process when instructing on the use of the strain index and to reduce sources of human error in timing and counting the number of exertions present in a job task. The program is available free of charge at http://ergocenter.tamu.edu/win-si.

9.6 Related Methods

There are relatively few methods that target the DUE as a distinct entity. The DUE components of Suzanne Rodgers's method based on localized muscle fatigue and the ACGIH TLV for HAL are probably most comparable. Rapid upper limb assessment (RULA) and rapid entire body assessment (REBA) are not comparable, since they incorporate data from the shoulder and torso. Risk-factor checklists could be considered related, but these suffer from either poor sensitivity or poor specificity in terms of hazard identification.

9.7 Standards and Regulations

The strain index has been identified as an acceptable method of job analysis in the ergonomics regulations promulgated by OSHA and adopted by Washington State. The strain index methods are consistent with the principles of job analysis mentioned in the ASC Z-365 draft standard.

9.8 Approximate Training and Application Time

Training time for only the strain index is typically one day. Application time varies with task complexity. Excluding observation or videotaping time, simple jobs can be analyzed in a few minutes, while complex jobs may take up to an hour.

9.9 Reliability and Validity

The predictive validity of the strain index has been demonstrated in pork processing, poultry processing, and two manufacturing settings. In terms of reliability, task-variable data and ratings have intraclass correlation coefficients (ICCs) between 0.66 and 0.84 for individuals and 0.48 to 0.93 for teams (Stevens, 2002). ICCs for strain index scores were 0.43 and 0.64 for the individuals and teams, respectively (Stevens, 2002). For hazard classification, KR-20 was 0.91 for the individuals and 0.89 for the teams (Stevens, 2002). Regarding test–retest reliability (specifically long-range stability), task ratings have stability coefficients between 0.83 and 0.93 for individuals and 0.68 to 0.96 for teams (Stephens, 2002). Stability coefficients for strain index scores were 0.7 and 0.84 for the individuals and teams, respectively (Stephens, 2002). For hazard classification, the stability coefficient was 0.81 for the individuals and 0.88 for the teams (Stephens, 2002).

9.10 Tools Needed

A video camera is recommended. If the recording is to be digitized for analysis using software, an appropriate computer interface is also necessary. Manual analysis is best performed using a stopwatch to measure time intervals and a lap counter or fingers to count exertions.

References

Knox, K. and Moore, J.S. (2001), Predictive validity of the strain index in turkey processing, *J. Occup. Environ. Med.*, 43, 451.

Moore, J.S. and Garg, A. (1995), The strain index: a proposed method to analyze jobs for risk of distal upper extremity disorders, *Am. Ind. Hyg. Assoc. J.*, 56, 443.

Moore, J.S. and Garg, A. (1997), Participatory ergonomics in red meat packing plant, Part 2: case studies, *Am. Ind. Hyg. Assoc. J.*, 58, 498–508.

Moore, J.S., Rucker, N.P., and Knox, K. (2001), Validity of generic risk factors and the strain index for predicting non-traumatic distal upper limb disorders, *Am. Ind. Hyg. Assoc. J.*, 62, 229–235.

Moore, J.S. (2002), Proposed biomechanical models for the pathogenesis of specific distal upper extremity disorders, *Am. J. Ind. Med.*, 41, 353–369.

Rucker, N.P. and Moore, J.S. (2002), Predictive validity of the strain index in manufacturing, *Appl. Occ. Env. Hyg.*, 17, 63.

Stevens, E. (2002), Inter-Rater Reliability of the Strain Index, Master's thesis, Texas A&M University, Bryan.

Stephens, J.-P. (2002), Test-Retest Repeatability of the Strain Index, Master's thesis, Texas A&M University, Bryan.

10

Posture Checklist Using Personal Digital Assistant (PDA) Technology

Karen Jacobs
Boston University

10.1 Background and Applications

Paper, pencil, and video cameras have prevailed as the de facto method of data entry for posture checklists by ergonomists and occupational health professionals (e.g., occupational therapists). This can severely burden the evaluator with large quantities of paperwork, compromise the accuracy of the information gained, and increase the associated costs of the delivery and planning of ergonomics services. Errors can occur upon completing the posture checklists at the point of data entry, while transcribing the data for digitization to a computer database, or while processing the information for tabulating scores and generating reports. Virtually all of the "repackaging" of data recorded on paper to electronic format for distribution to centralized or remote databases occurs at the evaluator's office, where it must be transcribed by hand or through an optical character recognition (OCR)-based system. Portable and mobile solutions are therefore needed, not only to facilitate electronic data entry at the point of service, but also to distribute it to various end users and devices, where these data can be stored or processed for report generation or integrated with other databases.

The handheld palm computers, referred to collectively as personal digital assistants (PDAs), are an emerging solution for remote data entry and management tasks. The market for health and rehabilitation applications for PDAs is set to explode. Two million of these devices were sold in the year 2000 alone, with projected sales to exceed 6 million by 2003 (Giusto, 2000). Originally developed a decade ago as a personal organizer featuring little more than an appointment calendar and address book, PDAs have rapidly evolved into true handheld personal computers (PCs). The boom in their popularity is directly related to their ability to deliver most, if not all, of the features of a PC with the added benefit of complete mobility and a pen interface instead of a keyboard, although a variety of keyboard solutions is also available. The pen interface allows users to interact with the PDA in a natural and familiar way by entering text, numbers, and graphics in "electronic ink" directly on the screen (Figure 10.1).

Most of the health and rehabilitation applications of the PDA have been directed at the physician (Sittig et al., 1998). The most popular medical applications are for patient-tracking systems, followed by formulary and prescription-printing systems that link and cross-reference Web-based pharmaceutical databases and that warn the prescribing physician of possible harmful drug interactions or contraindications. Other popular applications include portable electronic reference materials, decision-support tools, and billing systems.

A posture checklist found in a document developed by the Department of Defense (2002) has also been implemented on a PDA (Field Informatics, 2001). The checklist has been designed as an information guide for use by supervisors and end users. Typically, postures are documented in terms of the position of joints or body segments or duration from neutral position (Bohr, 1998). Indeed, the army computer workstation checklist (ACWC) included in this implementation evaluates the individual's posture while engaged in computer work. Typical measurements often used in describing posture — height, breadth, distance, curvature, and depth — are integrated into the interface between the user and the computer workstation.

10.2   Procedure

The ACWC (army computer workstation checklist) has nine sections that evaluate work area (desk, chair, footrest, monitor, document holder), work practice, input devices (keyboard, mouse, trackball), and lighting and glare. The user evaluates each of these areas by checking a "yes" or "no" response. If the answer is "no" to any of the questions, a potential problem exists.

In the PDA version, each of the nine areas has a separate drop-down window that cues the user to respond to questions with a "yes" or "no" response (Figure 10.2).

Data can also be easily input into the PDA using a digital camera (Figure 10.3).

Once the checklist is completed, any "no" responses automatically generate possible solutions, which are included in a report that can be printed on-site (Figure 10.4).

10.3   Advantages

- Measurements reflect actual working postures.
- Standards specified in the checklist are based on ANSI/HFES and ISO standards (ANSI/HFES 100-1988, 1988; ISO 9241-1, 1997; ISO 9241-2, 1992; ISO 9241-3, 1992; ISO 9241-4, 1998; ISO 9241-5, 1998; ISO 9241-6, 1999; ISO 9241-11, 1998) and Grandjean (1992).

FIGURE 10.1 Example of the PDA pen and keyboard input.

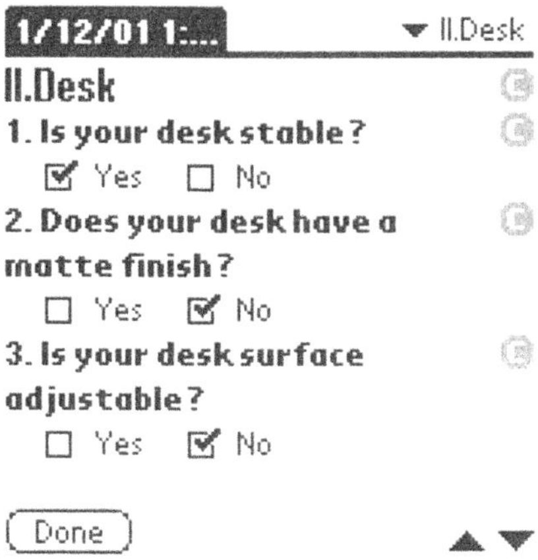

FIGURE 10.2   Example of the PDA drop-down window for the area "desk."

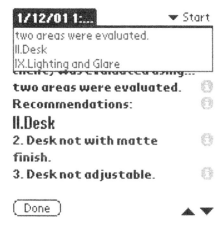

FIGURE 10.3   A digital camera that attaches to the PDA allows photographs to be directly inserted into the checklist and final report.

FIGURE 10.4   Example of PDA screen for the report of possible solutions generated from the checklist.

- It is easy to input data and produce summary analyses and reports.
- The program enters the data directly into a database.

10.4   Disadvantages

- No reliability or validity data are available.
- Standards specified in the checklist may be dated because they are based on the 1988 American National Standard for Human Factors Engineering of Visual Display Terminal Workstations, (ANSI/HFS Standard 100-1988).
- The checklist does not include psychosocial factors.

10.5   Approximate Training and Application Times

The initial training (self-learning) time of the new user is approximately 20 min. More time may be need for users who are unfamiliar with the PDA technology and the use of drop-down windows.

10.6   Reliability and Validity

No reliability or validity data are available. However, standards specified in the checklist are based on ANSI/HFES 100-1988 (1988), BSR/HFES 100 (2002), ISO 9241-1 (1997), ISO 9241-2 (1992), ISO 9241-3 (1992), ISO 9241-4 (1998), ISO 9241-5 (1998), ISO 9241-6 (1999), ISO 9241-11 (1998), Grandjean et al. (1983), and Grandjean (1992).

10.7   Tools Needed

The PDA version of the army checklist can be downloaded free from the Field Informatics Web site at http://www.fieldinformatics.com. The user will need a PDA with a 3.0 operating system (OS).

Acknowledgment

This author would like to acknowledge the contributions of information in this chapter by Dr. Serge Roy, Dr. Patrick Boissey, and Dr. Francois Galilee, formerly of Field Informatics LLC.

References

ANSI/HFS 100-1988 (1988), American National Standard for Human Factors Engineering of Computer Workstations, Human Factors and Ergonomics Society, Santa Monica, CA.

BSR/HFES 100 (2002), Draft American National Standard for Human Factors Engineering of Computer Workstations. Human Factors and Ergonomics Society, Santa Monica, CA

Bohr, P. (1998), Work analysis, in *Sourcebook of Occupational Rehabilitation*, King, P., Ed., Plenum Press, New York, p. 229–245.

Department of Defense (DOD) Ergonomics Working Group (2002), Creating the Ideal Computer Workstation: A Step-by-Step Guide, June 2002.

Giusto R. (2000), Mobility by the numbers, *Mobile Comput. Commun.*, 11, 22–23.

Grandjean, E., Hunting, W., and Pidermans, M. (1983), VDT workstation design: preferred settings and their effects, *Hum. Factors*, 25, 161–175.

Grandjean, E. (1992), *Ergonomics in Computerized Offices*, Taylor & Francis, New York.

ISO 9241-1 (1997), Ergonomic Requirements for Office Work with Visual Display Terminals (VDTs), Part 1: General Introduction, International Organization for Standardization, Geneva.

ISO 9241-2 (1992), Ergonomic Requirements for Office Work with Visual Display Terminals (VDTs), Part 2: Guidance on Task Requirements, International Organization for Standardization, Geneva.

ISO 9241-3 (1992), Ergonomic Requirements for Office Work with Visual Display Terminals (VDTs), Part 3: Visual Display Requirements, International Organization for Standardization, Geneva.

ISO 9241-4 (1998), Ergonomic Requirements for Office Work with Visual Display Terminals (VDTs), Part 4: Keyboard Requirements, International Organization for Standardization, Geneva.

ISO 9241-5 (1998), Ergonomic Requirements for Office Work with Visual Display Terminals (VDTs), Part 5: Workstation Layout and Postural Requirements, International Organization for Standardization, Geneva.

ISO 9241-6 (1999), Ergonomic Requirements for Office Work with Visual Display Terminals (VDTs), Part 6: Guidance on the Work Environment, International Organization for Standardization, Geneva.

ISO 9241-11 (1998), Ergonomic Requirements for Office Work with Visual Display Terminals (VDTs), Part 11: Guidance on Usability, International Organization for Standardization, Geneva.

Sittig, D.F., Kuperman, G.J., and Fiskio, J. (1998), Evaluating Physician Satisfaction Regarding User Interactions with an Electronic Medical Record System, technical report, Clinical Systems Research & Development, Partners Healthcare System Publication, Boston, MA, pp. 1–4.

Acknowledgment

This author would like to acknowledge the contributions of information in this chapter by Dr. Serge Roy, Dr. Patrick Boissey, and Dr. Francois Galilee, formerly of Field Informatics LLC.

11

Scaling Experiences during Work: Perceived Exertion and Difficulty

Gunnar Borg
Stockholm University

11.1 Background and Application

The concept of perceived exertion and the associated methods for measuring relevant variables were introduced to improve our understanding of physical work and its "costs" (Borg, 1962). The human sensory system can function as an efficient instrument to evaluate the workload by integrating many peripheral and central signals of strain. Psychophysical methods have been developed to complement the physiological methods. Because the way a person experiences the work and the situation is so fundamental to his/her adaptation, performance, and satisfaction, subjective assessments are also judged to be of value in themselves related to quality of life.

The main methods used to measure subjective experiences have used numerical rating scales (e.g., 1 to 6) anchored, in a symmetric way, with verbal expressions defining the scale points. In such cases, scale responses can only be ordered by rank. In psychophysical scaling, this is a great drawback. Ratio scaling methods have been developed for most kinds of sensory stimulus–response (S–R) functions (Stevens, 1975), and this functions well for relative S–R functions, but not for direct levels of intensity. Simple rating methods give direct levels, but they do not have good metric properties. Two new scaling methods have been developed. The first was initially developed for clinical diagnosis of overall perceived exertion, breathlessness, muscle fatigue, and pain. The second was developed for evaluation of most kinds of perceptions and feelings, including experiences of work tasks (of importance in human factors and ergonomics). Studies of effort and difficulty were originally applied in MMH (manual materials handling) and then for most other kinds of physical and mental work.

11.2 Procedure

11.2.1 The Borg RPE Scale®

Several principles and many experiments lay behind the construction of the Borg RPE (ratings (R) of perceived (P) exertion (E)) Scale (Borg, 1998). One principle was that the position of the verbal anchors on a numerical scale could be used to adjust the form of the growth function. A simple, symmetrical rating scale gave a negatively accelerating function with workload and heart rate during ergometer tests. A better approach would be to construct a scale that would increase linearly with oxygen consumption. By iterative trials, it was possible to construct a scale that grew linearly with workload and thus remained equidistant with regard to aerobic demands. By using 6 as the lowest number and 20 as the highest on the scale, a simple relation with heart rate for healthy middle-aged people was obtained. See Table 11.1

TABLE 11.1 The Borg RPE Scale®

6	No exertion at all
7	Extremely light
8	
9	Very light
10	
11	Light
12	
13	Somewhat hard
14	
15	Hard (heavy)
16	
17	Very hard
18	
19	Extremely hard
20	Maximal exertion

Source: Borg, G. (1998), *Borg's Perceived Exertion and Pain Scales*, Human Kinetics, Champaign, IL.
©Gunnar Borg, 1970, 1985, 1998.

and Figure 11.1.

11.2.2 Category-Ratio (CR) Scaling

A CR scale incorporates the best properties of a category-rating (C) scale for "absolute" levels of intensity and a ratio (R) scale for good metric properties. The most common CR scale is now the Borg CR10 scale. The fundamental principle behind the scale construction was that it should be a "ratio scaling" method (Stevens, 1975), even if the outcome only resulted in a "semiratio" scale (i.e., a scale that is not a real ratio scale but can function as such for descriptive purposes and predictions). Verbal anchors were used after careful determination of their "true" positions on a ratio scale. Quantitative semantics were used to determine the intensity associated with adjectives and adverbs. Since these can function partly as multiplicative constants, a first estimation could be obtained regarding "interpretation" (meaning and level) and preciseness (interindividual agreement). The range model was also an important principle, since most people and most modalities have about the same perceptual range. The size of this range (N) varies roughly according to a "magic exponent" 6 ± 1 in the equation:

$$N = a^{c-1}$$

where a is 2 and c is the number of categories. Thus if $c = 6$, then a minimal range is 1:32.

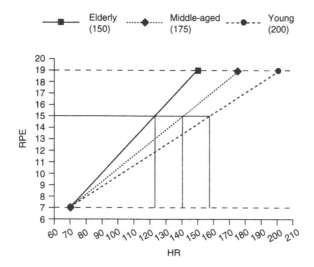

FIGURE 11.1 The relation between RPE and HR for work on a bicycle ergometer in three different age groups with different maximal HR (150, 175, and 200 beats/min).

A specific main anchor as a "fixed point" is 10 on the CR10 scale. For perceived exertion, the best anchor is the highest perception a person has previously experienced. "Maximal" perceived exertion and heaviness is also a good anchor for cross-modal comparisons. To avoid truncation effects, the scale is open at both ends, thus allowing the possibility of obtaining a rating higher or lower than the anchored end points. To obtain a congruence in meaning between numbers and verbal anchors, many experiments were performed with iterative trials to correct the positions of all anchors (Borg and Borg, 1994). Special effort was also devoted to the visual design of the scale (Borg and Borg, 2001).

The CR10 scale is shown in Table 11.2. The scale is a general intensity scale for most perceptual intensities, experiences, and emotions, not only perceived exertion and difficulty. The scale is a direct scale, providing a direct response that can be interpreted directly. It can be used in estimation studies as well as for production of intensities, thus allowing the possibility of two-way communication from the individual to the test leader and the test leader to the individual. That is not possible with the popular visual analogue (VAS) scale.

11.2.3 Administration of the Scales

To obtain reliable and valid results, scale administration should follow general psychometric principles, i.e., the test situation should be well prepared, and the test leader should be thoroughly familiar with the scaling methods and know how to test and evaluate. A written consent that the person is willing to participate may be needed.

The purpose of the test should be carefully explained to the subject. S/he must feel comfortable answering specific questions about feelings related to the work and the situation. Since there are so many different kinds of tasks and situations, and the purpose may either be to select or to change people or tasks, the explanation has to be well adapted to the specific purpose.

The definitions of questions/variables should be simple and clear according to common constitutional and contextual definitions so that the test administrator can convert these to valid operational definitions. It is fundamental that the administrator and respondent clearly understand what will be rated. For perceived exertion, this could be the overall perceived exertion or some specific symptom, such as breathlessness, muscle fatigue, pain, or other symptoms.

The test situation — for example, the presence of instruments or other people — can influence the participant. The person must be well motivated and feel that he/she can trust the test administrator. Even rather simple questions about perceived exertion and difficulty may be interpreted as sensitive.

TABLE 11.2 The Borg CR10 Scale

0	Nothing at all	
0.3		
0.5	Extremly weak	Just noticeable
0.7		
1	Very weak	
1.5		
2	Weak	Light
2.5		
3	Moderate	
4		
5	Strong	Heavy
6		
7	Very strong	
8		
9		
10	Extremely strong	"Maximal"
11		
•	Absolute maximum	Highest possible

Source: Borg, G. (1998), *Borg's Perceived Exertion and Pain Scales*, Human Kinetics, Champaign, IL.
©Gunnar Borg, 1982, 1998, 2003

Specific instructions should be given well before the test so that the respondent understands the scale and how the verbal anchors and numbers should be used. S/he should try to be as spontaneous and natural as possible and not underestimate or overestimate what is going to be rated, instead honestly expressing what s/he feels without considering what the difficulty may be in an "objective" sense. To accomplish this, the test administrator and the respondent must have a good rapport. The objective instruction should be subjectively checked by the test administrator to ensure comparable understanding across individuals to be evaluated.

A simple answer to the problem of how to evaluate a response cannot be given, since the scales are adapted for so many different kinds of situations and tasks. The test administrator must review the literature related to the specific tasks being evaluated (e.g., for perceived exertion [Borg, 1998]). In most situations, a rating above "moderate" or "somewhat strong" is too strong. "When it's hard, it's too hard." However, there may not be any "dangerous strain" in heavy work (performed in a correct way), even if a few tasks are "very heavy" (7 on CR10). Ergonomic improvements should diminish the difficulty of such tasks, which helps to avoid gender discrimination. The criterion of what is "too heavy and risky" also depends on duration without sufficient pauses. High-frequency, repetitive, monotonous work that endures for hours, e.g., work in the fast-checkout line is a kind of "insidious task," since it is not perceived to be especially hard. Even if it is only perceived to be of "light" to "moderate" intensity, it may in the long run be a risky job ("Cinderella" kind).

A good response protocol should be used for recording responses. It is also important to provide room for notations of special "qualitative" observations.

11.3   Advantages

- The RPE scale is easy to use, and the instruction is simple.
- Linear relations are obtained for work with high aerobic demands. The scale gives an individualized measure of exercise intensity.
- The CR10 scale can be used in a similar way as the RPE scale, but the CR10 scale has the advantage of determining "absolute" (level anchored) S–R functions.
- CR scales can be used for most kinds of experiences.
- A profile of symptoms can be obtained.

11.4   Disadvantages

- Responses obtained by the RPE scale do not reflect "true" growth functions.
- The RPE scale can only be used for perceived exertion and related symptoms.
- The CR10 scale is more complicated in its construction than the RPE scale. It is therefore more difficult to understand and requires more time for explanation, instruction, and training.

11.5   Example Output Table and Figure

Perceived exertion according to the RPE scale grows, like heart rate (HR), linearly with workload for aerobic work on a bicycle ergometer (see Figure 11.1). The form of the relation does not change with age, but the "absolute" levels do because of the decrease in maximum HR with age. When RPE is studied for different kinds of work, e.g., when cleaning is performed with mopping (rather light) and with swabbing (rather heavy) with increasing duration, e.g., 30 min, the function over time follows a very different course, with a rather level course for HR but an increasing function for RPE. The RPE function better reflects the increase in fatigue than the HR function (Winkel et al., 1983).

For short-term work in lifting objects, the growth function is positively accelerating when using the CR10 scale. A function for each individual can be obtained from a very light load to a heavy load. Predictions of functional capacity can then be made by extrapolations from submaximal ratings according to Figure 11.2. The extrapolation is possible by utilizing the knowledge of the numerical relations between

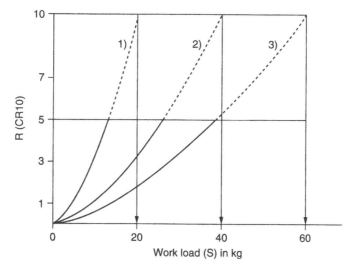

FIGURE 11.2   The figure shows the form of S–R functions for muscular effort and how functional capacity can be extrapolated from submaximal responses.

the anchors and the form of the S–R function.

Figure 11.3 shows a profile of symptoms for different kinds of physical or mental work. Several profiles can be obtained showing which factors are essential for specific tasks or groups of people. When the interest is focused on individual differences, e.g., in clinical diagnostics or tests of capacity to manage tasks of varying difficulty, these profiles will help to treat or select individuals. When the main problem is to find out weak points in products or situations, the profiles will help to identify critical factors that have to be improved. It is then important to be able to utilize one and the same scale for evaluations. There is seldom a need to have different scales for different tasks. This is true for physical as well as mental work and for "information ergonomics," including the design of signals or pictures, e.g., icons

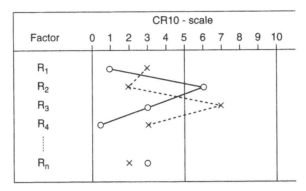

FIGURE 11.3 A profile of symptoms showing which "factors" are of critical importance for different persons or different products.

and symbols for navigation in computer work. Instructions and additional anchors are the tools to use for specifying what to rate.

11.6   Related Methods

Related methods use simple rating scales with weak metric properties (see p. 11-1), and interval scaling methods are represented by different partition scales, such as equisection or z-scales, where a specified range of intensities is divided into equal steps. The visual analogue scale (VAS) can be an interval scale. The relative ratio scales are the scales constructed by Stevens (1975) presented above. To this class may be added some methods that combine different ratio scales, e.g., with the help of cross-modal estimations or matchings (see Gescheider, 1997).

Most of these scales are constructed to cover a large intensity range. They can be used for both estimation and production of a desired workload. Some production methods are used to determine specific levels, such as an acceptable or preferred level, that the person can comfortably work at for a long time period (Mital et al., 1993).

11.7   Standard and Regulations

In 1981 the International Organization for Standardization (ISO) accepted the following principle with regard to work tasks and situations: "The work environment shall be designed and maintained so that physical, chemical and biological conditions have no noxious effect on people but serve to ensure their health, as well as their capacity and readiness to work. Account shall be taken of objectively measurable phenomena and of subjective assessments" (ISO, 1981). Consequently, it is important to agree on standardized methods for "subjective assessments." But currently there are no common standards in this area. Some methods are preferred or recommended by different organizations, but as yet there is no consensus. The use of the Borg RPE Scale is recommended by the American College of Sport Medicine (ACSM) guidelines for diagnostic assessments and prescriptions (Franklin, 2000) and by the American Heart Association (AHA) Science Advisory for both aerobic work and short-term resistance training (Pollock et al., 2000).

The use of the Borg CR10 scale is recommended by the International Ergonomics Association (IEA) technical committee on musculoskeletal disorders (TC13) for subjective assessments of force.

11.8   Approximate Training and Application Time

The Borg RPE scale is simple to apply. The instruction takes a couple of minutes, and most subjects immediately understand the scale and know how to rate overall perceived exertion or specific strain. The

CR-10 scale is more complicated in its construction, requiring more-detailed instruction and a follow-up to verify that the subject understands the scale. It may also be helpful to use training questions from other modalities: How sour is a lemon? (about 7); How black is a piece of black velvet? (about 9); How sweet is a ripe banana? (about 3).

11.9   Reliability and Validity

The reliability and validity of the RPE scale have been determined in intratests, parallel tests, and retests, with correlations above 0.90. Results for the CR10 scale are about the same.

The validity of the two scales is also very similar. More studies have been performed with the RPE scale showing good construct validity, concurrent validity, and predictive validity. Correlations between RPE and HR are about 0.80. Estimates of functional capacity obtained from ratings in ergometer tests have been correlated with field criteria, e.g., results from competitions in sports or salaries. Most correlations are between 0.50 and 0.70.

The validity of the form of the growth functions is a special problem. A main difference between the scales is that the RPE scale gives a linear relation to aerobic demands, while the CR10 scales gives positively accelerating functions (power functions with exponents about 1.6). The latter is more in congruence with other studies of effort and fatigue (Borg, 1998).

11.10   Tools Needed

The scales used should be those constructed and designed by Borg without any additional cues such as colors or pictures. The administration of the scales should follow the given rules. Responses can be given verbally, or the subject can use a calculator, point with a finger, or move a mouse.

References

Borg, G. (1962), Physical performance and perceived exertion, in *Studia Psychologica et Paedagogica*, series altera, Investigationes XI, Gleerup, Lund, Sweden.

Borg, G. and Borg, E. (1994), Principles and experiments in category-ratio scaling, No. 789 in Reports from the Department of Psychology, Stockholm University, Stockholm.

Borg, G. (1998), *Borg's Perceived Exertion and Pain Scales*, Human Kinetics, Champaign, IL.

Borg, G. and Borg, E. (2001), A new generation of scaling methods: level-anchored ratio scaling, *Psychologica*, 28, 15–45.

Franklin, B.A., Ed. (2000), *ACSM's Guidelines for Exercise Testing and Prescription*, 6th ed., Lippincott Williams & Wilkins, Philadelphia.

Gescheider, G.A. (1997), *Psychophysics: The Fundamentals*, 3rd ed., Lawrence Erlbaum, Mahwah, NJ.

International Organization for Standardization (1981), Ergonomic Principles in the Design of Work Systems, part 4.2, ISO 6385, ISO, Geneva.

Mital, A., Nicholson, A.S., and Ayoub, M.M. (1993), *A Guide to Manual Materials Handling*, Taylor & Francis, London.

Pollock, M., Franklin, B., Balady, G., Chaitman, B., Fleg, J., Fletcher, B., Limacher, M., Piña, I., Stein, R., Williams, M., and Bazzarre, T. (2000), Resistance exercise in individuals with and without cardiovascular disease, AHA science advisory, *Circulation 2000*, 101, 828–83.

Stevens, S.S. (1975), *Psychophysics: Introduction to Its Perceptual, Neural and Social Prospects*, Wiley, New York.

Winkel, J., Ekblom, B., Hagberg, M., and Jonsson, B. (1983), The working environment of cleaners: evaluation of physical strain in mopping and swabbing as a basis for job redesign, in *Ergonomics of Workstation Design*, Kvålseth, T.O., Ed., Butterworths, Woburn, MA.

12

Muscle Fatigue Assessment: Functional Job Analysis Technique

Suzanne H. Rodgers
Consultant in Ergonomics

12.1 Background

This muscle fatigue assessment method (MFA), also known as the functional job evaluation technique, was developed by Rodgers and Williams (1987) to characterize the discomfort described by workers on automobile assembly lines and fabrication tasks. When observing such workers, it was apparent that they were accumulating fatigue in some muscle groups as their shift progressed. Their perceived discomfort could not always be explained by biomechanical analyses of the job, but seemed to relate to their temporal work pattern. As task duration increased, some workers took shortcuts to get the efforts done more quickly than the standard required. From discussions, workers reported that they sped up their work to increase the recovery time for fatigued muscles after each effort cycle.

Since the workers appeared to be monitoring their fatigue, a method was sought that would estimate the quantity of accumulated fatigue in a task. Studies of physiological muscle fatigue at different effort levels and holding times provided the basis for this method (Monod and Scherrer, 1957; Rohmert, 1960a, 1960b). The frequency of muscle efforts determines how much recovery time is available between efforts. The amount of fatigue accumulated in a muscle during a task can be characterized by estimating how much time is needed from the isometric work/recovery time curves (Rohmert, 1973a) and comparing this to the actual time between efforts of the same intensity. If the cycle time is short, one can sum the deficit over a 5-min period and define the amount of fatigue accumulated. The greater the deficit, the more problematic the task is, especially if it is performed for more than 5 min.

To make the method easier to use, and to help prioritize between tasks when choices have to be made about which problem should be addressed first, the effort levels, effort durations (or holding times), and effort frequencies were reduced to three categories each. This was done by using multiple combinations of the three factors to calculate how much fatigue had accumulated. Four fatigue outcomes were chosen from the more detailed analysis, all calculated based on a continuous period of 5 min of work on the task. These were <30 sec (low), 30 to 90 sec (moderate), >90 sec to 3 min (high), and >3 min (very high). These outcomes are used to define the "priority for change" on the task. This is helpful when several tasks have been identified as needing improvements and one has to decide where to start. For the task analysis of each body part, the frequencies chosen were <1 per minute, 1 to 5 per minute, and >5 to 15 per minute. Holding times, or effort durations, were <6 sec, 6 to 20 sec, and 20 to 30 sec. Effort levels were chosen as percents of maximum strength in the postures used and were as follows: low = <40% max, moderate = 40 to <70% of maximum, and heavy = ≥70% of maximum. The effort intensity could be found by looking at a list of postural descriptors representing different degrees of risk when combined with force. Alternatively, the 10-point scale for large-muscle-group activity (Borg and Lindblad, 1976) could be used to let workers define the effort level for each body part. Multiplying the scale rating by 10 approximates the percent maximum muscle effort for the fatigue assessment rating.

The MFA method has been used to analyze more than 1000 jobs and tasks in more than 100 ergonomics team training courses between 1987 and 2002, and some modifications have resulted from this use. A fourth level for each of the factors has been added that defines when the method will underestimate the fatigue or injury risk. If any of the factor ratings are a 4, the analyst is told to automatically give the task the highest priority for change. A fourth factor has also been added to the analysis after the potential for fatigue has been determined. This factor is the total continuous task time before another activity is done. The total task time determines what some of the options can be to address the fatigue on a short-term basis, but it does not reduce the importance of making changes to reduce the tasks rated as high and very high priority for change. In 2002, Bernard altered the format by placing the table of "effort intensity" definitions into the body of the form and having a separate table for the "priority for change" score based on the three-number ratings when he placed the MFA method (also known as the Rodgers fatigue tool) on his Web site (Bernard and Rodgers, 2002).

12.2 Applications

This MFA method works best for evaluating production tasks having fewer than 12 to 15 repetitions per minute with the same muscle groups. It also can be used for studying some service and office jobs, but it will underestimate postural loads that are sustained continuously for more than 30 sec. It is not appropriate for use if fatigue is not likely to occur on a task, e.g., performing an occasional heavy lift. Any task that is beyond the capacity of more than half of the potential workforce should be fixed based on that very high effort level. Fatigue should only be a consideration if the effort is initially within reasonable guidelines.

The MFA method works best if all of the muscle groups are rated for each task, not just the ones that appear to be most involved in the heavier work. Some of the less heavily loaded muscles may have a combination of holding time and frequency of use that makes them more vulnerable to fatigue than the muscles that may be involved in short, heavy efforts. Also, when improvements to the tasks are identified, it is wise to rerate the task and all of the body parts again, because the proposed improvement may have shifted the load to another muscle group that now becomes the limiting one.

The MFA method can define which jobs might be appropriate for people to work on for a short term during initial return-to-work after an injury or illness. By rating all body parts on a task, those tasks that might exacerbate a muscle or joint problem can be separated from those tasks that should be acceptable for the injury or illness of concern during a short-term rehabilitation period. This reduces the need for general work restrictions and minimizes the chance of reinjury.

This method significantly underestimates accumulated fatigue in highly repetitive tasks where the muscle effort frequency is ≥30 per minute. High repetition represents almost continuous static efforts

TABLE 12.1 Muscle Fatigue Assessment Method Procedure

Procedure Step	How
1. Identify a problem job	Presence of injuries or complaints on job
	Difficulty of finding people who can do job
	Job reputation with workers/supervisors
2. Identify problem tasks on job	Ask workers/supervisors experts on job to rate difficulty, duration of task, frequency of doing task, and number of people exposed
	Review accident and injury/illness data
3. Select a task to analyze	Prioritize by worker and supervisory ratings in step 2.
4. Determine effort intensity levels for each body part	Use unedited videotape (4 to 6 min of continuous taping at a minimum) to study the task
	Ask workers to rate effort intensity using a psychophysical scale (Borg and Lindblad, 1976; Borg, 1998)
	Define effort level from definitions in Figure 12.1.
	Rate effort for right and left sides, if appropriate
	If more than one level of effort is present, (e.g., moderate and heavy), include both
5. Determine effort duration, in seconds, for each effort intensity for each body part	Use a stopwatch to time the seconds of continuous effort at a given effort intensity before a different effort intensity or relaxation occurs
6. Determine the frequency of efforts per minute at the same effort intensity for each body part	Count the number of new efforts at a given effort intensity over 1 min; if the task is very variable, measure the frequency over 5 min and divide by 5 to get the efforts per minute
7. Using the three-number rating generated from steps 4 to 6 above, determine the "priority for change" score. Put it in the last column for each body part.	Use Table 12.2 to obtain the "priority for change" rating from the three-number rating based on effort intensity, effort duration, and effort frequency for each body part.
8. Figure out how much the high and very high "priority for change" ratings would have to be changed to move them to a low priority for change	Start with the three-number rating and with any component that is rated a 3; figure out how much better the priority rating will be if the 3 is reduced to a 2; repeat until a low "priority for change" rating is found (see Table 12.2)
9. Identify why the three-number ratings are as they are for the high and very high "priorities for change"; develop several strategies to address the root causes	Use a problem-solving process to find the root causes for the risk factors identified by body part; keep asking "why" until root cause is identified
10. Rerate the task using all body parts to determine the impact of a suggested change on worker comfort or fatigue	Rate all body parts to be sure problem has not just shifted to another set of muscles.

for the active muscles because the time between contractions is too short for full relaxation and for adequate blood flow to be reestablished.

The MFA method is ideal for team evaluations of a task or job. Strategies for decreasing the risk exposure can be developed by defining the effort level through identification of postural and force-intensity risk factors as well as evaluating the pattern of work that either increases or decreases the risk of fatigue. Asking why the posture is there, why the force is so high, or why the holding time is so long for the effort can lead to new strategies for how to improve the task.

12.3 Procedure

Table 12.1 outlines the steps for using the MFA method shown in Figure 12.1 and Table 12.2. Table 12.3 shows the three-number ratings in order of increasing fatigue. Moving from bottom to top and right to left in Table 12.3 indicates reduced fatigue. The best improvements will move from a high or very high priority for change to a low priority for change. Moving up in the table within one priority indicates an improvement.

Job				Analyst			
Task				Date / /			

Region	Effort Level < 75% of All Workers Can Exert Effort - 4			Scores			Priority
	Light - 1	Moderate - 2	Heavy - 3	Effort	Dur	Freq	
Neck	Head turned partly to side, back or slightly forward	Head turned to side; head fully back; head forward about 20°	Same as Moderate but with force or weight; head stretched forward				
Shoulders	Arms slightly away from sides; arms extended with some support	Arms away from body, no support; working overhead	Exerting forces or holding weight with arms away from body or overhead	R L	R L	R L	R L
Back	Leaning to side or bending arching back	Bending forward; no load; lifting moderately heavy loads near body; working overhead	Lifting or exerting force while twisting; high force or load while bending				
Arms / Elbow	Arms away from body, no load; light forces lifting near body	Rotating arms while exerting moderate force	High forces exerted with rotation; lifting with arms extended	R L	R L	R L	R L
Wrists / Hands / Fingers	Light forces or weights handled close to body; straight wrists; comfortable power grips	Grips with wide or narrow span; moderate wrist angles, especially flexion; use of gloves with moderate forces	Pinch grips; strong wrist angles; slippery surfaces	R L	R L	R L	R L
Legs / Knees	Standing, walking without bending or leaning; weight on both feet	Bending forward, leaning on table; weight on one side; pivoting while exerting force	Exerting high force while pulling or lifting; crouching while exerting force	R L	R L	R L	R L
Ankles / Feet / Toes	Standing, walking without bending or leaning; weight on both feet	Bending forward, leaning on table; weight on one side; pivoting while exerting force	Exerting high force while pulling or lifting; crouching while exerting force; standing on tiptoes	R L	R L	R L	R L
Continuous Effort Duration	< 6 s 1	6 – 20s 2	20 – 30s 3	> 30 s 4 (Enter VH for Priority)			
Effort Frequency	< 1 / min 1	1 – 5 / min 2	> 5 – 15 / min 3	> 15 / min 4 (Enter VH for Priority)			

A

FIGURE 12.1 (A and B) Worksheet for muscle fatigue assessment method. (From Bernard, T. and Rodgers, S.H. [2002]; Rodgers, S.H. [1987, 1988, 1992, 1997]. With permission.)

Job: Carpet installation - rear compartment	Analyst S.H. Rodgers
Task: Carry carpet to car & install in rear compartment	Date 11/25/87

Region	Effort Level — Light - 1	Moderate - 2	Heavy - 3 < 75% of All Workers Can Exert Effort - 4	Effort	Dur	Freq	Priority
Neck	Head turned partly to side, back or slightly forward	Head turned to side; head fully back; head forward about 20°	Same as Moderate but with force or weight; head stretched forward	2	2	2	M
Shoulders	Arms slightly away from sides; arms extended with some support	Arms away from body, no support; working overhead	Exerting forces or holding weight with arms away from body or overhead	R 3 L 2	R 1 L 2	R 3 L 3	R H L H
Back	Leaning to side or bending arching back	Bending forward; no load; lifting moderately heavy loads near body; working overhead	Lifting or exerting force while twisting; high force or load while bending	3	2	2	H
Arms / Elbow	Arms away from body, no load; light forces lifting near body	Rotating arms while exerting moderate force	High forces exerted with rotation; lifting with arms extended	R 3 L 2	R 1 L 2	R 3 L 3	R H L H
Wrists / Hands / Fingers	Light forces or weights handled close to body; straight wrists; comfortable power grips	Grips with wide or narrow span; moderate risk angles, especially flexion; use of gloves with moderate forces	Pinch grips; strong wrist angles; slippery surfaces	R 3 L 2	R 1 L 2	R 3 L 3	R H L H
Legs / Knees	Standing, walking without bending or leaning; weight on both feet	Bending forward, leaning on table; weight on one side; pivoting while exerting force	Exerting high force while pulling or lifting; crouching while exerting force	R 3 L 2	R 1 L 1	R 3 L 3	R H L M
Ankles / Feet / Toes	Standing, walking without bending or leaning; weight on both feet	Bending forward, leaning on table; weight on one side; pivoting while exerting force	Exerting high force while pulling or lifting; crouching while exerting force; standing on tiptoes	R 3 L 2	R 1 L 1	R 3 L 3	R H L M
Continuous Effort Duration	< 6 s **1**	6 – 20s **2**	20 – 30s **3**	> 30 s **4** (Enter VH for Priority)			
Effort Frequency	< 1 / min **1**	1 – 5 / min **2**	> 5 – 15 / min **3**	> 15 / min **4** (Enter VH for Priority)			

B

FIGURE 12.1 (continued)

12.4   Example Using the Muscle Fatigue Analysis Method to Study Carpet Installation in Car Assembly

Workers on an assembly line had to install the rear-compartment carpet in cars at a frequency of one every minute. The task involved carrying rolled carpet (≈11 kg) from the supply rack to the car, placing

TABLE 12.2 "Priority for Change" Score from Three-Number Rating

Effort Level = 1			Effort Level = 2			Effort Level = 3		
Duration	Frequency	Priority	Duration	Frequency	Priority	Duration	Frequency	Priority
1	1	L	1	1	L	1	1	L
1	2	L	1	2	L	1	2	M
1	3	L	1	3	M	1	3	H
2	1	L	2	1	L	2	1	H
2	2	L	2	2	M	2	2	H
2	3	M	2	3	H	2	3	VH
3	1	L	3	1	M	3	1	VH
3	2	M	3	2	M	3	2	VH
3	3	—a	3	3	—a	3	3	—a

Note: Enter with the scores for effort level (top row) and for duration and frequency (columns within the section for effort level). A score of 4 for either effort level, duration, or frequency is automatically VH (very high). The "priority for change" from the table is low (L), moderate (M), high (H), or very high (VH).

a This combination of duration and frequency is not possible.

TABLE 12.3 Category Scores in the Order of Increasing Fatigue for Three-Number Rating (Effort, Continuous Effort Duration, and Frequency)

Low (L)	Moderate (M)	High (H)	Very High (VH)
111	123	223	323
112	132	313	331
113	213	321	332
211	222	322	
121	231		4xx
212	232		x4x
311	312		xx4
122			
131			
221			

Note: Level of fatigue increases as you move down each column or to the right.

TABLE 12.4 Task Evaluation for Muscle Fatigue Analysis

Problem job: Carpet installation	No. of people affected: 4
Problem task: Carrying and installing carpet	Minutes/shift: Up to 8 hours

it in the rear compartment, and pressing it down to conform to the car body. The assembler had to lean into the compartment when placing the carpet in the moving car and remain bent over while pulling the carpet into position and pressing it down. It usually took about 40 sec to complete the installation, leaving 20 sec for recovery and preparation for the next car.

Workers complained about the difficulty of the whole job, and indicated particular stress on their back, arms, shoulders, and hands. The whole job was analyzed using the MFA method. Table 12.4 summarizes the initial situation.

The observed postures included sustained bending into the compartment after the carpet was placed inside the car, with the right side of the body stretched forward to press the carpet on to the car frame on the other side and frequent pinch-grip tugs on the carpet to get it to fit properly. The left side of the body was holding on to the frame of the car and stabilizing the body during the extended reaches. Because the car was moving, the assembler had to walk as the job was done, so the duration of the load on each leg was reduced compared with the time the back was in the bent posture. The stress on the left leg and foot was less because the reaches were on the right side, and much of the body weight was on the right

TABLE 12.5 Body Parts with High Priority for Change Scores

Body Part	Three-Number Rating before Intervention (High Priority)	Three-Number Rating Goal after Intervention (Low Priority)	Strategy to Reduce Fatigue
R shoulder	313	212	Reduce reach, decrease force
L shoulder	223	221, 212	Reduce no. of forceful pushes
Back	322	221, 212	Decrease force, reduce reach, no.
R arm/elbow	313	212	Decrease force, reduce no.
L arm/elbow	223	212	Reduce no. and time
R wrist/hand	313	212	Reduce pinch grip, reach, no.
L wrist/hand	223	212	Reduce no. and time
R leg/knee	313	212	Reduce reach and no.
R ankle/foot	313	212	Reduce reach and no.

Note: Table 12.3 shows the category scores for the four levels of change priority (low, medium, high, very high).

leg during this time. The reaches, the forces required to press down the carpet with an extended reach, and the number of pressings done to get the carpet properly fitted all contributed to high stress on the back as well as the upper and lower body, as can be seen by the list of body parts where the priority for change is high (Table 12.5).

The analysis indicated that the best strategy to reduce fatigue required reducing the need for repetitive high press forces at full extension. The company was assigning the job to two teams of two assemblers who could alternate cars to obtain an extra minute of recovery time. However, this was still inadequate because of the need to carry the carpet for 8 to 10 sec before it was placed in the rear compartment (a 321 rating for the upper extremities, high priority, because of the carpet's bulkiness and the poor gripping situation). The assemblers had even tried to run the job by having each person do it alone so the time before repeating it would be 3 min and 20 sec instead of 80 sec, but this did not produce the required quality on the operation.

An interim solution was to provide a third team of two assemblers so each team had 2 min and 20 sec of recovery time between assemblies. The ultimate solution was to split the carpet and form it so it would lie flat in the compartment with fewer pressings required and less force. This also required less extended reaching. The smaller carpet was easier and lighter to carry, much easier to fit on the frame, and required fewer downward pressings. The supply of carpets could be moved closer to the start of the carpet installation station because the two halves could be staged on each side of the line and took less space than had the full carpet on one side of the line.

Rerating the job with the split carpet and with one team of two assemblers showed a reduction in the force, frequency, and in the carrying task duration of effort and intensity. The new ratings were 212s and 221s (low priority) in most cases. The effort time for the installation was reduced from 40 to 20 sec. This allowed adequate recovery time for the back and right-hand muscles that fell in the moderate "priority for change" category because of the type of grip used (312) and the need to work in a bent over position for more than 6 sec continuously (222). Complaints about the carpet installation job were rarely heard after the changes to the carpet were made. Although the cost of redesigning the carpet was significant, the savings in labor (that had been added because of the fatigue), improved quality of the installation, and reduced risk for injuries repaid the cost in a short time.

12.5 Advantages

- Fairly simple to use
- Worker cooperation needed to establish ratings
- Stimulates discussion about job while doing the ratings
- Not unidimensional; interactions are evaluated to estimate fatigue
- Evaluates all body muscle groups
- Identifies fatigue-producing patterns of work and shows how to improve them

- Prioritizes tasks for improvements
- Suggests several strategies for improving tasks during the analysis

12.6   Disadvantages

- Semiquantitative method that requires judgment
- Analyst has to gather job information on site
- Need to analyze tasks separately
- Focuses on muscle cycles instead of task cycles
- Less effective if done by one analyst rather than a team of people on the production floor

12.7   Related Methods

Several analysis techniques have been incorporated in the development of the MFA method. Muscle fatigue studies in France and West Germany (Monod and Scherrer, 1957; Rohmert, 1960a, 1960b, 1973a, 1973b) were used for the predictions of accumulating fatigue in the active muscles during 5 min of work. Muscle effort frequency at a given effort intensity identified whether the needed recovery time would be available in the pattern of work required by the task.

Determining the level of effort intensity borrows from other checklists on risk factors for musculoskeletal injury and illness (ANSI ASC Z-365, 2001; OSHA, 2000). The percent of maximum muscle effort for a muscle group uses the psychophysical method from the large-muscle-group activity scale developed by Borg and Lindblad (1976; Borg, 1998).

The problem-solving approach that takes the muscle fatigue assessment from a rating to a set of strategies for improving the task has its roots in two analysis systems: problem analysis from the Kepner–Tregoe problem-solving process (Kepner and Tregoe, 1965) and the FAST (functional analysis systems technique) diagramming method used in product development and organizational analyses (Bytheway, 1971; Caplan et al., 1991).

12.8   Standards and Regulations

American businesses can be cited for ergonomics problems under the General Duty Clause of the 1970 OSHA (Occupational Safety and Health Administration) Act (OSHA, 1970). Section 5 (a) (1) of this act says that "each employer shall furnish to each of his employees, employment and a place of employment which is free from recognized hazards that are causing or are likely to cause death or serious harm to his employees." Current information on the OSHA ergonomics program standard can be found on-line at www.osha.gov.

The ANSI (American National Standards Institute) guideline ASC Z-365 (ANSI, 2001) is in the final stages of approval.

The American Conference of Governmental Industrial Hygienists (ACGIH) is developing threshold limit values (TLVs) for hand activity level for acceptable repetitions for manual work and for lifting to provide weight limits based on frequency and total time of lifting (see www.acgih.org).

Several international standards (ILO and EU) incorporate ergonomics (see www.europa.eu.int, www.iso.org, www.cenorm.be, and www.perinorm.com) (Stuart-Buttle, 2003).

12.9   Approximate Training and Application Time

The basic rating method can be taught to industrial teams in 1 to 2 hours. A trained analyst can analyze a task in 15 to 30 min (Bernard, 2002).

12.10   Reliability and Validity

A major study of about 700 jobs in six automobile factories tested the analytical methods to see how well they predicted musculoskeletal injuries and illnesses. The study found the MFA method to be one of the best identifiers of good jobs, i.e., jobs that did not need improvements (Bernard and Bloswick, in preparation). It is very sensitive but not very specific in detecting the potential MSD (musculoskeletal disorder) risks (Bernard, 2002).

The MFA method is a good predictor of problem jobs that do not have significant biomechanical problems, especially where work rates are high and worker control over the work pattern is low, e.g., on paced assembly lines (Rodgers and Day, 2002). The method has been used by many teams in the automobile industry for over 5 years (Ford, 1995).

12.11   Tools Needed

Although more sophisticated measures can be made, the basic tools needed for this analysis are:
- Paper and pencil
- Videotapes showing continuous footage (unedited, at least 4 to 6 min) of job demands
- VCR (preferably four heads) and monitor
- Stopwatch
- Psychophysical scale (ten-point large-muscle-group activity)

References

ANSI (2001), Management of Work-Related Musculoskeletal Disorders, ASC Z-365, National Safety Council, www.nsc.org.

Bernard, T. (2000), personal communication, UAW/Ford study on MSDs.

Bernard, T.E. and Bloswick, D., Report on UAW/Ford Study of Job Analysis Methods to Evaluate the Risk of MSDs, in preparation.

Bernard, T. and Rodgers, S.H. (2002), Fatigue Analysis Tool, available on-line at www.hsc.usf.edu/~tbernard/ergotools.

Borg, G. (1998), *Borg's Perceived Exertion and Pain Scales*, Human Kinetics, Champaign, IL.

Borg, G. and Lindblad, I.M. (1976), The Determination of Subjective Intensities in Verbal Descriptors of Symptoms, no. 75, Institute of Applied Psychology, University of Stockholm, Stockholm.

Bytheway, C.W. (1971), The creative aspects of FAST diagramming, in *Proceedings of the SAVE Conference*, pp. 301–312.

Caplan, S.H., Rodgers, S.H., and Rosenfeld, H. (1991), A novel approach to clarifying organizational roles, in *Proceedings of the Human Factors Society*.

Ford Motor Company (1995), Ergonomics Training Program for Teams in Plants, Rodgers's fatigue analysis tool was one of three methods included in the training; the Union Ford Ergonomics Process: *Fitting Jobs to People*.

Kepner, C.H. and Tregoe, B.B. (1965), *The Rational Manager*, McGraw-Hill, New York.

Monod, H. and Scherrer, J. (1957), Capacité de travail statique d'un groupe musculaire synergique chez l'homme, *C. R. Soc. Biol. Paris*, 151, 1358–1362.

OSHA (1970), Occupational Safety and Health Act, www.osha.gov.

OSHA/DOL (2000), Part II: 29 CFR Part 1910, Ergonomics Program: Final Rule, Fed. Reg. 65(22): 68262–68870, November 14, 2000.

Rodgers, S.H. (1987), Recovery time needs for repetitive work, *Semin. Occup. Med.*, 2, 19–24.

Rodgers, S.H. (1988), Job evaluation in worker fitness determination, in *Worker Fitness and Risk Evaluations*, Himmelstein, J. and Pransky, G., Eds., Hanley and Belfus, Philadelphia.

Rodgers, S.H. (1992), A functional job analysis technique, in *Ergonomics*, Moore, J.S. and Garg, A., Eds., Hanley and Belfus, Philadelphia.

Rodgers, S.H. (1997), Work physiology: fatigue and recovery, in *Handbook of Human Factors and Ergonomics*, 2nd ed., Salvendy, G., Ed., John Wiley & Sons, New York, pp. 36–65.

Rodgers, S.H. and Day, D.E. (1987 to 2002), Ergonomics team training courses have used the Muscle Fatigue Assessment method to analyze production and service jobs, personal communication.

Rohmert, W. (1960a), Ermittlung vor Erhohlenspausen fur statische Arbeit des Menschen, *Int. Z. Einschl. Angew. Arbeitsphysiol.*, 18, 123–164.

Rohmert, W. (1960b), Zur Theorie der Erhohlenspausen bei dynamischen Arbeit, *Int. Z. Einschl. Angew. Arbeitsphysiol.*, 18, 191–212.

Rohmert, W. (1973a), Problems in determining rest allowances: part 1, use of modern methods to evaluate stress and strain in static muscular work, *Appl. Ergonomics*, 4, 91–95.

Rohmert, W. (1973b), Problems in determining rest allowances: part 2, determining rest allowances in different human tasks, *Appl. Ergonomics*, 4, 158–162.

Stuart-Buttle, C. (2003), Standards (chap. 1), in *Ergonomic Design for People at Work*, 2nd ed., Chengalur, S., Rodgers, S.H., and Bernard, T., Eds., John Wiley & Sons, New York.

13

Psychophysical Tables: Lifting, Lowering, Pushing, Pulling, and Carrying

Stover H. Snook
Harvard School of Public Health

13.1 Background and Application

A basic question when evaluating a manual handling task is: "How much should the worker be required to lift or lower or push or pull or carry?" Psychophysics is a particularly useful tool for answering that question. Psychophysics began over 165 years ago with the investigations of Ernst Heinrich Weber (1795–1878). Weber studied the sense of touch and found that a weight must be increased by a constant fraction of its value (about 1/40) for the perception of weight to be just noticeably different (JND), independent of the magnitude of the weight. He formulated Weber's law:

$$\text{delta } I/I = \text{constant}$$

where I is the weight (intensity) and delta I is the increment that the weight must be increased to be just noticeably different (Snook, 1999).

Gustav Theodor Fechner (1801–1887) expanded upon Weber's work, and in 1860 he proposed what is now known as Fechner's Law, which states that the strength of a sensation (S) is directly related to the intensity of its physical stimulus (I) by means of a logarithmic function:

$$S = k \log I$$

where the constant, k, is a function of the particular unit of measurement used.

Fechner's law is quite accurate in the middle ranges of stimuli and sensations, but it is inaccurate at the extremes. Stevens (1960) proposed that the relationship between stimuli and sensations was not a logarithmic one, but a power function

$$S = kI^n$$

that was more accurate throughout the entire range of stimuli and sensations.

When plotted in log–log coordinates, a power function is represented by a straight line, with the exponent, n, being equal to the slope of the line. Over the years, exponents have been experimentally determined for many types of stimuli: 3.5 for electric shock, 1.3 for taste (salt), 0.6 for loudness (binaural), and approximately 1.6 for the perception of muscular effort and force. Psychophysics has been used by Borg in developing ratings of perceived exertion (RPE), by Caldwell and associates in the development of effort scales, and by Snook and associates in the study of repetitive motion tasks (Snook, 1999).

About 35 years ago, the Liberty Mutual Research Center began to apply psychophysics to manual handling tasks to develop recommendations for use in reducing industrial low-back compensation claims. In these studies, the worker was given control of one of the task variables, usually the weight of the object being handled. All other variables such as repetition rate, size, height, distance, etc. were controlled. The worker then monitored his or her own feelings of exertion or fatigue and adjusted the weight of the object accordingly. It was believed that only the individual worker could (a) sense the various strains associated with manual handling tasks and (b) integrate the sensory inputs into one meaningful response (Snook, 1978).

Eleven separate experiments were conducted at the Liberty Mutual Research Center over a period of 25 years. Each experiment lasted 2 to 3 years. These experiments were unique in that they used realistic manual-handling tasks performed by industrial workers (male and female) over long periods of time (at least 80 hours of testing for each subject). Teams of three workers performed lifting, lowering, pushing, pulling, and carrying tasks. Physiological measurements of oxygen consumption and heart rate were recorded for comparison with psychophysical measurements. The experimental design also included 16 to 20 hours of training and conditioning, as well as a battery of 41 anthropometric measurements from each subject.

The results of the first seven experiments were combined and integrated into tables of maximum acceptable weights and forces for various percentages of the population (Snook, 1978). After four additional experiments, the tables were revised (Snook and Ciriello, 1991). The revised tables consist of the following:

Table 13.1: Maximum acceptable weight of lift for males (n = 68 subjects)
Table 13.2: Maximum acceptable weight of lift for females (n = 51 subjects)
Table 13.3: Maximum acceptable weight of lower for males (n = 48 subjects)
Table 13.4: Maximum acceptable weight of lower for females (n = 39 subjects)
Table 13.5: Maximum acceptable force of push for males (n = 63 subjects)
Table 13.6: Maximum acceptable force of push for females (n = 51 subjects)
Table 13.7: Maximum acceptable force of pull for males (n = 53 subjects)
Table 13.8: Maximum acceptable force of pull for females (n = 39 subjects)
Table 13.9: Maximum acceptable weight of carry (n = 38 male and 27 female subjects)

13.2   Procedure

Use the type of task (e.g., lifting) and the gender of the worker to identify the correct table.

For Table 13.1 through Table 13.4 (lifting and lowering):

1. Use the height of the task (e.g., floor level to knuckle height) to identify the correct set of columns (right, middle, or left) in the table.

2. Use the width of the box or object to identify the correct set of rows (upper, middle, or lower) in the table.
3. At the intersection of the correct sets of columns and rows, use the vertical distance of lift (or lower) to identify the correct 5 × 8 matrix (upper, middle, or lower).
4. Use the repetition rate (e.g., one lift every 2 min) and the percent of population to identify the correct value in the 5 × 8 matrix.

For Table 13.5 through Table 13.8 (pushing and pulling):

1. Determine the initial force and the sustained force separately.
2. Use the horizontal distance of movement (e.g., 15.2-m push) to identify the correct set of columns in the table.
3. Use the height of the hands above the floor to identify the correct set of rows in the table.
4. At the intersection of the correct sets of columns and rows, use the repetition rate and the percent of population to identify the correct value.

For Table 13.9 (carrying):

1. Use the distance of carry to identify the correct set of columns in the table.
2. Use the height of the hands above the floor to identify the correct set of rows in the table.
3. At the intersection of the correct sets of columns and rows, use the repetition rate and the percent of population to identify the correct value.

It is important to note that Table 13.1 through Table 13.4, and Table 13.9 are based on handling boxes with handles and handling boxes close to the body. The values in the tables should be reduced by approximately 15% when handling boxes without handles and by approximately 50% when handling smaller boxes with extended horizontal reaching between knee height and shoulder height.

It is also important to note that some of the weights and forces in Table 13.1 through Table 13.9 will exceed recommended physiological criteria when performed continuously for 8 hours or more. The recommended 8-hour criteria are approximately 1000 ml/min of oxygen consumption for males and 700 ml/min for females (NIOSH, 1981). Weights and forces that exceed these criteria appear in bold italics in Table 13.1 through Table 13.9.

Table 13.1 through Table 13.9 give maximum acceptable weights and forces for individual manual-handling tasks or components (e.g., lifting). Frequently, however, industrial tasks involve combinations with more than one component (e.g., lifting, carrying, and lowering). Each component of a combined task should be analyzed separately using the repetition rate of the combined task. The lowest maximum acceptable weight or force for the components represents the maximum acceptable weight or force for the combined task. However, since the physiological cost of the combined task will be greater than the cost for individual components, it should be recognized that some of the combined tasks may exceed recommended physiological criteria for extended periods of time (NIOSH, 1981).

13.3 Advantages

The major advantages and disadvantages of the psychophysical approach have been reviewed by Snook (1985) and Ayoub and Dempsey (1999). The advantages include:

- Capability to realistically simulate industrial work
- Capability to study very intermittent manual-handling tasks and very fast repetitive tasks (physiological methods have difficulty with intermittent tasks, and biomechanical methods have difficulty with fast repetitive tasks)
- Results that are consistent with the industrial engineering concept of a "fair day's work for a fair day's pay"
- Results that are very reproducible

TABLE 13.1 Maximum Acceptable Weight of Lift for Males (kg)

Width of Box [a] (cm)	Vertical Distance of Lift (cm)	Percent of Industrial Population	Floor Level to Knuckle Height								Knuckle Height to Shoulder Height								Shoulder Height to Arm Reach							
			One Lift Every								One Lift Every								One Lift Every							
			5 sec	9 sec	14 sec	1 min	2 min	5 min	30 min	8 hr	5 sec	9 sec	14 sec	1 min	2 min	5 min	30 min	8 hr	5 sec	9 sec	14 sec	1 min	2 min	5 min	30 min	8 hr
75	76	90	6	7	9	11	13	14	14	17	8	10	12	13	14	14	16	17	6	8	9	10	10	11	12	13
		75	9	11	13	16	19	20	21	24	10	14	16	18	18	19	21	23	8	10	12	14	14	14	16	17
		50	12	15	17	22	25	27	28	32	13	17	20	22	23	24	26	29	10	13	15	17	17	18	20	22
		25	15	18	21	28	31	34	35	41	16	21	24	27	27	28	32	35	11	16	18	21	21	22	24	27
		10	18	22	25	33	37	40	41	48	19	24	28	31	32	33	37	40	14	18	21	24	24	25	28	31
	51	90	6	8	9	12	13	15	15	17	8	11	13	15	15	16	18	19	6	8	9	12	12	12	14	15
		75	9	11	13	17	19	21	22	25	11	15	17	20	20	21	23	25	8	11	12	15	15	16	18	20
		50	13	15	18	23	26	28	29	34	14	19	21	25	25	26	29	32	10	14	16	19	20	20	23	25
		25	16	19	22	29	33	35	36	42	17	23	26	30	31	32	36	39	13	17	19	23	23	24	27	30
		10	19	22	26	34	38	42	43	50	20	26	30	35	36	37	41	45	15	19	22	27	27	29	32	35
	25	90	8	9	11	13	15	16	17	20	8	13	15	18	18	19	21	23	7	10	11	14	14	14	16	18
		75	11	13	15	19	22	24	24	28	11	17	20	23	24	25	27	30	10	13	15	18	18	18	21	23
		50	15	18	21	26	29	32	33	38	14	22	25	30	30	31	35	38	12	16	19	23	23	24	27	29
		25	18	22	26	33	37	40	41	48	17	27	30	36	36	38	42	46	15	20	22	28	28	29	32	35
		10	22	26	31	38	44	47	49	57	20	31	35	42	44	44	49	53	17	23	26	32	32	34	38	41
49	76	90	7	8	10	13	15	16	17	20	8	10	12	13	14	14	16	17	7	9	10	12	12	13	14	16
		75	10	12	14	19	22	24	24	28	10	14	16	18	18	19	21	23	9	11	13	16	16	17	19	21
		50	14	16	19	26	29	32	33	38	13	17	20	22	23	24	26	29	11	15	17	20	21	21	24	26
		25	17	20	24	33	37	40	41	48	16	21	24	27	27	28	32	35	13	18	20	25	25	26	29	31
		10	20	24	28	38	43	47	48	57	19	24	28	31	32	33	37	40	15	21	23	28	29	30	33	36
	51	90	7	9	10	14	16	17	18	20	8	11	13	15	15	16	18	19	7	9	11	14	14	14	16	18
		75	10	13	15	20	23	25	25	30	11	15	17	20	20	21	23	25	9	12	14	18	18	19	21	23
		50	14	17	20	27	30	34	34	40	14	19	21	25	25	26	29	32	12	15	18	23	23	23	24	29
		25	18	21	25	34	38	42	43	50	17	23	26	30	31	32	36	39	14	19	21	28	28	28	32	35
		10	21	25	29	40	45	49	50	59	20	26	30	35	36	37	41	45	16	22	25	32	32	34	37	41
	25	90	8	10	12	16	18	19	20	23	10	13	15	18	18	19	21	23	9	11	12	16	16	17	19	21
		75	12	15	17	23	26	28	29	33	13	17	20	23	24	25	27	30	11	14	16	21	21	22	25	27
		50	16	20	23	30	34	37	38	45	17	22	25	30	30	31	35	38	14	18	21	27	27	28	32	35
		25	21	25	29	38	43	47	48	56	20	27	30	36	36	38	42	46	16	22	25	33	33	34	38	42
		10	24	29	34	45	51	56	57	67	23	31	35	42	44	44	49	53	19	25	29	38	38	40	44	48

TABLE 13.1 Maximum Acceptable Weight of Lift for Males (kg) (continued)

Width of Box[a] (cm)	Vertical Distance of Lift (cm)	Percent of Industrial Population	Floor Level to Knuckle Height – One Lift Every								Knuckle Height to Shoulder Height – One Lift Every								Shoulder Height to Arm Reach – One Lift Every							
			5	9	14	1	2	5	30	8	5	9	14	1	2	5	30	8	5	9	14	1	2	5	30	8
			sec			min				hr	sec			min				hr	sec			min				hr
34	76	90	8	10	11	15	17	19	19	23	8	11	13	15	15	16	18	19	8	10	12	14	14	15	16	18
		75	12	14	17	22	25	28	28	33	11	15	17	20	20	21	23	25	10	14	16	18	19	19	22	24
		50	16	19	22	30	34	37	38	44	14	19	21	25	25	26	29	32	13	17	20	23	24	25	27	30
		25	20	24	28	37	42	47	47	55	17	23	26	30	31	32	36	39	16	21	24	28	29	30	33	36
		10	24	29	33	44	50	54	56	65	20	26	30	35	36	37	41	45	18	24	28	33	33	34	38	42
	51	90	9	10	12	16	18	20	20	24	9	12	14	17	17	18	20	22	8	11	13	16	16	17	18	20
		75	12	15	18	23	26	28	29	34	12	16	18	22	23	23	26	29	11	14	17	21	21	21	24	26
		50	17	20	24	31	35	38	39	46	15	20	23	28	29	30	33	36	14	18	21	26	27	28	31	34
		25	21	25	30	39	44	48	49	57	18	24	27	34	35	36	40	44	17	22	25	32	32	33	37	41
		10	25	30	35	46	52	57	58	68	21	28	32	40	40	42	46	51	19	26	29	37	37	39	43	47
	25	90	10	12	14	18	20	22	23	27	11	14	16	20	20	21	23	26	10	13	15	19	19	19	22	24
		75	15	18	21	26	30	32	33	38	14	18	21	26	27	28	31	34	13	17	20	24	25	26	29	31
		50	20	24	28	35	40	43	44	52	18	23	27	33	34	35	39	43	16	22	25	31	31	33	36	40
		25	26	30	35	44	50	54	55	65	21	28	32	40	41	42	47	52	20	26	30	37	38	39	44	46
		10	29	35	41	52	59	64	66	76	25	33	37	47	47	49	55	60	23	30	35	43	44	45	51	55

Note: Values in bold italics exceed 8-hour physiological criteria (see text).
[a] The dimension away from the body.

TABLE 13.2 Maximum Acceptable Weight of Lift for Females (kg)

Width of Box[a] (cm)	Vertical Distance of Lift (cm)	Percent of Industrial Population	Floor Level to Knuckle Height — One Lift Every								Knuckle Height to Shoulder Height — One Lift Every								Shoulder Height to Arm Reach — One Lift Every							
			5	9	14	1	2	5	30	8	5	9	14	1	2	5	30	8	5	9	14	1	2	5	30	8
			sec			min				hr	sec			min				hr	sec			min				hr
75	76	90	5	6	7	7	8	8	9	12	5	6	7	9	9	9	10	12	4	5	6	7	7	7	7	8
		75	7	8	9	9	10	10	11	14	6	7	8	10	11	11	12	14	5	6	7	8	8	8	8	10
		50	8	10	10	11	12	12	13	17	7	8	9	11	12	12	13	16	6	7	8	9	9	9	10	11
		25	9	11	12	13	14	14	15	21	8	9	10	13	14	14	15	18	7	7	9	10	10	10	11	13
		10	11	13	14	14	15	16	17	23	9	10	11	14	15	15	17	20	7	8	10	11	11	11	12	14
	51	90	6	7	8	8	9	9	10	14	6	7	8	9	10	10	11	13	5	6	7	7	7	7	8	9
		75	7	9	9	10	11	11	13	17	7	8	9	11	12	12	13	15	6	7	8	8	8	8	9	11
		50	9	10	11	12	13	14	15	21	9	9	11	13	14	14	15	17	7	8	9	9	9	10	10	13
		25	10	12	13	15	16	16	18	24	10	11	12	14	16	16	17	20	8	9	10	10	11	11	11	14
		10	11	14	15	17	18	18	20	27	11	12	14	16	17	17	19	22	9	10	11	12	13	13	14	16
	25	90	6	8	8	9	9	9	11	14	6	7	8	10	11	11	12	14	5	6	7	8	8	8	9	10
		75	8	10	11	11	12	13	13	18	7	8	9	12	13	13	14	17	6	7	8	9	9	9	10	12
		50	10	12	13	13	14	14	16	21	9	9	11	14	15	15	16	19	7	8	9	10	11	11	11	14
		25	11	14	15	15	16	17	19	25	10	11	12	16	17	17	19	22	8	9	10	12	12	12	12	14
		10	13	16	17	17	19	19	21	29	11	12	14	18	19	19	21	24	9	10	11	13	14	14	14	16
49	76	90	5	6	7	8	8	8	9	13	5	6	7	9	9	9	10	12	5	6	7	7	7	7	8	9
		75	7	8	9	10	10	10	12	16	6	7	8	10	11	11	12	14	6	7	8	8	8	8	9	11
		50	8	10	10	12	12	13	14	19	7	8	9	11	12	12	13	16	7	8	9	9	10	10	11	12
		25	9	11	12	14	15	15	17	22	8	9	10	13	14	14	15	18	8	9	10	11	11	11	12	14
		10	11	13	14	15	17	17	19	25	9	10	11	14	15	15	17	20	9	10	11	12	12	13	13	15
	51	90	6	7	8	9	10	10	11	15	6	7	8	9	10	10	11	13	5	6	7	7	7	7	8	9
		75	7	9	9	11	12	12	14	18	7	8	9	11	12	12	13	15	6	7	8	8	9	9	9	10
		50	9	10	11	13	15	15	16	22	9	9	11	13	14	14	15	17	7	8	9	10	10	11	11	14
		25	10	12	13	16	17	17	19	26	10	11	12	14	16	16	17	20	8	9	10	11	12	12	13	15
		10	11	14	15	18	20	20	22	30	11	12	14	16	17	17	19	22	9	10	11	13	14	14	15	17
	25	90	6	8	8	9	10	10	11	15	6	7	8	10	11	11	12	14	5	6	7	8	8	9	10	11
		75	8	10	11	12	13	14	14	19	7	8	9	12	13	13	14	17	6	7	8	9	10	10	11	13
		50	10	12	13	14	15	15	17	23	9	9	11	14	15	15	16	19	7	8	9	11	11	12	13	15
		25	11	14	15	16	18	18	20	27	10	11	12	16	17	17	19	22	8	9	10	12	13	13	15	17
		10	13	16	17	19	20	21	23	31	11	12	14	18	19	19	21	24	9	10	11	14	15	15	16	19

TABLE 13.2 Maximum Acceptable Weight of Lift for Females (kg) (continued)

Width of Box [a] (cm)	Vertical Distance of Lift (cm)	Percent of Industrial Population	Floor Level to Knuckle Height — One Lift Every									Knuckle Height to Shoulder Height — One Lift Every									Shoulder Height to Arm Reach — One Lift Every								
			sec			min				hr	sec			min				hr	sec			min				hr
			5	9	14	1	2	5	30	8	5	9	14	1	2	5	30	8	5	9	14	1	2	5	30	8
34	76	90	7	8	9	9	10	10	11	15	6	7	8	9	10	10	11	13	5	6	7	8	9	9	10	11
		75	8	10	11	12	13	13	14	19	7	8	9	11	12	12	13	15	6	7	8	9	10	10	11	13
		50	10	12	13	14	15	16	17	23	9	9	11	13	14	14	15	17	7	8	9	11	12	12	13	15
		25	12	14	15	17	18	18	20	27	10	11	12	14	16	16	17	20	8	9	10	12	13	13	15	17
		10	13	16	18	19	20	21	23	31	11	12	13	16	17	17	19	22	9	10	11	14	15	15	16	19
	51	90	7	9	9	11	12	12	13	18	8	8	9	10	11	11	12	14	7	7	8	9	10	10	11	12
		75	9	11	12	14	15	15	16	22	9	10	11	12	13	13	14	17	8	8	9	11	11	11	12	14
		50	11	13	14	16	18	18	20	27	10	11	13	14	15	15	17	19	9	10	11	12	13	13	14	17
		25	13	15	17	19	21	21	24	32	12	13	14	16	17	17	19	22	10	11	12	14	15	15	16	19
		10	14	18	19	22	24	24	27	36	13	14	16	18	19	19	21	24	11	12	14	15	16	16	18	21
	25	90	8	10	11	11	12	12	14	19	8	8	9	12	12	12	14	16	7	7	8	10	11	11	12	14
		75	10	12	13	14	15	15	17	23	9	10	11	13	14	14	16	18	8	8	9	12	12	12	14	16
		50	12	15	16	17	18	19	21	28	10	11	13	16	17	17	18	21	9	10	11	13	14	14	16	18
		25	14	17	19	20	22	22	24	33	12	13	14	18	19	19	21	24	10	11	12	15	16	16	18	21
		10	16	20	21	23	25	25	28	38	13	14	16	19	21	21	23	27	11	12	14	17	18	18	20	23

Note: Values in bold italics exceed 8-hour physiological criteria (see text).
[a] The dimension away from the body.

TABLE 13.3 Maximum Acceptable Weight of Lower for Males (kg)

Width of Box^a (cm)	Vertical Distance of Lower (cm)	Percent of Industrial Population	Knuckle Height to Floor Level — One Lower Every								Shoulder Height to Knuckle Height — One Lower Every								Arm Reach to Shoulder Height — One Lower Every							
			5 sec	9 sec	14 sec	1 min	2 min	5 min	30 min	8 hr	5 sec	9 sec	14 sec	1 min	2 min	5 min	30 min	8 hr	5 sec	9 sec	14 sec	1 min	2 min	5 min	30 min	8 hr
75	76	90	7	9	10	12	14	15	16	20	10	11	14	14	15	15	16	19	6	7	9	9	10	10	11	13
		75	10	13	14	18	20	22	22	29	13	15	16	18	18	21	21	26	9	10	12	12	14	14	14	18
		50	14	17	19	23	27	29	30	38	18	20	20	24	24	27	28	34	11	13	15	16	18	18	19	23
		25	17	21	24	29	33	36	37	47	21	25	25	29	29	34	34	42	14	16	19	19	22	23	23	28
		10	20	25	28	34	39	42	44	56	25	29	29	34	34	39	39	49	16	19	22	23	26	26	27	33
	51	90	8	10	11	13	15	16	17	21	11	12	14	15	17	17	18	22	7	8	9	10	12	12	12	15
		75	11	14	15	18	21	23	23	30	14	17	17	20	21	24	24	30	9	11	13	14	16	16	16	20
		50	14	18	20	24	28	30	31	40	19	21	21	25	27	31	31	38	12	14	16	18	21	21	21	26
		25	18	22	25	30	34	37	39	49	23	26	26	31	33	38	38	47	15	17	20	22	25	25	26	32
		10	21	26	29	36	41	44	46	58	27	31	31	36	38	44	44	55	17	20	24	26	30	30	30	37
	25	90	9	11	12	15	17	18	19	24	12	14	17	18	21	21	21	26	8	9	11	12	14	14	14	17
		75	13	16	17	21	24	25	26	34	17	20	23	24	28	28	28	35	11	13	15	16	19	19	19	24
		50	17	21	23	27	31	34	35	45	22	25	30	32	36	36	37	45	14	16	19	21	24	24	25	31
		25	21	26	29	34	39	42	44	56	27	31	37	39	44	44	45	56	17	20	24	26	30	30	30	38
		10	24	31	34	40	46	49	51	66	31	36	43	45	52	52	52	65	20	23	28	30	35	35	35	44
49	76	90	8	10	11	15	17	18	19	24	10	11	14	14	15	15	16	19	7	8	10	11	12	12	12	15
		75	12	15	16	21	24	26	26	34	13	16	16	18	18	21	21	26	10	11	14	15	17	17	17	21
		50	15	19	21	27	31	34	35	45	18	20	20	24	24	27	28	34	13	15	17	19	22	22	22	27
		25	19	24	26	34	39	42	44	56	21	25	25	29	29	34	34	42	16	18	21	23	27	27	27	33
		10	25	28	31	40	46	49	51	65	25	29	29	34	34	39	39	49	18	21	25	27	31	31	31	39
	51	90	9	11	12	15	17	19	19	25	11	12	14	15	17	17	18	22	8	9	10	12	14	14	14	17
		75	12	15	17	22	25	26	28	35	14	17	17	20	21	24	24	30	10	12	14	16	19	19	19	24
		50	16	20	22	29	33	35	37	47	19	21	21	25	27	31	31	38	14	16	18	21	24	24	25	31
		25	20	25	27	36	41	44	46	58	23	26	26	31	33	38	38	47	17	19	23	26	30	30	30	37
		10	23	29	32	42	48	51	54	68	27	31	31	36	38	44	44	55	19	22	26	30	35	35	35	44
	25	90	10	13	14	17	20	21	22	28	12	14	17	18	21	21	21	26	9	10	12	14	16	16	16	20
		75	14	18	19	24	28	30	31	40	17	20	23	24	28	28	28	35	12	14	17	19	22	22	22	28
		50	19	24	26	32	37	40	41	54	22	25	30	32	36	36	37	45	16	18	22	25	29	29	29	36
		25	23	29	32	40	46	49	51	65	27	31	37	39	44	44	45	56	20	23	27	31	35	35	36	44
		10	27	34	38	47	54	58	60	77	31	36	43	45	52	52	52	65	23	26	31	36	41	42	42	52

TABLE 13.3 Maximum Acceptable Weight of Lower for Males (kg) (continued)

Width of Box[a] (cm)	Vertical Distance of Lower (cm)	Percent of Industrial Population	Knuckle Height to Floor Level								Shoulder Height to Knuckle Height								Arm Reach to Shoulder Height							
			One Lower Every								One Lower Every								One Lower Every							
			5 sec	9 sec	14 sec	1 min	2 min	5 min	30 min	8 hr	5 sec	9 sec	14 sec	1 min	2 min	5 min	30 min	8 hr	5 sec	9 sec	14 sec	1 min	2 min	5 min	30 min	8 hr
34	76	90	10	12	13	17	19	21	21	27	11	12	14	15	17	17	18	22	9	10	12	12	14	14	14	18
		75	_**14**_	17	19	24	27	29	30	39	14	17	20	21	24	24	24	30	12	13	16	17	19	19	19	24
		50	_**18**_	_**23**_	25	32	36	39	40	51	19	21	25	27	31	31	31	38	15	17	21	22	25	25	25	31
		25	_**23**_	_**29**_	31	39	45	48	50	64	23	26	31	33	38	38	28	47	19	21	25	27	31	31	31	38
		10	_**27**_	_**34**_	_**37**_	46	53	57	59	75	27	31	36	38	44	44	44	55	22	25	30	31	36	36	36	45
	51	90	10	13	14	17	20	22	22	29	11	13	15	17	20	20	20	24	9	10	12	14	16	16	16	20
		75	14	18	20	25	28	30	32	40	15	18	21	23	27	27	27	33	12	14	17	19	22	22	22	27
		50	_**19**_	24	26	33	37	40	42	53	20	23	27	30	35	35	35	43	16	19	22	24	28	28	28	35
		25	_**24**_	_**30**_	33	41	47	50	52	67	24	28	33	37	42	42	43	53	20	23	27	30	34	34	35	43
		10	_**28**_	_**35**_	_**38**_	48	55	59	62	78	28	33	39	43	49	49	50	62	23	27	31	35	40	40	40	50
	25	90	12	15	16	20	23	24	25	32	13	15	18	20	23	23	23	29	11	12	15	16	19	19	19	23
		75	17	21	23	28	32	34	36	46	18	21	25	27	31	31	32	39	15	17	20	22	26	26	26	32
		50	_**23**_	28	31	37	42	46	47	60	23	27	32	35	41	41	41	51	19	22	26	29	33	33	33	41
		25	_**28**_	_**35**_	41	46	53	57	59	75	29	33	39	43	50	50	50	63	23	27	32	35	41	41	41	51
		10	_**33**_	_**41**_	_**45**_	54	62	67	70	89	33	39	46	51	58	58	59	73	27	31	37	41	47	47	48	59

Note: Values in bold italics exceed 8-hour physiological criteria (see text).

[a] The dimension away from the body.

TABLE 13.4 Maximum Acceptable Weight of Lower for Females (kg)

Width of Box^a (cm)	Vertical Distance of Lower (cm)	Percent of Industrial Population	Knuckle Height to Floor Level — One Lower Every								Shoulder Height to Knuckle Height — One Lower Every								Arm Reach to Shoulder Height — One Lower Every						
			5	9	14	1	2	5	30	8	5	9	14	1	2	5	30	8	5	9	14	1	2	5	30
			sec			min				hr	sec			min				hr	sec			min			
75	76	90	5	6	7	8	8	9	10	12	6	6	7	8	9	10	10	13	5	5	5	6	7	7	9
		75	6	8	9	10	10	11	11	14	7	8	8	10	11	12	12	15	5	6	6	7	8	8	11
		50	7	9	10	11	12	13	13	17	8	9	10	12	13	14	14	18	7	8	8	8	10	10	13
		25	9	11	12	14	14	15	17	20	9	11	11	13	15	17	17	21	8	9	9	10	11	12	15
		10	10	13	13	15	16	17	20	23	11	12	13	15	17	19	19	24	9	10	10	11	12	14	17
	51	90	6	7	8	9	10	10	10	14	7	8	8	9	10	11	11	14	6	6	6	7	8	8	10
		75	7	8	9	10	11	12	13	17	8	9	9	11	12	13	13	17	7	7	7	8	9	10	12
		50	8	10	11	12	14	14	15	20	10	11	11	13	15	16	16	20	8	8	9	9	11	11	15
		25	10	12	13	14	16	17	18	24	11	13	13	15	17	19	19	23	9	9	10	11	12	13	17
		10	11	13	14	16	18	19	20	27	13	15	15	17	19	21	21	26	10	12	12	12	13	15	19
	25	90	6	8	9	10	10	11	11	14	7	8	8	10	11	12	12	15	5	6	6	7	8	9	11
		75	8	10	10	11	12	12	13	17	8	9	9	12	13	15	15	19	7	7	8	9	10	11	13
		50	9	11	11	13	14	15	16	21	10	11	11	14	16	18	18	22	8	9	9	10	12	13	16
		25	11	13	13	15	17	17	19	25	11	13	13	16	19	20	20	26	9	10	10	12	13	15	19
		10	12	15	15	16	18	19	21	28	13	15	15	19	21	23	23	29	10	12	12	13	15	17	21
49	76	90	5	6	7	8	8	9	10	13	6	7	7	8	9	10	10	13	5	5	5	6	7	8	10
		75	6	8	8	10	10	11	12	16	7	8	8	10	11	12	12	15	5	6	6	8	9	9	12
		50	8	9	9	11	13	13	14	19	8	9	10	12	13	14	14	18	7	8	8	9	11	11	14
		25	9	11	11	13	15	16	17	22	9	11	11	13	15	17	17	21	8	9	9	11	12	13	16
		10	10	13	13	15	17	18	19	25	11	12	13	15	17	19	19	24	9	10	10	12	13	15	19
	51	90	6	7	7	9	10	10	11	15	7	8	8	9	10	11	11	14	5	5	6	7	8	9	11
		75	7	8	8	10	11	12	13	18	8	9	9	11	12	13	13	17	6	6	7	8	10	10	13
		50	8	10	11	12	13	14	15	22	10	11	11	13	15	16	16	20	7	8	8	10	11	13	16
		25	10	12	13	15	17	18	19	26	11	13	13	15	17	19	19	23	8	9	9	11	13	15	18
		10	11	13	14	17	19	20	22	29	13	15	15	17	19	21	21	26	9	10	10	12	13	16	21
	25	90	6	8	9	10	10	11	12	15	7	8	8	10	11	12	12	15	5	6	6	7	8	9	12
		75	8	10	11	12	13	13	14	19	8	9	9	12	13	15	15	19	6	7	7	9	10	12	14
		50	9	11	12	14	15	15	17	23	10	11	11	14	16	18	18	22	8	8	9	11	12	14	17
		25	11	13	14	16	18	19	20	27	11	13	13	16	19	20	20	26	9	9	10	13	14	16	20
		10	12	15	16	18	20	21	23	30	13	15	15	19	21	23	23	29	10	12	12	15	16	18	23

TABLE 13.4 Maximum Acceptable Weight of Lower for Females (kg) (continued)

Width of Box [a] (cm)	Vertical Distance of Lower (cm)	Percent of Industrial Population	Knuckle Height to Floor Level — One Lower Every								Shoulder Height to Knuckle Height — One Lower Every								Arm Reach to Shoulder Height — One Lower Every							
			5 sec	9 sec	14 sec	1 min	2 min	5 min	30 min	8 hr	5 sec	9 sec	14 sec	1 min	2 min	5 min	30 min	8 hr	5 sec	9 sec	14 sec	1 min	2 min	5 min	30 min	8 hr
34	76	90	6	8	9	9	10	11	12	15	7	8	8	9	10	11	11	14	6	6	7	8	9	9	9	12
		75	8	10	11	11	13	13	14	19	8	9	9	11	12	13	13	17	7	8	8	9	11	11	11	14
		50	**_10_**	12	13	14	15	16	17	23	10	11	11	13	15	16	16	20	8	9	10	11	13	14	14	17
		25	**_11_**	14	15	16	18	19	20	27	11	13	13	15	17	19	19	23	**_9_**	**_11_**	11	13	15	16	16	20
		10	**_13_**	**_16_**	17	18	20	21	23	30	12	14	15	17	19	21	21	26	**_11_**	**_12_**	13	14	16	18	18	23
	51	90	7	9	9	11	12	13	14	18	8	9	9	10	11	12	12	15	7	8	8	8	10	11	11	13
		75	**_9_**	11	11	13	15	16	17	22	9	11	11	12	14	15	15	19	8	9	10	10	12	13	13	16
		50	**_10_**	13	14	16	18	19	20	27	11	13	13	14	16	18	18	22	**_10_**	11	11	12	14	15	15	19
		25	**_12_**	**_15_**	16	19	21	22	24	31	13	15	15	17	19	21	21	26	**_11_**	**_13_**	13	14	16	18	18	22
		10	**_14_**	**_17_**	**_18_**	21	24	25	27	35	**_16_**	**_17_**	**_17_**	19	21	23	23	29	**_13_**	**_15_**	15	16	18	20	20	25
	25	90	8	10	10	11	13	13	14	19	8	9	9	11	12	13	13	17	7	8	8	9	11	12	12	15
		75	**_10_**	12	13	14	15	16	17	23	9	11	11	13	15	16	16	21	8	9	10	11	13	14	14	18
		50	**_12_**	14	15	17	19	20	21	28	11	13	13	16	18	20	20	25	**_10_**	11	11	14	15	17	17	21
		25	**_14_**	**_17_**	18	20	22	23	24	33	13	15	15	18	21	23	23	29	**_11_**	**_13_**	13	16	18	19	19	24
		10	**_15_**	**_19_**	20	22	25	26	28	37	**_15_**	**_17_**	**_17_**	21	23	26	26	32	**_13_**	**_15_**	15	18	20	22	22	28

Note: Values in bold italics exceed 8-hour physiological criteria (see text).

[a] The dimension away from the body.

TABLE 13.5 Maximum Acceptable Force of Push for Males (kg)

Height from Floor to Hands (cm)	Percent of Industrial Population	2.1-m Push							7.6-m Push							15.2-m Push							30.5-m Push					45.7-m Push					61.0-m Push			
		One Push Every							One Push Every							One Push Every							One Push Every					One Push Every					One Push Every			
		6 sec	12 sec	1 min	2 min	5 min	30 min	8 hr	15 sec	22 sec	1 min	2 min	5 min	30 min	8 hr	9 sec	14 sec	1 min	2 min	5 min	30 min	8 hr	1 min	2 min	5 min	30 min	8 hr	1 min	2 min	5 min	30 min	8 hr	2 min	5 min	30 min	8 hr
144	90	20	22	25	25	26	26	31	14	16	21	21	22	22	26	16	18	19	19	20	21	25	15	16	19	19	24	13	14	16	16	20	12	14	14	18
	75	26	29	32	32	34	34	41	18	20	27	27	28	28	34	21	23	25	25	26	27	32	19	21	25	25	31	16	18	21	21	26	16	18	18	23
	50	32	36	40	40	42	42	51	23	25	33	33	35	35	42	26	29	31	31	33	33	40	24	27	31	31	38	20	23	26	26	33	20	22	22	28
	25	38	43	47	47	50	51	61	27	31	40	40	42	42	51	31	35	37	37	40	40	48	28	32	37	37	46	24	27	32	32	39	23	27	27	34
	10	44	49	55	55	58	58	70	31	35	46	46	48	49	58	36	40	43	43	45	46	55	32	37	42	42	53	28	31	36	36	45	27	31	31	39
95	90	21	24	26	26	28	28	34	16	18	23	23	25	25	30	18	21	22	22	23	24	28	17	19	22	22	27	14	16	19	19	23	14	16	16	20
	75	28	31	34	34	36	36	44	21	23	30	30	32	32	39	24	27	28	28	30	30	36	21	24	28	28	35	18	21	24	24	30	18	21	21	26
	50	34	38	43	43	45	45	54	26	29	38	38	40	40	48	29	33	35	35	37	38	45	27	30	35	35	44	22	26	30	30	37	22	26	26	32
	25	41	46	51	51	54	55	65	31	35	45	45	48	48	58	35	40	42	42	45	45	54	32	36	42	42	52	27	31	36	36	45	27	31	31	38
	10	47	53	59	59	62	63	75	35	40	52	52	55	56	66	40	46	49	49	52	52	62	37	41	48	48	60	31	36	41	41	52	31	35	35	44
64	90	19	22	24	24	25	26	31	13	14	20	20	21	21	26	15	17	19	19	20	20	24	14	16	19	19	23	12	14	16	16	20	12	14	14	17
	75	25	28	31	31	33	33	40	16	19	26	26	27	28	33	19	21	24	24	26	26	31	18	21	24	24	30	16	18	21	21	26	15	18	18	22
	50	31	35	39	39	41	41	50	20	23	32	32	34	35	41	23	27	30	30	32	33	39	23	26	30	30	37	20	22	26	26	32	19	22	22	28
	25	38	42	46	46	49	50	59	25	28	39	39	41	41	50	28	32	36	36	39	39	47	28	31	36	36	45	24	27	31	31	39	23	26	26	33
	10	43	48	53	53	57	57	68	28	32	45	45	47	48	57	32	37	42	42	44	45	54	32	36	41	41	52	27	31	36	36	44	26	30	30	38

Initial Force (the force required to get an object in motion)

TABLE 13.5 Maximum Acceptable Force of Push for Males (kg) (continued)

Column groups are each headed "One Push Every". Units: sec, min, hr as indicated. Sustained Force (the force required to keep an object in motion).

Height from Floor to Hands (cm)	Percent of Industrial Population	2.1-m Push 6 sec	12 sec	1 min	2 min	5 min	30 min	8 hr	7.6-m Push 15 sec	22 sec	1 min	2 min	5 min	30 min	8 hr	15.2-m Push 9 sec	14 sec	1 min	2 min	5 min	30 min	8 hr	30.5-m Push 1 min	2 min	5 min	30 min	8 hr	45.7-m Push 1 min	2 min	5 min	30 min	8 hr	61.0-m Push 2 min	5 min	30 min	8 hr
144	90	10	13	15	16	18	18	22	8	9	13	13	15	16	18	8	9	11	12	13	14	16	8	10	12	13	16	7	8	10	11	13	7	8	9	11
	75	***13***	17	21	22	24	25	30	***10***	***13***	17	18	20	21	25	***11***	***13***	15	16	18	18	22	***11***	13	16	18	21	***10***	***11***	13	15	18	***9***	11	13	15
	50	***17***	22	27	28	31	32	38	***13***	***16***	22	23	26	27	32	***14***	***17***	***20***	20	23	24	28	***15***	***17***	20	23	28	***12***	***14***	17	19	23	***12***	14	16	19
	25	***21***	27	33	34	38	40	47	***16***	***20***	28	29	32	33	39	***17***	***20***	***24***	25	28	29	34	***18***	***21***	25	29	34	***15***	***18***	21	24	28	***15***	17	20	24
	10	***25***	***31***	38	40	45	46	54	***19***	***23***	32	33	38	39	46	***20***	***24***	***28***	29	33	34	40	***21***	***25***	29	33	39	***18***	***21***	24	28	33	***17***	20	23	28
95	90	10	13	16	17	19	19	23	8	10	13	13	15	15	18	8	***10***	11	12	13	13	16	8	10	12	13	16	7	8	9	11	13	7	8	9	11
	75	***14***	18	22	22	25	26	31	***11***	***13***	17	18	20	21	25	***11***	***13***	15	16	18	18	21	***11***	13	16	18	21	***9***	***11***	13	15	18	***9***	11	12	15
	50	***18***	23	28	29	33	34	40	***14***	***17***	22	23	26	27	32	***14***	***17***	19	20	23	23	28	***15***	***17***	20	23	27	***12***	***14***	17	19	23	***12***	14	16	19
	25	***22***	28	34	35	40	41	49	***17***	***21***	27	29	32	33	39	***17***	***21***	24	25	28	29	34	***18***	***21***	25	29	33	***15***	***18***	21	24	28	***15***	17	20	23
	10	***26***	***33***	40	41	46	48	57	***20***	***24***	32	33	37	38	45	***21***	***25***	28	29	32	33	40	***21***	***25***	29	33	39	***17***	***20***	24	27	32	***17***	20	23	27
64	90	10	13	16	16	18	19	23	8	10	12	13	14	15	18	8	***10***	11	11	12	13	15	8	9	11	13	15	7	8	9	11	13	7	8	9	10
	75	***14***	18	21	22	25	26	31	***11***	***13***	17	17	19	20	24	***11***	***13***	14	15	17	17	21	***11***	13	15	17	20	***9***	***11***	12	14	17	9	10	12	14
	50	***18***	23	28	29	32	33	39	***14***	***17***	21	22	25	26	31	***14***	***17***	19	19	22	22	27	***14***	***16***	19	22	26	***12***	***14***	16	18	22	12	14	15	18
	25	***22***	28	34	35	39	41	48	***17***	***21***	26	27	31	32	37	***18***	***21***	23	24	27	28	33	***17***	***20***	24	27	32	***14***	***17***	20	23	27	14	17	19	22
	10	***26***	***32***	39	41	46	48	56	***20***	***25***	30	32	36	37	44	***21***	***25***	27	28	31	32	38	***20***	***24***	28	32	37	***17***	***20***	23	26	31	16	19	22	26

Note: Values in bold italics exceed 8-hour physiological criteria (see text).

TABLE 13.6 Maximum Acceptable Force of Push for Females (kg)

Initial Force (the force required to get an object in motion)

Height from Floor to Hands (cm)	Percent of Industrial Population	2.1-m Push — One Push Every							7.6-m Push — One Push Every							15.2-m Push — One Push Every							30.5-m Push — One Push Every					45.7-m Push — One Push Every					61.0-m Push — One Push Every			
		6 sec	12 sec	1 min	2 min	5 min	30 min	8 hr	15 sec	22 sec	1 min	2 min	5 min	30 min	8 hr	9 sec	14 sec	1 min	2 min	5 min	30 min	8 hr	1 min	2 min	5 min	30 min	8 hr	1 min	2 min	5 min	30 min	8 hr	2 min	5 min	30 min	8 hr
135	90	14	15	17	18	20	21	22	15	16	16	16	18	19	20	12	14	14	14	15	16	17	12	13	14	15	17	12	13	14	15	17	12	13	14	15
	75	17	18	21	22	24	25	27	18	19	19	20	22	23	24	15	17	17	17	19	20	21	15	16	17	19	21	15	16	17	19	21	14	15	17	19
	50	20	22	25	26	29	30	32	21	23	23	24	26	27	29	18	20	20	20	22	23	25	18	19	21	22	25	18	19	21	22	25	17	18	20	22
	25	24	25	29	30	33	35	37	25	26	27	28	31	32	34	20	23	23	24	26	27	29	20	22	24	26	29	20	22	24	26	29	20	21	23	26
	10	26	28	33	34	38	39	41	28	30	30	31	34	36	38	23	26	26	26	29	31	32	23	25	27	29	33	23	25	27	29	33	22	24	26	29
89	90	14	15	17	18	20	21	22	14	15	16	17	19	19	21	11	13	14	14	16	16	17	12	14	15	16	18	12	14	15	16	18	12	13	14	16
	75	17	18	21	22	24	25	27	17	18	20	20	22	23	25	14	16	17	17	19	20	21	15	16	18	19	21	15	16	18	19	21	15	16	17	19
	50	20	22	25	26	29	30	32	20	21	23	24	27	28	30	16	19	20	21	23	24	25	18	20	21	23	26	18	20	21	23	26	18	19	20	23
	25	24	25	29	30	33	35	37	23	25	27	28	31	33	34	19	22	23	24	27	28	29	21	23	24	26	30	21	23	24	26	30	20	22	24	27
	10	26	28	33	34	38	39	41	26	28	31	32	35	37	39	22	24	26	27	30	31	33	24	26	28	30	33	24	26	28	30	33	23	25	26	30
57	90	11	12	14	14	16	17	18	11	12	14	14	16	16	17	9	11	12	12	13	14	15	11	12	12	13	15	11	12	12	13	15	10	11	12	13
	75	14	15	17	17	19	20	21	14	15	17	17	19	20	21	11	13	14	15	16	17	18	13	14	15	16	18	13	14	15	16	18	12	13	14	16
	50	16	17	20	21	23	24	25	16	18	20	20	23	23	25	14	15	17	18	19	20	21	15	17	18	19	22	15	17	18	19	22	15	16	17	19
	25	19	20	23	24	27	28	30	19	21	23	24	27	27	29	16	18	20	20	23	24	25	18	19	21	22	25	18	19	21	22	25	17	19	20	23
	10	21	23	26	27	30	31	33	22	23	26	27	30	31	33	18	20	22	23	25	26	28	20	22	23	25	28	20	22	23	25	28	19	21	23	25

TABLE 13.6 Maximum Acceptable Force of Push for Females (kg) (continued)

Sustained Force (the force required to keep an object in motion)

Height from Floor to Hands (cm)	Percent of Industrial Population	2.1-m Push							7.6-m Push							15.2-m Push							30.5-m Push					45.7-m Push					61.0-m Push			
		6 sec	12 sec	1 min	2 min	5 min	30 min	8 hr	15 sec	22 sec	1 min	2 min	5 min	30 min	8 hr	9 sec	14 sec	1 min	2 min	5 min	30 min	8 hr	1 min	2 min	5 min	30 min	8 hr	1 min	2 min	5 min	30 min	8 hr	2 min	5 min	30 min	8 hr
135	90	6	8	10	10	11	12	14	6	7	7	7	8	9	11	5	6	6	6	7	7	9	5	6	6	6	8	5	5	5	6	8	4	4	4	6
	75	9	12	14	14	16	17	21	9	10	11	11	12	13	16	7	8	9	9	10	11	13	7	8	9	9	12	7	8	8	8	11	6	6	6	9
	50	12	16	19	20	21	23	28	12	14	14	15	16	17	21	10	11	12	12	14	14	18	10	11	12	12	16	9	10	11	11	15	8	8	9	12
	25	16	20	24	25	27	29	36	15	17	18	18	20	22	27	12	14	15	16	17	18	22	13	14	15	15	21	11	13	13	14	19	10	10	11	15
	10	18	23	28	28	32	34	42	18	20	21	22	24	26	32	14	17	18	18	20	22	27	15	17	18	18	25	14	15	16	17	22	12	12	13	17
89	90	6	7	9	9	10	11	13	6	7	8	8	9	9	11	5	6	6	7	7	8	10	5	6	6	7	9	5	6	6	6	8	4	4	5	6
	75	8	11	13	13	15	16	19	9	10	11	11	13	13	17	7	8	9	10	11	11	14	6	9	9	10	13	7	8	8	9	12	6	6	7	9
	50	11	15	18	18	20	21	26	12	13	15	15	17	18	22	9	11	13	13	14	15	19	10	12	12	13	17	10	11	11	12	16	8	9	9	12
	25	14	18	22	22	25	27	33	15	17	19	19	21	23	28	12	14	16	16	18	19	24	13	15	15	16	22	12	14	14	15	20	11	11	12	15
	10	17	22	26	26	30	32	39	17	20	22	22	25	27	33	14	17	19	19	21	23	28	16	18	18	19	26	14	16	17	18	24	13	13	14	18
57	90	5	6	8	8	9	9	12	6	7	7	7	8	9	11	5	6	6	6	7	7	9	5	6	6	6	8	5	5	5	6	7	4	4	4	6
	75	7	9	11	12	13	14	17	8	10	11	11	12	12	15	7	8	9	9	10	10	13	7	8	8	9	12	7	7	8	8	11	6	6	6	8
	50	10	13	15	16	17	18	23	11	13	14	14	16	17	21	9	11	12	12	13	14	17	10	11	11	12	16	9	10	10	11	15	8	8	8	11
	25	12	16	19	20	22	23	29	14	17	18	18	20	21	26	12	14	15	15	17	18	22	12	14	14	15	20	11	13	13	14	19	10	10	11	14
	10	15	19	23	23	26	28	34	17	20	21	21	23	25	31	14	16	17	18	20	21	26	15	16	17	18	24	13	15	16	17	22	12	12	13	17

Note: Values in bold italics exceed 8-hour physiological criteria (see text).

TABLE 13.7 Maximum Acceptable Force of Pull for Males (kg)

Initial Force (the force required to get an object in motion)

Height from Floor to Hands (cm)	Percent of Industrial Population	2.1-m Pull — One Pull Every							7.6-m Pull — One Pull Every							15.2-m Pull — One Pull Every							30.5-m Pull — One Pull Every					45.7-m Pull — One Pull Every					61.0-m Pull — One Pull Every			
		sec 6	sec 12	min 1	min 2	min 5	min 30	hr 8	sec 15	sec 22	min 1	min 2	min 5	min 30	hr 8	sec 9	sec 14	min 1	min 2	min 5	min 30	hr 8	min 1	min 2	min 5	min 30	hr 8	min 1	min 2	min 5	min 30	hr 8	min 2	min 5	min 30	hr 8
144	90	14	16	18	18	19	19	23	11	13	16	16	17	18	21	13	15	15	15	16	17	20	12	13	15	15	19	10	11	13	13	16	10	11	11	14
	75	17	19	22	22	23	24	28	14	15	20	20	21	21	26	16	18	19	19	20	20	24	14	16	19	19	23	12	14	16	16	20	12	14	14	17
	50	20	23	26	26	28	28	33	16	18	24	24	25	26	31	19	21	22	22	24	24	29	17	19	22	22	27	15	16	19	19	24	14	16	16	20
	25	24	27	31	31	32	33	39	19	21	28	28	29	30	36	22	25	26	26	28	28	33	20	22	26	26	32	17	19	22	22	28	16	19	19	24
	10	26	30	34	34	36	37	44	21	24	31	31	33	33	40	24	28	29	29	31	31	38	22	25	29	29	37	20	22	25	25	31	18	21	21	27
95	90	19	22	25	25	27	27	32	15	18	23	23	24	24	29	18	20	21	21	23	23	28	16	18	21	21	26	14	16	18	18	23	13	16	16	19
	75	23	27	31	31	32	33	39	19	21	28	28	29	30	36	22	25	26	26	28	28	33	20	22	26	26	32	17	19	22	22	28	16	19	19	24
	50	28	32	36	36	39	39	47	23	26	33	33	35	35	42	26	29	31	31	33	33	40	24	27	31	31	38	20	23	27	27	33	20	23	23	28
	25	33	37	42	42	45	45	54	26	30	39	39	41	41	49	30	34	36	36	38	39	46	27	31	36	36	45	24	27	31	31	38	23	26	26	33
	10	37	42	48	48	51	51	61	30	33	43	43	46	47	56	33	38	41	41	43	44	52	31	35	40	40	50	27	30	35	35	43	26	30	30	37
64	90	22	25	28	28	30	30	36	18	20	26	26	27	28	33	20	23	24	24	26	26	31	18	21	24	24	30	16	18	21	21	26	15	18	18	22
	75	27	30	34	34	37	37	44	21	24	31	31	33	34	40	24	28	29	29	31	32	38	22	25	29	29	36	19	22	25	25	31	19	21	21	27
	50	32	36	41	41	44	44	53	25	29	37	37	40	40	48	29	33	35	35	37	38	45	27	30	35	35	43	23	26	30	30	37	22	26	26	32
	25	37	42	48	48	51	51	61	30	34	44	44	46	47	56	34	39	41	41	43	44	52	31	35	41	41	50	27	30	35	35	43	26	30	30	37
	10	42	48	54	54	57	58	69	33	38	49	49	52	53	63	38	43	46	46	49	49	59	35	39	46	46	57	30	34	39	39	49	29	34	34	42

TABLE 13.7 Maximum Acceptable Force of Pull for Males (kg) (continued)

Sustained Force (the force required to keep an object in motion)

Height from Floor to Hands (cm)	Percent of Industrial Population	2.1-m Pull One Pull Every							7.6-m Pull One Pull Every							15.2-m Pull One Pull Every							30.5-m Pull One Pull Every					45.7-m Pull One Pull Every					61.0-m Pull One Pull Every			
		6 sec	12 sec	1 min	2 min	5 min	30 min	8 hr	15 sec	22 sec	1 min	2 min	5 min	30 min	8 hr	9 sec	14 sec	1 min	2 min	5 min	30 min	8 hr	1 min	2 min	5 min	30 min	8 hr	1 min	2 min	5 min	30 min	8 hr	2 min	5 min	30 min	8 hr
144	90	8	10	12	13	15	15	18	8	10	11	12	12	12	15	7	8	9	9	10	11	13	7	8	9	11	13	6	7	8	9	10	6	6	7	9
	75	10	13	16	17	19	20	23	10	13	14	16	16	16	19	9	10	12	12	14	14	17	9	10	12	14	16	7	9	10	11	14	7	8	10	11
	50	13	16	20	21	23	24	28	13	16	17	19	20	20	23	11	13	14	15	17	17	20	11	13	15	17	20	9	11	12	14	17	9	10	12	14
	25	15	20	24	25	28	29	34	15	20	23	24	24	24	28	13	15	17	18	20	21	24	13	15	18	20	24	11	13	15	17	20	11	12	14	17
	10	17	22	27	28	32	33	39	17	22	26	27	27	27	32	14	17	19	20	23	24	28	15	17	20	23	27	12	14	17	19	23	12	14	16	19
95	90	10	13	16	17	19	20	24	8	10	13	14	16	16	19	9	10	12	12	14	14	17	9	10	12	14	17	7	9	10	12	14	7	9	10	12
	75	13	17	21	22	25	26	30	11	13	17	18	20	21	25	11	14	15	15	18	18	22	12	13	16	18	21	10	11	13	15	18	9	11	13	15
	50	16	21	26	27	31	32	37	13	17	21	22	25	26	31	14	17	19	19	22	23	27	14	17	19	22	26	12	14	16	19	22	12	14	16	18
	25	19	26	31	33	37	38	45	16	20	26	27	30	31	37	17	20	22	23	26	27	32	17	20	23	27	32	14	17	19	22	26	14	16	19	22
	10	22	29	36	37	42	43	51	18	23	29	31	34	36	42	19	23	26	27	30	31	37	19	23	27	31	36	16	19	22	25	30	16	19	21	25
64	90	11	14	17	18	20	21	25	9	11	14	15	17	17	20	9	11	12	13	15	15	18	9	11	13	15	18	8	9	11	12	15	8	9	11	12
	75	14	19	23	23	26	27	32	11	14	19	19	22	22	26	12	14	16	17	19	19	23	12	14	17	19	23	10	12	14	16	19	10	12	13	16
	50	17	23	28	29	32	34	40	14	18	23	24	27	28	33	15	18	20	21	23	24	28	15	18	21	24	28	13	15	17	20	23	12	14	16	20
	25	20	27	33	35	39	40	48	17	21	27	28	32	33	39	18	21	24	25	28	29	34	18	21	25	28	33	15	18	21	24	28	15	17	20	23
	10	23	31	38	40	45	46	54	19	24	31	32	37	38	45	20	24	27	28	32	33	39	21	24	28	32	38	17	20	24	27	32	17	20	23	27

Note: Values in bold italics exceed 8-hour physiological criteria (see text).

TABLE 13.8 Maximum Acceptable Force of Pull for Females (kg)

Initial Force (the force required to get an object in motion)

Height from Floor to Hands (cm)	Percent of Industrial Population	2.1-m Pull · 6 sec	12 sec	1 min	2 min	5 min	30 min	8 hr	7.6-m Pull · 15 sec	22 sec	1 min	2 min	5 min	30 min	8 hr	15.2-m Pull · 9 sec	14 sec	1 min	2 min	5 min	30 min	8 hr	30.5-m Pull · 1 min	2 min	5 min	30 min	8 hr	45.7-m Pull · 1 min	2 min	5 min	30 min	8 hr	61.0-m Pull · 2 min	5 min	30 min	8 hr
135	90	13	16	17	18	20	21	22	13	14	16	16	18	19	20	10	12	13	14	15	16	17	12	13	14	15	17	12	13	14	15	17	12	13	14	15
	75	16	19	20	21	24	25	26	16	17	19	19	21	22	24	12	14	16	16	18	19	20	14	16	17	18	20	14	16	17	18	20	14	15	16	18
	50	19	22	24	25	28	29	31	19	20	22	23	25	26	28	14	16	19	19	21	22	24	17	18	20	21	24	17	18	20	21	24	16	18	19	21
	25	21	25	28	29	32	33	35	21	23	25	26	29	30	32	16	19	21	22	25	26	27	19	21	23	24	27	19	21	23	24	27	19	20	22	25
	10	24	28	31	32	36	37	39	24	26	28	29	32	34	36	18	21	24	25	27	29	30	22	24	25	27	31	22	24	25	27	31	21	23	24	27
89	90	14	16	18	19	21	22	23	14	15	16	17	19	20	21	10	12	14	14	16	17	18	13	14	15	16	18	13	14	15	16	18	12	13	14	16
	75	16	19	21	22	25	26	27	17	18	19	20	22	23	25	12	15	17	17	19	20	21	15	16	18	19	21	15	16	18	19	21	15	16	17	19
	50	19	23	25	26	29	30	32	19	21	23	24	26	27	29	14	17	19	20	22	23	25	18	19	21	22	25	18	19	21	22	25	17	18	20	22
	25	22	26	29	30	33	35	37	22	24	26	27	30	31	33	16	20	22	23	26	27	28	20	22	24	25	29	20	22	24	25	29	20	21	23	26
	10	25	29	32	33	37	39	41	25	27	29	30	33	35	37	18	22	25	26	29	30	32	23	25	26	28	32	23	25	26	28	32	22	24	25	29
57	90	15	17	19	20	22	23	24	15	16	17	18	20	21	22	11	13	15	15	17	18	19	13	14	15	17	19	13	14	15	17	19	13	14	15	17
	75	17	20	22	23	26	27	28	17	19	20	21	23	24	26	13	15	17	18	20	21	22	16	17	18	20	22	16	17	18	20	22	15	16	18	20
	50	20	24	26	27	30	32	33	20	22	24	25	28	29	30	15	18	20	21	23	24	26	18	20	22	23	26	18	20	22	23	26	18	19	21	23
	25	23	27	30	31	35	36	38	23	25	27	29	32	33	35	17	21	23	24	27	28	30	21	23	25	27	30	21	23	25	27	30	21	22	24	27
	10	26	31	34	35	39	40	43	26	28	31	32	35	37	39	19	23	26	27	30	31	33	24	26	28	30	34	24	26	28	30	34	23	25	27	30

TABLE 13.8 Maximum Acceptable Force of Pull for Females (kg) (continued)

Sustained Force (the force required to keep an object in motion)

Height from Floor to Hands (cm)	Percent of Industrial Population	2.1-m Pull — One Pull Every							7.6-m Pull — One Pull Every							15.2-m Pull — One Pull Every							30.5-m Pull — One Pull Every					45.7-m Pull — One Pull Every					61.0-m Pull — One Pull Every			
		6 sec	12 sec	1 min	2 min	5 min	30 min	8 hr	15 sec	22 sec	1 min	2 min	5 min	30 min	8 hr	9 sec	14 sec	1 min	2 min	5 min	30 min	8 hr	1 min	2 min	5 min	30 min	8 hr	1 min	2 min	5 min	30 min	8 hr	2 min	5 min	30 min	8 hr
135	90	6	9	10	10	11	12	15	6	9	9	10	10	11	13	6	7	7	8	8	9	11	6	7	7	8	10	6	6	7	7	9	5	5	5	7
	75	8	12	13	14	15	16	20	8	11	12	13	14	15	19	7	9	10	11	12	14	19	8	9	10	10	14	8	9	9	9	12	7	7	7	10
	50	10	16	17	18	19	21	25	11	13	15	16	18	19	24	9	11	13	14	15	17	23	11	12	12	13	17	10	11	11	12	16	8	9	9	12
	25	13	19	21	21	23	25	31	13	16	18	18	20	22	29	11	14	15	16	18	20	27	13	15	15	16	21	12	13	14	14	19	10	11	11	15
	10	15	22	24	25	27	29	36	16	19	21	21	22	24	32	13	16	18	18	20	22	27	15	17	17	18	25	14	15	16	17	23	12	12	13	17
89	90	6	9	10	10	11	12	14	6	9	9	10	10	11	13	5	6	7	7	8	9	11	6	7	7	7	10	5	6	6	7	9	5	5	5	7
	75	8	12	13	13	15	16	19	8	11	12	12	13	15	17	7	8	9	10	11	12	14	8	9	9	10	13	7	8	8	9	12	6	7	7	9
	50	10	15	16	17	19	20	25	10	13	15	15	16	18	22	9	11	12	13	14	16	18	10	12	12	13	17	9	11	11	12	15	8	8	9	12
	25	12	18	20	21	23	24	30	13	16	18	18	20	22	27	11	13	15	15	17	18	22	12	14	14	15	21	11	13	13	14	19	10	10	11	15
	10	14	21	23	24	26	28	35	15	18	21	21	23	25	31	13	15	17	18	20	21	26	15	16	17	18	24	13	15	15	16	22	12	12	13	17
57	90	5	8	9	9	10	11	13	5	8	8	9	9	10	12	5	6	7	7	8	8	10	6	6	6	7	9	5	6	6	6	8	4	5	5	6
	75	7	11	12	12	13	14	18	7	10	11	11	12	13	16	7	8	9	10	11	11	13	7	8	9	9	12	7	8	8	8	11	6	6	6	9
	50	9	14	15	16	17	18	23	9	12	13	14	15	16	20	8	10	11	12	13	14	17	9	11	11	12	16	9	10	10	11	14	8	8	8	11
	25	11	17	18	19	21	22	27	11	15	16	17	19	20	24	10	12	14	14	16	17	21	11	13	13	14	19	11	12	12	13	17	9	9	10	13
	10	13	20	21	22	24	26	32	13	17	19	20	22	23	28	12	14	16	16	18	19	24	13	15	15	16	22	12	14	14	15	20	11	11	12	16

Note: Values in bold italics exceed 8-hour physiological criteria (see text).

TABLE 13.9 Maximum Acceptable Weight of Carry (kg)

Height from Floor to Hands (cm)	Percent of Industrial Population	2.1-m Carry — One Carry Every							4.3-m Carry — One Carry Every							8.5-m Carry — One Carry Every						
		6 sec	12 sec	1 min	2 min	5 min	30 min	8 hr	10 sec	16 sec	1 min	2 min	5 min	30 min	8 hr	18 sec	24 sec	1 min	2 min	5 min	30 min	8 hr
												Males										
111	90	10	14	17	17	19	21	25	9	11	15	15	17	19	22	10	11	13	13	15	17	20
	75	14	19	23	23	26	29	34	13	16	21	21	23	26	30	13	15	18	18	20	23	27
	50	19	25	30	30	33	38	44	17	20	27	27	30	34	39	17	19	23	24	26	29	35
	25	23	30	37	37	41	46	54	20	25	33	33	37	41	48	21	24	29	29	32	36	43
	10	27	35	43	43	48	54	63	24	29	38	39	43	48	57	24	28	34	34	38	42	50
79	90	13	17	21	21	23	26	31	11	14	18	19	21	23	27	13	15	17	18	20	22	26
	75	18	23	28	29	32	36	42	16	19	25	25	28	32	37	17	20	24	24	27	30	35
	50	23	30	37	37	41	46	54	20	25	32	33	36	41	48	22	26	31	31	35	39	46
	25	28	37	45	46	51	57	67	25	30	40	40	45	50	59	27	32	38	38	42	48	56
	10	33	43	53	53	59	66	78	29	35	47	47	52	59	69	32	38	44	45	50	56	65
												Females										
105	90	11	12	13	13	13	13	18	9	10	13	13	13	13	18	10	11	12	12	12	12	16
	75	13	14	15	15	16	16	21	11	12	15	15	16	16	21	12	13	14	14	14	14	19
	50	15	16	18	18	18	18	25	12	13	18	18	18	18	24	14	15	16	16	16	16	22
	25	17	18	20	20	21	21	28	14	15	20	20	21	21	28	15	17	18	18	19	19	25
	10	19	20	22	22	23	23	31	16	17	22	22	23	23	31	17	19	20	20	21	21	28
72	90	13	14	16	16	16	16	22	10	11	14	14	14	14	20	12	12	14	14	14	14	19
	75	15	17	18	18	19	19	25	11	13	16	16	17	17	23	14	15	16	16	17	17	23
	50	17	19	21	21	22	22	29	13	15	19	19	20	20	26	16	17	19	19	20	20	26
	25	20	22	24	24	25	25	33	15	17	22	22	22	22	30	18	19	21	22	22	22	30
	10	22	24	27	27	28	28	37	17	19	24	24	25	25	33	20	21	24	24	25	25	33

Note: Values in bold italics exceed 8-hour physiological criteria (see text).

- Capability to measure subjective variables such as pain, fatigue, and discomfort — variables that cannot be measured objectively
- Industrial application that is less costly and time consuming than most other methods
- Capability of exposing subjects to hazardous tasks without excessive risk

13.4   Disadvantages

The primary disadvantages of psychophysics include:

- Reliance on subjective judgments from subjects
- Results that may exceed recommended physiological criteria from manual-handling tasks with high repetition rates
- Apparent lack of sensitivity to the bending and twisting motions that are often associated with the onset of low-back pain

13.5   Examples

Use Table 13.1 through Table 13.9 to answer the following questions:

1. What is the maximum acceptable weight for 90% of males lifting a 34-cm-width box with handles for a distance of 76 cm from the floor once every minute? Answer: 15 kg.
2. What is the maximum acceptable initial force for 75% of females pushing a cart with an 89-cm-height handle for a distance of 15.2 m once every 5 min? Answer: 19 kg.
3. What is the maximum acceptable sustained force for item 2? Answer: 11 kg.
4. What is the maximum acceptable weight for 75% of males lifting a 34-cm-width box with handles for a distance of 76 cm between the floor and knuckle height, carrying it at a height of 79 cm for 8.5 m, and then lowering it back to the original height? The combined task is performed once every minute. Answer: 22 kg.

13.6   Related Methods

Biomechanical, physiological, intraabdominal pressure, epidemiological, and psychophysical methods have all been used in establishing guidelines for manual handling tasks. Several investigators have compared the different methods (Ayoub and Dempsey, 1999; Dempsey, 1998; Nicholson, 1989). There is evidence that a person incorporates both physiological and biomechanical stresses when making psychophysical judgments (Karwowski and Ayoub, 1986; Haslegrave and Corlett, 1995).

13.7   Standards and Regulations

In the United States, the "general duty clause" (Article 5.A.1) of the Occupational Safety and Health Act (OSHA) requires employers to provide employment and a place of work "free from recognized hazards that are causing or are likely to cause death or serious physical harm to his employees." Manual-handling tasks have long been associated with disorders of the lower back. In response, the National Institute for Occupational Safety and Health (NIOSH) used psychophysical, biomechanical, physiological, and epidemiological methods to develop the Work Practices Guide for Manual Lifting in 1981 and the revised NIOSH lifting equation in 1993 (NIOSH, 1981; Waters et al., 1993). The load constant for the revised NIOSH lifting equation (23 kg) is derived from the psychophysical tables; specifically, this is the maximum acceptable weight of lift between floor level and knuckle height for 75% of female workers under optimal conditions. Optimal conditions are defined as occasional lifting (once every 8 hours), small object (34-cm box), short lifting distance (25 cm), and good handles.

The NIOSH guidelines were developed specifically for lifting tasks, and they assume that lifting and lowering tasks have the same level of risk for low back injuries (Waters et al., 1993). The psychophysical tables do not make that assumption and can be used to directly evaluate all types of manual handling tasks (i.e., lifting, lowering, pushing, pulling, and carrying).

13.8   Approximate Training and Application Time

One hour should be sufficient for becoming familiar with the psychophysical tables. The application time is essentially the time required to measure the necessary weights, forces, distances, and sizes. Finding the correct value in the psychophysical tables should require no more than 1 or 2 min.

13.9   Reliability and Validity

A study by Marras et al. (1999) investigated the effectiveness of the 1981 NIOSH Work Practices Guide for Manual Lifting, the 1993 NIOSH lifting equation, and the psychophysical tables in correctly identifying jobs with high, medium, and low risk of low-back disorders. The study used a database of 353 industrial jobs representing over 21 million person-hours of exposure. The results indicated that all three methods were predictive of low back disorders, but in different ways. Table 13.10 depicts the percentage that

TABLE 13.10 Correct Identification of Jobs with High, Medium, and Low Risk of
Low-Back Disorders by Three Different Assessment Methods

	NIOSH 81	NIOSH 93	Psychophysical
High-Risk Jobs (Sensitivity)	10%	73%	40%
Medium-Risk Jobs	43%	21%	36%
Low-Risk Jobs (Specificity)	91%	55%	91%

Source: From Marras, W.S. et al. (1999), *Ergonomics*, 42, 229–245.

each method correctly predicted high-, medium-, and low-risk jobs.

The 1981 NIOSH Work Practices Guide underestimated the risk by predicting that most jobs were low risk (low sensitivity, high specificity). The 1993 NIOSH lifting equation overestimated the risk by predicting that most jobs were high risk (high sensitivity, moderate specificity). The psychophysical tables fell between the two (moderate sensitivity, high specificity).

Other studies have also concluded that recommendations based upon psychophysical results can reduce low-back disorders in industry (Snook et al., 1978; Liles et al., 1984; Herrin et al., 1986).

13.10   Tools Needed

A dynamometer or simple spring scale, a tape measure, and various straps for pulling tasks are needed for collecting the data necessary for using the psychophysical tables.

References

Ayoub, M.M. and Dempsey, P.G. (1999), The psychophysical approach to manual materials handling task design, *Ergonomics*, 42, 17–31.

Dempsey, P.G. (1998), A critical review of biomechanical, epidemiological, physiological and psychophysical criteria for designing manual materials handling tasks, *Ergonomics*, 42, 73–88.

Haslegrave, C.M. and Corlett, E.N. (1995), Evaluating work conditions for risk of injury: techniques for field surveys, in *Evaluation of Human Work*, Wilson, J.R. and Corlett, E.N., Eds., Taylor & Francis, London.

Herrin, G.D., Jaraiedi, M., and Anderson, C.K. (1986), Prediction of overexertion injuries using biomechanical and psychophysical models, *Am. Ind. Hyg. Assoc. J.*, 47, 322–330.

Karwowski, W. and Ayoub, M.M. (1986), Fuzzy modelling of stresses in manual lifting tasks, *Ergonomics*, 29, 237–248.

Liles, D.H., Deivanayagam, S., Ayoub, M.M., and Mahajan, P. (1984), A job severity index for the evaluation and control of lifting injury, *Hum. Factors*, 26, 683–693.

Marras, W.S., Fine, L.J., Ferguson, S.A., and Waters, T.R. (1999), The effectiveness of commonly used lifting assessment methods to identify industrial jobs associated with elevated risk of low-back disorders, *Ergonomics*, 42, 229–245.

NIOSH (1981), Work Practices Guide for Manual Lifting, DHHS (NIOSH) publication 81-122, National Institute for Occupational Safety and Health, Cincinnati.

Nicholson, A.S. (1989), A comparative study of methods for establishing load handling capabilities, *Ergonomics*, 32, 1125–1144.

Snook, S.H. (1978), The design of manual handling tasks, *Ergonomics*, 21, 963–985.

Snook, S.H. (1985), Psychophysical considerations in permissible loads, *Ergonomics*, 28, 327–330.

Snook, S.H. (1999), Future directions of psychophysical studies, *Scand. J. Work Environ. Health*, 25, 13–18.

Snook, S.H., Campanelli, R.A., and Hart, J.W. (1978), A study of three preventive approaches to low back injury, *J. Occup. Med.*, 20, 478–481.

Snook, S.H. and Ciriello, V.M. (1991), The design of manual handling tasks: revised tables of maximum acceptable weights and forces, *Ergonomics*, 34, 1197–1213.

Stevens, S.S. (1960), The psychophysics of sensory function, *Am. Scientist*, 48, 226–253.

Waters, T.R., Putz-Anderson, V., Garg, A., and Fine, L.J. (1993), Revised NIOSH equation for the design and evaluation of manual lifting tasks, *Ergonomics*, 36, 749–776.

14

Lumbar Motion Monitor

W. S. Marras
Ohio State University

W. G. Allread
Ohio State University

14.1 Background and Application

It is estimated that the prevalence of work-related back pain in the U.S., with at least one lost workday, is 4.6% (Guo et al., 1999). This suggests that nearly 1 in 20 employees is always afflicted with this disorder. Nationally, total indirect costs due to low back disorders (LBDs) are estimated at $40 billion to $60 billion (Cats-Baril, 1996). A variety of tools are available for assessing LBD risk. Many of these (e.g., static models, the NIOSH lifting equations) assume that motion is not a significant factor in injury causation or that all movements are slow and smooth. However, research (e.g., Bigos et al., 1986; Punnett et al., 1991) suggests that trunk movement plays an important role in LBD risk.

The lumbar motion monitor (LMM) was developed as a response to this need. It assesses the dynamic components of LBD risk in occupational settings, such as those requiring manual materials handling (MMH). The patented LMM (Figure 14.1) is a triaxial electrogoniometer that acts as a lightweight exoskeleton of the lumbar spine. It is placed on the back of an individual, directly in line with the spine, and is attached using harnesses around the pelvis and over the shoulders. The LMM uses potentiometers to measure the instantaneous position of the spine (as a unit), relative to the pelvis, in three-dimensional space. The position data are recorded on a computer using companion software, which also calculates the velocity and acceleration of the spine for the motion of interest.

The LBD risk model developed using the LMM was derived from over 400 repetitive MMH jobs (Marras et al., 1993). Trunk kinematic data, in addition to other workplace and personal factors, were recorded. This information was compared between "low-risk" jobs (those having no LBDs and no job turnover) and "high-risk" jobs (those having 12 or more LBDs annually per 100 full-time workers). Data analysis determined that five factors together determine the probability that a job measured using this method will be similar to those previously found to be "high risk." These factors include two workplace measures (the maximum external moment about the spine and the job's lift rate) and three

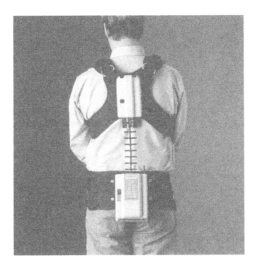

FIGURE 14.1 The lumbar motion monitor (LMM).

trunk-motion parameters (maximum sagittal flexion position, maximum lateral velocity, and average twisting velocity).

The lumbar motion monitor allows for data collection as employees are performing their actual jobs. It can be used in a wide variety of workplaces, including manufacturing environments, warehouses, and health-care facilities.

14.2 Procedure

Risk assessments using the LMM are derived using a four-step approach:

1. Placement of LMM on worker
2. Determination of a job's MMH components
3. Data collection
4. Analysis

Each is described below.

14.2.1 Placement of LMM on Worker

The LMM system is designed to accommodate a majority of individual body sizes and is adjustable to four lengths (extra small, small, medium, and large). It is important to ensure proper fit so that accurate trunk motions are measured. The appropriate size worn during data collection depends on a number of factors, primarily one's standing height and trunk length, and the amount of sagittal flexion required of the job. The adjustable harnesses worn with the LMM also help to ensure the correct fit.

14.2.2 Determination of a Job's MMH Components

To accurately assess a job's LBD risk level, it is important to correctly identify all job elements that have the potential to produce injury. This typically involves the lifting, lowering, pushing, pulling, and carrying tasks that are performed as part of the job's requirements. For jobs that require some sort of rotation, all tasks that comprise this rotation also must be included. The software used for data collection allows for eight tasks to be defined.

For jobs having only a few tasks, it may be helpful (especially during data interpretation) to define differences within a task that may exist, particularly if their physical nature varies. For example, in a job requiring a pallet to be loaded, categorizing the task as "place box on low level," "place box at medium level," etc. may assist the user in understanding more readily where the greatest LBD risk exists within the job.

14.2.3 Data Collection

The LMM software prompts users to structure jobs in a hierarchical manner. Before data collection can begin, the user is prompted to define the company within which the job is performed, the job itself, the tasks that make up the job, and the employees who will be doing the work. For later analysis purposes, the software allows the input of additional information within each of these categories (i.e., company address, department information, data related to the physical design of the workplace, employee age, gender, and anthropometric measurements), though this is not needed to assess LBD risk.

The primary goal of data collection is to gather information about a job that is fully descriptive of all required work. For example, if the job requires handling objects of many different weights or to/from a variety of locations, then data should be collected to represent these aspects of the job. The more data that are collected, the more likely it is to represent the requirements of the job. Previous research (Allread et al., 2000) found that, within the same object weights and lift locations, there was no additional reduction in data variability for a task after three cycles of the task and three employees performing the work were collected.

14.2.4 Analysis

The software provides numerous methods of evaluating the collected data, as detailed below.

14.2.4.1 Descriptive Information about Trunk Kinematics

This includes details of the positions, velocities, and accelerations of the three motion planes for each data trial collected. This information can be useful for general descriptions of the MMH performed or for comparisons with other tasks or jobs. This information also can be valuable for users who have formed hypotheses about, for example, which tasks require more trunk motions than others.

14.2.4.2 Probability of High-Risk Group Membership (LBD Risk)

Charts are produced that compare each of the five risk model factors (see Section 14.1) to the job database in order to determine the extent to which each is similar to a known group of other "high-risk" jobs. The chart shows an average of each of these five factor levels and calculates the overall "probability of high-risk group membership" (or LBD risk).

LBD risk can be computed in several ways. It can be determined for an individual employee, for a specific job task, or across an entire job that comprises several tasks. Risk also can be computed for a task or job that is averaged across two or more employees who have performed the activity while wearing the LMM.

These assessments allow the user to quantitatively ascertain which factors of the job (e.g., sagittal flexion, twisting velocity, lift rate) are most likely responsible for the level of risk produced. Also, for jobs with multiple tasks, it provides an assessment of which tasks are producing the most risk. These results can guide the user in making recommendations for improving the job (i.e., lowering LBD risk) from an ergonomics perspective.

The software also allows data to be exported into text files that can be analyzed using other applications. The database itself is stored in Microsoft Access® format for manipulation using that software, if desired.

14.3 Advantages

- Data gathered using the LMM are quantitative and allow three-dimensional trunk kinematics to be collected in real-world environments.
- The LBD risk model determines the extent to which the level of a particular risk factor, or the overall LBD risk level itself, is "too much."

- Risk levels are compared with a database of actual workplace factors and trunk motions previously found to have high and low LBD rates.
- The impact of job interventions can be assessed quickly.
- The risk model has been validated.

14.4   Disadvantages

- Use of the LMM requires the training of users.
- Data collection requires active involvement on the part of workers.
- Assessments typically require more data collection time than other tools.
- The LMM can come in contact with other equipment when worn in confined work spaces.
- The LBD risk model does not assess the potential risk of injury to other body parts.

14.5   Example Output

Figure 14.2, Figure 14.3, and Figure 14.4 illustrate outputs that can be derived from the LMM software. Figure 14.2 depicts the screen seen during data collection, which shows real-time trunk motions in the lateral, sagittal, and transverse planes. The vertical lines are user inputs, marking the beginning and ending of one cycle from a material-handling task. Sample kinematic output is shown in Figure 14.3. Following data collection, the computed positions, velocities, and accelerations can be viewed for each motion plane. Figure 14.4 shows a sample LBD risk model chart. The horizontal bars represent the magnitudes of each risk factor used in the model. The chart also shows the calculated risk probability value.

14.6   Related Methods

There are many different LBD risk assessment tools currently available, such as the NIOSH lifting equations (Waters et al., 1993), ACGIH lifting threshold limit values (ACGIH, 2002), and psychophysical tables (Snook and Ciriello, 1991), yet no others combine trunk kinematic factors with the more traditionally used measures, such as lifting frequency and load weight. Other methods are available to quantitatively assess trunk motions (e.g., Motion Analysis, Santa Rosa, CA), but these are difficult to use outside of the laboratory.

14.7   Approximate Training and Application Time

It takes approximately 8 hours to learn about the LMM, understand how to properly fit the device on users, and determine the methods by which to collect and analyze the data. The accompanying software developed for data gathering and risk assessment analysis is a Microsoft Windows-based application, which reduces learning time for those familiar with this format.

14.8   Reliability and Validity

The ability of the LMM to reliably measure trunk motions has been established by Marras et al. (1992). Their study found that the readings taken from the LMM, in all three planes of motion, were not significantly different from those determined using another reliable motion-analysis system.

The LBD risk model used with the LMM also has been prospectively validated (Marras et al., 2000) to ensure that it realistically reflects a job's injury risk. Here, MMH jobs were evaluated using the risk model both before and after significant ergonomic interventions were made. The jobs' low back injury rates also were determined before the intervention occurred and after a significant period of time had

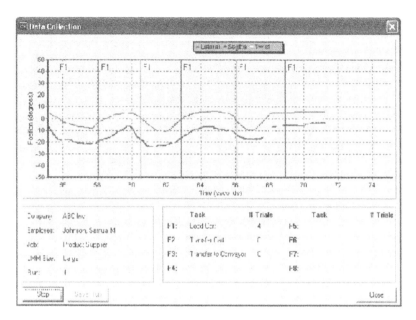

FIGURE 14.2 LMM software data collection screen.

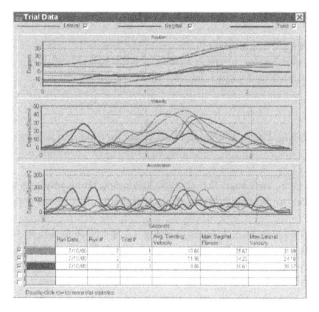

FIGURE 14.3 LMM software data view screen.

passed from when the job was modified. Changes in injury rates and risk assessments were analyzed relative to a comparison group of jobs in which no changes had been made. The results indicated that a statistically significant correlation existed between changes in the jobs' estimated LBD risk and changes in their actual low back incidence rates over the period of observation.

These data indicate that the LMM and the LBD risk model provide useful, reliable, and valid information for assessing a job's low back injury risk.

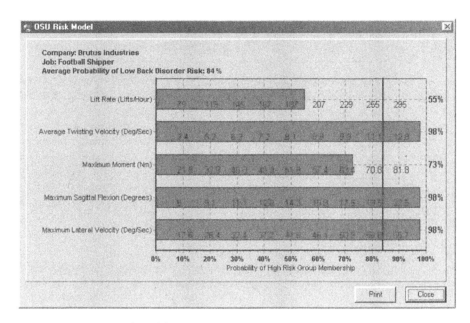

FIGURE 14.4 LMM software risk model output screen.

14.9   Tools Needed

LBD risk evaluation with the LMM requires the use of the Acupath™ system (LMM, harnesses, related electronic equipment, and a laptop computer loaded with the Ballet™ software). The user also will need a tape measure, scale or push-pull gauge, and data-recording forms.

References

ACGIH (2002), 2002 TLVs and BEIs: Threshold Limit Values for Chemical Substances and Physical Agents and Biological Exposure Indices, American Conference of Governmental Industrial Hygienists, Cincinnati.

Allread, W.G., Marras, W.S., and Burr, D.L. (2000), Measuring trunk motions in industry: variability due to task factors, individual differences, and the amount of data collected, *Ergonomics*, 43, 691–701.

Bigos, S.J., Spengler, D.M., Martin, N.A., Zeh, J., Fisher, L., and Nachemson, A. (1986), Back injuries in industry: a retrospective study, Part III: employee-related factors, *Spine*, 11, 252–256.

Cats-Baril, W. (1996), Cost of Low Back Pain Prevention, paper presented at the Low Back Pain Prevention, Control, and Treatment Symposium, St. Louis, MO, March 1996.

Guo, H.R., Tanaka, S., Halperin, W.E., and Cameron, L.L. (1999), Back pain prevalence in the U.S. industry and estimates of lost workdays, *Am. J. Public Health*, 89, 1029–1035.

Marras, W.S., Fathallah, F., Miller, R.J., Davis, S.W., and Mirka, G.A. (1992), Accuracy of a three dimensional lumbar motion monitor for recording dynamic trunk motion characteristics, *Int. J. Ind. Ergonomics*, 9, 75–87.

Marras, W.S., Lavender, S.A., Leurgans, S., Rajulu, S., Allread W.G., Fathallah, F., and Ferguson, S.A. (1993), The role of dynamic three dimensional trunk motion in occupationally related low back disorders: the effects of workplace factors, trunk position and trunk motion characteristics on injury, *Spine*, 18, 617–628.

Marras, W.S., Allread, W.G., Burr, D.L., and Fathallah, F.A. (2000), Prospective validation of a low-back disorder risk model and assessment of ergonomic interventions associated with manual materials handling tasks, *Ergonomics*, 43, 1866–1886.

Punnett, L., Fine, L.J., Keyserling, W.M., Herrin, G.D., and Chaffin, D.B. (1991), Back disorders and nonneutral trunk postures of automobile assembly workers, *Scand. J. Work, Environ., Health*, 17, 337–346.

Snook, S.H. and Ciriello, V.M. (1991), The design of manual handling tasks: revised tables of maximum acceptable weights and forces, *Ergonomics*, 34, 1197–1213.

Waters, T.R., Putz-Anderson, V., Garg, A., and Fine, L.J. (1993), Revised NIOSH equation for the design and evaluation of manual lifting tasks, *Ergonomics*, 36, 749–776.

15

The Occupational Repetitive Action (OCRA) Methods: OCRA Index and OCRA Checklist

Enrico Occhipinti
EPM-CEMOC

Daniela Colombini
EPM-CEMOC

15.1 Background and Applications

Occhipinti and Colombini (1996) developed the occupational repetitive action (OCRA) methods to analyze workers' exposure to tasks featuring various upper-limb injury risk factors (repetitiveness, force, awkward postures and movements, lack of recovery periods, and others, defined as "additionals"). The OCRA methods are largely based on a consensus document of the International Ergonomics Association (IEA) technical committee on musculoskeletal disorders (Colombini et al., 2001), and they generate synthetic indicators that also consider worker rotation among different tasks.

The OCRA index can be predictive of the risk of upper extremity (UE) work-related musculoskeletal disorders (WMSDs) in exposed populations. The OCRA index was the first, most analytical, and most reliable method developed. It is generally used for the (re)design or in-depth analysis of workstations and tasks (Colombini et al., 1998, 2002). The OCRA checklist, based on the OCRA index, is simpler to apply and is generally recommended for the initial screening of workstations featuring repetitive tasks (Occhipinti et al., 2000; Colombini et al., 2002).

Both OCRA methods are observational and are largely designed to be used by corporate technical specialists (occupational safety and health [OSH] operators, ergonomists, time and methods analysts, production engineers), who have proven in practice to be best suited to learning and applying the methods for prevention and also for improving production processes in general.

The methods have been applied in a wide cross-section of industries and workplaces. They target any jobs in manufacturing and the service sector that involve repetitive movements and/or efforts of the upper limbs (manufacture of mechanical components, electrical appliances, automobiles, textiles and clothing, ceramics, jewelry, meat and food processing). In Europe, it is estimated that these methods are currently used in more than 5,000 tasks that fit these categories, involving over 20,000 employees.

The methods are not suitable for assessing jobs that use a keyboard and mouse, or other computerized data-entry tools.

15.2 Procedures

15.2.1 General Aspects

The two assessment methods evaluate four main collective risk factors based on their respective duration:

1. Repetitiveness
2. Force
3. Awkward posture and movements
4. Lack of proper recovery periods

Other "additional factors" are also considered, such as mechanical, environmental, and organizational factors for which there is evidence of causal relationship with UE WMSDs. Each identified risk factor is properly described and classified to help identify possible requirements and preliminary preventive interventions. All factors contributing to the overall "exposure" are considered in a general and mutually integrated framework.

15.2.2 OCRA Definitions

Work (job) is composed of one or more tasks in one work shift:

- Within a single task, *cycles* are sequences of technical actions that are repeated over and over, always the same.
- Within each cycle, several *technical actions* can be identified. These are elementary operations that enable the completion of the cycle operational requirements (i.e., take, place, turn, push, pull, replace).

The suggested procedure for assessing the risk should be:

1. Pinpointing the repetitive tasks characterized by those cycles with significant durations
2. Finding the sequence of technical actions in a representative cycle of each task
3. Describing and classifying the risk factors within each cycle
4. Assembly of the data concerning the cycles in each task during the whole work shift, taking into consideration the duration and sequences of the different tasks and of the recovery periods
5. Brief and structured assessment of the risk factors for the job as a whole (exposure or risk index)

15.2.3 OCRA Risk Index

The OCRA index is the result of the ratio between the number of technical actions actually carried out during the work shift, and the number of technical actions which is specifically recommended. In practice, OCRA is defined as:

$$OCRA = \frac{\text{Overall number of technical actions carried out in the shift}}{\text{Overall number of technical actions recommended in the shift}}$$

The technical actions should not be identified as the joint movements. To make action frequency analysis more accessible, a conventional measurement unit has been chosen, the "technical action" of the upper limb. This definition is very similar to the method time measurement (MTM) elements (Barnes, 1968).

The overall number of actual technical actions (ATA) carried out within the shift can be calculated by organizational analysis (number of actions per cycle and number of actions per minute, with this last multiplied for the net duration of the repetitive task(s) analyzed to obtain ATA). The following general formula calculates the overall number of recommended technical actions (RTA) within a shift:

$$\text{Number of recommended technical actions} = \sum_{x=1}^{n} [\text{CF} \times (\text{Ffi} \times \text{Fpi} \times \text{Fci}) \times \text{Di}] \times \text{Fr} \times \text{Fd}$$

where

n = number of repetitive task/s performed during shift
ì = generic repetitive task
CF = frequency constant of technical actions (30 actions per minute)
Ff, Fp, Fc = multiplier factors with scores ranging from 0 to 1, selected for "force" (Ff), "posture" (Fp), and "additional elements" (Fc) in each of the n tasks
D = net duration in minutes of each repetitive task
Fr = multiplier factor for "lack of recovery period"
Fd = multiplier factor according to the daily duration of repetitive tasks

In practice, to determine the overall number of RTAs within a shift, proceed as follows:

1. For each repetitive task, start from a CF of 30 actions/min.
2. For each task, the frequency constant must be corrected for the presence and degree of the following risk factors: force, posture, and additional.
3. Multiply the weighted frequency for each task by the number of minutes of each repetitive task.
4. Sum the values obtained for the different tasks.
5. The resulting value is multiplied by the multiplier factor for recovery periods.
6. Apply the last multiplier factor that considers the total time spent in repetitive tasks.
7. The value thus obtained represents the total recommended number of actions (RTA) in the working shift.

The following sections briefly review the criteria and procedures involved in the determination of the OCRA index calculation variables. For additional details, refer to the handbook prepared by Colombini et al. (2002).

15.2.3.1 Action Frequency Constant (CF)

The literature, albeit not explicitly, supplies suggestions of "limit" action frequency values, and these range from 10 to 25 actions/movements per minute. On the basis of the above and given the practical considerations of the applicability of these proposals in the workplace, the action frequency constant (CF) is fixed at 30 actions per minute.

15.2.3.2 Force Factor (Ff)

Force is a good direct representation of the biomechanical commitment that is necessary to carry out a given technical action. It is difficult to quantify force in real working environments. To overcome this difficulty, one could use the Borg10-category scale for the rating of perceived exertion (Borg, 1982). Once the actions requiring exertion have been determined, operators will be asked to ascribe to each one (or homogeneous group) of them a progressive score from 1 to 10. The calculation of the average exertion weighted over time involves multiplying the Borg Scale score ascribed to each action by its percentage duration within the cycle. The partial results must then be added together.

When choosing the multiplier factor, it is necessary to refer to the average force value, weighted by cycle duration.

15.2.3.3 Postural Factor (Fp)

The description/assessment of the postures must be done over a representative cycle for each one of the repetitive tasks examined. This must be via the description of duration of the postures and/or movements of the four main anatomical segments (both right and left): shoulder, elbow, wrist, and hand.

For classification purposes, it is enough to see that, within the execution of every action, the joint segment involved reaches an excursion greater than 50% of joint range for at least one third of the cycle time. The longer the time, the higher is the score.

The presence of stereotypical movements can be pinpointed by observing those technical actions that are all equal to each other for at least 50% of cycle time or by a very short duration of the cycle (less than 15 sec). The presence of stereotypical movements increases scores for the joints involved.

All of these elements together lead to the design of a useful scheme to identify the values of the posture multiplier factor (Fp).

15.2.3.4 "Additional" Factor (Fc)

These factors are defined as additional not because they are of secondary importance, but because each one of them can be present or absent in the contexts examined. The list of these factors is not exhaustive and includes the use of vibrating tools; requirement for absolute accuracy; localized compressions; exposure to cold or refrigeration; the use of gloves that interfere with the required handling ability; objects handled have a slippery surface; sudden movements, "tearing" or "ripping" movements, or fast movements; repetitive impacts (e.g., hammering, hitting, etc.).

There are some factors (psychosocial) that are concerned with the individual sphere and cannot be included in methods considering a collective and occupational type of exposure. There are other factors, definable as organizational (working pace determined by machine, working on moving object), that should be taken into consideration.

For every additional factor indicated, variable scores can be assigned according to the type and duration.

15.2.3.5 "Recovery Periods" Factor (Fr)

A recovery period is a period during which one or more muscle-tendon groups are basically at rest. The following can be considered as recovery periods:

- Breaks, including the lunch break
- Visual control tasks
- Periods within the cycle that leave muscle groups totally at rest consecutively for at least 10 sec almost every few minutes

Using the indications supplied by some standards as a starting point, in the case of repetitive tasks, it is advisable to have a recovery period every 60 min, with a ratio of five work to one recovery. On the basis of this optimal distribution, it is possible to design criteria to evaluate the presence of risk in a concrete situation. The overall risk is determined by the overall number of hours at risk. For every hour without an adequate recovery period, there is a corresponding multiplier factor.

15.2.3.6 Duration Factor (Fd)

Within a working shift, the overall duration of tasks with repetitive and/or forced upper-limb movements is important to determine overall exposure. The index calculation model is based on scenarios where repetitive manual tasks continue for a good part (6 to 8 hours) of the shift.

15.2.3.7 Calculation of OCRA Exposure

Table 15.1 provides the necessary parameters for dealing with all of the multiplier factors and calculating the OCRA index. These results provide the basis for suggesting recommended technical actions in accord with the OCRA index.

TABLE 15.1 Calculation of OCRA Exposure Index

	Right Arm				Left Arm				
	A	B	C	D	A	B	C	D	Task/s
• Action frequency constant (actions/min)	30	30	30	30	30	30	30	30	CF

• Force factor (perceived effort)

Borg	0.5	1	1.5	2	2.5	3	3.5	4	4.5	5
Factor	1	0.85	0.75	0.65	0.55	0.45	0.35	0.2	0.1	0.01

A	B	C	D	A	B	C	D	Task(s)
								Ff

• Postural factor

Value	0–3	4–7	8–11	12–15	16
Factor	1	0.70	0.60	0.50	0.33

	A	B	C	D	A	B	C	D	Task(s)
Shoulder									(*) select lowest factor
Elbow									between elbow, wrist, and hand
Wrist									
Hand									Fp
(*)									

• Additional factors

Value	0	4	8	12
Factor	1	0.95	0.90	0.80

A	B	C	D	A	B	C	D	Task(s)
								Fc

X

• Duration of repetitive task

A	B	C	D	A	B	C	D	Task(s)

• No. recommended actions for repetitive task, and in total (partial result, without recovery factor)

								RIGHT	LEFT
α	β	γ	δ	α	β	γ	δ	(α+β+γ+δ)	(α+β+γ+δ)

• Factor referring to the lack of recovery periods (no. of hours without adequate recovery)

Hours	0	1	2	3	4	5	6	7	8	Fr
Factor	1	0.00	0.80	0.70	0.60	0.45	0.25	0.10	0	

• factor referring to overall duration of repetitive tasks

	RIGHT		LEFT
Minutes	<120	120–239	240–480	>480	Fd	=	$A_{RP} = \pi \times F_R \times F_D$	$A_{RP} = \pi \times F_R \times F_D$
Factor	2	1.5	1	0.5				

	RIGHT	LEFT		RIGHT	LEFT

$$\frac{\text{Total no. of technical actions observed in repetitive tasks}}{\text{Total no. recommended technical actions}} = \frac{\text{ATA}}{\text{RTA}} =$$

TABLE 15.2 Progressive Optimization of Task Using the OCRA Index

Right Limb	Actions/min	Actions/Shift	Force	Posture	Recovery Periods	OCRA Index
A	53.3	18,144	0.9	0.6	0.6	6.1
B	63.7		0.9	0.5		
A	**45**	14,472	0.9	0.6	0.6	4.9
B	**45**		0.9	0.5		
A	53.3	18,144	**1**	0.6	0.6	5.5
B	63.7		**1**	0.5		
A	53.3	18,144	0.9	0.6	**0.8**	4.5
B	63.7		0.9	0.5		
A	53.3	18,144	1	0.6	**0.8**	4.1
B	63.7		1	0.5		
A	**45**	14,472	1	0.6	**0.8**	3.3
B	**45**		1	0.5		
A	**45**	14,472	1	0.7	**1**	2.1
B	**45**		1	0.7		
Left Limb	**Actions/min**	**Actions/Shift**	**Force**	**Posture**	**Recovery Periods**	**OCRA Index**
A	40	12,864	0.8	0.5	0.6	5.4
B	40		0.9	0.5		
A	**35**	11,256	0.8	0.5	0.6	4.7
B	**35**		0.9	0.5		
A	40	12,864	**1**	0.5	0.6	4.4
B	40		**1**	0.5		
A	40	12,864	0.8	0.5	**0.8**	4
B	40		0.9	0.5		
A	**40**	12,864	1	0.5	**0.8**	3.3
B	**40**		1	0.5		
A	**35**	11,256	1	0.5	**0.6**	2.9
B	**35**		1	0.5		
A	**35**	11,256	1	0.7	**1**	1.7
B	**35**		1	0.7		

Note: Boldface items represent factors being optimized.

15.2.3.8 Example of How to Use the OCRA Index to Redesign Tasks/Workstations

Once the workstation has been analyzed using the OCRA index, and after checking for the presence of risk factors for the upper limbs, it is theoretically possible to use the same index to detect which risk factors should be dealt with to minimize the worker's exposure. Therefore, several versions of the OCRA index are described in which the different risk factors making up the index are gradually reduced. Table 15.2 proposes a summary of the OCRA indexes in which the optimization of each individual factor or set of factors is shown.

1. The initial OCRA values can be seen to be high: 6.1 for the right and 5.4 for the left. The job being analyzed comprises two alternating tasks (A and B) featuring: high-frequency actions (task A = 53.3 actions/min, task B = 63.7 actions/min, both involving the right limb); moderate use of force; high-risk hand posture; and inadequate distribution of recovery periods (there is an almost adequate rest period of 38 min, but concentrated into only two breaks).
2. After reducing the action frequency for both tasks A and B to 45 actions/min for the right and 35 for the left, the total number of actions in the shift is reduced to 14,472 for the right and 11,256 for the left. If the OCRA index is recalculated, the resulting values drop to 4.9 on the right and 4.7 on the left.
3. Introducing a reduction in the use of force, the OCRA index drops to 5.5 on the right and 4.4 on the left.

4. It is also possible to recalculate the OCRA index when nothing but the distribution of the recovery periods is optimized, in this case by dividing the available 38 min into four 9- to 10-min breaks. The resulting OCRA values drop to 4.5 on the right and 4 on the left.
5. By optimizing two factors simultaneously (use of force and distribution of recovery times), the OCRA indexes drop to 4.1 on the right and 3.3 on the left.
6. If three variables are optimized (the previous two, plus a reduction in the action frequency), the OCRA values drop to 3.3 on the right and 2.9 on the left.
7. If the objective is to reach the "risk absent" level, then further modifications have to be introduced, such as an improvement in hand posture and recovery times (with six 8-min breaks, obtained by increasing the 38-min rest factor to 48 min). The OCRA values then drop to 2.1 on the right and 1.7 on the left, values that can be regarded as being in the risk-absent area. Since it is not possible to increase the recovery periods, the action frequency can be reduced (still at a relatively high 45 actions/min) or, alternatively, the workers can rotate on low-risk tasks.

15.2.4 Classification of OCRA Index Results

The studies and experiments carried out until now (Occhipinti and Colombini, 2004) allow the identification of different exposure areas with key OCRA scores. By considering the trend of UE WMSDs in reference to working populations that are not exposed to specific occupational risks, it is possible to define the following OCRA index classification criteria and to indicate the consequent preventive actions to be adopted:

1. Index values ≤1.5 indicate full acceptability of the condition examined (green area or risk absent).
2. Index values between 1.6 and 2.2 (yellow/green area or not relevant risk) mean that exposure is still not relevant or not great enough to foresee significant excesses in the occurrence of UE WMSDs.
3. Index values between 2.3 and 3.5 (yellow/red area or very low risk) mean that exposure is not severe, but that there could be higher disease levels in the exposed groups with respect to a reference group of nonexposed. In these cases, it is advisable to introduce health surveillance, health education, and training and proceed to an improvement of working conditions.
4. Index values ≥3.6 (red area or medium risk up to 9.0, high risk ≥9.1) mean significant exposure levels. Working conditions must be improved, and close monitoring of all effects must be set up.

15.2.5 OCRA Checklist

The analysis system suggested with the checklist begins with the establishment of preassigned scores for each of the four main risk factors (recovery periods, frequency, force, posture) and for the additional factors. The sum total of the partial values obtained in this way produces a final score that estimates the actual exposure level.

The checklist describes a workplace and estimates the intrinsic level of exposure as if the workplace is used for the whole of the shift by one worker. This procedure makes it possible to find out quickly which workplaces in the company imply a significant exposure level (classified as absent, light, medium, and high). In the next stage, it is possible to estimate the exposure indexes for the operators considering their rotation through the different workplaces and applying the following formula:

$$(\text{score A} \times \%PA) + (\text{score B} \times \%PB) + \text{etc.}$$

where "score A" and "score B" are the scores obtained with the checklist for the various workplaces on which the same operator works, and %PA and %PB represent the percentage duration of the repetitive tasks within the shift.

Table 15.3 presents the contents of the checklist for each risk factor and the corresponding scores: the greater the risk, the higher is the score.

TABLE 15.3 The OCRA Checklist

TYPE OF WORK INTERRUPTION (WITH PAUSES OR OTHER VISUAL CONTROL TASKS) Choose one answer. It is possible to choose intermediate values.

0 - THERE IS AN INTERRUPTION OF AT LEAST 5 MINUTES EVERY HOUR IN THE REPETITIVE WORK (ALSO COUNT THE LUNCH BREAK).
1 - THERE ARE TWO INTERRUPTIONS IN THE MORNING AND TWO IN THE AFTERNOON (PLUS THE LUNCH BREAK), LASTING AT LEAST 7–10 MINUTES ON THE 7–8 HOUR SHIFT, OR AT LEAST FOUR INTERRUPTIONS PER SHIFT (PLUS THE LUNCH BREAK), OR FOUR 7–10 MINUTE INTERRUPTIONS IN THE 6-HOUR SHIFT.
3 - THERE ARE TWO PAUSES, LASTING AT LEAST 7–10 MINUTES EACH IN THE 6-HOUR SHIFT (WITHOUT LUNCH BREAK); OR, THREE PAUSES, PLUS THE LUNCH BREAK, IN A 7–8-HOUR SHIFT.
4 - THERE ARE TWO PAUSES, PLUS THE LUNCH BREAK, LASTING AT LEAST 7–10 MINUTES EACH OVER A 7–8 HOUR SHIFT (OR THREE PAUSES WITHOUT THE LUNCH BREAK), OR ONE PAUSE OF AT LEAST 7–10 MINUTES OVER A 6-HOUR SHIFT.
6 - THERE IS A SINGLE PAUSE, LASTING AT LEAST 10 MINUTES, IN A 7-HOUR SHIFT WITHOUT LUNCH BREAK; OR, IN AN 8-HOUR SHIFT THERE ONLY IS A LUNCH BREAK (THE LUNCH BREAK IS NOT COUNTED AMONG THE WORKING HOURS).
10 - THERE ARE NO REAL PAUSES EXCEPT FOR A FEW MINUTES (LESS THAN 5) IN A 7–8-HOUR SHIFT.

ARM ACTIVITY AND WORKING FREQUENCY WITH WHICH THE CYCLES ARE PERFORMED (IF NECESSARY, INTERMEDIATE SCORES CAN BE CHOSEN) Choose one answer (state whether left or right arm is involved the most).

0 - ARM MOVEMENTS ARE SLOW, AND FREQUENT SHORT INTERRUPTIONS ARE POSSIBLE (20 ACTIONS PER MINUTE).
1 - ARM MOVEMENTS ARE NOT TOO FAST, ARE CONSTANT AND REGULAR. SHORT INTERRUPTIONS ARE POSSIBLE (30 ACTIONS PER MINUTE).
3 - ARM MOVEMENTS ARE QUITE FAST AND REGULAR (ABOUT 40), BUT SHORT INTERRUPTIONS ARE POSSIBLE.
4 - ARM MOVEMENTS ARE QUITE FAST AND REGULAR, ONLY OCCASIONAL AND IRREGULAR SHORT PAUSES ARE POSSIBLE (ABOUT 40 ACTIONS PER MINUTE).
6 - ARM MOVEMENTS ARE FAST. ONLY OCCASIONAL AND IRREGULAR SHORT PAUSES ARE POSSIBLE (ABOUT 50 ACTIONS PER MINUTE).
8 - ARM MOVEMENTS ARE VERY FAST. THE LACK OF INTERRUPTIONS IN PACE MAKES IT DIFFICULT TO HOLD THE PACE, WHICH IS ABOUT 60 ACTIONS PER MINUTE.
10 - VERY HIGH FREQUENCIES, 70 ACTIONS PER MINUTE OR MORE. ABSOLUTELY NO INTERRUPTIONS ARE POSSIBLE

PRESENCE OF WORKING ACTIVITIES INVOLVING THE REPEATED USE OF FORCE IN THE HANDS-ARMS (AT LEAST ONCE EVERY FEW CYCLES DURING ALL THE TASK ANALYZED). More than one answer can be checked.

THIS WORKING TASK IMPLIES:

☐ THE HANDLING OF OBJECTS WEIGHING OVER 3 KG ☐ GRIPPING BETWEEN FOREFINGER AND THUMB AND LIFTING OBJECTS WEIGHING OVER 1 KG (IN PINCH) ☐ USING THE WEIGHT OF THE BODY TO OBTAIN THE NECESSARY FORCE TO CARRY OUT A WORKING ACTION ☐ THE HANDS ARE USED AS TOOLS TO HIT OR STRIKE SOMETHING	1 - ONCE EVERY FEW CYCLES 2 - ONCE EVERY CYCLE 4 - ABOUT HALF OF THE CYCLE 8 - FOR OVER HALF OF THE CYCLE

THE WORKING ACTIVITY REQUIRES THE USE OF INTENSE FORCE FOR:

☐ PULLING OR PUSHING LEVERS ☐ PUSHING BUTTONS ☐ CLOSING OR OPENING ☐ PRESSING OR HANDLING COMPONENTS ☐ USING TOOLS	4 - 1/3 OF THE TIME 6 - ABOUT HALF OF THE TIME 8 - OVER HALF OF THE TIME (*) 16 - NEARLY ALL THE TIME (*)

THE WORKING ACTIVITY REQUIRES THE USE OF MODERATE FORCE FOR:

☐ PULLING OR PUSHING LEVERS ☐ PUSHING BUTTONS ☐ CLOSING OR OPENING ☐ PRESSING OR HANDLING COMPONENTS ☐ USING TOOLS	2 - 1/3 OF THE TIME 4 - ABOUT HALF OF THE TIME 6 - OVER HALF OF THE TIME 8 - NEARLY ALL THE TIME

PRESENCE OF AWKWARD POSITIONS OF THE ARMS DURING THE REPETITIVE TASK.

☐ RIGHT   ☐ LEFT   ☐ BOTH   (mark the limb with greater involvement)

1 - THE ARM/ARMS ARE NOT LEANING ON THE WORKBENCH BUT ARE A LITTLE UPLIFTED FOR A LITTLE OVER HALF THE TIME   2 - THE ARMS HAVE NOTHING TO LEAN ON AND ARE KEPT NEARLY AT SHOULDER HEIGHT FOR ABOUT 1/3 OF THE TIME   4 - THE ARMS ARE KEPT AT ABOUT SHOULDER HEIGHT, WITHOUT SUPPORT, FOR OVER HALF OF THE TIME   8 - THE ARMS ARE KEPT AT ABOUT SHOULDER HEIGHT, WITHOUT SUPPORT, ALL THE TIME <div align="right">I___I A</div>
2 - THE WRIST MUST BEND IN AN EXTREME POSITION, OR MUST KEEP AWKWARD POSTURES (SUCH AS WIDE FLEXIONS OR EXTENSIONS, OR WIDE LATERAL DEVIATIONS) FOR AT LEAST 1/3 OF THE TIME   4 - THE WRIST MUST BEND IN AN EXTREME POSITION, OR MUST KEEP AWKWARD POSTURES (SUCH AS WIDE FLEXIONS OR EXTENSIONS, OR WIDE LATERAL DEVIATIONS) FOR OVER HALF OF THE TIME   8 - THE WRIST MUST BEND IN AN EXTREME POSITION ALL THE TIME <div align="right">I___I B</div>
2 - THE ELBOW EXECUTES SUDDEN MOVEMENTS (JERKING MOVEMENTS, STRIKING MOVEMENTS) FOR ABOUT 1/3 OF THE TIME   4 - THE ELBOW EXECUTES SUDDEN MOVEMENTS (JERKING MOVEMENTS, STRIKING MOVEMENTS) FOR OVER HALF OF THE TIME   8 - THE ELBOW EXECUTES SUDDEN MOVEMENTS (JERKING MOVEMENTS, STRIKING MOVEMENTS) NEARLY ALL THE TIME <div align="right">I___I C</div>
☐ GRIP OBJECTS, PARTS, OR TOOLS WITH FINGERTIPS WITH CONSTRICTED FINGERS (PINCH)   2 FOR ABOUT 1/3 OF THE TIME   ☐ GRIP OBJECTS, PARTS, OR TOOLS WITH FINGERTIPS WITH THE HAND NEARLY OPEN (PALMAR GRIP)   4 FOR OVER HALF THE TIME   8 ALL THE TIME   ☐ KEEPING FINGERS HOOKED <div align="right">I___I D</div>
PRESENCE OF IDENTICAL MOVEMENTS OF SHOULDER AND/OR ELBOW, AND/OR WRIST, AND/OR HANDS, REPEATED FOR AT LEAST 2/3 OF THE TIME (please cross 3 also if the cycle is shorter than 15 seconds) <div align="right">E 3</div>

PLEASE NOTE: use the highest value obtained among the four groups of questions (A,B,C,D) only once, and if possible add to that of the last question.

PRESENCE OF ADDITIONAL RISK FACTORS: only choose one answer per group of questions.

2 - GLOVES INADEQUATE TO THE TASK ARE USED FOR OVER HALF OF THE TIME (UNCOMFORTABLE, TOO THICK, WRONG SIZE, ETC.)
2 - VIBRATING TOOLS ARE USED FOR OVER HALF OF THE TIME
2 - THE TOOLS EMPLOYED CAUSE COMPRESSIONS OF THE SKIN (REDDENING, CALLOSITIES, BLISTERS, ETC.)
2 - PRECISION TASKS ARE CARRIED OUT FOR OVER HALF OF THE TIME (TASKS OVER AREAS SMALLER THAN 2 OR 3 MM)
2 - MORE THAN ONE ADDITIONAL FACTOR IS PRESENT AT THE SAME TIME AND, OVERALL, THEY OCCUPY OVER HALF OF THE TIME
3 - ONE OR MORE ADDITIONAL FACTORS ARE PRESENT, AND THEY OCCUPY THE WHOLE OF THE TIME (I.E.,............................)
1 - WORKING PACE SET BY THE MACHINE, BUT THERE ARE "BUFFERS" IN WHICH THE WORKING RHYTHM CAN EITHER BE SLOWED DOWN OR ACCELERATED
2 - WORKING PACE COMPLETELY DETERMINED BY THE MACHINE

Table 15.4 and Figure 15.1, Figure 15.2A, and Figure 15.2B report an example of mapping the risk for an assembly line, for individual departments, and for all of the workplaces within a company. Because the numerical values indicated in the checklist have been calibrated to the OCRA index multiplier factors, the final checklist value can be interpreted in terms of its correspondence to the OCRA values. Table 15.5 shows the checklist score and the corresponding OCRA index as recently updated (Occhipinti and Colombini, 2004).

15.3   Advantages

The advantages of the two methods are as follows:

OCRA index
- Provides a detailed analysis of the main mechanical and organizational determinants of the risk for UE WMSDs
- Linked with MTM analysis and subsequent task design: language easily understood by technicians
- Predicts (within set limits) health effects (UE WMSDs)

- Compares different work contexts (also pre/postintervention): can simulate different design or redesign solutions of the workplace and job organization
- Considers all the repetitive tasks involved in a complex (or rotating) job and estimates the worker's risk level

OCRA checklist
- Purely observational; easy and quick to use

- Produces scores related to exposure level (green, yellow, red, very red)

- Produces an "exposure map" in the production unit referred to the total population and to males/females separately
- Useful for setting priorities and planning job rotations, and for assessing previous exposures in relation to legal problems
- Considers all the repetitive tasks involved in a complex (or rotating) job estimating the worker's exposure level

15.4   Disadvantages

The disadvantages of the two methods are as follows:

OCRA index
- Can be time-consuming, especially for complex tasks and multiple task jobs
- Value of multiplier factors determined using nonhomogeneous approaches and data from the literature
- Initially difficult to learn the concept of "technical action" unless familiar with MTM analysis
- Does not consider all psychosocial factors related to the individual sphere
- Requires a video camera for performing the analysis in slow motion

OCRA checklist
- Allows only a preliminary analysis of the main risk determinants with a preset overestimation
- Allows only an estimation of exposure per risk area (green, yellow, red, very red) and not a precise risk evaluation (as for the OCRA index)
- If observers are not well trained, there is possibility of misclassifying the risk factors
- Does not consider all psychosocial factors related to the individual sphere
- Not useful for analytical design or redesign of tasks and workplaces (for that purpose the OCRA index is preferable)

15.5   Related Methods

The OCRA methods are based on and enlarge upon the indications contained in the IEA technical committee on musculoskeletal disorders document entitled "Exposure assessment of upper limb repetitive movements: a consensus document" (Colombini et al., 2001).

Where the frequency of the technical actions of the upper limbs is analyzed, in the OCRA index approach, there is a specific connection with the concepts envisioned in the motion time measurement (MTM) method (Barnes, 1968). In the case of the OCRA checklist, there are similarities with a proposal, put forward by a group of researchers from the University of Michigan, that was incorporated in the hand-activity level (HAL) proposed by the ACGIH (2000).

TABLE 15.4 Example of Procedures for Analyzing Exposure to Repetitive Movements Using the OCRA Checklist: Analytical Results of Single Checklists in an Assembly Line

N° chk-list	WORK PLACE	recovery	frequency	force	site	shoulder	wrist	elbow	hand	stereotypy	tot. posture score	additional	checklist score	shift type	similar work stations no.	tot.	female	male
ASSEMBLY LINE A																		
1–2	1	3	6	2	right	2	0	0	6	0	6	3	20	3	2	6	6	0
3–4	2	3	7	1	right	2	1	0	8	3	11	1	23	3	2	6	6	0
5–6	3	3	7	1	right	1	0	0	7	3	10	1	22	3	2	6	6	0
7–8	4	3	4	2	right	2	0	0	4	0	4	1	14	3	2	6	6	0
9	5	3	3	1	right	1	0	0	3	3	6	1	14	3	1	3	3	0
10	6	3	3	0	right	1	0	0	0	3	4	1	11	3	1	3	2	1
11	7	3	6	0	right/left	2	0	0	6	3	9	1	19	3	1	3	3	0
12–13–1	8	3	6	2	right/left	1	2	0	2	3	5	1	17	3	2	6	6	0
15	9	3	0	0	right/left	3	0	0	0	2	5	1	9	3	1	3	3	0
16	10	3	2	0	right	3	2	0	0	3	6	1	12	3	1	3	3	0
17	11	3	3	0	right/left	4	2	2	4	3	7	1	14	3	1	3	3	0
18	12	3	1	0	right	0	0	0	0	3	3	2	9	3	1	3	1	2
19	13	3	6	4	right/left	1	2	2	3	3	6	1	20	3	1	3	1	2
20	14	3	4	1	right	2	0	0	3	3	5	3	16	3	1	3	0	3
21	15	3	1	2	right	4	0	0	1	3	7	1	14	3	1	3	0	3
22–23	16	3	1	4	right	4	4	0	4	3	7	5	20	3	1	3	0	3
25	17	1	1	0	right	1	0	0	1	1	2	1	5	3	1	3	1	2
26	18	3	0	0	right/left	1	0	0	0	3	4	2	9	3	1	3	2	1
27–28	19	3	4	4	right/left	1	2	0	4	3	7	1	19	3	2	6	5	1
1–E23	20	3	1	2	right	2	0	2	0	3	5	1	12	3	1	3	3	0
average		2,9	3,3	1,3		1,9	0,8	0,3	2,7	2,6	6,0	1,5	15			78	60	18

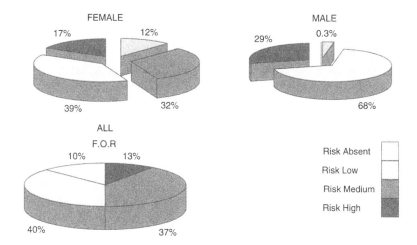

FEMALE — 12%, 17%, 39%, 32%

MALE — 29%, 0.3%, 68%

ALL F.O.R — 13%, 10%, 40%, 37%

Risk Absent
Risk Low
Risk Medium
Risk High

FIGURE 15.1 Results of the final scores of the checklist in a production department in total and for gender.

15.6 Standards and Regulations

The European Council Directive 89/331/EEC, "Introduction of measures to encourage improvements in the safety and health of workers at work," has been incorporated in the legislation of all the European state members. This directive requires employers to undertake a "risk assessment." Specifically, the directive states that "the employers shall … evaluate the risks to the safety and health of workers…; subsequent to this evaluation and as necessary, the preventive measures and the working and production methods implemented by the employer must assure an improvement in the level of protection." The

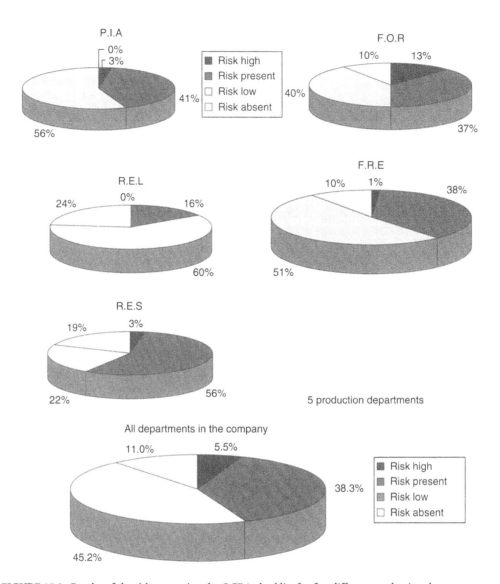

FIGURE 15.2   Results of the risk map using the OCRA checklist for five different production departments.

TABLE 15.5   Checklist Score and OCRA Index

Checklist Score	OCRA Index	Exposure Level
≤7.5	2.2	no exposure
7.6–11	2.3–3.5	very low exposure
11.1–14.0	3.6–4.5	light exposure
14.1–22.5	4.6–9.0	medium exposure
≥22.6	≥9.1	high exposure

OCRA methods, at different levels, are tools tailored for the assessment and management of risks associated with UE WMSDs.

Another European directive, 98/37/EEC and relevant modifications, sets forth essential safety and ergonomic requirements in the design, construction, and marketing of new machines. The directive has prompted the European Committee for Standardization (CEN) to develop a large number of technical standards aimed at verifying compliance with these requirements. Among these standards, those belonging to the EN-1005 series concern the use of manual force on machinery. Standard 1005-5, still in the draft stage, concerns manual activities featuring low force and high frequency. The current draft is largely based on the evaluation procedures in the OCRA method.

15.7 Approximate Length of Training and Application

Both methods, being closely related, generally require 2 days of training time. Follow-up sessions to ensure the training efficacy (8 hours) are highly recommended. The application time for the OCRA index is dependent on the complexity of the task/job. For a task with a cycle time of 30 seconds it takes about 30 to 45 min to complete the analysis. The analysis of a generic task/workplace using the checklist takes about 10 to 15 min.

15.8 Reliability and Validity

Based on existing studies, Occhipinti and Colombini (2004) report that the OCRA index is highly associated with the prevalence of UE WMSDs in the exposed populations. In particular, the following linear regression equation can be used to predict the expected prevalence of persons affected by UE WMSDs (with 95% confidence limit):

$$\text{UE WMSDs\%} = (2.39 \pm 0.27) \times \text{OCRA index}$$

The association expressed by the regression equation shows $R^2 = 0.92$ and is statistically very significant ($p < 0.00001$).

Occhipinti et al. (2000) reported a very high concordance and a close association between the OCRA index and the checklist scores when the two methods are applied to the same work contexts by two different experts. No formal studies on inter- and intraobserver variability exist yet. However, empirical data strongly suggest that the reliability and validity are highly dependent upon the training and expertise of the analyst.

15.9 Tools Needed

Both methods can be carried out using just a pen and paper. The OCRA index method, however, often requires the use of a video camera that allows films to be viewed in slow motion. The checklist, by definition, is completed directly at the workplace. Both methods have specialized software available for loading and processing the data and results (Colombini et al., 2002).

References

ACGIH (2000), Threshold Limit Values for Chemical Substances in the Work Environment, American Conference of Governmental Industrial Hygienists, Cincinnati, pp. 119–121.

Barnes, R.M. (1968), *Motion and Time Study: Design and Measurement of Work*, 6th ed., Wiley, New York.

Borg, G.A.V. (1982), A category scale with ratio properties for intermodal and interindividual comparison, in *Psychophysical Judgement and the Process of Perception*, Geissler, H.G. and Petrold, P., Eds., VEB Deutscher Verlag der Wissenschaften, Berlin, pp. 25–34.

Colombini, D., Grieco, A., and Occhipinti, E. (1998), Occupational musculoskeletal disorders of the upper limbs due to mechanical overload, *Ergonomics*, 41 (special issue).

Colombini, D., Occhipinti, E., Delleman, N., Fallentin, N., Kilbom, A., and Grieco, A. (2001), Exposure assessment of upper limb repetitive movements: a consensus document, in *International Encyclopaedia of Ergonomics and Human Factors*, Karwowski, W., Ed., Taylor & Francis, London.

Colombini, D., Occhipinti, E., and Grieco, A. (2002), *Risk Assessment and Management of Repetitive Movements and Exertions of Upper Limbs*, Vol. 2, Elsevier Science, New York.

Occhipinti, E. and Colombini, D. (1996), Alterazioni muscolo-scheletriche degli arti superiori da sovraccarico biomeccanico: metodi e criteri per l'inquadramento dell'esposizione lavorativa, *Med. Lav.*, 87, 491–525.

Occhipinti, E. and Colombini, D. (2004), Metodo OCRA: aggiornamento dei valori di riferimento e dei modelli di previsione di UL-WMSDs in popolasioni lavozative esposte a movimenti e sfors rifetuti degli arti superiori, *Med. Lav.*, in press.

Occhipinti, E., Colombini, D., Cairoli, S., and Baracco, A. (2000), Proposta e validazione preliminare di una checklist per la stima dell'esposizione lavorativa a movimenti e sforzi ripetuti degli arti superiori, *Med. Lav.*, 91, 470–485.

16

Assessment of Exposure to Manual Patient Handling in Hospital Wards: MAPO Index (Movement and Assistance of Hospital Patients)

Olga Menoni
EPM-CEMOC
Milan, Italy

Maria Grazia Ricci
EPM-CEMOC
Milan, Italy

Daniela Panciera
EPM-CEMOC
Milan, Italy

Natale Battevi
EPM-CEMOC
Milan, Italy

16.1 Background and Application

Previous research has relied on two main criteria to assess musculoskeletal injury risk exposure: those based on epidemiological investigations and those based on analysis of the potential biomechanical overload of the lumbar disks. Data from epidemiological studies point unequivocally to the existence of a relationship between the type and number of maneuvers involved in manual movement of patients and the occurrence of certain acute and chronic disorders of the lumbar spine (Bordini et al., 1999; Colombini et al., 1999a, 1999b). Biomechanical studies briefly refer to measurements of lumbar loads during lifting or handling movements with noncooperative patients. Gagnon (1986) calculated 641 kg as the maximum load acting on the lumbar disks when lifting a patient weighing 75 kg from sitting to standing position. Garg et al. (1991) calculated 448 kg as the average load on disk L5/S1 when moving a patient from bed to wheelchair. In a recent study, Ulin and Chaffin (1997) calculated a disk load of 1020 kg when moving a noncooperative patient weighing 95 kg.

These and other studies showed how manual handling of patients often produces a disk load exceeding the values defined as tolerable (about 275 kg in women and 400 kg in men), which roughly correspond to the concept of "action limit" (NIOSH, 1981). Other studies (Dehlin, 1976; Magora, 1970; Stobbe et al., 1988; Takala, 1987; Winkelmolen et al., 1994) also correlate the injury risk due to manual patient handling with:

- Degree of patient's disability
- Type of transfer operation performed
- Daily frequency of lifting operations
- Unsuitability of beds or absence of equipment (patient lifting aids)

Menoni et al. (1999) developed the MAPO (movement and assistance of hospital patients) method practical tool for analysis and intervention and prevention. The MAPO exposure index was calculated on 440 different types of wards in hospital and residences (nursing home) for both acute and long-term patients (with few exceptions) and for 6400 nurses exposed to manual patient handling. This method is not applicable in accident and emergency settings, in operating theaters, or to physiotherapy.

16.2 Procedures

16.2.1 General Aspects

It is necessary to identify the following main factors which, together, characterize occupational exposure:

- Patient care load produced by the presence of disabled patients
- Type and degree of motor disability of the patients
- Structural aspects of the working environment and the wards
- Equipment installed
- Training of staff on the specific topic

To calculate the MAPO index, we used a data-recording worksheet (see Figure 16.1) that consists of two parts:

1. Part I is completed during an interview with the head nurse. The interviewer collects all information concerning organizational and training aspects
2. Part II is completed during an on-site inspection. This section of the worksheet is specifically designed to facilitate the analysis of environmental and equipment aspects and to assess specific subsidiary maneuvers.

16.2.1.1 Disabled Patient/Operator Ratios (NC/Op and PC/Op)

For a description of the care load, the following information needs to be collected:

- Number and type of workers (operators) employed in the unit and the number assigned to manual patient handling divided into three shifts (Op)
- Type of patients usually encountered in manual handling (noncooperative [NC] or partially cooperative [PC])

It is necessary to know the average number of disabled patients present in the unit. Disabled patients are further classified, on the basis of residual motor capacity and current illness, into totally noncooperative (NC) and partially cooperative (PC). By totally noncooperative (NC), we mean a patient who is unable to use the upper and lower limbs and who therefore has to be fully lifted in transfer operations.

DATA COLLECTION SHEET FOR THE MAPO INDEX

HOSPITAL: UNIT: Medicine
TOTAL NO. STAFF ENGAGED ON PATIENT TRANSFER OVER 3 SHIFTS:
Morning 4   Afternoon 3   Night 2   Total no. operators | 9 |

TYPE OF PATIENT:

DISABLED (D)   22   (Average No.)

Noncooperative patients (NC) No. 16   Partially cooperative patients (PC) No. 6
WHICH MANUAL (D) PATIENT TRANSFER OPERATIONS ARE PERFORMED?:
☒ - bed-wheelchair   ☒ - bed-trolley   ☐ - to pillow   ☒ - other ....
....

WHEELCHAIR AND COMMODES:

FEATURES AND INADEQUACY SCORE OF WHEELCHAIRS (Wh) AND/ OR COMMODES (com)	Score	TYPE OF WHEELCHAIR OR COMMODE							No. of wheelchairs	10	
		A ☐ Wh ☐ com No. 3	B ☐ Wh ☐ com No. 1	C ☐ Wh ☐ com No. 3	D ☐ Wh ☐ com No. 3	E ☐ Wh ☐ com No.	F ☐ Wh ☐ com No.	G ☐ Wh ☐ com No.	
Poor maintenance									
Malfunctioning brakes	1	X	X		X				
Armrest not extractable	1		X		X				
Footrest not extractable									
Backrest cumbersome	1	X	X		X				Total wheelchair score
Width exceeding 70 cm	1		X						
Column score No. (Wh or com) × sum of scores		6	4	0	9				19

Mean score (MSWh) = total wheelchair score/no. wheelchairs | 1.9 | MSWh

ARE WHEELCHAIRS SUFFICIENT IN NUMBER? (at least 50% of total no. of disabled patients)
☐ YES   ☒ NO

LIFTING DEVICES: MANUAL No. | 1 | ELECTRIC No. | 1 |

ARE LIFTING DEVICES NORMALLY USED?   ☒ YES
If yes, for which operations?: _____ it's only used for electric type _____

If no, why?   ☒ not suitable for the unit's requirements
                                     ☐ lack of training
                                     ☒ often broken
                                     ☐ use too time-consuming
                                     ☐ not enough space for use

ARE MANUAL PATIENT LIFTING OPERATIONS COMPLETELY ELIMINATED BY USE OF LIFTING DEVICES?
☐ YES   ☒ NO

FIGURE 16.1   Data collection sheet for MAPO index. Example of application.

OTHER AIDS AVAILABLE:

GLIDEBOARDS	TRANSFER DISCS	ROLLERS	ERGONOMIC BELTS	SLIDING SHEETS
No. _____	No. _____	No. __1__	No. _____	No. _____

FOR WHICH OPERATIONS ARE THESE AIDS USED?

☒ bed/wheelchair transfer   ☐ bed/trolley transfer   ☐ moving in bed   ☐ other_____

STRUCTURAL FEATURES OF ENVIRONMENT

BATHROOMS (centralized or individual in rooms):

FEATURES AND INADEQUACY SCORE OF BATHROOMS WITH SHOWER/BATH centr. = centralized indiv. = individual	Score	A ☐ centr. ☐ indiv. No. 1	B ☐ centr. ☐ indiv. No. 2	C ☐ centr. ☐ indiv. No.	D ☐ centr. ☐ indiv. No.	E ☐ centr. ☐ indiv. No.	F ☐ centr. ☐ indiv. No.	G ☐ centr. ☐ indiv. No.	
									Total no. bathrooms
Free space inadequate for use of aids	2	X	X						|_3_|
Door opening inward (not outward)									
No shower									
No fixed bath									Total score
Door width less than 85 cm	1	X							for
Nonremovable obstacles	1								bathroom:
Column score (No. bathrooms × sum of scores)		3	4						7

Mean score bathrooms (**MSB**) = total score bathrooms/total no. bathrooms: |_2.33_| MSB

TOILETS (WC) (centralized or individual in rooms):

FEATURES AND INADEQUACY SCORE OF TOILETS centr. = centralized indiv. = individual	Score	A ☐ centr. ☐ indiv. No. 1	B ☐ centr. ☐ indiv. No. 7	C ☐ centr. ☐ indiv. No.	D ☐ centr. ☐ indiv. No.	E ☐ centr. ☐ indiv. No.	F ☐ centr. ☐ indiv. No.	G ☐ centr. ☐ indiv. No.	
Free space insufficient to turn wheelchair round	2	X	X						Total nno. toilets (WC)
Door opening inward (not outward)									|_8_|
Height of WC insuffient (below 50 cm)	1	X							
WC without side grips	1		X						
Door width less than 85 cm	1								Total WC
Space at side of WC less than 80 cm	1		X						score
Column score (No. toilets × sum of scores)		3	28						31

Mean score toilets (**MSWC**) = total WC score/no. WC: |_3.87_| MSWC

No. WARDS: 10

FIGURE 16.1   Data collection sheet for MAPO index. Example of application. (continued)

		TYPE OF WARDS							
FEATURES AND INADEQUACY SCORE OF WARDS	Score	No. 2 Wards No. 4 beds	No. 8 Wards No. 2 beds	No. __ Wards No. __ beds	No.__ Wards No. __ beds	No. __ Wards No. __ beds	No. __ Wards No. __ beds	No. __ Wards No. __ beds	
Space between beds or between bed and wall less than 90 cm	2	X	X						Total no. wards
Space at foot of bed less than 120 cm	2	X								10	
Presence of nonremovable obstacles									
Fixed beds with height less than 70 cm									
Bed unsuitable: needs to be partially lifted	1								
Side flaps inadequate									
Space between bed and floor less than 15 cm	2								
Beds with two wheels or without wheels									Total ward
Height of armchair seat less than 50 cm	0.5	X							score
Column score (No. wards × sum of scores)		9	16						25

Mean score wards (MSW) = total score wards/total no. wards | 2.5 | MSW

MEAN ENVIRONMENT SCORE = MSB + MSWC + MSW = |8, 7| MSE

Presence of height-adjustable beds: ☐ YES ☒ NO

If yes, no. (in unit) _____ ☐ with three sections ☐ manual ☐ electric

Space between bed and floor less than 15 cm: ☐ YES ☐ NO

STAFF TRAINING IN MANUAL LOAD HANDLING

☒ not given (2)

☐ included in training course (0.75)

☐ given only via training course on use of aids (1)

☐ only via information brochures (1)

FIGURE 16.1 Data collection sheet for MAPO index. Example of application. (continued)

By partially cooperative (PC), we mean a patient who has residual motor capacity and who is therefore only partially lifted.

The choice of dividing the disabled patients into totally noncooperative (NC) and partially cooperative (PC), which is also well proven in the literature, is derived from the evidence indicating a different biomechanical overload on the lumbar spine in relation to the various types of maneuvers performed.

16.2.1.2 Lifting Factor (LF)

Assessment of patient lifting devices combines two aspects: a sufficient number compared with the number of totally noncooperative patients, and their adequacy compared with the needs of the unit. By "sufficient number," we mean the presence of one lifting device for every eight totally noncooperative patients (NC).

STRUCTURAL FEATURES OF ENVIRONMENT

NUMBER OF DISABLED PATIENTS/OPERATORS RATIO					
No. noncooperative patients (NC) 16	mean	no. operators	(OP) 9 =	1.77	mean NC/OP
No. partially cooperative patients (PC) 6	mean	no. operators	(OP) 9 =	0.66	mean PC/OP

LIFTING DEVICE FACTOR (LF)	VALUE OF LF	
Lifting devices ABSENT or	4	
INADEQUATE + INSUFFICIENT	2	|_2_| LF
Lifting devices INSUFFICIENT or INADEQUATE	0.5	
Lifting devices ADEQUATE and SUFFICIENT		

MINOR AIDS FACTOR (AF)	VALUE OF AF	
Minor aids ABSENT or INSUFFICIENT	1	|_1_| AF
Minor aids SUFFICIENT and ADEQUATE	0.5	

WHEELCHAIR FACTOR (WF)							
Mean wheelchair score (MSWh)							|_1_| WF
	0.5–1.33		1.34–2.66		2.67–4		
Numerically sufficient	YES	NO	YES	NO	YES	NO	
VALUE OF WF	0.75	1	1.12	1.5	1.5	2	

ENVIRONMENT FACTOR (EF)				
Mean environment score (MSE)	0–5.8	5.9–11.6	11.7–17.5	
VALUE OF EF	0.75	1.25	1.5	|_1.25_| EF

TRAINING FACTOR	TF FACTOR	
Adequate training	0.75	
Only information	1	|_2_| TF
No training	2	

MAPO EXPOSURE INDEX

$$\text{MAPO INDEX} = [(|1.77| \times |2|) + (|0.66| \times |1|)] \times |1| \times |1.25| \times |2| = 10.56$$

$$\qquad\quad \underset{\text{NC/OP}}{} \quad \underset{\text{LF}}{} \quad \underset{\text{PC/OP}}{} \quad \underset{\text{AF}}{} \quad \underset{\text{WF}}{} \quad \underset{\text{EF}}{} \quad \underset{\text{TF}}{}$$

FIGURE 16.1 Data collection sheet for MAPO index. Example of application. (continued)

We consider "inadequate for the needs of the unit" a lifting device that:

- Cannot be used for the type of patient normally present in the department
- Is in a poor state of repair (often broken)
- Cannot be used due to the environmental features of the wards and/or bathrooms

The value assigned to the lifting factor (LF) varies from 0.5 to 4, as seen in Table 16.1. In brief, the values assigned are based on the features of sufficient number and/or adequacy described above, having first and foremost estimated a standard lifting frequency (an NC patient is usually moved at least four times per day), which produces the maximum obtainable score for the parameter: LF = 4.

TABLE 16.1 Values Assigned to Lifting Factor (LF)

Features of Lifting Device	LF Value
Absent or inadequate + insufficient	4
Insufficient or inadequate	2
Sufficient + adequate	0.5

16.2.1.3 Minor Aids Factor (AF)

We consider as minor aids equipment that reduces the number of or the overload produced by certain operations to partially move the weight of the patient (sliding sheet, transfer disc, roller, ergonomic belt). They are considered to be present when the unit is equipped with one sliding sheet plus at least two of the other aids mentioned.

A reducing value (0.5) was assigned to the relative factor, considering that the presence of these aids reduces the number of the said operations. When minor aids are not present or are insufficient, the value assigned is 1.

16.2.1.4 Wheelchair Factor (WF)

Assessment of wheelchairs and/or commodes considers two aspects in an integrated manner: sufficient number compared with the number of disabled patients, and the presence of ergonomic requirements. By sufficient number, we mean the presence of a number of wheelchairs equal to at least half of the disabled patients in the unit. This choice was based on the observation that some totally noncooperative (bedridden) or partially cooperative patients do not use wheelchairs. Assessment of the ergonomic requirements is made by assigning a value of 1 to each type of wheelchair/commode identified during on-site inspection for absence of each of the following features:

- Arm rests, which should be removable
- Back rest, which should not be cumbersome
- Equipped with reliable brakes
- Width, which should not exceed 70 cm

From the sum of the "inadequacy" score for each type of wheelchair, multiplied by the number of wheelchairs (with the same features), the total score for each type of wheelchair is obtained (column score, see Figure 16.1). From the sum of the various column scores, divided by the total number of wheelchairs, the mean wheelchair score (MSWh) is obtained, which is therefore an assessment of the ergonomic suitability of all wheelchairs/commodes present in the unit.

It is thus possible to define the wheelchair factor (WF) value by combining the two aspects assessed (number and ergonomic requirements), as shown in Table 16.2. The value of this factor varies from 0.75 to 2 because, on the basis of preliminary observations, the presence of inadequate or numerically insufficient wheelchairs or commodes leads to at least double the frequency of patient transfer operations that produce a biomechanical overload of the lumbar spine.

16.2.1.5 Environment Factor (EF)

Only the structural aspects of the environment that can cause an increase or a decrease in the transfer movements overloading the lumbar spine were considered. Three sections in the data recording worksheet were reserved for this purpose (see Figure 16.1), which cover analysis of bathrooms, toilets, and wards.

TABLE 16.2 Values Assigned to Wheelchair Factor (WF)

	Mean Wheelchair Score					
	0–1.33		1.34–2.66		2.67–4	
Numerically sufficient	Yes	No	Yes	No	Yes	No
Value of wheelchair factor (WF)	0.75	1	1.12	1.5	1.5	2

TABLE 16.3 Scores Assigned to Structural Features Recorded in Bathrooms, Toilets, and Wards

Structural Features	Score
Bathrooms:	
Free space inadequate for use of aids	2
Door width less than 85 cm	1
Nonremovable obstacles	1
Toilets:	
Free space insufficient to turn wheelchair round	2
Height of wheelchair insufficient (below 50 cm)	1
Wheelchair without side grips	1
Door width less than 85 cm	1
Space at side of wheelchair less than 80 cm	1
Wards:	
Space between beds less than 90 cm	2
Space at foot of bed less than 120 cm	2
Bed unsuitable: needs to be partially lifted	1
Space between bed and floor less than 15 cm	2
Armchairs unsuitable (height of seat less than 50 cm)	0.5

For each section, we identified a number of inadequacy features with scores as shown in Table 16.3. The highest scores (1 or 2) were assigned to environmental aspects that, if inadequate, oblige the operators to perform a greater number of patient transfer maneuvers. The lowest score (0.5) was assigned to the presence of furniture (e.g., armchairs) that prevents the partially cooperative patient from using any residual motor capacity, so that the operator has to lift the patient.

For each section — bathrooms (B), toilets (WC), wards (W) — the procedure is the same as for wheelchairs, i.e., calculating the mean score (MS) of "inadequacy" of the sections (MSB, MSWC, MSW). The sum of the mean scores of the three sections constitutes the mean environment score (MSE), which is divided into three categories of equidistant range to express low, average, and high inadequacy, as shown in Table 16.4.

The environment factor (EF) value varies from 0.75 to 1.5. On the basis of preliminary observations, it was possible to establish that total absence of ergonomic requirements in environment structures leads to an increase of about 1.5 in the number of maneuvers producing a biomechanical overload of the lumbar spine.

16.2.1.6 Training Factor (TF)

The last determining factor contributing to a definition of the exposure index is the specific training of operators. Experience in verifying the efficacy of training enabled minimum requirements to be defined for specific training adequacy based on the following features:

- Training course lasting 6 hours divided into a theoretical section and practical exercises on techniques for partially lifting patients that produce the least overload
- Practical exercises on the correct use of equipment

Where training had these features, it was noted, via on-site observations even though these were not performed systematically, that the number of movements producing an overload of the lumbar spine decreased considerably and that the remainder were performed in a "less overloading" manner. For these reasons, a reducing value of 0.75 was assigned to the cases of adequate training. Where training was limited simply to giving information (verbal or via leaflets), no significant reduction was observed in the

TABLE 16.4 Values Attributed to Environment Factor (EF)

Degree of inadequacy	Low	Average	High
Mean environment score (MSE)	0–5.8	5.9–11.6	11.7–17.5
Environment factor value (EF)	0.75	1.25	1.5

number of movements producing overload. Therefore a training factor of 1 was assigned. In the cases where no type of training was given, the frequency–severity of the overloading movements was doubled (training factor 2).

16.2.2 General Features of the Calculation Model of MAPO Index

MAPO synthetic exposure index is calculated using the following expression:

$$\text{MAPO} = [(\text{NC/Op} \times \text{LF}) + (\text{PC/Op} \times \text{AF})] \times \text{WF} \times \text{EF} \times \text{TF}$$

In the above expression, the relationships between disabled patients and operators (NC/Op and PC/Op) are, in view of the previous observations, of primary importance and are a function of the frequency of the lifting and/or transfer operations objectively demanded of the operators in the unit under study.

The ratios NC/Op and PC/Op are "weighted" with respect to "lifting" and "minor aids" factors, respectively, in order to assess the potential biomechanical overload produced by transfer operations according to the presence/absence and suitability of the aids under study. The other factors (WF, EF, TF) act as multipliers (negative or positive) of the general exposure level (increase–decrease in frequency or overload in manual patient handling operations).

In the calculation model, the lifting factor (LF) is a multiplier or a reduction factor only in noncooperative patients, whereas the minor aids factor (AF) is related only to partially cooperative patients. This approach was chosen to rationalize the model, even though in reality both types of aids refer to the total number of disabled patients. The other factors (wheelchairs, environment, and training) were correlated both with noncooperative and partially cooperative patients, since if they are inadequate they can produce an increase in the frequency–severity of movement–transfer operations of disabled patients.

16.2.3 Classification of MAPO Index Results

The studies and experiments carried out so far have made it possible to identify different exposure levels (green, yellow, and red) with key MAPO scores. By considering the trend of odds ratio in reference to negligible levels of exposure, it is possible to define the following MAPO index classification criteria and to indicate the consequent preventive actions to be adopted.

The green band corresponds to an index level between 0 and 1.5, where risk is negligible save in the case of the previously described exceptions (ratio of noncooperative patients/operators >0.25 without lifting devices). Within this range, the prevalence of low back pain appears identical to that of the general population (3.5%).

The yellow "alert" band falls within a range of index values between 1.51 and 5. In this range, low back pain can have an incidence 2.5 times higher than in the green band. At this level, it is necessary to perform a medium- and long-term intervention plan to address the issues of health surveillance, aid equipment, and training.

The red band, with exposure index above 5, corresponds to certain and always higher risk, where low back pain can have an incidence up to 5.6 times the expected incidence. In this case, a short-term intervention plan must be performed to address the issues of health surveillance, aid equipment, training, and environment.

16.3 Advantages

- Allows identification of three action levels in accordance with the well-known traffic light model (green, yellow, red), which is of great practical value
- Provides detailed analysis of the main risk determinants for low back pain in nurses
- Facilitates comparison of different wards
- Allows pre- and postintervention plan comparison, thus making it possible to simulate different types of intervention

- Enables simple and quick analyses

16.4 Disadvantages

- MAPO is not an individual index but represents a risk level of the analyzed ward.
- MAPO is not applicable in emergency wards.
- In some specific situations, it is possible to have a residual risk when the MAPO value is under 1.5.
- The environment factor does not consider the ergonomic features of beds.
- Sometimes in nursing homes, the value of the wheelchair factor is inadequate.

16.5 Related Methods

The only method related to the MAPO index was proposed by Stobbe in 1988. Stobbe identified two exposure levels — high and low risk — that were correlated with the frequency of patient lifting (high being more than five liftings per operator and per shift, low being less than three liftings). The author used an interview with head nurses to quantify the mean number of lifting maneuvers for the ward operators. Only two types of maneuvers were considered: bed–wheelchair and toilet–wheelchair.

16.6 Standards and Regulations

In Europe, a special impulse was given to the assessment of risk involved in manual handling of loads by the European Council Directive 90/269/EEC, which defines what is meant by manual load handling and the relevant obligations of the employer. In the spirit of the legislation, the primary objective is to eliminate risk (check of automation possibility or complete operations mechanization).

The MAPO method is also a tool to guide prevention measures, both in choice of priorities and also for facilitating the relocation of operators who are judged not "sufficiently fit."

For assessment of environmental ergonomic requirements (bath, toilet, and wards), MAPO takes into account the laws and regulations on architectural barrier demolition.

As regards the lifting factor of the MAPO index, reference is made to the standard EN ISO 10535.

16.7 Approximate Training and Application Time

The MAPO method requires, on the whole, 12 hours of training time. A follow-up of training efficacy (4 hours) is highly recommended.

It normally takes about 45 min for a trained operator to evaluate the risk of a single ward using the MAPO index. A preliminary meeting with head nurses (of all the wards) is required to obtain the information needed to calculate the MAPO index.

16.8 Reliability and Validity

A study of the association between exposure (MAPO index) and low back pain was carried out with a multicenter study in 23 hospitals and nursing homes, 234 wards, and 3400 nurses (Menoni et al., 1999). The analysis was conducted with odds ratio (logistic regression analyses) and incidence rate ratio (Poisson regression).

Results reported by Battevi et al. (1999) show that at index level between 0 and 1.5, the prevalence of low back pain appears identical to that of the general population (3.5%). Over the range of MAPO index between 1.51 and 5, the incidence of low back pain is 2.5 times higher than the general population. When the MAPO index is above 5, there incidence of low back pain can be up to 5.6 times than general population.

16.9 Tools Needed

Just a pen, a paper, and a tape measure are necessary.

References

Battevi, N. et al. (1999), Application of the synthetic exposure index in manual lifting of patients: preliminary validation experience, *Med. Lav.*, 90, 256–275.

Bordini, L. et al. (1999), Epidemiology of musculo-skeletal alterations due to biomechanical overload of the spine in manual lifting of patients, *Med. Lav.*, 90, 103–116.

Colombini, D. et al. (1999a), Acute low back pain caused by manual lifting of patients in hospital ward: prevalence and incidence data, *Med. Lav.*, 90, 229–243.

Colombini, D. et al. (1999b), Preliminary epidemiological data on clinical symptoms in health care workers with tasks involving manual lifting of patients in hospital wards, *Med. Lav.*, 90, 201–228.

Dehlin, O. (1976), Back symptoms in nursing aides in a geriatric hospital, *Scand. J. Rehab. Med.*, 8, 47–52.

Gagnon, M. (1986), Evaluation of forces on the lumbo-sacral joint and assessment of work and energy transfers in nursing aides lifting patient, *Ergonomics*, 29, 407–421.

Garg, A. (1991), A biomechanical and ergonomics evaluation of patient transferring tasks: bed to wheelchair and wheelchair to bed, *Ergonomics*, 34, 289–312.

Magora, A. (1970), Investigation of the relation between low back pain and occupation work history, *Ind. Med. Surg.*, 39, 31–37.

Menoni, O. et al. (1999), Manual handling of patients in hospital and one particular kind of consequent diseases: acute and/or chronic spine alterations, *Medicina Lavoro*, 90, 99–436.

NIOSH (1981), Work Practices Guide for Manual Lifting, technical report 81-122, National Institute for Occupational Safety and Health, U.S. Department of Health and Human Services, Washington, D.C.

Stobbe, T.J. et al. (1988), Incidence of low back injuries among nursing personnel as a function of patient lifting frequency, *J. Saf. Res.*, 19, 21–28.

Takala, E.P. (1987), The handling of patients on geriatric wards, *Appl. Ergonomics*, 18, 17–22.

Ulin, S.S. and Chaffin, D.B. (1997), A biomechanical analysis of methods used for transferring totally dependent patients, *SCI Nurs.*, 14, 19–27.

Winkelmolen, G., Landeweerd, J.A., and Drost, M.R. (1994), An evaluation of patient lifting techniques, *Ergonomics*, 37, 921–932.

Psychophysiological Methods

17

Psychophysiological Methods

Karel A. Brookhuis
University of Groningen

Various methods of measuring physiology used in the medical field are increasingly being borrowed for human factors and ergonomics purposes to study operators in workplaces with respect to workload or, more specifically, mental workload. There are many reasons why the measurement of operators' mental workload earns great interest these days, and will increasingly enjoy this status in the near future. First, the nature of work has largely changed, or at least been extended, from physical (e.g., measured by muscle force exertion, addressed in this section) to cognitive (e.g., measured in brain activity, also covered in this section), a trend that has not reached ceiling yet. Second, accidents in workplaces of all sorts are numerous and costly, and they are seemingly ineradicably and in fact largely attributable to the victims themselves, human beings. Third, human errors related to mental workload, in the sense of inadequate information processing, are the major causes of the majority of accidents (Smiley and Brookhuis, 1987).

While both low and high mental workload (e.g., as reflected in heart rate parameters [Smiley and Brookhuis, 1987]) are undoubtedly basic conditions for the occurrence of errors, an exact relationship between mental workload and accident causation is not easily established, let alone measured in practice. De Waard and Brookhuis (1997) discriminated between underload and overload, the former leading to reduced alertness and lowered attention (e.g., reflected in eye parameters), the latter to distraction, diverted attention, and insufficient time for adequate information processing. Both factors have been studied in relationship to operator state; however, the coupling to error occurrence is not via a direct link (see also Brookhuis et al., 2002). Criteria for *when* operator state is below a certain threshold, leading to erroneous behavior, should be established. Only then can accidents and mental workload (high or low) be related, in conjunction with the origins such as information overload (e.g., measured by blood pressure or galvanic skin response [Brookhuis et al., 2002]), fatigue (e.g., reflected in the electroencephalogram [Brookhuis et al., 2002]), or even a factor such as alcohol (e.g., measured with a breathalyzer [Brookhuis et al., 2002]). The operator's working environment and the work itself will only gain in complexity, at least for the time being, with the rapid growth in complex electronic applications for control and management. And last but not least, aging plays a role in the interest of the measurement of operators' mental workload these days, and will increasingly do so in the near future (the "gray wave").

As long as 30 years ago, Kahneman (1973) defined mental workload as directly related to the proportion of the mental capacity an operator expends on task performance. The measurement of mental workload is the specification of that proportion (O'Donnell and Eggemeier, 1986; De Waard and Brookhuis, 1997) in terms of the costs of the cognitive processing, which is also referred to as mental effort (Mulder, 1980). Mental effort is similar to what is commonly referred to as doing your best to achieve a certain target level, to even "trying hard" in case of a strong cognitive processing demand, reflected in several physiological measures. The concomitant changes in effort will not show easily in work performance measures because operators are inclined to cope actively with changes in task demands, for instance in traffic, as drivers do by adapting their driving behavior to control safety (Cnossen et al., 1997). However, the changes in effort will be apparent in self-reports of the drivers and, *a fortiori*, in the changes in certain physiological measures such as activity in certain brain regions as well as heart rate and heart rate variability (cf. De Waard, 1996).

Mulder (1986) discriminates between two types of mental effort, i.e., the mental effort devoted to the processing of information in controlled mode (computational effort) and the mental effort needed to apply when the operator's energy state is affected (compensatory effort). Computational effort is exerted to keep task performance at an acceptable level, for instance, when task complexity level varies or secondary tasks are added to the primary task. In case of (ominous) overload, extra computational effort could forestall safety hazards. Compensatory effort takes care of performance decrement in case of, for instance, fatigue up to a certain level. Underload by boredom, affecting the operator's capability to deal with the task demands, might be compensated as well. In case effort is exerted, be it computational or compensatory, both task difficulty and mental workload will be increased. Effort is a voluntary process under control by the operator, while mental workload is determined by the interaction of operator and task. As an alternative to exerting effort, the operator might decide to change the (sub)goals of the task. Adapting driving behavior as a strategic solution is a well-known phenomenon. For example, overload because of an additional task, such as looking up telephone numbers while driving, is demonstrated to be reduced by lowering vehicle speed (see De Waard et al., 1998, 1999).

Basically there are three global categories of measurement distinguished in this field: measures of task performance, subjective reports, and physiological measures (see also Eggemeier and Wilson, 1991; Wierwille and Eggemeier, 1993; Brookhuis, 1993). The first and the most widely used category of measures is based on techniques of direct registration of the operator's capability to perform a task at an acceptable level, i.e., with respect to an acceptably low accident likelihood. Subjective reports of operator performance are of two kinds: observer reports, which are mostly given by experts, and self-reports by the operators. The value of observer reports is by virtue of strict protocols that limit variation as produced by personal interpretation; the value of self-reports is mainly by virtue of validation through multiple applications in controlled settings. Well-known examples of the latter are the NASA task-load index (NASA-TLX) (Hart and Staveland, 1988) and the rating scale of mental effort (RMSE) (Zijlstra, 1993). Finally, physiological measures are the most natural type of workload index, since, by definition, work demands physiological activity. Both physical and mental workload have, for instance, a clear impact on heart rate and heart rate variability (Mulder, 1980, 1986, 1988, 1992; Brookhuis et al., 1991), on galvanic skin response (Boucsein, 1992), on blood pressure (Rau, 2001), and on respiration (Mulder, 1992; Wientjes et al., 1998). Mental workload might increase heart rate and decrease heart rate variability at the same time (Mulder et al., Chapter 20, this volume). Other measures of major interest are event-related phenomena in the brain activity (Kramer, 1991; Noesselt et al., 2002) and environmental effects on certain muscles (Jessurun, 1997).

In this section, i.e., the next nine chapters of the handbook (Chapter 18 through Chapter 26), the methodology of measuring a number of relevant physiological parameters is elaborated. These include the cardiovascular parameters of heart rate and heart rate variability; the electrocortical parameters of frequency shifts in the electroencephalogram and event-related potentials; galvanic skin responses; blood pressure responses; respiration rate; magnetic reflections of activities in the brain; eyelid movements; and muscle activity.

The section starts with an overview of a very old method, the measurement of electrical phenomena in the skin (Chapter 18). Measurement techniques include galvanic skin response (GSR), skin potential, peripheral autonomic surface potentials, etc., all with the aim of studying electrodermal activity. The latter can be regarded as a psychophysiological indicator of arousal, stress–strain processes, and emotion. The measurement of electrodermal activity is used to investigate orienting responses and their habituation, for studying autonomic conditioning, for determining the amount of information-processing capacity needed during a task, and for determining the arousal/stress level, especially in situations evoking negative emotions. It has also been used for measuring workload and mental strain, specifically emotional strain; increases in certain types of electrodermal activity indicate readiness for action.

Chapter 19 addresses electromyography (EMG), i.e., muscle function through analysis of the electrical signals emanated during muscular contractions. EMG is commonly used in ergonomics and occupational health research because it is noninvasive, allowing convenient measures of physical effort during movements as well as physiological reactions caused by mentally controlled processes.

In Chapter 20, heart rate is the central topic. Heart rate is derived from the electrocardiogram (ECG), which reflects the (electrical) activity of the heart. For the assessment of mental effort, the ECG itself is not of interest; rather, the time duration between heartbeats is the interesting information. During task performance, operators have to expend (mental) effort, which is usually reflected in increased heart rate and decreased heart rate variability when compared with resting situations. The general cardiovascular response pattern that is found in many mental-effort studies can be characterized by an increase of heart rate and blood pressure and a decrease of variability in heart rate and blood pressure in all frequency bands. This pattern is comparable with a defense reaction and is predominantly found in laboratory studies using short-lasting tasks requiring challenging mental operations in working memory.

Chapter 21 describes the technique for detecting hypovigilance, sleepiness, or even sleep, by ambulatory EEG (and EOG) recording, enabling the investigator to obtain a second-to-second measure of manifest sleepiness (and sleep). The measurement technique is relatively old, much used, reliable, and well accepted by the research community.

Chapter 22 describes mental chronometry using the event-related potential (ERP), which is derived as a transient series of voltage oscillations in the brain that can be recorded from the scalp in response to discrete stimuli and responses. Some ERP components, usually defined in terms of polarity and latency with respect to discrete stimuli or responses, have been found to reflect a number of distinct perceptual, cognitive, and motor processes, thereby proving useful in decomposing the processing requirements of complex tasks. ERPs are being used to study aspects of cognition that are relevant to human factors and ergonomics research, including such topics as vigilance, mental workload, fatigue, adaptive aiding, stressor effects on cognition, and automation.

Chapter 23 is a companion to Chapter 22. Both chapters describe the measurement of neural activity within the brain, but Chapter 23 focuses on the use of a noninvasive outside-the-head technique that uses magnetic fields to monitor neural activity (deep) within the brain. The corresponding recordings are known as a magnetoencephalogram (MEG), which is supplemented by a comparable technique called magnetic resonance imaging (MRI). In recent years, these techniques have gained considerable interest in neurophysiological applications because they enable the analysis of the substrate of specific cognitive processes. Recent attempts to employ these methods in the diagnosis of certain neurological diseases seem to be successful.

Chapter 24 describes techniques for evaluating workload by measuring ambulatory blood pressure. This type of ambulant technique can assess interactions — behavior, emotion, and activation — with workload under real work conditions. Carryover effects of workload on activities, behavior, strain after work, and recovery effects during rest can also be measured. This implies an enhancement of the load–strain paradigm from short-term effects (fatigue, boredom, vigilance, etc.) to long-term effects of work (disturbed recovery processes after work, cardiovascular health diseases, diabetes mellitus, depression, etc.).

Chapter 25 is about alertness monitoring. Certain measures of ocular psychophysiology have been identified for their potential to detect minute-to-minute changes in the drowsiness and hypovigilance that are associated with lapses of attention and diminishing alertness during performance. A measure of slow eyelid closure, referred to as percentage of closure (PERCLOS), correlated highly with lapses in visual vigilance performance. The technique is increasingly being used to monitor operators in their working environment (e.g., professional drivers).

Finally, Chapter 26 describes the measurement of respiration in applied research. Respiratory measurement is a potentially powerful asset, as it seems closely related to a variety of important functional psychological dimensions, including response requirements and appraisal patterns. Respiratory measures can also provide valuable supplementary information to alternative measures (subjective measures and other measures of operator workload) and in cases where the task environment is stressful or potentially hazardous.

The selection criteria for the chapters included in this section were: (a) nonintrusiveness and (b) proved effects in relation to mental work conditions. Measurement of most of the included physiological parameters is relatively easy, or at least feasible, in the working environment. However, some measures of brain activity in the working environment are difficult (event-related potentials in the electroencephalogram) or even impossible (magnetic phenomena within the cortex), at least for the time being. Nevertheless, all of these measurement techniques are relevant within the context of this section.

References

Boucsein, W. (1992), *Electrodermal Activity*, Plenum Press, New York.

Brookhuis, K.A. (1993), The use of physiological measures to validate driver monitoring, in *Driving Future Vehicles*, Parkes, A.M. and Franzén, S., Eds., Taylor & Francis, London, pp. 365–377.

Brookhuis, K.A., De Vries, G., and De Waard, D. (1991), The effects of mobile telephoning on driving performance, *Accident Anal. Prev.*, 23, 309–316.

Brookhuis, K.A., Van Winsum, W., Heijer, T., and Duynstee, L. (1999), Assessing behavioural effects of in-vehicle information systems, *Transp. Hum. Factors*, 1, 261–272.

Brookhuis, K.A., de Waard, D., and Fairclough, S.H. (2002), Criteria for driver impairment, *Ergonomics*, 45, 433–445.

Cnossen, F., Brookhuis, K.A., and Meijman, T. (1997), The effects of in-car information systems on mental workload: a driving simulator study, in *Simulators and Traffic Psychology*, Brookhuis, K.A., de Waard, D., and Weikert, C., Eds., Centre for Environmental and Traffic Psychology, Groningen, the Netherlands, pp. 151–163.

De Waard, D. (1996), The Measurement of Drivers' Mental Workload, Ph.D. thesis, Traffic Research Centre, University of Groningen, Haren, the Netherlands.

De Waard, D. and Brookhuis, K.A. (1997), On the measurement of driver mental workload, in *Traffic and Transport Psychology*, Rothengatter, J.A. and Carbonell Vaya, E., Eds., Pergamon, Amsterdam, pp. 161–171.

De Waard, D., Van der Hulst, M., and Brookhuis, K.A. (1998), The detection of driver inattention and breakdown, in *Human Factors in Road Traffic II, Traffic Psychology and Engineering*, Santos, J., Albuquerque, P., Pires da Costa, A., and Rodrigues, R., Eds., University of Minho, Braga, Portugal.

De Waard, D., Van Der Hulst, M., and Brookhuis, K.A. (1999), Elderly and young drivers' reaction to an in-car enforcement and tutoring system, *Appl. Ergonomics*, 30, 147–157.

Eggemeier, F.T. and Wilson, G.F. (1991), Performance-based and subjective assessment of workload in multi-task environments, in *Multiple-Task Performance*, Damos, D.L., Ed., Taylor & Francis, London, pp. 207–216.

Hart, S.G. and Staveland, L.E. (1988), Development of NASA-TLX (task load index): results of experimental and theoretical research, in *Human Mental Workload*, Hancock, P.A. and Meshkati, N., Eds., North-Holland, Amsterdam.

Jessurun, M. (1997), Driving through a Road Environment, Ph.D. thesis, Traffic Research Centre, University of Groningen, Haren, the Netherlands.

Kahneman, D. (1973), *Attention and Effort*, Prentice-Hall, Englewood Cliffs, NJ.

Kramer, A.F. (1991), Physiological metrics of mental workload: a review of recent progress, in *Multiple-Task Performance*, Damos, D.L., Ed., Taylor & Francis, London, pp. 279–328.

Mulder, G. (1980), The Heart of Mental Effort, Ph.D. thesis, University of Groningen, Groningen, the Netherlands.

Mulder, G. (1986), The concept and measurement of mental effort, in *Energetics and Human Information Processing*, Hockey, G.R.J., Gaillard, A.W.K., and Coles, M.G.H., Eds., Martinus Nijhoff Publishers, Dordrecht, the Netherlands, pp. 175–198.

Mulder, L.J.M. (1988), Assessment of Cardiovascular Reactivity by Means of Spectral Analysis, Ph.D. thesis, University of Groningen, Groningen, the Netherlands.

Mulder, L.J.M. (1992), Measurement and analysis methods of heart rate and respiration for use in applied environments, *Biol. Psychol.*, 34, 205–236.

Noesselt, T., Hillyard, S.A., Woldorff, M.G., Schoenfeld, A., Hagner, T., Jäncke, L., Tempelmann, C., Hinrichs, H., Heinze, H.J. (2002), Delayed striate cortical activation during spatial attention, *Neuron*, 35, 575–687.

O'Donnell, R.D. and Eggemeier, F.T. (1986), Workload assessment methodology, in *Handbook of Perception and Human Performance*, Boff, K.R., Kaufman, L., and Thomas, J.P., Eds., Vol. II, 42, Cognitive Processes and Performance, Wiley, New York, pp. 1–49.

Rau, R. (2001), Objective characteristics of jobs affect blood pressure at work, after work and at night, in *Progress in Ambulatory Assessment*, Fahrenberg, J. and Myrtek, M., Eds., Hogrefe and Huber, Seattle, WA, pp. 361–386.

Smiley, A. and Brookhuis, K.A. (1987), Alcohol, drugs and traffic safety, in *Road Users and Traffic Safety*, Rothengatter, J.A. and de Bruin, R.A., Eds., Van Gorcum, Assen, the Netherlands, pp. 83–105.

Wientjes, C.J.E., Grossman, P., and Gaillard, A.W.K. (1998), Influence of drive and timing mechanisms on breathing pattern and ventilation during mental task performance, *Biol. Psychol.*, 49, 53–70.

Wierwille, W.W. and Eggemeier, F.T. (1993), Recommendation for mental workload measurement in a test and evaluation environment, *Hum. Factors*, 35, 263–281.

Zijlstra, F.R.H. (1993), Efficiency in Work Behavior: A Design Approach for Modern Tools, Ph.D. thesis, Delft University of Technology, Delft University Press, the Netherlands.

18

Electrodermal Measurement

Wolfram Boucsein
University of Wuppertal

18.1 Background and Applications

Electrodermal activity (EDA) is a common term for all electrical phenomena in the skin. The term comprises galvanic skin response (GSR), skin potential, peripheral autonomic surface potentials, etc. It can be measured with or without the application of an external voltage to the skin, although the use of a 0.5-V constant external voltage is most common. Electrodermal recording dates back to the late 19th century and has been widely used as a psychophysiological method (Boucsein, 1992). It is easy to measure and interpret.

Since EDA is generated by sweat gland activity, its central-nervous origin lies solely in the sympathetic branch of the autonomic nervous system. Therefore, EDA can be regarded as a psychophysiological indicator of arousal, stress–strain processes, and emotion that is not influenced by the parasympathetic branch of the autonomic nervous system. The phasic part of EDA depends to a great extent on the number of sweat gland ducts that are momentarily filled with sweat, thus providing electric shunts between the skin surface and the well-conducting deeper tissue. Tonic EDA, on the other hand, depends mostly on the degree of moisture in the upper epidermal layers as a consequence of phasic electrodermal phenomena. Since these so-called electrodermal responses (EDRs) will moisten the upper layers, their count can be used as an index of tonic EDA as well.

In laboratory settings, EDA has been used for investigating orienting responses and their habituation, for autonomic conditioning, for determining the amount of information-processing capacity needed during a task, and for measuring the arousal/stress level, especially in situations evoking negative emotions (Boucsein, 1992). As a method in human factors and ergonomics, EDA has been used for determining workload, mental strain, and, most specifically, the amount of emotional strain (Boucsein and Backs, 2000). In the context of task-related information processing, an increase in phasic EDA indicates orienting or directing one's attention toward a stimulus, whereas an increase in tonic EDA indicates readiness for action (Figure 18.1).

18.2 Procedure

18.2.1 Recording Sites

Although sweat gland activity appears more or less at the whole body surface, palms and soles have been found to be the most active sites, especially when emotional components of stress or workload are involved. If ergonomic or physical reasons prohibit recording from these sites, EDA can be taken from the inner aspect of the foot adjacent to the sole (see Boucsein, 1992, Figure 28). Two electrodes are attached in a 3- to 4-cm distance. No previous treatment of the sites is required, except for an inactive reference site in the case of skin potential measurement, which is, however, uncommon in ergonomics.

18.2.2 Electrodes and Electrolyte

Electrodermal recording is performed with 6-mm-diameter disk electrodes with a sintered silver/silver chloride surface and a cylindrical plastic chamber for the electrolyte. They are attached to the skin with double-sided adhesive collars, the hole of which fits the chamber diameter. An isotonic electrolyte jelly (0.9% sodium chloride in a neutral ointment) is used to prevent unwanted interactions between skin and electrolyte. The whole system should be allowed to stabilize for 10 to 15 min prior to EDA recording. After the measurement, electrodes should be cleaned with warm water without mechanically destroying the electrode surface.

18.2.3 Recording Devices

The most commonly used DC (direct current) skin conductance measurement requires a special coupler between skin electrodes and bioamplifier (Figure 18.2). For safety reasons, the 0.5-V constant voltage should be generated from battery power and the signal fed into the amplifier via optical fibers. It is recommended that future recording devices make use of AC (alternating current) measurement techniques in order to avoid electrode polarization and to allow for a level-independent measure of phase angle (Schaefer and Boucsein, 2000). Skin potential can be recorded by a bioamplifier without a coupler, but this technique requires an electrodermally inactive recording site, and the results are not so easy to interpret (Boucsein, 1992). Amplifiers should be the high-resistance type and allow for time constants of at least 10 sec.

18.2.4 Signal Storage and Evaluation

Simultaneous recording of electrodermal level and response characteristics requires at least 16-bit resolution when digitally stored. A sampling rate of 20 Hz is sufficient, since the phasic EDA changes are relatively slow. Constant-voltage DC measurement produces tonic and phasic conductance values (expressed in μS = micro-Siemens, which is identical to μmho) that are labeled, respectively, skin conductance level (SCL) and skin conductance response (SCR). For SCR, in addition to the response amplitude, rise time and recovery time are also evaluated (Figure 18.3), and in the case of a distinct stimulation, the SCR latency is also measured (i.e., the time from stimulus onset to response onset). Instead of the SCL, the number of SCRs per minute that exceed a minimum change in amplitude can also be counted as a measure for tonic EDA (labeled NS.SCR frequency, i.e., frequency of nonspecific SCRs, or more generally labeled NS.EDRs, i.e., nonspecific electrodermal responses). The sensitivity of this measure for emotional strain has been repeatedly demonstrated in ergonomics (see Boucsein and Backs, 2000, Table 1.4).

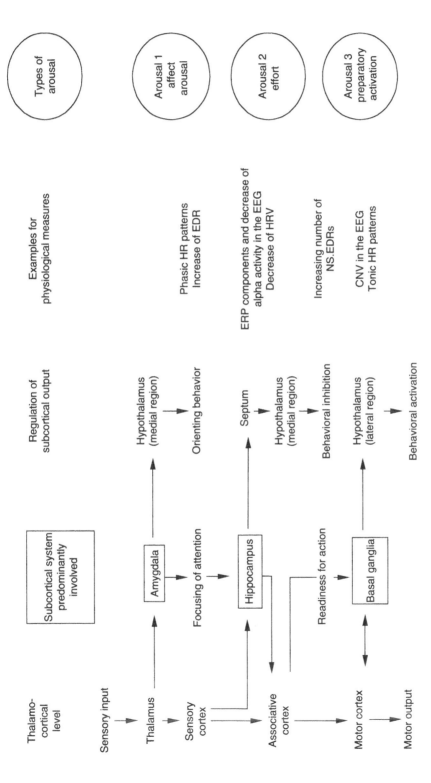

FIGURE 18.1 A three-arousal model for the use of psychophysiology in ergonomics. CNV, contingent negative variation; EEG, electroencephalogram; EDR, electrodermal response; ERP, event-related potential; HR, heart rate; HRV, heart rate variability; NS.EDRs, nonstimulus-specific electrodermal responses. (Adapted from Boucsein, W. and Backs, R.W. [2000], *Engineering Psychophysiology: Issues and Applications*, Backs, R.W. and Boucsein, W., Eds., Lawrence Erlbaum Associates, Mahwah, NJ. Copyright © 2000 by Lawrence Erlbaum Associates. Used by permission of the publisher.)

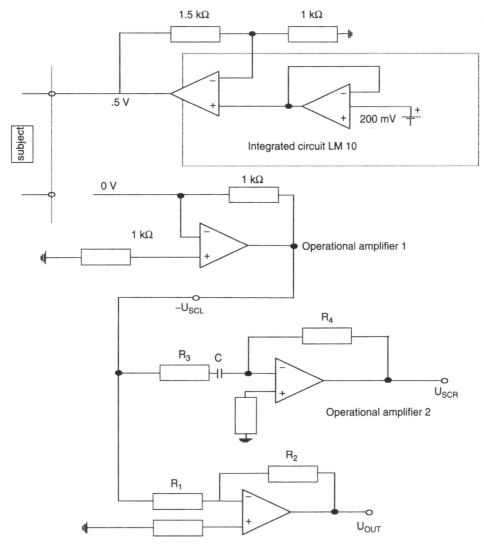

FIGURE 18.2 EDA coupler (upper part), electrodes attached to the subject's skin (upper left side), and a set of bioamplifiers (lower part). The integrated circuit LM 10 provides a highly constant voltage of 0.5 V. The first operational amplifier converts the current that flows through the subject's skin to a negative voltage, $-U_{SCL}$, that is proportional to the SCL, with a sensitivity of 0.5 V/µS. The upper exit, U_{SCR}, in the lower part of the figure provides an amplified SCR, whereas the lower exit, U_{OUT}, provides the total SCL.

18.2.5 Precautions

Various possibilities for generating artifacts have to be taken into account, especially in nonlaboratory environments. High ambient temperatures enhance EDA, whereas increases in humidity may have the opposite effect. Excessive sweating can lead to detachment of the adhesive electrode collars from the skin, but it does not necessarily correlate with increased EDA. Recording artifacts may result from detachment of or pressure on electrodes, body movements, or speaking. Especially, taking a deep breath can produce EDR-like changes in the signal. Therefore, it is recommended to avoid possible artifacts as much as practicable or at least take records of respiratory activity (Schneider et al., 2003). For sources of artifacts and their treatment, see Boucsein (1992).

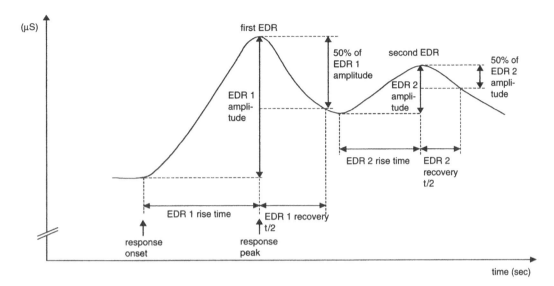

FIGURE 18.3 A typical electrodermal response (EDR) with its parameters (rise time, amplitude, recovery time) as measured by the application of an external voltage. Recovery t/2 refers to falling back to half of the amplitude. In the event that a second EDR starts prior to reaching this point, which is not the case in the example shown, EDR recovery t/2 can not be obtained without extrapolation.

18.3 Advantages

- An easy-to-measure and interpretable physiological signal
- Pure measure of the sympathetic branch of the autonomic nervous system
- Demonstrated sensitivity for workload and emotional strain

18.4 Disadvantages

- Need of a specific coupler for recording
- Artifact proneness in nonlaboratory settings
- Indiscriminately reactive/sensitive to any sympathetic activity

18.5 Example of Electrodermal Measurement in Human Factors/Ergonomics

The following example is taken from a study performed in the European Patent Office in The Hague, The Netherlands (Boucsein and Thum, 1996, 1997). Ambulatory assessment of electrodermal activity was used to monitor stress–strain processes during 8 hours of highly demanding computer work under different work/rest schedules. Eleven patent examiners performed their complex computer task under two different break regimes in counterbalanced order: a 7.5-min break after each 50 min of work on one day and a 15-min break after each 100 min of work on the other day.

18.5.1 Recording

By means of an ambulatory monitoring system (VITAPORT 1, see Jain et al., 1996), EDA was continuously recorded as skin conductance throughout the workday. Standard silver/silver chloride electrodes were filled with an isotonic electrolyte jelly and attached to the medial side of the right foot, as described by

Boucsein (1992, Fig. 28), by means of double-sided adhesive collars. The adhesiveness of the EDA electrodes was improved by using a skin-friendly glue for attaching the collars to the skin. The sites were pretreated with cleansing alcohol to further increase the adhesiveness of the collars. The recording sites were marked with a felt pen to ensure identical electrode positions for the second day. A protection ring of an adhesive collar was used for the marking, in order to prevent the glue from being spread on the skin area to which the electrode jelly would be applied. The glue was given an opportunity to dry for 5 to 10 min until it became threadlike by touch. The electrodes were additionally fixed with a skin-friendly adhesive tape. The electrode leads were fixed by tape above the ankle and carried between the subject's legs and pants to the waist belt, where the registration system was worn. The technique used in this study was successful in preventing electrode detachment, even through stair climbing over several floors several times a day. To reduce the risk of mechanical artifacts, the subjects were instructed to wear slippers or sandals that would not exert pressure on the electrodes.

18.5.2 Data Storage and Analysis

The EDA channel of the registration system performed an online preevaluation of the recorded skin conductance that allowed for a storage rate as low as 8 Hz with a resolution of 16 bits. After the end of each 9-hour recording, the signal was read out to a personal computer and analyzed by a specially in-house-developed EDA evaluation program (EDA_PARA, see Schaefer, 1995) that allowed for the detection and elimination of artifacts. The resolution was 0.01 uS for each evaluation period of 2 min; NS.SCR frequency was automatically detected and stored for further computations. The 2-min evaluation periods were selected from the recordings at the beginning and after the termination of each break. During these periods, the subjects were instructed to sit relaxed in their chairs in order to avoid as many moving artifacts as possible.

Figure 18.4 depicts differences in NS.SCR frequency between the 2-min periods after the break and the 2-min periods at the beginning of the break at three identical times of day for the two work/rest schedules (i.e., 7.5-min break after each 50 min of work versus 15-min break after each 100 min of work). Analysis of variance yielded a significant interaction between the kind of work/rest schedule and the time of day. The results suggest that the long-break schedule actually increases emotional strain during the

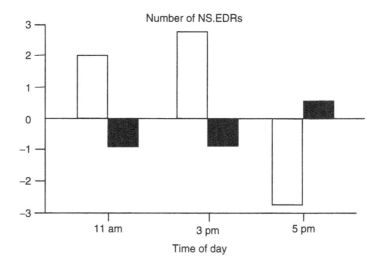

FIGURE 18.4 Differences in tonic EDA (expressed as NS.EDRs before the break subtracted from those after the break) under two different work/rest schedules at three time points during the working day. White: long work and long rest break; black: short work and short rest break.

11 A.M. and the 3 P.M. break. This is in accordance with the results obtained earlier with overly long system response times (Boucsein, 2000). In contrast, the 5 P.M. break during the long-break schedule was associated with a reduction of emotional strain, resulting in a recommendation to prefer short rest breaks until the early afternoon and thereafter prolong the breaks.

18.6   Standards

Standards for recording and interpreting EDA are comprehensively described by Boucsein (1992). Recording devices must meet criteria for medical safety.

18.7   Approximate Training and Application Times

The use of psychophysiological methodology requires special training of several months in a laboratory that has expertise in combining psychological and physiological approaches. Ambulatory monitoring as applied in the above example will add another month of training. Basic knowledge of the special EDA evaluation software can be acquired within two days.

18.8   Reliability and Validity

Short-time reliabilities for electrodermal responses (i.e., within a single recording session on the same day) are at least 0.8 and often exceed 0.9, whereas after one year the reliability coefficient drops to 0.6, which is quite common for psychophysiological measurements. The NS.SCR frequency as used in the above example is a more reliable measure of tonic EDA than the SCL, yielding test–retest correlations of 0.76 after one year compared with 0.61 for SCL (Schell et al., 2002). Validity coefficients between EDA and emotion strength can exceed 0.9 in laboratory sessions (Boucsein, 1992) but are not quantitatively available from applied studies. The NS.SCR frequency is regarded as a valid indicator for the strength of an emotion and for the description of the course of emotional strain under both laboratory and field conditions, as opposed to physical strain, which is more adequately reflected in cardiovascular measures such as heart rate and blood pressure (Boucsein, 1992, 2001). Quantitative estimates of validity are available from the application of EDA in the field of detection of deception, where 68 to 86% correct classifications for guilty subjects were reported (Boucsein, 1992).

18.9   Costs and Tools Needed

At the time of writing, the cost of the successor of the system that was used in the above example was about $10,000 for hardware and software, excluding a laptop for setup and data evaluation. For laboratory studies, nonportable recording devices can be used, including a specific EDA coupler that costs about $2,000. The EDA-evaluation software described here is available for $400. Electrodes, adhesive collars, and electrode jelly add up to approximately $1,000 per study with 10 to 12 subjects. However, electrodes can be used for several studies if they are treated with care.

18.10   Related Methods

Electrodermal measurement is one of the noninvasive methods for psychophysiological assessment in human factors/ergonomics. Other frequently used measures in this domain are heart rate, heart rate variability, body temperature, eye blink rate, eye movements, electromyogram, electroencephalogram, and endocrine measures. For an overview on these methods, see Boucsein and Backs (2000) or Boucsein (2001).

References

Boucsein, W. (1992), *Electrodermal Activity*, Plenum Press, New York.

Boucsein, W. (2000), The use of psychophysiology for evaluation of stress-strain processes in human computer interaction, in *Engineering Psychophysiology: Issues and Applications*, Backs, R.W. and Boucsein, W., Eds., Lawrence Erlbaum Associates, Mahwah, NJ, pp. 289–309.

Boucsein, W. (2001), Psychophysiological methods, in *International Encyclopedia of Ergonomics and Human Factors*, Vol. III, Karwowski, W., Ed., Taylor & Francis, London, pp. 1889–1895.

Boucsein, W. and Backs, R.W. (2000), Engineering psychophysiology as a discipline: historical and theoretical aspects, in *Engineering Psychophysiology: Issues and Applications*, Backs, R.W. and Boucsein, W., Eds., Lawrence Erlbaum Associates, Mahwah, NJ, pp. 3–30.

Boucsein, W. and Thum, M. (1996), Multivariate psychophysiological analysis of stress-strain processes under different break schedules during computer work, in *Ambulatory Assessment*, Fahrenberg, J. and Myrtek, M., Eds., Hogrefe and Huber, Seattle, WA, pp. 305–313.

Boucsein, W. and Thum, M. (1997), Design of work/rest schedules for computer work based on psychophysiological recovery measures, *Int. J. Ind. Ergonomics*, 20, 51–57.

Jain, A., Martens, W.L.J., Mutz, G., Weiss, R.K., and Stephan, E. (1996), Towards a comprehensive technology for recording and analysis of multiple physiological parameters within their behavioral and environmental context, in *Ambulatory Assessment*, Fahrenberg, J. and Myrtek, M., Eds., Hogrefe and Huber, Seattle, WA, pp. 215–235.

Schaefer, F. (1995), EDR_PARA: An Interactive Computer Program for the Evaluation of Electrodermal Recordings, unpublished report, University of Wuppertal, Wuppertal, Germany.

Schaefer, F. and Boucsein, W. (2000), Comparison of electrodermal constant voltage and constant current recording techniques using phase angle between alternating voltage and current, *Psychophysiology*, 37, 85–91.

Schell, A.M., Dawson, M.E., Nuechterlein, K.H., Subotnik, K.L., and Ventura, J. (2002), The temporal stability of electrodermal variables over a one-year period in patients with recent-onset schizophrenia and in normal subjects, *Psychophysiology*, 39, 124–132.

Schneider, R., Schmidt, S., Binder, M., Schaefer, F., and Walach, H. (2003), Respiration-related artifacts in EDA recordings: introducing a standardized method to overcome multiple interpretations, *Psychol. Rep.*, 93, 907–920.

19

Electromyography (EMG)

Matthias Göbel
Berlin University of Technology

19.1 Background and Application

Electromyography (EMG) studies muscle function through analysis of the electrical signals emanated during muscular contractions.

Electrical potentials evoked during voluntary muscular contraction were observed long before the biochemical function of muscular force generation was explored (Matteucci, 1844; Piper, 1912). Despite the simplicity of voltage-measuring electrodes attached to the surface of the skin, it required advanced semiconductor technology to reveal significant information from the complex and noisy electrode signal. During the last few decades, more sophisticated electronic instrumentation and more powerful analysis methods permitted the use of EMG in various areas of research.

Motor activity is initiated by commands generated in the anterior horn of the spinal cord and transmitted along alpha motor neurons to the periphery. Each muscle fiber consists of multiple chains of contractile sarcomeres (actin-myosin-filament units) that create the force of muscle action. The composite of cell body, connecting neuron, and muscle fiber cluster build a motor unit in which all muscle fibers are activated synchronously.

The local neuron of each motor unit chemically activates the muscle fibers connected at its myoneural junction by a cellular depolarization (amplitude $\approx$100 mV, duration 2 to 14 msec). Spreading along the membrane of the muscle, this muscle action potential stimulates the sarcomeres to contract. Electrolytes in tissues and skin conduct electrical potentials, which makes it technically possible to track the muscle activity by measuring the local voltage of electrodes inserted into the muscle or fixed at the skin surface.

Muscle force is controlled by the number of motor units recruited and by variation of the discharge rate of each motor unit (5 to 50/sec). Thus, larger wire electrodes as well as surface electrodes detect the (algebraic) sum of different motor-unit potentials around their pickup zone (Basmajian and DeLuca, 1985). The larger the electrode dimensions and the larger the distance between muscle fibers and electrode, the more motor units will be detected. Asynchronous firing of the different motor units and variation of firing rate and motor-unit recruitment builds an interference pattern at the electrode (Figure 19.1). The EMG signal is, hence, described as a train of quasi-randomly shaped spikes varying in amplitude and duration without an identifiable sequence (Kramer et al., 1972). In spite of its noisy character, significant information can be processed from the EMG signal.

There is a correlation between the number and intensity of generated spikes and the muscular contraction force. Thus, the average EMG signal intensity increases with muscle contraction. Besides this most widely used category of information, various types of signal processing and pattern analysis allow a deeper insight into muscle activation and exertion. Anyway, it is important to recognize that EMG does not measure force, joint position, or anything else other than voltage, representing local muscle recruitment.

In ergonomics EMG measures are used for workplace and tool design as well as for scheduling of work process based on examination of:

- Muscle load (static and dynamic)
- Local muscle fatigue due to overload
- Muscle timing and coordination
- Motor-unit recruitment pattern, explaining low-level muscle fatigue and mentally induced strain

In most cases, EMGs are complemented by measures of external load, body postures, or joint movements to aid in interpretation.

Recordings of EMG can be performed either by needle electrodes inserted into the muscle or by surface electrodes taped on the skin over the target muscle. Needle electrodes are used mostly for medical and rehabilitation purposes, extracting detailed information on muscle innervation. Surface EMG is commonly used in ergonomics and occupational health due to its noninvasiveness, thus allowing convenient measures during movements. However, the more indirect measurement of surface EMG complicates application and processing because only the muscles located beneath the skin surface are accessible, and cross-talk from adjacent muscles may interfere. Furthermore, low signal reception and large signal variability complicate undisturbed evaluation and proper scaling.

19.2 Procedure

19.2.1 Equipment

For each muscle, the electrode signal is passed to a preamplifier, making the signal suitable for further processing. Subsequent filtering passes frequencies related to muscle activity and cuts electrical disturbance (50 or 60 Hz, sometimes called hum), noise, and artifacts. After being checked, the signal is processed for related information (e.g., averaged intensity) and then stored, or, inversely, the signal is stored in a raw form and then filtered and processed. Finally, the processed signal is statistically evaluated in conjunction with task conditions or other control measures.

19.2.2 Selection of Muscles

Due to the large number of muscles involved for movements, an appropriate selection of accessible muscles is required. Thus, the investigator using electromyography must have a good understanding of the anatomy of the human body.

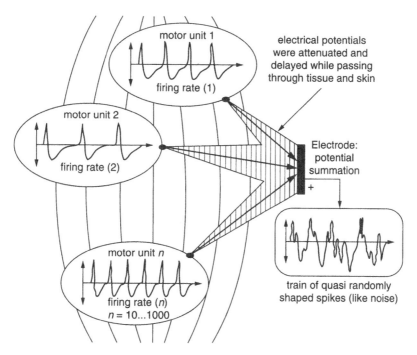

FIGURE 19.1 EMG signal composition using surface electrodes.

19.2.3 Electrode Configuration and Placement

Electrodes are usually 5- to 10-mm-sized discs coated with or made of silver/silver chloride that are connected to skin by a conductive gel and fixed by adhesive tape. Fixation may require some routine to fill the interspace with gel while leaving the adhesive collar free of gel.

To improve electrical and mechanical contact, the skin surface should be cleaned with alcohol before fixing the electrodes. Cables should be fixed without inner tension by adhesive tape. Electrodes are placed on the belly of the muscle; bony landmarks or motor-point finders may help in placement.

The basic (monopolar) EMG measure uses a single active electrode placed on the muscle and a ground electrode placed above bones or passive tissue. An improved directivity with reduced cross-talk from other muscles can be obtained by a bipolar electrode arrangement using two active electrodes (about 2 cm apart from each other) and a differential amplification (Gath and Stalberg, 1976). With this most widely applied arrangement, only the signal difference between both active electrodes is amplified, and common signal portions (from muscles located farther away) are suppressed. A third, tripolar electrode arrangement uses three active electrodes (or a concentric three-pole electrode) and a double differentiation. This setup provides an even better directivity than the bipolar arrangement.

19.2.4 Setting of Amplification, Filtering, and Storage

The electrode signal has a very low level (typically 0.1 to 5 mV) and high impedance (20,000 to 1 million ohms). This requires at first a strong amplification with high impedance input ($>10^8$ ohms) and low noise; amplifiers for bipolar and tripolar electrode arrangements should provide a common-mode rejection of >80 dB. To avoid migration of noise and hum into the low-level electrode cables, active electrodes with built-in amplifiers can be used.

The large variation of signal intensity requires an individual gain adjustment. Signal filtering is required to cut high-frequency noise, low-frequency motion artifacts, and hum from power lines. In the raw EMG, most of the signal is between 20 and 200 Hz (depending on electrode size and placement), with lesser components up to 1000 Hz and down to 2 to 3 Hz. Because the high-frequency components increasingly contain noise but do not add significant information, a low-pass filter is inserted, usually set at 500 to 600 Hz. The charge gradient at the electrode–electrolyte interface alters its capacitance during movements, so low-frequency interferences (motion artifacts) should be cut by a 10- to 20-Hz high-pass filter. Hum from power lines, particularly caused by scant electrode contact, can be suppressed by a 50/60-Hz notch filter. There is a long debate about appropriate filter settings (dependent on electrode design and placement and on subjects measured), but because no specific information is to be found within particular frequency intervals of the stochastic raw EMG, stronger filtering affects the results only marginally.

19.2.5 Signal Control

Due to the stochastic character of EMG signals, it cannot be discriminated easily from noise and other interference. A manual signal control (like an oscilloscope) is recommended, in particular to detect scant electrode leads. Testing is performed by checking signal variation during voluntary contraction and relaxation. A spectral analytic view may help to identify disturbance and noise overlay.

19.2.6 Signal Processing

The most widely used type of processing is to compute the average signal intensity that corresponds directly to muscular activation level (Figure 19.2). Basically, this is performed by rectification (or root-mean-square computation) of the raw EMG with subsequent signal integration or low-pass filtering. Like filter setting, processing options differ subtly, but the selection of integration interval (or cutoff frequency) is very important because it affects temporal resolution and noise. (For more details, see Luczak and Göbel, 2000.) Measures of muscular activation level are then interpreted according to movement or work sequences, to work conditions, or correlated to movement parameters. Amplitude distribution can provide further information on static and peak activation. Isometric contractions with constant load can be checked for muscular fatigue if increased muscle activation is observed with time.

A second type of processing focuses on the frequency spectrum of the raw EMG. As the muscle fatigues, propagation velocity of the action potentials decreases due to the accumulation of acid metabolites. This leads to a shift and a compression of the EMG frequency spectrum toward lower frequencies (Lindström and Magnusson, 1977). There are also numerous other signal analysis approaches, mostly focusing on motor-unit recruitment characteristics (e.g., Forsman et al., 1999).

19.2.7 Scaling

The amplitude of the detected motor-unit action potential depends on many factors, such as the electrode size, the distance between active muscle fiber and the detection site, etc. Hence, absolute EMG amplitude varies strongly between subjects and measures (up to a factor 100). However, data need to be presented in a common format for comparisons. Therefore, a second measure is used as a scaling reference. The

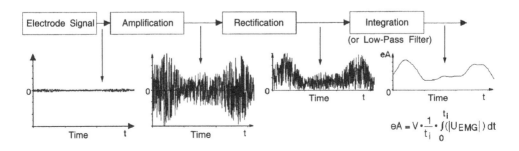

FIGURE 19.2 Recovering the intensity function from a surface EMG.

most widely used technique involves the scaling of EMG amplitude as a percentage of the value measured during maximum voluntary contraction (constituting an individual reference). Another variant refers to the EMG signal amplitude when exerting a predefined force (constituting a physical reference). Both types of reference lack the reliability to voluntarily control force of a specific muscle (typical variance ±20%).

Complementary to amplitude scaling, time intervals can be standardized for pattern correlation, but this approach also has similar difficulties with reliability.

19.2.8 Measurement

During data collection, the most crucial point is to ensure suitable electrode contact. As the hours pass, electrode gel diffuses into skin, and sweating can dissolve electrode adhesion. Further, signal amplitudes need to be checked for possible overmodulation of electronics.

19.3 Advantages

EMG provides continuous and quantitative measured data with a high temporal resolution and only marginal interference with task execution. It allows the detection of muscle fatigue in its early stages and thus can be used as an objective measure, even for not deliberately accessible motor reactions. Multi-channel EMG serves as a practical means of identifying muscular bottlenecks.

19.4 Disadvantages

Using surface EMG, only muscles located directly beneath the skin can be accessed. A proper analysis is only feasible for single muscles of individuals who are not too obese. Careful calibration, instrumentation, data treatment, and interpretation are required. Preparation of EMG measures is somewhat lengthy (15 to 30 min for experienced ergonomists), and interpretation of the data requires additional information (e.g., workplace conditions or simultaneous measures of the work positions). Calibration lacks reliability due to nonlinear characteristics, which would require the user to define an individual calibration function.

19.5 Example

The following example (Göbel, 1996) shows a subtle type of analysis: movement coordination is, at least partially, controlled by sensory feedback of different modalities. To optimize motor control tasks and tools, knowledge about the use of sensory feedback would be very helpful. Although such a measure is not directly accessible, it can be explored by EMG. Due to the inevitable delay of human information processing, effective feedback loops must produce oscillatory components. Its frequency depends on the delay of the information-processing loop, but it is always higher than the frequency of external movements. Thus, the activity of sensory feedback can be estimated by the higher frequency components of the EMG intensity function (Figure 19.3).

Using this method to analyze the learning phase of a manual assembly task, it could be shown that (monosynaptic and polysynaptic) reflex activity correlated negatively with execution speed during the initial work periods (up to 8 to 12 hours total work duration), but this effect would disappear with higher degrees of experience (Figure 19.4, upper side). In contrast to this, the longer haptic and visual feedback loop did not affect working speed during the initial work periods, but showed a strong positive correlation during the later learning phases (Figure 19.4, lower side). It can be concluded that sensory feedback is incorporated first for monosynaptic and then for polysynaptic reflexes. Visual and haptic feedback follow again with delay. As experience increases, all types of sensory feedback activity abet execution speed.

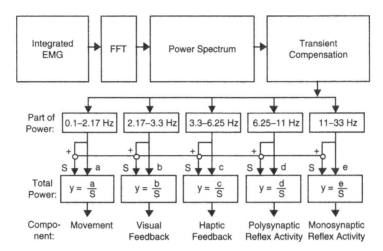

FIGURE 19.3 Signal-processing structure to estimate the activity of sensory feedback on movement control (simplified).

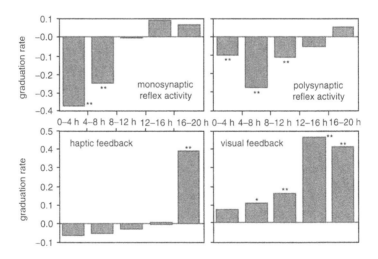

FIGURE 19.4 Correlation between feedback control and execution speed during a learning phase for a manual-assembly task (ten subjects during five subsequent periods, each of 4-hour duration; mean cycle duration 22 sec; *: $p < 0.05$; **: $p < 0.01$)

19.6   Related Methods

EMG is a subset of psychophysiological measures, a noninvasive technique for detecting physiological reactions caused by mentally controlled processes. Like other psychophysiological measures (e.g., heart-rate variability, skin resistance, eye blink frequency, electroencephalography, etc.), the indirect assessment of biological reactions complicates data detection and interpretation. EMG used for analysis of muscle load also represents a physiological measure, such as oxygen consumption or heart rate frequency.

Besides the application of EMG in ergonomics and human factors research, another large domain is established in clinical EMG for medical analysis of muscle and motor control functions.

19.7 Standards and Regulations

Because of the electrical connection to human subjects, equipment must ensure security against electrical shocks. EMG hardware has to fulfill the safety and control regulations for medical equipment, and thus only certified equipment must be used.

Many attempts have been made to standardize EMG recordings (Winter et al., 1980; Zipp, 1982). However, due to the sheer number of factors influencing the information content of the detected EMG (electrode size, type, configuration and location, filter frequencies and characteristics, signal processing, normalization, etc.), as well as the fact that parameter setting depends on subjects, on experimental layout, and on each other, no general standardization in ergonomics has yet been established. Consequently, detailed documentation of EMG recording and processing conditions is required, thus restricting comparison of results and making reports somewhat laborious.

19.8 Approximate Training and Application Times

Although measurements themselves can be performed quite quickly, extensive experience is required for definition of experimental setup and auxiliary measures as well as for muscle selection, electrode placement, filter selection, calibration, and data interpretation. Thus, while measures for demonstration purposes can be performed very quickly, experience requirements increase dramatically with precision and detail of measurements (up to years).

While hardware and basic software equipment are available for almost any type of application, software for data interpretation often needs to be created or modified according to the experimental setup.

19.9 Reliability and Validity

Despite the extensive use of EMG for the last several decades, little work has been completed that examines reliability issues. Reliability of direct measures is very low between subjects and between electrode settings due to differences in muscle geometry, subcutaneous tissue, and electrode location. With normalization, maximum reliability coefficients range between 0.8 and 0.9 for submaximal contractions and between 0.5 and 0.8 for maximal contractions (Viitasalo and Komi, 1975; Yang and Winter, 1983). Within-subject reliability is far superior if electrodes are not replaced (0.9 to 0.99, reported by Lippold, 1952). In general, across-muscle comparisons are precluded because of differences in muscle and body tissues constituencies.

Validity depends mainly on electrode location according to muscle site and possible separation conflicts.

19.10 Tools Needed

Investment	Consumptions
• Electrodes and cables	• Adhesive electrode tape and electrode gel or disposable electrodes
• Preamplifiers	
• Hardware filter units	• Adhesive tape for cable fixation
• Digital-to-analog converter	
• Computer or other storage medium	
• Recording and control software	
• Data analysis software	

Although a large variety of mobile and stationary products is available, one should be aware that most products are designed only for one type of electrode configuration (commonly bipolar arrangement).

References

Basmajian, J.V. and DeLuca, C.J. (1985), *Muscles Alive: Their Functions Revealed by Electromyography*, Williams & Wilkins, Baltimore, MD.

Forsman, M., Kadefors, R., Zhang, Q., Birch, L., and Palmerud, G. (1999), Motor-unit recruitment in the trapezius muscle during arm movements and in VDU precision work, *Int. J. Ind. Ergonomics*, 24, 619–630.

Gath, I. and Stalberg, E.V. (1976), Techniques for improving the selectivity of electromyographical recordings, *IEEE Trans. BME*, 23, 467–472.

Göbel, M. (1996), Electromyographic evaluation of sensory feedback for movement control, in *Proceedings of the XIth International Ergonomics and Safety Conference*, July 1996, Zurich.

Kramer, H., Kuchler, G., and Brauer, D. (1972), Investigations of the potential distribution of activated skeletal muscles in man by means of surface electrodes, *Electromyogr. Clin. Neurophysiol.*, 12, 19–24.

Lindström, L. and Magnusson, R. (1977), Interpretation of myoelectric power spectra: a model and its application, *Proc. IEEE*, 65, 653–662.

Lippold, O.C.J. (1952), The relation between integrated action potentials in a human muscle and its isometric tension, *J. Physiol.*, 117, 492–499.

Luczak, H. and Göbel, M. (2000), Signal processing and analysis in application, in *Engineering Psychophysiology*, Backs, R.W. and Boucsein, W., Eds., Lawrence Erlbaum, Mahwah, NJ, pp. 79–110.

Matteucci, C. (1844), *Traites des Phenomenes Electropysiologiques*, Paris.

Piper, H. (1912), *Electrophysiologie menschlicher Muskeln*, Springer, Berlin.

Viitasalo, J.H. and Komi, P.V. (1975), Signal characteristics of EMG with special reference to reproducibility of measurements, *Acta Physiol. Scand.*, 93, 531–539.

Winter D.A., Rau, G., Kadefors, R., Broman, H., and DeLuca, C.J. (1980), Units, Terms, and Standards in the Reporting of EMG Research, a report by the ad hoc committee of the International Society of Electrophysiology and Kinesiology.

Yang, J.F. and Winter D.A. (1983), Electromyography reliability in maximal and submaximal isometric contractions, *Arch. Phys. Med. Rehabil.*, 64, 417–420.

Zipp, P. (1982), Recommendations for the standardization of lead positions in surface electromyography, *Eur. J. Appl. Physiol.*, 50, 55–70.

20

Estimating Mental Effort Using Heart Rate and Heart Rate Variability

Lambertus (Ben) J.M.
Mulder
University of Groningen

Dick de Waard
University of Groningen

Karel A. Brookhuis
University of Groningen

20.1 Background and Applications

Heart rate can be derived from the electrocardiogram (ECG), which reflects the (electrical) activity of the heart. For the assessment of mental effort, it is not the ECG itself that is of interest but, rather, the time durations between heartbeats. In this respect, heart rate (HR) is the number of heart beats in a certain time period (usually a minute), while mean heart period or interbeat interval (IBI) is the average time duration of the heart beats in that period. Heartbeats have varying time durations, resulting in IBI time series with characteristic patterns and frequency contents. This is called heart rate variability (HRV). During task performance, subjects have to expend mental effort, which is usually reflected in increased HR and decreased HRV when compared with resting situations.

    The normal rhythm of the heart is controlled by membrane processes of the cardiac sinoatrial node, which are modulated by innervation from both the sympathetic and parasympathetic branches of the autonomous nervous system (Berntson et al., 1997; Levy, 1977). Levy (1977) showed a clear nonlinear relation between sympathetic and parasympathetic activity at one side and HR at the other, while Berntson et al. (1994) indicated that the relationship between the two autonomous branches and heart period was much more linear. Van Roon (1998) showed this same phenomenon using a baroreflex simulation model in which the heart rate control part was described by the well-known equations of Rosenblueth and Simeone (1934).

HR is mainly controlled by nuclei in the brain stem, guided by the hypothalamus and (prefrontal) cortical structures. Two different control modes have to be distinguished (Van Roon, 1998):

1. Parasympathetic (vagal) and sympathetic output to the heart. This is directly controlled by the hypothalamus via brain-stem nuclei; Porges (1995) labeled these, respectively, as vagal tone and sympathetic tone.
2. Mediation of baroreflex activity. In this case, incoming information of the baroreceptors into the nucleus tractus solitarius serves as input to the brain-stem nuclei, determining vagal and sympathetic activity. The role of the hypothalamus in this mode is the mediation of baroreflex gain factors to the autonomic system branches.

Distinction between these two modes of cardiac control is important with respect to interpretation of HR and HRV changes during mental effort and stress.

HRV can be related to changes in autonomic control. Two main sources of variability can be distinguished: respiratory sinus arrhythmia and spontaneous fluctuations mainly related to short-term blood pressure control (baroreflex). Respiratory sinus arrhythmia (Porges, 1995) is a reflection of the respiratory pattern into the HR pattern. During inspiration, vagal control to the heart is diminished (vagal gating), resulting in an increased HR, while during expiration this vagal suppression disappears, resulting in an increased HR (Porges et al., 1982; Grossman et al., 1991). According to this mechanism, and related to the relatively high frequency of normal breathing, it is believed that respiratory sinus arrhythmia is mainly vagally determined. The second source of variation is strongly related to baroreflex control and to the "eigenrhythm" of this control loop. In many cardiovascular variables, including HRV, a characteristic 10-sec rhythm (0.10-Hz component) can be found that is modulated by baroreflex gain. Wesseling and Settels (1985) considered this type of variation as random fluctuations within the baroreflex. By using spectral analysis, Mulder (1992) distinguished three different frequency bands: a low-frequency area between 0.02 and 0.06 Hz; a mid-frequency band between 0.07 and 0.14 Hz; and a high-frequency band between 0.15 and 0.40 Hz. Other authors (Berntson et al., 1997) do not use the low-frequency band and indicate the (sometimes extended) mid-frequency area as the low-frequency band. Some authors (Pagani et al., 1986) consider the high-frequency area as completely vagally determined and consider the power in the low-frequency band as an index for sympathetic activity. The latter is an overly simplistic view because of the large effects of vagal control on the low frequencies.

The general cardiovascular response pattern that is found in many mental-effort studies can be characterized by an increase of HR and blood pressure and a decrease of HRV and blood pressure variability in all frequency bands. This pattern is comparable with a defense reaction and is mainly found in laboratory studies using short-lasting tasks that require challenging mental operations in working memory. An important empirical finding is that the mid-frequency band is the most sensitive measure for variations in mental effort (Mulder and Mulder, 1987). Simulation studies (Van Roon, 1998) indicated that this could be attributed to the fact that two effects occur at the same time: a decreased vagal activation and an increased sympathetic activation.

20.2 Procedure

A number of steps have to be taken in order to obtain artifact-free HR data and to compute the spectral power-band values.

20.2.1 Measuring the ECG

The ECG can be measured with three electrodes at the chest. A good lead yielding a relative high R-wave and a suppressed T-wave (most ideal for detecting R-peaks) is a precordial bipolar one with electrodes on position V6 and the other at the sternum. Derivations using electrodes at extremities (arms, legs) should be avoided for reasons of artifact sensitivity. For registrations of long duration, self-adhesive electrodes that are used in child cardiology are recommended. The source signal of the ECG is at the millivolt level, implying that the signal is relatively insensitive to 50- or 60-Hz disturbances. However, if such disturbances do occur, low-pass filtering with a cut-off frequency of about 35 Hz and an attenuation of 20 to 40 dB (factor 10 or 100) at 50 Hz can easily solve such problems.

20.2.2 Sampling and R-peak Detection

Hardware detection of R-peaks in the ECG and registration of the R-wave event times at an accuracy of 1 msec is the most effective and straightforward method of data acquisition and can be considered as the preferred technique. Mulder (1992) describes such a device that can obtain an accuracy of 1 msec, while detection errors (i.e., missing R-peaks or additional triggering at the T-wave) are very seldom seen. Alternatively, the same features can be implemented in software for off-line application. In this case, the ECG has to be sampled at least at 400 Hz in order to obtain a detection accuracy of 2 to 3 msec. In situations where small HRV changes have to be measured (e.g., vagal blockade studies), such an accuracy level may be critically low. Often ECG is measured in combination with other signals, such as respiration, finger blood pressure, and EEG. In such cases, a lower sampling rate than necessary for adequate R-peak detection is often chosen.

20.2.3 Artifact Detection and Correction

Detection and correction of artifacts in the measured R-peak event series is very critical because of the high sensitivity of spectral power measures for missing or additional events. Mulder (1992) describes that one missing R-peak in a time segment of 100 sec increases computed HRV measures by 100%. Artifacts can have different causes. With subjects in a normal resting, sitting position, no more than 10% of the artifacts have technical backgrounds, i.e., most problems are related to physiological origins, varying from occasionally occurring extra systoles to specific short-lasting patterns of sudden vagal activation or artifacts related to respiration (sighing). Inexperienced users may not be aware of the fact that subjects can show extra systoles or other conduction problems at the heart without having medical indications of heart problems. Automatic detection and correction of such "artifacts" is very difficult if not impossible, implying that when the number of artifacts is too large, the subject should be excluded from the study. For this reason, data from about 1 in 20 subjects has to be skipped. In the field, e.g., in a moving vehicle, artifacts due to electrical disturbances and/or vibration are more common, but still restricted. The loss of data in on-road studies is estimated to be between 5 and 15% (e.g., de Waard et al., 1995; Steyvers and de Waard, 2000).

While automatic detection and correction is not always successful and satisfactory, detection and classification with visual support is more adequate, although time-consuming. Correctly detected and classified artifacts can be corrected relatively easily by using linear interpolation (Mulder, 1992).

20.2.4 Spectral Procedures

Spectral analysis serves to decompose the IBI time series in different rhythms, e.g., for distinguishing the 0.10-Hz component from the respiratory-related components in HRV. As a matter of fact, the variance (power) in each of the three earlier defined frequency bands has to be determined.

Several techniques and procedures are used for spectral analysis of the corrected IBI data in order to obtain HRV spectral power values (Berntson et al., 1997; Berger et al., 1986; Mulder, 1992). A variety of spectral procedures can be applied, each with its own advantages and restrictions. The results are not really dependent on the way the spectra are computed. However, serious differences can be expected when spectra of HR and IBI are compared. This is discussed in the next section.

20.2.5 HR, IBI, or Normalized Values?

Although intuitively it might be expected that spectra of HR and IBI produce the same results, this is not the case. When HRV is decreased in a certain measuring period, this will be reflected in lower values for spectra of both HR and IBI. However, this does not occur in a linear way because of the inverse relationship between HR and IBI. De Boer et al. (1984) and Mulder (1992) described particularly the strong dependency on mean HR. Mulder (1992) concluded that the likelihood of finding statistically significant results in an experiment on mental effort would be greater if IBI measures were taken instead of HR measures. This is caused by the expected concurrent decrease of IBI and HRV in such situations. In our opinion, such a difference in expected results between HR and IBI measures is undesirable. Therefore, Mulder (1992) proposed to use normalized values. This means that for the calculation of heart rate variability, the values in the original time series of either HR or IBI are divided by its mean values in the analysis segment at hand. This corresponds to taking the coefficient of variation instead of computing the standard deviation. Mulder (1992) showed that this transformation prevents the strange discrepancies between HR and IBI outcomes. For this reason, we prefer to use normalized power values. The obtained power values have the unit "squared modulation index."

20.2.6   Logarithmic Transformation

Bendat and Piersol (1986) showed that spectral power values have a chi-squared statistical distribution. For this reason, Van Roon (1998) argued that taking the log values of these power values is the most appropriate transformation, both on theoretical and practical grounds, in order to obtain a more normal distribution that is suitable for statistical analysis. Therefore, we propose to take the natural logarithm for all spectral band values before statistical analysis.

20.3   Disadvantages

HR and HRV are used as indicators of mental effort: the higher the invested effort, the higher is HR and the lower is HRV. However, because of the complex relationship of HR with baroreflex blood pressure control and with the autonomous nervous activity, there are several reasons why such a simple starting point as indicated above does not hold under all circumstances. The most stable results, according to this viewpoint, are obtained when short-duration laboratory tasks are performed with relative high demands on working memory.

Practical disadvantages and restrictions of the method are its sensitivity for artifacts in the obtained IBI series and its sensitivity for changes in the respiratory pattern. With respect to the latter, the effects of speaking and sighing on HRV are large. Artifact correction is time consuming. Moreover, the need for using electrodes and the necessity of having registration apparatus can be considered obtrusive. However, nowadays sports medicine wristwatches are available that can register HR from beat-to-beat in a reliable way at an accuracy level that is adequate for human factors studies (Van de Ven, 2002).

20.4   Example

In an advanced driving simulator, 22 participants completed test rides while driving over different types of roads. Among these were roundabouts, quiet dual carriageways, and urban roads. During the test rides, participants' heart rates were registered. (For a full report, see de Waard et al. [1999].) From recorded interbeat intervals, a moving-average heart rate and the power in the 0.10-Hz band of heart rate variability were calculated with aid of the so-called profile technique (e.g., Mulder, 1992). With this technique, averages are calculated, in this case using 30 sec of data as input. Then the calculation window is moved with a step of 10 sec, and again averages are calculated. The resulting moving averages are displayed in Figure 20.1.

The ride through different road environments is clearly reflected in both heart rate and the 0.10-Hz component. Driving on a high-density road (roundabout) is more effortful — heart rate is higher and variability is reduced — than driving over a dual carriageway. Very illustrative is the additional information that the 0.10-Hz component of HRV gives. When looking at the ride through the first built-up area, heart rate is relatively high compared with driving on the dual carriageway, and it remains high during

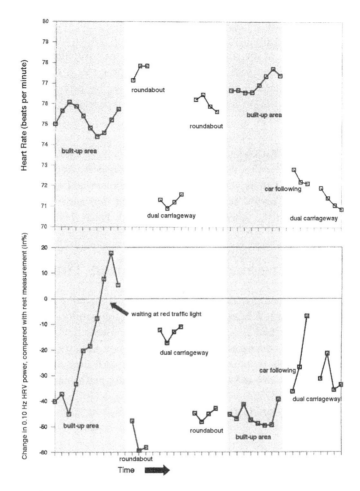

FIGURE 20.1 Average heart rate (top panel) and the 0.10-Hz component of HRV (lower panel) during test rides in a driving simulator (N = 22). Time is displayed in steps of 10 sec. HRV is displayed compared with a resting level of 0. (Adapted from De Waard, D. [1996], The Measurement of Drivers' Mental Workload, Ph.D. thesis, University of Groningen, Groningen, the Netherlands. With permission.)

that part of the ride. Heart rate variability, however, is only low (i.e., high effort) during the first half of the ride through the city. And that part actually *is* the only driving part, as during the second part participants were waiting for a red traffic light to turn green. Obviously, the waiting itself is not an effort-demanding task, which is clearly reflected in HRV.

20.5 Related Methods

There are several related methods, such as subjective scales and questionnaires, for obtaining indications of invested effort (Zijlstra and Van Doorn, 1985; Hart and Staveland, 1988). However, such methods should be considered complementary and not as replacements for HRV measures because there are several reasons why subjective and psychophysiological indices might differ. There are strong links of HR and HRV with blood pressure changes and with respiratory patterns. For this reason, it is informative to measure respiration and (finger) blood pressure from beat-to-beat in situations where this is possible. Having additional information about the respiratory pattern at an individual level helps interpretation of HRV tremendously. This holds for speaking and sighing as well as for shifts in main respiratory rate. For instance, when respiration rate increases during task performance, there will be a decreased HRV in the high-frequency band simply for this reason (Angelone and Coulter, 1964). Moreover, when main respiratory rate comes within the mid-frequency band, this will lead to an enormous increase in HRV.

Measuring finger blood pressure gives the possibility of obtaining systolic and diastolic blood pressure values from beat to beat. Because of the strong dependency, via the baroreflex, of blood pressure and HR, the insight into what is going on with the control of ANS processes is considerably increased. In particular, with respect to interpretation of HR changes in relation to the two mechanisms presented (either direct hypothalamic control or the occurrence of a defense-type reaction), the availability of blood pressure indices is essential.

20.6   Standards and Regulations

There are no standards and regulations with respect to measurement and analysis of HR and HRV, although there are two extended review studies that give insight into and recommendations for the application of such techniques in the world of cardiology (Task Force, 1996) and psychophysiology (Berntson et al., 1997).

20.7   Approximate Training and Application Times

With the right measuring and registration devices, an adequate data-analysis package, as well as software for data transformation into the right format, the training for using the mentioned measuring and analysis techniques does not require more than 2 to 3 hours. However, experience has taught that understanding the backgrounds of HRV as well as the necessary data-transformation and artifact-correction algorithms requires more time. Several weeks or months of working with such data may be necessary to fully understand what is going on with HR and HRV during mental task performance.

20.8   Reliability and Validity

Reliability and repeatability of obtained HR and HRV in short-duration mental-loading laboratory tasks results are usually high, but this does not hold for use in practical settings. The reason for this may be connected to uncertainty about what subjects are doing in terms of cognitive activities on the one hand, and emotional/motivational factors as well as compensatory effort on the other. According to the two main mechanisms of HR regulation mentioned in the introduction, this can have important implications. Furthermore, fluctuations in breathing pattern, for instance during speaking, have large effects on HRV.

With respect to diagnostics, different positions are taken. Gaillard and Kramer (2000) consider HRV nondiagnostic as an index for mental effort because differences in types of mental operations cannot be distinguished. Mulder et al. (2000), however, do consider the same index diagnostic because a clear distinction can be made between attention-demanding and nonattention-demanding mental work.

It has to be realized, however, that the sensitivity of the measure is not very high, making it difficult to distinguish between levels of task load and related invested effort. Moreover, using the index on an individual level requires several repetitions in order to obtain adequate statistical reliability.

The validity aspect, finally, has to be related to the HR regulation mechanisms mentioned in the introduction. If only the second mechanism, i.e., a defense-type reaction, is in play, we consider the HRV index as very valid for invested mental effort. However, if compensatory mechanisms are becoming more important, interpretation of HR and HRV changes in terms of mental effort is not sufficiently reliable.

20.9   Tools Needed

Data collection requires three electrodes, an amplifier, an R-wave triggering device (either in hardware or software), and a data-collection device that saves the data either as samples of the ECG or as R-peak-event time points. The latter is the most efficient. Several types of equipment are available, ranging from a wristwatch or another portable event-collection device to a full-blown electrophysiological recording apparatus. For data analysis, in principle, two methods are followed. The first is transformation of the nonequidistant series of events into an equidistant series of samples (Berger et al., 1986) while using a standard package for spectral analysis. The second includes a direct Fourier transform of the events series (Mulder, 1992). In both cases, adequate artifact correction is required additionally. There are some software packages that combine the necessary processing steps (e.g., Carspan, see Mulder [1992]), including data preprocessing of other cardiovascular time series.

References

Angelone, A. and Coulter, N.A. (1964), Respiratory-sinus arrhythmia: a frequency dependent phenomenon, *J. Appl. Physiol.*, 19, 479–482.

Bendat, J.S. and Piersol, A.G. (1986), *Random Data: Analysis and Measurement Procedures*, 2nd ed., John Wiley & Sons, New York.

Berntson, G.G., Cacioppo, J.T., Quigley, K.S., and Fabro, V.T. (1994), Autonomic space and psychophysiological response, *Psychophysiology*, 31, 44–61.

Berntson, G.G. et al. (1997), Heart rate variability: origins, methods, and interpretive caveats, *Psychophysiology*, 34, 623–648.

Berger, R.D., Akselrod, S., Gordon, D., and Cohen, R.J. (1986), An efficient algorithm for spectral analysis of heart rate variability, *IEEE Trans. Biomed. Eng.*, 33, 900–904.

De Boer, R.W., Karemaker, J.M., and Strackee, J. (1984), Comparing spectra of a series of point events, particularly for heart rate variability data, *IEEE Trans. Biomed. Eng.*, 31, 384–387.

De Waard, D. (1996), The Measurement of Drivers' Mental Workload, Ph.D. thesis, University of Groningen, Groningen, the Netherlands.

De Waard, D., Jessurun, M., Steyvers, F.J.J.M., Raggatt, P.T.F., and Brookhuis, K.A. (1995), Effect of road layout and road environment on driving performance, drivers' physiology and road appreciation, *Ergonomics*, 38, 1395–1407.

De Waard, D., Van Der Hulst, M., and Brookhuis, K.A. (1999), Elderly and young drivers' reaction to an in-car enforcement and tutoring system, *Appl. Ergonomics*, 30, 147–157.

Gaillard, A.W.K. and Kramer, A.F. (2000), Theoretical and methodological issues in psychophysiological research, in *Engineering Psychophysiology: Issues and Applications*, Backs, R.W. and Boucsein, W., Eds., Lawrence Erlbaum Associates, Mahwah, NJ, pp. 31–58.

Grossman, P., Karemaker, J., and Wieling, W. (1991), Prediction of tonic parasympathetic cardiac control using respiratory sinus arrhythmia: the need for respiratory control, *Psychophysiology*, 28, 201–216.

Hart, S.H. and Staveland, L.E. (1988), Development of NASA-TLX (task load index): results of empirical and theoretical research, in *Human Mental Workload*, Hancock, P.A. and Meshkati, N., Eds., North-Holland, Amsterdam, pp. 139–183.

Levy, M.N. (1977), Parasympathetic control of the heart, in *Neural Regulation of the Heart*, Randall, W.C., Ed., Oxford University Press, New York, pp. 95–129.

Mulder, L.J.M. (1992), Measurement and analysis methods of heart rate and respiration for use in applied environments, *Biol. Psychol.*, 34, 205–236.

Mulder, L.J.M. and Mulder, G. (1987), Cardiovascular reactivity and mental workload, in *The Beat-by-Beat Investigation of Cardiovascular Function*, Rompelman, O. and Kitney, R.I., Eds., Oxford University Press, Oxford, pp. 216–253.

Mulder, G., Mulder, L.J.M., Meijman, T.F., Veldman, J.B.P., and Van Roon, A.M. (2000), A psychophysiological approach to working conditions, in *Engineering Psychophysiology: Issues and Applications*, Backs, R.W. and Boucsein, W., Eds., Lawrence Erlbaum Associates, Mahwah, NJ, pp. 139–159.

Pagani, M. et al. (1986), Power spectral analysis of heart rate and arterial pressure variabilities as a marker of sympathovagal interaction in man and conscious dog, *Circ. Res.*, 59, 178–193.

Porges, S.W. (1995), Orienting in a defensive world: mammalian modifications of our evolutionary heritage — a polyvagal theory, *Psychophysiology*, 32, 301–318.

Porges, S.W., McCabe, P.M., and Yongue, B.G. (1982), Respiratory-heart-rate interactions: psychophysiological implications for pathophysiology and behavior, in *Perspectives in Cardiovascular Psychophysiology*, Cacioppo, J. and Petty, R., Eds., Guilford Press, New York, pp. 223–259.

Rosenblueth, A. and Simeone, F.A. (1934), The interrelations of vagal and accelerator effects on the cardiac rate, *Am. J. Physiol.*, 265, H 1577–1587.

Steyvers, F.J.J.M. and de Waard, D. (2000), Road-edge delineation in rural areas: effects on driving behaviour, *Ergonomics*, 43, 223–238.

Task Force of the European Society of Cardiology and the North American Society of Pacing and Electrophysiology (1996), Heart rate variability: standards of measurement, physiological interpretation, and clinical use, *Circulation*, 93, 1043–1065.

Van de Ven, T. (2002), Getting a Grip on Mental Workload, Ph.D. thesis, Catholic University, Nijmegen, the Netherlands.

Van Roon, A.M. (1998), Short-term Cardiovascular Effects of Mental Tasks: Physiology, Experiments and Computer Simulations, Ph.D. thesis, University of Groningen, Groningen, the Netherlands.

Wesseling, K.H. and Settels, J.J. (1985), Baromodulation explains short-term blood pressure variability, in *The Psychophysiology of Cardiovascular Control*, Orlebeke, J.F., Mulder, G., and van Doornen, L.P.J., Eds., Plenum Press, New york, pp. 69–97.

Zijlstra, F.R.H. and Van Doorn, L. (1985), The Construction of a Subjective Effort Scale, Technical University of Delft, Delft, the Netherlands.

21

Ambulatory EEG Methods and Sleepiness

Torbjörn Åkerstedt

National Institute for Psychosocial Factors and Health

21.1 Background and Applications

Sleepiness has been identified as one of the major causes of accidents in transport and industry (Dinges, 1995). The causes mainly involve sleep loss, long time awake, work at the circadian trough of physiological activation and alertness, and monotony. Indeed, drugs including alcohol, sedatives, hypnotics, antihistamines, and others can also be included in the list.

The concept of sleepiness usually refers to signs that indicate a tendency toward sleep (Dement and Carskadon, 1982). This commonsense conception involves subjective as well as behavioral and physiological components. Most of the existing knowledge about sleepiness has been based on self-rating techniques. Such techniques do, however, have limitations with respect to validity and with respect to the possibility to capture moment-to-moment fluctuations. Performance measures present similar problems and can hardly be said to indicate sleepiness per se (although certainly the consequences of it). Other, more physiologically oriented techniques include evoked potentials, pupillography, and critical flicker fusion. None of these, however, can easily be used to monitor sleepiness continuously. For this purpose, only polysomnographical methods would work, i.e., registration and analysis of electroencephalography (EEG), electrooculography (EOG), and electromyography (EMG).

The EEG represents the sum of the electrical brain activity that can be recorded on the scalp via surface or needle electrodes. In animal studies, electrodes implanted deep into the brain are also used. Normally, the alert brain presents a high-frequency (16 to 50 Hz) irregular pattern, since different structures of the brain are differentially active. When alertness falls, the frequency of the EEG falls and the amplitude increases as more and more neurons are synchronized to fire simultaneously by the thalamus. This is essentially the rationale behind the use of the EEG as an indicator of sleepiness.

The first EEG descriptions of the process of falling asleep were accomplished by Loomis et al. (1935a, 1935b, 1937). They showed that relaxed subjects lying with closed eyes showed a predominant alpha (8 to 12 Hz) activity and responded to environmental stimuli. When alpha started to break up, however, the subjects ceased to respond. Further progression showed that the EEG frequency decreased into the theta (4 to 8 Hz) and, later, the delta (0 to 2 Hz) range, i.e., sleep proper. Bjerner et al. (1949) showed that exceptionally long reaction times were associated with alpha blocking downward toward theta, and they interpreted this as "transient phenomena of the same nature as sleep."

Not only closure of the eyelid, but also the movement of the eye, is related to EEG changes. Usually, slow (0.1 to 0.6 Hz) eye movements start to appear when the alpha activity breaks up and the subject begins to "drift off" into sleep stage 1 (Kuhlo and Lehman, 1964). As sleep is more firmly established, the slow eye movements begin to disappear, although they may sometimes remain for a while in sleep stage 2.

It should be emphasized that the above studies have used relaxed and closed-eyes conditions, in which case the presence of alpha signals alertness. If the eyes are open, however, this is reversed (Daniel, 1967), and poor detection performance covaries with increased alpha and theta activity (O'Hanlon and Beatty, 1977). Furthermore, individuals falling asleep while performing a task show increased alpha activity before the event and a "terminal" theta burst when the neck muscle tonus drops and the subject's head "nods off" (Torsvall and Åkerstedt, 1988). Slow eye movements also increase with the increase in alpha activity.

Alpha and theta activity, as well as slow eye movements, also increase gradually with increased sleep loss and proximity to the circadian trough, but effects are not seen until one has reached levels of subjective sleepiness indicating some effort to stay awake (Åkerstedt and Gillberg, 1990). Recently, it has been demonstrated that different 1-Hz bands react differently to sleep loss and proximity to the circadian trough (Aeschbach et al., 1997), i.e., there seems to be a homeostatic and circadian type of sleepiness.

Using the approaches indicated above, a number of studies have focused on polysomnographical sleepiness in different work settings. Lecret and Pottier (1971) showed that time with alpha activity (filtered) increased with rural and uneventful driving. O'Hanlon and Kelley (1977) showed increased alpha activity and lane drifting when driving was monotonous. Studies of shift workers have shown that train drivers show increased alpha and theta activity (and increased slow eye movements) toward the late parts of the night (Torsvall and Åkerstedt, 1987). Truck drivers behave in much the same way (Kecklund and Åkerstedt, 1993). One quarter of process operators show actual sleep during night shifts (Torsvall et al., 1989). Mitler et al. (1997), however, found very little changes in alpha and theta activity in long-distance truck driving despite self reports of sleepiness and videotape evidence of frequent bouts of sleepiness. Aircrew show increased alpha and theta activity during long-haul flights (Gundel et al., 1995). Recently there has been a number of studies showing increased theta/alpha activity in driving simulators as a function of sleep loss and alcohol, a trend that was counteracted by breaks, naps, noise, etc. (Reyner and Horne, 1997, 1998; Landström et al., 1998).

21.2   Procedure

The procedure of preparing a polysomnographical sleep recording is extensively described in the literature, most central of which is the "gold standard" manual by Rechtschaffen and Kales (1968), and the *Principles and Practice of Sleep Medicine* (Kryger et al., 2000). The procedure uses part of the so-called ten-twenty system for electrode positioning. It starts with the measurement of the distance between the nasion (the recession between the eyes where the nose bone starts to protrude from the skull) and the inion (the recession where the skull meets the neck). The distance between the midpoint of the frontal part of one ear to the other also should be measured. Using these measures, the ideal electrode positions are then identified. For sleep, C3 and A2 (mastoid behind the left ear) or C4 and A1 (mastoid behind the left ear) are used. For sleepiness measures, the derivation O2-P4 is often preferred. In general, alpha activity is most pronounced in occipital or parietal derivations. Frontal positions often contain large artifacts from eye movements and therefore may be difficult to use. The exact positioning of the electrodes are quite important when using a bipolar derivation, since the amplitude of the recording will increase with increasing distance between the electrodes, up to a point.

After identification with a marker pen, the electrode sites are cleaned with spirit and acetone (to remove fat). A proper rubbing is necessary for a good recording. A silver–silver chloride cup electrode is applied on the site and fastened using liquid collodium or a self-adhesive ring. EOG electrodes are placed as a horizontal pair (at the outer canthi of each eye) and as a vertical pair (above and below the eye). If only few channels are available, one might get away with only one pair placed obliquely, i.e., one electrode at the outer canthus, slightly above the midline of the eye, and the other below the midline. This permits the recording of eye blinks (vertical) as well as horizontal eye movements. Frequently, self-adhesive electrodes can be used for the EEG recording, mainly because of the high-voltage signal (0.5 mV).

When the electrodes have been applied, the conductive medium, electrode jelly, is injected through the hole at the top of the electrode using a blunt hypodermic needle. For good contact, one can abrade the skin somewhat using the blunt syringe. However, care must be taken not to draw blood, since this will change the electrical properties of the electrode. The resistance of the applied electrodes should be below 5 kOhm. At higher levels, artifacts will be pronounced and create a poor signal-to-noise ratio. The level should be checked with an impedance meter and, where unsatisfactory, more jelly should be applied or the skin abraded a bit more. The EOG, with its high-voltage signal (0.5 mV), can accept higher levels of impedance than the EEG with its 0.01- to 0.05-mV level for alpha and theta activity.

In the next step, the leads from the electrodes are collected in a "pony tail" so as not to interfere with movement of the head. At several points, the tail should be secured with surgical tape to the neck in order to prevent movement artifacts. When applying the tape, it is important to leave enough slack in the leads to allow for normal movement of the head.

The leads are then connected with the receptacle of a small, portable EEG recorder, which is placed in its carrying pouch (on the belt, for example). The recorder, which has been preprogrammed for certain types of recordings, is then started and calibrated. This usually involves several rounds of "look right, look left," "tense your cheek muscles," "close your eyes," and "look straight ahead."

Each calibration exercise should be identified with the event marker of the recorder. These recordings are later used to identify particular patterns in the EEG and to remove artifacts. The subject is then released to go about his normal activities or to wait for the experiment to start. In most cases, the subject will be instructed to press the event marker upon going to sleep and upon rising or when starting and ending a work shift, depending on the design of the study. These events are then used to unequivocally identify the critical segments of the recording. In most studies, the subject will also be supplied with a small diary in which to note times of going to bed, of rising, of starting and ending work, etc. The diary is also used to record sleepiness, stress, and other states every 2 to 3 hours, as well as any important events, such as having to turn off the recorder, etc.

The subject typically returns 24 hours later to have the electrodes and recorder removed. However, the subject often will have received instructions on how to remove the electrodes (acetone on a wad of cotton), so logistics can be simplified and the recorder returned or picked up with efficient timing. The logistics are often important, since the recorders are expensive and time is of essence. At the lab, the contents of the recorder are dumped into a computer for later processing. However, an immediate eyeing of the recording is important to establish whether the data will be usable or whether a new recording will have to be arranged.

21.3 Analysis

The analysis of the recorded data is done on-screen, often with the help of spectral analysis. The scorer divides each 20-sec window into 4-sec segments and scores for presence of sleepiness (alpha activity, theta activity, slow eye movements). At the same time, all artifacts are removed from the material to make a later quantitative spectral analysis possible. One also needs to be prepared to score the recording into the traditional sleep stages (Rechtschaffen and Kales, 1968), since full-blown sleep may well occur. Essentially, sleep is divided into stage 1 (alpha activity has disappeared, slow eye movements occur, theta activity is present), stage 2 (theta activity with sleep spindles, i.e., 14- to 16-Hz activity longer than 0.5 sec, together with K-complexes — fast, high-amplitude waves), stage 3 (delta activity 20 to 50% of the epoch and with an amplitude >75 µV), stage 4 (delta activity >50%), and REM sleep (theta activity, rapid eye movements, sharply reduced EMG activity).

Computerized analysis is mainly some type of spectral analysis, which has replaced simpler techniques using analog or digital filters. Spectral analysis often uses the fast Fourier transform (FFT). The resulting spectrum (amplitude or power) can be integrated or averaged across the frequencies of interest. It divides the EEG activity in frequency bands depending on the amount of activity in each band. The bandwidth depends on the sampling interval, e.g., a 4-sec interval permits a resolution into 1/4 bands. One can present the data as total power, expressed as squared microvolts per hertz. Frequently, one prefers to integrate (sum) power across several bands, such as the alpha (8 to 12 Hz), theta (4 to 8 Hz), 16 to 30 Hz, and delta (0.5 to 4 Hz) intervals. Usually, each band is expressed in percent of the mean across the study or in relation to the value in a controlled situation at the start. Often, for example, the alpha band is dichotomized into a high and a low portion or the alpha and theta band summed. As yet there is no absolute level that is used. It is therefore relatively difficult to compare the results of different conditions.

21.4 Advantages and Disadvantages

The main advantage of the ambulatory EEG (and EOG) recording is that the investigator obtains a second-to-second measure of manifest sleepiness (and sleep). No other method can provide that. The absence of such changes cannot, however, be taken to indicate low (latent) sleepiness, since a variety of factors can mask the appearance of the sleepy EEG. To obtain a dependable measure of latent "deep" sleepiness, one needs some sort of controlled low-stimulation condition introduced (see Section 21.6, Related Methods).

One problem is also that the experimenter usually does not know what the ambulatory subject does every minute around the clock when out of the laboratory. A section of extreme drowsiness or even sleep could be completely innocuous if it occurs during a break, but fatal if it occurs while operating equipment. Thus some sort of frequent diary notation by the subject is invaluable.

An obvious problem is also the abundance of artifacts due to movement or electrical interference from the environment. Movement artifacts can be eliminated through meticulous application of electrodes and leads. Electrical interference can often be filtered out. On an individual level, weak alpha producers may be difficult to analyze.

21.5 Example

Figure 21.1 shows the EEG/EOG pattern of a subject with severe sleepiness performing a test with open eyes. The first seconds show beta activity with relatively normal eye blinks (the triangular forms). This is followed by an increase in alpha activity together with obvious eye closure and slow rolling eye movements. Fifteen seconds later, the dozing-off episode is terminated, beta activity reappears, and eye blinks return.

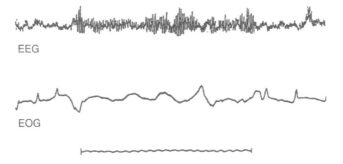

FIGURE 21.1 EEG and EOG recording during an episode of nodding off in a subject performing a test.

21.6   Related Methods

There is no alternative method for measuring continuous fluctuations of alertness, but there are three or four related ones. They are all based on the idea of analyzing the EEG during a well-controlled period of relaxation with open or closed eyes. The advantage is, of course, that external interference is reduced and that latent sleepiness is brought to the surface through the relaxed and soporific situation.

The first is the multiple sleep latency test, which simply involves lying down on a bed in a dark room for 20 min with the instruction to relax and to permit oneself to fall asleep (Carskadon et al., 1986). The test is terminated when the first three epochs of stage 1 sleep have appeared. Normal values in alert individuals are 15 to 20 min, while 5 min constitutes an indicator of pathology or presence of severe sleepiness because of sleep loss, late-night work, or drug intake. The test is usually repeated four to five times during daytime, and the mean is used as a score. Sleep-onset REM may suggest narcolepsy or depression. This test has the disadvantage that subjects with a good ability to relax may obtain spuriously short latencies.

The maintenance-of-wakefulness test uses the same rationale but involves an instruction to stay awake for as long as possible (Mitler et al., 1982). This test is increased in duration to 60 min.

The Karolinska drowsiness test is run for 10 min with open eyes and 5 min with closed eyes (Åkerstedt and Gillberg, 1990). The amount of time with sleepiness indications is used as a measure, and the time with clear alpha or theta activity and/or slow eye movements in the EOG is summed. This test has the advantage that it comes relatively close to sedentary work tasks.

Another interesting method is called the "alpha attenuation test" and is based on a ratio between the EEG alpha power with eyes open and with eyes closed, recorded alternately in 1-min intervals. Since alpha power (spectral content) decreases with sleepiness when the eyes are closed and increases with the eyes open, the ratio will increase with increasing sleepiness (Stampi and Stone, 1995).

21.7   Tools Needed

The primary tool for this type of measurement is a small, lightweight, ambulatory EEG recorder. Older tape-recorder types of systems have been replaced by hard disk or solid-state minicomputers that can record for 24 hours on a large number of channels. Two frequently used instruments are the Embla from Flaga (Iceland) and Vitaport (The Netherlands). The former offers 17 channels; the second uses a modular system that can add on series of amplifiers. Both can record any type of electrophysiological measures, including respiration, light, physical activity, etc. Several other types of equipment are used for sleep analysis in the home, but we are not aware of them being used for true ambulatory wake monitoring.

Apart from the recording equipment, a personal computer is needed for downloading the recorded signals. The software for manual sleep analysis and spectral analysis is supplied by the manufacturers of the recorders, but the user may need to make modifications. We are using the programming package LabView to produce customized analysis packages.

EEG recording also requires the consumables such as electrodes (silver–silver chloride or other types), spirits, and acetone for cleaning application sites.

21.8   Application and Training Time

The application time for the experienced lab specialist is between 15 and 20 min for the minimum number of electrodes. Learning to apply electrodes takes only a few hours, but one needs 10 to 20 applications with follow-up of recording quality before performance is reasonably reliable. Scoring the EEG, on the other hand, takes a few months of training with close monitoring of quality by an experienced scorer. Furthermore, repeated quality checks are necessary.

21.9 Standards and Regulations

There are no standards developed specifically for continuous EEG recording for sleepiness, although there are standards for sleep recording (Rechtschaffen and Kales, 1968). The established literature cited above should be used for guidance.

21.10 Reliability and Validity

Formal reliability has not been established for ambulatory EEG methods of sleepiness, since reliability refers to the precision with which a method measures what it is intended to measure. This definition is rather meaningless, since the methods define exactly what they are intended to measure, and reliability is thus essentially perfect. Another approach is to evaluate the repeatability of the measure, but since the EEG is constantly changing, this approach also lacks meaning.

With respect to validity as an indicator of sleepiness, a number of studies cited above have established a relation between subjective sleepiness and sleepiness-impaired performance. Essentially, purposeful interaction with the environment is simply not feasible when the EEG is dominated by alpha or theta activity with slow eye movements.

References

Aeschbach, D.R., Postolache, T.T., Jackson, M.A., Giesen, H.A., and Wehr, T.A. (1997), Dynamics of the human EEG during prolonged wakefulness: evidence for frequency-specific circadian and homeo-static influences, *Neurosci. Lett.*, 239, 121–124.

Åkerstedt, T. and Gillberg, M. (1990), Subjective and objective sleepiness in the active individual, *Int. J. Neurosci.*, 52, 29–37.

Bjerner, B. (1949), Alpha depression and lowered pulse rate during delayed actions in a serial reaction test, *Acta Physiol. Scand.*, 19, 1–93.

Carskadon, M.A., Dement, W.C., Mitler, M.M., Roth, T., Westbrook, P.R., and Keenan, S. (1986), Guidelines for the multiple sleep latency test (MSLT): a standard measure of sleepiness, *Sleep*, 9, 519–524.

Daniel, R.S. (1967), Alpha and theta EEG in vigilance, *Perceptual Motor Skills*, 25, 697–703.

Dement, W.C. and Carskadon, M.A. (1982), Current perspectives on daytime sleepiness: the issues, *Sleep*, 5, 56–66.

Dinges, D.F. (1995), An overview of sleepiness and accidents, *J. Sleep Res.*, 4, 4–14.

Gundel, A., Drescher, J., Maass, H., Samel, A., and Vejvoda, M. (1995), Sleepiness of civil airline pilots during two consecutive night flights of extended duration, *Biol. Psychol.*, 40, 131–141.

Kecklund, G. and Åkerstedt, T. (1993), Sleepiness in long distance truck driving: an ambulatory EEG study of night driving, *Ergonomics*, 36, 1007–1017.

Kryger, M.H., Roth, T., and Dement, C.W. (2000), *Principles and Practice of Sleep Medicine*, W.B. Saunders, Philadelphia, 1336 pp.

Kuhlo, W. and Lehman, D. (1964), Das Einschlaferleben und seine neurophysiologischen Korrelate, *Arch. Psychiat. Z. Ges Neurol.*, 205, 687–716.

Landström, U., Englund, K., Nordström, B., and Åström, A. (1998), Laboratory studies of a sound system that maintains wakefulness, *Perceptual Motor Skills*, 86, 147–161.

Lecret, F. and Pottier, M. (1971), La vigilance, facteur de sécurité dans la conduite automobile, *Travail Humain*, 34, 51–68.

Loomis, A.L., Harvey, E.N., and Hobart, G. (1935a), Further observations on the potential rhythms of the cerebral cortex during sleep, *Science*, 82, 198–200.

Loomis, A.L., Harvey, E.N., and Hobart, G. (1935b), Potential rhythms of the cerebral cortex during sleep, *Science*, 81, 597–598.

Loomis, A.L., Harvey, E.N., and Hobart, G. (1937), Cerebral states during sleep, as studied by human brain potentials, *J. Exp. Psychol.*, 21, 127–144.

Mitler, M., Gujavarty, K., and Broman, C. (1982), Maintenance of wakefulness test: a polysomnographic technique for evaluating treatment efficacy in patients with excessive somnolence, *Electroencephalogr. Clin. Neurophysiol.*, 1982, 658–661.

Mitler, M.M., Miller, J.C., Lipsitz, J.J., Walsh, J.K., and Wylie, C.D. (1997), The sleep of long-haul truck drivers, *N. Engl. J. Med.*, 337, 755–761.

O'Hanlon, J.F. and Beatty, J. (1977), Concurrence of electroencephalographic and performance changes during a simulated radar watch and some implications for the arousal theory of vigilance, in *Vigilance*, Mackie, R.R., Ed., Plenum Press, New York, pp. 189–202.

O'Hanlon, J.F. and Kelley, G.R. (1977), Comparison of performance and physiological changes between drivers who perform well and poorly during prolonged vehicular operation, in *Vigilance*, Mackie, R.R., Ed., Plenum Press, New York, pp. 87–111.

Rechtschaffen, A. and Kales, A. (1968), A Manual of Standardized Terminology, Techniques and Scoring System for Sleep Stages of Human Subjects, U.S. Department of Health, Education and Welfare, Public Health Service, Bethesda, MD.

Reyner, L.A. and Horne, J.A. (1997), Suppression of sleepiness in drivers: combination of caffeine with a short nap, *Psychophysiology*, 34, 721–725.

Reyner, L.A. and Horne, J.A. (1998), Evaluation of "in-car" countermeasures to sleepiness: cold air and radio, *Sleep*, 21, 46–50.

Stampi, C. and Stone, P. (1995), A new quantitative method for assessing sleepiness: the alpha attention test, *Work Stress*, 9, 368–376.

Torsvall, L. and Åkerstedt, T. (1987), Sleepiness on the job: continuously measured EEG changes in train drivers, *Electroencephalogr. Clin. Neurophysiol.*, 66, 502–511.

Torsvall, L. and Åkerstedt, T. (1988), Extreme sleepiness: quantification of EOG and spectral EEG parameters, *Int. J. Neurosci.*, 38, 435–441.

Torsvall, L., Åkerstedt, T., Gillander, K., and Knutsson, A. (1989), Sleep on the night shift: 24-hour EEG monitoring of spontaneous sleep/wake behavior, *Psychophysiology*, 26, 352–358..

22

Assessing Brain Function and Mental Chronometry with Event-Related Potentials (ERP)

Arthur F. Kramer
University of Illinois

Artem Belopolsky
University of Illinois

22.1 Background and Applications

The event-related potential (ERP) is a transient series of voltage oscillations in the brain that can be recorded from the scalp in response to discrete stimuli and responses. Specific ERP components, usually defined in terms of polarity and minimum latency with respect to a discrete stimulus or response, have been found to reflect a number of distinct perceptual, cognitive, and motor processes, thereby proving useful in decomposing the processing requirements of complex tasks (Fabiani et al., 2000).

ERPs have a long history of being used to examine aspects of cognition that relate to a number of issues of relevance to human factors and ergonomics, including vigilance, mental workload, fatigue, adaptive aiding, stressor effects on cognition, and automation. (See Kramer and Weber [2000] and Byrne and Parasuraman [1996] for reviews of this literature.) For example, Kramer et al. (1983) found that increasing difficulty on a primary task resulted in decreases in the amplitude of a late positive component of the ERP, the P300, elicited by a secondary task. In a later study, Sirevaag et al. (1987) found reciprocity in the amplitude of P300s elicited by primary and secondary tasks. As the primary task became more difficult, P300s elicited by primary task events increased in amplitude concurrently with decreases in the amplitude of P300s elicited by secondary task events. This pattern of P300 reciprocity is consistent with resource tradeoffs predicted by multitask processing models (Wickens and Hollands, 2000). Other studies have employed ERPs to examine changes in performance as a function of time on task (Humphrey et al., 1994), in on-line adaptive communication algorithms (Farwell and Donchin, 1988), as a means to examine drug effects on human cognition (Ilan and Polich, 2001), as well as other human factors issues.

22.2 Procedure

The recording and analysis of ERPs require a number of steps with branching points (i.e., choices with regard to methods of recording and analyses) at many of the steps.

22.2.1 Step 1: Design of the Experimental Paradigm

Given that many ERP components have a relatively small signal-to-noise ratio, ERP paradigms often need to be designed such that a number of trials can be recorded in each of the conditions of interest. Exceptions to the need for multiple trials include the later ERP components such as the P300 and slow wave, which can be resolved with few trials (Humphrey and Kramer, 1994). There is also a need to isolate the electrodes and recording device from other electrical sources. Finally, ERPs are most useful as a means to decompose aspects of cognition in well-designed paradigms in which perceptual, cognitive, and motor processes can be systematically manipulated.

22.2.2 Step 2: Preparing the Subject for ERP Recording

This step involves the application of electrodes to the subject's or operator's scalp, the selection of appropriate gains and filters for the amplifiers, and ensuring that "clean" signals (i.e., electrical signals uncontaminated by electrical line noise, movement artifact, drift, etc.) are being recorded. Monitoring of the electrical signals during the conduct of the experiment is also prudent to ensure data quality.

22.2.3 Step 3: Preparation of the ERP Data for Analysis

This step often entails additional filtering based on the characteristics of the ERP components that one is interested in as well as screening for eye movements, which can contaminate the electrical activity recorded from the brain. Trials on which eye movements occur can either be rejected or adjusted to reduce the influence of electrical activity associated with saccades from the activity recorded from the brain. Techniques are applied to enhance the signal-to-noise ratio of the ERP components. In many cases, this involves signal averaging, although other approaches such as spatial filtering and autoregressive modeling have also been employed with some success (Gratton, 2000; Gratton et al., 1988).

22.2.4 Step 4. Component Definition and Pattern Recognition

Prior to statistical analysis, it is important to determine how ERP components are to be resolved and measured. Perhaps the simplest approach is to examine components in restricted time windows and define them in terms of measures of base-to-peak amplitude, peak-to-peak amplitude, or area. However, while straightforward, this approach makes it difficult to deconvolve temporally overlapping components. This can be accomplished with techniques such as principal components analysis, Vector filtering, and wavelet analysis (Effern et al., 2000; van Boxtel, 1998).

22.2.5 Step 5. Data Analysis

ERP components can be analyzed with a variety of traditional univariate and multivariate procedures in an effort to examine the sensitivity of specific components to experimental manipulation and system parameters. Journals such as *Psychophysiology, Biological Psychology,* and *Electroencephalography and Clinical Neurophysiology* can be consulted for examples of the application of particular statistical methods of data analysis.

22.3   Advantages

Perhaps the primary advantage of ERP recording is the wealth of knowledge that we currently possess concerning the functional significance of different ERP components. This knowledge can be used to decompose the information-processing requirements of different tasks and system configurations. That is, ERPs can and have been used to decompose, in a temporally precise manner, the information-processing activities that transpire between the time stimuli impinge on sensory receptors until an individual produces an action, whether it is an eye movement, a vocalization, or the skilled movement of the hands or feet. (See Kramer [1991] and Rugg and Coles [1995] for a summary of this literature.)

Another important advantage of ERPs is that, in many cases (i.e., for a number of ERP components), the brain regions from which a component is generated are known. This knowledge enables the researcher to capitalize on the extensive neuropsychological literature that maps particular cognitive functions to neuroanatomical circuits. For example, on the basis of ERP and other neouroimaging data, Just et al. (2003) have recently proposed a resource-based model of mental workload that predicts multitask performance deficits on the basis of whether distinct or overlapping brain regions will be used in the performance of particular tasks.

ERPs can also be obtained in the absence of operator actions and performance. This is especially useful in vigilance situations in which few operator actions are required as well as in highly automated systems in which operators serve primarily as system monitors or supervisors.

22.4   Disadvantages

Given problems with motion artifacts and electrical noise, it is preferable to record ERPs from individuals who are not ambulatory. This clearly reduces the number of situations in which ERPs can be utilized. Nevertheless, there are still ample professions for which ERP recording is possible (e.g., pilot, driver, office worker, process control operator, etc.).

ERPs require a discrete stimulus or response. Therefore, situations in which it is either difficult to record ERPs from stimuli that already appear in a task or system (e.g., the appearance of a new track on an air-traffic-control display) or it is unfeasible to introduce signals from which ERPs can be recorded (e.g., in a secondary task) are not easily amenable to ERP recording.

Finally, ERP recording, analysis and interpretation require relatively substantial training both in terms of the recording procedures as well as in the relevant psychological and physiological literatures.

22.5   Example

Figure 22.1 presents a set of average ERPs recorded from eight different scalp sites from ten young adult volunteers in an experiment conducted to examine the attentional requirements of visual marking. Visual marking is a phenomenon that entails focusing attention on one set of stimuli while ignoring other physically interspersed stimuli. The ERPs shown in Figure 22.1 were recorded from the presentation of a set of stimuli that subjects were instructed to ignore (gap condition), memorize for later report (memory condition), or determine whether one of the stimuli changed (neutral condition). The ERPs presented in Figure 22.1 were averaged across the ten volunteers after having been filtered and screened for potential artifacts. Several different ERP components can be identified in the waveforms: a P100 at approximately 150 ms poststimulus (the stimulus is indicated by the vertical line at 0 ms on the *x*-axis), an N100 at approximately 200 ms poststimulus, and a later positive-going waveform called the P300. As can be seen at several different electrode sites (e.g., P3 and O1), the three different conditions are clearly discriminated

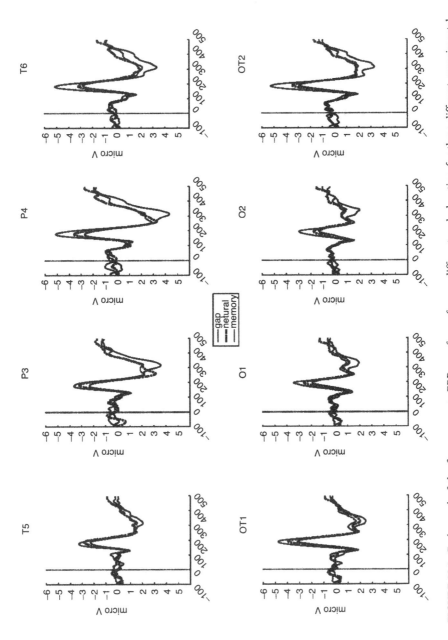

FIGURE 22.1 Each panel of the figure presents ERP waveforms from a different scalp location for three different experimental conditions. The x-axis indicates the time, relative to the presentation of a search display (at 0 msec), of different waveform components. The y-axis provides a measure of the amplitude of ERP components in microvolts.

by the N100 component. Given that N100 reflects the allocation of spatial attention, these data suggest that attention is indeed distributed differently during the three conditions.

22.6   Related Methods

The most closely related method is electroencephalographic recording (EEG). ERPs are the transient electrical activity, elicited by specific stimuli or responses, within the EEG signal. Unlike ERPs however, EEG activity is usually analyzed in the frequency rather than the time domain and tends not to be related to specific stimuli and/or responses (although it can be). Like ERPs, EEG can also be recorded in laboratory, simulator, and field environments.

Other techniques that enable an assessment of central nervous system activity include positron emission tomography (PET) and functional magnetic resonance imaging (fMRI). PET and fMRI are used to image functional activity in the brain, often as an individual is performing a specific task. Both of these techniques involve inferring changes in neuronal activity from changes in blood flow or metabolic activity in the brain (Reiman et al., 2000). In PET, cerebral blood flow and metabolic activity are measured on the basis of clearance of radionuclides from cortical tissues. These radionuclides, which are either inhaled or injected, decay by the emission of positrons that combine with electrons to produce gamma rays, which are detected by a series of sensors placed around the head. Each PET image, which is acquired over an interval of anywhere from 1 to 45 min, depending on the nature of the radionuclide employed in a study, represents all of the brain activity during the integration period. These PET images are then coregistered with structural scans, often obtained from MRIs, to indicate the location of the functional activity. fMRI is similar to PET in that it provides a map of functional activity of the brain. However, fMRI activity can be obtained more quickly (within a few seconds), does not depend on the inhalation or injection of radioactive isotopes, and can be collected in the same system as the structural information. The blood-oxygen-level-dependent technique (BOLD) of fMRI uses the perturbation of local magnetic fields due to changes in the oxygen content of blood during increased blood flow to image functional brain activity (Belliveau et al., 1991; Ogawa and Lee, 1990). While both PET and fMRI provide excellent spatial resolution for brain activity related to different aspects of cognition, neither technique can be used outside of the laboratory.

22.7   Standards and Regulations

There are no regulations that govern the collection and analysis of ERP data. However, a number of guidelines have been published concerning ERP data recording, analysis, and presentation (see Picton et al., 2000).

22.8   Approximate Training and Application Times

The time required to learn how to record and analyze ERP data varies depending on whether you want to learn the basics (a couple of months) or instead become knowledgeable about the basis of ERP signals (an advanced degree). Application times for the ERP electrodes also vary from approximately 15 min for a few electrode sites to up to 45 min for a large array of electrodes.

22.9   Reliability and Validity

The validity of different ERP components as metrics of specific cognitive constructs (e.g., attention, aspects of language processing, aspects of memory, etc.) has been convincingly demonstrated through a series of converging operations (Kramer, 1991; Fabiani et al., 2000).

Reliability of ERP components has been established through extensive replications of a variety of effects, which have involved the mapping of ERP component latency and amplitude changes to specific experimental manipulations. More formal reliability analyses have also been performed with a subset of ERP components. For example, Fabiani et al. (1987) evaluated the reliability of P300 amplitude and latency measures in a series of simple oddball studies in which individuals were instructed to respond to a subset of stimuli. The split-half reliability was 0.92 for P300 amplitude and 0.83 for P300 latency. The test–retest reliability assessed over a period of several days was 0.83 for P300 amplitude and 0.63 for P300 latency.

22.10 Tools Needed

ERP recording requires sensors (electrodes); amplifiers; a computer system for data collection, analysis, and stimulus presentation; as well as software for data collection and analysis. There are many complete systems available on the market for EEG/ERP recording. However, many of these systems are designed for clinical use and therefore do not provide the flexibility often needed for research. A few examples of systems that are designed for ERP research include Neuroscan (http://www.neuro.com/medsys/), BIOPAC (http://www.biopac.com/newsletter/september/erp.htm), Biologic Systems (http://www.blsc.com/neurology/explorerep/index.html), and Nicolet (http://www.nicoletbiomedical.com/tr_intuition.shtml).

References

Belliveau, J.W., Kennedy, D.N., Mckinstry, R.C., Buchbinder, B.R., Weisskoff, R.M., Cohen, M.S., Vevea, J.M., Brady, T.J., and Rosen, B.R. (1991), Functional mapping of the human visual cortex by magnetic resonance imaging, *Science*, 254, 716–719.

Byrne, E.A. and Parasuraman, R. (1996), Psychophysiology and adaptive automation, *Biol. Psychol.*, 42, 249–268.

Effern, A., Lehnertz, K., Grunwald, T., Fernandez, G., David, P., and Elger, C.E. (2000), Time adaptive denoising of single trial event-related potentials in the wavelet domain, *Psychophysiology*, 37, 859–865.

Fabiani, M., Gratton, G., and Coles, M.G.H. (2000), Event-related brain potentials, in *Handbook of Psychophysiology*, Cacioppo, J., Tassinary, L., and Bertson, G., Eds., Cambridge University Press, New York, pp. 53–84.

Fabiani, M., Gratton, G., Karis, D., and Donchin, E. (1987), Definition, identification, and reliability of measurement of the P300 component of the event-related brain potential, in *Advances in Psychophysiology*, Ackles, P., Ed., JAI Press, New York, pp. 1–78.

Farewell, L.A. and Donchin, E. (1988), Talking off the top of your head: toward a mental prosthesis utilizing event-related brain potentials, *Electroencephalogr. Clin. Neurophysiol.*, 70, 510–523.

Gratton, G. (2000), Biosignal processing, in *Handbook of Psychophysiology*, Cacioppo, J., Tassinary, L., and Bertson, G., Eds., Cambridge University Press, New York, pp. 900–923.

Gratton, G., Coles, M.G.H., and Donchin, E. (1988), A procedure for using multi-electrode information in the analysis of components of the event-related potential, *Psychophysiology*, 26, 222–232.

Humphrey, D. and Kramer, A.F. (1994), Towards a psychophysiological assessment of dynamic changes in mental workload, *Hum. Factors*, 36, 3–26.

Humphrey, D., Kramer, A.F., and Stanny, R. (1994), Influence of extended-wakefulness on automatic and non-automatic processing, *Hum. Factors*, 36, 652–669.

Ilan, A.B. and Polich, J. (2001), Tobacco smoking and event-related brain potentials in a Stroop task, *Int. J. Psychophysiol.*, 40, 109–118.

Just, M.A., Carpenter, P.A., and Miyake, A. (2003), Neuroindices of cognitive workload: neuroimaging, pupillometric, and event-related potential studies of brain work, *Theor. Issues Ergonomic Sci.*, 4, 56–88.

Kramer, A.F. (1991), Physiological measures of mental workload: a review of recent progress, in *Multiple Task Performance*, Damos, D., Ed., Taylor & Francis, London, pp. 279–328.

Kramer, A.F. and Weber, T. (2000), Application of psychophysiology to human factors, in *Handbook of Psychophysiology*, Cacioppo, J., Tassinary, L., and Bertson, G., Eds., Cambridge University Press, New York, pp. 794–814.

Kramer, A.F., Wickens, C.D., and Donchin, E. (1983), An analysis of the processing demands of a complex perceptual-motor task, *Hum. Factors*, 25, 597–622.

Ogawa, S. and Lee, T.M. (1990), Magnetic resonance imaging of blood vessels at high fields: *in vivo* and *in vitro* measurements and image simulation, *Magn. Resonance Med.*, 16, 9–18.

Picton, T.W., Bentin, S., Berg, P., Donchin, E., Hillyard, S.A., Johnson, R., Miller, G.A., Ritter, W., Ruchkin, D.S., Rugg, M.D., and Taylor, M.J. (2000), Guidelines for using human event-related potentials to study cognition: recording standards and publication criteria, *Psychophysiology*, 37, 127–152.

Reiman, E.M., Lane, D., Van Petten, C., and Bandetinni, P.A. (2000), Positron emission tomography and functional magnetic resonance imaging, in *Handbook of Psychophysiology*, 2nd ed., Cacioppo, J.T., Tassinary, L.G., and Berntson, G.G., Eds., Cambridge University Press, New York, pp. 85–118.

Rugg, M. and Coles, M.G.H. (1995), *Electrophysiology of Mind: Event-related Brain Potentials and Cognition*, Oxford University Press, New York.

Sirevaag, E., Kramer, A.F., Coles, M.G.H., and Donchin, E. (1987), Resource reciprocity: an event-related brain potentials analysis, *Acta Psychol.*, 70, 77–90.

Van Boxtel, G.J.M. (1998), Computational and statistical methods for analyzing event-related potential data, *Behav. Res. Methods, Instrum., and Comput.*, 30, 87–102.

Wickens, C.D. and Hollands, J.G. (2000), *Engineering Psychology and Human Performance*, 3rd ed., Prentice-Hall, Upper Saddle River, NJ.

23

MEG and fMRI

Hermann Hinrichs
University of Magdeburg

23.1 Background and Applications

The neural activity of the brain generates electrical currents that, in accord with general physical laws, can be monitored outside the head by means of their resulting electrical and magnetic fields. The corresponding recordings are known as the electroencephalogram (EEG) and magnetoencephalogram (MEG), respectively. Whereas EEG recordings have been used for many years to evaluate clinical conditions such as epilepsy, the technologies that permit MEG recordings to be made are relatively recent. Due to the intrinsic noise present in the sensors and amplifiers used to obtain MEG records, only the relatively strong fields produced by the simultaneous firing of large ensembles of similarly oriented neurons can be detected.

In addition, the increased metabolic activity produced by local regions of increased neuronal activity leads to local increases in the blood supply to these neurons. The increase in oxygenated hemoglobin can be detected by appropriate magnetic resonance imaging (MRI) techniques. Because these local modulations in blood supply are a consequence of the execution of cerebral functions, this MRI variant is called functional MRI (fMRI).

In recent years, these techniques have gained increasing importance in neuroscientific applications such as the analysis of the neural substrate of specific cognitive processes. Very recently, there have also been attempts to employ these methods in the diagnosis of certain neurological diseases. MEG and fMRI are to some extent complementary. While MEG can monitor neural functions with excellent temporal resolution (in the range of msec) but only poor spatial resolution, fMRI can localizes activations with substantial spatial resolution (a few mm) but only coarse temporal resolution (a few seconds). By using the two methods in combination, one can therefore perform precise temporal–spatial analyses of higher neural functions.

Because the usefulness of MEG for medical diagnosis is still under debate, the number of MEG laboratories is still limited (some 50 at the time of writing), and MEG applications are still largely restricted to pure scientific investigations. FMRI, on the other hand, is being utilized in many more than 100 clinical and research institutes worldwide as well as in applied studies.

Both fMRI and MEG are best suited for the analysis of cognitive and other brain functions that are restricted to limited brain areas. FMRI is equally sensitive throughout all regions of the brain, with the exception of some regions close to caverns and ventricles. In contrast, MEG is primarily sensitive to neural activity in superficial brain regions that are oriented tangentially with respect to the skull. Processes in the center of the brain and/or with radial orientation are almost undetectable by the MEG.

Some neuroscientific research areas amenable to investigation with MEG and fMRI are:

- Movement-related brain activation
- Memory processes (encoding and retrieval)
- Visual perception, attention, and selection
- Auditory perception, attention, and selection
- Language production and processing
- Perception of music
- Learning and brain plasticity with respect to cognitive functions

Some clinical applications of these techniques include the localization of functionally important brain areas in the context of neurosurgery, the localization of epileptic foci based on specific areas of neural spiking (with MEG supplementing EEG for this purpose), and the estimation of the impact of certain brain lesions on higher neural functions.

In some investigations, there will be no canonical basis for preferring one of the two methods discussed here, and the choice will depend on the system that is available. However, if an analysis of temporal responses is desired, the MEG will be preferred. If the desired emphasis is on the spatial specificity of responses, fMRI will be preferred. As previously indicated, maximal information will be derived when the methods can be combined. It should be noted, though, that some subjects or patients cannot be examined by either of the methods due to technical or safety considerations (for instance, the presence of ferromagnetic inserts).

23.2 Basic Mechanisms

23.2.1 MEG

The brain's magnetic fields are approximately 10^{-8} smaller than the earth's magnetic field. Conventional magnetic measuring devices (such as fluxgates) cannot be used to monitor field amplitudes this low. Instead, superconducting quantum interference devices (SQUIDs) must be employed. These sensors exploit the quantum mechanical "Josephson-effect" (Josephson, 1962), which can be observed when certain materials are in a superconducting state. In order to achieve this state, these devices need to be maintained at a temperature of 4 K (−269°C) by liquid helium. Modern MEG systems record the signal from as many as 150 to 300 SQUIDs spread equally over the head surface (with no physical contact needed), thereby providing an intersensor distance of approximately 2 to 3 cm. In order to eliminate external magnetic noise (for instance, noise caused by moving iron objects like cars), MEG experiments must be conducted in a dedicated room that is magnetically shielded by multiple layers of mu-metal, an iron alloy especially developed for magnetic shielding. This shielding can have a total weight of several tons.

To further suppress external noise, special variants of the MEG sensors have been developed. These new variants are known as gradiometers, in contrast to the original magnetometers (for details see Hämäläinen et al., 1993; Lounasmaa et al., 1996).

23.2.2 fMRI

When the spin tilt of protons is aligned by a strong magnetic field, and that spin tilt is momentarily perturbed by a brief electromagnetic pulse, the protons will emit a burst of radio-frequency (RF) energy as they return to their initial aligned state. MRI, in general, relies on an analysis of the RF energy produced during the return or relaxation period. The characteristics of the emitted RF signal will depend on the character and sequence of the perturbing pulses and on the intensity of the aligning field. If the strength of the aligning field is given a spatial gradient (by the superimposition of a separate gradient field), the emitted signals will have an RF signature that is dependent on their location in the gradient. Since the overall strength of an emitted signal is dependent upon the density of the protons generating that signal, the strength of the emitted signal bearing a particular RF signature will allow one to determine the proton density (and thus tissue characteristics) at the spatial location that is the source of that signal. (For details, see Jezzard et al. [2001] or Turner et al. [1998].) In this manner, a map of tissue densities throughout the brain can be derived. In the case of fMRI, a relaxation signal time constant (which is designated $T2^*$) is of special interest. This time constant characterizes the exponential decay of the RF burst emitted by the excited protons (see above). $T2^*$ values vary according to the oxygenation level of the venous blood at particular locations. This oxygenation level is in turn dependent on the level of neural activity at a location, because there is an overcompensating increase in the blood supply in regions with enhanced metabolism. The change in the RF signal that reflects this change in oxygenated hemoglobin is known as the blood-oxygenation-level-dependent (BOLD) response. BOLD responses onset about 4 sec after an increase in metabolic activity and have a total duration of about 15 sec. FMRI maps the spatial distribution of these BOLD responses.

To generate the huge static magnetic field required for MRI and fMRI (typically 1.5 tesla, although systems up to 7 tesla have been built), the system requires correspondingly strong currents. These can only be provided by superconduction. Therefore, MRI systems require permanent cooling using liquid helium. In addition, fast and powerful amplifiers and receivers are required to set up the magnetic gradients, generate short RF pulses with arbitrary shapes, and record the emitted RF echoes. RF shielding of the MRI room is mandatory to prevent interference. See Figure 23.1 for a typical MRI scanner.

MRI and fMRI images are organized into slices of a few millimeters thickness, each made up of a rectangular matrix of voxels (i.e., volume elements), which represent the signal from a small rectangular volume of tissue. The number of voxels per slice is typically 256×256 to 1024×1024 for structural scans, and 64×64 to 128×128 for functional scans. The geometric region covered by the slices is called the field of view (FOV).

23.3 Procedure

23.3.1 MEG

At the beginning of the measurement period, a helmetlike gantry holding the sensors is placed over the subject's head. Coils fixed on the head provide weak magnetic sources at known anatomical sites. These sources are used to establish an anatomical coordinate system, which is needed to assign the measured magnetic fields to appropriate anatomical structures. After this coordinate system has been established, the MEG is recorded continuously or in a stimulus-locked/epoch-based manner, depending on the experimental design. The evoked magnetic fields are sampled at a rate of some hundred Hertz.

MRI device   MEG device

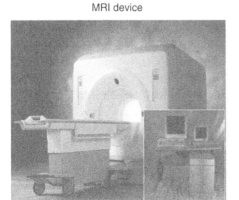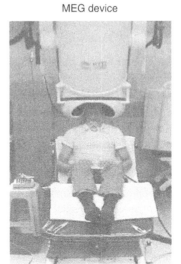

FIGURE 23.1 Typical examples of MRI scanner and MEG device. The computer system attached to the MRI image indicates the console system needed to control the imaging process, located outside the shielded MRI room.

23.3.2 fMRI

The subjects recline on a movable gantry, which is shifted into the bore of the magnet. Usually, fMRI experiments require at least two steps: (1) a structural scan to acquire an image of the anatomical structure at high resolution (taking some 10 min) and (2) repeated (up to some hundred) functional scans that acquire the task-specific BOLD image at a low structural resolution. Depending on the type of experiment, the entire process can take up to 1 hour. The BOLD signals are typically acquired with a sampling frequency of 0.1 to 1 Hz. During subsequent processing, the sequence of functional images is statistically analyzed (see below) and combined with the structural scan (a process called coregistration) to allow for a precise anatomical localization of the functional results.

23.4 Experimental Setup

During data recording, a sequence of visual or auditory or other sensory stimuli is presented to the subject, who has to process them according to a predefined task. For example, a sequence of words may have to be encoded, or geometric figures classified, or there may be a sequence of tactile stimuli that require no overt response. Often, the same or similar stimuli are presented repeatedly with different task requirements. Generally, differences between the BOLD responses in the different experimental conditions are evaluated to determine what brain regions are specifically activated by a particular task or a particular type of sensory input. In order to access the generality of the obtained results, several subjects (typically 10 to 20) are examined using the same experimental manipulation.

In clinical applications, certain phasic events like epileptic MEG or EEG signal patterns can be used as triggers for the data-extraction process. Signal-averaging techniques can then be used to extract specific characteristics of the signals or images associated with those events.

23.4.1 Data Analysis

23.4.1.1 MEG

Since the strength of the task or stimulus-specific–response waveforms are normally at least one order of magnitude smaller than the ongoing background activity, event-locked epochs are initially averaged separately for each subject, each channel (sensor), and each task or condition. Segments containing artifacts (due to subject movements, eye movements, or other sources of distortion) are rejected or corrected (see, for example, Nolte and Curio, 1999). Next, the grand average waveforms (waveforms averaged over all subjects) are scanned for components, i.e., temporal peaks and troughs. These typically have a duration of 50 to 100 msec. Time and amplitude measures are determined for each component, task, and subject and then subjected to a statistical analysis to ascertain if there are any significant differences between the different tasks or conditions. Finally, significantly differing components are localized with respect to their intracerebral origin by the application of source analysis techniques. Equivalent current dipoles have been introduced in this context as a model for depicting confined sources of neural activity. Unfortunately, this localization technique is inexact. For further information, see Dale and Sereno, 1993; Scherg et al., 1999; and Fuchs et al., 1999.

Significant sources can be anatomically localized if the sensor coordinates can be described in terms of an MRI coordinate system. Various coregistration techniques have been proposed for this purpose.

Various approaches for analyzing continuous MEG recordings have been proposed. Among these are the estimation of nonlinear parameters using algorithms from chaos theory (see for instance Kowalik and Elbert, 1994), coherence measures to identify synchronized oscillations (Singh et al., 2002), and Fourier spectral analyses (Hari et al., 1997). The details of these methods are beyond the scope of this chapter.

23.4.1.2 fMRI

The continuously recorded fMRI/BOLD data are sorted according to the different stimulus or task conditions during acquisition. As with MEG, the resulting data sets are compared using standard statistical comparisons. Potential movement-related errors and other distortions are corrected by processing algorithms. The resulting functional images show significant differences in activation levels at the level of the individual voxels. In contrast to MEG, fMRI usually allows for single-subject statistical analyses because of the better signal-to-noise ratio obtainable in fMRI experiments. Comprehensive evaluation packages are available; one popular version is the statistical parametric mapping (SPM) package, which is freely provided by the imaging group of the Welcome institute in London, U.K. There are also advanced applications techniques to deconvolve temporally overlapping BOLD responses (see for instance Hinrichs et al., 2000).

23.4.2 Combined Analysis

There is still no canonical method to combine MEG and fMRI analysis. One approach is to run both analyses individually and then attribute sources found in the MEG to the nearest fMRI activation peaks. Alternatively, fMRI results can be included in the MEG source analysis by emphasizing those brain areas where significant fMRI activations were found as potential source locations.

23.5 Example

In a recent study, we conducted an experiment focusing on visual spatial attention processes (Noesselt et al., 2002). Subjects were presented sequences of short bilateral stimuli including a central fixation cross. In repeated runs, they had to direct their attention to either side of the screen and detect occasionally occurring target patterns. In separate sessions, fMRI images and MEG signals were acquired during this task. As seen in Figure 23.2, the fMRI showed a clear attention-specific activation in various areas of the occipital cortex, including V1, the primary visual cortical receiving area. MEG records revealed a strong attention-related component that could also be attributed to V1. By looking at the timing of the MEG signals, it could be shown that the V1 activation in this case did not reflect a direct modulation by the sensory input but, rather, reflected a delayed feedback to V1 from higher visual cortical areas.

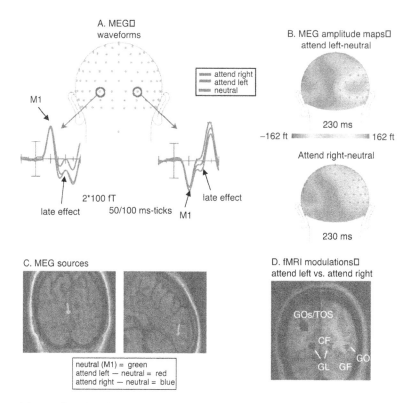

FIGURE 23.2 (A) Stimulus-related average MEG waveforms at two of the 148 MEG sensor sites. Separate averages are shown for the attend-left, attend-right, and neutral conditions. (B) Topographical distribution of the difference amplitudes at latency 230 msec. (C) Best-fit equivalent dipole sources for these difference amplitudes. In addition, the dipole fitted for the attend left condition at 80 ms is shown. It reflects the primary/sensory response to the stimulus (thus indicating visual area V1). (D) Attention-related fMRI modulations. Pixels shown in red (figure presented here in black and white) had significantly larger BOLD signals during attend-left than attend-right conditions, and pixels shown in blue had the reverse. Significance level was $p < 0.05$, corrected for spatial extent. Abbreviations: calcarine fissure (CF = visual area V1), lingual gyrus (GL), fusiform gyrus (GF), inferior occipital gyrus (GOi), superior occipital gyrus/transverse occipital gyrus (Gos/TOS). Note the activation in the calcarine fissure (CF), which corresponds to the dipolar source found in the MEG (Figure 23.2C). (Adapted from Noesselt, T. et al. [2002], *Neuron*, 35, 575–687.)

23.6 Standards

With respect to system calibration, phantoms with known physical properties are available for both methods. On the MEG side, technical current dipoles contained in a confined saline solution are used to check the precision of the sensors. For fMRI, similar volumes filled with saline solution plus some chemical additives allow to control for potential spatial distortions as well as intensity inhomogeneities.

Widely accepted normative databases are still missing for both modalities.

23.7 Required Training

Routine predefined protocols can be performed by technicians after a training period of a few weeks. For more sophisticated applications, at least one trained full-time engineer or physicist should be available (this training requires approximately 6 months). In case of clinical applications, the cooperation of a dedicated physician is required for an appropriate interpretation of the data. In neuroscientific applications, it is important that investigators have a sound background in experimental design as well as a neurophysiological education. The specific training that is needed will depend to some extent on the specifics of the investigation to be conducted.

23.8 Testing Times

The time required to conduct an experimental run will depend on the nature of the investigation, but this time is typically about 1.5 hours, with another 15 min needed for preparation of the subject.

23.9 Reliability and Validity

23.9.1 MEG

Even with thorough shielding and postprocessing to reduce residual artifacts, artifacts cannot be avoided in routine MEG recordings. In the raw signals, these distortions will normally easily be recognized by an experienced rater. The potential impact of artifacts will depend on the signal-to-noise ratios. To avoid misinterpretation of average MEG waveforms, the following measures are usually taken:

1. Careful monitoring during recording
2. Automatic artifact detection and rejection
3. A check of the obtained waveforms for reasonable morphology and topography

In the case of subsequent source analysis, the investigator should know that while this method can provide reasonable solutions when carefully applied, it is far from foolproof and may in some cases yield seriously misleading results. In summary, the reliability and validity of the MEG method mainly relies on the user's experience.

23.9.2 fMRI

Given that the MRI scanner is under continuous maintenance, the raw images should be acceptable in most cases. However, artifacts due to discontinuities in magnetic susceptibility (which might occur in the vicinity of ventricles, intracranial cavities, and internal or external nonbiological parts) can lead to local structural distortions and signal loss. These effects are easily detected but not readily corrected. Errors can also be introduced during statistical analyses. If nonexperts run these analyses, invalid results may be obtained due to inappropriate setting of the various parameters that guide the evaluation. But even if the formal fMRI handling is correct, erroneous conclusions may be drawn from the data obtained using inappropriate experimental designs. In summary, the reliability and validity of the fMRI method also mainly relies on the user's experience.

23.10 Costs

23.10.1 Investment

The investment required for an MEG system will be about $2 million. A 1.5-tesla MRI scanner is available for about $1.5 million. Of course, actual costs will depend on the details of the system configuration (for MEG, the number of channels/sensors, analysis software, etc.; for fMRI, the type of gradient field, number of RF channels, etc.).

23.10.2 Running Costs

23.10.2.1 MEG

These costs are mainly determined by the consumption of liquid helium. Assuming approximately 100 l/week, the total amount comes to approximately $30,000/year. Energy consumption (electronics and air conditioning) add up to an additional $2,000, depending on the type of MEG system and the local energy rates.

23.10.2.2 fMRI

The total costs/year are composed of the fractions for electrical energy, cooling (electronics and room temperature), water, and liquid helium. Overall, about $30,000/year is needed, but the actual cost will depend substantially on the kind and frequency of usage, type of system, and on the local rates for water, energy, etc.

23.10.3 Maintenance

23.10.3.1 MEG

In many institutes, maintenance is primarily provided by local staff, supported by the system manufacturer. However, maintenance contracts are also available. These typically cost about $50,000/year, but costs vary substantially depending on the details of the device and of the contract.

23.10.3.2 fMRI

Maintenance contracts are quite usual for these devices. The annual rate is in the range of $100,000/year.

23.11 Tools Needed

No special tools are needed to use these systems under normal conditions, except for the devices used to present visual, auditory, or other stimuli to the subjects and record any subject behavioral responses. These systems must be compatible with the strong magnetic field produced by MRI scanners and the sensitive RF systems of these scanners, or with the extreme magnetic sensitivity of the MEG. Their are vendors who can supply this specialized equipment.

Most suppliers provide data-analysis tools as an integral part of the system. However, for sophisticated analyses and/or for extraordinary experimental designs, special programs are available from commercial and noncommercial groups.

23.12 Related Methods

23.12.1 MEG

The EEG in principle reflects the same neural activity as the MEG. In contrast to its magnetic counterpart, EEG is equally sensitive to tangential and radial sources. However, in contrast to MEG, the topographical distribution of EEG potentials is determined not only by intraneural currents, but also by volume currents distributed across the entire volume of the brain. Consequently, the spatial resolution of the EEG is clearly less than that of MEG. Nevertheless, many laboratories acquire data with both types of systems simultaneously in order to gather as much information as possible. Using specialized hardware, the EEG can even be acquired during MRI. This allows for paradigms such as neural spike-triggered imaging.

23.12.2 fMRI

Positron-emission-tomography (PET) with ^{15}O-labeled water as a tracer can be used to monitor neural activation in a manner similar to fMRI. PET has the advantage that its results more directly reflect modulations in the regional cerebral blood and thus are less prone to artifacts. On the other hand, because PET requires radioactive tracers, it can only be repeated for a few runs with a given subject. For the same reason, subjects should not undergo a PET examination more than once a year. In addition, the spatial resolution of PET is substantially lower than the resolution of fMRI.

Near-infrared spectroscopy recently has gained some interest because it allows for a direct monitoring of certain metabolic processes in the brain. Infrared light from light-emitting diodes (LED) is transmitted through the skull into the brain. After being diffracted by the cortical tissue, some part of the light is transmitted back through the skull and picked up by phototransistors. The amount of diffraction, and thus the intensity of the returned light, is modulated by certain metabolic processes tied to neural activity. However, while this method can directly reveal neural activity, its spatial resolution and depth sensitivity are extremely poor.

References

Dale, A. and Sereno, M. (1993), Improved localization of cortical activity by combining EEG and MEG with MRI cortical surface reconstruction: a linear approach, *J. Cognitive Neurosci.*, 5, 162–176.

Fuchs, M., Wagner, M., Kohler, T., and Wischmann, H.A. (1999), Linear and nonlinear current density reconstructions, *J. Clin. Neurophysiol.*, 16, 267–95.

Hämäläinen, M., Hari, R., Ilmoniemi, R.J., Knuutila, J., and Lounasmaa, O.V. (1993), Magnetoencephalography: theory, instrumentation, and applications to noninvasive studies of the working human brain, *Rev. Modern Phys.*, 65, 413–497.

Hari, R., Salmelin, R., Makela, J.P., Salenius, S., and Helle, M. (1997), Magnetoencephalographic cortical rhythms, *Int. J. Psychophysiol.*, 26, 51–62.

Hinrichs, H., Scholz, M., Tempelmann, C., Woldorff, M.G., Dale, A.M., and Heinze, H.J. (2000), Deconvolution of event-related fMRI responses in fast-rate experimental designs: tracking amplitude variations, *J. Cognitive Neurosci.*, 12, 76–89.

Jezzard, P., Matthews, P.M., and Smith, S.M. (2001), *Functional MRI: An Introduction to Methods*, Oxford University Press, New York.

Josephson, B.D. (1962), Possible new effects in superconducting tunnelling, *Phys. Lett.*, 1, 251–253.

Lounasmaa, O.V., Hamalainen, M., Hari, R., and Salmelin, R. (1996), Information processing in the human brain: magnetoencephalographic approach, *Proc. Natl. Acad. Sci. U.S.A.*, 93, 8809–8815.

Kowalik, Z.J. and Elbert, T. (1994), Changes of chaoticness in spontaneous EEG/MEG, *Integr. Physiol. Behav. Sci.*, 29, 270–282.

Noesselt, T., Hillyard, S.A., Woldorff, M.G., Schoenfeld, A., Hagner, T., Jäncke, L., Tempelmann, C., Hinrichs, H., and Heinze, H.J. (2002), Delayed striate cortical activation during spatial attention, *Neuron*, 35, 575–687.

Nolte, G. and Curio, G. (1999), The effect of artifact rejection by signal space projection on source localization accuracy in MEG measurements, *IEEE Trans. Biomed. Eng.*, 46, 400–408.

Scherg, M., Bast, T., and Berg, P. (1999), Multiple source analysis of interictal spikes: goals, requirements, and clinical value, *J. Clin. Neurophysiol.*, 16, 214–224.

Singh, K.D., Barnes, G.R., Hillebrand, A., Forde, E.M., and Williams, A.L. (2002), Task-related changes in cortical synchronization are spatially coincident with the hemodynamic response, *Neuroimage*, 16, 103–114.

Turner, R., Howseman, A., Rees, G.E., Josephs, O., and Friston, K. (1998), Functional magnetic resonance imaging of the human brain: data acquisition and analysis, *Exp. Brain Res.*, 123, 5–12.

24

Ambulatory Assessment of Blood Pressure to Evaluate Workload

Renate Rau
University of Technology, Dresden

24.1   Background and Applications

Although strain is indicated among the criteria affecting mood and behavior, performance, as well as physiological effort, the majority of strain analyses have only considered psychological parameters (supported by a wide range of questionnaires). One of the most striking developments in the field of strain analyses over the past decades is the advance in ambulatory assessment techniques. The progress in ambulant techniques has permitted the assessment of behavioral, emotional, and activational interaction with workload under real work conditions. Moreover, carryover effects of workload on activities, behavior, and strain after work as well as on recovery effects during the night can be measured. This means an enhancement of the load–strain paradigm from short-term effects, such as fatigue, boredom, vigilance, etc., to the long-term effects of work, such as disturbed recovery processes after work, and to long-term effects, such as cardiovascular health diseases, diabetes mellitus, depression, etc.

The focus of this chapter will be on the use of ambulatory blood pressure monitoring (ABPM) for assessing strain effects during and after work in relation to workload. ABPM is one form of ambulatory assessment developed for clinical purposes in the first place. However, since about 1980, this technique has also been used for the investigation of psychosocial work characteristics related to the risk of cardiovascular disease. The majority of the work-related ABPM studies were based on the job-demand/decision-latitude model proposed by Karasek and coworkers (Karasek, 1979; Karasek and Theorell, 1990). Data reported by Theorell et al. (1991), Pickering (1991), Rau (1996), Schnall et al. (1998), and Belkic et al. (1998) indicate that sympathetic nervous activity can be increased during the exposure to high job strain, which is a combination of high demands at work with low decision latitude or control. There is considerable evidence linking the exposure to "job strain" with hypertension (e.g., Schnall et al., 1990; Theorell et al., 1993; Van Egeren, 1992; Pickering, 1993; Schnall et al., 1998; Peter and Siegrist, 2000; Rau et al., 2001). Besides the studies that were done to investigate the impact and/or direct effects of high job strain on blood pressure changes or on blood pressure status, the effects of social support (Unden et al., 1991), of perceived control (Gerin et al., 1992), of effort–reward imbalance (Siegrist, 1998), and of sequential and hierarchical completeness of tasks (Hacker, 1994; Rau, 2001) were additionally tested.

24.2   Procedure

Ambulatory blood pressure monitoring means the use of portable recorders for automatic, repeated, noninvasive registration of arterial blood pressure. Parameters that are evaluated are the level of systolic and diastolic and sometimes also the arterial mean blood pressure as well as the dynamic of blood pressure.

Evaluating workload by means of ABPM involves the following steps:

1. Procure an ABPM device
2. Select an appropriate work analysis by objective methods (e.g., Task Diagnosis Survey, Pohlandt et al., 1999; Job Exposure Matrix, Fredlund et al., 2000) and/or by subjective methods of analysis (e.g., Job Content Questionnaire, Karasek et al., 1998)
3. Develop a prequestionnaire about normal daily activities (necessary for programming measurement intervals in BP monitor)
4. Maintain a diary on handheld computer or paper with questions for:
   - Time of answering (record the time using the handheld computer)
   - Setting (where are you, current activity, body position, alone/not alone, etc.)
   - Current status (recommended observations include the perceived mental and physical load, mood, perceived control)
5. Prepare the BP monitor with subject number, date, and with measurement intervals
6. Fit the BP monitor to the subject (additional recommendation is fitting an activity sensor)
7. Instruct the subject about BP measurement, possibility of interruption, self-induced BP measurement, use of handheld computer or diary, how to sleep with equipment
8. Subject can start with daily routine, e.g., with work
9. Remove BP monitor and diaries after the time of ambulant monitoring (e.g., after 24 hours)
10. Transfer data into computer

Hypertension is a disease defined by abnormally elevated blood pressure, which acts in arteries as an injuring mechanical force, which in turn can contribute to the increase in cardiovascular risks such as arteriosclerosis, myocardial infarction, or stroke. BP status can be determined by the evaluation of the averaged daytime blood pressure recorded by ABPM.

There is evidence that the workplace plays an important etiological role in developing (essential) hypertension. The empirical data is strongest with respect to exposure to job strain (which is the combination of high psychological demands and low decision latitude) and elevated ambulatory blood pressure at work (Belkic et al., 2001). Studies investigating the effects of job strain on BP status reported increased odds of hypertension in employees dependent on job strain, particularly in men (e.g., Landsbergis et al., 1994: odds ratio 2.9, 95% confidence interval 1.3 to 6.6, n = 262 men; Tsutsumi et al., 2001: odds ratio 1.18, 95% confidence interval 1.05 to 1.32, n = 3187 men).

BP readings describe the actual individual strain effect on workload because the BP reacts quickly to situational changes. This reaction can also be a conditional reaction on the type of load, and this condition can be pathological. An example for such a conditional reaction is that professional drivers with ischemic heart disease showed a higher diastolic BP reactivity to a visual-avoidance task than nondrivers (Emdad et al., 1998). This reactivity pattern is assumed to be conditioned by driving experience and via cerebral pathways that contribute to the development of hypertension (Belkic et al., 1994). Examples for the study of short-term effects in BP on situational changes during work were reported for social support and workload (e.g., Theorell, 1992; Melamed et al., 1998).

Results of the Framingham study (Kannel et al., 1971; Stokes et al., 1989; Fiebach et al., 1989) repeatedly disclosed that systolic blood pressure predicts cardiovascular heart disease (e.g., myocardial infarction). There are many studies that suggest that job strain acts, in part, to cause cardiovascular diseases through the mechanism of elevated blood pressure (van Egeren, 1992; Theorell et al., 1991, 1993; Schnall et al., 1992; Rau et al., 2001). Furthermore, the risk of pathological changes of the cardiovascular system rises with the frequency and the height of situational BP peaks (Schmidt, 1982). For example, situational BP changes were evaluated regarding work-related noise (Van Kempen et al., 2002).

If job strain plays a role in the development of sustained hypertension, it should be possible to demonstrate that job strain can increase the blood pressure not only during work, but also during leisure time and night. The study of such a carryover effect of job strain on nocturnal BP is important, since nocturnal BP is independently associated with end organ damage above the risk associated with daytime BP (Verdecchia et al., 1997). Furthermore, it has also been shown that a reduced nocturnal reduction rate (no or only small BP drops at night) is associated with target organ involvement. Work-related effects on nocturnal BP were reported by Theorell et al. (1991), Pickering et al. (1991), van Egeren (1992), Fox et al. (1993), Pickering (1997), Schnall et al. (1998), and Rau et al. (2001). Their data indicate that BP may increase at work in high-strain jobs (high-strain definition based on the job-demand/control model by Karasek [1979]) and remain elevated at night, thus hampering rewind.

The labels "leisure time" and "time at home" have been often used relatively undifferentiated in literature about work strain and carryover effects after work. Since the activities after work consist both of stressing activities and of passive relaxation, the labels "leisure time" and "time at home" have to be defined in future studies.

24.3   Requirements for ABPM Evaluation

The assessment of blood pressure values is only possible with information about the body position and motor activity at the time of blood pressure measurement. However these are the minimum information details. Ambulatory blood pressure monitoring should be combined with psychological data monitoring by paper-and-pencil diary or, better, by questionnaires on a handheld computer. Thus, the assessment of strain by ABPM includes the ambulatory measured blood pressure, the information about the momentary setting, and, if possible, ratings about the momentary psychological state (e.g., perceived mental load, perceived control, mood, motivation, etc.).

Both, the ABPM and the ambulant psychological data monitoring require:

- Determination of the sampling method
- Determination of data segmentation
- Data reduction

24.3.1   Determination of the Sampling Method

The method used can be a time, an event, or a combination of time and event sampling. A typical example of time sampling in job-strain studies is the automatic recording of blood pressure at 15-min intervals during the daytime and at 30- or 60-min intervals at nighttime. Guidelines for ABPM indicate that measurement intervals of 15 to 20 min at daytime and 30 min at nighttime suffice for a valid assessment of blood pressure level and variability (Pickering, 1991; Palatini et al., 1994). Whereas in the early ABPM studies a standardized segment of 6:00 A.M. to 10:00 P.M. was used to define the daytime and the interval from 10:00 P.M. to 6:00 A.M. was defined as nighttime, it is now common to define daytime and nighttime by the subject's own schedule, i.e., by when the subject goes to bed and wakes up. The sole use of event sampling is not common for ABPM, but the combination of time and event sampling has been used most frequently. In addition to the time-sampling plan described above, participants can also be asked to take a measure of blood pressure and/or to answer the diary/handheld computer questions by themselves when starting and finishing work, when changing the activity at work, when starting and finishing household duties, and/or when going to bed, etc. Other types of events for event-sampling methods are emotions (Myrtek et al., 2001; Triemer and Rau, 2001) or stressful situations (Perrez and Reicherts, 1996).

24.3.2 Determination of Data Segmentation

Segmentation is necessary to distinguish relatively homogeneous data units within the whole course of ambulatory monitoring (Fahrenberg, 1996). Data can be separated by different criteria, depending on the aim of the study. For example, such criteria can be the main activities (e.g., work time, time for the journey to and from work, time for household duties, leisure time, nighttime) or, more generally, from the beginning to the end of selected time segments. Another criterion can be an event, e.g., an emotion. Then the starting time or number of blood pressure measures before this event and the ending time or number of blood pressure measures after the event determine the segment. The precondition of segmentation is that there is information about the setting for all blood pressure measurements.

24.3.3 Data Reduction

Ambulatory assessment technology produces an enormous data set that must be reduced before further evaluation. The required data-reduction strategies should be specified before the data acquisition. For example, use only mean values and values of variance of blood pressure of a defined segment, or use only data without physical impact, or use the median of psychological ratings of the defined segments, etc.

24.4 Advantages

ABPM allows the repeated measurement of BP in combination with a simultaneous record of workload as well as the subject's strain experience and behavior. In addition, work-strain-related effects on BP recovery during the nighttime can be assessed.

24.5 Disadvantages

ABPM can affect the daily activity (e.g., subject avoids sports activities, although such activity is possible with the monitor) and the sleep quality of a few subjects.

24.6 Example

The objective of this study was to explore the effect of objectively measured exposure to job strain on the cardiovascular reactions at work, after work, and during sleep following a working day by measuring heart rate and blood pressure.

24.6.1 Background

The combination of high psychological job demands and low decision latitude (high job strain) has been associated with an increased risk of coronary heart disease.

24.6.2 Method

A sample group of 105 women was tested during 24 hours of a real working day by means of a computerized diary and ambulant monitoring of blood pressure (see Figure 24.1). By using the task diagnosis survey (Pohlandt et al., 1999) for an objective analysis of work, workplaces were classified according to their potential of exposing employees to high vs. low job strain. From the total sample, 41 participants were exposed to low strain and 17 to high strain. The other 47 women were either exposed to low or to high job strain.

24.6.3 Result

High job strain was related to both systolic and diastolic blood pressure levels during daily life. High job strain was associated with elevated blood pressure at work and at night and also during the time between work and night (see Figure 24.2). However, it was not the leisure time that showed a significant relationship to job strain, but the time, which includes further burden like household duties, child care, or care of elderly and ill, etc. Furthermore, more women of the high-job-strain group reported a disturbed relaxation ability and problems with falling asleep.

24.6.4 Conclusion

The exposure to job strain in the present study was independently assessed by external observers. The results support the demand for a change in designing workplaces toward healthy jobs.

24.7 Related Methods

Related methods include ambulatory assessment of other biological parameters (e.g., electrocardiogram, accelerometry) and of psychological data. For a good overview, see Fahrenberg and Myrtek (1996, 2001).

24.8 Standards and Regulations

Beside the standards related to mental workload (ISO 10075-1/-2; ISO/CD 10075-3), there are special standards for the diagnosis, procedure, and techniques of ABPM.

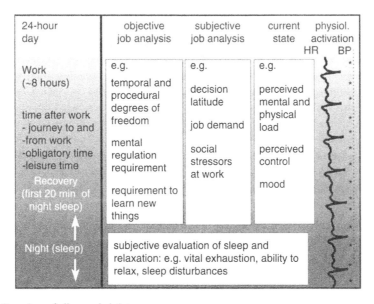

FIGURE 24.1 Overview of all recorded data.

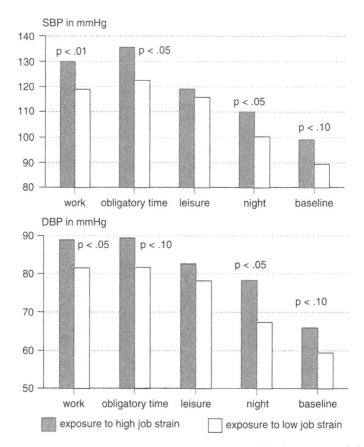

FIGURE 24.2 Relationship between systolic (SBP) and diastolic (DBP) blood pressure, and the exposure to job strain in women working in white-collar professions.

24.8.1   Standards for Diagnosis Related to BP

Because blood pressure measured by ABPM is systematically lower than surgery or clinic measurements in hypertensive and normotensive people, the classification levels of what is a "normal" ABPM reading had to be adjusted downward compared with office readings (Parati et al., 1998; Staessen et al., 1993; Lefevre and Aronson, 2001). The American Society of Hypertension (O'Brien et al., 2000) and the British Hypertension Society (Myers et al., 1999) have issued recommendations for interpreting ambulatory blood pressure measurements as follows. Normal readings for adults are considered a daytime-averaged systolic blood pressure (SBP) of less than 135 mmHg and a daytime-averaged diastolic blood pressure (DBP) of less than 85 mmHg. Thus a daytime-averaged blood pressure reading equal to or greater than 135 to 140 mmHg (systolic) over 85 to 90 mmHg (diastolic) is considered borderline elevated, and greater than 140 mmHg (systolic) over 90 mmHg (diastolic) is considered (probably) abnormally elevated.

24.8.2   Standards for Ambulatory BP Devices

Ambulatory BP devices should be independently evaluated for accuracy according to the two most widely used protocols of the British Hypertension Society (BHS protocol) and/or of the Association for the Advancement of Medical Instrumentation (AAMI standard). Use only calibrated equipment (O'Brien, 2001; O'Brien et al., 2001).

24.8.3 Standards for BP Measurement Variables that Should Be Controlled

Accurate measurement of blood pressure depends on a number of factors. The most important factors that should be considered in ABMP are as follows:

24.8.3.1 Age and Body-Mass Index

There is a linear rise in SBP with age and a concurrent increase in DBP and mean arterial BP until about 50 years of age (Franklin et al., 1997). Furthermore, SBP and DBP rise with increasing body-mass index (BMI = weight [kg]/length [m^2]) (Kannel et al., 1967; Sowers, 1998). Therefore, all statistical analyses must be controlled for age and body-mass index (e.g., consider age and BMI as covariables or constitute groups that are homogeneous in age and BMI).

24.8.3.2 Body Position and Physical Activity

In mental workload studies, data confounded by motor activity are, as a rule, excluded from further data analyses (Pickering, 1991; Gellman et al., 1990). All discussions of results based on comparison of data segments in different body positions have to account for these confounding effects of motor activity.

24.8.3.3 Emotional Episodes

Because emotional episodes can influence cardiovascular activity (James et al., 1986; Shimomitsu and Theorell, 1996; Triemer and Rau, 2001), data units associated with emotional episodes should be excluded. Information from the participants about the episode are needed (e.g., information from the self-report ratings of the momentary state) to define the duration of measurements or the number of measures that should be excluded.

24.8.3.4 Sex

All studies that evaluate nocturnal recovery of blood pressure have to exclude the participants who have had sex during the night. The background is that sexual intercourse produces an extreme increase in blood pressure (Falk, 2001) followed by a fast unwinding within a few minutes after orgasm. Consequently, all parameters of blood pressure describing nocturnal recovery are affected by these changes.

24.8.3.5 Tobacco Use

Involving tobacco users in a study requires control procedures (e.g., including the variable in regression analyses), since both smoking and nonsmoking tobacco users were found to have a significantly elevated ambulatory daytime DBP (Bolinder and de Faire, 1998).

24.8.3.6 Antihypertensive Medication

Antihypertensive medicine includes all medicine influencing the cardiovascular system. Strain analyses to evaluate workload should exclude all potential participants who use antihypertensive drugs or drugs that influence the cardiovascular system, as well as all participants diagnosed as hypertensive (with the obvious exception of a study comparing normal subjects versus hypertensive subjects).

24.8.3.7 Renal Diseases

All potential participants suffering from renal diseases should be excluded from data recording.

24.8.3.8 Size of Cuff

The relationship between arm circumference and blood pressure cuff size has an impact on accurate blood pressure measurement. Using too small a cuff relative to arm circumference leads to an overestimation of blood pressure (e.g., Linfors et al., 1984); using too large a cuff may underestimate blood pressure (Graves, 2001). Table 24.1 shows cuff sizes recommended by the American Heart Association (Frohlich et al., 1988) and by the British Hypertension Society (O'Brien et al., 1997).

TABLE 24.1 Recommended Cuff Sizes

American Heart Association		British Hypertension Society [a]	
Upper Arm Circumference	Recommended Cuff Size	Upper Arm Circumference	Recommended Cuff Size
24–32 cm	13 cm × 24 cm	17–26 cm	10 cm × 18 cm
32–42 cm	17 cm × 32 cm	26–33 cm	12 cm × 26 cm
>42 cm	20 cm × 42 cm	>33 cm	12 cm × 40 cm

[a] All dimensions have a tolerance of ±1 cm; the center of the cuff bladder is over the brachial artery.

24.8.3.9 Gender

Analyses should include gender as an additional independent variable, since studies according to job strain and blood pressure show different results for men and women (Pickering, 1991).

24.9   Approximate Training and Application Times

The investigator should be experienced in the principles of traditional blood pressure measurement, monitor functioning (including effect of cuff size), and the interpretation of ABPM readings (e.g., determining BP status, impact of physical load, etc.). In addition, the main variables that have an influence on blood pressure should be known, and the investigator should be able to control for these variables (e.g., use a diary protocol by handheld computer or paper and pencil, record the age, BMI, CHD family history, smoking behavior, etc.).

Fitting the BP monitor on a subject takes about 15 to 20 min. Subjects should be instructed about the measurement procedure, including the frequency of inflation and deflation, the need for automatic repetition of the measurement in the event that a measure failed, and how the subject can deflate the cuff. The instruction for measurement (keep the arm steady, without motion during measurement; engage in normal activities between measurements) should be given in spoken and written form. One or more test readings should be done to train the subject.

Removal of the BP monitor should be done by the investigator, with the immediate next step being a follow-up interview about the activities and events during the course of measurement (interview time is about 30 min).

A good overview of training, application, and (clinical) interpretation of ABPM is given by the British Hypertension Society (O'Brien et al., 2001).

24.10   Tools Needed

Required is a device for ambulatory measurement that has been validated independently according to the BHS or AAM standard.

Recommended is a handheld computer for recording psychological data and setting information. The device should include a clock and have a wide screen in order to display questions, ratings, and multiple-choice items.

References

Belkic, K.L., Schnall, P.L., Landsbergis, P.A., Schwartz, J.E., Gerber, L.M., Baker, D., and Pickering, T.G. (2001), Hypertension at the workplace: an occult disease? The need for work site surveillance, *Adv. Psychosomatic Med.*, 22, 116–138.

Belkic, K., Emdad, R., and Theorell, T. (1998), Occupational profile and cardiac risk: possible mechanisms and implications for professional drivers, *Int. J. Occup. Med. Environ. Health*, 11, 37–57.

Belkic, K., Savic, C., Theorell, T., Rakic, L., Ercegovac, D., and Djordjevic, M. (1994), Mechanisms of cardiac risk factors among professional drivers, *Scand. J. Work Environ. Health*, 20, 73–86.

Bolinder, G. and de Faire, U. (1998), Ambulatory 24-h blood pressure monitoring in healthy, middle-aged smokeless tobacco users, smokers, and nontobacco users, *Am. J. Hypertension*, 11, 1153–1163.

Emdad, R., Belkic K., Theorell, T., Cizinsky, S., Savic, C., and Olsson, K. (1998), Psychophysiologic sensitization to headlight glare among professional drivers with and without cardiovascular disease, *J. Occup. Health Psychol.*, 3, 147–160.

Fahrenberg, J. (1996), Ambulatory assessment: issues and perspectives, in *Ambulatory Assessment: Computer-Assisted Psychological and Psychophysiological Methods in Monitoring and Field Studies*, Fahrenberg, J. and Myrtek, M., Eds., Hogrefe and Huber, Seattle, WA, pp. 3–20.

Fahrenberg, J. and Myrtek, M., Eds. (1996), *Ambulatory Assessment: Computer-assisted Psychological and Psychophysiological Methods in Monitoring and Field Studies*, Hogrefe and Huber, Seattle, WA.

Fahrenberg, J. and Myrtek, M., Eds. (2001), *Progress in Ambulatory Assessment*, Hogrefe and Huber, Seattle, WA.

Falk, R.H. (2001), The cardiovascular response to sexual activity: do we know enough? *Clin. Cardiol.*, 24, 271–275.

Fiebach, N.H., Hebert, P.R., Stampfer, M.J., Colditz, G.A., Willett, W.C., Rosner, B., Speizer, F.E., and Hennekens, C. (1989), A prospective study of high blood pressure and cardiovascular disease in women, *Am. J. Epidemiol.*, 130, 646–654.

Fox, M.L., Dwyer, D.J., and Ganster, D.C. (1993), Effects of stressful job demands and control on physiological and attitudinal outcomes in a hospital setting, *Acad. Manage. J.*, 36, 289–318.

Franklin, S.S., Gustin, W., Wong, N.D., Weber, M., and Larson, M. (1997), Hemodynamic patterns of age-related changes in blood pressure: Framingham heart study, *Circulation*, 96, 308–315.

Fredlund, P., Hallqvist, J., and Diderichsen, F. (2000), Psykosocial yrkesexponeringsmatris: en uppdatering av ett klassifikationssystem för yrkesrela terade psykosociala exponeringar, Arbete och Hälsa, 2000, 11, Arbetslivsinstitutet, Stockholm.

Frohlich, E.D., Grim, C., Labarthe, D.R., Maxwell, M.H., Perloff, D., and Weidman, W.H. (1988), Recommendations for human blood pressure determination by sphygmomanometers: report of a special task force appointed by the steering committee, American Heart Association, *Hypertension*, 11, 209A–222A.

Gellman, M., Spitzer, S., Ironson, G., Llabre, M., Saab, P., DeCarlo Pasin, R., Weidler, D.J., and Schneiderman, N. (1990), Posture, place, and mood effects on ambulatory blood pressure, *Psychophysiology*, 27, 544–551.

Gerin, W., Pieper, C., Marchese, L., and Pickering, T.G. (1992), The multidimensional nature of active coping: differential effects of effort and enhanced control on cardiovascular reactivity, *Psychosomatic Med.*, 54, 707–719.

Graves, J.W. (2001), Prevalence of blood pressure cuff sizes in a referral practice of 430 consecutive adult hypertensives, *Blood Pressure Monitoring*, 6, 17–20.

Hacker, W. (1994), Action regulation theory and occupational psychology: review of German empirical research since 1987, *German J. Psychol.*, 18, 91–120.

ISO (1991), Ergonomic Principles Related to Mental Work-load: General Terms and Definitions, ISO 10075, International Organization for Standardization, Geneva.

ISO (1996), Ergonomic Principles Related to Mental Work-load, Part 2: Design Principles, ISO 10075-2, International Organization for Standardization, Geneva.

ISO (2001), Ergonomic Principles Related to Mental Workload: Measurement and Assessment of Mental Workload, ISO/CD 10075-3:2001, committee draft, International Organization for Standardization, Geneva.

James, G.D., Yee, L.S., Harshfield, G.A., Blank, S.G., and Pickering, T.G. (1986), The influence of happiness, anger, and anxiety on the blood pressure of borderline hypertensives, *Psychosomatic Med.*, 48, 502–508.

Kannel, W.B., Brand, M., Skinner, J.J., Jr., Dawber, T.R., and McNamara, P.M. (1967), The relation of adiposity to blood pressure and development of hypertension: the Framingham study, *Ann. Intern. Med.*, 67, 48–59.

Kannel, W.B., Gordon, T., and Schwartz, M.J. (1971), Systolic versus diastolic blood pressure and risk of coronary heart disease: the Framingham study, *Am. J. Cardiol.*, 27, 335–345.

Karasek, R. (1979), Job demands, job decision latitude, and mental strain: implications for job redesign, *Administrative Sci. Q.*, 24, 285–307.

Karasek, R.A. and Theorell, T. (1990), *Healthy Work*, Basic Books, New York.

Karasek, R., Brisson, C., Kawakami, N., Houtman, I., and Bomgers, P. (1998), The job content questionnaire (JCQ): an instrument for internationally comparative assessments of psychosocial job characteristics, *J. Occup. Health Psychol.*, 3, 322–355.

Landsbergis, P., Schnall, P., Warren, K., Pickering, T.G., and Schwartz, J.E. (1994), Association between ambulatory blood pressure and alternative formulations of job strain, *Scand. J. Work Environ. Health*, 20, 349–363.

Lefevre, F.V. and Aronson, N. (2001), Ambulatory Blood Pressure Monitoring for Adults with Elevated Office Blood Pressure, Blue Cross and Blue Shield Association, government sponsorship contract 290-97-001-5, http://www.cms.hhs.gov/mcac/8b1-e3.pdf.

Linfors, E.W., Feussner, J.R., Blessing, C.L., Starmer, C.F., Neelon, F.A., and McKee, P.A. (1984), Spurious hypertension in the obese patient: effect of sphygmomanometer cuff size on prevalence of hypertension, *Arch. Int. Med.*, 144, 1482–1485.

Melamed, S., Kristal-Boneh, E., Harari, G., Froom, P., and Ribak, J. (1998), Variation in the ambulatory blood pressure response to daily work load: the moderating role of job control, *Scand. J. Work Environ. Health*, 24, 190–196.

Myers, M.G., Haynes, R.B., and Rabkin, S.W. (1999), Canadian hypertension society guidelines for ambulatory blood pressure monitoring, *Am. J. Hypertension*, 12, 1149–1157.

Myrtek, M., Zanda, D., and Aschenbrenner, E. (2001), Interactive psychophysiological monitoring of emotions in students' everyday life: a replication study, in *Progress in Ambulatory Assessment: Computer-assisted Psychological and Psychophysiological Methods in Monitoring and Field Studies*, Fahrenberg, J. and Myrtek, M., Eds., Hogrefe and Huber, Seattle, pp. 399–414.

O'Brien, E. (2001), State of the market in 2001 for blood pressure measuring devices, *Blood Pressure Monitoring*, 6, 171–176.

O'Brien, E., Waeber, B., Parati, G., Staessen, G., and Myers, M.G. (2001), Blood pressure measuring devices: recommendations of the European Society of Hypertension, *Brit. Med. J.*, 322, 531–536.

O'Brien, E., Coats, A., Owens, P., Petrie, J., Padfield, P.L., Littler, W.A., de Swiet, M., and Mee, F. (2000), Use and interpretation of ambulatory blood pressure monitoring: recommendations of the British Hypertension Society, *Brit. Med. J.*, 320, 1128–1134.

O'Brien, E., Petrie, J., Littler, W.A., de Swiet, M., Padfield, P.D., and Dillon, M.J. (1997), *Blood Pressure Measurement: Recommendations of the British Hypertension Society*, BMJ Books, London.

Palatini, P., Mormino, P., and Canali, C. (1994), Factors affecting ambulatory blood pressure reproducibility: results of the HARVEST trial, *Hypertension*, 23, 211–216.

Parati, G., Omboni, S., Staessen, J., Thijs, L., Fagard, R., Ulian, L., and Mancia, G. (1998), Limitations of the difference between clinic and daytime blood pressure as a surrogate measure of the "white-coat" effect, *J. Hypertension*, 16, 23–29.

Perrez, M. and Reicherts, M. (1996), A computer-assisted self-monitoring procedure for assessing stress-related behavior under real-life conditions, in *Ambulatory Assessment: Computer-assisted Psychological and Psychophysiological Methods in Monitoring and Field*, Fahrenberg, J. and Myrtek, M., Eds., Hogrefe and Huber, Seattle, WA, pp. 51–67.

Peter, R. and Siegrist, J. (2000), Psychosocial work environment and the risk of coronary heart disease, *Int. Arch. Occup. Environ. Health*, 73, S41–S45.

Pickering, T. (1997), The effects of occupational stress on blood pressure in men and women, *Acta Physiol. Scand. Suppl.*, 640, 125–128.

Pickering, T.G. (1991), *Ambulatory Monitoring and Blood Pressure Variability*, Science Press, London.

Pickering, T.G. (1993), Applications of ambulatory blood pressure monitoring in behavioral medicine, *Ann. Behavioral Med.*, 15, 26–32.

Pickering, T.G., Gary, D.J., Schnall, P.L., Schlussel, Y.R., Pieper, C.F., Gerin, W., and Karasek, R.A. (1991), Occupational stress and blood pressure, in *Women, Work, Health*, Frankenhaeuser, M., Lundberg, U., and Chesney, M., Eds., Plenum Press, New York, pp. 171–186.

Pohlandt, A., Hacker, W., and Richter, P. (1999), Das Tätigkeitsbewertungssystem TBS, in *Handbuch psychologischer Arbeitsanalyseverfahren: ein praxisorietierter Überblick*, Dunckel, H., Hrsg., (Schriftenreihe Mensch, Technik, Organisation, Band 14, S. 515-530), vdf Hochschulverlag an der ETH Zürich, Zürich.

Rau, R. (2001), Objective characteristics of jobs affect blood pressure at work, after work and at night, in *Progress in Ambulatory Assessment*, Fahrenberg, J. and Myrtek, M., Eds., Hogrefe and Huber, Seattle, WA, pp. 361–386.

Rau, R. (1996), Psychophysiological assessment of human reliability in a simulated complex system, *Biol. Psychol.*, 42, 287–300.

Rau, R., Georgiades, A., Lemne, C., de Faire, U., and Fredrikson, M. (2001), Psychosocial work characteristics and perceived control in relation to cardiovascular rewind at night, *J. Occup. Health Psychol.*, 6, 171–181.

Schmidt, T.H. (1982), Die Situationshypertonie als Risikofaktor, in *Essentielle Hypertonie*, Vaitl, D., Hrsg., Springer, Berlin.

Schnall, P.L., Schwartz, J.E., Landsbergis, P.A., Warren, K., and Pickering, T.G. (1998), A longitudinal study of job strain and ambulatory blood pressure: results from a three-year follow-up, *Psychosomatic Med.*, 60, 697–706.

Schnall, P.L., Schwartz, J.E., Landsbergis, P.A., Warren, K., and Pickering, T.G. (1992), The relationship between job strain, alcohol and ambulatory blood pressure, *Hypertension*, 19, 488–494.

Schnall, P.L., Pieper, C., Schwartz, J.E., Karasek, R.A., Schlussel, Y., Devereux, R.B., Ganau, A., Alderman, M., Warren, K., and Pickering, T.G. (1990), The relationship between "job strain," work place diastolic blood pressure, and left ventricular mass index, *J.A.M.A.*, 263, 1929–1935.

Shimomitsu, T. and Theorell, T. (1996), Intraindividual relationships between blood pressure level and emotional state, *Psychotherapy Psychosomatics*, 65, 137–144.

Siegrist, J. (1998), Berufliche Gratifikationskrisen und Gesundheit: ein soziogenetisches Modell mit differentiellen Erklärungschancen, in *Gesundheits- oder Krankheitstheorie? Saluto- versus pathogenetische Ansätze im Gesundheitswesen*, Margraf, J., Siegrist, J., and Neumer, S., Hrsg., Springer, Berlin.

Sowers, J.R. (1998), Obesity and cardiovascular disease, *Clin. Chem.*, 44, 1821–1825.

Staessen, J.A., O'Brien, E.T., Atkins, N., and Amery, A.K. (1993), Short report: ambulatory blood pressure in normotensive compared with hypertensive subjects, ad hoc working group, *J. Hypertension*, 11, 1289–1297.

Stokes, J., III, Kannel, W.B., Wolf, P.A., D'Agostino, R.B., and Cupples, L.A. (1989), Blood pressure as a risk factor for cardiovascular disease: the Framingham study, 30 years of follow-up, *Hypertension*, 13, I13–I18.

Theorell, T. (1992), The psycho-social environment stress and coronary heart disease, in *Coronary Heart Disease, Epidemiology*, Marmott, M. and Elliott, P., Eds., Oxford University Press, Oxford, pp. 256–273.

Theorell, T., Ahlberg-Hulten, G., Jodko, M., Sigala, F., and Torre, B. (1993), Influence of job strain and emotion on blood pressure in female hospital personnel during work hours, *Scand. J. Work Environ. Health*, 19, 313–318.

Theorell, T., de Faire, U., Johnson, J., Hall, E., Perski, A., and Stewart, W. (1991), Job strain and ambulatory blood pressure profiles, *Scand. J. Work Environ. Health*, 17, 380–385.

Triemer, A. and Rau, R. (2001), Stimmungskurven im Arbeitsalltag, eine Feldstudie, *Z. Differentielle Diagnostische Psychologie*, 22, 42–55.

Tsutsumi, A., Kayaba, K., Tsutsumi, K., and Igarashi, M. (2001), Association between job strain and prevalence of hypertension: a cross-sectional analysis in a Japanese working population with a wide range of occupations: the Jichi Medical School cohort study, *Occup. Environ. Med.*, 58, 367–73.

Unden, A.L., Orth-Gomer, K., and Elofsson, S. (1991), Cardiovascular effects of social support in the work place: twenty-four-hour ECG monitoring in men and women, *Psychosomatic Med.*, 53, 50–60.

Van Egeren, L.F. (1992), The relationship between job strain and blood pressure at work, at home, and during sleep, *Psychosomatic Med.*, 54, 337–343.

Van Kempen, E.E, Kruize, H., Boshuizen, H.C., Ameling, C.B., Staatsen, B.A., and de Hollander, A.E. (2002), The association between noise exposure and blood pressure and ischemic heart disease: a meta-analysis, *Environ. Health Perspect.*, 110, 307–317.

Verdecchia, P., Schillaci, G., Borgioni, C., Ciucci, A., Gattobigio, R., Guerrieri, M., Comparato, E., Benemio, G., and Porcellati, C. (1997), Altered circadian blood pressure profile and prognosis, *Blood Pressure Monitoring*, 2, 347–352.

25

Monitoring Alertness by Eyelid Closure

Melissa M. Mallis
NASA Ames Research Center

David F. Dinges
University of Pennsylvania

25.1 Background and Applications

Objective identification of moment-to-moment changes in alertness and drowsiness is complicated by two factors:

1. Alertness and drowsiness are the neurobiological products of an interaction between the endogenous circadian pacemaker and the homeostatic need for sleep (Van Dongen and Dinges, 2000), resulting in nonlinear changes in endogenous fatigue, alertness, and vigilance over time.
2. Objective biobehavioral signs of alertness and drowsiness often require intrusive physiological monitoring or probed performance tasks, which complicate their utility in operational environments such as transportation, where continuous attention is demanded.

The development of biobehaviorally based technologies that objectively provide on-line measures of alertness and drowsiness has recently undergone intensified activity due to demands imposed by the somewhat unrealistic expectation that humans can work and maintain performance continuously throughout the 24-hour day. Technology developments in the area of continuous detection of fatigue, while people perform (e.g., driving), have focused on sensors and monitors that can detect drowsiness and hypovigilance. Advances in electronics, optics, data-acquisition systems, algorithm developments, and many other areas have made it far more likely that the goal of an affordable, unobtrusive drowsiness detection system will be incorporated and implemented in various operational environments in the near future.

The need to develop such objective alertness and drowsiness detection technologies is critical because many experiments have shown that subjective estimates are unreliable, and an individual cannot accurately determine when he or she will experience uncontrollable sleepiness that results in serious vigilance lapses (Dinges, 1989). If objective, on-line drowsiness detection technologies can be shown to validly detect drowsiness, then they may offer the potential for a more reliable warning of performance impairing drowsiness, before fatigue leads to a catastrophic outcome. Valid on-line detection of drowsiness and hypovigilance could become a key component in fatigue management (Dinges and Mallis, 1998). Such technology could potentially provide individuals with immediate information on their fatigue levels, particularly fatigue associated with drowsiness and involuntary microsleeps. Subsequently, through the delivery of informational feedback, users could then take appropriate countermeasures when drowsiness is detected.

In recent years, a number of measures of ocular psychophysiology have been identified for their potential to detect minute-to-minute changes in drowsiness and hypovigilance associated with lapses of attention during performance. To determine the scientific validity and reliability of a subset of these prototypical technologies in detecting drowsiness during performance, a double-blind, controlled laboratory study of six technologies that claimed to detect drowsiness was performed at the University of Pennsylvania (Dinges et al., 1998). This study revealed that a measure of slow eyelid closure — referred to as percentage of slow eyelid closure (PERCLOS) (Wierwille et al., 1994; Wierwille and Ellsworth, 1994) — correlated highly with visual vigilance performance lapses in all subjects across 42 hours of wakefulness (Dinges et al., 1998). PERCLOS, which involved video based scoring of slow eyelid closures by trained observers, not only had the highest coherence of the technologies tested, but it also correlated higher with performance lapses than did subjects' own ratings of their sleepiness. Consequently, the results of this experiment, in conjunction with the work of Wierwille et al. (1994), suggested that PERCLOS had the potential to detect fatigue-induced lapses of attention. However, this would only be the case once PERCLOS scoring algorithm used by human observers was automated in a computer algorithm, interfaced to provide informational feedback on alertness and drowsiness levels, and scientifically validated in a controlled laboratory experiment.

25.2 Procedure

An automated PERCLOS scoring algorithm was developed based on the results of the validation study conducted by Dinges and Mallis (1998) as described above, which demonstrated that PERCLOS has the potential to detect fatigue-induced lapses of attention. The automated system involves monitoring slow eyelid closures using an infrared, retinal reflectance, on-line PERCLOS monitor (CoPilot®) that yields an estimate of PERCLOS. Based on the calculated PERCLOS values, the individual being monitored receives informational feedback on alertness and drowsiness. The current feedback is an advisory tone that is reinforced by a visual gauge (Grace, 2001) that encourages the individual being monitored to use an effective fatigue countermeasure. The CoPilot was developed specifically for use in commercial operations involving nighttime driving in a trucking environment (Grace, 2001).

The CoPilot is capable of measuring eye closure under low light conditions using charge coupled device (CCD) imaging technology and a dedicated PC/104 with PCI Bus computer platform. PERCLOS values, based on the recorded eye closure measurements, are calculated in real time. The PERCLOS monitor records eye closure based on the physiological property that the human retina reflects different amounts of infrared light at varying wavelengths. The retina reflects 90% of the incident light at 850 nm; however, at 950 nm, the intensity of reflection is sharply reduced due to the absorption of light by water molecules.

The monitor utilizes two separate cameras, situated at a 90-degree angle to one another, but both focus on the same point being the eye. The image is passed through a beam splitter that transmits or reflects the image onto the lenses of each camera. In order to isolate the correct wavelengths of light, one camera is outfitted with an 850-nm filter, and the other has a 950-nm filter. The 850-nm filter yields a "bright-eye" camera image (i.e., distinct glowing of the individual's pupils, or the red-eye effect), as seen in panel A of Figure 25.1. The 950-nm filter yields a "dark-eye" (dark pupil) image, as seen in panel B of Figure 25.1. Thus the two images are identical except for the brightness of the pupils in the image. A third image enhances the bright eyes by calculating the difference of the two images, as seen in panel C of Figure 25.1 (Grace et al., 1999). The user's eyes are identified in this third image by applying a threshold. The threshold is determined adaptively by examining the average brightness in each video frame. The retinal image sizes are measured, and the results are used to calculate PERCLOS values. This process is repeated several times per second (Grace, 2001).

Panel A   Panel B   Panel C

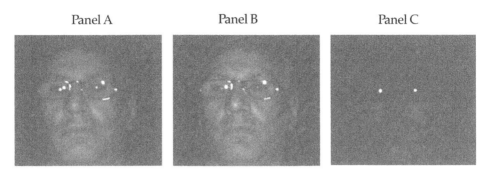

FIGURE 25.1 The three images obtained by PERCLOS are shown. The "bright-eye" image (Panel A) and the "dark-eye" image (Panel B) are essentially identical except for the glowing pupils in the "bright-eye" image. The difference image (Panel C) eliminates all image features except for the bright pupils.

The automated, on-line PERCLOS system, the CoPilot, is an example of a technology that appears to objectively measure alertness and drowsiness through the calculation of PERCLOS values. The CoPilot has undergone an initial controlled laboratory trial that demonstrated that the automated on-line measure of slow eyelid closures (PERCLOS) in fatigued individuals has a robust relationship to lapses of attention, an indicator of reduced alertness in sleepy individuals (Dinges et al., 2000). Additionally, the CoPilot is currently being tested in field studies for reliability and utility in the management and prevention of drowsy driving.

25.3 Advantages

The identification of drowsiness and alertness through an on-line, near-real-time, automated slow eyelid closure system that measures PERCLOS, using video images of the eyes, is unobtrusive to the user. In other words, an automated slow eyelid closure system does not come in direct contact with the user. Ideally, the system would objectively determine alertness levels during performance (e.g., driving) based on preset thresholds in the device's validated algorithm, thus bypassing the need to rely on subjective estimates. If the required validity and reliability of such a monitor could be established, the system could be deployed in operational environments that involve attention-demanding tasks and require an operator to remain stationary with limited physical activity during performance. Incorporation of an alertness feedback interface into an automated slow eyelid closure sensor system can provide the operator with information on increasing likelihood of hypovigilance, allowing the user to take an appropriate action such as a biologically effective fatigue countermeasure (e.g., napping, caffeine) and thus preventing a catastrophic incident or accident due to drowsiness or reduced alertness.

25.4 Disadvantages

There are potential disadvantages to even well validated on-line, alertness-monitoring systems. First, they may not work in all situations. For example, the use of an on-line automated slow eyelid closure system, very likely, will require a restricted field of view. Consequently, if the user is operationally required to move around frequently and not remain seated in one position, the system cannot capture the user's eyes with the use of a single camera array. Detection of alertness and drowsiness in an operational environment in which the user is mobile would require either an array of cameras or a modified system that would be mounted to the head of the user, making it obtrusive and restricting the individual's overall field of view. Second, the use of an automated on-line drowsiness system that relies on slow eyelid closures as the input variable is not ideal in low humidity environments, where users are likely to close their eyes slowly (and keep them closed over a period of time) in an attempt to moisten the eyes. An automated slow eyelid closure system, based on video images only, cannot differentiate between eyes closed due to fatigue or eyes closed due to a rewetting of the eyes, which can potentially result in false positives. Finally, even validated and effective on-line, alertness-monitoring systems cannot be used safely without some instruction on the proper interpretation of the device's feedback (information on alertness provided through an interface). Misuse would consist of an operator in a safety sensitive occupation assuming that the information from the system prevents errors, fatigue, or uncontrolled sleep attacks. Information, even when presented as compelling alerts, alarms, or warnings, does not reduce fatigue as effectively as recovery sleep, naps, caffeine, and other biological interventions. Rather, information from an on-line drowsiness monitor is intended to inform the operator of one's increasing behavioral impairment (e.g., hypovigilance) and to reinforce perception of the need to engage in a biologically effective countermeasure. Hence, without proper use, on-line fatigue and alertness monitors can be misused (e.g., a driver who continues driving when the system repeatedly warns of drowsiness because he/she reasons that the warnings will keep him/her awake).

25.5 Standards and Regulations

Currently, there are no federally mandated standards or regulations concerning drowsiness-detection devices. The U.S. Department of Transportation has developed its own requirement since it is one of the fastest growing operational environments for drowsiness detection. They have indicated an expectation that any device or system claiming to provide on-line monitoring of a driver's alertness or drowsiness must undergo validation against a scientifically established standard for drowsiness and its effects on performance relevant to driving (Rau, 1999). However, given the rapid increase in the development of drowsiness-detection devices (and their proposed use in fatigue management programs), government agencies and the scientific community will need to work collaboratively to establish minimum standards for scientific/engineering criteria, practical/implementation criteria, and legal/policy criteria (Dinges and Mallis, 1998). Hearings held by the U.S. National Transportation Safety Board (NTSB) on such technologies revealed substantial differences of opinion among highway-use stakeholders (e.g., trucking companies, drivers, law enforcement, safety advocates, accident investigators, etc.) regarding the usefulness of such technologies, especially if the drowsiness-detection devices store the alertness information in memory for access by one or more of these entities.

25.6 Approximate Training and Application Times

Other than teaching operators to interpret feedback information indicating drowsiness (eg., stop driving and engage in a fatigue countermeasure), no training time is required to use the automated version of the PERCLOS system. Since its primary development for use is in the driving environment, it has been designed to be small and easily installed on the dashboard, just to the right of the steering wheel, with two degrees of freedom for adjustment. Drivers can then adjust the unit as one would adjust a rearview mirror until they can see their reflection in the face of the unit (Grace, 2001).

25.7 Reliability and Validity

To be safe and effective, any system designed to detect loss of alertness, drowsiness, or hypovigilance on-line must be demonstrated to be scientifically valid. The biobehavioral variable it measures, and the algorithm it uses to provide alertness and drowsiness feedback, must actually reflect a substantial portion of the fatigue-induced variance in performance (Dinges and Mallis, 1998). Failure to meet this basic criterion makes even the most technically sophisticated monitor useless or dangerous. In other words, accurate detection of loss of performance capability is the goal (i.e., accurate detection of hypovigilance with minimal misses and false alarms). Inventors of such systems often fail to appreciate that there are two levels of validation. The first level of validation concerns selection of a biobehavioral parameter(s) that is consistently affected by drowsiness. Contrary to widespread beliefs and casual claims, this level of validation can only be established in a controlled laboratory scientific trial (ideally double blind) in which alertness is manipulated across a wide range in multiple individuals, and in which the biobehavioral marker selected for monitoring changes in close relationship to the performance effects of waxing and waning fatigue (Dinges et al., 1998). Assuming validity solely based on a belief or claim (regardless of how authoritative) that a measure (e.g., specific algorithms for eye blinks, head dips, etc.) reflects drowsiness is inadequate for validation. The second level of validation concerns the specificity of the biobehavioral parameter measured. Once an automated device has been developed that claims to record alertness and drowsiness (e.g., measures slow eyelid closures), it is critical that it be demonstrated to work, recording what it claims to record, and to do so consistently, thus demonstrating engineering validity and reliability. For example, a device claiming to measure slow eyelid closures should be measuring only slow eyelid closures and not eye blinks or other parameters. Unfortunately, because such systems are often invented and developed by engineers, their focus is exclusively on this second level of validity with little appreciation for the fact that it is meaningless if it fails to include a valid biobehavioral marker of alertness and drowsiness in the first place (level 1 validity).

Scientific validity and reliability should first be established across a dynamic range of performance (from alert to drowsy) to ensure that the technology and associated algorithm measure a meaningful fatigue-induced change. It is also important that it be easy to use and that it identify all (or nearly all) fatigue- or drowsiness-related events in fatigued individuals (sensitivity) without false alarms (specificity). If the automated device has high sensitivity but low specificity in detecting drowsiness, it may give too many false alarms, adversely affecting the compliance of the user. In contrast, if the device has a low sensitivity but high specificity, it will give few false alarms, but it may miss too many fatigue and drowsiness events.

25.8 Example

Following an extensive laboratory test of the validity of six drowsiness-detection devices for detecting hypovigilance previously described in Dinges et al. (1998), an early generation of an automated PERCLOS device was developed by researchers at Carnegie Mellon University and tested in the trucking research environment. Mallis and colleagues (2000) conducted a study at Carnegie Mellon University that implemented a first-generation automated PERCLOS device, with drowsiness and alertness feedback, in a full fidelity truck simulator to evaluate the biobehavioral responses of truck drivers during a simulated nighttime drive. The automated PERCLOS system was mounted on the dashboard of the truck driving simulator. The system was designed with three levels of interaction (feedback) to the driver (visual gauge, audible tone, and an odor vs. a voice command). The study established that feedback from the device, when PERCLOS values were above a threshold, improved driving variability and driver alertness, especially when drivers were very drowsy (Mallis et al., 2000).

This study demonstrated that it was possible to interface an on-line, automated drowsiness-detection technology in a simulator environment with some modification and improvement, thus leading to the development of the CoPilot (Grace, 2001). It is currently being deployed along with other on-line fatigue management technologies in a pilot field study of long-haul trucks being conducted by the U.S. Department of Transportation, the Federal Motor Carrier Safety Administration, and the Trucking Research Institute of the American Trucking Associations Foundation. The study seeks to determine how truck drivers react to having a variety of devices that inform them of their alertness and sleep need.

25.9 Related Methods

Although the focus of this chapter has been on slow eye closures as a valid measurement of drowsiness, there are many other drowsiness detection devices in varying stages of development (Mallis, 1999). Independent of whether a device is being developed by the government, academia, or private sector, all drowsiness detection devices should be held to the same standards of validity and reliability detailed above. It must be scientifically valid (measuring a biobehavioral marker of hypovigilance due to drowsiness) and reliable (working at all times and for all users). While a drowsiness detection device may prevent a catastrophic incident or accident, it is no substitute for establishing functional capabilities of an individual before an individual begins work.

References

Dinges, D.F. (1989), The nature of sleepiness: causes, contexts and consequences, in *Perspectives in Behavioral Medicine: Eating, Sleeping and Sex*, Stunkard, A. and Baum, A., Eds., Lawrence Erlbaum, Hillsdale, NJ.

Dinges, D.F. and Mallis, M.M. (1998), Managing fatigue by drowsiness detection: can technological promises be realized? in *Managing Fatigue in Transportation*, Hartley, L., Ed., Elsevier Science, Oxford, pp. 209–229.

Dinges, D.F., Mallis, M., Maislin, G., and Powell, J.W. (1998), Evaluation of Techniques for Ocular Measurement as an Index of Fatigue and the Basis for Alertness Management, final report for the U.S. Department of Transportation, Report DOT HS 808, National Highway Traffic Safety Administration, Washington, D.C.

Dinges, D.F., Price, N., Maislin, G., Powell, J.W., Ecker, A., Mallis, M., and Szuba, M. (2002), Prospective Laboratory Re-Validation of Ocular-Based Drowsiness Detection Technologies and Countermeasures (Subtask A), final report for the U.S. Department of Transportation, National Highway Traffic Safety Administration, Washington, D.C.

Grace, R. (2001), Drowsy Driver Monitor and Warning System, in Proceedings International Driving Symposium on Human Factors in Driving Assessment, Training and Vehicle Design, Snowmass, CO.

Grace, R., Byrne, V.E., Legrand, J.M., Gricourt, D.J., Davis, R.K., Staszewski, J.J., and Carnahan, B. (1999), A Machine Vision Based Drowsy Driver Detection System for Heavy Vehicles, Report FHWA-MC-99-136, U.S. Department of Transportation, National Highway Traffic Safety Administration, in *Ocular Measures of Driver Alertness: Technical Conference Proceedings*, Herndon, VA, pp. 75–86.

Mallis, M.M. (1999), Evaluation of Techniques for Drowsiness Detection: Experiment on Performance-Based Validation of Fatigue-Tracking Technologies, Ph.D. thesis, Drexel University, Philadelphia.

Mallis, M., Maislin, G., Konowal, N., Byrne, V., Bierman, D., Davis, R., Grace, R., and Dinges, D.F. (2000), Biobehavioral Responses to Drowsy Driving Alarms and Alerting Stimuli, Report DOT HS 809 202, final report for the U.S. Department of Transportation, National Highway Traffic Safety Administration, Washington, D.C.

Rau, P.S. (1999), A Heavy Vehicle Drowsy Driver Detection and Warning System: Scientific Issues and Technical Challenges, Report FHWA-MC-99-136, U.S. Department of Transportation, National Highway Traffic Safety Administration, in *Ocular Measures of Driver Alertness: Technical Conference Proceedings*, Herndon, VA, pp. 24–30.

Van Dongen, H.P.A. and Dinges, D.F. (2000), Circadian rhythms in fatigue, alertness and performance, in *Principles and Practice of Sleep Medicine*, 3rd ed., Kryger, M.H., Roth, T., and Dement, W.C., Eds., W.B. Saunders, Philadelphia, pp. 391–399.

Wierwille, W.W. and Ellsworth, L.A. (1994), Evaluation of driver drowsiness by trained raters, *Accident Anal. Prev.*, 26, 571–581.

Wierwille, W.W., Ellsworth, L.A., Wreggit, S.S., Fairbanks, R.J., and Kirn, C.L. (1994), Research on Vehicle-based Driver Status/Performance Monitoring: Development, Validation, and Refinement of Algorithms for Detection of Driver Drowsiness, Final Report DOT HS 808 247, National Highway Traffic Safety Administration, Washington, D.C.

26

Measurement of Respiration in Applied Human Factors and Ergonomics Research

Cornelis J.E. Wientjes
NATO Research and Technology Agency

Paul Grossman
Freiburg Institute for Mindfulness Research

26.1 Background and Applications

Respiratory measurement may be a potentially powerful asset in certain applied studies. Respiration appears to be closely coupled to a variety of important functional psychological dimensions, such as response requirements and appraisal patterns (Boiten et al., 1994), as well as to subtle variations in mental effort investment (Wientjes et al., 1998). Yet, perhaps the most striking feature of some respiratory measures is the degree to which they are linked to various dimensions of emotion, affect, and mood (Boiten et al., 1994; Nyklíek et al., 1997), as well as to the experience of psychosomatic symptoms (Grossman and Wientjes, 1989; Wientjes and Grossman, 1994).

Respiratory measures can only be cursorily discussed here. Much more extensive treatment of the topic can be found in Harver and Lorig (2000) and in Wientjes (1992). Briefly, there are two major types of respiratory assessment to be considered: (a) assessment of how depth and frequency of breathing contribute to ventilation and (b) measurement of parameters associated with gas exchange. Depth of breathing is usually expressed in terms of tidal volume (i.e., the volume displaced per single breath), and frequency in terms of respiration rate (the number of breaths per minute). The volume of air displaced per minute is expressed as minute ventilation, which is the product of tidal volume and respiration rate and typically reflects metabolic activity. In addition to these basic parameters, respiratory measurement often includes a more sophisticated evaluation of the breathing cycle in terms of the duration of the phases of the breathing cycle (inspiratory time and expiratory time), total cycle time, mean inspiratory flow rate (tidal volume/inspiratory time), and duty cycle time (see Wientjes, 1992). Measurement of gas exchange includes assessment of the volume or quantity of oxygen (VO_2) consumed per time unit, and of the quantity of carbon dioxide (VCO_2) produced. These measures can be used to calculate energy expenditure. An important final aspect of respiratory regulation should be added. While ventilation is normally in tune with metabolic requirements, the equilibrium can, under certain conditions, break down. Ventilation in excess of metabolic demands (i.e., hyperventilation) causes more CO_2 to be eliminated from the body per time unit than is produced by metabolic processes. As a consequence, arterial partial CO_2 (PCO_2) levels decrease below normal values, which results in a state of hypocapnia. The occurrence of hyperventilation and hypocapnia can be determined by measuring the PCO_2 at the end of normal expiration ($P_{et}CO_2$). If properly measured, $P_{et}CO_2$ may be considered to be a valid approximation of arterial PCO_2 (Wientjes, 1992).

26.2 Applications

Respiratory measures can provide valuable supplementary information to subjective measures among investigations assessing task or systems demands, operator workload, and stressful or potentially hazardous aspects of task environments (Wientjes, 1992; Wientjes et al., 1998). By virtue of its easy assessment, respiration rate is unquestionably the most popular respiratory measure used in this context. It has been suggested that this measure may be more sensitive to variations in workload or task difficulty than heart rate (Brookings et al., 1996). Although respiration rate may, indeed, sometimes be a convenient measure of workload and stress, a word of caution is necessary. Because alterations in respiration rate may often be secondary, or compensatory, to respiratory volume changes (Wientjes et al., 1998), simultaneous assessment of volumetric measures is strongly recommended.

Wientjes et al. (1998) have elaborated many details of the respiratory response to varying mental task demands in a laboratory study. In addition, several ambulatory studies have assessed respiratory responses, including tidal volume, respiration rate, minute ventilation, and $P_{et}CO_2$, to mental workload and stress. Harding (1987) identified highly demanding flight profiles on the basis of minute ventilation and $P_{et}CO_2$ among jet fighter pilots flying high-performance aircraft. Bles et al. (1988) and Bles and Wientjes (1988) found that exposure to vestibular stimulation (ship movements) induced hyperventilation among navy personnel. Brookings et al. (1996) reported that respiration rate was one of the most sensitive measures to differentiate between different levels of air-traffic-control workload, while Wientjes et al. (1996), also assessing air-traffic-control workload, found that transcutaneous PCO_2 (as an estimate of arterial PCO_2) differentiated between high workload episodes and episodes characterized by stress. In a series of within-subject studies among astronauts and cosmonauts during long-duration (135-day) simulated space flight, Wientjes et al. (1996, 1997) assessed the cumulative effects of isolation, operator workload, and fatigue on tidal volume, respiration rate, minute ventilation, heart rate, and blood pressure. They found that minute ventilation and systolic blood pressure were elevated during stressful episodes, in particular among the crewmates that were responsible for critical aspects of the missions. Wilhelm and Roth (1996, 1998), on the basis of ambulatory measurement of minute ventilation, provide an innovative procedure aimed at characterizing ambulatory behavioral or emotional heart rate responses that dissociate from overall change in metabolic activity. This corresponds to the emotional excess tachycardia and systemic overperfusion first demonstrated by Obrist (1981). This additional heart rate measure was also used among military personnel during exposure to a realistic stressor task (crossing of a rope bridge across an 80-m-deep ravine) (Fahrenberg and Wientjes, 2000). Although, as yet, only few ambulatory studies have assessed $P_{et}CO_2$, this type of monitoring would seem to be very promising and can document the extent of clinical improvement and synchrony of change between self-reported anxiety and changes in respiration (Grossman et al., 1985).

Several studies have used respiratory measures not as their primary target responses, but in order to obtain approximations of energy expenditure (e.g., Wertheim, 1998) or as estimates for the computation of additional heart rate (see above). Respiratory measures can also be employed as important control parameters in studies measuring cardiac vagal tone by means of respiratory sinus arrhythmia (e.g., Grossman et al., 1991; Grossman and Kollai, 1993).

26.3 Measurement

In the past decades, several approaches to respiratory measurement have been developed that provide a reasonably valid quantitative estimation without the interference introduced by spontaneous breathing, a problem that is typical of the more intrusive classical measurement techniques (Wientjes, 1992). Typically, the separate motions of the rib cage and the abdomen are measured, mostly by means of inductive respiratory plethysmography (Inductotrace™, Ambulatory Monitoring, Inc., Ardsley, NY; or Respitrace™, Vivometrics, Inc., Ventura, CA). This approach is based on a model that was introduced by Konno and Mead (1967), who argued that the volume of air that is displaced during each breathing cycle can be calculated on the basis of the circumference changes of the rib cage and the abdominal compartments (see Figure 26.1). A number of different calibration techniques have been developed that allow the estimation of the rib cage and abdominal gains by means of multiple regression or other procedures (e.g., Banzett et al., 1995; Brown et al., 1998; Groote et al., 2001; Leino et al., 2001; Millard, 2002; Reilly et al., 2002; Sackner, 1996; Strömberg, 2001; Strömberg et al., 1993). A recently developed ambulatory device, LifeShirt™ (Vivometrics, Inc., Ventura, CA), incorporates Respitrace, accelerometry, ECG, and oximetry or $P_{et}CO_2$ into a complete system that allows long-term monitoring and analysis of all major volumetric and timing parameters of respiration, as well as autonomic and cardiac measures, physical activity, and posture.

Although the most commonly used equipment for the breath-by-breath assessment of $P_{et}CO_2$ is the infrared gas analyzer (capnograph), a few studies have employed ambulatory measurement of transcutaneous PCO_2 (Wientjes, 1992).

26.4 Procedure

Monitoring and analysis of respiration requires consideration and preparation of a number of different stages. Because human factors and ergonomic research and evaluation often utilize monitoring under real-world conditions, we will describe a procedure employing the only presently commercially available, nonintrusive ambulatory system that allows estimation of both volumetric and timing components of respiration, the LifeShirt (see Figure 26.2). An incorporated microprocessor also permits registration of electronic diary information. This system, of course, can be used, as well, in laboratory environments and can be optionally fitted with a portable $P_{et}CO_2$ monitor.

26.4.1 Step 1: Research Design

Respiratory parameters have proved to be well suited as measures of physical, mental, and emotional workload and seem to reflect systematic variations in motivation and effort. Studies aimed at any of these dimensions may be appropriate candidates for respiratory measurement. Because inductive plethysmography employs thoracic and abdominal displacements for assessment of all respiratory parameters, it is best to minimize the likelihood of extreme postural movements in order to reduce artifacts. However, normal movement and postural adjustments do not greatly disturb estimation, and lager postural shifts that may cause artifacts can be eliminated from the registration during data analysis.

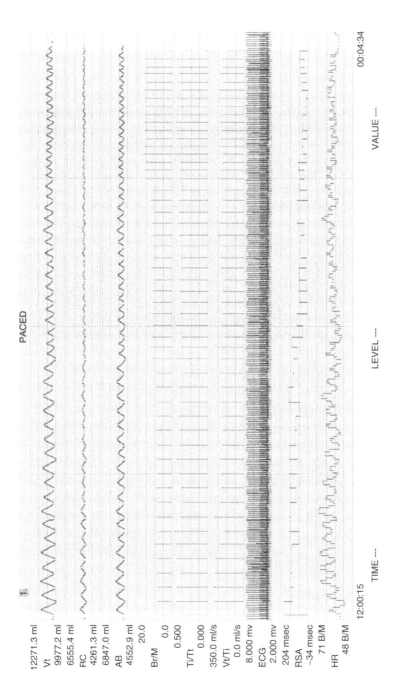

FIGURE 26.1 Vivologic™ screenshot of selected analyzed respiratory and cardiac parameters taken from a paced-breathing condition at three rates of respiration. From top to bottom, tidal volume (milliliters of air per breath), rib-cage contribution to tidal volume (RC); abdominal contribution (AB); respiration rate (breaths per minute, Br/M); respiratory timing (inspiration time/total cycle time, Ti/Tt); inspiratory flow rate (tidal volume/Ti); the raw electrocardiographic signal (ECG); peak–valley respiratory sinus arrhythmia (RSA, parasympathetically mediated respiratory fluctuations in heart rate); and momentaneous heart rate (HR). Note how the different respiratory parameters diverge, even with mere voluntary in rate and volume. Also note that RSA magnitude (reflected in both of the lowest two traces) is related to respiration rate and volume.

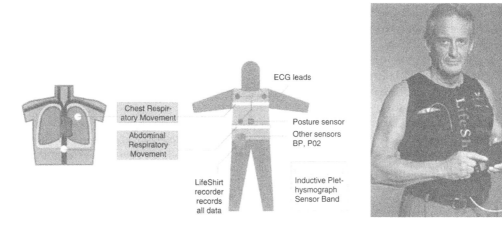

FIGURE 26.2 Left, schematic representation of the LifeShirt monitoring system; right, an example of the actual monitoring jacket and data-acquisition and electronic-diary unit.

26.4.2   Step 2: Preparing the Subject for Registration

The appropriately sized LifeShirt Lycra vest is fitted to the subject (see Figure 26.2, "snug" respiratory bands), ECG electrodes are applied, all cables are connected, and the unit is started. The subject is briefly instructed about the use of the monitor, employment of the optional electronic diary, and calibration procedure. Calibration, employing a disposable 800-ml breathing bag and nose clip, is performed in whatever postures will be assumed during actual monitoring. Initial on-line check of signals from the microprocessor display is made to ensure reliable registration.

26.4.3   Step 3: Physiological Monitoring

Respiratory signals are continuously sampled at 50 Hz (ECG at 1000 Hz and two-dimensional accelerometry at 10 Hz) during the measurement period. Data are stored on removable media, and the current battery duration is 24 hours. Battery and storage media are easily replaceable by subjects during longer recordings. Vests can be comfortably worn for 24-hour periods or longer, with some discomfort under extremely warm conditions.

26.4.4   Step 4: Data Acquisition and Analysis

Data are downloaded onto a PC hard drive via a conventional flashcard reader. Analysis of data is performed via a propriety program (VivoLogic, Vivometrics, Inc., Ventura, CA). Both original waveform traces and up to 40 different derived measures are displayed on a strip-chart-type screen for interactive assessment (parameters of volume, timing, mechanics, thoraco-abdominal coordination, derivatives, cardiac function, posture, and physical activity; see Figure 26.1). Data can be exported for further signal processing or statistical analyses as the original traces, minute values, averages for a specific segment, or breath-by-breath values.

26.5   Advantages

As has been noted by Gaillard and Kramer (2000), physiological measures are particularly valuable in applied studies of complex task or systems demands and effort investment among human operators. In these situations, performance-based methods or subjective methods, although indispensable, may often fail to provide sufficient insight in how the available resources are managed by the operator. While this is true for a range of physiological measures, the potential supplemental value of employing respiration is that respiratory parameters provide a range of information regarding dimensions of mental effort, physical workload, and emotional strain. Because minute ventilation, the product of respiration rate and tidal volume, is a measure of metabolic activity under normal circumstances, respiratory quantification permits relatively accurate and continuous assessment of energy expenditure during different types of mental and physical activity. At the same time, respiratory parameters, like inspiratory flow rate, also seem to reflect motivation and effort, whereas parameters of $P_{et}CO_2$ and variability of tidal volume and respiration rate yield information about levels of acute and chronic anxiety as well as physical pain (Wilhelm et al., 2001). Indeed, when a sufficient number of parameters are considered, respiration is capable of distinguishing between a wide variety of physical and mental performance dimensions (Grossman and Wientjes, 2001). Just this multifaceted interplay of metabolic, cognitive, and emotional dimensions of breathing parameters may serve to make respiratory monitoring either desirable or complicated, depending upon the researcher's needs and interests. If one is interested in disentangling these different dimensions, respiration assessment can provide a particularly valuable tool.

26.6 Disadvantages

For many applied research purposes, respiration may seem an unlikely choice as a target response system. Regulation of respiration is accomplished via an intricate interplay of brainstem, metabolic, and volitional influences, and these are not always easy to unravel. In addition, respiration may not be convenient to measure when one is only interested in monitoring oxygen consumption, $P_{et}CO_2$, or CO_2 production. Different types of recording equipment, many unfamiliar to the conventional researcher, may sometimes be required to assess different aspects of respiratory behavior. As previously mentioned, respiration is not a simple, one-dimensional phenomenon, but respiratory responses typically vary in a complex manner along dimensions of volume and timing, and of pattern and intensity. Given these qualifications, researchers potentially interested in respiratory assessment should ask themselves whether respiratory quantification could best address the questions they have in mind.

26.7 Example

Wientjes et al. (1998) describe an investigation in which respiratory parameters tracked different levels of mental load while quietly sitting. Demands of a memory-comparison task were varied by means of feedback and monetary incentive to yield three conditions of progressively increasing mental demands and motivation. Respitrace and $P_{et}CO_2$ monitoring were performed. Monotonic increases of both minute ventilation and inspiratory flow rate mirrored intensity of task demands. Respiration rate and $P_{et}CO_2$ did not vary between task conditions, but tidal volume displayed monotonic increases. These data indicate that minute ventilation and inspiratory flow rate seem to sensitively reflect variations in mental load due to task difficulty and/or motivation. Respiratory sinus arrhythmia changes indicated task-related effects upon cardiac vagal tone (Grossman et al., 2002). Additionally, varying mental demand in this manner appears to have a small but discernible and systematic effect upon metabolic activity, as indexed by minute ventilation variations. Further unreported findings on individual differences in respiratory regularity might suggest differential levels of task-related anxiety.

26.8 Related Methods

Respiration is closely related to other physiological processes, perhaps most apparently to the cardiovascular system (Grossman, 1983). Of considerable current interest is respiratory sinus arrhythmia (also called high-frequency heart-rate variability), frequently employed as a noninvasive index of cardiac vagal tone. Respiratory sinus arrhythmia is defined as fluctuations in heart rate that are in phase with breathing: heart rate typically accelerates during inspiration and decelerates during expiration. Measurement of respiration is necessary to adequately quantify respiratory sinus arrhythmia. Furthermore, because variations in respiration rate and tidal volume can dramatically influence respiratory sinus arrhythmia magnitude without affecting cardiac vagal tone, respiration may severely confound the relationship between vagal tone and respiratory sinus arrhythmia (Grossman et al., 1991; Grossman and Kollai, 1993). Therefore, respiratory variations must be carefully considered whenever vagal tone is estimated using respiratory sinus arrhythmia. Similarly, changes in respiration may exert significant effects upon other physiological functions, such as brain activity or skin resistance (e.g., Stern and Anschel, 1968).

26.9   Standards and Regulations

There are no standard or regulations for the registration and analysis of respiration, except for clinical medical protocols. However, a useful summary of the most important practical guidelines is provided by Wientjes (1992).

26.10   Approximate Training and Application Times

Inevitably, like mastering most complex skills, a considerable investment in time, effort, and resources is needed to sufficiently familiarize oneself with the underlying physiology, the measurement, and the analysis of respiratory responses. Under expert supervision, a student or laboratory assistant can learn the basics necessary for carrying out measurements in a few weeks. However, developing the advanced expert knowledge that is necessary to supervise serious respiratory research is a major undertaking that can take much more time.

Application of the sensors, carrying out the calibration procedure, and checking the quality of the signals can, in particular when employing the user-friendly Respitrace and LifeShirt systems, vary from 10 to 30 min.

26.11   Reliability and Validity

The reliability and validity of various calibration methods for Respitrace have, in the last decade, received much attention from investigators (e.g., Banzett et al., 1995; Brown et al., 1998; Groote et al., 2001; Leino et al., 2001; Millard, 2002; Reilly et al., 2002; Sackner, 1996; Strömberg, 2001; Strömberg et al., 1993).

An important potential setback that needs to be addressed is that Respitrace bands respond not only to movements associated with respiratory activity, but also to postural changes and other movements. Some investigators have used simple filters to remove movement-related artifacts during on-line processing of the respiratory signals (e.g., Anderson and Frank, 1990). This somewhat crude approach has received criticism from Wilhelm and Roth (1998), who have argued that respiratory data analysis is too complex to be performed completely automatically.

The LifeShirt system employs advanced digital filtering of the respiratory signal, and concurrent estimation of postural changes through accelerometry permits both automatic and manual adjustments for movement-related artifacts. Additionally, the accompanying physiological analysis program allows for visual inspection and correction of artifact-prone segments.

Measurement of $P_{et}CO_2$ may also pose serious validity problems that need to be taken into account. The CO_2 waveforms should be very carefully inspected before inclusion in the analyses. Breaths with a low percentage of end-tidal air (mostly shallow breaths) can be recognized by the fact that the CO_2 waveform does not reach a horizontal plateau but, rather, increases steeply and then breaks off suddenly (Gardner et al., 1986). Automatic identification and exclusion of such breaths is still a contentious issue.

26.12   Tools Needed

Useful information on Respitrace and LifeShirt can be obtained from http://www.vivometrics.com/site/index.html. Information on an alternative device for inductive respiratory plethysmography, Inductotrace, can be obtained at http://www.ambulatory-monitoring.com/. Information on an alternative ambulatory monitoring system, which can be equipped with a Respitrace module for cardiorespiratory measurements, can be obtained from http://www.rrz.uni-koeln.de/phil-fak/psych/diagnostik/ VITAPORT/.

References

Anderson, D.E. and Frank, L.B. (1990), A microprocessor system for ambulatory monitoring of respiration, *J. Ambulatory Monitoring*, 3, 11–20.

Banzett, R.B., Mahan, S.T., Garner, D.M., Brughera, A., and Loring, S.H. (1995), A simple and reliable method to calibrate respiratory magnetometers and Respitrace, *J. Appl. Physiol.*, 79, 2169–2176.

Bles, W. and Wientjes, C.J.E. (1988), Well-being, task performance and hyperventilation in the tilting room: effects of visual surround and artificial horizon (Dutch), IZF Report 1988–30, TNO Institute for Perception, Soesterberg, the Netherlands.

Bles, W., Boer, L.C., Keuning, J.A., Vermeij, P., and Wientjes, C.J.E. (1988), Seasickness: dose-effect registrations with HMS Makkum (Dutch), IZF Report 1988–5, TNO Institute for Perception, Soesterberg, the Netherlands.

Boiten, F.A., Frijda, N.H., and Wientjes, C.J.E. (1994), Emotions and respiratory patterns: review and critical analysis, *Int. J. Psychophysiol.*, 17, 103–128.

Brookings, J.B., Wilson, G.F., and Swain, C.R. (1996), Psychophysiological responses to changes in workload during simulated air traffic control, *Biol. Psychol.*, 42, 361–377.

Brown, K., Aun, C., Jackson, E., Mackersie, A., Hatch, D., and Stocks, J. (1998), Validation of respiratory inductive plethysmography using the qualitative diagnostic calibration method in anaesthetized infants, *Euro. Respir. J.*, 12, 935–943.

Fahrenberg, J. and Wientjes, C.J.E. (2000), Recording methods in applied environments, in *Engineering Psychophysiology*, Backs, R.W. and Boucsein, W., Eds., Lawrence Erlbaum Associates, Mahwah, NJ.

Gaillard, A.W.K. and Kramer, A.F. (2000), Theoretical and methodological issues in psychophysiological research, in *Engineering Psychophysiology*, Backs, R.W. and Boucsein, W., Eds., Lawrence Erlbaum Associates, Mahwah, NJ.

Gardner, W.N., Meah, M.S., and Bass, C. (1986), Controlled study of respiratory responses during prolonged measurement in patients with chronic hyperventilation, *Lancet*, i, 826–830.

Groote de, A., Paiva, M., and Verbandt, Y. (2001), Mathematical assessment of qualitative diagnostic calibration for respiratory inductive plethysmography, *J. Appl. Physiol.*, 90, 1025–1030.

Grossman, P. (1983), Respiration, stress and cardiovascular function, *Psychophysiology*, 20, 284–300.

Grossman, P., DeSwart, H., and Defares, P.B. (1985), A controlled study of breathing therapy for hyperventilation syndrome, *J. Psychosomatic Res.*, 29, 49–58.

Grossman, P., Karemaker, J., and Wieling, W. (1991), Respiratory sinus arrhythmia, cardiac vagal tone and respiration: within- and between-individual relations, *Psychophysiology*, 30, 486–495.

Grossman, P. and Kollai, M. (1993), Prediction of tonic parasympathetic cardiac control using respiratory sinus arrhythmia, *Psychophysiology*, 28, 201–216.

Grossman, P. and Wientjes, C.J.E. (1989), Respiratory disorders; asthma and hyperventilation syndrome, in *Handbook of Clinical Psychophysiology*, Turpin, G., Ed., Wiley, Chichester, U.K.

Grossman, P. and Wientjes, C.J.E. (2001), How breathing adjusts to mental and physical demands, in *Respiration and Emotion*, Haruki, Y., Homma, I., Umezawa, A., and Masaoka, Y., Eds., Springer, New York.

Grossman, P., Wilhelm, F.H., and Wientjes, C.J.E. (2002), On the ambiguity of high-frequency heart-rate variability (respiratory sinus arrhythmia) as a measure of cardiac vagal tone during mental tasks: further evidence of RSA as a respiratory rate– and volume-dependent phenomenon, *Psychophysiology*, 39, S64.

Harding, R.M. (1987), Human Respiratory Responses during High Performance Flight, AGARDograph 312, NATO Advisory Group for Aerospace Research and Development (AGARD), Neuilly-sur-Seine, France.

Harver, A. and Lorig, T.S. (2000), Respiration, in *Handbook of Psychophysiology*, 2nd ed., Cacioppo, J.T., Tassinari, L.G., and Berntson, G.G., Eds., Cambridge University Press, Cambridge, U.K.

Konno, K. and Mead, J. (1967), Measurement of the separate volume changes of rib cage and abdomen during breathing, *J. Appl. Physiol.*, 22, 407–422.

Leino, K., Nunes, S., Valta, P., and Takala, J. (2001), Validation of a new respiratory inductive plethysmograph, *Acta Anaesthesiol. Scand.*, 45, 104–111.

Millard, R.K. (2002), Key to better qualitative diagnostic calibrations in respiratory inductive plethysmography, *Physiol. Meas.*, 23, N1–8.

Nyklíek, I., Thayer, J.F., and van Doornen, L.J.P. (1997), Cardiorespiratory differentiation of musically induced emotions, *J. Psychophysiol.*, 11, 304–321.

Obrist, P.A. (1981), *Cardiovascular Psychophysiology: A Perspective*, Plenum Press, New York.

Reilly, B.P., Bolton, M.P., Lewis, M.J., Houghton, L.A., and Whorwell, P.J. (2002), A device for 24 hour ambulatory monitoring of abdominal girth using inductive plethysmography, *Physiol. Meas.*, 23, 661–670.

Sackner, M.A. (1996), A simple and reliable method to calibrate respiratory magnetometers and Respitrace, *J. Appl. Physiol.*, 81, 516–517.

Stern, R.M. and Anschel, C. (1968), Deep inspirations as a stimulus for responses of the autonomic nervous system, *Psychophysiology*, 5, 132–141.

Strömberg, N.O.T. (2001), Error analysis of a natural breathing calibration method for respiratory inductive plethysmography, *Med. Biol. Eng. Comput.*, 39, 310–314.

Strömberg, N.O.T, Dahlbäck, G.O., and Gustafsson, P.M. (1993), Evaluation of various models for respiratory inductance plethysmography calibration, *J. Appl. Physiol.*, 74, 1206–1211.

Wertheim, A.H. (1998), Working in a moving environment, *Ergonomics*, 41, 1845–1858.

Wilhelm, F.H, Gevirtz, R., and Roth, W.T. (2001), Respiratory dysregulation in anxiety, functional cardiac, and pain disorders: assessment, phenomenology, and treatment, *Behav. Modification*, 25, 513–545.

Wilhelm, F.H. and Roth, W.T. (1996), Ambulatory assessment of clinical anxiety, in *Ambulatory Assessment*, Fahrenberg, J. and Myrtek, M., Eds., Hogrefe and Huber, Göttingen, Germany.

Wilhelm, F.H. and Roth, W.T. (1998), Using minute ventilation for ambulatory estimation of additional heart rate, *Biol. Psychol.*, 49, 137–150.

Wientjes, C.J.E. (1992), Respiration in psychophysiology: measurement issues and applications, *Biological Psychol.*, 34, 179–203.

Wientjes, C.J.E., Gaillard, A.W.K., and ter Maat, R. (1996), Measurement of transcutaneous PCO_2 among air traffic controllers: differential responding to workload and stress, *Biol. Psychol.*, 43, 250–251.

Wientjes, C.J.E. and Grossman, P. (1994), Overreactivity of the psyche or the soma? Interindividual associations between psychosomatic symptoms, anxiety, heart rate, and end/tidal partial carbon dioxide pressure, *Psychosomatic Med.*, 56, 533–540.

Wientjes, C.J.E, Grossman, P., and Gaillard, A.W.K. (1998), Influence of drive and timing mechanisms on breathing pattern and ventilation during mental task performance, *Biol. Psychol.*, 49, 53–70.

Wientjes, C.J.E., Veltman, J.A., and Gaillard, A.W.K. (1996), Cardiovascular and respiratory responses during complex decision-taking task under prolonged isolation, in *Advances in Space Biology and Medicine*, Vol. 5, Bonting, S.L., Ed., JAI Press, Greenwich, CT.

Wientjes, C.J.E., Veltman, J.A., and Gaillard, A.W.K. (1997), Cardiovascular and respiratory responses during simulation of a 135-day space flight: a longitudinal within-subject study, *Biol. Psychol.*, 46, 79–80.

Behavioral and
Cognitive Methods

27

Behavioral and Cognitive Methods

Neville A. Stanton
Brunel University

27.1   Introduction

Behavioral and cognitive methods have their original foundation in the psychological disciplines. The methods presented in this section of the handbook generally provide information about the perceptions, cognitive processes, and (potential) response(s) of individuals, although this information may be aggregated from many people. This information can be presented in many different forms and guises, including human errors, human tasks, task times, goals and subgoals, decisions, workload, and user preferences. Overarching and integrating these data are general psychological models of human performance. Three such models have dominated human factors and ergonomics over the past three decades: Norman's (1986) model of action, Wickens's (1992) model of multiple attentional resources in information processing, and Neisser's (1976) model of the perceptual cycle in human ecology. Each of these models is briefly discussed.

Norman's (1986) model of action presents a description of human activity in two distinct phases: execution (where human action brings about changes in the world) and evaluation (where the changes in the world are evaluated). Both of these phases are linked by goals, which define the purpose of the activity. In the seven-stage model, a goal is translated into intention, which is in turn translated into a sequence of actions, which are finally executed. Feedback on the effects of the action are perceived by the sensory systems, those perceptions are then interpreted, and interpretations evaluated and related to the goal. Norman has used the model to illustrate the extent to which system design supports or fails to support the execution and evaluation phases of human activity. This model of action may be linked to ergonomics methods; for example, hierarchical task analysis (HTA) provides a methodology of representing human and system goals.

Wickens's (1992) model of multiple attentional resources in information processing builds upon the standard stage-based process model (i.e., where information is perceived via the sensory systems, presented to a decision-making system that draws on memory, from which actions are decided upon and executed. Wickens extends the model to propose a theory of multiple pools of attentional resources in relation to different information-processing demands: speech and text utilize a verbal information-processing code and draw upon a different pool of attentional resources than tones and pictures, which utilize a spatial processing code. Wickens argues that when the attentional resources assigned to the verbal processing code are exhausted, workload demands may be increased further by using the alternative spatial information processing code through the presentation of tones or pictures (although these pools are not wholly mutually exclusive). The idea of *demands* and *resources* provides a useful conceptual framework for ergonomics. Demands and resources could come from the task, the device, and the user. For example, user resources (e.g., knowledge, experience, and expertise) and demands (e.g., user goals and standards) interact with task demands (e.g., task goals and standards) and task resources (e.g., instruction manuals and training). This interaction is mediated by demands (e.g., device complexity) and resources (e.g., clarity of the user-interface, which could reduce demands) of the device being operated. This is a familiar concept in discussions of task workload and situational awareness, and it is implied that demand–resource imbalance can occur as both task underload and task overload, both of which are detrimental to task performance.

Neisser (1976) puts forward a view of how human thought is closely coupled with a person's interaction with the world. He argued that knowledge of how the world works (e.g., mental models) leads to the anticipation of certain kinds of information, which in turn directs behavior to seek out certain kinds of information and provide a ready means of interpretation. During the course of events, as the environment is sampled, the information serves to update and modify the internal, cognitive schema of the world, which will again direct further search. Arguably, the product constructs elicited by the repertory-grid methodology draw upon these schemas. If, as schema theory predicts, action is directed by schema, then faulty schemas or faulty activation of schemas will lead to erroneous performance. As Table 27.1 shows, this can occur in at least three ways. First, we can select the wrong schema due to misinterpretation of the situation. Second, we can activate the wrong schema because of similarities in the trigger conditions. Third, we can activate schemas too early or too late. The error prediction methodologies are based upon trying to identify what can go wrong from a description of what should happen.

Hancock and Dias (2002) identify a major stumbling block for a unified ergonomics theory as the cover-all requirement: to predict all manner of systems interaction in a wide variety of contexts. It should come as no surprise that context specificity in theoretical development is the current order of the day. Despite all of these problems, Hancock and Dias are optimistic about the integration of concepts in ergonomics. As an example, they cite the intersection between the field-of-safe-travel research and the work on situational awareness. The former comes from a tradition of ecological psychology (e.g., Neisser , 1976), whereas the latter comes from information-processing theory (e.g., Wickens, 1992). They argue that there is more uniting these concepts than separating them, as both theories are concerned with people's perceptions, interpretations, and predictions of the world.

Past development of methods comes from a variety of processes based upon enhancements, modifications, and combinations of existing methods or transferring methods from one domain and adapting it to another. Examples of enhancements and modifications include the developments to hierarchical task analysis and developments to the systematic human-error reduction and prediction approach. Examples of combining methods include the combination of state–space diagrams, hierarchical task analysis, and transition matrices in order to produce task analysis for error identification. Examples of moving an analysis method from one domain and adapting it to another includes the application of repertory-grid methodology from personality-questionnaire development to the field of product-evaluation constructs and the application of the critical-path analysis methodology from project management to multimodal cognitive modeling.

The methods in this section have been classified into four categories: general analysis methods, cognitive task analysis methods, error analysis methods, and workload and situational analysis methods. An overview of the methods follows.

27.2   General Analysis Methods

Five of the methods presented in this section of the handbook address general aspects of human factors. These methods are observation (Chapter 28), interviews (Chapter 29), verbal protocol analysis (Chapter 30), repertory grids (Chapter 31), and focus groups (Chapter 32).

Observation of activities, in combination with interviews of people performing those activities, are the core approaches in the ergonomist's toolkit. These approaches feed data into many other methods, such as hierarchical tasks analysis, cognitive task analysis, verbal protocol analysis, human error identification methodologies, and workload analyses. Yet observation is also fraught with many practical and methodological difficulties. The main problems with observation as a data collection technique include the effect the observer has on the observed, and the representativeness of the people and tasks observed. Overcoming these problems requires planning, piloting of the methods, careful selection of the participants and tasks, familiarization of participant with the observer(s), and studies of reliability and validity of the data. The three main ways that observational data are collected are direct observation (i.e., watching live performance of the task), indirect observation (i.e., watching and/or hearing a recording of the task), and participant observation (i.e., participating in the task oneself). These methods can be used in combination with each other in order obtain a broader understanding of the task. Observational data can take many forms, such as time and error data, descriptions and frequencies of activities, verbatim protocols, and behavioral narratives.

In a similar manner to observation, interviews can take many forms, from completely unstructured (i.e., just random questions asked as they occur to the interviewer), to semistructured (i.e., following an agenda of questions, but deviating off the path to following interesting lines of enquiry, before returning to the agenda items), to completely structured (i.e., a verbal questionnaire where the interviewer simply speaks the written questions and records the responses). Obviously, the type of approach adopted will depend upon the purpose for which the interview is being conducted. The semistructured approach is recommended for the purposes of a product evaluation interview. An agenda allows the interviewer to follow a purposeful course, but with the flexibility to pursue relevant additional lines of questioning as they arise. A structure of the questioning is proposed as starting with open questions, following up with probing questions, and finishing with closed questions.

Verbal protocol analysis uses data from verbal transcripts to analyze the content therein. These transcripts could come from protocols gathered from recordings of live performance of the task. The transcript can be coded and analyzed at various levels of detail, from individual words, to phrases, to sentences, to themes. If the protocols are to come from "think aloud" scenarios, the participant may have to be instructed and trained in the activity of protocol generation first. The process of generating verbal protocols while simultaneously performing the task places additional demand on the participant, which needs to be taken into account. Nonetheless, verbal protocols from novice and expert task performers can provide a useful source of data. Encoding the verbalizations requires the development of a mutually exclusive and exhaustive categorization scheme, which should first be tested for reliability before it is used to analyze the transcripts. Once reliability has been established, the scheme can be used with some confidence. Any changes in the scheme or tasks require that the reliability be reestablished.

The repertory-grid technique is a method used to elicit personal constructs from people. It is a grounded theory approach, relying upon the participant to generate the data set, rather than imposing a top-down framework within which the person responds. Based upon personal construct theory developed in the field of personality psychology, the repertory-grid methodology has three main stages. First, the pool of artifacts from which the constructs will be based are defined. Second, triads (i.e., groups of three artifacts) are drawn from the pool, and participants are asked to state what makes two of the artifacts the same and different from the third. This results in a construct that should have a logical opposite. The process is repeated until all constructs and/or all combination of triads are exhausted. Third, participants are asked to rate all of the artifacts against the constructs, and the matrix is analyzed. The result provides a set of personal constructs for product evaluation. If the process is repeated with a large set of individuals and a large set of artifacts, then some common constructs are likely to emerge. The process is likely to be of most use at the level of classes of artifact to examine the different sorts of constructs that people hold, rather than looking for commonality across everything.

Focus groups are an extension of the individual interview and have a long tradition in market research and product design. Focus groups have been found to be useful when gathering people together to talk about individual and collective reactions to events. Talking in a group has both positive and negative connotations. On the positive side, there is safety in numbers. People might be more willing to discuss their reactions with others sharing similar points of view. It may also be possible to draw a consensus of feeling about the event under discussion. On the negative side, people may feel inhibited by the presence of others if the issue is of a sensitive nature or if they do not agree with the general consensus. In addition, Cooper and Baber (Chapter 32) identify a major limitation in product design, which is that people are not very good a imagining themselves using devices that currently do not exist. Therefore, they have developed the idea of scenario-based focus groups. This method requires the focus group to develop a scenario — using task steps, story-boards, or role-play — based on an existing product that is in some way related to the conceptual product under evaluation. This can serve as an outline script for the new product. In this way, the consideration of the new product can be integrated into their everyday activities. Then they can question the potential advantages and disadvantages offered by the new product and consider how they might use it. The benefit of the group approach is that ideas can be developed by more than one person, and individuals can help each other see how the device might be used.

27.3   Cognitive Task Analysis Methods

Four of the methods presented in this section of the handbook address cognitive task analysis and allocation of system function. These methods are hierarchical task analysis (Chapter 33), an extension of hierarchical task analysis known as allocation of function methodology (Chapter 34), the critical decision method (Chapter 35), and applied cognitive work analysis (Chapter 36).

Hierarchical task analysis (HTA) was initially developed in response to the need for greater understanding of cognitive tasks. With greater degrees of automation in industrial work practices, the nature of worker tasks were changing in the 1960s. It has been argued that because these tasks involved significant cognitive components (such as monitoring, anticipating, predicting, and decision making), a method of analyzing and representing this form of work was required. Traditional approaches tended to focus on observable aspects of performance, whereas HTA sought to represent system goals and plans. Hierarchical task analysis describes a system in terms of goals and subgoals, with feedback loops in a nested hierarchy. Its enduring popularity can be put down to two key points. First, it is inherently flexible: the approach can be used to describe any system. Second, it can be used to many ends: from person specification, to training requirements, to error prediction, to team performance assessment, and to system design. HTA also anticipated the interest in cognitive task analysis by more that 15 years.

One extension of HTA is presented in the form of a cognitive allocation of function methodology. Numerous approaches have been taken to the allocation of system functions over the years The approach presented in this section develops from allocation of the system subgoals in HTA. While many have used HTA to describe human activity, there is no reason why it cannot be extended to describe general system activity (i.e., the activities of both human and machine subsystems). Many system descriptions are indeed HTA-like in their representation (e.g., Diaper and Stanton, 2003), so that this is a very natural progression. In the method of allocation of function presented, there is a differential between four basic types of subgoal allocation: human only, computer only, shared but human in charge, shared but computer in charge. Working through the HTA, each of the subgoals can be allocated as appropriate, depending upon the skills and technology available. In allocation of system function, four criteria are applied: the job satisfaction of the person doing the work, the potential for human error, the potential effect on the situational awareness of the person doing the work, and the resource implications of the allocation. The process is supposed to be reiterative and requires constant review and redescription.

The critical decision method (CDM) is an update and extension of the critical incident technique. The critical decision method structures the interview in an incident analysis by asking the interviewee to review critical decision points as the event unfolded. The investigation begins with detailing the incident timeline and then looking at decision points along that timeline. A series of questions are then presented to probe each decision, such as:

- What information cues were attended to?
- What were the situation assessments?

- What information was considered?
- What options were considered?
- What basis was used to select the final option?
- What goals were to be achieved?

The approach can generate large amounts of data that have to be carefully managed. The structure of the questioning should help the analyst identify conflicts and contradictions between and within interviewee's responses. The structured and exhaustive approach is very thorough. The critical decision method usually requires four sweeps of the event at deeper and deeper levels of analysis to be completed. This is inevitably time-consuming, and the accuracy of the evidence is wholly dependent on a person's recall memory. Correlating the evidence with other sources can increase confidence in the reliability of the data.

The applied cognitive work analysis (ACWA) method is distinct from CDM in that it is based on the analysis of demands and constraints that are inherent in the task domain. ACWA has typically been used in process-control and command-and-control environments. The methodology aims to approach design holistically, considering all aspects of the system, including the organizational structure, procedures, training program, automation, database design, and sensors. The systems-based, sociotechnical, ecological approach offers a formative approach to design. This offers a considerable improvement to the design process over the reiterative cycle of build-then-test that is often seen in interface design. The basic steps in the process include development of a functional abstraction network, determining cognitive work requirements, identifying information relationship requirements, representing design requirements, and presenting design concepts. The analysis is driven by a comprehensive analysis of the context and constraints within which the work is to be performed.

27.4   Error Analysis Methods

Two of the methods presented in this section focus on the prediction of human error. These methods are the systematic human error reduction and prediction approach (SHERPA) (Chapter 37) and task analysis for error identification (TAFEI) (Chapter 38).

Systematic human error reduction and prediction approach (SHERPA) is based, in part, on HTA as a description of normative, error-free behavior. The SHERPA analyst uses this description as a basis to consider what can go wrong in task performance. At the core of SHERPA is a task and error taxonomy. The idea is that each task can be classified into one of five basic types. The error taxonomy is continually under revision and development, so it is best considered as a work in progress. SHERPA uses hierarchical task analysis together with an error taxonomy to identify credible errors associated with a sequence of human activity. In essence, the SHERPA technique works by indicating which error modes are credible for each task step in turn, based upon an analysis of work activity. This indication is based upon the judgment of the analyst, and it requires input from a subject-matter expert to be realistic. In its favor, SHERPA, like HTA, has been applied in a diverse set of industries to a wide range of tasks, including control room tasks, maintenance tasks, transportation tasks, and command-and-control tasks. Research comparing SHERPA with other human error identification methodologies suggests that it performs better than most in a wide set of scenarios.

Task analysis for error identification (TAFEI) is an error identification method that is based upon a theory of human–product interaction called rewritable routines. The idea of rewritable routines is that they are transitory, either becoming completely overwritten or modified. From this theory of human–product interaction, TAFEI has developed as a method for predicting, representing, and analyzing the dialogue between people and products. TAFEI has two forms of output. The first form of output is predicted errors from human interaction with a device, based on the analysis of transition matrices. The second form of output is a model of task flow based on mapping human action onto state–space diagrams. The task flow model has been used as part of an analytical prototyping procedure to assess a virtual product. Validation research suggests that the modeling is reasonably realistic, and the TAFEI certainly outperforms heuristic evaluations.

SHERPA and TAFEI are inherently different in the way they work. SHERPA is a divergent error-prediction method: it works by associating up to ten error modes with each action. In the hands of a novice, it is typical for there to be an overinclusive strategy for selecting error modes. The novice user could use a play-safe-than-be-sorry approach and predict many more errors than are likely to occur. This could be problematic; "crying wolf" too many times might ruin the credibility of the approach. TAFEI, by contrast, is a convergent error-prediction technique: it works by identifying the possible transitions between the different states of a device and uses the normative description of behavior (provided by the HTA) to identify potentially erroneous actions. Even in the hands of a novice, the technique seems to prevent the individual generating too many false alarms, certainly no more than they do using heuristics. In fact, by constraining the user of TAFEI to the problem space surrounding the transitions between device states, it should exclude extraneous error prediction. That said, SHERPA is a simpler technique to use, as it does not require an explicit representation of machine activity.

27.5   Workload and Situational Analysis Methods

The last four methods presented in this section of the handbook address mental workload, modeling task time, and situational awareness. The methods presented include the NASA task load index and the situation-awareness rating technique (Chapter 39), multiple resources time-sharing model (Chapter 40), critical path analysis (Chapter 41), and the situational awareness global assessment technique (Chapter 42).

There is a range of methods available for the measurement of mental workload, and it is argued that this field has become increasingly important with more emphasis on cognitive task demands. Mental workload is defined as a multidimensional concept incorporating task and performance demands together with operator skill and attention. Both mental overload and mental underload are associated with performance decrements, and task design is challenged with keeping workload within an optimal performance zone, where workload is neither too high nor too low. Measures of mental workload include measures of primary and secondary task performance, physiological measures (see also Section II on psychophysiological measures, edited by Karel Brookhuis), and subjective measures.

The multiple resources time-sharing model (MRTSM) has been developed from Wickens's (1992) multiple resources theory into a practical approach for predicting workload in situations where multiple tasks are performed concurrently. The model distinguishes between perceptual modalities, processing stages, processing codes, and responses. The methodology for predicting multiple task performance can be used for heuristic and computational evaluation. Through a worked example, Wickens (Chapter 40) demonstrates how overload can be predicted. This offers a considerable extension of multiple-resource theory, developing it into a predictive methodology. In its present form, the method is not able to predict mental underload on tasks.

The multimodal critical path analysis (mmCPA) method has its roots in two traditions. CPA is based in project management literature, whereas multimodality of people is based in a human factors literature. Traditional methods for modeling human response time are constrained because they do not represent multimodality. For example, the keystroke-level model (KLM) method offers a simple additive method for calculating response times in computing tasks. However, this traditional approach does not take account of multiple attentional resource theory, which proposes that tasks can be performed in parallel if they use different modalities and draw upon different resources. A fundamental difference in performance is found between tasks performed in parallel and those performed in series. If two or more tasks occupied the same modality, then they must be performed in series, but if they occupied different modalities they could be performed in parallel. The mmCPA uses this approach to model task time with claims for greater accuracy than the KLM model.

The situational awareness global assessment technique (SAGAT) measures three levels of awareness (i.e., perception of elements, comprehension of the situation, and prediction of future status) by presentation of recall-probe questions when the task is interrupted. The recall probes are developed using a HTA-type technique so that operator goals can be elicited. From this, questions at each of the three levels of awareness can be developed. In an aviation example, questions about perception of the elements might

TABLE 27.1 Analysis of the Behavioral and Cognitive Methods by Form of Data Output

	Methods	Output						
		Time	Errors	Tasks	Goals	Decisions	Workload	Usability
General Analysis Methods	Observation	▓	▓	░				
	Interview		░	░	░	░	░	
	Verbal Protocol	░	░	░		░	░	
	Rep. Grids							░
	Focus Groups		░	░	░	░	░	▓
Cognitive Task Analysis	HTA			░	▓	░		
	Function Alloc.			░	░		░	
	CDM	░		░	░	▓	░	
	ACWA			░	░	░	░	▓
Error Analysis	SHERPA		▓	░	░	░		
	TAFEI		▓	░	░	░		░
Workload and Situation Analysis	Workload			░	░	░	▓	
	MRTSM			░	░	░	▓	
	CPA	▓		░	░	░	░	░
	SAGAT		░	░	░	░	░	░

include air speed, altitude, engine revolutions, fuel status, location, and heading; whereas questions about comprehension of the situation might include time and distance with fuel available, tactical status of threats, and mission status; and questions about project of future status might include potential aircraft conflicts. In the case of a task simulation, the technique requires stopping the simulation while the questions are presented. This process is undertaken several times at random points throughout the experimental session. Guidelines for the administration of the SAGAT queries and data analysis are presented by Jones and Kaber in Chapter 42.

27.6 Conclusions

In conclusion, the methods presented in this section of the handbook provide a variety of types of data about human performance. As indicated in Table 27.1, the coverage of the methods vary but are complementary. Most projects would require using multiple methods in order to provide a comprehensive picture of cognitive and behavioral demand and constraints.

As illustrated in Table 27.1, the 15 methods in this section vary in terms of their output and area of focus. The darker shading is intended to represent the primary form of output, and the lighter shading is intended to represent secondary forms of output. Some methods have very general coverage, in which cases a primary form of output is not identified.

As with all ergonomics methods, selection of the appropriate set of methods requires that the analyst undertake careful definition of the purpose of the analysis. Most studies are likely to involve a combination of methods. This has the advantage of cross-fertilization and validation between data from a variety of sources. Application of methods will benefit from a pilot study to determine whether the methods are likely to yield the kind, and form, of data required.

References

Diaper, D. and Stanton, N.A. (2003), *Task Analysis in Human–Computer Interaction*, Lawrence Erlbaum Associates, Mahwah, NJ.

Hancock, P.A. and Dias, D.D. (2002), Ergonomics as a foundation of science of purpose, *Theor. Issues Ergonomics Sci.*, 3, 115–123.

Neisser, U. (1976), *Cognition and Reality: Principles and Implications of Cognitive Psychology*, Freeman, San Francisco.

Norman, D.A. (1986), Cognitive engineering, in *User Centered System Design*, Norman, D.A. and Draper, S.W., Eds., Lawrence Erlbaum Associates, Hillsdale, NJ.

Wickens, C.D. (1992), *Engineering Psychology and Human Performance*, Harper Collins, New York.

28
Observation

Neville A. Stanton
Brunel University

Christopher Baber
University of Birmingham

Mark S. Young
University of New South Wales

28.1 Background and Applications

Observation of people interacting with a device to perform a task provides a way of capturing data on errors and performance time as well as providing some insight into the ease or difficulty with which the task is performed (Baber and Stanton, 1996; Stanton, 1999; Stanton and Young, 1999). There are many and varied observational techniques, which fall into three broad categories: direct, indirect, and participant observation (Drury, 1995). Ideally, the participants would be representative end users of the system being analyzed, but this is not always possible.

Observation can be very useful for recording physical task sequences or interactions between workers. Indeed, Baber and Stanton (1996) have already considered the potential of observation as a technique for usability evaluation, and they provide helpful guidelines for those conducting research with these methods. When we observe something, what do we see and how should we ensure that other people will see the same thing? If our observations are being conducted as part of the design process, then it becomes important to collect observations that can be interpreted and applied by other members of the design team. Even if our observations are conducted to get a "feel" for what is going on, we ought to consider how to communicate this "feeling." We might decide to employ generalizations about a product, such as "most people find the product easy to use," or we might prefer to provide some form of anecdotal evidence, such as "one user of a ticket vending machine was observed rolling a five pound note into a small tube and attempted to insert this tube into the coin slot." We should aim to present a record that can be interpreted as unambiguously as possible, based on a set of observations that can be unequivocally defined. We may wish to record some commentary made by the person on their action. This commentary could consist of a description of what they are doing, an account of their planning and intentions, a justification of what they have done, etc., all of which fall under the general heading of verbal protocol (see Walker on Verbal Protocol Analysis, Chapter 30), or could consist of discussions with another co-user of the product (see Cooper and Baber on Focus Groups, Chapter 32).

A further problem arises from knowing what to observe and how often to observe it. While we might wish to limit our observations to the use of a particular piece of equipment, we are still left with the issue of who will be using it and where they will be using it. If we conduct our observations in a usability laboratory, can we be sure that it is sufficiently similar to the domain in which the object will be used? For instance, in one set of studies reported to us, the researchers found that performance times in the usability laboratory were consistently faster than in an office, and that perceived levels of workload for the same tasks were rated as higher in the usability laboratory than in the office. These differences were thought to stem partly from participants' perceptions of the usability assessment, and partly from the excising of the task from its normal work environment. This latter phenomenon led to problems of pacing the work, of planning tasks within the context of work routine, and of the feeling of being watched.

Clearly, watching somebody can have a bearing on their performance. We may try to be as unobtrusive as possible, but the presence of an observer can lead people to engage in behavior additional to their normal activity. For instance, observation could lead people to demonstrate a knowledge of how a product ought to be used, rather than how they actually use it. This is especially true if people are removed from the normal settings in which they use the product. Alternatively, one could attempt to observe behavior covertly, i.e., by using hidden cameras. However, this raises issues of the ethics of conducting observations. Obviously, if the project involves a field study of several hundred people, it would be difficult to canvas all of the people to obtain their consent. A final type of observation that may be useful is known as participant observation. In this approach, the observer actually performs the task or work under consideration. For some studies, this may require a considerable period of training, e.g., when conducting participant observation of the behavior of pilots in aircraft, and it may be easier to obtain similar subjective information from interviewing experts. Actually performing a task yourself can provide some useful insight into potential problems that users may encounter. If one is studying the design of automatic teller machines, a simple form of participant observation would be to use a series of different machines, noting the aspects of machine use that were problematic. This could then form the basis of an observation schedule for a study. Alternatively, if one is interested in the process by which products are designed, actually sitting in and contributing to a design meeting is an invaluable form of participant observation.

28.2   Procedure and Advice

Observation seems perhaps, at first glance, to be the most obvious way of collecting performance data on people to inform user-centered design. Essentially, the method simply requires one to observe users performing tasks. However, this belies the complexity of potentially interacting and confounding variables. The observational method begins with a scenario: the observer should present the participant with the device and a list of tasks to perform. The observer can then sit back and record aspects of human–device interaction that are of interest. Typical measures are execution times and any errors observed. This information can be integrated into the design process for the next generation of devices. The wise observer will draw up an observation sheet for use in data collection prior to commencing. Filling in cells on a table is quicker and easier than writing prose while your participant is performing a task. Video observation can be a valuable tool, particularly with the computer-assisted analysis techniques that are now available. These can greatly reduce data collection and analysis time.

One of the main concerns with observation is the intrusiveness of the observational method. It is well known that the behavior of people can change purely as a result of being watched. Observing people affects what they do; people observed can bias the results as they might perform an unrepresentative range of tasks; and the way in which the data are recorded could compromise the reliability and validity of the observations. Overcoming these potential problems requires careful preparation and piloting of the observational study.

1. Determine what activities are to be observed.
2. Specify the characteristics and size of the sample population to ensure that they are representative of the population the results will generalize (e.g., experts or novices, males or females, older or younger people, etc.).
3. Decide what aspects of performance are required. Are you interested in thoughts (which can be elicited through verbal protocols), errors (which can be written down), speed of performance (which can be timed), or behavior (which can be recorded on a precoded observation sheet)?
4. Decide how the reliability of the data is to be checked. The observational data are useless unless you can be sure they are correct. There are two ways of checking reliability: either have two observers and compare recordings, or videotape the activities and conduct a reliability analysis on a sample of the videotape by comparing it with the direct observation.
5. When conducting the observation study, it is worth spending some time beforehand with the person being observed to get them used to you and the idea of being watched. This can help reduce some of the bias.

Some forms of observation can be conducted in the absence of people actually using products. For instance, one could infer the volume of traffic in a public place from carpet wear. One could also infer the use of functions on a machine from wear on buttons. For example, a drink dispenser had handwritten labels stuck on the coffee and hot chocolate buttons but a near-pristine label for tea, because the tea was awful. One could also infer patterns of use from signs of damage or wear and tear. For example, on a food mixer, a set of scratches and marks where the bowl was placed attested to the difficulty of actually putting the bowl onto its holder. Finally, one also could infer problems from the presence of additional labels on machine, e.g., ticket vending machines with "do not insert money first" (which shows a problem but could be misread). These "observations" rely on inferences about the states in which products operate, i.e., things that are done by the product. Such states can be recorded in terms of their frequency and duration, and additional temporal data can be obtained, such as fraction of total performance time, interkey-press times, etc. However, these data may be difficult to interpret without an indication of the actions performed by a user. These data can be defined in similar terms, but focus on human behavior rather than product states. Finally, another major problem for observation is that one cannot infer causality from simple observation. That is, the data recorded must be a purely objective record of what actually happened, without any conjecture as to why.

Observation of live performance is, in many ways, rich because it allows a potential to interact with the people performing tasks, etc. However, this level of interaction will depend on the degree of "naturalness" one wishes. For instance, asking questions of participants may produce more information, but it can also alter or otherwise interfere with their task performance. Videotaping the task performance is useful in that it provides an opportunity for reviewing the performance, possibly allowing for more than one form of analysis, and it can be edited for presentations. However, it is vital that the analysis be conducted with the aim of collecting some data. Software packages are available that allow analysts to sample a range of activities from recorded or live performance.

28.3 Advantages

- Provides objective information that can be compared and ratified by other means
- Can be used to identify individual differences in task performance
- Gives real-life insight into human–machine interaction

28.4 Disadvantages

- Very resource-intensive, particularly during analysis
- Effect on observed party
- Lab vs. field trade-off (i.e., control vs. ecological validity)
- Does not reveal any cognitive information

28.5   Examples

The simplest form of recording is a frequency count of specific events, i.e., product state changes or human actions. Table 28.1 shows a simple frequency count based on a study involving a prototype electronic book. The product was a nonfunctioning mock-up, requiring an experimenter to perform physical actions, such as turn a page, in response to "user" commands. The commands were issued via buttons. The aim of the study was to assess the ease with which a user could guess the functions of the buttons. An illustration is shown in Table 28.1.

In order for these counts to serve as data, it is necessary to consider the relationship between actions and states, i.e., to say which buttons were pressed to produce the states. This required another frequency count. One could, of course, combine the counts into a single table. What these data do not tells us is the speed with which the product could be used. (In this study, it was felt that timing the use of a mock-up would not produce realistic data.) However, the point to note is that the actions were defined as unambiguously as possible prior to observation. Table 28.2 shows data collected during a field observation of a ticket vending machine. Here, the actions are both timed and categorized. The aim of categorizing actions was twofold: to simplify recording and to produce information relating to a specific research question, in this case, how successful were users of the machines (S = success, R = success with repetition, A = abort).

In another example, observations have been made of performance on a radio-cassette machine. For this study, a task list was constructed comprising the following tasks:

1. Switch on
2. Adjust volume
3. Adjust bass
4. Adjust treble
5. Adjust balance

TABLE 28.1   Frequency Count for Electronic Book Study

Turn to contents page	| | |
Turn to index	|
Turn to next page	ⲙⲏⳒ ⲙⲏⳒ |
Turn to previous page	| |

TABLE 28.2   Sample of Field Observations

Change Given			Exact Money		
Subject	Time	Outcome	Subject	Time	Outcome
1	16	S	8	35	A
2	23	S	9	56	S
3	48	A	10	21	R

6. Choose a new preset station
7. Choose a new station using the seek function and store it
8. Choose a new station using a manual search and store it
9. Insert cassette
10. Find the next track on the other side of cassette
11. Eject cassette
12. Switch off

For each task, time and error data were collected on two occasions (called T1 and T2 in Table 28.3) for two participants. One participant was an expert in using the device, and the other participant was a novice. The total number of errors and the average response times of both participants are shown in Table 28.3.

There will be instances in which a number of items in a product will be used in combination, and the analyst may be interested in describing the frequency with which the products are used along with the orders in which they are used. From these data, one might decide to reposition items within a product, placing the most frequently associated controls together. This form of representation can be achieved through the use of link analysis. In link analysis, one indicates the number of times an item is selected and indicates, via lines, the frequency with which other items are selected next. Figure 28.1 illustrates the link analysis for a proposed redesign of an electron-microscopy workstation using link analysis.

TABLE 28.3 Observational Study of a Novice and Expert

Task[a]	Errors Observed	Error Frequency @ T1	Error Frequency @ T2	Time @ T1 (sec) mean	Time @ T2 (sec) mean
1	Pressed knob instead of turning	—	1	5.37	5.27
2		—	—	4.66	4.67
3		—	—	10.03	6.77
4		—	—	9.64	4.39
5	Adjusted fade instead of balance	1	1	5.26	4.94
6		—	—	6.5	4.38
7	Pressed scan instead of seek	1	—	18.19	8.92
	Failed to store	1	—		
8	Failed to store	1	1	22.5	9.8
9		—	—	3.21	4.63
10	Did not push reverse button hard enough	—	—	28.09	16.25
	Pressed fast forward only	1	1		
11		—	—	4.4	2.96
12		—	—	3.17	2.9

[a] Tasks correspond to the numbered list above this table.

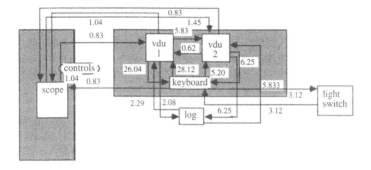

FIGURE 28.1 Link analysis of electron-microscopy workstation.

In the final example, the number of control-room activities during the periods of observation have been divided by the duration of the observation to produce an activity rate (i.e., the mean number of activities per minute), as shown in Table 28.4. The purpose of this analysis was to evaluate the workload of some control-room engineers. For details on workload analysis, see Chapter 39.

The variation in the activity rate on days 1 and 3, compared with days 2, 4, and 5, is due to failures or breakdowns on the busier days. When a failure or breakdown occurs, the engineer is involved in twice as many manual activities. The mean number of activities by type divided into three time periods is shown in Table 28.5.

Taken together, these analyses present a picture of an engineer who is engaged in activities every minute or two. When not performing these activities, the engineer is monitoring the automatic system, checking paperwork, and communicating with colleagues.

28.6   Related Methods

Observation, like the interview, is a core method that relates to many other methods. Examples of related methods include link/layout analysis, content analysis, and hierarchical task analysis.

28.7   Standards and Regulations

In assessing the usability of a device and the extent to which it supports working practice, observation would seem to be a necessary part of the design and development process. Standards and regulations are not especially prescriptive on exactly which methods should be employed, but observational trials with end-user participants seems to follow the spirit of fulfilling the usability framework requirements of ISO 9241.

28.8 Approximate Training and Application Times

A study reported by Stanton and Young (1998, 1999) suggests that it is relatively easy to train users to apply observational techniques. For example, in their study of radio-cassette machines, training in the observational method took only 45 min. Newly trained observers took approximately 30 to 40 min to collect response-time data and populate the table with data from the participants in the radio-cassette study. It can take a great deal more time to analyze the data and interpret the findings.

TABLE 28.4 Frequency Rates of Control-Room Activities

Monitoring Period	Minutes of Monitoring	Number of Activities	Activity Rate (mean activities per min)
Day 1: 07:00 to 10:00	180	127	0.70
Day 2: 07:00 to 10:00	180	282	1.57
Day 3: 07:00 to 10:00	180	159	0.88
Day 4: 07:00 to 10:00	180	294	1.63
Day 5: 07:00 to 10:00	180	315	1.75
Total	**900**	**1177**	**1.31**

TABLE 28.5 Types of Control-Room Activities

Activity Type	07:00 to 08:00	08:00 to 09:00	09:00 to 10:00
Phone calls	7.6 (5.9)	8.0 (3.0)	7.0 (2.6)
Manual control	48.2 (27.7)	58.0 (30.2)	33.8 (11.8)
Other	32.8 (27.0)	25.8 (10.7)	14.2 (3.5)

Note: Mean frequency of activities (with standard deviation).

TABLE 28.6 Reliability and Validity Data for Observation

	Time Data	Error Data
Interanalyst reliability	0.209	0.304
Intraanalyst reliability	0.623	0.89
Validity	0.729	0.474

28.9   Reliability and Validity

The reliability and validity data for observation are based upon a study reported by Stanton and Young (1999) and are presented in Table 28.6.

It is desirable that more studies collect data on reliability and validity so that generalizations about the effectiveness of techniques can be made.

28.10   Tools Required

The simplest, cheapest, and easiest to operate of all observation tools are pen and paper. There are also products on the market that allow quick and efficient analysis of video data. (Indeed, at least one of these products can also be used on handheld computers for real-time observation.) If we remember that the data collection concerns the frequency and timing of events, one could imagine software that uses key presses to stand for specific events, with each key press being time-stamped. This is the basis of packages called The Observer and Drum. Without going into too much detail, these packages have the potential to allow the analyst to control the rate at which video plays, so the tape can be fast-forwarded to allow quick capture of rare events observed over long periods. These products also will perform basic statistical analysis of the frequency and duration of specified events. This allows analysis of videotapes to be performed almost in real time.

References

Baber, C. and Stanton, N.A. (1996), Observation as a technique for usability evaluations, in *Usability in Industry*, Jordan, P.W., Thomas, B., Weerdmeester, B.A., and McClelland, I.L., Eds., Taylor & Francis, London, pp. 85–94.

Drury, C.G. (1995), Methods for direct observation of performance, in *Evaluation of Human Work: A Practical Ergonomics Methodology*, 2nd ed., Wilson, J. and Corlett, E.N., Eds., Taylor & Francis, London, pp. 45–68.

Stanton, N.A. (1999), Direct observation, in *The Methods Lab: User Research for Design*, Aldersley-Williams, H., Bound, J., and Coleman, R., Eds., Design for Ageing Network, London.

Stanton, N.A. and Young, M. (1998), Is utility in the mind of the beholder? A review of ergonomics methods, *Appl. Ergonomics*, 29, 41–54.

Stanton, N.A. and Young, M. (1999), *A Guide to Methodology in Ergonomics: Designing for Human Use*, Taylor & Francis, London.

29

Applying Interviews to Usability Assessment

Mark S. Young
University of New South Wales

Neville A. Stanton
Brunel University

29.1 Background and Applications

The interview is one of the original methods for gathering general information, and it has been popularly applied across a range of fields. Common perceptions of interviews are in employment and in the questioning of witnesses to a crime, but they can yield results in any situation where a person's opinion or perspective is sought. Examples of more diverse uses are in mental health or for presurgical psychological screening (see Memon and Bull, 1999).

In the context of usability evaluations, interviews are intended to elicit users' or designers' views about a particular task or system. They are truly multipurpose, even within the usability context. Applications include task analysis for human reliability assessment (Kirwan, 1990), predesign information gathering (Christie and Gardiner, 1990), and collecting data on product assessment after a user trial (McClelland, 1990).

One major advantage of interviews is the high degree of ecological validity. If you want to find out what a person thinks of a device, you simply ask him/her. Researchers agree that the flexibility of the interview is also a great asset (Kirwan and Ainsworth, 1992; Sinclair, 1990), in that particular lines of inquiry can be pursued if desired. Furthermore, the interview technique is very well documented, with an abundance of literature on this method. Finally, the main advantage of an interview is its familiarity to the respondent as a technique. This, combined with the face-to-face nature, is likely to elicit more information, and probably more accurate information.

29.2 Procedure

While the popular assumption may be to use interviews in market research once a product is in existence, interviews can also be exploited at any stage in the design process. In usability evaluations, a user trial is implied before carrying out an interview. Thus, a partial prototype of the product under test should be available.

The procedure described in this chapter follows this precedent, using the interview in a manner similar to the "cooperative evaluation" technique cited by Christie et al. (1990). Ideally, in any interview scenario, access to the actual user population is most desirable. If end users are available, then the output is likely to be more revealing by using these people as interviewees. However, in the absence of actual users, potential designers can consider interviewing other colleagues, although in this case the potential bias in the results must be acknowledged.

If a product is in development, designers can use a prototype to create a user-trial scenario on which to base the interview. Two members of the design team collaborate, one acting as the user and one acting as the interviewer. The user performs a series of tasks on the device and is then interviewed about the usability issues involved.

Interviewing takes many forms, from the completely unstructured interview to the formally planned, structured interview. For the current approach, particularly with untrained interviewers, a semistructured interview format is recommended. This has proved effective in other domains, such as personnel selection (e.g., Wright et al., 1989). A useful structure for usability evaluations is based on the Ravden and Johnson (1989) checklist. Although this checklist was designed for assessing usability of human–computer interfaces, it can be applicable to interaction with other devices. By using the sections of the checklist as a scaffold for the interview, one can maintain flexibility while ensuring thoroughness in covering all aspects of usability. Each section title is presented in turn, and relevant questions about the device under scrutiny are posed.

The interviewee should be granted an exhaustive user trial with the device under analysis and then interviewed for his/her thoughts. Each section title of the checklist should be used as a prompt for asking questions (e.g., "Let's talk about visual clarity — did you think information was clear and well organized?" etc.). It should be noted that the structure is just the bones upon which to build an interview; it is by no means fixed and should not be viewed as a script for asking questions. It is more of an agenda to ensure that all aspects are covered. The interviewer should direct the questioning from open questions ("What did you think of this aspect?") through probing questions ("Why do you think that?") to more closed ones ("Is this a good thing?"). It may be useful to keep a protocol sheet in hand as a prompt for this (the flowchart in Figure 29.1 would serve this purpose). The idea is that the interviewer opens a line of inquiry with an open question and then follows it up. When the line of inquiry is exhausted, s/he moves to another line of inquiry. By doing this for every aspect of the device, one can be sure of having conducted a thorough interview. It is helpful to have a prepared data sheet for filling in responses during the interview.

Interviews are, of course, adaptive by nature, and if the interviewer feels that any particular section is irrelevant to the particular device s/he is studying, s/he is free to exclude it. The professional wisdom of the interviewer can be an advantage for this technique.

29.3   Advantages

- Technique is familiar for most respondents.
- Flexibility: information can be followed up on-line.
- Structured interview offers consistency and thoroughness.

29.4   Disadvantages

- Necessitates a user trial.
- Analysis is time-consuming.
- Demand characteristics of situation may lead to misleading results.

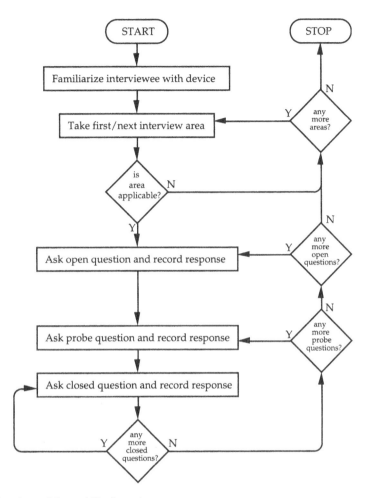

FIGURE 29.1 Flowchart of the usability interview process.

29.5 Example

The following is an example of interview output based on an analysis of using a typical in-car radio-cassette machine:

Section 1: Visual clarity. Information displayed on the screen should be clear, well organized, and easy to read.
- There is a certain amount of visual clutter on the LCD.
- Writing (labeling) is small but readable.
- Abbreviations are ambiguous (e.g., DX/LO; ASPM ME-SCAN).

Section 2: Consistency. The way the system looks and works should be consistent at all times.
- Tuning buttons (especially Scan and Seek functions) present inconsistent labeling.
- Moded functions create problems in knowing how to initiate the function.

Section 3: Compatibility. The way the system looks and works should be compatible with user expectations.
- Four functions on "On/Off" switch makes it somewhat incompatible with user's expectations.
- Autoreverse function could cause cognitive compatibility problems.

Section 4: Informative Feedback. Users should be given clear, informative feedback on where they are in the system.

- Tactile feedback is poor, particularly for the "On/Off" switch.
- Operational feedback poor when programming a preset station.

Section 5: Explicitness. The way the system works and is structured should be clear to the user.
- Novice users may not understand station programming without instruction.
- Resuming normal cassette playback after FF or RWD is not clear.

Section 6: Appropriate Functionality. The system should meet the needs and requirements of users when carrying out tasks.
- Rotating dial is not appropriate for front/rear fader control.
- Prompts for task steps may be useful when programming stations.

Section 7: Flexibility and Control. To suit the needs and requirements of all users, the interface should be sufficiently flexible in structure, information presentation, and in terms of what the user can do.
- Users with larger fingers may find controls to be clumsy.
- Radio is inaudible while winding cassette; this is inflexible.

Section 8: Error Prevention and Correction. The system should be designed to minimize the possibility of user error; users should be able to check their inputs and to correct errors.
- There is no "undo" function for stored stations.
- Separate functions would be better initiated from separate buttons.

Section 9: User Guidance and Support. Informative, easy-to-use, and relevant guidance and support should be provided.
- Manual is not well structured; relevant sections are difficult to find.
- Instructions in the manual are not matched to the task.

Section 10: System Usability Problems.
- There are minor problems in understanding function of two or three buttons.
- Treble and bass controls are tiny.

Section 11: General System Usability.
- Best aspect: This radio is *not* mode-dependent.
- Worst aspect: Ambiguity in button labeling.
- Common mistakes: Adjusting balance instead of volume.
- Recommended changes: Substitute pushbutton operation for "On/Off" control.

29.6 Related Methods

Interviews are very closely related to questionnaires, as well as being linked with observation.

Questionnaires can take many forms, but they can be thought of as an extreme form of structured interview. Indeed, a formally structured interview is essentially just an administered questionnaire. The obvious advantage of a questionnaire is that it can be completed on paper by the participant, thus enabling huge samples of data collection with relatively minimal effort on the part of the experimenter. Disadvantages, however, concern the inflexibility of questionnaires and the inability to pursue interesting lines of inquiry or follow up on answers that may be unclear.

Observations of a user trial can be used in conjunction with the posttrial interview output to corroborate (or otherwise) the benefits and problems with the product's design. Where the interview reveals subjective opinion and perception of product usability, the observations demonstrate actual errors and performance times in using the device.

29.7 Standards and Regulations

ISO 9241-11 (1988) defines usability as "the extent to which a product can be used by specified users to achieve specified goals with effectiveness, efficiency and satisfaction in a specified context of use." The interview can certainly help in developing an understanding of the extent to which users of a product feel that the device has met these criteria.

29.8   Approximate Training and Application Times

In a study examining the suitability of interviews as a usability evaluation technique, Stanton and Young (1999) trained a group of engineering students in the semistructured interview method. Engineering students were chosen for their relevance in the design process (i.e., the technique will ultimately be used by engineers and designers). The same study also assessed 11 other usability methods.

Combined training and practice times were relatively low compared with other more complex methods (such as hierarchical task analysis or task analysis for error identification). The training session lasted just under an hour, which was actually comparable with most of the other methods studied. Only the very simple techniques (such as heuristics, checklists, and questionnaires) were drastically quicker to train. However, real savings were observed in practice times. On average, participants spent just over 40 min practicing the technique in an example scenario. The reason for this was likely due to the fact that interviews are a familiar method to most people, and so the training and practice simply had to focus on the specifics of the semistructured process.

Execution times, on the other hand, were rather longer than anticipated. On average, the engineering students spent approximately 1 hour conducting the interview. It might be argued that this was due to their unfamiliarity with the procedure, but on a retest two weeks later, this time increased slightly (though nonsignificantly). Naturally, the time taken on an interview will reflect the complexity of the product under test and the amount of information gathered. It follows that a speedy interviewer is not necessarily producing efficient output.

29.9   Reliability and Validity

In their study, Stanton and Young (1999) measured predictive validity, intrarater reliability, and interrater reliability.

Predictive validity was assessed using signal-detection theory. By examining the hit ratio (percentage of predictions that were actually observed) and the false alarm ratio (percentage of predictions that were not observed), it is possible to arrive at a single figure that essentially represents how accurate the predictions were. A statistic of 0.5 indicates that the hit ratio and false alarm ratio are equal; greater than 0.5 means that the hit ratio exceeds the false alarm ratio. A result of 1.0 is perfect accuracy.

Predictive validity of the engineers' output on their first attempt rated 0.488. This did not improve on their second attempt two weeks later, when the score was 0.466. On the basis of these results, validity of the interview as a usability tool is rather disappointing.

Intrarater reliability was measured by simple correlation coefficients. Again, the result was disappointing: a correlation coefficient of 0.449, which failed to reach significance. This leads to the conclusion that interviewers are not consistent over time in eliciting responses from their participants.

Interrater reliability, on the other hand, was assessed using the kurtosis statistic. Being a numerical measure of the shape of a distribution, kurtosis gives an indication of how tightly grouped the responses were. A result of zero means a perfect normal distribution; greater than zero indicates less spread; less than zero means a flatter distribution. Good interrater reliability, then, will be reflected in a high positive value of the kurtosis statistic. For the first test session, the kurtosis statistic turned out to be −1.66. For session two, this improved to 0.362. Once again, these results are less than encouraging for the fate of the interview as a usability tool.

Nevertheless, since the interview does not take long to learn or apply, it does have potential in a resource-limited environment. This is also reflected in the moderately good reception it received by the participants in this experiment. Attempts to reduce the drain on resources by using engineers as interviewees apparently failed, so potential analysts are well advised to target end users for their views.

29.10 Tools Needed

The interview is a relatively simple tool, and can largely be conducted as a pen-and-paper exercise. Some preparation on the part of the interviewer will prove beneficial. For a semistructured interview, this would include a *pro forma* or a checklist of the main headings to be covered, as well as a data sheet for recording responses. Audio recording equipment is also highly recommended as a means of keeping an accurate transcript of the interview.

References

Christie, B. and Gardiner, M.M. (1990), Evaluation of the human-computer interface, in *Evaluation of Human Work: A Practical Ergonomics Methodology*, Wilson, J.R. and Corlett, E.N., Eds., Taylor & Francis, London, pp. 271–320.

Christie, B., Scane, R., and Collyer, J. (1990), Evaluation of human–computer interaction at the user interface to advanced IT systems, in *Evaluation of Human Work: A Practical Ergonomics Methodology*, 2nd ed., Wilson, J.R. and Corlett, E.N., Eds., Taylor & Francis, London, pp. 310–356.

International Organization for Standardization (1988), ISO 9241–11: Ergonomic requirements for office work with visual display terminals (VDTs) — Part II: Guidance on usability.

Kirwan, B. (1990), Human reliability assessment, in *Evaluation of Human Work: A Practical Ergonomics Methodology*, 2nd ed., Wilson, J.R. and Corlett, E.N., Eds., Taylor & Francis, London, pp. 921–968.

Kirwan, B. and Ainsworth, L.K. (1992), *A Guide to Task Analysis*, Taylor & Francis, London.

McClelland, I. (1990), Product assessment and user trials, in *Evaluation of Human Work: A Practical Ergonomics Methodology*, 2nd ed., Wilson, J.R. and Corlett, E.N., Eds., Taylor & Francis, London, pp. 249–284.

Memon, A. and Bull, R., Eds. (1999), *Handbook of the Psychology of Interviewing*, Wiley, Chichester, U.K.

Ravden, S.J. and Johnson, G.I. (1989), *Evaluating Usability of Human–Computer Interfaces: A Practical Method*, Ellis Horwood, Chichester, U.K.

Sinclair, M.A. (1990), Subjective assessment, in *Evaluation of Human Work: A Practical Ergonomics Methodology*, 2nd ed., Wilson, J.R. and Corlett, E.N., Eds., Taylor & Francis, London, pp. 69–100.

Stanton, N.A. and Young, M.S. (1999), *A Guide to Methodology in Ergonomics: Designing for Human Use*, Taylor & Francis, London.

Wright, P.M., Lichtenfels, P.A., and Pursell, E.D. (1989), The structured interview: additional studies and a meta-analysis, *J. Occup. Psychol.*, 62, 191–199.

30

Verbal Protocol Analysis

Guy Walker
Brunel University

30.1 Background and Applications

The purpose of verbal protocol analysis is to make "valid inferences" from the content of discourse (Weber, 1990). In human factors applications, this discourse is a written transcript gained from individuals thinking aloud as they perform a task. The valid content of this written transcript can be found either within individual words, word senses, phrases, sentences, or themes. The analysis proceeds by extracting this valid content and categorizing it according to a defined categorization scheme. Thus, verbal protocol analysis is a means of data reduction, of keeping the content derived from verbal transcripts manageable in size and theoretically valid. This enables relevant concepts or interrelationships to be analyzed and inferred.

Verbal protocol analysis has found use within human factors research as a means of gaining insight into the cognitive underpinnings of complex behaviors. Applications are as diverse as steel-melting furnaces (Bainbridge, 1974), usability of the Internet (Hess, 1999), and driving (Walker et al., 2001). In these human factors settings it has been shown to be a good exploratory method, and careful experimental design can help to optimize reliability and validity. Providing that verbalizations are concerned with the content or outcomes of thinking (and thinking aloud can provide this information), then psychological processes can be inferred even if individuals are not generally self-aware of them. Within the context of exploring hypotheses and conducting studies in naturalistic settings, verbal protocol analysis can be extremely useful and has much to offer.

30.2 Advantages

- Verbalizations provide a rich data source in quantity and content.

- The process lends itself well to examining behaviors in naturalistic settings.
- Protocol analysis is especially good at analyzing sequences of activities.
- Content and outcomes of thinking can provide an insight into cognitive processes.
- Experts can often provide excellent verbal data.

30.3   Disadvantages

- Data collection can be time-consuming.
- Data analysis can be very time-consuming.
- Providing a verbal commentary can change the nature of the task, especially if certain processes are not normally verbalized, such as skill-based or automatic behaviors.
- There are theoretical issues concerned with verbal reports not necessarily correlating with knowledge used in task enactment.
- High task demands often lead to reduced quantity of verbalizations and therefore loss of resolution.

30.4   Procedure

Practitioners of protocol analysis are keen to point out that there are not necessarily any hard and fast rules for the design of a protocol analysis paradigm. The procedure given below is a generic template for distilling themes from verbal protocols within human factors applications, and will provide a good starting point from which other add-ins can be employed (discussed below) to suit the particular scenario.

30.4.1   Data Collection Phase

30.4.1.1   Step 1: Devise Scenario

This will normally involve, for instance, some set tasks, a particular task scenario, and the operation of one type or different types of equipment or system. For example, in the study by Walker et al. (2001), car drivers were required to drive their own vehicles around a predetermined test route. The scenario will be based upon a hypothesis, and observational methodologies or task analysis methods can be useful at this stage (see Stanton and Young, 1999; Annett and Stanton, 2000).

30.4.1.2   Step 2: Instruct and Train the Participant

Standard instructions and training should be provided to the participant. These often take the form of telling the participant what things they should be talking about and, most importantly, informing them to keep talking even if what is being said does not seem to make much sense to them. The experimenter should also demonstrate the method to the participant, showing them the desired form and content of verbalizations. Furthermore, it is helpful if a brief practice can be undertaken, with the experimenter able to interject in the participant's verbal commentary to offer advice and feedback.

30.4.1.3   Step 3: Record the Scenario

As a minimum requirement, some means of recording audio with a time index should be sought. Digital recording products such as MiniDisc™ are particularly useful, as are portable computers. It can often be helpful to simultaneously record video to back up the verbal commentary. A particularly good method is to collect the data digitally via a laptop computer, and to use any proprietary software audio and video player to transcribe it. Aside from audio and video data, other data of interest can be gathered simultaneously, such as eye-tracking data or system telemetry that will objectively inform on exactly how the system is being used. This can often be a useful counterpoint to the verbal data supplied by the participant.

30.4.2 Data Reduction Phase/Content Analysis

30.4.2.1 Step 4: Transcribe the Verbalizations

After collecting the verbal data, it then needs to be transcribed verbatim into written form. A spreadsheet can be devised to achieve this. The rate of verbalizations for relatively fast-paced tasks can easily reach 130 words per minute. In Figure 30.1, a time index of 2-sec increments would suffice for this rate of verbalization. A good technique for accurate and rapid time-indexing is to pause the audio recording after hearing a section of speech (such as a brief sentence or phrase), note the time, then subtract 2 sec and type in the verbalization at this new time point.

30.4.2.2 Step 5: Encode the Verbalizations

30.4.2.2.1 Decide Whether to Encode Words, Word Senses, Phrases, Sentences, or Themes

After transcribing the verbal commentary into written form, it then has to be categorized (Weber, 1990). The experimenter must make a decision at this point. Will words, word senses, phrases, sentences, or themes be encoded?

- Words: encoding of the occurrence of discrete words
- Word senses: the encoding of words with multiple meanings (e.g., to be "hard up" vs. "a hard surface," or "tent peg" vs. "off the peg," etc.)
- Phrase: the encoding of phrases that constitute a semantic unit (e.g., "Windsor Great Park," "London Orbital Motorway," "park and ride," etc.)
- Sentence: the encoding of what particular sentences refer to, or express (e.g., sentences expressing positive vs. negative sentiments about protocol analysis)
- Theme: the encoding of the meaning of phrases and sentences into shorter thematic units or segments (Weber, 1990), e.g., "the engine's at 3000 rpm, nice and smooth." The thematic segments might be as follows. Here the user is referring to the *behavior* of a piece of *equipment*, that is being *perceived* and *comprehended* by the operator, and the *information* is being *gained* from equipment *instruments*.

A protocol analysis based on themes provides the richest and most flexible source of data and can be recommended as a starting point.

30.4.2.2.2 Establish a Conceptual Framework for the Encoding Scheme

The themes then have to be encoded according to some rationale determined by the research question. This involves attempting to ground the encoding scheme according to some established theory or approach. For example, the encoding scheme could be based on established theories of mental workload, or cognitive control, or situational awareness. This helps to ensure some degree of construct validity, inasmuch as the encoding scheme is measuring verbalizations according to a theoretical background. The theoretical background will of course be determined by the particular research question.

TIME mm	ss	VERBALIZATIONS
14	0.52	
14	0.54	I'm operating control x
14	0.56	
14	0.58	Now I'm looking at display y for a reading z
15	0.00	
15	0.02	
15	0.04	

FIGURE 30.1 Transcription sheet.

30.4.2.2.3 Devise Encoding Instructions

The next step is to draw up highly defined written instructions for the encoding scheme. This is a good exercise in terms of tightly setting the encoding criteria. Given the length of time it can take to encode data of this sort, these instructions should be constantly referred to, and this in turn will help to ensure intrarater reliability (within raters). Of course, these same instructions will be used for when interrater reliability (between raters) has to be established later.

30.4.2.2.4 Complete Encoding

An example of an encoding worksheet is shown in Figure 30.2. Another decision has to be made at this point to determine whether the encoding categories should be mutually exclusive or exhaustive. Typically, for an analysis based on themes, mutual exclusivity need not be applied, and the theme can fit into as many encoding categories as defined from the written encoding instructions. Under this scheme, the encoding is "exhaustive." As such, whenever a theme meets the definitions described in the encoding instructions, the number *1* is entered in the relevant encoding box.

It is worth mentioning at this point that various computer programs exist to assist in encoding verbal transcripts. These are typically very good at analyses couched in determining the occurrence of words or phrases and/or the structural analysis of texts. Obviously, one of the simplest ways of counting the occurrence of discrete words is to use the *find* function in any word-processing package. Examples of computer protocol analysis packages include General Enquirer™, TextQuest™, and WordStat™, to name a few. Although computer packages exist for more qualitative analyses, it is fair to say that, in general, computer software tends to be weaker with theme-based analyses. Nevertheless, there can be a considerable time saving combined with improved encoding reliability through the use of computer software. The particular research question will determine whether its use is appropriate or not.

30.4.2.3 Step 6: Devise Other Data Columns

After transcribing the verbal data against a time index, and having encoded the themes, the final part of the worksheet consists of other data columns, as shown in Figure 30.3. This is an opportunity to note any mitigating circumstances that may have occurred during the trial, and that may have affected the

ENCODING								
Group 1			Group 2			Group 3		
Category1	Category 2	Category 3	Category1	Category 2	Category 3	Category1	Category 2	Category 3
1				1		1	1	
	1					1		
1	1			1	1	1		
						1	1	1
	1				1			

FIGURE 30.2 Encoding sheet.

MITIGATING EVENTS	STAGE OF TASK ENACTMENT
	Inputting command
System responding slowly	
Figure z on screen	
	Executing command

FIGURE 30.3 Other simultaneous information columns.

verbal report. At this stage, it also may also be helpful to note different stages of task enactment, or to tie up telemetry or eye-tracker data with the verbal report, using separate worksheet columns as required. For example, in the study by Walker et al. (2001), the structure of encoding could also be analyzed with reference to the type of road that the driver was driving upon, thus providing extra resolution within the analysis.

30.4.2.4   Step 7: Establish Inter- and Intrarater Reliability

After completing the encoding process (although in the context of a pilot run, it is sufficient only to encode and measure a smaller subset of verbalizations), the reliability of the encoding scheme has to be established. For interrater reliability between different raters, this is especially important when using themes, as themes rely on shared meaning, and an analysis of reliability will indicate how well this meaning is in fact shared. Intrarater reliability (within raters) will help to measure any potential drifting in encoding performance over time.

In protocol analysis, reliability is established through reproducibility. In other words, an independent rater or raters need to encode previously encoded analyses (or a subset of previously encoded analyses). They do this in a blind condition, unaware of the previous rater's encoding performance. The independent raters make use of the same categorization instructions that the original rater employed before beginning their own encoding. The dual encoding of the same analysis can then be analyzed using basic correlational statistics, such as Pearsons or Spearmans. These same methods can also be used to gain a measure of intrarater reliability. More-specialized protocol analysis reliability techniques, such as Krippendorff's alpha (Krippendorff, 1980), are also available.

30.4.2.5   Step 8: Perform Small-Scale Pilot Study

Having reached this point, the protocol analysis procedure should now be put to the test within the context of a small pilot study or pilot run. This will demonstrate whether the verbal data collected are useful, whether the encoding system works, and whether inter- and intrarater reliability are satisfactory. This offers an opportunity to refine the procedure further before returning to step 1 and conducting the final complete study.

30.4.2.6   Step 9: Analyze Structure of Encodings

Having conceptually grounded the encoding scheme by relying on established theories or constructs, and having established intrarater reliability through the use of encoding instructions and inter-rater reliability by employing independent encoders, it is now time to analyze the results of the content analysis.

From this point the analysis will proceed contingent upon the research question at hand, but all analyses will need to sum the responses given in each encoding category, and this is achieved simply by adding up the frequency of occurrence noted in each category. Obviously, for a more fine-grained analysis, the structure of encodings can be analyzed contingent upon events that have been noted in the "other data" column(s) of the worksheet, or in light of other data that have been collected simultaneously.

30.5 Example of Protocol Analysis Recording, Transcription, and Encoding Procedure for an On-Road Driving Study

This digital video image (Figure 30.4) is taken from the study reported by Walker et al. (2001) and shows how the protocol analysis was performed with normal drivers. The driver in Figure 30.4 is providing a concurrent verbal protocol while being simultaneously videotaped. The driver's verbalizations and other data gained from the visual scene are transcribed into the transcription sheet in Figure 30.5, which illustrates the 2-sec incremental time index, the actual verbalizations provided by the driver's verbal commentary, the encoding categories, the events column, and the protocol structure.

In this study, three encoding groups were defined: behavior, cognitive processes, and feedback. The behavior group defined the verbalizations as referring to the driver's own behavior (OB), behavior of the vehicle (BC), behavior of the road environment (RE), and behavior of other traffic (OT). The cognitive processes group was subdivided into perception (PC), comprehension (CM), projection (PR), and action execution (AC). The feedback category offered an opportunity for vehicle feedback to be further categorized according to whether it referred to system or control dynamics (SD or CD) or to vehicle instruments (IN). The cognitive processes and feedback encoding categories were couched in relevant theories in order to establish a conceptual framework. The events column was for noting road events from the simultaneous video log, and the protocol structure was color-coded according to the road type being traveled upon. In this case, the shade corresponds to a motorway and would permit further analysis of the structure of encoding contingent upon road type. The section frequency counts simply sum the frequency of encoding for each category for that particular road section.

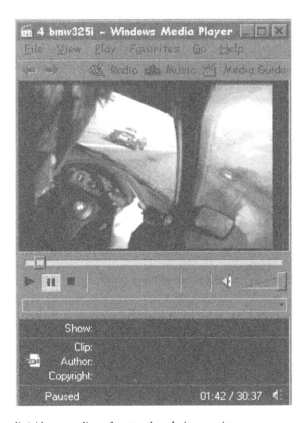

FIGURE 30.4 Digital audio/video recording of protocol analysis scenario.

TIME (mm: ss)	VERBALIZATIONS	\<BEHAV\> OB	BC	RE	OT	\<COG\> PC	CM	PR	AC	\<F/B\> SD	CD	IN	EVENTS
01:34	70mph, 5th gear	1				1					1	1	Glances at gear lever
01:36	2800 rpm	1				1						1	
01:38	that's quite smooth	1					1						
01:40	he's slowing down			1	1								Other car crossing from lane 3 over
01:42	don't know what's wrong with him	1				1		1					to hard shoulder in front of driver
01:44													
01:46													
01:48													
01:50													
01:52													
01:54													
01:56													
01:58													
02:00													
02:02													
02:04	it's all clear ahead			1	1								
02:06													
02:08	chap behind has eased off a bit luckily			1	1	1							
02:10													
02:12	make my intention clear that I'm going right	1							1				Indicating right
02:14	so I'll stick to the right side of this slip lane	1	1			1							
02:16													
02:18													
02:20	bit worried about overtaking him	1				1		1					Passing other vehicle
02:22													
02:24													
02:26													
Section Frequency Counts		4	3	1	5	5	5	0	1	0	1	2	
02:28													

FIGURE 30.5 Transcription and encoding sheet.

30.6 Related Methods

Verbal protocol analysis is related to observational methods, and often the modification of this method and the addition of an ongoing verbal protocol will turn an observational study into a satisfactory protocol analysis paradigm. Observational methods and/or task analysis can be helpful in discerning the scenario (and hypotheses) to be analyzed via verbal protocol analysis. Eye tracking and various other system telemetry can be a useful adjunct to the method, providing additional information to correlate or contrast with the verbal data.

30.7 Standards and Regulations

ISO 9241 defines usability as follows: "the extent to which a product can be used by specified users to achieve specified goals with effectiveness, efficiency and satisfaction in a specified context of use."

The British Standards Institute (for example) has around 200 standards that relate directly to ergonomics and the user-centered design of artifacts and systems. These standards include BS EN ISO 10075-2:2000, Ergonomic principles related to mental work load: design principles; BS EN ISO 13407:1999, Human-centred design processes for interactive systems; and BS EN ISO 11064-1:2001, Ergonomic design of control centres: principles for the design of control centres.

The European commission helps to set various proposals that form the basis of health and safety law in individual member states. The basis of much of this law is the assessment of risk. In the U.K., the health and safety executive deals with guidance, approved codes of practice (ACOP), and regulations. In particular, approved codes of practice specify exactly what reasonably practicable steps should be taken given the nature of a specific task, the equipment that is used, within the environment in which it is being used. These steps ensure that relevant health and safety legislation is complied with.

Protocol analysis is an ideal way of analyzing the user, the task, the equipment that is used, within the environment in which it is being used. Protocol analysis can be an effective means of analyzing the outcome of that interaction with reference to specified goals, objectives, and standards.

30.8 Approximate Training and Application Times

The method is easy to train, and the setup and organization of an experimental scenario is comparable with any other user trial of this kind (such as observational methods). However, analysis can be very lengthy. If transcribed and encoded by hand, 20 min of verbal data at around 130 words per minute can take in the region of 6 to 8 hours to transcribe and encode. In order to establish reliability, an independent rater or raters need to be employed along with the associated time taken to reanalyze the transcript.

30.9 Reliability and Validity

Given a theoretically grounded encoding scheme, and provided that the same encoding instructions are used, reliability is reassuringly high, even using a theme-based analysis. For example, Walker et al. (2001) employed two independent raters and established interrater reliability at Rho = 0.9 for rater 1 and Rho = 0.7 for rater 2. Intrarater reliability during the same study was also reassuringly high, being in the region of Rho = 0.95. In other studies using protocol analysis, and employing a wide range of reliability techniques, correlations of the order of 0.8 are not unusual.

In terms of validity, there is still a degree of debate in terms of the relationship between verbalizations and the content of cognition. Self-awareness of psychological processes is widely regarded as poor. However, verbalizations reflecting the *content* and *outcomes* of thinking are argued to posses theoretical validity. As such, cognitive processes and concepts can be inferred from the participants' thinking aloud on the content and outcome of their thinking. Validity is also aided by ensuring some degree of construct validity by providing a conceptual framework for the encoding scheme. With these provisos in place, and as an exploratory method, the use of protocol analysis is justified.

30.10 Tools Needed

As a minimum, protocol analysis requires an experimental scenario and some means of recording audio against a time index measured in seconds. More efficient means of conducting the analysis can employ digital audio/video recording techniques. Software-based video players allow direct storage access to audio/video data (as opposed to sequential access with tape recorders) and are the most convenient choice when transcribing verbalizations. A spreadsheet package such as Microsoft's Excel™ is also required, but the use of the specialized protocol analysis software packages (mentioned above) is a matter of discretion, depending on the specific nature of the analysis paradigm.

References

Annett, J. and Stanton, N.A. (2000), *Task Analysis*, Taylor & Francis, London.

Bainbridge, L. (1974), A summary of the cognitive processes of operators controlling the electricity supply to electric-arc steel-making furnaces, in *The Human Operator in Process Control*, Edwards, E. and Lees, F.P., Eds., Taylor & Francis, London, pp. 146–158.

Hess, B. (1999), Graduate student cognition during information retrieval using the World Wide Web: a pilot study, *Comput. Educ.*, 33, 1–13.

Krippendorff, K. (1980), *Content Analysis: An Introduction to Its Methodology*, Sage, London.

Stanton, N.A. and Young, M.S. (1999), *A Guide to Methodology in Ergonomics*, Taylor & Francis, London.

Walker, G.H., Stanton, N.A., and Young, M.S. (2001), An on-road investigation of vehicle feedback and its role in driver cognition: implications for cognitive ergonomics, *Int. J. Cognitive Ergonomics*, 5, 421–444.

Weber, R.P. (1990), *Basic Content Analysis*, Sage, London.

Wilson, J.R. and Corlett, N.E. (1995), *Evaluation of Human Work: A Practical Ergonomics Methodology*, Taylor & Francis, London.

31

Repertory Grid for Product Evaluation

Christopher Baber
University of Birmingham

31.1 Background and Applications

Subjective responses constitute a significant component in the evaluation of software and other products. There are a variety of techniques for obtaining responses from users relating to issues such as whether they found a product easy to use or whether they enjoyed using the product. However, a possible criticism of such approaches is that the evaluator is effectively putting words into the mouths of the respondents, i.e., by having predefined terms, the respondents might be unable to express precisely how they feel. Furthermore, it can be very useful to have an idea of *how* users might think about a given product, particularly during the initial phases of design. A technique for eliciting user's concepts relating to a product, therefore, might provide some flexibility in defining rating terms and also provide an insight into how people think about the product. One such technique is the repertory grid.

Basically, the repertory-grid method involves participants selecting one item from a set of three and stating how it differs from the others. The feature or aspect that defines the difference is assumed to be a significant component of the person's concept of these items. Originally, the technique was used as a means of studying a patient's interaction with other people, e.g., by examining responses to authority or to attachment (Kelly, 1955). The "items" in these studies were often descriptions, names, or pictures of other people. As patients selected one from three, the analyst was able to develop a sense of the key components, or attributes, that the patient assigned to people in general. The key components were termed "constructs," in that they reflected ways in which the respondent construed the world. The outcome of the technique then helped the analyst to treat the patient, e.g., by pointing out and addressing particular components.

For human interaction with products, it is possible to observe collections of responses or attributes among users (Sinclair, 1995; Baber, 1996). For example, when speaking of technology acceptance, we might distinguish between the "early adopters," who often respond to high-tech products in a strong and favorable manner, with the "laggards," who will wait until a technology is mature before committing themselves to purchasing it. Asking people to compare sets of high-tech products, therefore, might allow us to determine which of the features of the products are most likely to attract the early adopters or the laggards. Alternatively, asking people to consider a set of Web pages might allow the analyst to determine which features are most attractive and which features link to concepts such as ease of use.

There are several approaches to analyzing a repertory grid. (See Fransella and Bannister [1977] for a review.) Many approaches rely on complex statistical analysis, such as principal-components analysis and other factor-analysis techniques. However, there seem to me to be two problems associated with the use of complex statistics for repertory grid. The first is simply the need to access the knowledge (and probably software) to both run and interpret the analysis. The second, and I feel more significant, problem relates to the issue of whether repertory grids are individual or collective. In other words, most of the early work on repertory grids focused on individual responses, with the acceptance that any set of concepts represented one person's view of the world. In order to perform complex statistics, one usually requires a set of responses, i.e., it is necessary to pool the results of several respondents. My concern with this is that it is not clear how the "constructs" of an individual can be combined with those of someone else in order to produce collective constructs. Consequently, the technique reported here is designed to be used with individuals and assumes no statistical knowledge. Indeed, the technique is a variation on that used in the original work by Kelly (1955) and modified by Coshall (1991).

31.2 Procedure

There are 14 steps in the repertory-grid analysis as follows.

31.2.1 Step 1: Define a Set of Items for Comparison

At this stage, it is sufficient to select items that share some common feature. Thus, one might select a set of Web pages or mobile telephones or kettles. A simple rule of thumb is that the items ought to look as though they belong together, so that when you begin asking the participant to generate constructs, they will not feel as if the items are a random jumble of unrelated things. I have also found that it is very important to emphasize *why* the items have been grouped. For example, in one study we examined "wearable technology" and included wristwatches, GPS (global positioning system) units, and head-mounted displays. Without the overall term "wearable technology," respondents might not have noticed the link between the items. Bearing in mind that participants will work with sets of three items, it is worth considering how many comparisons you expect people to make. For example, if you only have three items then respondents can only perform one comparison, whereas if you have six items they can perform some 20 comparisons, and with ten items they can perform some 136 comparisons. As a rule of thumb, if you have ten items or more, it is probably wise to think about reducing your set. A further point to note is that you must randomize the order of triads; at least make sure that all items appear at least three times during the study.

31.2.2 Step 2: Brief the Participant

The instructions need not be too complex, but they should emphasize that you want the participant to select one item and to give a single word (or short phrase) to·justify that selection. It is not necessary, at this stage, to point out that the words represent "constructs." You might also encourage participants to answer as quickly as possible and to inform them that a choice *must* be made.

31.2.3 Step 3: Present a Set of Three Items (Triad)

The items can be presented as real objects or as pictures or as verbal descriptions (although I have found that the latter can be confusing for participants). It is useful to combine a standard description with the presentation of each item, e.g., by giving the item a name or providing a short description. Respondents are then requested to "decide which two items are most similar, and how the third item differs from them," or to "say which item is the odd one out, and why." The most important part of this stage is that the respondent provides a reason for the selection. It might be useful for the analyst and respondent to agree on an acceptable term; ideally the term should be a single word or short phrase. This represents the "construct" for that triad.

31.2.4 Step 4: Repeat Step 3

Continue presenting triads until the respondent is unable to generate new constructs. I think that after the tenth presentation, people's attention is starting to flag, and by the 15th presentation they are becoming very bored indeed. Thus, it might not be possible to exhaust all possible triads. For this reason, it is very important to ensure that the triads have been randomized. (See the comment in step 1 relating to showing each item three times.)

31.2.5 Step 5: Construct a Repertory-Grid Table

At this point, you might invite the participant to have a rest and a drink. By hand, draw up a table (see Table 31.1) with the items across the top and the constructs down the right-hand side. It is useful to leave a fairly large margin on the right of the paper, after the construct listing, for the analysis.

31.2.6 Step 6: Define Contrasts

Call back the participant and ask him/her to define opposites for each construct. Write these next to the constructs. The idea is that the construct/contrast represents a continuum for a particular idea or concept.

31.2.7 Step 7: Relate Constructs to Items

For each construct, ask the participant to state whether that construct fits an item or not. This is a binary forced-choice response, i.e., the participant can only say yes or no. Where there is some doubt, leave the space blank and continue with the other constructs. Return to the blank spaces at the end of the session. For a "yes" response, enter *1* on the table, and for a "no" response, enter *0*.

31.2.8 Step 8: Review the Repertory Grid Table

Make sure that the participant agrees with the scoring of the constructs and items.

TABLE 31.1 Outline Repertory Grid Table

	Item 1	Item 2	Item 3	Item n	Construct	Contrast	Score
	1	0	0	1	A	a	2
	1	1	0	0	B	b	4
	1	1	1	0	C	c	3
Sum	3	2	1	1	—	—	—
Template	1	1	0	0	—	—	—

31.2.9 Step 9: Perform First Pass

Again, it is wise to dismiss the participant, perhaps arranging to meet later in the day or on the following day. This analysis is, I suppose, comparable with factor analysis in that one is looking for item groupings but does not require statistics. The first task is to define a template, which will be used to determine the membership of the group. For each item, sum the number of 1s and 0s in each item column and enter this value in the appropriate cell at the foot of the table. Thus, in Table 31.1, the sum for item 1 is 2. Convert these numbers into a template, which is a pattern of 1s and 0s, i.e., decide on a cut-off point and assign 0 to all numbers below this point and 1 to all numbers above this point. This stage is somewhat arbitrary, but as a rule of thumb, look to have equal numbers of 1s and 0s. In Table 31.1, the cut-off point would be 1, and all values of 1 or 0 would be assigned 0 and all values >1 would be assigned 1.

31.2.10 Step 10: Compare the Template with the Construct

Take the pattern of 1s and 0s in the template, and overlay this on each construct's line. Where the template matches the line, then add 1 to the score. If fewer than half are matches, then report a negative score and go to step 11. Enter the final total in the "score" column and proceed to step 12.

31.2.11 Step 11: Reflection

Sometimes, the construct is not sufficiently matched to the template. In this case, it might be useful to "reflect" (or reverse) the construct/contrast. Thus, in Table 31.1, construct A produces a score of –2. Reflecting this, to use "a" as the construct (and A as the contrast) would lead to a score of 2.

31.2.12 Step 12: Define Groups

The method assumes binomial distribution of responses, i.e., a statistical majority is required before a group can be defined. For four items, this majority is four, but for eight items, this might be seven or eight. The appendix at the end of this chapter describes the definition of these majorities. The grouping is analogous to the factor that arises from factor analysis.

31.2.13 Step 13: Name Identified Factors

The participants are asked to provide names for the factors that have been identified.

31.2.14 Step 14: Discuss Products

The factors, and their names, are then used to discuss the products.

31.3 Advantages

- Procedure is structured and comprehensive.
- Manual analysis does not require statistics but gives "sensible" results.

31.4 Disadvantages

- Analysis can be tedious and time-consuming for large sets of items.
- Procedure does not always produce factors.

31.5 Example of Repertory Grid

In this section, an example of using repertory grid to assess five Web sites will be presented. The Web sites are fictitious, and the aim of the example is to illustrate the method.

A participant, with little experience of using the Internet, was shown screen dumps of five Web pages that were related to e-commerce. This would require around ten triads to exhaust the set of items. The Web pages {A, B, C, D, E} were presented in triads in the following order: ABC, CDE, ACE, BDE, ABD. After only five presentations, the participant was unable to generate more constructs (although notice that each Web page has been presented at least three times). The respondent was then asked to assign a score 1 or 0 to each Web page, in terms of the constructs. Table 31.2 shows this analysis.

From Table 31.2, it is apparent that three of the constructs produce negative scores. Table 31.3 was produced by reflecting these constructs.

From the appendix at the end of this chapter, it is clear that, for five items, a majority of five is required for membership in the group. From Table 31.3, one can see that two constructs meet this criterion, and so form a factor:

Factor 1 <name>: Clear/Cluttered, Easy controls/Hard to see controls

These constructs are then removed and the analysis is repeated, as shown in Table 31.4.

TABLE 31.2 Initial Repertory Grid for Example

	A	B	C	D	E	Construct	Contrast	Score
	1	0	1	1	0	Colorful	Dull	−3
	0	1	1	0	1	Banks	Shops	−3
	0	1	0	1	0	Clear	Cluttered	5
	0	1	0	1	0	Easy controls	Hard to see controls	5
	1	0	0	0	1	Too many words	Brief	−4
Score	2	3	2	3	2			
Template	0	1	0	1	0			

TABLE 31.3 Repertory Grid following Reflection

	A	B	C	D	E	Construct	Contrast	Score
	0	1	0	0	1	Dull	Colorful	3
	1	0	0	1	0	Shops	Banks	3
	0	1	0	1	0	Clear	Cluttered	5
	0	1	0	1	0	Easy controls	Hard to see controls	5
	0	1	1	1	0	Brief	Too many words	4
Score	1	4	1	4	1			
Template	0	1	0	1	0			

TABLE 31.4 Second-Pass Repertory Grid (following Reflection)

	A	B	C	D	E	Construct	Contrast	Score
	0	1	0	0	1	Dull	Colorful	5
	0	1	1	0	1	Banks	Shops	4
	1	0	0	0	1	Too many words	Brief	3
Score	1	2	1	0	3			
Template	0	1	0	0	1			

From Table 31.4, a further factor is identified:

Factor 2 <name>: Dull/Colorful.

The analysis is repeated until no more constructs arise. In this example, we have two factors so far. The participant is then recalled and asked to provide names for the factors.

31.6   Related Methods

As mentioned in the introduction, there are many techniques that have been designed for knowledge elicitation and for generating subjective responses. Having said this, there are few techniques that are able to combine both into a single method like repertory grid.

31.7   Standards and Regulations

Most standards relating to product evaluation require some assessment of user response to a product. Repertory grid can be a useful first step in defining appropriate dimensions for scaling such responses.

31.8   Approximate Training and Application Times

Stanton and Young (1999) propose that it takes about 30 min to train people to use the repertory grid, with a further 60 min of practice. In my experience, people can learn to become proficient users within 2 to 3 hours, although further practice is often very useful.

31.9   Reliability and Validity

Stanton and Young (1999) report data on the interrater and intrarater reliability of repertory grid together with its validity. From their analysis, they show that repertory grid has reasonable validity (0.533) and intra-rater reliability (0.562), but poor interrater reliability (0.157). This last result might not be too surprising, in that the analysis is very much participant dependent and (it would appear) also prone to variation in the application by the analyst. Consequently, I would be cautious of claiming that the technique produces generalizable data, but suggest that its strength lies in the ability of the technique to elicit sensible groupings of constructs from participants.

31.10   Tools Needed

The method can be applied using pen and paper and makes use of the table of binomial distributions in Table 31.A1 in the appendix.

References

Baber, C. (1996), Repertory grid and its application to product evaluation, in *Usability Evaluation in Industry*, Jordan, P.W., Thomas, B., Weerdmeester, B.A., and McClelland, I., Eds., Taylor & Francis, London, pp. 157–166.

Coshall, J.T. (1991), An appropriate method for eliciting personal construct subsystems from repertory grids, *Psychologist*, 4, 354–357.

Fransella, F. and Bannister, D. (1977), *A Manual for Repertory Grid Technique*, Academic Press, London.

Kelly, G.A. (1955), *The Psychology of Personal Constructs*, Norton, New York.

Sinclair, M.A. (1995), Subjective assessment, in *The Evaluation of Human Work*, 2nd ed., Wilson, J.R. and Corlett, E.N., Eds., Taylor & Francis, London, pp. 69–100.

Stanton, N.A. and Young, M.S. (1999), *A Guide to Methodology in Ergonomics*, Taylor & Francis, London.

Appendix

The values in Table 31.A1 represent significance levels, i.e., p-values. In order for there to be significant agreement, p must be less than 0.05. Read the table by selecting the value for N, i.e., the number of items being compared, and then read along the row of p values until you find a value where $p > 0.05$. From the top of the table, read the N – x number to give you the score at which a construct is defined as belonging to a factor. For example, if you have five items, then N = 5. Reading along this row, you see that only one value of p is less than 0.05, i.e., 0.031, and that this value is in the column N – 0. Therefore, for five items, you need a score of 5 – 0 = 5 for the construct to be selected. To take another example, assume that you have eight items. Reading along the row N = 8, you see two values less than 0.05, i.e., 0.002 and 0.02. Therefore, for eight items, constructs with scores of N – 0 and N – 1 (or 7 and 8) are acceptable. Finally, for ten items, I usually assume that significant agreement is obtained for N – 0, N – 1, and N – 2 (although, of course, for N – 2, $p = 0.055$).

TABLE 31.A1 Table of Binomial Distributions

N	N – 0	N – 1	N – 2	N – 3
5	0.031	0.188	0.500	0.812
6	0.018	0.109	0.344	0.656
7	0.008	0.062	0.277	0.500
8	0.004	0.035	0.145	0.363
9	0.002	0.020	0.090	0.254
10	0.001	0.011	0.055	0.172

32

Focus Groups

Lee Cooper
University of Birmingham

Christopher Baber
University of Birmingham

32.1 Background and Applications

An intuitively simple way of evaluating software and other products is to ask people what they think of them. The group interview, or "focus group," is a well-established research technique that is used throughout the product design community. A focus group is a "carefully planned [group] discussion designed to obtain perceptions on a defined area of interest in a permissive nonthreatening environment" (Krueger, 1988). O'Donnell et al. (1991) elaborate:

> Normally ... [focus groups] consist of 8–12 members with a leader or moderator.... Members are a sample of the customers or end users ... organised into relatively homogeneous groups. These are more likely to produce the desired exchange of information between members rather than a flow of information from individual members to the leader.

Focus groups have a long history in market research. In one study from the 1940s, researchers evaluated listeners' emotional reactions to radio programs. Participants were provided with buttons that they pressed during the program to indicate positive and negative reactions. After responding to the radio programs, participants took part in a focus group to give reasons for their individual and collective responses (Merton, 1987). In a more recent study, O'Donnell et al. (1991) used focus groups to survey peoples' opinions of a domestic heating control system. During the interview, the moderator took the group through each feature of the task, asking for group comments on what was happening or what they thought he should be thinking to handle the task. Possible modifications to the system's interface were inferred from these comments.

While asking people what they want and need of a system is straightforward if the system is an incremental improvement on an existing product, it is less straightforward if the system has no obvious precursor (Sato and Salvador, 1999). This is because most people cannot imagine themselves acting habitually in ways other than they currently do: "In 1980, few executives could picture themselves at a keyboard any more than a fashionable girl of 1955 could imagine herself well dressed without bouffant petticoats" (Ireland and Johnson, 1995). According to Norman (1998), "Users have great trouble imagining how they may use new products, and when it comes to entirely new product categories — forget it." This chapter details a scenario-based focus group procedure that addresses these problems.

A scenario is a postulated sequence of events — a story. With regard to product design, a scenario refers to the sequence of events that outline a person's interaction with a particular product. For example, the scenario featured in Figure 32.1 details the events associated with a client using an ATM (automated teller machine). People in every age, in every place, and in every society have used stories to consider and communicate their experiences (Barthes, 1977), so it is not surprising that scenarios are proving useful in evaluating products. The focus group procedure that is detailed in this chapter encourages interviewees to think about and discuss stories that detail how they use existing products before evaluating a prototype product. Specifically, they are asked to use these scenarios to illustrate what is good and bad about existing products that are directly or indirectly related to a prototype product. The interviewees are then asked to use these scenarios as a context within which to evaluate the prototype product.

This procedure is useful in that it encourages interviewees to consider how they actually use existing products *and* that these products are not ideal. Norman (1998) argues that participants in focus groups have great trouble imagining how they may use new products because much of peoples' current behavior is not conscious: "we are primarily aware of our acts when they go wrong or when we have difficulty." Furthermore, people have difficulty anticipating the potential benefits of new products because they cannot see any problems with existing products and services.

Second, this procedure restricts the scope of the evaluation task, which might enhance the interviewees' ability to complete the undertaking. The idea that constraints might enhance such tasks seems paradoxical

---

The use case begins when the Client inserts an ATM card. The system reads and validates the information on the card.

System prompts for PIN. The Client enters PIN. The system validates the PIN.

System asks which operation the client wishes to perform. Client selects "Cash withdrawal."

System requests amounts. Client enters amount.

System requests type. Client selects account type (checking, savings, credit).

The system communicates with the ATM network to validate account ID, PIN, and availability of the amount requested.

The system asks the client to withdraw the card. Client withdraws card.

System dispenses the requested amount of cash.

System prints receipt.

FIGURE 32.1   A scenario used by Krutchen. (From Krutchen, P. [1999], *The Rational Unified Process: An Introduction*, Addison-Wesley, Reading, MA.)

given the widely held belief that such activities require as few restraints as possible. For example, Macaulay (1996) argues that the purpose of a focus group is to allow a group of users to talk in a free-form discussion with each other. However, concentrating only on a small subset of scenarios provides a "constraining framework." Such frameworks can enhance people's ability to notice or recognize the unexpected (Perkins, 1981). Indeed, Finke et al. (1992) argue that such constraints can enhance peoples' ability to participate in a variety of design tasks.

Finally, and perhaps most importantly, the scenario-based procedure provides the interviewees with a naturalistic communication tool. Scenarios seem to allow people from different backgrounds to discuss products by providing a common and perhaps natural sense-making mechanism. For example, Lloyd (2000) notes the unstructured use of scenarios by engineers, who would tell stories to each other to develop a common understanding of a particular product. Moreover, people without any design experience also seem to use scenarios informally when discussing systems. For example, there is considerable urban mythology associated with the use of fake ATMs to steal money. The story goes that a gang of thieves sites a dummy ATM in a shopping center. As people unsuccessfully attempt to withdraw money, the fake ATM records their personal identity numbers and card details. The gang then uses this information to access peoples' accounts. In this scenario, the validity of the claim (i.e., is this technically possible) may be less important than the opportunity to discuss the security risks posed by ATMs. In a similar vein, Payne (1991) discusses the spontaneous generation of scenarios by participants asked to explain how ATMs work.

32.2 Procedure

There are three steps in the scenario-based focus group procedure.

32.2.1 Step 1: Scenario Generation

Ask interviewees to develop a set of usage scenarios to exemplify distinctive and typical usage situations for an existing product that are related (directly or indirectly) to the product that is to be evaluated. The scenarios could be illustrated using the task steps approach (see Figure 32.1) or through storyboards or through role play. Each approach leads the group to focus on different aspects of their interactions with the products, and each approach encourages attention to different product characteristics. Thus, for example, the use of task steps tends to concentrate on the functionality of the product, whereas the use of role play tends to focus on the social contexts in which the product might be used.

32.2.2 Step 2: Claims Analysis

Ask the interviewees to identify features of these products that they consider significant. One approach to performing this activity is to use large sheets of paper on which the group will write a title to define a particular feature. For instance, a personal medical device might be assigned the feature "it knows all about my medical history." Having written this feature on the top of the sheet, the group then generates comments about the feature, e.g., they might mention issues of privacy or completeness of the records, etc. For each comment, ask the participants to determine whether it is good and desirable or whether it is bad or undesirable. It might be useful to divide the paper into two columns headed "positive" and "negative." Encourage the group to put the sheets of paper onto the walls around the room in which the focus group is being held, and to periodically return to the sheets to add or edit the comments.

32.2.3 Step 3: Evaluation

Introduce the interviewees to alternative product concepts. These concepts might be the result of design team activities or might arise from previous focus groups. Ask the group to rescript their scenarios, replacing the existing product with the new product and evaluate how its features would emphasize or deemphasize the pros and cons identified earlier. For example, our hero might thus be presented with a new ATM that features a biometric security facility. The interviewee would then rewrite the scenario to include the ATM requesting the client to provide a fingerprint as identification. The interviewee might thus judge the system to be beneficial in that it eases the cognitive load on the ATM user.

32.3   Advantages

- Group interviews allow researchers to survey a large number of opinions quickly.
- Group interviews can also overcome two of the problems associated with individual interviews.

Kelly (2001) argues that interviews are problematic for two reasons:

One reason is the same factor that prevents you from learning your meat loaf tastes like sawdust. Your dinner guests are too polite to tell you the unvarnished truth, too wrapped up in trying to give you the expected answer. How's the meat loaf? "Fine," they say. ("Delicious," if they care about you or think it will make you happy.) How many people volunteer that they're having a lousy day? It's human nature to put a bright face on a dismal situation.… A second reason for the "fine" response is that your guests don't know or can't articulate the "true" answer. Maybe the meat loaf needs more salt or less onion. The problem is that your customers may lack the vocabulary or the palate to explain what's wrong, and especially what's *missing*.

In short, people are frequently unwilling or unable to discuss particular issues directly. However, in focus groups, interviewees can prompt one another to explore the gaps in their thinking and thus compensate for individual failings (O'Donnell et al., 1991). Moreover, the interviewees can also provide support for one another to say things that they might otherwise be unwilling to discuss with an interviewer (Lee, 1993). For example, Morgan (1988, in Lunt and Livingstone, 1996) used group interviews to study the grieving process of widows and argued that under such circumstances, the groups can take on the character of private self-help groups or confessionals. Indeed, Lunt and Livingstone (1996) argue that focus groups could "be understood not by analogy to the survey, as a convenient aggregate of individual opinion, but as a simulation of these routine but relatively inaccessible communicative contexts that can help us discover the processes by which meaning is socially constructed through everyday talk." In their opinion the former view of focus groups predominates contemporary practice. As such, scenarios may be key in establishing the focus group as a theoretically significant approach rather than as a convenient (or contaminated) source of singular belief.

32.4   Disadvantages

- The analysis of the data derived from group interviews can be time consuming.

32.5   Example of Focus Groups

The following case study illustrates a scenario-based focus group that followed the above procedure. The topic of the discussion was an electronic wallet. Electronic wallets intend to replace the traditional wallet as electronic money replaces traditional money. The rationale is that the wallet is inherently associated with personal finance and will therefore provide a natural platform for electronic commerce and banking (Cooper et al., 1999). For instance, AT&T's electronic wallet provides a container for "electronic purses." Electronic purses are similar to conventional credit cards but contain an embedded computer chip that can store money in the form of electronic cash. AT&T's wallet also allows its users to access information relating to any transactions they have made using the electronic money that is stored in their purses. The objective of the following case study was to evaluate a particular electronic wallet prototype

Four individual groups of six interviewees were asked to develop scenarios exemplifying how they currently use their wallets and electronic banking products such as automated teller machines (ATMs) and Internet banking products. These products were chosen as they were regarded as being most closely associated with electronic wallets. The interviewees were given pens, pencils, and paper and instructed to illustrate their usage scenarios. The participants were then asked to identify "features" of these situations that they considered significant. The participants were then asked to elaborate both what is good or desirable (the pros) and what is bad or undesirable (the cons) about these features as a group. They were then introduced to an electronic wallet prototype and instructed to evaluate how the features of the new technology would emphasize or deemphasize the pros and cons identified earlier. Two of the features — storage and access — that were identified by the participants are reported below.

32.5.1 Storage

Many of the interviewees (18/24) identified "storage" as an important feature of wallet use. For example, that "you can store most any small thing in a wallet" was regarded as a pro. However, that "you can store only a limited amount of small things in a wallet/purse" was regarded as a con. The perceived inability of the prototype electronic wallet to store objects other than a few banknotes was anticipated to be a real problem. While the participants most closely associated the wallet with the action of storing financial objects such as cash and bank cards, they also revealed in their scenarios that they used it to stash a multitude of other objects. In fact, the interviewees, prompted by the storage scenario in Figure 32.2, discussed hoarding rather unusual items, including plasters, textile samples, and teeth! Furthermore, several of the interviewees went on to say that they only really carried a wallet to store sentimental things, such as pictures of their children. Indeed, as the discussion moved away from talking about wallets as financial artifacts, it became obvious that they were as much an emotive artifact as anything else.

A friend of mine gave me a press cutting at a party.

I didn't want to lose it so I went to find my jacket and my purse.

I put the cutting into my purse for safe keeping.

Pros
- Safe place.
- I won't lose it.
- I will have it with me at most times.

Cons
- It takes up space in my purse.
- It will get creased.

FIGURE 32.2 A participant's storage scenario.

32.5.2 Access

The prototype fared better with regard to "access." The majority of the interviewees (20/24) identified the accessibility as an important feature of banking systems. Although the only constraint placed on the participants was to consider a "banking system," all of the participants produced ATM usage scenarios. For example, that "you can use an ATM at any time" to recover information was regarded extremely positively. However, that "you have to know where an ATM is located (sometimes in an unfamiliar place)" was regarded very negatively. As previously, the participants used these and other ATM pros and cons to evaluate the features of the concept electronic wallet prototype that they were presented with. Clearly, the portability of the electronic wallet maintains peoples' ability to access financial information anytime. Moreover, the electronic wallet would allow them in the future to access financial information "any place." Furthermore, the portability of the concept electronic wallet meant that they did not have to search for the system beyond their bag or pocket.

32.6 Related Methods

The focus group's closest relative is the individual interview.

32.7 Standards and Regulations

Most standards relating to product evaluation require some assessment of user response to a product. Focus groups can be a useful first step in defining appropriate dimensions for scaling such responses.

32.8 Approximate Training and Application Times

Focus groups require extensive planning and rely on an adept moderator. Moreover, conducting the actual interviews and analyzing the results can be time consuming.

32.9 Reliability and Validity

In discussing interviews, Robson (1993) notes:

> Asking people directly about what is going on is an obvious short cut in seeking answers to our research questions.... The lack of standardization that it implies inevitably raises concerns about reliability. Biases are difficult to rule out. There are ways of dealing with these problems but they call for a degree of professionalism that does not come easily.

32.10 Tools Needed

The focus group that is used as an example in this article required only a few pens, pencils, drawing pads, and a tape recorder to note what was said.

References

Barthes, R. (1977), *Image-music-text*, Fontana, London.

Cooper, L., Johnson, G., and Baber, C. (1999), A run on Sterling: personal finance on the move, in *Third International Symposium on Wearable Computers*, IEEE Computer Society, Los Alamitos, CA.

Finke, R., Ward, T., and Smith, S. (1992), *Creative Cognition: Theory, Research and Applications*, MIT Press, Cambridge, MA.

Ireland, C. and Johnson, B. (1995), Exploring the future in the present, *Design Manage. J.*, 6, 57–64.

Kelly, T. (2001), *The Art of Innovation: Lessons in Creativity from IDEO, America's Leading Design Firm*, Harper Collins, London.

Krueger, R. (1988), *Focus Groups: A Practical Guide for Applied Research*, 3rd ed., Sage, London.

Krutchen, P. (1999), *The Rational Unified Process: An Introduction*, Addison-Wesley, Reading, MA.

Lee, R.M. (1993), *Doing Research on Sensitive Topics*, Sage, London.

Lloyd, P. (2000), Storytelling and the development of discourse in the engineering design process, *Design Studies*, 21, 357–373.

Lunt, P. and Livingstone, S. (1996), Rethinking the focus group in media and communications research, *J. Commun.*, 46, 79–98.

Macaulay, L. (1996), *Requirements Engineering*, Springer, London.

Merton, R. (1987), The focussed interview and focus groups: continuities and discontinuities, *Public Opinion Q.*, 51, 550–566.

Norman, D. (1998), *The Invisible Computer*, MIT Press, Cambridge, MA.

O'Donnell, P.J., Scobie, G., and Baxter, I. (1991), The use of focus groups as an evaluation technique in HCI, in *People and Computers VI: Proceedings of the HCI '91 Conference*, Diaper, D. and Hammond, N., Eds., Cambridge University Press, Cambridge, U.K., pp. 211–224.

Payne, S.J. (1991), A descriptive study of mental models, *Behav. Inf. Technol.*, 10, 3–22.

Perkins, D. (1981), *The Mind's Best Work*, Harvard University Press, Cambridge, MA.

Robson, C. (1993), *Real World Research*, 12th ed., Blackwell Scientific, Oxford, U.K.

Sato, S. and Salvador, T. (1999), Playacting and focus troupes: theatre techniques for creating quick, intense, immersive, and engaging focus group sessions, *Interactions*, Sep./Oct., 35–41.

33

Hierarchical Task Analysis (HTA)

John Annett
University of Warwick

33.1 Background and Applications

Hierarchical task analysis (HTA) was developed at the University of Hull in response to the need to analyze complex tasks, such as those found in the chemical-processing and power-generation industries (Annett et al., 1971). The training of process-control operators was a matter of concern because the time-and-motion methods of task analysis, developed originally for the routine repetitive manual operations used in manufacturing industries, did little justice to the skills needed in modern automated industries, which involve less physical activity combined with a high degree of cognitive skill and knowledge on the part of the operator.

HTA analyzes not actions per se but *goals* and *operations*, the means of attaining goals. Complex tasks are decomposed into a hierarchy of operations and suboperations with the aim of identifying those that are likely to fail due to poor design or lack of expertise and thus to propose solutions that might involve redesigning the task or providing special training.

Operations are specified by the conditions under which the goal becomes active, known as the *input*; the means by which the goal is attained, known as *actions*; and the indications of goal attainment, known as *feedback*. Operations can be decomposed into constituent suboperations grouped together as a *plan*. There are four main types of plan: a simple sequence of operations or *routine procedures*, a conditional sequence involving a *decision* such that the appropriate action depends on a specific pattern of inputs, a *time-shared procedure* when two goals must be attained at the same time, and an *unordered procedure* where all subgoals must be attained but order is unimportant.

Operations can be decomposed to whatever level of detail is required by the purpose of the analysis, but a general rule is to stop when the probability of failure of an operation times the cost of failure (p × c) is acceptable and that a remedy for actual or potential failure can be offered.

HTA has been widely used in the process-control and power-generation industries and in military applications (Kirwan and Ainsworth, 1992; Ainsworth and Marshall, 1998) and has been adapted for use in most human factor and human–computer interaction (HCI) applications, including training (Shepherd, 2002), design (Lim and Long, 1994), error and risk analysis (Baber and Stanton, 1994), and the identification and assessment of team skills (Annett et al., 2000).

33.2 Procedure

HTA is a very flexible tool that can be used in a variety of ways, but the following seven-step procedural guidelines will meet most cases.

33.2.1 Step 1: Decide on the Purpose of the Analysis

HTA is not simply a record of how a task is normally done; rather it is also a means of identifying the sources of actual or potential performance failure and proposing remedies (see step 7). These may be in the form of redesign of the equipment, changing the way the task is carried out, or optimizing the use of personnel or the content or style of training. Thus the output of an HTA is a report that addresses the original question, such as a modified equipment design or operating procedure, a recommended training syllabus or training medium, a risk/hazard assessment, etc.

33.2.2 Step 2: Determine Task Goals and Performance Criteria

The "owners" of the task and stakeholders such as designers, managers, supervisors, instructors, and operators should agree on the goals, the organizational values, and the outputs desired. Most importantly, they should agree on objective performance criteria. This step may require close questioning and even negotiation between stakeholders, since people can sometimes be unclear about their goals and values and acceptable costs or even resist the urge to say what they really want.

33.2.3 Step 3: Identify Sources of Task Information

It is desirable to make use of as many sources as possible. These include documentation, such as drawings and manuals for maintenance and operating procedures; expert opinion from designers, managers, instructors, and operators; and records of plant or operator performance, including accident and maintenance data. Direct observation is often helpful for initial orientation and for checking opinion, but the more varied the task and the greater the involvement of cognitive processes (as opposed to physical activity), the less useful it is as the primary source of data.

33.2.4 Step 4: Acquire Data and Draft Decomposition Table/Diagram

It is usually best to start at the top, i.e., with top-level goals, asking in turn how each subgoal is attained. It is also important to ask not only what *should* happen but what *might* happen and especially *what can go wrong* and what would be the consequences of failure to attain any goal or subgoal. The decomposition table and diagram (see Section 33.5 for an example of HTA application) should reveal the overall structure of the task, including significant plans such as long procedures, critical decision rules, dual tasks, etc. All operations at the same level of analysis should be (a) mutually exclusive and (b) account completely for the superordinate operation from which they derive.

33.2.5 Step 5: Recheck Validity of Your Decomposition with Stakeholders

Extracting unambiguous information is seldom easy, and stakeholders need to be assured that the analysis is consistent with the facts, constraints, and values associated with the task and its context. It may be necessary to revisit the analysis, or parts of it, on several occasions in order to resolve ambiguities. It is particularly important to establish objective performance criteria associated with high-level goals and critical suboperations, since this is the only means by which the analysis can be validated by evidence of problems identified and solved.

33.2.6 Step 6: Identify Significant Operations in the Light of the Purpose of the Analysis

A significant operation is one that fails the $p \times c$ criterion. The reason for failure may be obvious upon inspection of the details of the operation, but it is helpful to consider failures related to input, to action and plans, and to feedback. Inputs may be physically obscure, such as an illegible instrument, or conceptually difficult, such as a rare fault pattern. The appropriate action may be problematic for reasons ranging from the physical (cannot reach the control) to the conceptual (not knowing what to do). Plans, while composed of simple elements, may involve long procedures, complex decisions, or division of attention and effort between two or more simultaneous demands. Feedback essential to correct performance may suffer the problems of any other type of perceptual input, but may be especially disruptive if subject to delay.

33.2.7 Step 7: Generate and Test Hypothetical Solutions to the Performance Problems Identified in the Analysis

Having identified the likely source(s) of unsatisfactory performance, plausible solutions, based on current theory and best practice, are presented. These may be related to the design of the task and the equipment, personnel use, procedures or training, and other forms of support, depending on the purpose of the analysis established in step 1. Usually the type of solution, e.g., to modify the design of the equipment or to construct a training syllabus, is predetermined, but the analyst should not refrain from drawing attention to alternative solutions where these may offer advantages.

33.3 Advantages

- As a generic method, HTA is adaptable to a wide range of purposes.
- Tasks can be analyzed to any required level of detail, depending on the purpose.
- When used correctly, HTA provides an exhaustive analysis of the problem addressed.

33.4 Disadvantages

- Requires handling by an analyst well trained in a variety of methods of data collection and in relevant human-factors principles.

- Requires full collaboration of relevant stakeholders.
- Requires time in proportion to the complexity of the task and the depth of the analysis.

33.5 Example: The Analysis of Antisubmarine Warfare Team Skills

Because HTA has been used for a wide variety of purposes, and because the documentation of any nontrivial example would take too much space, the following example simply illustrates how each of the seven recommended steps were followed in one specific case. Further detail of this example is provided in Annett et al. (2000). A wide range of examples can be found in Shepherd (2002).

33.5.1 Step 1: Purpose of the Analysis

The aim was to identify and measure team skills critical to successful antisubmarine warfare (ASW). This analysis focuses on operations depending critically on team interaction.

33.5.2 Step 2: Determine Task Goals and Performance Criteria

Naval doctrine specifies goals for the warfare team of a ship as "float, move, fight." When escorting a highly valued unit (e.g., a troopship), safe arrival at the designated time and place and successfully countering all threats are the top-level goals. A variety of *objective criteria* are available, including geographic location, proportion of threats correctly identified within specified time limits, defensive attacks successfully executed, etc.

33.5.3 Step 3: Identify Information Sources

The primary source was a senior ASW instructor, supported by manuals specifying warfare doctrine and standard operating procedures. The analyst was also able to observe teams operating in a simulator and to make use of electronic records of events occurring in exercises. The latter proved to be better at telling what happened than how or why it happened. At a later stage, video recordings were made of teams operating in prearranged scenarios in a simulator. These served as checks on the information provided by the expert and also as examples of data that could be used to measure team skills (see step 7).

33.5.4 Step 4: Draft Decomposition Diagram/Table

The questioning of subject-matter experts focused on goals, such as, "How do you…?" and failures, such as, "What happens if you do not…?" Especially important are questions such as, "How would you know if 'x' had been correctly carried out?" Examples of classic disasters collected as anecdotes from experienced naval staff were valuable pointers to sources of team failure.

A small section of the analysis is illustrated in diagrammatic form in Figure 33.1. The clear rectangles show how part of the overall operation is decomposed into suboperations. The shaded rectangles provide the immediate context. The plans are indicated in square brackets […]. The four types of plan are represented by the symbols [1 + 2], meaning that both subgoals 1 and 2 are active simultaneously, i.e., do 1 *and* 2; [1 > 2], meaning that subgoal 1 must be attained and then subgoal 2, i.e., do 1 *then* 2; [1/2], meaning that subgoal 1 or subgoal 2 is active, i.e., do 1 *or* 2, depending on a rule; [1:2], meaning that subgoals 1 and 2 are both active but not simultaneously and in no specified order. Another way of representing a plan is to specify the algorithm in words such as "do 1 then 2," or "do 1 or 2 depending on condition x," etc. The tabular form (Table 33.1) contains notes on teamwork and criterion measures relating to the clear rectangles.

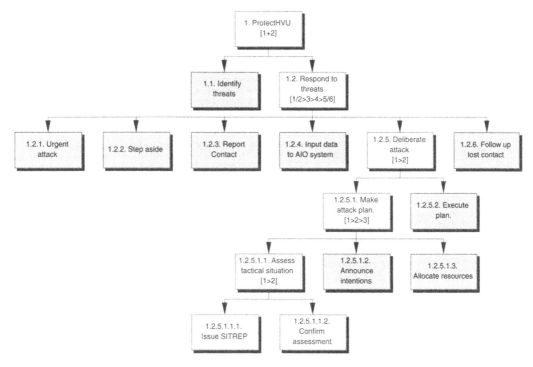

FIGURE 33.1 Partial analysis of an antisubmarine warfare team task.

33.5.5 Step 5: Recheck Validity of Decomposition with Stakeholders

It is rarely possible to avoid misunderstandings of the task structure on the basis of a single interview. At least two iterations of the decomposition were required before the stakeholders, and the analyst, were satisfied that this was an accurate representation of the task goals and the various means of attaining them.

33.5.6 Step 6: Identify Significant Operations

Referring to step 1, the aim of the analysis was to identify and measure team skills critical to successful team performance. The depth of the analysis was therefore determined by the lowest level at which various team activities could be associated with critically important team products. An example is seen in operation 1.2.5.1.1.1 (Table 33.1), where discussion between team members, each of whom hold partial information on the nature of the threat, is a highly desirable means of minimizing ambiguity of interpretation of the tactical situation.

33.5.7 Step 7: Generate and Test Hypothetical Solutions to Performance Problems

The aim of the analysis was to identify and measure different ways in which poor team skills could lead to unsatisfactory performance. The analysis identified a number of critical operations in which various types of team interaction could lead to failure to achieve team goals. Five types of critical team behavior were identified: send information, receive information, discuss, collaborate, and synchronize. (Note that different categories could have been used; these happened to coincide with a particular theory of teamwork adopted by the analysts.) The performance of ASW teams was observed during a standard exercise and scored as the percentage of satisfactorily executed team operations in each of the five categories.

TABLE 33.1 Tabular Form of Selected ASW Team Operations

Protect highly valued unit (HVU) [1+2]	*Goal:* Ensure safe and timely arrival of HVU *Teamwork:* Principal Warfare Officer (PWO) in unit gaining initial contact with threat assumes tactical command and follows standard operating procedures in this role *Plan:* Continues to monitor threats [1] while responding to identified threat *Criterion measure:* Safe and timely arrival of HVU
Respond to threats [1/2>3>4>5/6]	*Goal:* Respond to threat according to classification *Teamwork:* PWO selects response based on information provided by other team members *Plan:* If threat is immediate (e.g., torpedo) go to urgent attack [1.2.1], else execute 2, 3, 4, and 5 or 6 *Criterion measure:* Appropriate response with minimal delay
1.2.5: Deliberate attack [1>2]	*Goal:* Get weapon in water within 6 min *Teamwork:* See further breakdown below *Plan:* Make attack plan, then execute *Criterion measure:* Time elapsed since classification and/or previous attack
1.2.5.1: Make attack plan [1>2>3]	*Goals:* Plan understood and accepted by team *Teamwork:* Information regarding tactical situation and resources available from team members to PWO *Plan:* Assess tactical situation; announce intentions; allocate resources *Criterion measure:* Accurate information provided
1.2.5.1.1: Assess tactical situation [1>2]	*Goal:* Arrive at correct assessment of tactical situation *Teamwork:* PWO must gather all relevant information by up-to-date status reports from own team and sensors and other friendly forces *Plan:* Issue situation report (SITREP), then confirm assessment *Criterion measures:* Correct assessment; time to make an assessment
1.2.5.1.1.1: Issue SITREP	*Goal:* To ensure whole team is aware of threat situation and to provide an opportunity for other team members to check any omissions or errors in tactical appreciation *Teamwork:* PWO issues SITREP at appropriate time; all team members check against information they hold *Criterion measure:* All team members have accurate tactical information
1.2.5.1.1.2: Confirm tactical assessment	*Goal:* Construct an accurate assessment of the threat and of resources available to meet it *Teamwork:* Final responsibility lies with the PWO, but information provided by and discussion with other team members are essential to identify and resolve any inconsistencies *Criterion measure:* Accurate assessment in light of information and resources available

Validation studies by the Royal Navy School of Education and Training Technology, Portsmouth, to establish how well the method predicts overall team success are under way at the time of writing.

33.6 Related Methods

HTA has been widely used as a basis for the investigation of a variety of problems, for example as the first step in the TAFEI (task analysis for error identification) method for hazard and risk assessment (Baber and Stanton, 1994), in SHERPA (systematic human error reduction and prediction approach) for predicting human error (Baber and Stanton, 1996), in MUSE (method for usability engineering) usability assessment (Lim and Long, 1994), the sub-goal template (SGT) method for specification of information requirements (Ormerod et al., 1998), and the TAKD (task analysis for knowledge-based descriptions) method for the capture of task knowledge requirements in HCI (Johnson et al., 1984).

33.7 Standards and Regulations

HTA does not purport to provide any standard measure and has not, as yet, been incorporated into specific standards.

33.8 Approximate Training and Application Times

There are no formal training courses in HTA, and hence training time cannot be estimated. A study by Patrick et al. (2000) gave students a few hours training with not entirely satisfactory results on the analysis of a very simple task, although performance improved with further training. A survey by Ainsworth and Marshall (1998) found that the more experienced practitioners produced more-complete and acceptable analyses. Clearly the application time is entirely dependent on the complexity of the task, the depth of the analysis, and the ease with which data can be acquired.

33.9 Reliability and Validity

Reliability depends principally on the care taken in data collection. Attention is drawn especially to cross-checking of information in step 5. Validity depends on whether the analyst correctly addresses the questions asked (step 1) and provides effective solutions (step 7). These factors are not easily assessed quantitatively. Although HTA is widely used, data from a sufficient sample have never been collected, and it would take a major research project to do so. In the example given above, further trials are currently being conducted by the Royal Navy.

33.10 Tools Needed

HTA can be carried out using only pencil and paper. In the example given above, interviews were tape-recorded and the diagrammatic and tabular forms were constructed on a laptop computer using a general-purpose software idea development and planning tool called Inspiration™. Video recording was used in step 7 for sampling team behavior in critical operations.

References

Ainsworth, L. and Marshall, E. (1998), Issues of quality and practicality in task analysis: preliminary results from two surveys, *Ergonomics*, 41, 1604–1617; reprinted in Annett and Stanton (2000).

Annett, J., Duncan, K.D., Stammers, R.B., and Gray, M. (1971), *Task Analysis*, Her Majesty's Stationery Office, London.

Annett, J., Cunningham, D.J., and Mathias-Jones, P. (2000), A method for measuring team skills, *Ergonomics*, 43, 1076–1094.

Annett, J. and Stanton, N.A., Eds. (2000), *Task Analysis*, Taylor & Francis, London, pp. 79–89, 114–135.

Baber, C. and Stanton, N.A. (1994), Task analysis for error identification, *Ergonomics*, 37, 1923–1941.

Baber, C. and Stanton, N.A. (1996), Human error identification techniques applied to public technology: predictions compared with observed use, *Appl. Ergonomics*, 27, 119–131.

Johnson, P., Diaper, D., and Long, J. (1984), Tasks, skills and knowledge: task analysis for knowledge-based descriptions, in *Interact '84: First IFIP Conference on Human-Computer Interaction*, Shackel, B., Ed., Elsevier, Amsterdam, pp. 23–27.

Kirwan, B. and Ainsworth, L. (1992), *A Guide to Task Analysis*, Taylor & Francis, London.

Lim, K.Y. and Long, J. (1994), *The MUSE Method for Usability Engineering*, Cambridge University Press, Cambridge, U.K.

Ormerod, T.C., Richardson, J., and Shepherd, A. (1998), Enhancing the usability of a task analysis method: a notation and environment for requirements, *Ergonomics*, 41, 1642–1663; reprinted in Annett and Stanton (2000).

Patrick, J., Gregov, A., and Halliday, P. (2000), Analysing and training task analysis, *Instructional Sci.*, 28, 51–79.

Shepherd, A. (2002), *Hierarchical Task Analysis*, Taylor & Francis, London.

34

Allocation of Functions

Philip Marsden
University of Huddersfield

Mark Kirby
University of Huddersfield

34.1 Background and Applications

Allocation of system function (Chapanis, 1970), function allocation (Hollnagel, 1999), or sociotechnical allocations (Clegg et al., 2000) are names given to the process in which members of a design team decide whether to allocate jobs, tasks, system functions, or responsibility to human or automated agents in sociotechnical work environments. Although a variety of tools, techniques, and automated aids are now available to assist the function allocation process, consideration of the relative merits of people and machines remains at the heart of the process. Whereas originally, comparison of the contrasting abilities of people and machines concentrated purely on task-related performance factors, such as the ability to process large volumes of data or speed of processing, contemporary schemes have been broadened to consider wider social issues, such as job satisfaction, user engagement, and empowerment of system stakeholders (Kirby, 1997).

Numerous examples of function allocation in practice can be found in the human factors literature. The first and perhaps best-known case study can be found in a research report edited by Fitts (1951). Fitts pioneered the allocation-of-function approach in the early 1950s as part of an investigation into the role of automation in air-traffic control. More recently, an insightful method for allocation of system function has been described by Price et al. (1982) in the context of nuclear power generation. Marsden (1991) advocated use of an approach similar to that described by Price et al. in the context of manned space flight, while Strain et al. (1997) described an approach called Organizational Requirements Definition for Information Technology (ORDIT) to help designers allocate system function during the design of new warships for the British Royal Navy. These are just a few examples of the diverse range of applications that have been submitted to a formal allocation-of-function process. For a fuller description of each example, and others not mentioned, the reader is referred to Fallon et al. (1997) and an article in the *IJHCS* (Anon., 2000).

It is commonly assumed that automation confers many benefits on complex and dynamic real-world systems and that the advantages of its use far outweigh any disadvantages. While this may be true, it is also the case that automation can create special problems for the human component of a highly automated system. There are a number of important limitations to the design and implementation of automated aids. The question of allocating function to humans or machines has been of interest to human factors for over four decades. It is highly appropriate to apply the paradigm to the problem of determining which function to allocate to the operator and which functions to allocate to the automated systems. Singleton (1989) argues that optimal allocation depends upon technological capability and the feasibility of human tasks The first step is to separate system functions into discrete categories. This serves as a basis for allocation. Functions are either allocated to the humans and/or to automation. These allocations can be validated by task analysis (see Chapter 33 for details of methods) and technological assessment. If the validation outcomes are satisfactory, the functions are transformed into design activities.

Allocation-of-function methods include tables of relative merit (TRM), psychometric approaches, computational aids, and the hypothetical-deductive model (HDM). The TRM approach is perhaps in its most well-known form as the Fitts list (1951). This list is continually being updated, for example the Swain list (1980). The TRM method employs the task-dichotomy approach: tasks that machines are good at and humans are poor at, and vice versa. Essentially all of the approaches characterize the differences in abilities between humans and machines. When these differences have been determined, decisions can be made to form prescriptions for the design of systems. In an extensive review, Marsden (1991) concluded that more formal and balanced approaches to allocation of function (such as HDM) offer a significant advance over the TRM approach. The HDM (Price, 1985) consists of five main stages:

1. Specification, in which the system requirements are clarified
2. Identification, in which system functions are identified and defined in terms of the inputs and outputs that characterize the various operations
3. Hypothesization of solutions, in which hypothetical design solutions are advanced by various specialist teams
4. Testing and evaluation, in which experimentation and data gathering are undertaken to check the utility of functional configuration for the overall design
5. Optimization of design, in which design iterations are made to correct errors

The present chapter attempts to bring together strands apparent in a number of different schemes in an effort to provide a generic overview of the function-allocation procedure. In what follows, we shall outline a sociotechnical systems approach to function allocation in which particular attention is given to the broader requirements of the stakeholder community.

34.2 Procedure

The procedure involves three main steps: task analysis, stakeholder analysis, and analysis of alternative allocations of function.

34.2.1 Step 1: Task Analysis

The simplest suitable method for step 1 is hierarchical task analysis (HTA). This starts with the identification of the purpose of the system, then logically considers which tasks need to be performed to achieve that purpose. Each task is related to a goal. For each task, the analyst needs to identify which subtasks need to be carried out to achieve that goal. The task model can be developed in greater detail by breaking subtasks down into sub-subtasks in the same way.

If stakeholders do not agree on the purpose of the system, more than one model should be considered.

It is important that the upper levels of the hierarchy be solution independent, i.e., the analyst must try and indicate what tasks need to be carried out, not how they will be carried out. For example, if a goal requires a person to go through an identity check, the subtask should be called "go through an identity check" and not "enter password." The latter assumes a particular way of carrying out the task and preempts the allocation of function, prematurely ruling out alternative allocations of function and creative solutions.

The analyst should stop developing the detailed levels of the model when it becomes impossible to break the tasks down further without having a solution in mind. Further development of the model should be postponed until decisions have been made about allocation of function.

Other approaches to task analysis can be used instead of hierarchical task analysis, as long as a model is produced that lends itself to annotation with different types of allocation of function at step 3.

34.2.2 Step 2: Stakeholder Analysis for Allocation of Function

It is important to carry out this step without thinking in detail about the design of a computer system; otherwise, premature decisions might be made about allocation of function. The aim is to identify the sources of job satisfaction and dissatisfaction of any stakeholder whose job might conceivably be affected by changes in computer systems at work. The analysis is better conducted by observing stakeholders at work than by asking stakeholders questions directly. For allocation of function, the most relevant parts of stakeholder analysis are:

- Identifying both the current knowledge and skills of the stakeholders and their potential to gain new knowledge and skills, e.g., through training programs
- Considering which of the following aspects of work are important to the stakeholders:
  Having a challenging job to do
  Opportunity to exercise specialist skills
  Development of new skills
  Keeping the job as simple as possible
  Removing tedious aspects of work
  Having a variety of work to do
  Enjoying interaction with other people
  Avoiding interaction with other people
  Working as a member of a team
  Working alone
  Status gained from doing a specialist job
  Status gained from being a source of important information
  Knowing what is going on
  Pride in contributing to a successful product or service
  Challenge of dealing with difficult problems

Where stakeholders share views, this should be noted. Where there is variance in views, the extent of disagreement should also be noted.

34.2.3 Step 3: Analyzing Alternative Allocations

34.2.3.1 Consider Relative Capabilities of Human and Computer

The first part of the analysis involves consideration of the relative capabilities of the human and the computer. While working through the task hierarchy, allocate tasks according to whether the human or computer can perform the task best, or whether human and computer are likely to produce a better result working together. During this step, the analyst should bear in mind both the current and potential skills of the humans. If a task is to be shared between human and computer, it is important to consider whether the human or the computer should be in control. A simple way to record allocations of function is to annotate the task model as follows:

- H: human only
- H-C: shared between human and computer, human in control
- C-H: shared between human and computer, computer in control
- C: computer only

More-complex classifications of shared human–computer tasks can be used if desired.

34.2.3.2 Review Impact of Allocation

Review the potential impact of this allocation of function on job satisfaction and task performance by considering:

- The extent to which it is compatible with or conflicts with the job satisfaction criteria identified in step 2, bearing variances in mind
- The potential for human error, particularly where the task is safety critical or there are security issues
- The need for the human to be sufficiently involved in the task to stay alert and have sufficient knowledge of the situation to be able to act when needed
- Any likely change in the cost and use of resources, including time if speed of task completion is important

34.2.3.3 Explore Alternative Allocation

Explore an alternative allocation of function by changing some of the annotations and repeating the step 3b review for the alternative model. Consider as many alternatives as time allows. Compare the different alternatives and make a choice. It is preferable to involve the stakeholders in making a choice. The rationale for the choice should be recorded for future reference. When requirements change or amendments to a computer system are considered in the future, it is useful to understand why the original allocation of function was chosen.

34.2.3.4 Add Detail to Model

Where H-C or C-H has been allocated, it is useful to add more detail to the task model. These shared tasks should be broken down into more detail until the lowest levels of the hierarchy contain only H or C allocations, thus indicating how the task is shared between human and computer.

34.3   Example

The list below shows a hierarchical task analysis, where the hierarchy is indicated by indentation. (A graphical representation of a hierarchy can be used if preferred.) The context for this example concerns a small brewery deciding whether it has the resources to take on additional orders for beer. The top level of the hierarchy, task 1, indicates purpose. Tasks 1.1 to 1.4 are the tasks that need to be carried out to achieve that purpose.

34.3.1   Example of a Tabular HTA for Decision-Support System in a Brewery Context

    1 Check the desirability of trying to meet a potential increase in demand:
       1.1 Forecast demand
          1.1.1 Review regular sales
          1.1.2 Review demand from pub chains
          1.1.3 Review potential demand from one-off events
       1.2 Produce provisional resource plan

1.2.1 Calculate expected demand for each type of beer
1.2.2 Make adjustment for production minima and maxima
1.3 Check feasibility of plan
1.3.1 Do materials explosion of ingredients
1.3.2 Do materials explosion of casks and other packaging
1.3.3 Check material stocks
1.3.4 Calculate materials required
1.3.5 Negotiate with suppliers
1.3.6 Check staff availability
1.3.7 Check ability to deliver beer to customers
1.4 Review potential impact
1.4.1 Review impact of plan on cash flow
1.4.2 Review impact of plan on staff
1.4.3 Review impact on customer relations
1.4.4 Review impact on supplier relations

In the brewery case, the analyst took into account stakeholders' needs to exercise craftsmanship and other specialist skills, their wish to maintain control of planning heuristics, and their needs to promote cohesive team work. The analyst opted for a high degree of allocation to the human. A greater amount of computer support would have better addressed a need for quick recalculation of plans when under stress, but this was in conflict with the needs most associated with the stakeholders' sources of job satisfaction. There is some allocation to the computer and to the computer in support of the human where tedious bookwork could be reduced in a way that makes the task quicker to perform and less stressful. The final allocation of function is outlined in the next section.

34.3.2 Example of Function Allocations Based on Stakeholder Analysis of the Sociotechnical System

1 Check the desirability of trying to meet a potential increase in demand:
1.1 Forecast demand **H**
1.1.1 Review regular sales **H**
1.1.2 Review demand from pub chains **H**
1.1.3 Review potential demand from one-off events **H**
1.2 Produce provisional resource plan **H-C**
1.2.1 Calculate expected demand for each type of beer **H-C**
1.2.2 Make adjustment for production minima and maxima **C**
1.3 Check feasibility of plan **H-C**
1.3.1 Do materials explosion of ingredients **H-C**
1.3.2 Do materials explosion of casks and other packaging **C**
1.3.3 Check material stocks **H-C**
1.3.4 Calculate materials required **C**
1.3.5 Negotiate with suppliers **H**
1.3.6 Check staff availability **H**
1.3.7 Check ability to deliver beer to customers **H**
1.4 Review potential impact **H**
1.4.1 Review impact of plan on cash flow **H**
1.4.2 Review impact of plan on staff **H**
1.4.3 Review impact on customer relations **H**
1.4.4 Review impact on supplier relations **H**

34.4   Advantages

- The process provides a structure for automation decisions.
- The process ensures that automation decisions are traceable.
- The process helps ensure that the system user will not be inappropriately delegated the role of supervisor of automated systems.

34.5   Disadvantages

- The process can be costly in large-scale systems.
- The process requires a degree of expertise and familiarity with the human factor.
- Access to and involvement with system stakeholders is essential.

34.6   Related Methods

- Task analysis methodologies
- Sociotechnical systems analysis
- Stakeholder analysis

34.7   Standards and Regulations

The requirement to complete a formal allocation-of-function process during systems design is given prominence in ISO 13407:1999, Human centred design processes for interactive systems. This standard was given the status of a British standard in 2000 (BS EN 13407:1999, BSI, [2000]). Specifically, the standard states:

> Whatever the design process and allocation responsibilities and roles adopted, the incorporation of a human centred approach is characterized by the following: (a) the active involvement of users and a clear understanding of user and task requirements, (b) an appropriate allocation of function between users and technology, (c) the iteration of design solutions, (d) multidisciplinary design.

To clarify what is meant by the term "appropriate allocation of function," the standard (BS EN 13407:1999, 2000) states:

> The decision … [to allocate a function] … should be based on many factors, such as the relative capabilities and limitations of humans versus technology in terms of reliability, speed, accuracy, strength, flexibility of response, financial cost, the importance of successful or timely accomplishment of tasks and user well being. They should not simply be based on determining which functions the technology is capable of performing and then simply allocating the remaining functions to users … the resulting human functions should form a meaningful set of tasks.

In addition to the above, particular industries are required as part of their licensing agreement to carry out and demonstrate appropriate function allocation. These additional industry-specific requirements are of course important, but they are beyond the scope of this particular chapter.

34.8   Approximate Training and Application Times

Allocation-of-function practitioners must have several core skills. Analysts should be familiar with each of the background methodologies used and should have some experience of dealing with the human (as opposed to the automation) factor. A design team comprising technical staff, target users, and human-factors specialists can best achieve appropriate allocations. Simple allocations, such as the example provided above, can be achieved relatively quickly assuming users of the system are available. Function allocation in complex systems involves a more drawn-out procedure in which allocations may be made over a period of months, and in some cases years, due to the iterative nature of design evolution in complex systems.

34.9 Reliability and Validity

While most practitioners recognize that automation decisions need to be supported, in principle, by consideration of the function allocation case of support, there has been considerable debate surrounding the validity of the concept in practice. Several authors, for example, have concluded that allocation of function methodology is deeply flawed (e.g., Jordan, 1963; Fuld, 2000). In the words of Wirstad (1979), "Although the principle is clear, function allocation has never worked in practice." Objections to inclusion of function allocation in the design process take several forms. However, two main problems seem to recur throughout the literature:

1. While function allocation provides a useful theory for automation decisions, the approach typically prescribes methods, or incorporates practices, in an idealized form that fails to correspond to requirements in a practical context.
2. There is no real evidence that misallocation has ever contributed to system failure. A variant of this objection centers on the notion that theoreticians have misrepresented the principle of automating all feasible functions (the automate-or-liquidate philosophy).

Irrespective of the position taken in relation to this debate, it is important to recognize that, for the present at least, evidence of function allocation decisions is often required for development of systems involving a significant loss potential.

34.10 Tools Needed

No special tools are needed for allocation of system functions, although several automated aids have been developed to facilitate the process. What is perhaps most important is the need to create design documentation that can be audited to justify particular allocations.

References

Anon. (2000), Allocation of function, *Int. J. Hum Comput. Stud.*, 52 (special issue).

BSI (2000), Human Centred Design Process for Interactive Systems, BS EN ISO 13407:1999, British Standards Institute, London.

Chapanis, A. (1970), Human factors in systems engineering, in *Systems Psychology*, Greene, K.B., Ed., McGraw-Hill, New York.

Clegg, C.W., Older Gray, M., and Waterson, P.E. (2000), The charge of the byte brigade and socio-technical response, *Int. J. Hum.-Comput. Stud.*, 52 (special issue), 235–251.

Fallon, E.F., Hogan, M., Bannon, L., and McCarthy, J., Eds. (1997), *ALLFN '97, Proceeding of the First International Conference on Allocation of Functions*, Vols. 1–2, National University of Ireland, Galway, Ireland.

Fitts, P.M. (1951), *Human Engineering for an Effective Air Navigation and Traffic Control System*, Ohio State University Research Foundation, Columbus.

Hollnagel, E. (1999), From function allocation to function congruence, in *Coping with Computers in the Cockpit*, Dekker, S. and Hollnagel, E., Eds., Ashgate, Aldershot, U.K.

Kirby, M.A.R. (1997), Designing computer systems to support human jobs, in *ALLFN '97, Proceeding of the First International Conference on Allocation of Functions*, Vol. 2, National University of Ireland, Galway, Ireland, pp. 177–184.

Marsden, P. (1991), Allocation of System Function, report prepared for European Space Agency, Human Reliability Associates Ltd., Parbold, Wigan, U.K.

Price, H. E. (1985), The allocation of system functions, *Hum. Factors*, 27, 33–45.

Price, H.E., Maisano, R.E., and van Cott, H.P. (1982), The Allocation of Functions in Man–Machine Systems: a Perspective and Literature Review, NUREG CR-2623, Nuclear Regulatory Authority, Washington, D.C.

Singleton, W.T. (1989), *The Mind at Work*, Cambridge University Press, Cambridge, U.K.

Strain, J., Eason, K., Preece, D., and Kemp, L. (1997), New technology and employment categories in warships: a socio-technical study of skills and job allocation in the Royal Navy, in *ALLFN '97, Proceeding of the First International Conference on Allocation of Functions*, Vol. 1, National University of Ireland, Galway, Ireland, pp. 89–100.

Swain, A.D. (1980), *Design Techniques for Improving Human Performance in Production*, Albuquerque, NM.

35

Critical Decision Method

Gary Klein
Klein Associates Inc.

Amelia A. Armstrong
Klein Associates Inc.

35.1 Background and Application

The naturalistic study of expert decision making requires an in-depth analysis of complex cognitive tasks. It involves the exploration of the use of perceptual cues, the development of expert knowledge, and the evolution of expert strategies. This type of exploration is most effective when experts are studied in the context of their complex environments. Cognitive Task Analysis (CTA) methods for eliciting and analyzing the aspects of expertise needed to perform complex tasks have been evolving over the past two decades.

CTA methods must be highly adaptable for two reasons. First, the data collection conditions can vary from one project to the next, and therefore the methods have to be flexible. Second, CTA methods are often used to make new discoveries, which mean that the phenomena of interest are not well known at the outset of the research. Consequently, the CTA methods being used must be adjustable to take advantage of what has been learned so that they can better focus on the critical issues.

The method was developed more than 15 years ago (Klein et al., 1986) and continues to be useful for conducting naturalistic decision-making research. We have codified the procedures for conducting Critical Decision Method (CDM) projects, but at the same time we, and others in our field, continue to learn how to adapt the CDM by changing the way it is executed and by synthesizing it with complementary data-collection tools and approaches.

35.2 Procedure: Overview of the Critical Decision Method

One approach to cognitive task analysis is the CDM (Klein et al., 1989; Hoffman et al., 1998). The CDM was developed as an extension of the critical incident technique (CIT) (Flanagan, 1954) and uses in-depth interviews to gather retrospective accounts of challenging incidents. The CDM is a semistructured interviewing technique for investigating phenomena that rely on subtle cues, knowledge, goals, expectancies, and expert strategies.

The CDM does not use a strict protocol of interview questions. It is structured by a set of interview phases or "sweeps" that examine the incident in successively greater detail. A typical CDM session requires approximately 2 hours to move through each of the four interview sweeps:

1. Identify a complex incident that has the potential to elicit discoveries about cognitive phenomena.
2. Create a detailed incident timeline that shows the sequence of events.
3. Deepen strategies for managing the decision points embedded in the timeline.
4. Probe with what-if queries to elicit potential expert/novice differences.

35.3 Advantages

- Elicitation of real incidents. Capturing incidents within an expert's experience that required complex cognitive behavior and thought allows the researcher to identify influences and strategies that might not be included in even a very realistic task simulation.
- In-depth iterative structure. The four-sweep structure allows for an iterative approach to data collection so that the final sweeps of the interview can deepen on the issues that surface during the initial detailing.
- Efficiency. As Hoffman (1987) has shown, the use of critical incidents is a highly efficient means of cognitive task analysis (CTA). Subtle aspects of expertise are brought into play, along with the routine aspects of performance that serve as background.
- Informed cognitive probes. The cognitive probes and what-if queries used in sweeps three and four of the CDM have been utilized for years and deemed to be fruitful in the research-and-development environment to capture processes such as recognitional decision making (Klein, 1998).

35.4 Disadvantages

- Uncertain reliability. Given that CDM methodologies elicit retrospective incidents, concerns of data reliability have been raised due to evidence of memory degradation over time for critical events and details of those events. Taynor et al. (1987) did not find evidence of distortion due to memory loss, but we are still careful not to treat verbal protocols as exact depictions of incidents. We take what experts say as sources of hypotheses, rather than as ground truth, and seek replication across experts' reports of similar types of incidents as a way of establishing reliability.
- Resource intensive, small data set. The CDM interviews are more demanding than traditional surveys or structured interviews. These costs appear to be more than balanced, however, by the richness of the data obtained from each interview.
- Sophisticated methodology requires training. Utilizing the CDM methodology requires a high level of expertise and training. Effective use of the CDM also requires knowledge of the cognitive processes/phenomena being investigated, along with domain familiarization.

TABLE 35.1 Research Team Observation Notes

Scenario Start: 13:23.15

Time	Team	Event
13:24.19	Investigators	Dispatched to the Comm Center by DCA
13:42.26	Investigators	Report fire in Comm Center
13:44.59	DCA	Orders direct attack on Comm Center
13:48.35	DCA	Attempts to reach investigation team no. 1 about Comm Center status — no answer
13:50.14	Investigators	Comm Center door is hot and jammed shut. Scuttle is hot — need to find another way in
13:51.30	Attack teams	Use exothermic torch to gain access
13:51.44	Attack teams	Report access gained
13:55.49	Attack teams	Fire out in Comm Center — reflash watch set
14:00.39	Support team	Overhaul of fire in Comm Center complete

Elapsed Time: 37:24.

35.5 Example: An Adaptation in Practice

The following example illustrates a cross-method adaptation of CDM. It exemplifies the pairing of real-time observation and postobservation interviewing within the domain of shipboard firefighting. This example represents a single effort of many people who aided in the design of a real-time damage-control decision-support system built for the U.S. Navy (Miller et al., 2002). There were three primary approaches to data collection:

1. Observe and record team communication streams
2. Observe and probe human–system interactions
3. Conduct postobservation subject-matter expert interviews

35.5.1 Background and Scenario

Live-fire scenarios were conducted over a one-week period in September 2001 aboard the ex-*U.S.S. Shadwell*, the world's largest ship-fire research complex.[1] Scenarios simulated the wartime detonation of a medium-sized warhead to the ship's starboard side, which caused significant live-fire damage and progressive flooding. Twenty-five sailors participated in the live-fire tests and were manned as the damage control assistant (DCA), the casualty coordinator, console operators, team leaders, investigators, or attack team members.

35.5.2 Observation of Scenario Events

Sailors were stationed at their assigned locations throughout the ship. Members of the research team were placed in these same locations in an effort to capture multiple viewpoints on a single scenario. As scenario events unfolded, the research team recorded time-stamped communication streams that illustrated team roles and functions, courses of action implemented, and the order and duration of scenario events (Table 35.1).

---

[1] The Shadwell program is housed under the Damage Control-Automation for Reduced Manning (DC-ARM) program directed by the Naval Research Laboratory (NRL).

35.5.3   Observations of Human–System Interactions

As researchers observed scenario events and recorded time-stamped accounts of the incident, others were stationed in the control room observing human–system interactions and probing for key points of feedback. The systems being tested were built to help characterize damage, monitor personnel, track the progression of damage and the movement of personnel, and assist in building overall situation awareness.

35.5.4   Postobservation Interviews

When live-fire tests were complete, researchers convened into pairs and began interviewing key scenario "players." A member of each interviewing team had firsthand knowledge of the scenario and held notes similar to those in Table 35.1. Based on scenario events, 20-min interviews were conducted with the key personnel. A sample probe and response is listed below.

[Researcher] "You made repeated attempts to contact investigators during the scenario and often waited minutes for a response. What were you thinking during those periods? What were your biggest concerns?"

[DCA] "I'd say my biggest concern was that I didn't know where my men were. Several minutes passed and I'd heard nothing. Our comms were down and visibility was low — the black, tar-like smoke had filled most passageways before we could establish boundaries. At this point my investigators were my most important resource. I need to know where they are and what they see, otherwise, I'm blind. In this case, I relied on their tracking patterns to determine their location. I knew they'd circle the decks and search opposite sides and ends of the ship. Team one would search port side, aft compartments. Team two would search starboard side, forward compartments. Based on how much time had passed and my knowledge of the ship, I estimated their location and progress."

35.6   Related Methods

35.6.1   Adapting Critical Decision Method to Address Critical Needs

We have made generally two types of CDM methodological adaptations: within-method adaptations that modify the way we conduct the interviews, and cross-method adaptations that synthesize the CDM with other related methods.

35.6.2   Within-Method Adaptations

We have identified several adaptations of the traditional CDM method that we are calling within-method adaptations, i.e., adaptations of the four traditional CDM sweeps. There are two sets of notable within-method variations of CDM: variations of timeline usage and adaptations of CDM cognitive probes.

35.6.2.1   Variations of Timeline Usage

Establishing an incident timeline remains a significant sweep within the CDM methodology in order to understand the sequence of events and to establish the critical decision points that will be focused upon in subsequent sweeps. Establishing a timeline for an incident is key when the incident is complex, if the domain places emphasis on the sequence or timing of events, or when there is a need to resolve data inconsistencies. There are circumstances, however, when establishing a timeline is not as useful, e.g., if the incident is not straightforward or does not include twists and turns of situation events, if there are aspects of the domain/incident that are more important than the time element, or if time to conduct the CDM interview is limited.

35.6.2.2 Adaptations of Cognitive Probes

There is a set of cognitive probes that we have found to be effective for successfully focusing on critical incidents. These probes elicit information about situation assessments, situational cues, expert strategies, goals of incident players, and critical decisions and judgments. The cognitive probes should be adapted for the domain and for the research question. Thus, CDM probes can be adapted to fill out work domain hierarchies, task hierarchies, or to pinpoint/elicit information about specific cognitive processes.

35.6.3 Cross-Method Adaptations

35.6.3.1 Postobservation CDM

In a number of cases, we have been able to observe training exercises. In these cases, we have not needed to conduct the follow-up CDM interviews from scratch. Because we have observed the incident and the decision making during the incident, we can spend the interview time on more event-specific probes. This adaptation also mitigates the problem of interviewee memory degradation. The CDM in-depth interview is conducted immediately following an observed event, and CDM interview probes serve as cues for the recall of recent exercise events.

The adaptation also allows the CDM interviewer and exercise interviewee to begin the CDM session with established common ground, or shared exposure to a common event. This shared exposure allows for time-saving method abbreviations in that it allows the interviewer to skip the first two CDM sweeps that involve the eliciting of a critical incident and the construction of an incident timeline. The critical incident in this case is the observed exercise/event, which, if tracked correctly, becomes the ideal interview scenario.

Another advantage of the postobservation CDM adaptation is that it allows an interviewer to track both individual and team interactions. The experiences of the decision makers are not studied in the "vacuum" of a single isolated incident, but rather in a complex exercise with observable team interactions.

35.6.3.2 Additional Cross-Method Critical Decision Method Adaptations

We are currently exploring how the knowledge audit (Klein and Militello, in press), simulation-based methods, and additional cued-recall methods (Murphy and Awe, 1985) can be combined with the CDM to create useful adaptations that address clear methodological needs. We provide a snapshot of the adaptations we are considering below, with brief descriptions, explanations of when the adaptations are most appropriately used, and declarations of their inherent value to researchers.

35.6.3.2.1 *Knowledge Audit + CDM*

The knowledge audit (Hutton and Militello, 1996; Klein and Militello, in press) is an applied cognitive task analysis method that is best used for examining expert/novice differences and for unpacking the nature of expertise in a particular domain or task. The knowledge audit employs a set of probes designed to describe types of domain knowledge or skills and elicit appropriate examples. It focuses on the categories of knowledge and skills that distinguish experts from others, such as metacognition, mental models, perceptual cues and patterns, analogues, and declarative knowledge. The knowledge audit is best used when an interviewee's expertise is limited or far removed, or when interviewees have trouble recalling challenging and complex incidents. We have found that when we use the knowledge audit prior to the CDM, the knowledge audit probes often trigger recall of critical incidents that can be probed in depth. This arrangement allows the use of the knowledge audit as a survey of the types of relevant expertise, along with the CDM, which provides the details of an incident account.

35.6.3.2.2 *Simulation-Based Methods + CDM*

Simulation can be best described as the "technical mimicking" of a real-life system or operation, or of innovative technical concepts intended for real-time implementation. Simulation comes in many forms, such as Web-based gaming, scenario playbacks or replays, or intelligent interactive simulation. Simulation-based methods can be extended by adding CDM probes at discrete points during an exercise or an interactive simulation. These probes can elicit a user's understanding of the events being played or simulated and of the technology that supports those events. CDM probing can thereby expand the usefulness of a simulation for system and usability testing.

35.6.3.2.3 *Cued-Recall Techniques + CDM*

Using observation as a primer for in-depth interviewing was one cued-recall technique appropriate for melding with CDM methodology. There are additional cued-recall techniques that may be well suited for CDM adaptations. The most promising is a method conducted by a research team at Swinburne University (Omodei et al., 1998). The method is informally known as video-based incident recall and uses real-time video feed of critical incidents to cue an interviewee's recall during post-event in-depth interviewing. The value of this method is clear in that it significantly decreases the probability that incident memory degradation will occur and is an excellent alternative to observation if observation is not possible.

35.7   Standards, Regulations, and Adaptations

Whenever possible, the CDM should be conducted in its entirety, i.e., as a four-sweep, 2-hour, in-depth interview process. Resource and domain constraints, however, do not always allow for CDM to be conducted in such a fashion. In order to ensure high data quality given resource and domain constraints, we have generated methodological modifications or variations of the critical decision method that require less time and complexity. In the remainder of the chapter, we describe a few of these adaptations, including when and where they can be used most effectively.

35.8   Training and Application

Klein Associates has been using the CDM for more than 15 years to "get inside the heads" of the experts we study. During that time, we have worked to enhance the methodology's strengths and address its weaknesses. We have completed over 100 CTA efforts, in excess of 60 domains, and for a variety of applications, such as the design of systems, instruction/training, and for conducting consumer research. When completed in full, the methodology provides in-depth critical incident accounts that illustrate expert cognitive processes in the context of an expert's environment, providing insightful data that can be widely applied.

We have employed a variety of approaches to training the CDM, including a shadowing/mentorship program, audio and video analysis of CDM interviews and critical incidents, on-the-job training, and literature reviews. Developing the interviewing and investigation skills needed to capture in-depth, cognitive accounts of critical incidents is an evolutionary process and often requires years of training and practice to perfect. In the next section, we describe some lessons we have learned for modifying or streamlining the method when required by conditions.

35.9   Tools

As discussed, the CDM is structured by a set of interview "sweeps" that examine a complex incident in successively greater detail. Tools in the form of cognitive probes and listening cues have been developed for each CDM sweep and have proved to be insightful, even in the face of methodological adaptations or sweep abbreviations. Tools, tips, and sample probes are listed below for each CDM interview sweep (Hoffman et al., 1998).

35.9.1   Identifying an Incident

Ask the interviewee, "Can you think of a time when your skills as an expert (specific task) were challenged? Or when your skills really made a difference?"

Listen for an incident that fits your research goals, or that highlights the processes you are studying. Make sure the incident reflects the key role your interviewee played.

35.9.2 Creating a Detailed Timeline

Ask the interviewee to give a quick run-through of the incident.

Listen for areas to probe further. Identify decision points, gaps in the story, conceptual leaps, errors, or shifts in situation assessment.

When the interviewee mentions, "I just knew x would happen," or "It was just a gut feeling to choose that COA [course of action]," **flag** those points for further probing.

35.9.3 Deepening on Decision Points

Ask key probes that investigate flags you have noted in creating the timeline.

Probes for investigating decision points and shifts in situation assessment may include:

"What was it about the situation that let you know what was going to happen?"

"What were your overriding concerns at that point?"

"How would you summarize the situation at that point?"

Probes for investigating cues, expert strategies, and goals may include:

"What were you noticing at that point?"

"What information did you use in making this decision?"

"What knowledge did you have that was absolutely necessary?"

"What were you hoping/intending to accomplish at this point?"

35.9.4 Probing with "What-If" Queries

Ask about other alternatives that the interviewee may have considered.

Ask if someone else, perhaps with lesser experience, might have taken the same position. Determine what influence the interviewee's experience had on the chosen course of action.

Acknowledgments

We would like to thank Laura Militello and Beth Crandall for their comments on a draft of this chapter. We would also like to thank Patrick Tissington, Georgina Fletcher, and Margaret Crichton of Robert Gordon University; Jon Holbrook of NASA; Neelam Naikar of the Defense Science Technology Organization; and Glenn Elliott of the Swinburne University of Technology for their contributions to this chapter.

This article was prepared through participation in the Advanced Decision Architectures Collaborative Technology Alliance sponsored by the U.S. Army Research Laboratory under Cooperative Agreement DAAD 19-01-2-0009; Michael Strub was the contracting office's technical representative.

References

Flanagan, J.C. (1954), The critical incident technique, *Psychol. Bull.*, 51, 327–358.

Hoffman, R.R. (1987), The problem of extracting the knowledge of experts from the perspective of experimental psychology, *AI Mag.*, 8, 53–67.

Hoffman, R.R., Crandall, B.W., and Shadbolt, N.R. (1998), Use of the critical decision method to elicit expert knowledge: a case study in cognitive task analysis methodology, *Hum. Factors*, 40, 254–276.

Hutton, R.J.B. and Militello, L.G. (1996), Applied cognitive task analysis (ACTA): a practitioner's window into skilled decision making, in *Engineering Psychology and Cognitive Ergonomics: Job Design and Product Design*, Vol. 2, Harris, D., Ed., Ashgate, Aldershot, U.K., pp. 17–23.

Klein, G. (1998), *Sources of Power: How People Make Decisions*, MIT Press, Cambridge, MA.

Klein, G. and Militello, L. (in press), The knowledge audit as a method for cognitive task analysis, in *How Professionals Make Decisions*, Brehmer, B., Lipshitz, R., and Montgomery, H., Eds., Lawrence Erlbaum Associates, Mahwah, NJ.

Klein, G.A., Calderwood, R., and Clinton-Cirocco, A. (1986), Rapid decision making on the fireground, in *Proceedings of the Human Factors and Ergonomics Society 30th Annual Meeting*, Vol.1, The Human Factors Society, Santa Monica, CA, pp. 576–580.

Klein, G.A., Calderwood, R., and MacGregor, D. (1989), Critical decision method for eliciting knowledge, *IEEE Trans. Syst., Man, Cybern.*, 19, 462–472.

Miller, T.E., Armstrong, A.A., Wiggins, S.L., Brockett, A., Hamilton, A., and Schieffer, L. (2002), Damage Control Decision Support: Reconstructing Shattered Situation Awareness, final technical report, Contract N00178-00-C-3041, prepared for Naval Surface Warfare Center, Dahlgren, VA, Klein Associates Inc., Fairborn, OH.

Murphy, M.R. and Awe, C.A. (1985), Aircrew Coordination and Decision Making: Performance Ratings of Video Tapes Made during a Full Mission Simulation, paper presented at the 21st Annual Conference on Manual Control, Ohio State University, Columbus.

Omodei, M., Wearing, A., and McLennan, J., Eds. (1998), *Head-mounted Video Recording: a Methodology for Studying Naturalistic Decision Making*, Ashgate, Aldershot, U.K.

Taynor, J., Crandall, B., and Wiggins, S. (1987), The Reliability of the Critical Decision Method, Contract MDA903-86-C-0170 for the U.S. Army Research Institute Field Unit, Alexandria, VA, Klein Associates Inc., Fairborn, OH.

36

Applied Cognitive Work Analysis (ACWA)

W.C. Elm
Aegis Research Corporation

E.M. Roth
Roth Cognitive Engineering

S.S. Potter
Aegis Research Corporation

J.W. Gualtieri
Aegis Research Corporation

J.R. Easter
Aegis Research Corporation

36.1 Background and Applications

Many critical jobs (e.g., air traffic controller, power plant operator, medical operating room staff) involve complex knowledge and cognitive activity such as monitoring, situation assessment, planning, deciding, anticipating, and prioritizing. Cognitive task analysis (CTA) methods provide a means to explicitly identify the requirements of cognitive work so as to be able to specify ways to improve individual and team performance (be it through new forms of training, user interfaces, or decision aids). A broad overview of CTA methods and applications can be found in Schraagen et al. (2000).

There are two complementary approaches to performing a CTA. One approach relies on an analysis of the application domain to uncover the demands inherent in domain tasks. These are usually based on some form of goal–means decomposition. That is, the domain application is analyzed in terms of the goals or functions that need to be achieved for success and the means that are available to achieve those goals (cf. Woods and Hollnagel, 1987; Rasmussen et al., 1994; Vicente, 1999). From this analysis, one can derive an assessment of the range and complexity of tasks facing the user. This provides the basis for specification of the content and format of displays and controls. A second complementary approach employs interview and observation techniques to analyze how people actually go about performing the task (either in the actual task environment or in a simulated task environment). This approach enables discovery of the knowledge and strategies that domain practitioners utilize to cope with domain demands (Hoffman et al., 1998; Militello and Hutton, 1998; Potter et al., 2000). In practice, a CTA involves a combination of both approaches and relies on interviews and observations of domain experts to elicit the core knowledge that forms the basis of the CTA.

Vicente (1999) and Rasmussen et al. (1994) present an extensive methodology for analysis and modeling of cognitive work called cognitive work analysis (CWA). CWA supports the derivation of display design requirements, information requirements, human–automation function-allocation decision, and operator knowledge and skill requirements based on successive and iterative analyses of system and task constraints. A distinguishing characteristic of CWA is that it focuses on identifying the complexities and constraints of the work domain that serve to shape and constrain the behavior of domain practitioners.

Applied cognitive work analysis (ACWA) falls within the broad class of CWA methods. It provides a practical, step-by-step approach that links the demands of the domain as revealed by the cognitive analysis to the elements of a decision-support system. A detailed description of ACWA can be found in Elm et al. (2003). Several case studies of systems developed using the ACWA methodology are found in Potter et al. (2003).

ACWA has been used across a wide range of domains, from classically designed process control to so-called intentional domains such as military command and control (e.g., Roth et al., 2001; Potter et al., 2003). In each case, it has developed decision-support concepts that, in hindsight, appear intuitively obvious (as an ideal decision-support system should), yet remained undiscovered prior to the application of ACWA.

36.2   Procedure

The ACWA approach is a structured, principled methodology that systematically moves from analyzing the demands of a domain to identifying visualizations and decision-aiding concepts that will provide effective support. The five steps in this process are:

1. Using a functional abstraction network (FAN) model to capture the essential domain concepts and relationships that define the problem-space confronting the domain practitioners
2. Overlaying cognitive work requirements (CWR) on the functional model as a way of identifying the cognitive demands/tasks/decisions that arise in the domain and require support
3. Identifying the information/relationship requirements (IRR) for successful execution of these cognitive work requirements
4. Specifying the representation design requirements (RDR) to define the shaping and processing of the information/relationships that are to be presented to the practitioner(s)
5. Developing presentation design concepts (PDC) to explore techniques to implement these representation requirements into the syntax and dynamics of presentation forms in order to produce effective support.

In the ACWA analysis and design approach, each step is associated with a design artifact that captures the results. These design artifacts form a continuous design thread that provides a principled, traceable link from cognitive analysis to design. While steps in the ACWA process are presented as if they are performed in a strictly sequential order, in practice the process is much more parallel, opportunistic, and iterative in nature.

36.2.1   Representing the Way the World Works — Building a Functional Abstraction Network

ACWA begins with a function-based goal–means decomposition of the domain. This technique has its roots in the formal, analytic goal–means decomposition method pioneered by Rasmussen and his colleagues as a formalism for representing cognitive work domains as an abstraction hierarchy (e.g., Rasmussen et al., 1994; Vicente, 1999). A work-domain analysis is conducted to understand and document the goals to be achieved in the domain and the functional means available for achieving them. The objective of this functional analysis is a structured representation of the functional concepts and their relationships to serve as the context for the information system to be designed. This abstraction network is intended to approximate an "experts' mental model" of the domain. This includes knowledge of the system's characteristics and the purposes or functions of the specific entities. The result of this phase is a functional abstraction network (FAN) — a multilevel recursive means–ends representation of the structure of the work domain.

The FAN specifies the domain objectives and the functions that must be available and satisfied in order to achieve their goals. In turn, these functions may be abstract entities that need to have other, less abstract functions available and satisfied in order that they might be achieved. This creates a decomposition network of objectives or purposes that are linked together from abstract goals to specific means to achieve these goals. For example, in the case of engineered systems, such as a process plant, functional representations are developed that characterize the purposes for which the engineered system has been designed and the means structurally available for achieving those objectives. In the case of military command-and-control systems, the functional representations characterize the functional capabilities of individual weapon systems, maneuvers, or forces and the higher-level goals related to military objectives.

The work-domain analysis is performed based on extensive interactions with expert practitioners in the domain and includes face-to-face interviews with the experts, watching the experts work in the domain, verbal protocol techniques, and other CTA methods (cf. Potter et al., 2000). The FAN can be initialized from training materials, observations, interviews, or whatever is the richest initial source of domain information. Once initialized, it becomes a working hypothesis to be refined with each additional piece of domain knowledge acquired.

36.2.2 Modeling Decision Making — Deriving Cognitive Work Requirements

With the FAN representation of the work domain's concepts and how they are interrelated as the underlying framework, it is possible to derive the cognitive demands for each part of that domain model. The ACWA methodology refers to these as cognitive work requirements (CWR), meaning all types of recognition, decision making, and problem-solving reasoning activities required of the domain practitioner.

Based on the underlying premises of the ACWA methodology, these CWRs center around elements of the FAN (e.g., goals or functional process represented in the FAN). Examples include monitoring for goal satisfaction and resource availability, planning and selection among alternative means to achieve goals, and controlling functional processes (initiating, tuning, and terminating) to achieve goals (Roth and Mumaw, 1995). By organizing the specification of operator cognitive requirements around nodes in the FAN, rather than organizing requirements around predefined task sequences (as in traditional approaches to task analysis), the representation helps ensure a consistent, decision-centered perspective.

The cognitive demands that are captured at this step of the analysis constitute another intermediate design artifact that captures an essential part of the design: the explicit identification of the cognitive demands placed on the domain practitioner by the domain itself. They constitute an explicit enumeration of the cognitive tasks to be supported by the decision-support system. Since many of the abstract functional nodes in the FAN are "nontraditional" representations, it is also common to see CWRs that are unlike any elements explicitly visible in documentation or training materials.

36.2.3 Capturing the Means for Effective Decision Making — Identifying Information Requirements

The next step in the process is to identify and document the information required for each decision to be made. Information/relationship resources (IRRs) are defined as the set of information elements necessary for successful resolution of the associated CWR. The focus of this step in the methodology is on identifying the ideal and complete set of information for the associated decision making. It is important to note that identifying IRRs is focused on satisfying the decision requirements and is *not* limited by data availability in the current system. In cases where the required data are not directly available, ACWA provides a rationale for obtaining that data (e.g., pulling data from a variety of previously stovepiped databases, adding additional sensors, or creating "synthetic" values). This is a critical change from the typical role that human-factors engineers have had in the past (designing an interface after the instrumentation has been specified).

36.2.4 Linking Decision Requirements to Aiding Concepts — Developing and Documenting a "Model of Support"

This step in the ACWA process develops the specification of the display concept and how it supports the cognitive tasks and is captured in representation design requirements (RDR) for the eventual development of presentation design concepts (PDC). The RDR defines the goals and scope of the information representation in terms of the cognitive tasks it is intended to support (and thus a defined target region of the FAN). It also provides a specification of the supporting information required to support the cognitive tasks. An RDR is another span of the bridge that helps to link the decisions within the work domain to the visualization and decision-support concepts intended to support those decisions. The RDR is not only a compilation of information developed earlier, it has the added value of providing a more complete description of the behaviors and features needed to communicate the information effectively as well as allocating the information/relationship resources across the entire set of displays within the workspace. When done correctly, it is still in the form of a "requirement" and not an implementation. This artifact becomes a key transition between the cognitive system engineer, the system developer, and the system (effectiveness) tester.

36.2.5 Developing Presentations — Instantiating the Aiding Concept as Presentation Design Concepts

From the RDR's specification of how information is to be represented within the decision-support system, the next step of the ACWA process is the explicit development of presentation design concepts (PDCs) for the decision-support system. (A similar process is used for the design of auditory, visual, or other senses' presentations of the RDR's specification.) This final step requires an understanding of human perception and its interaction with the various presentation techniques and attributes. As such, it requires considerable skill and ability beyond cognitive work analysis. The actual design of a revolutionary aiding concept is probably one of the largest "design gaps" that is needed to be bridged within the ACWA process. One of the major design challenges is to create visualizations (sometimes referred to as representational aids or ecological interfaces) that reveal the critical information/relationship resources and constraints of the decision task through the user interface in such a way as to capitalize on the characteristics of human perception and cognition (Woods, 1995; Vicente, 2002).

With the RDR as a guide, the sketches, proposals, and brainstorming concepts can all be resolved back against the display's intent and requirements. The issues of how it is perceived can best be addressed with empirical testing of prototypes, often requiring considerable tuning and adjustment to achieve the representational capabilities specified in the RDR.

36.3 Advantages

- Requiring analysts to identify and represent high-level domain goals and functions as part of the FAN promotes development of novel visualizations of nonphysical abstractions corresponding to how expert domain practitioners think (or should think) about the domain for more effective support of individual and collaborative decision making/planning.
- Organizing operator cognitive requirements around nodes in the FAN, rather than organizing requirements around predefined task sequences (as in traditional approaches to task analysis), results in decision-support systems that have a decision-centered perspective, and are thus able to support performance in unanticipated situations as well as in expected situations.

• Providing a step-by-step set of linked processes from cognitive analysis to design ensures traceability of design elements to the cognitive requirements they are intended to support.

36.4 Disadvantages

• Requires comprehensive analysis and documentation of domain demands and decision-support system requirements
• Requires training in the FAN and ACWA methodology

36.5 Examples of ACWA Artifacts for a "Relative Combat Power Display"

Table 36.1 and Figure 36.1 are examples of ACWA artifacts that were created as part of the process of developing a "relative combat power" display intended to support military commanders in generating and evaluating courses of action in combat situations (Potter et al., 2003). Figure 36.2 shows the resulting display that provides a visualization of relative combat power of red and blue forces under different assumptions of when, where, and what forces will meet.

36.6 Related Methods

The ACWA methodology has its roots in the formal cognitive work analysis method pioneered by Rasmussen and colleagues (e.g., Rasmussen et al., 1994; Vicente, 1999; Woods and Hollnagel, 1987).

36.7 Standards and Regulations

Currently, there are no standards or regulations with respect to cognitive task analysis or cognitive work analysis methods.

36.8 Approximate Training and Application Times

Currently, training is primarily achieved in an apprenticeship mode, although books and book chapters that explain the methodology are becoming available (Vicente, 1999; Elm et al., 2003). There are also multiday workshops offered on the ACWA method, as well as semester college-level courses offered on cognitive task analysis.

Application times vary depending on the scope of the support system or training system to be developed (cf. Potter et al., 2000).

36.9 Reliability and Validity

Data on the reliability and validity of CTA methods in general and CWA methods in particular are limited. One recent study (Bisantz et al., 2002) compared the results of two independent teams performing a CWA of very similar domains and found significant overlap in the results, increasing confidence in the reliability of the approach.

36.10 Tools Needed

While there are ongoing efforts to develop software tools to support work domain analysis and the ACWA methodology, no special tools are required to perform the analyses.

TABLE 36.1 Representation Design Requirements for the "Choose Combat Power" Decision-Support Concept

Representation Design Requirements
Context
This display is intended for combat applications where the mission objective is to engage and defeat an enemy. It should assist the commander in choosing combined combat power and managing the space/time tradeoff to control the battle space. Specifically, it is to assist in deciding: (1) at what location to engage the enemy, (2) when to engage the enemy, and (3) what combat resources to deploy in order to maximize the potential to defeat the enemy.

Decision and Information Requirements	Visualization Requirements
CWR P6-1 — "Determine the point in time when the enemy will reach a specified point in space, and monitor the enemy's combat power over time." **IRR P6-1.1** — "Expected arrival time of enemy combat resources at the specified point in space" (i.e., the lead unit as well as other follow-on units). **IRR P6-1.2** — "Estimated measure of combined enemy combat power at the specified point in space, beginning at the arrival time of the first enemy unit and extending through follow-on units." **CWR P6-2** — "Choose among the friendly combat resources that can bring their combat power to bear at the specific point in space and time." **IRR P6-2.1** — "The time required for selected friendly combat resources to reach the specified point in space." **IRR P6-2.2** — "Estimated measure of combined combat power of the selected friendly combat power resources once they reach the specified point in space." **CWR P6-3** — "Estimate the potential to defeat the enemy after the application of the chosen power at the specific point in time and space." **IRR P6-3.1** — "Measure of combat power ratio of friendly to enemy combat power beginning with the arrival of the first unit (friendly or enemy) over time." **IRR P6-3.2** — "Indication of combat power ratios required to defeat the enemy under different battle conditions (i.e., doctrinal/procedural referent information)." **IRR P6-3.3** — "Location of alternative resources of both friendly and enemy combat power that could be brought to bear." **IRR P6-3.4** — "The time required to bring to bear the combat power of these alternative friendly and enemy combat resources." **IRR P6-3.5** — "Measures of cumulative combat power of both friendly and enemy resources as additional friendly and enemy resources are selected (over a specified window in time)."	1. Visualize the time required for enemy unit(s) to reach location designated by commander, and visualize cumulative enemy combat power at that location as a function of time. 2. Visualize the time required for friendly "combat power" forces to bring their combat power to bear on the designated point (in space and time) and visualize cumulative friendly combat power at that location as a function of time. 3. Visualize relative friendly to enemy combat power ratio at the designated point in space as a function of time, and compare to combat power ratio required to defeat enemy under different battle conditions. 4. Visualize uncertainty of estimates of "combat power ratio." 5. Visualize changes in enemy and friendly combat power and combat power ratio as selected enemy and/or friendly units are added or removed.

Source: From Potter, S.S. et al. (2003), Case studies: applied cognitive work analysis in the design of innovative decision support, in *Handbook for Cognitive Task Design*, Hollnagel, E., Ed., Lawrence Erlbaum Associates, London.

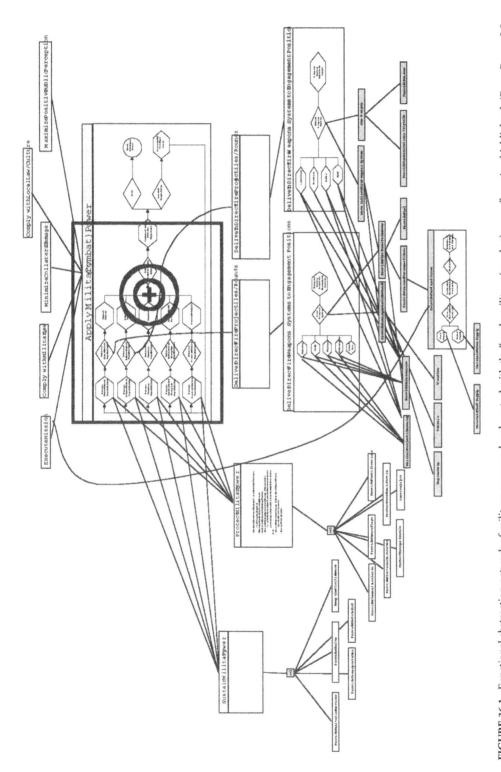

FIGURE 36.1 Functional abstraction network of military command and control with the "apply military (combat) power" portion highlighted. (From Potter, S.S. et al. [2003], Case studies: applied cognitive work analysis in the design of innovative decision support, in *Handbook for Cognitive Task Design*, Hollnagel, E, Ed., Lawrence Erlbaum Associates, London. With permission.)

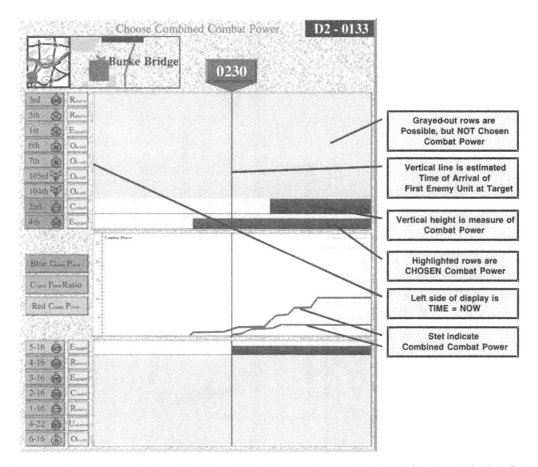

FIGURE 36.2 "Choose combat power" decision aid showing combat power of Red and Blue forces at the time that the 2nd and 4th Blue battalions and 5th Red battalion reach the "choke point" south of the Burke Bridge. (From Potter, S.S. et al. [2003], Case studies: applied cognitive work analysis in the design of innovative decision support, in *Handbook for Cognitive Task Design*, Hollnagel, E., Ed., Lawrence Erlbaum Associates, London. With permission.)

References

Bisantz, A.M., Burns, C.M., and Roth, E.M. (2002), Validating methods in cognitive engineering: a comparison of two work domain models, in *Proceedings of the Human Factors and Ergonomics Society 46th Annual Meeting*, Human Factors Society, Santa Monica, CA, pp. 521–527.

Burns, C.M., Barsalou, E., Handler, C., Kuo, J., and Harrigan, K. (2000), A work domain analysis for network management, in *Proceedings of the LEA 2000/HFES 2000 Congress*, Vol. 1, HFES, Santa Monica, CA, pp. 469–472.

Elm, W.C., Potter, S.S., Gualtieri, J.W., Roth, E.M., and Easter, J.R. (2003), Applied cognitive work analysis: a pragmatic methodology for designing revolutionary cognitive affordances, in *Handbook for Cognitive Task Design*, Hollnagel, E., Ed., Lawrence Erlbaum Associates, London, pp. 357–382.

Hoffman, R.R., Crandall, B., and Shadbolt, N. (1998), Use of the critical decision method to elicit expert knowledge: a case study in the methodology of cognitive task analysis, *Hum. Factors*, 40, 254–277.

Militello, L. and Hutton, R. (1998), Applied cognitive task analysis (ACTA): a practitioner's tool kit for understanding cognitive task demands, *Ergonomics, Special Issue*: Task Analysis, 41, 1618–1641.

Potter, S.S., Gualtieri, J.W., and Elm, W.C. (2003), Case studies: applied cognitive work analysis in the design of innovative decision support, in *Handbook for Cognitive Task Design*, Hollnagel, E., Ed., Lawrence Erlbaum Associates, London, pp 653–677.

Potter, S.S., Roth, E.M., Woods, D.D., and Elm, W. (2000), Bootstrapping multiple converging cognitive task analysis techniques for system design, in *Cognitive Task Analysis*, Schraagen, J.M., Chipman, S.F., and Shalin, V.L., Eds., Lawrence Erlbaum Associates, Mahwah, NJ, pp. 317–340.

Roth, E.M., Lin, L., Kerch, S., Kenney, S.J., and Sugibayashi, N. (2001), Designing a first-of-a-kind group view display for team decision making: a case study, in *Linking Expertise and Naturalistic Decision Making*, Salas, E. and Klein, G., Eds., Lawrence Erlbaum Associates, Mahwah, NJ, pp. 113–135.

Roth, E.M. and Mumaw, R.J. (1995), Using cognitive task analysis to define human interface requirements for first-of-a-kind systems, in *Proceedings of the Human Factors and Ergonomics Society 39th Annual Meeting*, Human Factors and Ergonomics Society, Santa Monica, CA, pp. 520–524.

Rasmussen J., Pejtersen, A.M., and Goodstein, L.P. (1994), *Cognitive Systems Engineering*, John Wiley & Sons, New York.

Schraagen, J.M.C., Chipman, S.F., and Shalin, V.L., Eds. (2000), *Cognitive Task Analysis*, Lawrence Erlbaum Associates, Mahwah, NJ.

Vicente, K.J. (1999), *Cognitive Work Analysis: Towards Safe, Productive, and Healthy Computer-based Work*, Lawrence Erlbaum Associates, Mahwah, NJ.

Vicente, K. (2002), Ecological interface design: progress and challenges, *Hum. Factors*, 44, 62–78.

Woods, D.D. (1995), Toward a theoretical base for representation design in the computer medium: ecological perception and aiding human cognition, in *Global Perspectives on the Ecology of Human–Machine Systems*, Flach, J., Hancock, P., Caird, J., and Vicente, K., Eds., Lawrence Erlbaum Associates, Hillsdale, NJ, pp. 157–188.

Woods, D.D. and Hollnagel, E. (1987), Mapping cognitive demands in complex problem-solving worlds, *Int. J. Man-Machine Stud.*, 26, 257–275.

37

Systematic Human Error Reduction and Prediction Approach (SHERPA)

Neville A. Stanton
Brunel University

37.1 Background and Applications

The systematic human error reduction and prediction approach (SHERPA) was developed by Embrey (1986) as a human-error prediction technique that also analyzes tasks and identifies potential solutions to errors in a structured manner. The technique is based upon a taxonomy of human error, and in its original form it specified the psychological mechanism implicated in the error. The method is subject to ongoing development, which includes the removal of this reference to the underlying psychological mechanism.

In general, most of the existing human-error prediction techniques have two key problems (Stanton, 2002). The first of these problems relates to the lack of representation of the external environment or objects. Typically, human-error analysis techniques treat the activity of the device and the material with which the human interacts in only a passing manner. Hollnagel (1993) emphasizes that human reliability analysis (HRA) often fails to take adequate account of the context in which performance occurs. Second, there tends to be a good deal of dependence made upon the judgment of the analyst. Different analysts, with different experience, may make different predictions regarding the same problem (called interanalyst reliability). Similarly, the same analyst may make different judgments on different occasions (intraanalyst reliability). This subjectivity of analysis can weaken the confidence that can be placed in any predictions made. The analyst is required to be an expert in the technique as well as the operation of the device being analyzed if the analysis has a hope of being realistic.

SHERPA was originally designed to assist people in the process industries (e.g., conventional and nuclear power generation, petrochemical processing, oil and gas extraction, and power distribution) (Embrey, 1986). An example of the application of SHERPA applied to the procedure for filling a chorine road tanker can be found in Kirwan (1994). A recent example of SHERPA applied to oil and gas exploration can be found in Stanton and Wilson (2000). The domain of application has broadened in recent years to include ticket machines (Baber and Stanton, 1996), vending machines (Stanton and Stevenage, 1998), and in-car radio-cassette machines (Stanton and Young, 1999).

37.2   Procedure

There are eight steps in the SHERPA analysis, as follows:

37.2.1   Step 1: Hierarchical Task Analysis (HTA)

The process begins with the analysis of work activities, using hierarchical task analysis (HTA). HTA (Annett, Chapter 33, this book; Shepherd, 2001; Kirwan and Ainsworth, 1992) is based upon the notion that task performance can be expressed in terms of a hierarchy of goals (what the person is seeking to achieve), operations (the activities executed to achieve the goals), and plans (the sequence in which the operations are executed). The hierarchical structure of the analysis enables the analyst to progressively redescribe the activity in greater degrees of detail. The analysis begins with an overall goal of the task, which is then broken down into subordinate goals. At this point, plans are introduced to indicate in which sequence the subactivities are performed. When the analyst is satisfied that this level of analysis is sufficiently comprehensive, the next level can be scrutinized. The analysis proceeds downward until an appropriate stopping point is reached. (For a discussion of the stopping rule, see Annet [Chapter 33, this book] and Shepherd [2001].)

37.2.2   Step 2: Task Classification

Each operation from the bottom level of the analysis is taken in turn and is classified from the error taxonomy into one of the following types:

- Action (e.g., pressing a button, pulling a switch, opening a door)
- Retrieval (e.g., getting information from a screen or manual)
- Checking (e.g., conducting a procedural check)
- Selection (e.g., choosing one alternative over another)
- Information communication (e.g., talking to another party)

37.2.3   Step 3: Human-Error Identification (HEI)

This classification of the task (step 2) then leads the analyst to consider credible error modes associated with that activity, using the error taxonomy in Table 37.1. For each credible error (i.e., those judged by a subject-matter expert to be possible), a description of the form that the error would take is given, as seen in Table 37.1.

37.2.4   Step 4: Consequence Analysis

Considering the consequence of each error on a system is an essential next step, as the consequence has implications for the criticality of the error.

TABLE 37.1 Taxonomy of Credible Errors

	Action Errors		Checking Errors		Retrieval Errors		Communication Errors		Selection Errors
A1	Operation too long/short	C1	Check omitted	R1	Information not obtained	I1	Information not communicated	S1	Selection omitted
A2	Operation mistimed	C2	Check incomplete	R2	Wrong information obtained	I2	Wrong information communicated	S2	Wrong selection made
A3	Operation in wrong direction	C3	Right check on wrong object	R3	Information retrieval incomplete	I3	Information communication incomplete		
A4	Operation too little/much	C4	Wrong check on right object						
A5	Misalign	C5	Check mistimed						
A6	Right operation on wrong object	C6	Wrong check on wrong object						
A7	Wrong operation on right object								
A8	Operation omitted								
A9	Operation incomplete								
A10	Wrong operation on wrong object								

37.2.5 Step 5: Recovery Analysis

If there is a later task step at which the error could be recovered, it is entered next. If there is no recovery step, then "None" is entered.

37.2.6 Step 6: Ordinal Probability Analysis

An ordinal probability value is entered as either low, medium, or high. If the error has never been known to occur, then a low (L) probability is assigned. If the error has occurred on previous occasions, then a medium (M) probability is assigned. Finally, if the error occurs frequently, then a high (H) probability is assigned. The assigned classification relies upon historical data and/or a subject-matter expert.

37.2.7 Step 7: Criticality Analysis

If the consequence is deemed to be critical (i.e., it causes unacceptable losses), then a note of this is made. Criticality is assigned in a binary manner. If the error would lead to a serious incident (this would have to be defined clearly before the analysis), then it is labeled as critical (denoted thus: !). Typically, a critical consequence would be one that would lead to substantial damage to plant or product and/or injury to personnel.

37.2.8 Step 8: Remedy Analysis

The final stage in the process is to propose error reduction strategies. These are presented in the form of suggested changes to the work system that could have prevented the error from occurring or, at the very least, reduced the consequences. This is done in the form of a structured brainstorming exercise to propose ways of circumventing the error or to reduce the effects of the error. Typically, these strategies can be categorized under four headings:

1. Equipment (e.g., redesign or modification of existing equipment)
2. Training (e.g., changes in training provided)
3. Procedures (e.g., provision of new, or redesign of old, procedures)

4. Organizational (e.g., changes in organizational policy or culture)

Some of these remedies may be very costly to implement. Therefore, they needed to be judged with regard to the consequences, criticality, and probability of the error. Each recommendation is analyzed with respect to four criteria:

1. Incident prevention efficacy (the degree to which the recommendation, if implemented, would prevent the incident from occurring)
2. Cost effectiveness (the ratio of the cost of implementing the recommendation to the cost of the incident × the expected incident frequency)
3. User acceptance (the degree to which workers and organization are likely to accept the implementation of the recommendation)
4. Practicability (the technical and social feasibility of recommendation)

This evaluation then leads to a rating for each recommendation.

37.3 Advantages

- Structured and comprehensive procedure, yet maintains usability
- Taxonomy prompts analyst for potential errors
- Encouraging validity and reliability data
- Substantial time economy compared with observation
- Error reduction strategies offered as part of the analysis, in addition to predicted errors

37.4 Disadvantages

- Can be tedious and time-consuming for complex tasks.
- Extra work is involved if HTA is not already available.
- Does not model cognitive components of error mechanisms.
- Some predicted errors and remedies are unlikely or lack credibility, thus posing a false economy.
- Current taxonomy lacks generalizability.

37.5 Example of SHERPA Output Based on Programming a VCR

The process begins with the analysis of work activities, using hierarchical task analysis. HTA (see Chapter 33) is based upon the notion that task performance can be expressed in terms of a hierarchy of goals (what the person is seeking to achieve), operations (the activities executed to achieve the goals), and plans (the sequence in which the operations are executed). An example of HTA for the programming of a videocassette recorder is shown in Figure 37.1.

For the application of SHERPA, each task step from the bottom level of the analysis is taken in turn. First, each task step is classified into a type from the taxonomy:

- Action (e.g., pressing a button, pulling a switch, opening a door)
- Retrieval (e.g., getting information from a screen or manual)
- Checking (e.g., conducting a procedural check)
- Information communication (e.g., talking to another party)
- Selection (e.g., choosing one alternative over another)

This classification of the task step then leads the analyst to consider credible error modes associated with that activity, as shown in step three of the procedure.

For each credible error (i.e., those judged by a subject-matter expert to be possible), a description of the form that the error would take is given as illustrated in Table 37.2. The consequence of the error on system needs to be determined next, as this has implications for the criticality of the error. The last four

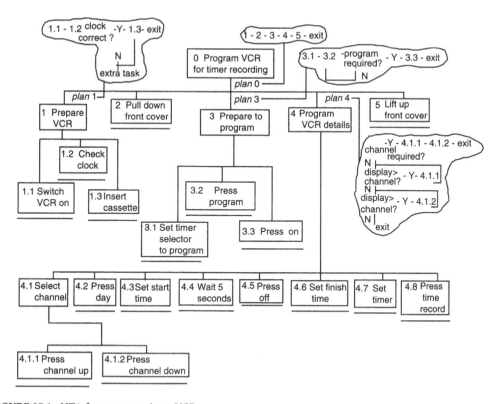

FIGURE 37.1 HTA for programming a VCR.

steps consider the possibility for error recovery, the ordinal probability of the error (high, medium, or low), its criticality (either critical or not critical), and potential remedies. Again, these are shown in Table 37.2.

As Table 37.2 shows, there are six basic error types associated with the activities of programming a VCR. These are:

1. Failing to check that the VCR clock is correct
2. Failing to insert a cassette
3. Failing to select the program number
4. Failing to wait
5. Failing to enter programming information correctly
6. Failing to press the confirmatory buttons

The purpose of SHERPA is not only to identify potential errors with the current design, but to guide future design considerations. The structured nature of the analysis can help to focus the design remedies on solving problems, as shown in the column labeled "remedial strategy." As this analysis shows, quite a lot of improvements could be made. It is important to note, however, that the improvements are constrained by the analysis. The analysis does not address radically different design solutions, i.e., those that may remove the need to program at all.

37.6 Related Methods

SHERPA relies heavily upon hierarchical task analysis (HTA), which must be conducted before SHERPA can be carried out. The taxonomic approach is rather like a human version of a hazard-and-operability study. Kirwan (1994) has argued that more accurate predictions of human error are produced by using

TABLE 37.2 SHERPA Description

Task Step	Error Mode	Error Description	Consequence	Recovery	P[a]	C[b]	Remedial Strategy
1.1	A8	Fail to switch VCR on	Cannot proceed	Immediate	L	—	Press of any button to switch VCR on
1.2	C1	Omit to check clock	VCR clock time may be	None	L	!	Automatic clock setting and
	C2	Incomplete check	incorrect				adjust via radio transmitter
1.3	A3	Insert cassette wrong way	Damage to VCR	Immediate	L	!	Strengthen mechanism
	A8	around	Cannot record	Task 3	L	—	On-screen prompt
		Fail to insert cassette					
2	A8	Fail to pull down front cover	Cannot proceed	Immediate	L	—	Remove cover to programming
3.1	S1	Fail to move timer selector	Cannot proceed	Immediate	L	—	Separate timer selector from programming function
3.2	A8	Fail to press PROGRAM	Cannot proceed	Immediate	L	—	Remove this task step from sequence
3.3	A8	Fail to press ON button	Cannot proceed	Immediate	L	—	Label button START TIME
4.1.1	A8	Fail to press UP button	Wrong channel selected	None	M	!	Enter channel number directly from keypad
4.1.2	A8	Fail to press DOWN button	Wrong channel selected	None	M	!	Enter channel number directly from keypad
4.2	A8	Fail to press DAY button	Wrong day selected	None	M	!	Present day via a calendar
4.3	I1	No time entered	No program recorded	None	L	!	Dial time in via analogue
	I2	Wrong time entered	Wrong program recorded	None	L	!	clock
							Dial time in via analogue clock
4.4	A1	Fail to wait	Start time not set	Task 4.5	L	—	Remove need to wait
4.5	A8	Fail to press OFF button	Cannot set finish time	—	—	—	Label button FINISH TIME
4.6	I1	No time entered	No program recorded	None	L	!	Dial time in via analogue
	I2	Wrong time entered	Wrong program recorded	None	L	!	clock
							Dial time in via analogue clock
4.7	A8	Fail to set timer	No program recorded	None	L	!	Separate timer selector from programming function
4.8	A8	Fail to press TIME RECORD button	No program recorded	None	L	!	Remove this task step from sequence
5	A8	Fail to lift up front cover	Cover left down	Immediate	L	—	Remove cover to programming

[a] Probability (see step 6).
[b] Criticality (see step 7).

multiple methods, so SHERPA could be used in conjunction with task analysis for error identification (TAFEI) (see Chapter 38). Our research suggests that more-accurate predictions are also found by pooling the data from multiple analysts using the same method.

37.7 Standards and Regulations

The Management of Health and Safety at Work (MH&S@W) regulations (1992) require employers to undertake a "risk assessment." Specifically, the regulations state that "employers should undertake a systematic general examination of their work activity and that they should record the significant findings of that risk assessment." SHERPA offers a means of achieving this analysis and ensuring that the regulations are met (Stanton, 1995).

For product design, both ISO (International Organization for Standardization) and BS (British Standards) allude to the need for error reduction. ISO 9241 requires standards of usability to ensure "the extent to which a product can be used by specified users to achieve specified goals with effectiveness, efficiency and satisfaction in a specified context of use." This definition was intended for the design of software products, but it is generic enough to apply to any kind of product.

Similarly, BS 3456 (1987) is concerned with the safety of household and similar electrical appliances. Part of the standard is quoted below to illustrate that it covers both abnormal and normal product use.

BS 3456 (1987) Safety of Household and Similar Electrical Appliances
Part 101: General Requirements
19.1 Appliances shall be designed so that the risk of fire and mechanical damage impairing safety or protection against electrical shock as a result of **abnormal or careless use** is obviated as far as possible.
20.2 Moving parts of motor-operated appliances shall, as far as is compatible with the use and working of the appliance, be so arranged or enclosed as to provide, **in normal use**, adequate protection against personal injury.

The focus on use of the product would seem to indicate a clear role for human-error prediction in product design and development so that the final version of the product entering the marketplace can be optimized to meet the requirements of the standard.

37.8 Approximate Training and Application Times

Based on the example of the application to the radio-cassette machine, Stanton and Young (1998) report training times of around 3 hours. (This estimate is doubled if training in hierarchical task analysis is included.) It took an average of 2 hours and 40 min for people to evaluate the radio-cassette machine using SHERPA.

37.9 Reliability and Validity

Kirwan (1992) reports that SHERPA was the most highly rated of five human-error prediction techniques by expert users. Baber and Stanton (1996) report a concurrent validity statistic of 0.8 and a reliability statistic of 0.9 in the application of SHERPA by two expert users for prediction of errors on a ticket vending machine. Stanton and Stevenage (1998) report a concurrent validity statistic of 0.74 and a reliability statistic of 0.65 in the application of SHERPA by 25 novice users for prediction of errors on a confectionery vending machine. Stanton and Young (1999) report a concurrent validity statistic of 0.2 and a reliability statistic of 0.4 in the application of SHERPA by eight novice users for prediction of errors on a radio-cassette machine. These results suggest that reliability and validity are highly dependent upon the expertise of the analyst and the complexity of the device being analyzed (Stanton and Baber, 2002).

37.10 Tools Needed

At its simplest, SHERPA can be conducted with just a pen and paper. This can become slightly more sophisticated with the use of a computerized spreadsheet or table on a computer. The latter has the advantage of making the process less tedious when reorganizing the material. Finally, some companies offer specialist software for conducting the analysis. These labor-saving systems also offer prompts to aid novice users (Bass et al., 1995).

References

Baber, C. and Stanton, N.A. (1996), Human error identification techniques applied to public technology: predictions compared with observed use, *Appl. Ergonomics*, 27, 119–131.
Bass, A., Aspinal, J., Walter, G., and Stanton, N.A. (1995), A software toolkit for hierarchical task analysis, *Appl. Ergonomics*, 26, 147–151.

Embrey, D.E. (1986), SHERPA: A Systematic Human Error Reduction and Prediction Approach, paper presented at the International Meeting on Advances in Nuclear Power Systems, Knoxville, TN.

Hollnagel, E. (1993), *Human Reliability Analysis: Context and Control*, Academic Press, London.

Kirwan, B. (1990), Human reliability assessment, in *Evaluation of Human Work: a Practical Ergonomics Methodology*, Wilson, J.R. and Corlett, E.N., Eds., 2nd ed., Taylor & Francis, London, pp. 921–968.

Kirwan, B. (1992), Human error identification in human reliability assessment. II. Detailed comparison of techniques, *Appl. Ergonomics*, 23, 371–381.

Kirwan, B. (1994), *A Guide to Practical Human Reliability Assessment*, Taylor & Francis, London.

Shepherd, A. (2001), *Hierarchical Task Analysis*, Taylor & Francis, London.

Stanton, N.A. (1995), Analysing worker activity: a new approach to risk assessment? *Health Safety Bull.*, 240, 9–11.

Stanton, N.A. (2002), Human error identification in human computer interaction, in *The Human-Computer Interaction Handbook*, Jacko, J. and Sears, A., Eds., Lawrence Erlbaum Associates, Mahwah, NJ.

Stanton, N.A. and Baber, C. (2002), Error by design: methods to predict device usability, *Design Stud.*, 23, 363–384.

Stanton, N.A. and Stevenage, S.V. (1998), Learning to predict human error: issues of acceptability, reliability and validity, *Ergonomics*, 41, 1737–1756.

Stanton, N.A. and Wilson, J. (2000), Human factors: step change improvements in effectiveness and safety, *Drilling Contractor*, Jan./Feb., 46–41.

Stanton, N.A. and Young, M. (1998), Is utility in the mind of the beholder? A review of ergonomics methods, *Appl. Ergonomics*, 29, 41–54.

Stanton, N.A. and Young, M. (1999), What price ergonomics? *Nature*, 399, 197–198.

38

Task Analysis for Error Identification

Neville A. Stanton
Brunel University

Christopher Baber
University of Birmingham

38.1 Background and Applications

Task analysis for error identification (TAFEI) is a method that enables people to predict errors in the use of a device by modeling the interaction between user and device. It assumes that people use devices in a purposeful manner, such that the interaction can be described as a "cooperative endeavor," and it is by this process that problems arise. Furthermore, the technique makes the assumption that actions are constrained by the state of the product at any particular point in the interaction, and that the device offers information to the user about its functionality. Thus, the interaction between users and devices progresses through a sequence of states. At each state, the user selects the action most relevant to his/her goal, based on the system image.

The foundation for the approach is based on general systems theory. This theory is potentially useful in addressing the interaction between subcomponents in systems (i.e., the human and the device). It also assumes a hierarchical order of system components, i.e., that all structures and functions are ordered by their relation to other structures and functions, and that any particular object or event comprises lesser objects and events. Information regarding the status of the machine is received by the human part of the system through sensory and perceptual processes and converted to physical activity in the form of input to the machine. The input modifies the internal state of the machine, and feedback is provided to the human in the form of output. Of particular interest here is the boundary between humans and machines, as this is where errors become apparent. We believe that it is essential for a method of error prediction to examine explicitly the nature of the interaction.

The theory draws upon the ideas of scripts and schema. We can imagine that a person approaching a ticket-vending machine might draw upon a "vending machine" or a "ticket kiosk" script when using a ticket machine. From one script, the user might expect the first action to be "insert money," but from the other script, the user might expect the first action to be "select item." The success, or failure, of the interaction would depend on how closely they were able to determine a match between the script and the actual operation of the machine. The role of the comparator is vital in this interaction. If it detects differences from the expected states, then it is able to modify the routines. Failure to detect any differences is likely to result in errors. Following Bartlett's (1932) lead, the notion of schema is assumed to reflect a person's "effort after meaning," arising from the active processing (by the person) of a given stimulus. This active processing involves combining prior knowledge with information contained in the stimulus. While schema theory is not without its critics (see Brewer [2000] for a review), the notion of an active processing of stimuli clearly has resonance with our proposal for rewritable routines. The reader might feel that there are similarities between the notion of rewritable routines and some of the research on mental models that was popular in the 1980s. Recent developments in the theory underpinning TAFEI by the authors have distinguished between global prototypical routines (i.e., a repertoire of stereotypical responses that allow people to perform repetitive and mundane activities with little or no conscious effort) and local, state-specific, routines (i.e., responses that are developed only for a specific state of the system). The interesting part of the theory is the proposed relationship between global and local routines. It is our contention that these routines are analogous to global and local variables in computer programming code. In the same manner as a local variable in programming code, a local routine is overwritten (or rewritable in our terms) once the user has moved beyond the specific state for which it was developed. (See Baber and Stanton [2002] for a more detailed discussion of the theory.)

Examples of applications of TAFEI include prediction of errors in boiling kettles (Baber and Stanton, 1994; Stanton and Baber, 1998), comparison of word processing packages (Stanton and Baber, 1996; Baber and Stanton, 1999), withdrawing cash from automatic teller machines (Burford, 1993), medical applications (Baber and Stanton, 1999; Yamaoka and Baber, 2000), recording on tape-to-tape machines (Baber and Stanton, 1994), programming a menu on cookers (Crawford et al., 2001), programming videocassette recorders (Baber and Stanton, 1994; Stanton and Baber, 1998), operating radio-cassette machines (Stanton and Young, 1999), recalling a phone number on mobile phones (Baber and Stanton, 2002), buying a rail ticket from the ticket machines in the London Underground (Baber and Stanton, 1996), and operating high-voltage switchgear in substations (Glendon and McKenna, 1995).

38.2 Advantages

- Structured and thorough procedure
- Sound theoretical underpinning
- Flexible, generic methodology

38.3 Disadvantages

- Not a rapid technique, as hierarchical task analysis (HTA) and state–space diagrams (SSDs) are prerequisites
- Requires some skill to perform effectively
- Limited to goal-directed behavior

38.4 Procedure and Advice

Procedurally, TAFEI comprises three main stages. First, hierarchical task analysis (HTA, see Chapter 33) is performed to model the human side of the interaction. Of course, one could employ any technique to describe human activity. However, HTA suits our purposes for the following reasons:

1. It is related to goals and tasks.
2. It is directed at a specific goal.
3. It allows consideration of task sequences (through "plans").

As will become apparent, TAFEI focuses on a sequence of tasks aimed at reaching a specific goal. Next, SSDs are constructed to represent the behavior of the artifact. Plans from the HTA are mapped onto the SSD to form the TAFEI diagram. Finally, a transition matrix is devised to display state transitions during device use. TAFEI aims to assist the design of artifacts by illustrating when a state transition is possible but undesirable (i.e., "illegal"). Making all illegal transitions impossible should facilitate the cooperative endeavor of device use. An indication of the steps is provided in Figure 38.1.

For illustrative purposes of how to conduct the method, a simple, manually operated, electric kettle is used in this example. The first step in a TAFEI analysis is to obtain an appropriate HTA for the device, as shown in Figure 38.2. As TAFEI is best applied to scenario analyses, it is wise to consider just one specific goal, as described by the HTA (e.g., a specific, closed-loop task of interest) rather than the whole design. Once this goal has been selected, the analysis proceeds to constructing state–space diagrams (SSDs) for device operation.

A SSD essentially consists of a series of states through which the device passes, from a starting state to the goal state. For each series of states, there will be a current state and a set of possible exits to other

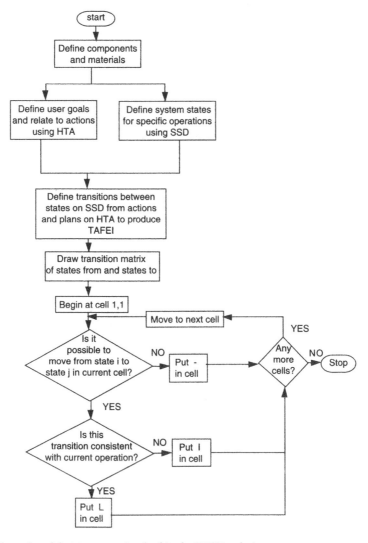

FIGURE 38.1 The series of decision stages involved in the TAFEI technique.

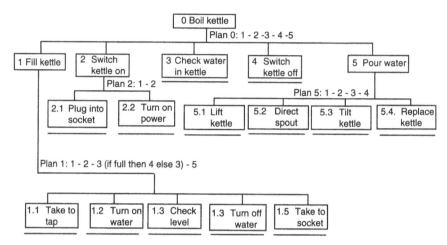

FIGURE 38.2 Hierarchical task analysis.

states. At a basic level, the current state might be "off," with the exit condition "switch on" taking the device to the state "on." Thus, when the device is "off," it is "waiting to…" some action (or set of actions) that will take it to the state "on." It is very important to have, upon completing the SSD, an exhaustive set of states for the device under analysis. Numbered plans from the HTA are then mapped onto the SSD, indicating which human actions take the device from one state to another. Thus the plans are mapped onto the state transitions. (If a transition is activated by the machine, this is also indicated on the SSD, using the letter M on the TAFEI diagram.) This results in a state–space TAFEI diagram, as shown in Figure 38.3. Potential state-dependent hazards have also been identified.

The most important part of the analysis from the point of view of improving usability is the transition matrix. All possible states are entered as headers on a matrix (see Table 38.1). The cells represent state transitions (e.g., the cell at row 1, column 2 represents the transition between state 1 and state 2) and are then filled in one of three ways. If a transition is deemed impossible (i.e., you simply cannot go from this state to that one), a "—" is entered into the cell. If a transition is deemed possible and desirable (i.e., it progresses the user toward the goal state, a correct action), this is a legal transition and "L" is entered into the cell. If, however, a transition is both possible but undesirable (a deviation from the intended path, an error), this is termed illegal, and the cell is filled with an "I." The idea behind TAFEI is that usability can be improved by making all illegal transitions (errors) impossible, thereby limiting the user to only performing desirable actions. It is up to the analyst to conceive of design solutions to achieve this.

The states are typically numbered, but in this example the text description is used. The character L denotes all of the error-free transitions, and the character I denotes all of the errors. Each error has an associated character (i.e., A to G) for the purposes of this example. These characters represent error descriptions and design solutions, as described in Table 38.2.

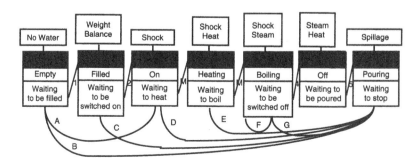

FIGURE 38.3 State–space TAFEI diagram.

TABLE 38.1 Transition Matrix

					To State			
		Empty	Filled	On	Heating	Boiling	Off	Pouring
	Empty	—	L (1)	I (A)	—	—	—	I (B)
	Filled		—	L (2)	—	—	—	I (C)
From	On			—	L (M)	—	—	I (D)
State	Heating					L (M)	—	I (E)
	Boiling					I (F)	L (4)	I (G)
	Off							L (5)
	Pouring							

TABLE 38.2 Error Descriptions and Design Solutions

Error	Transition	Error Description	Design Solution
A	1 to 3	Switch empty kettle on	Transparent kettle walls and/or link to water supply
B	1 to 7	Pour empty kettle	Transparent kettle walls and/or link to water supply
C	2 to 7	Pour cold water	Constant hot water or autoheat when kettle placed on base after filling
D	3 to 7	Pour kettle before boiled	Kettle status indicator showing water temperature
E	4 to 7	Pour kettle before boiled	Kettle status indicator showing water temperature
F	5 to 5	Fail to turn off boiling kettle	Auto cutoff switch when kettle boiling
G	5 to 7	Pour boiling water before turning kettle off	Auto cutoff switch when kettle boiling

Obviously, the design solutions in Table 38.2 are just illustrative and would need to be formally assessed for their feasibility and cost.

What TAFEI does best is enable the analyst to model the interaction between human action and system states. This can be used to identify potential errors and consider the task flow in a goal-oriented scenario. Potential conflicts and contradictions in task flow should come to light. For example, in a study of medical imaging equipment design, Baber and Stanton (1999) identified disruptions in task flow that made the device difficult to use. TAFEI enabled the design to be modified and led to the development of a better task flow. This process of analytical prototyping is key to the use of TAFEI in designing new systems. Obviously, TAFEI can also be used to evaluate existing systems. There is a potential problem that the number of possible states that a device can hold could overwhelm the analyst. Our experience suggests that there are two possible approaches. First, only analyze goal-oriented task scenarios. The process is pointless without a goal, and HTA can help focus the analysis. Second, the analysis can be nested at various levels in the task hierarchy, revealing more and more detail. This can make each level of analysis relatively self-contained and not overwhelming. The final piece of advice is to start with a small project and build up from that position.

38.5 Example

The following example of TAFEI was used to analyze the task of programming a videocassette recorder. The task analysis, state–space diagrams, and transition matrix are all presented. First of all, the task analysis is performed to describe human activity, as shown in Figure 38.4. Next, the state–space diagrams are drawn, as shown in Figure 38.5. From the TAFEI diagram, a transition matrix is compiled, and each transition is scrutinized, as shown in Table 38.3.

Thirteen of the transitions in Table 38.3 are defined as "illegal." These can be reduced to a subset of six basic error types:

1. Switch VCR off inadvertently
2. Insert cassette into machine when switched off

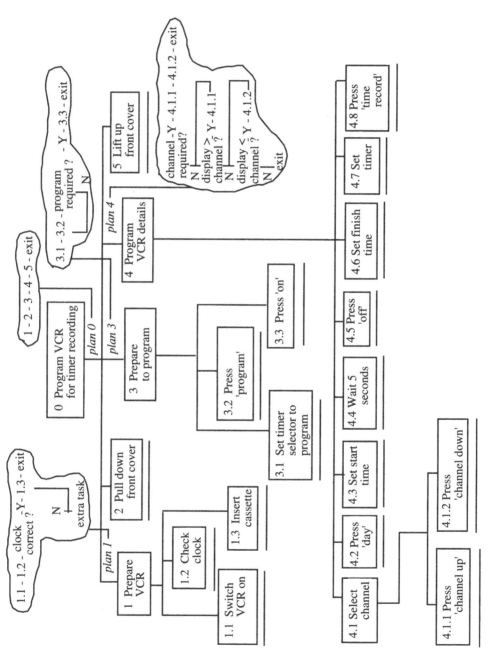

FIGURE 38.4 Hierarchical task analysis.

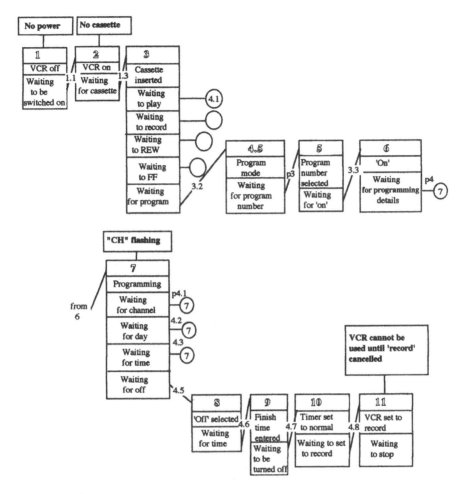

FIGURE 38.5 TAFEI description.

3. Program without cassette inserted
4. Fail to select program number
5. Fail to wait for "on" light
6. Fail to enter programming information

In addition, one legal transition has been highlighted because it requires a recursive activity to be performed. These activities seem to be particularly prone to errors of omission. These predictions then serve as a basis for the designer to address the redesign of the VCR. A number of illegal transitions could be dealt with fairly easily by considering the use of modes in the operation of the device, such as switching off the VCR without stopping the tape and pressing "play" without inserting the tape.

38.6 Related Methods

TAFEI is related to HTA, which provides a description of human activity. Like SHERPA, TAFEI is used to predict human error with artifacts. Kirwan and colleagues recommend that multiple human-error-identification methods can be used to improve the predictive validity of the techniques. This is based on the premise that one method may identify an error that another one misses. Therefore, using SHERPA and TAFEI in combination may be better than using either alone. We have found that the use of multiple analysts similarly improves performance of a method. This is based on the premise that one analyst may identify

TABLE 38.3 Transition Matrix

		To state:									
		1	2	3	4.5	5	6	7	8	9	10	11
From state:	1	-	L	I	-	-	-	-	-	-	-	-
	2	L	-	L	I	-	-	-	-	-	-	-
	3	L	-	-	L	-	-	-	-	-	-	-
	4.5	I	-	-	-	L	I	-	-	-	-	-
	5	I	-	-	-	-	L	I	I	-	-	-
	6	I	-	-	-	-	-	L	I	-	-	-
	7	I	-	-	-	-	-	I	L	-	-	-
	8	I	-	-	-	-	-	-	-	L	-	-
	9	I	-	-	-	-	-	-	-	-	L	-
	10	I	-	-	-	-	-	-	-	-	-	L
	11	-	-	-	-	-	-	-	-	-	-	-

TABLE 38.4 Reliability and Validity Data for TAFEI

	Novices [a]	Experts [b]
Reliability	$r = 0.67$	$r = 0.9$
Validity	SI = 0.79	SI = 0.9

[a] From Stanton, N.A. and Baber, C. (2002), *Design Stud.*, 23, 363–384.
[b] From Baber, C. and Stanton, N.A. (1996), *Appl. Ergonomics*, 27, 119–131.

an error that another one misses. Therefore using SHERPA or TAFEI with multiple analysts may perform better than one analyst using SHERPA or TAFEI alone (Stanton, 2002).

38.7 Standards and Regulations

As with SHERPA, both national and international standards allude to the need for error reduction. In product design, the standards of usability outlined in ISO 9241 require that devices be used efficiently, without causing user frustration. This implies that errors should be reduced to the bare minimum.

Similarly, regulations and standards concerned with the safety of household and similar electrical appliances require that the user be safe from injury in both abnormal and normal product use. This means that human-error identification techniques could be used to anticipate user errors in normal use and the errors themselves as illustrations of abnormal use. Under both conditions, the user should be kept safe. Demonstration of the application of these procedures may help inform analyses of product liability and litigation, as the outcomes suggest what could be reasonably anticipated in the design of a product or procedure.

38.8 Approximate Training and Application Times

A study reported by Stanton and Young (1999) suggests that observational techniques are relatively easily trained and applied. In that study, which involved radio-cassette machines, training in the TAFEI method took approximately 3 hours. Newly trained people who applied the method in the radio-cassette study took approximately 3 hours to predict the errors.

38.9 Reliability and Validity

There are two studies that report on the reliability and validity of TAFEI for both expert and novice analysts. These data are reported in Table 38.4.

38.10 Tools Needed

There is currently no software available to undertake TAFEI, although there are software packages to support HTA.

References

Baber, C. and Stanton, N.A. (1994), Task analysis for error identification: a methodology for designing error-tolerant consumer products, *Ergonomics*, 37, 1923–1941.

Baber, C. and Stanton, N.A. (1996), Human error identification techniques applied to public technology: predictions compared with observed use, *Appl. Ergonomics*, 27, 119–131.

Baber, C. and Stanton, N.A. (1999), Analytical prototyping, in *Interface Technology: The Leading Edge*, Noyes, J.M. and Cook, M., Eds., Research Studies Press, Baldock, U.K..

Baber, C. and Stanton, N.A. (2002), Task analysis for error identification: theory, method and validation, *Theor. Issues Ergonomics Sci.*, 3, 212–227.

Bartlett, F.C. (1932), *Remembering: A Study in Experimental and Social Psychology*, Cambridge University Press, Cambridge, U.K.

Brewer, W.F. (2000), Bartlett's concept of the schema and its impact on theories of knowledge representation in contemporary cognitive psychology, in *Bartlett, Culture and Cognition*, Saito, A., Ed., Psychology Press, London, pp. 69–89.

Burford, B. (1993), *Designing Adaptive ATMs*, M.Sc. thesis, University of Birmingham, Birmingham, U.K.

Crawford, J.O., Taylor, C., and Po, N.L.W. (2001), A case study of on-screen prototypes and usability evaluation of electronic timers and food menu systems, *Int. J. Hum.–Comput. Interact.*, 13, 187–201.

Glendon, A.I. and McKenna, E.F. (1995), *Human Safety and Risk Management*, Chapman & Hall, London.

Kirwan, B. (1994), *A Guide to Practical Human Reliability Assessment*, Taylor & Francis, London.

Stanton, N.A. (2002), Human error identification in human computer interaction, in *The Human Computer Interaction Handbook*, Jacko, J. and Sears, A., Eds., Lawrence Erlbaum Associates, Mahwah, NJ, pp. 371–383.

Stanton, N.A. and Baber, C. (1996a), A systems approach to human error identification, *Saf. Sci.*, 22, 215–228.

Stanton, N.A. and Baber, C. (1996b), Task analysis for error identification: applying HEI to product design and evaluation, in *Usability Evaluation in Industry*, Jordan, P.W., Thomas, B., Weerdmeester, B.A., and McClelland, I.L., Eds., Taylor & Francis, London, pp. 215–224.

Stanton, N.A. and Baber, C. (1998), A systems analysis of consumer products, in *Human Factors in Consumer Products*, Stanton, N.A., Ed., Taylor & Francis, London, pp. 75–90.

Stanton, N.A. and Baber, C. (2002), Error by design, *Design Stud.*, 23, 363–384.

Stanton, N.A. and Young, M. (1999), *A Guide to Methodology in Ergonomics: Designing for Human Use*, Taylor & Francis, London.

Yamaoka, T. and Baber, C. (2000), Three-point task analysis and human error estimation, in *Proceedings of the Human Interface Symposium 2000*, Tokyo, pp. 395–398.

39

Mental Workload

Mark S. Young
University of New South Wales

Neville A. Stanton
Brunel University

39.1 Background and Applications

Mental workload (MWL) is a pervasive concept throughout the ergonomics and human factors literature (e.g., Sanders and McCormick, 1993; Singleton, 1989), and it is a topic of increasing importance. As modern technology in many working environments imposes greater cognitive demands upon operators than physical demands (Singleton, 1989), the understanding of how MWL impinges on performance is critical.

There is no universally accepted definition of MWL, although an analogy is often drawn with physical load (Schlegel, 1993). In this sense, MWL can broadly comprise two components: stress (task demands) and strain (the resulting impact upon the individual).

Attentional resource theories form a useful basis for describing MWL. These theories assume that individuals possess a finite attentional capacity that can be allocated to one or more tasks. Essentially, mental workload represents the proportion of resources required to meet the task demands (Welford, 1978). If demands begin to exceed capacity, the skilled operator either adjusts his/her strategy to compensate (Singleton, 1989) or performance degrades.

An alternative perspective takes into account the level of operator skill and the extent to which cognitive processing is automatic. Gopher and Kimchi (1989) reviewed evidence that MWL in real-world tasks is determined by the balance of automatic and controlled processing involved. This is consistent with the attentional-resources approach, as automaticity releases attentional resources for other tasks, with a resulting decrease in MWL.

Thus, we see that MWL is a multidimensional concept, determined by characteristics of the task (e.g., demands, performance) and of the operator (e.g., skill, attention) (Leplat, 1978). Collating this information enabled Young and Stanton (2001) to propose an operational definition of MWL:

> The mental workload of a task represents the level of attentional resources required to meet both objective and subjective performance criteria, which may be mediated by task demands, external support, and past experience.

In this definition, the level of attentional resources is assumed to have a finite capacity, beyond which any further increases in demand are manifest in performance degradation. Performance criteria can be imposed by external authorities, or they may represent the internal goals of the individual. Examples of task demands are time pressure and complexity. Support can be in the form of peer assistance or technological aids. Finally, past experience can influence mental workload via changes in skill or knowledge.

In the applied arena, MWL is a core area for research in virtually every field imaginable. Two areas that have received particular attention in recent years are driving and automation.

Driver MWL can be affected by a number of factors (Schlegel, 1993) that are either external to the individual (e.g., traffic, road situation) or internal (e.g., age, experience). These can also interact. For example, different levels of traffic do not affect the skilled driver, while high traffic increases the workload for the unskilled driver (Verwey, 1993). In addition, different elements of the driving task (e.g., vehicle control and guidance, navigation) can impose varying levels of MWL. For instance, steering appears to be a significant source of workload in vehicle control (Young and Stanton, 1997), while tuning a car radio is actually one of the most demanding in-car tasks (Dingus et al., 1989). However, there is also some evidence of MWL homeostasis in driving, in that the driver may increase demands when the task is easy (e.g., by driving faster) and reduce them if more difficulty is introduced (Zeitlin, 1995).

It may seem like a contradiction, but automated systems can both reduce and increase MWL. For instance, it has been observed (Hughes and Dornheim, 1995) that glass cockpits in commercial aircraft have relieved workload by automating flight procedures and reducing display clutter. Increased trust in automation also serves to relieve MWL, as the operator feel a lower burden while monitoring the system (Kantowitz and Campbell, 1996). However, the same cockpit systems have increased workload in other areas, such as increased confusion as to the operating mode. This can lead to mental underload during highly automated activities (such as cruise flight) but mental overload during more critical operations (such as take-off and landing) (Parasuraman et al., 1996). Others have predicted that future systems could increase complexity or excessively reduce demands in both aircraft and cars (Labiale, 1997; Lovesey, 1995; Roscoe, 1992; Verwey, 1993). Mental overload and mental underload are therefore very real possibilities, and both are equally serious conditions that can lead to performance degradation, attentional lapses, and errors (Wilson and Rajan, 1995).

39.2   Measurement Procedures

The measurement of MWL is as diverse a topic as its theoretical counterpart, with many techniques available. Indeed, researchers in applied domains tend to favor the use of a battery of measures to assess workload rather than any one measure (Gopher and Kimchi, 1989; Hockey et al., 1989). This use of multidimensional devices to assess workload seems sensible given the previous discussion of workload as a multidimensional concept. The main categories of workload measures are:

1. Primary task performance measures
2. Secondary task performance measures
3. Physiological measures
4. Subjective ratings

39.2.1   Primary and Secondary Tasks

Performance measures on primary and secondary tasks are widely used in workload assessment. The basic premise is that a task with higher workload will be more difficult, resulting in degraded performance compared with a low workload task. Of course, though, following from attentional-resources theory, an increase in difficulty (workload) may not lead to performance deficits if the increase is still within the capacity of the operator. Thus a secondary task, designed to compete for the same resources as the primary task, can be used as a measure of spare attentional capacity. According to the definition of MWL proposed above, the level of MWL in a task can be directly inferred from measures of attentional capacity. In the secondary-task technique, participants are instructed to maintain consistent performance on the primary task while attempting to perform the secondary task when their primary task demands allow them to. Differences in workload between primary tasks are then reflected in performance of the secondary task. Researchers wishing to use a secondary task are well advised to adopt discrete stimuli (Brown, 1978) that occupy the same attentional resource pools as the primary task (e.g., if the primary task is driving, the secondary task should also be visual–spatial, with a manual response). This ensures that the technique really is measuring spare capacity and not an alternative resource pool.

39.2.2 Physiological Measures

Physiological measures of MWL are many and varied. For instance, various researchers have used respiration and heart rate (Roscoe, 1992), heart rate variability (Jorna, 1992), electrodermal response (Helander, 1978), eye movements and pupillary responses (Backs and Walrath, 1992; Itoh et al., 1990), and event-related potentials (Kramer et al., 1996) as indices of mental effort. Many physiological measures have been reliably associated with mental effort (e.g., Helander, 1978). However, it is generally recommended that physiological measures only be applied if they are unobtrusive and reliable (Fairclough, 1993) and in conjunction with other measures of workload (e.g., Backs and Walrath, 1992).

39.2.3 Subjective Ratings

Many authors claim that the use of subjective ratings may well be the only index of "true" MWL (e.g., Hart and Staveland, 1988). Subjective MWL scores are sensitive to perceived difficulty, the presence of automation (Liu and Wickens, 1994), concurrent activities, and demand for multiple resources (Hockey et al., 1989). One particular advantage of subjective measures is that they are sensitive to changes in effort when such effort maintains primary task performance at stable levels (Hockey et al., 1989). Subjective ratings materialize in many forms, but these can be reduced to two categories: unidimensional and multidimensional. Unidimensional measures tend to be simpler to apply and analyze, but they offer only a general workload score. In contrast, multidimensional measures provide some diagnostics for identifying the sources of MWL, but these procedures are typically more complex.

39.3 Advantages

- Primary task: Direct index of performance; effective in measuring long periods of workload and performance, overload, and individual differences in resource competition.
- Secondary task: Can discriminate between tasks when no differences are observed in primary performance; useful for quantifying short periods of workload, spare attentional capacity, and even automaticity.
- Physiological: Continuous monitoring of data; increased sensitivity in measurement; does not interfere with primary task performance.
- Subjective: Easy to administer and analyze; provides an index of perceived strain; multidimensional measures can determine the source of MWL.

39.4 Disadvantages

- Primary task: Cannot distinguish between tasks if they are within the attentional capacity of the operator; can be unfeasible or uneconomical to measure in real-world conditions; not a reliable measure in isolation.
- Secondary task: Only sensitive to gross changes in MWL; can be intrusive, particularly at low levels of primary task workload; must be carefully designed in order to be a true measure of spare attentional capacity.

- Physiological: Easily confounded by extraneous interference; physically obtrusive equipment; difficult to obtain and analyze data.
- Subjective: Can only be administered posttask, thus influencing reliability for long task durations; metacognitive limitations can cloud accurate reporting; difficult to make absolute comparisons between participants

39.5 Examples of Mental Workload Measurement

By their very nature, primary and secondary task measures of MWL will be tailored specifically to suit the application being studied. Physiological and subjective measures, though, are more generic and can be used across most domains. The following examples are based on some of our previous research into vehicle automation and driver MWL. The results were obtained using a driving simulator.

In terms of primary task measures, the simulator recorded data on a range of variables, including speed, lateral position, and distance from the vehicle in front (headway). In order to evaluate performance on these variables, some measure of stability along these variables was needed. For speed and headway, this was achieved using the standard error around the regression line (as used by Bloomfield and Carroll [1996]). For lateral control, it was considered that instability measures would not be an appropriate reflection of driving performance on a road that includes both curved and straight sections. Instead, simple measures of lane excursions were used to assess lateral control, with the assumption being that good driving performance is rewarded with fewer lane excursions. The total number of lane excursions and the time spent out of lane were the dependent variables for lateral control.

The secondary task used in the driving simulator consisted of a rotated-figures task (as used by Baber [1991]), presented in the lower left corner of the screen (Figure 39.1). Each stimulus was a pair of stick figures (one upright; the other rotated through 0°, 90°, 180°, or 270°) holding one or two flags. Pairs were presented in a random order. The flags were simple geometrical shapes, either squares or diamonds. The task was to make a judgment as to whether the figures were the same or different, based on the flags they were holding. Responses were made via buttons attached to the steering column stalks, and brief visual feedback was provided before presentation of the next stimulus. The subsidiary-task technique was used in an attempt to measure spare attentional capacity. Participants were thus instructed to attend to the secondary task only when they had time to do so. The secondary task was visual–spatial, requiring a manual response and, as such, was intended to occupy the same attentional resource pools as driving. This ensured that the task was indeed a measure of spare mental capacity (based on multiple-resources

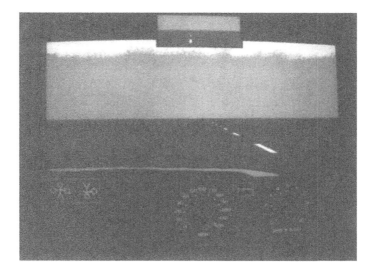

FIGURE 39.1 Screenshot of the driving simulator.

theory) and not some alternative cognitive resource. It was found in our studies that error rate remained quite constant across all trials (at around 5%), and therefore the dependent variable we chose to associate with the secondary task was simply the number of correct responses. Higher frequencies implied increased spare capacity, thus lower MWL.

Physiological measures were not taken in many of our experiments, as it was felt that the collection and interpretation of such data with respect to MWL was fraught with difficulty. Given that three of the four major methods of MWL assessment were already being used (i.e., primary task, secondary task, and subjective measures), it was felt that the complications involved with collection and interpretation of physiological data for MWL measurement outweighed the potential benefits. However, in one study, heart rate (HR) was recorded as a pure measure of physiological arousal, for which conclusions are more readily drawn than for MWL. Physiological measures of MWL, such as heart rate variability, are indirect and depend on a number of assumptions. Arousal, however, is directly reflected in HR and is not open to other explanations.

Subjective MWL was assessed with the NASA-task load index (TLX) (Hart and Staveland, 1988). The NASA-TLX was selected over other subjective MWL measures, e.g., the Cooper-Harper Scale (Cooper and Harper, 1969) and the subjective workload assessment technique (SWAT) (Reid and Nygren, 1988), for a number of reasons. First, a multidimensional technique was preferred over unidimensional measures for its utility in assessing the components of MWL that characterize the experimental task. Of the multidimensional measures, TLX and SWAT seem to be the most widely used (e.g., Hendy et al., 1993). The choice of TLX was made because it is more acceptable to participants (Hill et al., 1992) and is more sensitive to MWL differences than SWAT, particularly at low workload levels (Nygren, 1991). Hart and Staveland (1988) claim that their procedure is practically and statistically superior to SWAT because the independent components of TLX provide additional diagnostic information unavailable in SWAT. Finally, and most importantly, the TLX is far easier to administer than its counterparts. This is particularly true in the light of research suggesting that the weighting procedure of TLX is superfluous and can be omitted without compromising the measure (Hendy et al., 1993; Hill et al., 1992; Nygren, 1991). It is thus more acceptable to participants and increases the likelihood of genuine responses (Hill et al., 1992). The dependent variables are then simply the raw scores from each of the six scales and the arithmetic mean of these scores, which constitutes overall workload (OWL).

The rating scales of the NASA-TLX are: mental demand (MD), physical demand (PD), temporal demand (TD), and the individual's perceived level of performance (PE), effort (EF), and frustration (FR). Each dimension is rated on a visual analogue scale, with five-point steps between 0 and 100. Participants are given definitions of the rating scales to assist them in making their assessments (see Table 39.1). In our studies, MWL was measured using a secondary task as well as the TLX. It was thus important that participants were instructed to rate only the primary driving task and not the combined demands of the primary and secondary tasks.

39.6 Related Methods

MWL is often associated with situation awareness (SA), which is measured using a subjective rating scale, similar to the TLX. The situation-awareness rating technique (SART) was constructed by researchers interested in the performance of fighter pilots (Taylor and Selcon, 1991). SART consists of 14 items, each rated on a visual analogue scale. Four items relate to perceived demand on attentional resources; five assess how much attention the participant is supplying to the situation; and four measure how well the participant believes s/he understands the situation. There is also a scale for rating overall situational awareness.

39.7 Standards and Regulations

ISO 9241-11 (1998) defines usability as follows: "the extent to which a product can be used by specified users to achieve specified goals with effectiveness, efficiency and satisfaction in a specified context of use."

TABLE 39.1 NASA-TLX Rating Scale Definitions

Title	Endpoints	Descriptions
Mental demand	low/high	How much mental and perceptual activity was required (e.g., thinking, deciding, calculating, remembering, looking, searching, etc.)? Was the task easy or demanding, simple or complex, exacting or forgiving?
Physical demand	low/high	How much physical activity was required (e.g., pushing, pulling, turning, controlling, activating, etc.)? Was the task easy or demanding, slow or brisk, slack or strenuous, restful or laborious?
Temporal demand	low/high	How much time pressure did you feel due to the rate or pace at which the tasks or task elements occurred? Was the pace slow and leisurely or rapid and frantic?
Performance	good/poor	How successful do you think you were in accomplishing the goals of the task set by the experimenter (or yourself)? How satisfied were you with your performance in accomplishing these goals?
Effort	low/high	How hard did you have to work (mentally and physically) to accomplish your level of performance?
Frustration level	low/high	How insecure, discouraged, irritated, stressed, and annoyed versus secure, gratified, content, relaxed, and complacent did you feel during the task?

Source: Hart, S.G. and Staveland, L.E. (1988), Development of NASA-TLX (task load index): results of empirical and theoretical research, in *Human Mental Workload*, Hancock, P.A. and Meshkati, N., Eds., North-Holland, Amsterdam, pp. 138–183.

The British Standards Institute has around 200 standards that relate directly to ergonomics and the user-centered design of artifacts and systems. These standards include BS EN ISO 10075-2:2000, Ergonomic principles related to mental workload. There are similar national standards in other countries.

39.8 Approximate Training and Application Times

Primary and secondary task techniques tend to be specific to the application they are measuring, so training is somewhat irrelevant. Domain expertise will inform the researchers about the types of measures to employ. However, a search of the literature will reveal typical methods of primary and secondary tasks (e.g., Schlegel, 1993; Baber, 1991).

Physiological measures represent the most difficult end of the spectrum, as a deep understanding of psychophysiology is necessary to employ and interpret physiological methods correctly. In addition, training in using the physiological monitoring hardware is necessary, and this can take a few hours if the user is unfamiliar with such equipment. Finally, there is a measure of statistical expertise involved in filtering and processing the raw data that physiological tools output.

Subjective techniques are probably the easiest MWL measure to apply without prior experience, since they are essentially simple questionnaires. It is recommended that researchers familiarize themselves with the original derivation of the chosen technique, e.g., the NASA-TLX (Hart and Staveland, 1988), since this will improve understanding of the rating scales and confirm their relevance for the application. The source papers will also explain the scoring methods in more detail (although, as stated above, current opinion is that the raw scores are sufficient for the TLX).

39.9 Reliability and Validity

Because primary and secondary task measures are specific to the application being studied, it is not possible to comment on reliability and validity for these techniques. Researchers are best advised to use a battery of techniques, selecting at least one measure from each of the MWL categories (primary task, secondary task, physiological, and subjective) as far as this is feasible. In this way, some degree of consistency between the measures can be checked. Reliability of the secondary task can be optimized by ensuring that the task is designed to occupy the same attentional-resource pools as the primary task. Critics might argue that this would increase the intrusiveness of the secondary task. However, if the secondary task draws upon separate pools, it can no longer be relied upon as a true measure of spare capacity.

Physiological measures are supported by a good deal of research, although their validity as a MWL measure can vary between techniques. The abundance of literature is too great to summarize here, but recent opinion suggests that heart rate variability (HRV) is one of the better techniques for measuring MWL. The relative ease of recording is an added advantage for HRV. It should be noted that in the example given in this chapter, simple heart rate was used as a measure of arousal rather than workload. Although MWL can be related to stress (and hence would influence arousal), it is generally accepted that HRV is a more reliable index of MWL.

As for the subjective methods, both the NASA-TLX and SWAT were derived via rigorous testing and factor-analysis methods. Their extensive application in MWL studies testifies to their utility, although this does not necessarily advocate for their validity. However, there are certainly a number of follow-up studies on the TLX (e.g., Hendy et al., 1993; Hill et al., 1992; Nygren, 1991) that confirm its reliability as a MWL metric, even in the absence of the lengthy weighting procedure. Reliability of subjective scales can be greatly enhanced by anchoring the scales, preferably by asking participants to rate a simple baseline task before proceeding with the experimental conditions. Naturally, if within-subject designs are used, all measures will be relative anyway, so there is less need for an anchor task.

In our own studies, we found that the TLX significantly correlated with the secondary task, with $r^2 \approx 0.4$. The remaining variability suggested that, although the techniques were measuring the same underlying construct, they may have been accessing different aspects of that construct.

39.10 Tools Needed

The tools required for primary and secondary task measurement will entirely depend on the application. Ideally, data collection for both will be computerized, such as in our driving simulator. It is recognized, though, that this might not always be practicable in field experiments. A separate laptop computer can provide a useful and accurate means of presenting a secondary task. Failing that, a series of laminate cards displaying the secondary task figures can be used, with the experimenter noting the results manually on a crib-sheet.

Physiological measures require complex (and often expensive) monitoring equipment. However, advances in sports technology mean that heart monitors are now readily available and relatively simple to use. In our experiment, we used a Polar Vantage NV heart rate monitor, a nonintrusive chest belt developed for sports science applications. It was generally quite reliable, although the plethora of electrical equipment in the simulator laboratory sometimes caused interference. Dedicated processing and analysis software is also available, which is invaluable in filtering the files for spurious peaks caused by technical artifacts. When filtering data, though, care should be taken to remove only the nonphysiological data without altering genuine interbeat intervals.

Finally, subjective measures are mostly pen-and-paper exercises. For the TLX, a sheet should be created with visual analogue scales for each of the six dimensions, marked with 20 divisions (it is also useful for participants if there are major and minor ticks along the scale, particularly to highlight the midpoint). A copy of the rating scale definitions should be given to participants for reference when completing the TLX. The analysis process can, of course, be automated by presenting the questionnaire on a computer. If this is the case, researchers should be careful to ensure that reliability has not been affected.

References

Baber, C. (1991), *Speech Technology in Control Room Systems: A Human Factors Perspective*, Ellis Horwood, Chichester, U.K.

Backs, R.W. and Walrath, L.C. (1992), Eye movements and pupillary response indices of mental workload during visual search of symbolic displays, *Appl. Ergonomics*, 23, 243–254.

Bloomfield, J.R. and Carroll, S.A. (1996), New measures of driving performance, in *Contemporary Ergonomics 1996*, Robertson, S.A., Ed., Taylor & Francis, London, pp. 335–340.

British Standards Institute (2000), BS EN ISO 10075-2: Ergonomic principles related to mental workload. Design principles.

Brown, I.D. (1978), Dual task methods of assessing work-load, *Ergonomics*, 21, 221–224.

Cooper, G.E. and Harper, R.P. (1969), The use of pilot rating in the evaluation of aircraft handling, Report ASD-TR-76-19, National Aeronautics and Space Administration, Moffett Field, CA.

Dingus, T.A., Antin, J.F., Hulse, M.C., and Wierwille, W.W. (1989), Attentional demand of an automobile moving-map navigation system, *Transp. Res.-A*, 23, 301–315.

Fairclough, S. (1993), Psychophysiological measures of workload and stress, in *Driving Future Vehicles*, Parkes, A.M. and Franzen, S., Eds., Taylor & Francis, London, pp. 377–390.

Gopher, D. and Kimchi, R. (1989), Engineering psychology, *Ann. Rev. Psychol.*, 40, 431–455.

Hart, S.G. and Staveland, L.E. (1988), Development of NASA-TLX (task load index): results of empirical and theoretical research, in *Human Mental Workload*, Hancock, P.A. and Meshkati, N., Eds., North-Holland, Amsterdam, pp. 138–183.

Helander, M. (1978), Applicability of drivers' electrodermal response to the design of the traffic environment, *J. Appl. Psychol.*, 63, 481–488.

Hendy, K.C., Hamilton, K.M., and Landry, L.N. (1993), Measuring subjective mental workload: when is one scale better than many? *Hum. Factors*, 35, 579–601.

Hill, S.G., Iavecchia, H.P., Byers, J.C., Bittner, A.C., Zaklad, A.L., and Christ, R.E. (1992), Comparison of four subjective workload rating scales, *Hum. Factors*, 34, 429–439.

Hockey, G.R.J., Briner, R.B., Tatersall, A.J., and Wiethoff, M. (1989), Assessing the impact of computer workload on operator stress: the role of system controllability, *Ergonomics*, 32, 1401–1418.

Hughes, D. and Dornheim, M.A. (1995), Accidents direct focus on cockpit automation, *Aviation Week Space Technol.*, Jan. 30, 52–54.

International Organization for Standardization (1998), ISO 9241–11: ergonomic requirements for office work with visual display terminals (VDTs). Part II: Guidance on usability.

Itoh, Y., Hayashi, Y., Tsuki, I., and Saito, S. (1990), The ergonomic evaluation of eye movement and mental workload in aircraft pilots, *Ergonomics*, 33, 719–733.

Jorna, P.G.A.M. (1992), Spectral analysis of heart rate and psychological state: a review of its validity as a workload index, *Biol. Psychol.*, 34, 237–257.

Kantowitz, B.H. and Campbell, J.L. (1996), Pilot workload and flightdeck automation, in *Automation and Human Performance: Theory and Applications*, Parasuraman, R. and Mouloua, M., Eds., Lawrence Erlbaum Associates, Mahwah, NJ, pp. 117–136.

Kramer, A.F., Trejo, L.J., and Humphrey, D.G. (1996), Psychophysiological measures of workload: potential applications to adaptively automated systems, in *Automation and Human Performance: Theory and Applications*, Parasuraman, R. and Mouloua, M., Eds., Lawrence Erlbaum Associates, Mahwah, NJ, pp. 137–162.

Labiale, G. (1997), Cognitive ergonomics and intelligent systems in the automobile, in *Ergonomics and Safety of Intelligent Driver Interfaces*, Noy, Y.I., Ed., Lawrence Erlbaum Associates, Mahwah, NJ, pp. 169–184.

Leplat, J. (1978), Factors determining work-load, *Ergonomics*, 21, 143–149.

Liu, Y. and Wickens, C.D. (1994), Mental workload and cognitive task automaticity: an evaluation of subjective and time estimation metrics, *Ergonomics*, 37, 1843–1854.

Lovesey, E. (1995), Information flow between cockpit and aircrew, *Ergonomics*, 38, 558–564.

Nygren, T.E. (1991), Psychometric properties of subjective workload measurement techniques: implications for their use in the assessment of perceived mental workload, *Hum. Factors*, 33, 17–33.

Parasuraman, R., Mouloua, M., Molloy, R., and Hilburn, B. (1996), Monitoring of automated systems, in *Automation and Human Performance: Theory and Applications*, Parasuraman, R. and Mouloua, M., Eds., Lawrence Erlbaum Associates, Mahwah, NJ, pp. 91–115.

Reid, G.B. and Nygren, T.E. (1988), The subjective workload assessment technique: a scaling procedure for measuring mental workload, in *Human Mental Workload*, Hancock, P.A. and Meshkati, N., Eds., North-Holland, Amsterdam.

Roscoe, A.H. (1992), Assessing pilot workload: why measure heart rate, HRV and respiration? *Biol. Psychol.*, 34, 259–287.

Sanders, M.S. and McCormick, E.J. (1993), *Human Factors in Engineering and Design*, McGraw-Hill, New York.

Schlegel, R.E. (1993), Driver mental workload, in *Automotive Ergonomics*, Peacock, B. and Karwowski, W., Eds., Taylor & Francis, London, pp. 359–382.

Singleton, W.T. (1989), *The Mind at Work: Psychological Ergonomics*, Cambridge University Press, Cambridge, U.K.

Taylor, R.M. and Selcon, S.J. (1991), Subjective measurement of situational awareness, in *Designing for Everyone: Proceedings of the 11th Congress of the International Ergonomics Association*, Quéinnec, Y. and Daniellou, F., Eds., Taylor & Francis, London, pp. 789–791.

Verwey, W.B. (1993), How can we prevent overload of the driver? in *Driving Future Vehicles*, Parkes, A.M. and Franzen, S., Eds., Taylor & Francis, London, pp. 235–244.

Welford, A.T. (1978), Mental work-load as a function of demand, capacity, strategy and skill, *Ergonomics*, 21, 151–167.

Wilson, J.R. and Rajan, J.A. (1995), Human-machine interfaces for systems control, in *Evaluation of Human Work: A Practical Ergonomics Methodology*, Wilson, J.R. and Corlett, E.N., Eds., Taylor & Francis, London, pp. 357–405.

Young, M.S. and Stanon, N.A. (1997), Automotive automation: investigating the impact on drivers' mental workload, *Int. J. Cognitive Ergonomics*, 1(4), 325–336.

Young, M.S. and Stanton, N.A. (2001), Mental workload: theory, measurement, and application, in *International Encyclopedia of Ergonomics and Human Factors*, Vol. 1, Karwowski, W., Ed., Taylor & Francis, London, pp. 507–509.

Zeitlin, L.R. (1995), Estimates of driver mental workload: a long-term field trial of two subsidiary tasks, *Hum. Factors*, 37, 611–621.

40

Multiple Resource Time Sharing Models

Christopher D. Wickens
*University of Illinois at
Urbana-Champaign*

40.1 Background and Applications

The multiple-resource model predicts the degree of interference between two time-shared tasks, i.e., it predicts the loss in performance of one or both tasks carried out concurrently, relative to their single-task baseline measures. As such, it is a model of workload effects on performance, in which the sources of workload are multiple task demands. The model we describe here is based upon multiple-resource theory (Navon and Gopher, 1979; Wickens, 1980, 2002), which posits that three factors are important in predicting how well (or poorly) a task will be performed when time-shared with another:

1. The difficulty, or demand for resources, of each single task component (e.g., driving in traffic is more resource demanding than driving on an open road)
2. The allocation of those limited resources between two time-shared task (e.g., whether driving is emphasized at the expense of using in-vehicle technology, or the converse)
3. The extent to which the two tasks demand common or separate attentional resources (e.g., an in-vehicle visual display will demand more common resources with driving than will an in-vehicle auditory display)

As the above example suggests, separate resources are defined by auditory versus visual processing. In addition, synthesis of dual-task research (Wickens and Hollands, 2000) also suggests that separate resources are defined by spatial (analogue) versus verbal (linguistic) processing; by perception and working memory versus responding; and by focal versus ambient vision. Thus any task can be represented by a set of demands on either dichotomous level of one or more of these four dimensions. Any *pair of* tasks can then be represented by the degree to which they *share* common levels on each dimension (e.g., both auditory) *and* by their combined demand for resources. The amount of shared resources and the combined demand predict the total interference between tasks. Then the resource-allocation policy between tasks (the extent to which one is favored and the other neglected) determines how this interference (dual-task decrement) is apportioned between them.

40.2 Procedure

The computational version of the multiple-resource model (Wickens, 2002b) involves the following steps:

40.2.1 Step 1: Code Time-Shared Tasks

Each task that will be time-shared is coded by the extent to which it depends on separate resources defined by the four dichotomous dimensions mentioned above, as shown in Table 40.1. Demand levels (including 0) within each resource can take on simple integer values, with greater demand implying greater value. As an example, a very simple conversational task would be coded as: Perception: audition-verbal (= 1), Working memory: verbal (= 1), Response: verbal (= 1). A very demanding vigilance task (detecting weapons in x-rayed luggage) would be coded as: Perception: vision-focal-spatial (= 3), Working memory (= 0), Response: vocal or manual (= 1). Thus each task spawns what is called a *demand vector*. This vector has two important properties for computing interference:

1. The average level of demand across all resources involved. For the simple conversation task, this is 1.0; for the vigilance task it is 3.0.
2. Which resources within Table 40.1 are loaded.

40.2.2 Step 2: Calculate Total Demand Score

The total resource demand for two time-shared tasks is summed to predict a total demand score component. In the above case of time sharing conversation with vigilance, this would be 1 + 3 = 4. The larger this score, the greater is the amount of interference.

40.2.3 Step 3: Calculate Resource-Conflict Score

A resource-conflict score component is computed based on the extent to which the two tasks demand overlapping resources within the four-dimensional model. Since there are four dichotomous dimensions, each task can compete with each other task for common levels of zero, one, two, three, or all four

TABLE 40.1 Resource Coding

	Processing Stage	
Perception	Cognition (Working Memory)	Response
Vision		
Ambient		
Focal	Spatial	Spatial (manual)
Spatial		
Verbal	Verbal	Verbal (vocal)
Audition		
Spatial		
Verbal		

Note: Ambient vision can only be employed for spatial processing; the distinction between spatial and verbal working memory is independent of the perceptual modality and is independent of the distinction between spatial and verbal responses. Also, while perception and cognition define different stages, they do not define different resources.

TABLE 40.2 A Simplified Two-Dimensional Conflict Matrix

			Task A			
			P/C		Response	
			Spatial	Verbal	Spatial	Verbal
	P/C	Spatial	0.7	0.5	0.5	0.3
		Verbal	0.5	0.7	0.3	0.5
	Response	Spatial	0.3	0.3	0.8	0.6
Task B		Verbal (speech)	0.3	0.5	0.6	1.0

Note: P/C = perceptual/cognitive.

dimensions. The multiple-resource model further predicts that the amount of interference between task elements (elements in the demand vector) grows with the number of shared-resource features with the other tasks. A multiple-resource *conflict matrix* is employed to calculate this. Because of the high complexity of such a matrix with the four-dimensional model, it will not be presented here (for details, see Wickens, 2002b). Instead, a simple two-dimensional model will be employed. Here, assume that the two dichotomous dimensions are (1) whether the task is spatial or verbal and (2) the level of its demands on perceptual-cognitive versus response resources. This defines the simple conflict matrix shown in Table 40.2. Higher values within each cell dictate greater conflict.

There are several features to note about the matrix. First, all values are bounded between 1.0 (maximum conflict) and 0 (no conflict). Second, each cell that shares an additional resource between its rows and columns increments the amount of conflict by 0.2. As a consequence, those cells on the negative diagonal (defining identical resources between two tasks) involve the greatest conflict. Those off of the negative diagonal involve less conflict. Third, there is somewhat greater conflict between the response component of two tasks than between the perceptual/cognitive components. One very intuitive reason for this asymmetry is the simple fact that the mouth cannot produce different vocal responses for two different tasks at the same time (hence the maximum conflict value of 1.0 is placed in the bottom right cell for verbal [speech] responses). A more subtle reason is the long-standing research finding in psychology that response selection tends to act as a "bottleneck" in time-shared tasks (Welford, 1967; Pashler, 1998).

In computing the conflict, the two task-demand vectors derived in step 1 are placed across the rows and columns of the table. Any cell in the matrix that has nonzero values in both its row and column entry will contribute to a resource-conflict score by an amount equal to the value in the cell. The sum of such conflict values across all cells determines the total resource-conflict component.

40.2.4 Step 4: Calculate Total Interference Score

The sum of the total demand component and the resource conflict component determines the total dual task interference score.

40.2.5 Step 5: Apportion Interference Score

This interference score can be apportioned to one task or the other, or both, as a function of how the operator is inferred to prioritize the two tasks.

40.2.6 Extending the Model

The simple model in Table 40.2 can readily be extended in two ways. First, more resource dichotomies can be added. For example, if it is desired to include the differences between auditory and visual presentation, then the perceptual/cognitive entries get expanded from two to four (since either spatial or verbal information can be presented auditorily or visually). Still, the algorithm of incrementing interference by 0.2 each time a shared resource is created can be applied (see Wickens, 2002). Second, certain "special circumstances" may require adjustment of the values in certain cells within the conflict matrix. Most importantly, the value in any cell characterizing visual conflict between two tasks must be elevated to the extent that the two sources are separated. In particular, any sources requiring focal vision that are separated by more than around 4 degrees of visual angle (such that the fovea cannot access both at once) will have a conflict value set to 1.0.

40.2.7 Applying the Model

In applying the model, a few points are important:

1. Any language-based task, or any task involving symbolic meaning, is classified as "verbal."
2. Demand levels should be kept simple, at low values. When there is doubt, the value within any level of the demand matrix can be set at either 0 or 1.
3. There is no firm basis for establishing the relative weighting between the demand and the resource-conflict components of the model. The maximum value of each will be application dependent. A useful heuristic is to scale each component relative to the maximum possible value in a given model application.
4. The absolute level of interference is of less importance than the relative level comparing two (or more) dual-task conditions or time-sharing interfaces.

40.3 Advantages

- Captures known empirical phenomena that influence dual-task performance in many multitask environments (e.g., vehicle control)
- Based upon theory
- Simple (arithmetic) in its computations
- Relatively robust to simplifications (e.g., equal-demand coding)
- Flexible in its applications (e.g., number of resources, qualitatively different tasks)

40.4 Disadvantages

- Requires some modeling expertise to code conflict matrix (commercial software is not available)
- Requires domain expertise to estimate demand values
- Model output does not translate into direct absolute measure of dual-task performance; but rather it yields a relative measure of task interference between different dual task combinations
- Has received only limited validation
- Does not account for all multitask phenomena, in particular task switching and "cognitive tunneling"; model assumes operator is "trying" to time-share tasks

40.5 Example

To provide a very simple example of such a computation, consider the model shown in Table 40.3, which postulates only two resources: perceptual cognitive (P/C) versus response (R). Consistent with the discussion of step 3, this matrix shows greater conflict within a stage (the negative diagonal) than across stages (the positive diagonal). Furthermore, consistent with single-channel bottleneck models of processing, it portrays the inability to respond to two tasks at once (1.0) and the greater capacity to time share the perceptual–cognitive aspects of a pair of tasks (0.80).

TABLE 40.3 A Simplified Two-Resource Conflict Matrix

		Task A	
		P/C	R
	P/C	0.80	0.60
Task B	R	0.60	1.00

Note: P/C = perceptual/cognitive; R = response.

TABLE 40.4 Predicted Total Interference Values Resulting from Six Task Combinations

Tasks	Demand Component	Conflict Component	Total Interference
AA	1 + 1 = 2	0.8 + 0 + 0 + 0 = 0.8	2.8
BB	1 + 1 = 2	0.8 + 1 + 0.6 + 0.6 = 3.0	5.0
CC	1.5 + 1.5 = 3	0.8 + 1 + 0.6 + 0.6 = 3.0	6.0
AB	1 + 1 = 2	0.8 + 0 + 0.6 + 0 = 1.4	3.4
AC	1 + 1.5 = 2.5	0.8 + 0 + 0.6 + 0 = 1.4	3.9
BC	1 + 1.5 = 2.5	0.8 + 1 + 0.6 + 0.6 = 3.0	5.5

Note: Demand component represents task difficulty; conflict component represents resource conflicts.

Now consider three tasks. Task A involves pure, demanding monitoring, so its vector of demands across the two resources is [2,0]. Task B involves standard information transmission (e.g., a tracking task) involving perception and response [1,1]. Task C is also a tracking task, but it has an incompatible control, so that control movement must be reversed from the expected direction in order to correct an error. Because display-control compatibility is found to influence the difficulty of response selection (Wickens and Hollands, 2000), its demands are [1,2]. Table 40.4 shows the predicted interference patterns resulting from the two components of the formula (total demand and resource conflict) across the six possible dual-task combinations reflected by the different two-way combinations of the three tasks, A, B, and C. The demand component is computed by summing the average demand across all resources within a task, across both tasks. The conflict component is computed by summing the conflict matrix components of all cells that are demanded by both tasks. Here, total interference assumes equal weighting between the two components.

40.6 Related Methods

The Boeing timeline analysis (Parks and Boucek, 1989) analyzes task interference as a function of tasks overlapping in time, but it does not consider resource conflict or demand values. MicroSaint/Windex is a multiple-resource approach (North and Riley, 1989; Laughery, 1989) that is somewhat more complex than that described above, but provides similar calculations. Task networking methods (Laughery and Corker, 1997) such as MIDAS provide outputs on multiple resources and account for task switching. The IMPRINT modeling tool (Laughery and Corker, 1997) provides table lookup of demand values for many tasks. Some of these methods are reviewed and contrasted by Sarno and Wickens (1995).

40.7 Standards and Regulations

Certain organizations, such as the U.S. Federal Aviation Administration (FAA) impose workload standards in certifying new equipment (FAA Federal Air Regulation 25.1523 Appendix D). However, they do not specify methods whereby this should be done.

40.8   Approximate Training and Application Time

Application time is approximately 5 hours for a simple dual-task problem. Training time decreases to the extent that the user has greater familiarity with cognitive task analysis and with the domain of application. An application to driving performance was validated by Horrey and Wickens (2003).

40.9   Reliability and Validity

Few studies have been conducted to validate the method. The most exhaustive of these was carried out by Sarno and Wickens (1995), in which several operators performed simulated flight tasks, and the measures of task interference were predicted by (1) a full multiple-resource model (similar to that presented here) along with models involving (2) only resource-demand values, (3) only resource-conflict values, and (4) neither (pure timeline model). Models of types 1, 2, and 3 were all validated to account for over half of the variance in decrement, across 16 different interface conditions. It was found that including a resource-conflict component was more important than including a resource-demand component.

References

Horrey, W.J. and Wickens, C.D (2003), Multiple resouce modeling of task interference in vehicle control, hazard awareness and in-vehicle task performance, in Proceedings of the Second International Driving Symposium on Human Factors in Driver Assessment, Training, and Vehicle Design, Park City, UT, pp. 7–12.

Laughery, K.R. (1989), Micro SAINT: a tool for modeling human performance in systems, in *Applications of Human Performance Models to System Design*, Defense Research Series, Vol. 2, McMillan, G.R., Beevis, D., Salas, E., Strub, M.H., Sutton, R., and Van Breda, L., Eds., Plenum Press, New York, pp. 219–230.

Laughery, K.R. and Corker, K. (1997), Computer modeling and simulation, in *Handbook of Human Factors and Ergonomics*, 2nd ed., Salvendy, G., Ed., Wiley, New York, pp. 1375–1408.

Navon, D. and Gopher, D. (1979), On the economy of the human processing systems, *Psychol. Rev.*, 86, 254–255.

North, R.A. and Riley, V.A. (1989), W/INDEX: a predictive model of operator workload, in *Applications of Human Performance Models to System Design*, Defense Research Series, Vol. 2, McMillan, G.R., Beevis, D., Salas, E., Strub, M.H., Sutton, R., and Van Breda, L., Eds., Plenum Press, New York, pp. 81–90.

Parks, D.L. and Boucek, G.P., Jr. (1989), Workload prediction, diagnosis, and continuing challenges, in *Applications of Human Performance Models to System Design*, McMillan, G.R., Beevis, D., Salas, E., Strub, M.H., Sutton, R., and Van Breda, L., Eds., Plenum Press, New York, pp. 47–64.

Pashler, H.E. (1998), *The Psychology of Attention*, MIT Press, Cambridge, MA.

Sarno, K.J. and Wickens, C.D. (1995), The role of multiple resources in predicting time-sharing efficiency, *Int. J. Aviation Psychol.*, 5, 107–130.

Welford, A.T. (1967), Single channel operation in the brain, *Acta Psychol.*, 27, 5–21.

Wickens, C.D. (1980), The structure of attentional resources, in *Attention and Performance VIII*, Nickerson, R., Ed., Lawrence Erlbaum Associates, Hillsdale, NJ, pp. 239–257.

Wickens, C.D. (2002), Multiple resources and performance prediction, *Theor. Issues Ergonomics Sci.*, 3, 159–177.

Wickens, C.D. and Hollands, J. (2000), *Engineering Psychology and Human Performance*, 3rd ed., Prentice-Hall, Upper Saddle River, NJ.

41

Critical Path Analysis for Multimodal Activity

Christopher Baber
University of Birmingham

41.1 Background and Applications

The idea of using time as the basis for predicting human activity has its roots in the early 20th century, specifically in the "scientific management" of Fredrick Taylor (although the idea of breaking work into constituent parts and timing these parts can be traced to the Industrial Revolution in the late 18th century). The basic idea of such approaches was to simplify work and then seek ways of making the work as efficient as possible, i.e., to reduce the time taken for each task step and, as a consequence, to reduce the overall time for the activity. Obviously, such an approach is not without problems. For example, Taylor faced Presidential Select Committee hearings in the U.S. when workers rioted or went on strike in response to the imposition of his methods. At a more basic level, there is no clear evidence that there is "one best way" to perform a sequence of tasks, and people often are adept in employing several ways. Thus, while the timing of task steps can be seen as fairly straightforward, the combination of the task steps into meaningful wholes is problematic.

In recent years, the increase in human–computer interaction (HCI) has prompted researchers to seek techniques that will allow modeling of the interaction between user and computer in order to determine whether a proposed design will be worth developing. One such set of techniques involves breaking activity into discrete tasks and then defining times for these tasks. Combining the tasks into sequences would then result in a prediction of overall time for the sequence. This is basically the approach used by the keystroke-level model (see Section 41.5, Related Methods).

Researchers have been investigating approaches that will allow them to combine discrete tasks in more flexible ways. One such approach draws on critical path analysis (CPA), which is a project management tool that is used to calculate the combination of tasks that will most affect the time taken to complete a job. (See Harrison [1997] or Lockyer and Gordon [1991] for more-detailed descriptions of CPA as a project management technique.) Any change in the tasks on the "critical path" will change the overall job completion time (and changes in tasks off the critical path, within limits, can be accommodated without problem). In the version presented in this chapter, the critical path is defined both in terms of time (so that a task will need to be completed before a subsequent task can begin) and modality (so that two tasks sharing the same modality must be performed in series).

Among the earliest studies that employed critical path analysis in HCI were those reported by Gray et al. (1993) and Lawrence et al. (1995). In this study, a telephone company wanted to reequip its exchanges with new computer equipment. Critical path analysis was used to investigate the relationship between computer use and other activities in call handling. It was shown that computer use did not lie on the critical path, so investment in such equipment would not have improved performance.

41.2   Procedure

There are seven steps in critical path analysis, as follows:

41.2.1   Step 1: Define Tasks

This could take the form of a task analysis, or it could be a simple decomposition of the activity into constituent tasks. Thus, the activity of "accessing an automated teller machine" might consist of the following nine task steps: (1) retrieve card from wallet; (2) insert card into ATM; (3) recall PIN; (4) wait for screen to change; (5) read prompt; (6) type in digit of PIN; (7) listen for confirmatory beep; (8) Repeat steps 6 and 7 for all digits in PIN; (9) wait for screen to change.

41.2.2   Step 2: Define Tasks in Terms of Input and Output Sensory Modality

Define tasks in terms of input and output sensory modality: manual (left or right hand), visual, auditory, cognitive, and speech. There will also be times associated with various system responses. Table 41.1 relates task steps to modality. The table might require a degree of judgment from the analyst, e.g., some task steps might require more than modality or might not easily fit into the scheme. However, taking the dominant modality usually seems to work.

41.2.3   Step 3: Construct Chart Showing Task Sequence and Dependencies between Tasks

As mentioned above, dependency is defined in terms of time, i.e., a specific task needs to be completed before another task can commence, and modality, i.e., two tasks in the same modality must occur in series. Figure 41.1 shows a chart for the worked example. The example takes the task sequence up to the

TABLE 41.1 Relating Task Steps to Modality

Task Step	Manual-L	Manual-R	Speech	Auditory	Visual	Cognitive	System
Retrieve card	X	X					
Insert card		X					
Recall PIN						X	
Screen change							X
Read prompt					X		
Type digit		X					
Listen for beep				X			
Screen change							X

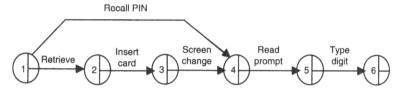

FIGURE 41.1 Initial part of CPA chart.

first digit being entered, for reasons of space (the other four digits will need to be entered, with the user pausing for the "beep" prior to the next digit, and the final screen change will occur for the sequence to be completed). In this diagram, an action-on-arrow approach is used. This means that each node is linked by an action, which takes a definable length of time. The nodes are numbered, and also have spaces to insert earliest start time and latest finish time (see step 5).

41.2.4   Step 4: Assign Times to the Tasks

Table 41.2 provides a set of times for the example. Appendix A provides a larger set of data. The diagram shown in Figure 41.1 can be redrawn in the form of a table, which helps in the following steps (see Table 41.2).

41.2.5   Step 5: Calculate Forward Pass

Begin at the first node of Figure 41.1 and assign an earliest start time of 0. The finish time for task from this node will be 0 + the duration of the task step; in this case, "retrieve card" takes 500 msec, so the earliest finish time will be 500 msec. Enter these values into Table 41.2 and move to the next node. The earliest finish time of one task becomes the earliest start time (EST) for the next task. A simple rule is to calculate EST on the forward pass. When more than one task feeds into a node, take the highest time. Repeat the steps until you reach the last node.

41.2.6   Step 6: Calculate Backward Pass

Begin at the last node and assign a latest finish time (in this case, the time will equal the earliest finish time). To produce the latest start time, subtract the task duration from the latest finish time. The time on the connection becomes the latest finish time (LFT) for that task. When more than one task feeds into a node, take the lowest time. Repeat the steps until you reach the first node.

41.2.7   Step 7: Calculate Critical Path

The critical path consists of all nodes that have zero difference between EST and LFT. In the example in Table 41.3, the task step on "recall PIN" has a nonzero float, which means that it can be started up to 320 msec into the other tasks without having an impact on total task performance. It is possible to

TABLE 41.2   Critical-Path Calculation Table: Forward Pass

Task Step	Duration	Earliest Start	Latest Start	Earliest Finish	Latest Finish	Float
Retrieve card	500 msec	0	—	500	—	—
Insert card	350 msec	500	—	850	—	—
Recall PIN	780 msec	0	—	780	—	—
Screen change	250 msec	850	—	1100	—	—
Read prompt	350 msec	1100	—	1450	—	—
Type digit	180 msec	1450	—	1630	—	—
Wait for beep	100 msec	1630	—	1730	—	—

TABLE 41.3 Critical-Path Calculation Table

Task Step	Duration	Earliest Start	Latest Start	Earliest Finish	Latest Finish	Float
Retrieve card	500 msec	0	0	500	500	0
Insert card	350 msec	500	500	850	850	0
Recall PIN	780 msec	0	320	780	1100	320
Screen change	250 msec	850	850	1100	1100	0
Read prompt	350 msec	1100	1100	1450	1450	0
Type digit	180 msec	1450	1450	1630	1630	0
Wait for beep	100 msec	1630	1630	1730	1730	0

perform the calculations using commercial software, such as Microsoft Project™. (Note, however, that Microsoft Project works in terms of days, hours, and months rather than milliseconds or seconds, which can produce some misleading calculations unless you set all of the parameters appropriately.) Alternatively, you can perform the calculations using Microsoft Excel™ (see Appendix B).

41.3   Advantages

- Structured and comprehensive procedure
- Can accommodate parallelism in user performance
- Provides reasonable fit with observed data

41.4   Disadvantages

- Can be tedious and time consuming for complex tasks
- Modality can be difficult to define
- Can only be used for activities that can be described in terms of performance times
- Times not available for all actions
- Can be overly reductionistic, particularly for tasks that are mainly cognitive in nature

41.5   Related Methods

The earliest, and most influential, model of transaction time was the keystroke-level model (Card et al., 1983). The keystroke-level model (KLM) sought to decompose human activity into unit tasks and to assign standard times to each of these unit tasks. Transaction time was calculated by summing all standard times. KLM represents a particular approach to HCI, which can be thought of as reducing humans to engineering systems, i.e., with standardized, predictable actions that can be assigned standard times. KLM has proved to be effective at predicting transaction time within acceptable limits of tolerance, e.g., usually within 20% of the mean time observed from human performance (Card et al., 1983; Olson and Olson, 1990). However, there are a number of criticisms that have been leveled at KLM, including the following:

1. KLM assumes expert performance, where the definition of an expert is a person who uses the most efficient strategy to perform a sequence of unit tasks and who works as fast as possible without error.
2. KLM ignores flexibility in human activity.
3. KLM ignores other unit-task activity or variation in performance.
4. KLM assumes that unit tasks are combined in series, i.e., that performance is serial and that there is no parallel activity.

The first criticism has been the subject of much discussion. Experts are users with a wide repertoire of methods and techniques for achieving the same goal, rather than people programmed with a single efficient procedure. Thus, a technique that reduces performance to a simple, linear description will obviously miss the variability and subtlety of human performance. Furthermore, nonexpert users will typically exhibit a wide variety of activity, and the notion that this activity can be reduced to "one best way" is questionable.

The main response to the second criticism is that the approach seeks to produce "engineering approximations" of human performance rather than a detailed description (Card et al., 1983). As such, the approach can be considered as a means of making task analysis "dynamic" (in the sense that times can be applied to unit tasks in order to predict the likely performance time of a sequence of such unit tasks). This shifts the debate from the utility of KLM per se and onto the inherent reductionism of task-analysis techniques. Recent discussions of human–computer interaction have tended to focus on the broad range of issues associated with the context of HCI, and these have argued against descriptions that focus too narrowly on one person using one computer. It is proposed that a requirement of user-modeling techniques ought to be that they can adequately reflect that range of activities that a user performs, giving the context of work. Consequently, KLM might be too narrowly focused on one user performing one task using one computer (following one best way of working), and alternative methods should be developed to rectify these problems.

The third criticism has been the subject of less debate, although there have been attempts to capture performance variation. Researchers have examined how systems respond to definable variability in performance. For example, speech-recognition systems can be defined by their recognition accuracy, and it is important to know how variation in recognition accuracy can influence system efficiency. Rudnicky and Hauptmann (1991) have used Markov models to describe HCI, working from the assumption that dialogues progress through a sequence of states, and that each state can be described by its duration. By varying state-transition parameters, it is possible to accommodate variation in the recognition accuracy of speech recognizers. Ainsworth (1988) employs a slightly different technique to the same end. His work models the impact of error correction and degradation of recognition accuracy on transaction time. We have used unit-task network models (specifically MicroSaint™) to investigate error correction and the effects of constraint on speech-based interaction with computers (Hone and Baber, 1999). Examination of the issues surrounding the combination of unit times for prediction of human performance raises questions concerning the scheduling of unit tasks and the coordination of activity. It also leads to concerns over how unit tasks might be performed in parallel (which relates to the fourth criticism).

41.6 Standards and Regulations

While there are no standards that specifically relate to modeling of human performance, recent ISO standards relating to human-centered system design propose that design targets can be useful aids in evaluating design, e.g., ISO 9241, ISO 13407, and ISO 9126. One way in which to define such targets is to produce a model describing performance of potential users, and then to use the predictions from the model as a benchmark for subsequent analysis.

41.7 Approximate Training and Application Times

There are no published figures defining the training times for the method. An undergraduate engineering student would be expected to develop a grasp of the basic principles of the technique within a couple of hours. As with most techniques, practice will improve both the comprehension of the method and also the speed with which it can be applied. In general, the most time-consuming aspect of the approach is the initial task analysis. The actual calculation of times is relatively easy; a calculation of 20 nodes would typically take around 30 min. It is possible to reduce this time still further by using software. There are commercially available project management packages that can be used to calculate critical paths. (It is worth noting, however, that these software packages tend to use hours as the smallest time unit, and that running calculations in which the figures are milliseconds may not lead to sensible or valid results.) If the description has more than 20 nodes (as a rule of thumb), the use of CPA and the resulting calculations and model might well be unduly complex.

41.8   Reliability and Validity

Baber and Mellor (2001) compared predictions using critical path analysis with the results obtained from user trials; they found that the fit between observed and predicted values had an error of less than 20%. This suggests that the approach can provide robust and useful approximations of human performance.

41.9   Tools Needed

The only tools needed are pencil and paper. Having said this, it is useful to have a collection of approximate times for generic tasks in hand prior to starting CPA.

References

Ainsworth, W. (1988), Optimization of string length for spoken digit input with error correction, *Int J. Man Machine Stud.*, 28, 573–581.

Baber, C. and Mellor, B.A. (2001), Modelling multimodal human–computer interaction using critical path analysis, *Int. J. Hum. Comput. Stud.*, 54, 613–636.

Card, S.K., Moran, T.P., and Newell, A. (1983), *The Psychology of Human–Computer Interaction*, LEA, Hillsdale, NJ.

Gray, W.D., John, B.E., and Atwood, M.E. (1993), Project Ernestine: validating a GOMS analysis for predicting and explaining real-world performance, *Hum.–Comput. Interact.*, 8, 237–309.

Harrison, A. (1997), *A Survival Guide to Critical Path Analysis*, Butterworth-Heinemann, London.

Hone, K.S. and Baber, C. (1999), Modelling the effect of constraint on speech-based human computer interaction, *Int. J. Hum. Comput. Stud.*, 50, 85–105.

ISO (1998), Ergonomics of Office Work with VDTs: Guidance on Usability, ISO 9241, International Standards Organization, Geneva.

ISO (1999), Human-Centred Design Processes for Interactive Systems, ISO 13407, International Standards Organization, Geneva.

ISO (2000), Software Engineering: Product Quality, ISO 9126, International Standards Organization, Geneva.

Lawrence, D., Atwood, M.E., Dews, S., and Turner, T. (1995), Social interaction in the use and design of a workstation: two contexts of interaction, in *The Social and Interactional Dimensions of Human-Computer Interaction*, Thomas, P.J., Ed., Cambridge University Press, Cambridge, U.K., pp. 240–260.

Lockyer, K. and Gordon, J. (1991), *Critical Path Analysis and Other Project Network Techniques*, Pitman, London.

Olson, J.R. and Olson, G.M. (1990), The growth of cognitive modelling in human–computer interaction since GOMS, *Hum.–Comput. Interact.*, 3, 309–350.

Rudnicky, A.I. and Hauptmann, A.G. (1991), Models for evaluating interaction protocols in speech recognition, *Proceedings of CHI '91*, ACM Press, New York, pp. 285–291.

Appendix A: Defining Standard Times for Unit Tasks

The times reported in this appendix have been collated from several sources, primarily Card et al. (1983) and Olson and Olson (1990).

Select and Drag Objects

Prior to moving a cursor or using a keyboard, the hand might need to be moved from its resting place to the interaction device. Values for this task range from 214 to 400 msec, and in this work a standard time of 320 msec is used.

One could use a standard time of 1100 msec to describe cursor positioning on a standard visual display screen. However, this is not without its problems, i.e., it fails to take into account the relationship between the size of the target and the distance that the cursor needs to move. This relationship is generally described using Fitts's Law:

$$\text{Movement Time} = a + b \log_2 (2A/W)$$

The equation essentially describes a straight line (i.e., in terms $y = mx + c$), and the values of a and b represent the intercept and slope of the line, respectively. If one could determine values for a and b for different interaction devices, then the equation could be used predictively. For example, values of –107 for a and 223 for b have been derived for a mouse, and 75 for a and 300 for b for a trackball. Using these values, assume that a cursor move 100 mm across the screen onto an icon of 20-mm width. Therefore, for the mouse the movement time should be: $-107 + 223 \log_2 (200/20) = 385$ msec; for the trackball the movement time should be: $75 + 300 \log_2 (200/20) = 1245$ msec.

Dragging objects can be considered in the same terms (the object being dragged, after all, simply becomes a big cursor). Values of 135 for a and 249 for b have been derived for a mouse dragging an object, and –349 for a and 688 for b for a trackball. Applying these data to the example above gives movement times of 1275 msec for the mouse and 1126 msec for the trackball.

Data Entry

The timing of data entry varies according to several factors, e.g., the interaction device used, the user's level of expertise, etc. For keyboard entry, values can range from 80 to 750 msec, depending on the level of expertise and task complexity. For the purposes of this work, a value of 200 msec is used as the standard time. For speech recognition, a value of 100 msec per phoneme, together with 50 msec for checking response, would be sufficient to describe data entry to a 100% accurate speech recognizer, i.e., entering a digit would take 350 msec. However, speech recognition can vary in terms of accuracy, and so the following equation is used to counter this problem:

$$\text{Time(RA)} = \text{time(100\%)} + \{[(100 - RA)/100] \text{ time(100\%)}\}$$

Thus, for the example above, if the recognition accuracy was only 75%, then the predicted time would be:

$$350 + \{[(100 - 75)/100]350\} = 437.5 \text{ msec}$$

Cognition and Perceptual Activities

Unit tasks are assumed to be initiated and verified by mental operators that take 50 msec. Typically, these mental operators are placed on either side of a main task. A general cognitive operator of 1350 msec accounts for the cognitive processes involved in complex tasks. While this is used as the standard time, it is possible that some choice tasks will have a standard time of 1760 msec, and some simple problem-solving tasks will have a standard time of 990 msec.

Perceptual responses require the user to switch attention from one part of a visual display to another (a standard 320 msec), recognition of familiar words or complex objects (314 to 340 msec), and reading of specific information, i.e., short textual descriptions (1800 msec). For auditory processing, a time of 2700 msec is proposed.

Deriving New Standard Times

Any unit tasks that do not have standard times can be derived by using the difference between the sum of known standard times and the observed transaction time. Thus, if one observed a transaction time of 3000 msec and predicted a time of 2500 msec + X (where X is the unknown value), then the standard time for X will be 500 msec.

Appendix B: Calculating Critical Path Using Solver in Excel

This section presents a way of using the Solver function in Excel to perform the calculations. This has the advantage of being faster than manual calculation, particularly for large networks, and of minimizing calculation error (although the result depends, of course, on accurate data entry).

Sketch a chart depicting the relationship between unit tasks. The chart has lettered nodes, with the links between nodes representing unit tasks and their duration. In Excel, create a table with the following headings: Activity, Duration, Start, End, Allowed, Float. This will be used for the calculation. The Activity column requires the name of each unit task, and the Duration column requires the standard times that have been employed in the model.

The Start column requires the name of the node to the left of the unit task. For example, task one starts with node A. Thus, the Task 1 cell will include +A_ to indicate that this node will be added to the equation. The End column applies the same logic. For example, task 1 ends with node B and thus has +B_ in the relevant cell. In the first cell of the Allowed column, enter an equation describing End – Start time. For example, if the Start column is in column C of the Excel spreadsheet, and the End column is in column D, the equation would be "= +C1–D1." Select this equation and drag over the remaining cells in the Allowed column. In the first cell of the Float column, use an equation describing "allowed – duration," i.e., the Duration column might be B in the Excel spreadsheet and the Allowed column might be E, and so the equation would be "= +E1–B1." Select this equation and drag over the remaining cells in the Float column.

The calculations are performed using the "solver" tool in the "tools" menu. This requires that the elements in the equation be defined. Create a separate table with two columns: one called Node and one called Endtime. In the Node column, enter the letter for each node in the network (using the format letter and underscore, e.g., A_). In the Endtime column, enter 0 in the first cell and leave the remaining cells blank. Each cell in the Endtime column needs to be named as elements in the equation. Use the "insert" menu, with the "name" and "define" menu items. This calls up a dialogue box that allows naming of cells. For the cell containing "0," use the name "A_." For the next cell, use the name "B_," and so on. Thus, if you have six nodes in the critical path network, the last cell in the Endtime column will be named "F_." This cell will be used as the Outcome variable in the equation. Use the "solver" menu item from the "tools" menu to call up a dialogue box and enter the following information. The "target cell" will be the name of the last cell in the Endtime column (in this example, it will be F_). The "changing" field will be all the cells in the Endtime column (in this example, A12:A18). The "constraints" will be that the Endtime and Float columns must exceed zero (in this instance, A12:A18> = 0, and F1:F6> = 0). Finally, select "options" and ensure that the "assume linear model" and "use automatic scaling" boxes are checked. Select "OK" and "solve." The equation should run and produce calculated times in the Allowed and Float columns, together with values in the Endtime column. Values of 0 in the Float column indicate unit tasks that lie on the critical path, and the final cell in the Endtime column indicates the total transaction time.

42

Situation Awareness Measurement and the Situation Awareness Global Assessment Technique

Debra G. Jones
SA Technologies, Inc.

David B. Kaber
North Carolina State University

42.1 Background and Application

Situation awareness (SA) can be formally defined as the "perception of the elements within a volume of time and space (Level 1), the comprehension of their meaning (Level 2), and the projection of their status in the near future (Level 3)" (Endsley, 1995). Because SA is an inferred mental construct, it is somewhat elusive to measure. The most common measures of SA are based on subjective perceptions or ratings of SA (e.g., the Situation Awareness Rating Technique [Taylor, 1990]), which, although easy to administer, have many shortcomings that can distort a participant's ability to reliably report their SA. For example, participants do not know what they do not know about a work environment, and subjective assessments at the end of a test trial are often affected by performance outcomes and memory decay (Jones, 2000). One alternative method for measuring SA is the Situation Awareness Global Assessment Technique (SAGAT). It provides an unbiased assessment of individual SA by directly querying operators regarding their current knowledge of the various elements in an environment and comparing their responses to the actual state of the environment.

The SAGAT involves temporarily stopping, or freezing, operator activity (usually in a simulation) and administering a battery of questions that target the operator's dynamic information needs (i.e., SA requirements) with respect to the domain of interest (Endsley, 2000). Although some research has identified the technique as being intrusive or disruptive to operator performance (e.g., Pew, 1995), other empirical work has shown that SAGAT stops do not interfere with performance and that task freezes can last as long as 5 to 6 min without concern for operator memory decay (Endsley, 1995). Endsley (1995) also observed that when SAGAT queries are appropriately designed and administered, there is no biasing of operator attention.

42.2 Procedure

42.2.1 Developing SAGAT Queries

The foundation of a successful SAGAT data-collection effort rests on the efficacy of the queries. Before queries can be developed, the operator's SA requirements must be defined. This task can be accomplished through a goal-directed task analysis (GDTA). Goal-directed task analysis is very similar to hierarchical task analysis (HTA), a methodology commonly used in human–computer interaction studies as a basis for interface design. The GDTA seeks to uncover operator goals in a particular domain, the decisions that must be made to achieve these goals, and the dynamic information requirements needed to support the decisions. (For more information on GDTA, see Endsley [1993].)

The SA requirements cover all aspects of perception, comprehension, and projection in the task domain and, consequently, form the basis for the SAGAT queries. In order to create a thorough yet concise query set, questions should be designed in such a way that an operator's response will encompass knowledge of multiple SA requirements. Further, the wording of the queries should be compatible with the operator's frame of reference and appropriate to the language of the domain.

42.2.2 Selecting SAGAT Queries

SAGAT queries can be presented in their entirety at each freeze, or the query set can be customized for the specific needs of the experimental design. The selection of which SAGAT queries to administer during a trial depends on several factors, including the objective of the study, the time available to answer the queries, the phase of the task at the time of a stop, and any limitations of the test bed. For example, if a flight simulation is not able to reflect real-time weather issues, questions related to the weather may not be appropriate.

Although not all questions have to be asked at each stop, care should be taken to avoid narrowing the question set excessively, which might allow operators to prepare for specific queries. Asking only questions that are high priority at a particular point in the simulated scenario should also be avoided so as not to cue operators to specific events and, thus, alter SA and performance. When time constraints prevent asking the entire set of queries, the SAGAT can involve a random subset at each freeze to prevent participants from preparing for specific questions.

Although SAGAT queries are randomized at each stop, the first question typically involves presenting participants with a map of the work environment (complete with appropriate boundaries and reference points) and asking them to fill in the locations of the various elements in the scenario. That is, the question is typically a recall or memory test. This map usually forms the basis for the other questions administered during the stop. By restricting subsequent questions to the elements in the situation of which participants are aware, an operator is not penalized more than once for a lack of awareness of specific information. Additionally, insight can be gained regarding the participant's understanding of the situation at that moment in time without unduly cueing participants about elements in the situation of which they are unaware.

In the event that a map of the task environment cannot be presented as part of an initial query, general questions on the status of the task or system can be posed with the option for participants to skip questions to which they cannot respond. In addition, if there are various phases to a task and certain SAGAT queries are only relevant to a particular phase, participants can be queried as to the phase of the task, and all subsequent questions can focus on operator comprehension and projection related to the specific phase (cf. approach used by Kaber et al., 2002a).

42.2.3   Implementing the SAGAT

Although no hard-and-fast rules govern SAGAT implementation, several guidelines have been suggested (e.g., Endsley, 2000):

1. The timing of SAGAT stops should be randomly determined and not occur only at times of increased activity.
2. A SAGAT stop should not occur within the first 3 to 5 min of an experimental trial (in order to enable the participant to develop a thorough understanding of what is going on in the scenario prior to being quizzed).
3. SAGAT stops should not occur within 1 min of each other.
4. Multiple SAGAT stops can be incorporated into a single scenario (as many as three stops within 15 min have been used with no adverse effects [Endsley, 1995]).
5. Over the course of an experiment, 30 to 60 samplings should be collected per SA query (across subjects and trials in a within-subjects experimental design) for each experimental condition (Endsley, 2000).

Other dependent measures (e.g., performance or workload measures) can also be incorporated into the same trial as the SAGAT. Note, however, that including trials without SAGAT stops allows for post hoc statistical analysis to ensure that the stops did not impact operator performance or workload and to accurately describe an effect if one did occur.

42.2.4   Administering the SAGAT

Prior to the first experimental trial, the participant should be fully briefed on SAGAT methodology, examples of the queries that will be asked, and the manner in which the queries will be presented. Additionally, several training trials should be provided in which SAGAT stops are administered. (Generally, three to five training trials are adequate [Endsley, 2000].) The participant should be comfortable with the procedure and understand the questions before the experiment begins in order to avoid any confusion as to the methodology or meaning of the questions during actual experimental testing.

The SAGAT is implemented by temporarily freezing a simulation at predetermined random times and blanking task displays (or simply covering them if blanking is not possible). A set of queries is then administered to the participant. When more than one participant is taking part in a study, all participants should be given the SAGAT battery at the same time, but no communication should be allowed between the participants. No outside information should be available to the participants during data collection that might influence their responses to SAGAT queries. Once a participant completes the SAGAT test battery or a predetermined quiz time interval has elapsed, the simulation is resumed from the exact point at which it was frozen for the SAGAT queries.

In order to accurately evaluate participant responses to SAGAT queries, the correct answers to the queries must be recorded at the same time the participant is completing the SAGAT battery. When possible, the data on the actual state of the system should be recorded directly by the simulator. For queries whose answers cannot be collected using a simulation computer (e.g., because it involves a higher-order SA question), a subject-matter expert, who is fully versed in the domain and the specific test scenario, should be used to supply the correct answers.

42.2.5 Analyzing Data

Responses to SAGAT queries are scored as either correct or incorrect, based on acceptable tolerance bands. Thus, the data are binomial and can be analyzed using methods appropriate for binomial data (e.g., the chi-square test). Because SAGAT data typically violates the assumptions of an analysis of variance (i.e., normality, linearity, etc.), applying a transformation to the SAGAT response measure (e.g., Y' = arcsine[Y]) may be necessary before a valid F-test can be conducted.

When evaluating SAGAT data, each query is typically analyzed separately. For example, the percentage of correct responses to Query 1 under Condition 1 should be compared with the percent correct responses to Query 1 under Condition 2. Generally, queries are not aggregated and then compared (i.e., all queries under Condition 1 vs. all queries under Condition 2). Because SAGAT data provide diagnostic information on an individual query basis, combining the data might mask trade-offs in SA between conditions (e.g., Jones and Endsley, 2000). However, other research has successfully aggregated results on SAGAT queries within levels of SA (i.e., perception, comprehension, or projection) in an effort to provide a more comprehensive assessment of a particular aspect of operator SA at a particular point in time (i.e., at a specific stop) (Endsley and Kaber, 1999). With this approach, correlation analyses are used to ensure consistency between query response patterns at the selected level of SA (e.g., operator comprehension). The objective is to ensure interquery reliability so that results on multiple SAGAT queries can be used to define operator SA.

42.3   Advantages

- Objectively measures SA; no subjective judgments required.
- Directly measures SA; no inferences based on performance or behavior.
- Maximizes data collection (i.e., numerous repeated measures) in minimal time.
- Provides diagnostic information on specific elements of SA as well as composite representations of all levels of operator SA.
- Collects SA information throughout activities; avoids pitfalls associated with collecting data after the fact.

42.4   Disadvantages

- Requires extensive preparation; delineating SA requirements through GDTA is time consuming and can be tedious.
- Requires access to simulation facilities.
- Requires ability to stop and restart simulation.
- May be intrusive to performance, depending upon task circumstances and the individual.
- Not well suited to actual operations.

42.5   Example

Figure 42.1 and Figure 42.2 show examples of a computerized presentation of two SAGAT queries developed by Endsley and Kiris (1995) based on a TRACON air traffic control simulation. A partial list of the SAGAT queries developed for this domain is shown in Table 42.1.

In analyzing response data to these queries, the number of correct and incorrect answers is initially counted for each participant across stops on a query-by-query basis (see Table 42.2). For this example, let us assume that ten aircraft were present in the simulation scenario at the point in time at which the first SAGAT stop (Stop 1) occurred. If Participant 1 correctly recalled the position of only eight of the aircraft, the score for that participant/stop/query would be eight out of ten. Many queries are simply either correct/incorrect and thus would have binary scores of 0 or 1 only (see scores for Queries 2 and

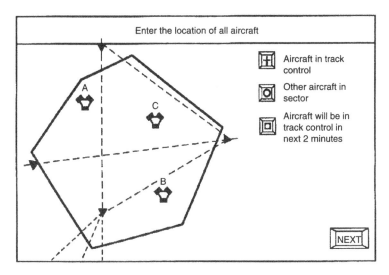

FIGURE 42.1   Query 1: Sector map for TRACON air traffic control.

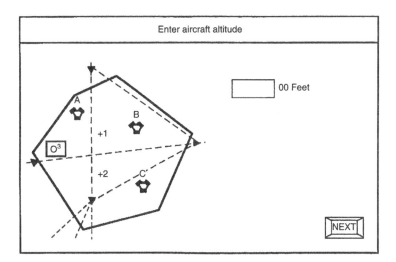

FIGURE 42.2   Additional query on TRACON simulation.

TABLE 42.1   SAGAT Queries for Air Traffic Control (TRACON)

1. Enter the location of all aircraft (on the provided sector map): aircraft in track control, other aircraft in sector, aircraft that will be in track control in next 2 min
2. Enter aircraft call sign (for aircraft highlighted of those entered in Query 1)
3. Enter aircraft altitude (for aircraft highlighted of those entered in Query 1)
4. Enter aircraft ground speed (for aircraft highlighted of those entered in Query 1)
5. Enter aircraft heading (for aircraft highlighted of those entered in Query 1)
6. Enter aircraft's next sector (for aircraft highlighted of those entered in Query 1)
7. Which pairs of aircraft have lost or will lose separation if they stay on their current (assigned) courses?
8. Which aircraft have been issued assignments (clearances) that have not been completed?
9. Did the aircraft receive its assignment correctly?
10. Which aircraft are currently conforming to their assignments?

Source: Endsley, M.R. and Kiris, E.O. (1995), *Situation Awareness Global Assessment Technique (SAGAT) TRACON Air Traffic Control Version User's Guide*, Texas Tech University Press, Lubbock. Reprinted with permission.

TABLE 42.2 Example Data: Participant

Condition 1	Stop 1	Stop 2	Stop 3	Stop 4	Stop 5	Total
Query 1	8/10	2/4	5/5	6/6	3/4	0.828
Query 2	1	0	1	1	1	0.8
Query 3	1	0	1	0	1	0.6
Etc.	—	—	—	—	—	—

TABLE 42.3 Example Data: Query Summed across Participants

Condition 1	Participant 1	Participant 2	Participant 3	Total
Query 1	0.828	0.765	1	0.864
Query 2	0.8	0.5	0.9	0.733
Query 3	0.6	0.7	1	0.767
Etc.	—	—	—	—

3 in Table 42.2). Once scores have been determined for each query and stop, the percentage of correct responses to a specific query is calculated across all stops.

Next, the scores are averaged across all participants for a single query (within an experimental condition), as seen in Table 42.3. These scores are then ready for comparison with other experimental conditions using an appropriate statistical technique.

42.6 Related Methods

For situations where an evaluation of SA is desired but work stoppages are not an option, a methodology utilizing real-time probes has shown promise. Real-time probes can be created using the same methodology used to develop SAGAT queries; the probes are administered verbally and do not require a halt in operator activity. Initial research has demonstrated that this metric has the potential to provide accurate assessments of operator SA (Jones and Endsley, 2000). However, correlation analyses of real-time probe results and measures of workload in early studies warrant further investigation before real-time probes are accepted as a valid metric of SA (Jones and Endsley, 2000).

42.7 Standards and Regulations

No standards or regulations exist with respect to the SAGAT or other metrics of SA.

42.8 Approximate Training and Application Times

Training time for the actual administration of the SAGAT is minimal. Participants simply need to be presented with example queries at appropriate times in a training trial. However, training time for researchers to learn methods to delineate SA requirements (e.g., GDTA) and transforming these requirements into queries may be somewhat more extensive.

42.9 Reliability and Validity

Numerous studies have been performed to assess the validity of the SAGAT (e.g., Endsley, 1995), and in each case, the evidence indicates that the method is indeed a valid metric of SA. The SAGAT has been shown to have a high degree of reliability (e.g., Endsley and Boldstad, 1994), to possess sensitivity to condition manipulations (Endsley, 2000), and to be effective across a variety of domains, including the following: air traffic control (Endsley et al., 2000); infantry operations (Matthews et al., 2000); commercial aviation (Endsley and Kiris, 1995; Kaber et al., 2002a); and teleoperations (Kaber et al., 2000; Kaber et al., 2002b). Further, the SAGAT has been found to be a method for assessing SA without influencing participant behavior despite the potentially intrusive nature of the methodology (i.e., interrupting simulations to query participants) and without falling prey to problems associated with recall (Endsley, 2000).

42.10   Tools Needed

The SAGAT can be administered via pencil and paper or by developing a program on a personal computer (PC). Collecting SAGAT data using a PC simplifies both query randomization and data analysis. That is, the computer can randomize query presentation automatically, score participant responses, and write the results into an output file.

References

Endsley, M.R. (1993), A survey of situation awareness requirements in air-to-air combat fighters, *Int. J. Aviation Psychol.*, 3, 157–168.

Endsley, M.R. (1995), Measurement of situation awareness in dynamic systems, *Hum. Factors*, 37, 65–84.

Endsley, M.R. (2000), Direct measurement of situation awareness: validity and use of SAGAT, in *Situation Awareness Analysis and Measurement*, Endsley, M.R. and Garland, D.J., Eds., Lawrence Erlbaum Associates, Mahwah, NJ, pp. 147–173.

Endsley, M.R. and Boldstad, C.A. (1994), Individual differences in pilot situation awareness, *Int. J. Aviation Psychol.*, 4, 241–264.

Endsley, M.R. and Kaber, D.B. (1999), Level of automation effects on performance, situation awareness and workload in a dynamic control task, *Ergonomics*, 42, 462–492.

Endsley, M.R. and Kiris, E.O. (1995), *Situation Awareness Global Assessment Technique (SAGAT) TRACON Air Traffic Control Version User's Guide*, Texas Tech University Press, Lubbock.

Endsley, M.R., Sollenberger, R., Nakata, A., and Stein, E. (2000), Situation Awareness in Air Traffic Control: Enhanced Displays for Advanced Operations, DOT/FAA/CT-TN00/01, Federal Aviation Administration William J. Hughes Technical Center, Atlantic City, NJ.

Jones, D.G. (2000), Subjective measures of situation awareness, in *Situation Awareness Analysis and Measurement*, Endsley, M.R. and Garland, D.J., Eds., Lawrence Erlbaum Associates, Mahwah, NJ, pp. 113–128.

Jones, D.G. and Endsley, M.R. (2000), Can real-time probes provide a valid measure of situation awareness? in *Human Performance, Situation Awareness, and Automation: User-Centered Design for the New Millennium*, Kaber, D.B. and Endsley, M.R., Eds., SA Technologies, Atlanta, GA.

Kaber, D.B., Onal, E., and Endsley, M.R. (2000), Design of automation for telerobots and the effect on performance, operator situation awareness and subjective workload, *Hum. Factors Ergonomics Manufacturing*, 10, 409–430.

Kaber, D.B., Endsley, M.R., Wright, M.C., and Warren, H.L. (2002a), The Effects of Levels of Automation on Performance, Situation Awareness, and Workload in an Advanced Commercial Aircraft Flight Simulation, Grant NAG-1-01002, NASA Langley Research Center, Hampton, VA.

Kaber, D.B., Wright, M.C., and Hughes, L.E. (2002b), Automation-State Changes and Sensory Cueing in Complex Systems Control, Award N0001401010402, Office of Naval Research, Arlington, VA.

Matthews, M.D., Pleban, R.J., Endsley, M.R., and Strater, L.G. (2000), Measures of infantry situation awareness for a virtual MOUT environment, in *Human Performance, Situation Awareness, and Automation: User-Centered Design for the New Millennium*, Kaber, D.B. and Endsley, M.R., Eds., SA Technologies, Atlanta, GA.

Pew, R.W. (1995), The state of situation awareness measurement: circa 1995, in *Experimental Analysis and Measurement of Situation Awareness*, Garland, D.J. and Endsley, M.R., Eds., Embry-Riddle Aeronautical University Press, Daytona Beach, FL, pp. 7–16.

Taylor, R.M. (1990), Situational awareness rating technique (SART): the development of a tool for aircrew systems design, in Situational Awareness in Aerospace Operations, AGARD-CP-478, NATO-AGARD, Neuilly-sur-Siene, France, pp. 3/1–3/17.

Team Methods

43

Team Methods

Eduardo Salas
University of Central Florida

The benefits of teams in organizations are undeniable in this day and age. While the trend toward teams in a variety of settings can be traced back decades, the current environment is even more receptive to team-based systems at a number of levels. Savoie (1998) reported that teams have risen dramatically, with reports of "team presence" by workers rising from 5% in 1980 to 50% in the mid 1990s. In fact, an American Society for Quality Control (ASQC) survey (1993) of over 1200 organizations reported that more than 80% of workers were a part of some work team within their organization, and over 84% of workers surveyed were a part of multiple teams throughout their organization (Fiore et al., 2001; Stough et al., 2000). With the influx of teams in the 1980s, managers and workers assumed that, with a collected group of people, there are more skills and resources available to solve complex problems. However, as the use of teams increased, it became evident that they are not automatically effective and that putting a group of workers together does not necessarily make a team. Therefore, the potential benefits of teams often were not realized. While there are numerous benefits to be gained by the implementation of teams, there are also a number of reasons teams fail. These failures can often be traced back to errors in team implementation and training.

As a consequence, research into understanding the factors contributing to team effectiveness and their training has increased over the past 20 years (see Salas and Cannon-Bowers, 2000). Industry globalization, emerging technology, safety issues, efforts to reduce human error, and higher productivity goals are increasingly pulling teams to the forefront of business, industry, aviation, medicine, and the military and other government organizations. Furthermore, the increasingly complex and competitive environments within industries have led to organizational interest in team topics (e.g., team performance measurement and training). In order to properly optimize team effectiveness, however, all the components that might affect team performance must be understood. It is important to remember that teamwork is a multidimensional and dynamic phenomenon that is sometimes deceiving and difficult to observe and capture (Cannon-Bowers et al., 1995). The process of managing team performance requires a deep understanding of the team competencies, the communication and task requirements, the team environments, and the team objectives and mission. Team training, task analysis, and performance measurement are necessary methods for organizations to optimize team functioning. And to accomplish all of that, methods are needed. This section of the handbook provides a related set of tools for both researchers and practitioners addressing some of these requirements.

While originating mostly at the individual level, team training has increasingly become more scientific (e.g., theory driven) and multilevel (e.g., individual- and team-level training interventions) (Cannon-Bowers et al., 1993; Cannon-Bowers et al., 1995; Tannenbaum et al., 1996). In addition, a number of different tools, methods, and strategies for training have evolved. However, it is important to point out that teamwork, team performance, and team performance measurement cannot be conducted haphazardly. Rather, it must be done systematically, using the tools and strategies gleaned from the team research done over the last three decades.

This section of the handbook begins with an overview of selected methods of team training (Chapter 44). The chapter covers the structure, tools, methods, and strategies for implementing and evaluating team training. The structure of team training, developed by Salas and Cannon-Bowers (1997), is explored as a means for understanding the nature and demands of team training. Teamwork knowledge, skills/ behaviors, and attitudes (i.e., KSAs) are discussed to provide insight into what teams think, do, and feel (Salas et al., 1992). A number of principles (e.g., regarding skill development, outcome measures, simulation, and guided practice) for properly integrating team training into an organization are provided, and some myths commonly held regarding training are addressed. The next three chapters (Chapters 45, 46, and 47) provide an overview of different training techniques. Chapter 45 discusses a more specific instance of team training in distributed simulation. The discussion uses "distributed mission training," a major distributed training program, to demonstrate how simulations can be used to train distributed teams. This approach can be used to apply technology to training large, multidisciplinary teams in conducting operations and decision making. While this approach has previously been primarily used by the military, the use of distributed simulation team training is increasingly finding its way into other disciplines (e.g., medicine, manufacturing, product design). Chapter 46 covers another simulation option: synthetic task environments (STE) for team training. Specifically, this chapter outlines a technique using STEs and uninhabited air vehicles (UAVs), designed for three interdependent distributed team members, that were investigated for the U.S. military. This technique is particularly good for training team behavioral and cognitive processes in complex settings. Communication protocols (e.g., dynamic data exchange and high-level architecture) are recommended, and software appropriate for the synthetic environments is described. Chapter 47 covers the event-based training (EBAT) procedure, a tool incorporated into training programs and simulations. EBAT entails the use of embedded trigger events within the training environment. Key aspects of this training technique include the practice and feedback components. The steps for implementing EBAT are discussed, and some applications for implementing and debriefing EBAT are discussed. This training strategy has been used effectively in a number of circumstances (Goldsmith and Johnson, 2002; Oser et al., 1999; Salas et al., 2002).

Chapter 48 focuses on team building, a method of improving team role clarification and fostering team cohesion. While team building is not the same thing as team training, it can be a useful addition to training strategies for improving certain team KSAs. In general, team training can be described as the systematic use of tools and methods derived from theory (e.g., shared-mental-model theory), while team building focuses on organizational environments and structures (Tannenbaum et al., 1992) by working to clarify the roles and responsibilities of team members. The components of team building (e.g., role clarification, goal setting, problem solving) are discussed. Although team building is a useful tool, there are varying degrees of support for this technique within the research community.

The next nine chapters (Chapter 49 through Chapter 57) cover assessment techniques related to teams. Chapter 49 discusses the difficulty of measuring team cognition and shared mental models and instead provides a comprehensive technique for measuring team knowledge of the task. The authors refer to the collective phenomenon of team cognition instead of often-used but sometimes ambiguous terms such as "shared mental models." As in other chapters, a research example of the procedure using UAVs for the military is provided. Steps for measuring team knowledge are explained. This technique for knowledge elicitation utilizes several established tools (e.g., pathfinder, Dearholt and Schvaneveldt [1990]; KNOT [knowledge network organizing tool] software). Chapter 50 outlines team communication analysis and discusses the utilization of team communication analysis in the study of team performance. As stated earlier, communication is a vital component of team performance and has been the focus of numerous studies. Frequency of communication and patterns of communication are discussed as approaches to studying team communication. The steps used to capture and analyze communication are explored, and different methods of pattern identification (e.g., lag–sequential and Markov chain) are cited. Chapter 51 provides an example of team decision-making procedures in team training as a method of measuring team situational awareness through the task-mutual-awareness questionnaire. The researchers show the relationships between task, team, and workload awareness. Examples of the survey are provided to give readers a clearer picture of the types of questions included, as well as an example of the scoring. Chapter

52 also covers a team knowledge assessment technique called "team decision requirement exercise," which attempts to make decision requirements explicit to teams while they are training. This assessment of teamwork is a method for training teams. The goal of this exercise is to make explicit the decisions that teams formulate. The authors provide steps for conducting the team decision requirement exercise, as well as advice for implementation (e.g., common errors, suggested changes, and why it is difficult). Chapters 53 and 54 provide information on two techniques for assessing team behavioral performance: TARGET (targeted acceptable responses to generated events) and BOS (behavioral observation scales), respectively. TARGET is based on "trigger events" within a scenario that requires team members to perform the behaviors of interest. The measurement is based on a checklist developed by subject matter experts (SMEs). BOS focuses on observable teamwork and often uses different measurement scales. Chapter 54 describes the ten major steps of BOS. Chapter 55 discusses team situation assessment (SA), a technique that attempts to evaluate how teams react to unexpected events by assessing team cognition and coordination. Team SA has an impact on team adaptability and should be incorporated into training. Chapter 56 addresses team task analysis, which is used in several steps of team training (covered in Chapter 44). Team task analysis includes a variety of methods for data collection and involves the identification of tasks relating to the job the team must accomplish. Ultimately, tasks are translated into KSAs or KSAOs (knowledge, skills, attitudes, and other competencies). The chapter provides further details on how to do this. Chapter 57 provides information about team workload analysis, which emphasizes the importance of workload in team performance. The workload measure focused on in this chapter is the NASA TLX (task load index), a popular validated measure used for individuals as well as teams.

Lastly, Chapter 58 outlines a technique for assessing the relationships in a network, which can inform researchers and practitioners about any constraints that may exist in an organization regarding relationships between entities (e.g., teams or individuals). Social network analysis, as outlined in the chapter, has four steps. The evaluation of networks also can be used to assess team performance efficiency.

This section of the handbook attempts to give the reader an overview of the different tools and techniques involved in team performance and team training. There are numerous methods, tools, and strategies that can be applied to team performance, measurement, and training. This section includes several of the most widely used and most tested methods. Each of the chapters in this section provides a piece of information to help in solving the puzzle of team performance. Ergonomists and human-factors professionals will find this section to be a useful toolbox.

References

ASQC (1993), Gallup Organization Survey, American Society for Quality Control, Milwaukee, WI.

Cannon-Bowers, J.A., Salas, E., and Converse, S. (1993), Shared mental models in expert team decision making, in *Individual and Group Decision Making: Current Issues*, Castellan, N.J., Jr., Ed., Lawrence Erlbaum Associates, Hillsdale, NJ, pp. 221–246.

Cannon-Bowers, J.A., Tannenbaum, S.I., Salas, E., and Volpe, C.E. (1995), Defining team competencies: implications for training requirements and strategies, in *Team Effectiveness and Decision Making in Organizations*, Guzzo, R.A. and Salas, E., Eds., Jossey-Bass, San Francisco, pp. 333–380.

Dearholt, D.W. and Schvaneveldt, R.W. (1990), Properites of Pathfinder networks, in *Pathfinder Associative Networks: Studies in Knowledge Organization*, Schvaneveldt, R.W., Ed., Ablex, Norwood, NJ, pp. 1–30.

Fiore, S.M., Salas, E., and Cannon-Bowers, J.A. (2001), Group dynamics and shared mental model development, in *How People Evaluate Others in Organizations*, London, M., Ed., Lawrence Erlbaum Associates, Mahwah, NJ, pp. 309–336.

Goldsmith T.E. and Johnson P.J. (2002), Assessing and improving evaluation of aircrew performance, *Int. J. Aviation Psychol.*, 12, 223–240.

Oser, R.L., Gualtieri, J.W., Cannon-Bowers, J.A., and Salas, E. (1999), Training team problem solving skills: an event-based approach, *Comput. Hum. Behav.*, 15, 441–462.

Salas, E. and Cannon-Bowers, J.A. (1997), Methods, tools, and strategies for team training, in *Training for a Rapidly Changing Workplace: Applications of Psychological Research*, Quiñones, M.A. and Ehrensstein, A., Eds., American Psychological Association, Washington, D.C., pp. 249–279.

Salas, E. and Cannon-Bowers, J.A. (2000), The anatomy of team training, in *Training and Retraining: A Handbook for Business, Industry, Government, and the Military*, Tobias, S. and Fletcher, J.D., Eds., Macmillan Reference, New York, pp. 312–335.

Salas, E., Dickinson, T.L., Converse, S.A., and Tannebaum, S.I. (1992), Toward an understanding of team performance and training, in *Teams: Their Training and Performance*, Swezey, R. and Salas, E., Eds., Ablex, Norwood, NJ, pp. 3–29.

Salas, E., Oser, R.L., Cannon-Bowers, J.A., and Daskarolis-Kring, E. (2002), Team training in virtual environments: an event-based approach, in *Handbook of Virtual Environments: Design, Implementation, and Applications, Human Factors and Ergonomics*, Stanney, K.M., Ed., Lawrence Erlbaum Associates, Mahwah, NJ, pp. 873–892.

Savoie, E.J. (1998), Tapping the power of teams, in *Theory and Research on Small Groups: Social Psychological Applications to Social Issues*, Vol. 4, Tindale, R.S., Heath, L., et al., Eds., Plenum Press, New York, pp. 229–244.

Stough, S., Eom, S., and Buckenmyer, J. (2000), Virtual teaming: a strategy for moving your organization into the new millennium, *Ind. Manage. Data Syst.*, 100, 370–378.

Tannenbaum, S.I., Beard, R.L., and Salas, E. (1992), Team building and its influence on team effectiveness: an examination of conceptual and empirical developments, in *Issues, Theory, and Research in Industrial/Organizational Psychology*, Kelley, K., Ed., Elsevier Science Publishers, Amsterdam.

Tannenbaum, S.I., Salas, E., and Cannon-Bowers, J.A. (1996), Promoting team effectiveness, in *Handbook of Workgroup Psychology*, West, M., Ed., John Wiley & Sons, Sussex, U.K., pp. 503–529.

44

Team Training

Eduardo Salas
University of Central Florida

Heather A. Priest
University of Central Florida

44.1 Background and Applications

Teams are of interest to a number of organizations in such fields as aviation, the military, and industry (Guzzo and Dickson, 1996). Over the past 20 years, the research community has invested a great deal of resources into the study of teams in complex environments (Salas and Cannon-Bowers, 2001). For example, research concerning teams in aviation arose through the examination of accidents and incidents in commercial and military flights. It is well documented that 60 to 80% of mishaps in aviation can be traced back to human error (NASA, 2002). One solution to this problem, as suggested by NASA, is training for aircrews or teams involved in teamwork. Because of its heavy reliance on teams (e.g., cockpit crews, ground crews), aviation has been at the forefront of team research from the beginning. This has led to some notable successes, such as crew or cockpit resource management (CRM). The military has also been interested in team research and training for decades (Salas et al., 1995). For example, due to incidents in the Persian Gulf in the late 1980s and early 1990s, teamwork and team training has risen to the forefront of military research in the U.S. and abroad (Cannon-Bowers and Salas, 1998). Taken together, much progress has been made in these areas. There are tools, methods, and strategies that must be applied in order to successfully design and deliver team training.

All teams are not created equal (Blickensderfer et al., 2000). In order to learn more about teams and training, researchers and practitioners had to agree on what exactly they were talking about. The accepted definition designates a team (as opposed to a group) as consisting of two or more people, dealing with multiple information resources, who work to accomplish some shared goal. Other characteristics of teams include meaningful task interdependency, coordination among team members, specialized member roles and responsibilities, and intensive communication. What follows applies to these kinds of teams.

44.2 Procedure

44.2.1 Structure of Team Training

The overarching goal of team training is to develop competencies to allow effective synchronization, coordination, and communication between team members (see Figure 44.1). The application of what we have learned over the past two decades can best be organized around four critical elements of team training, outlined by Salas et al. (2001). The following steps are extracted from that work.

44.2.1.1 Step 1: Determining the Team Training Requirements

Applications of tools are needed to design effective team interventions. Some, such as task analysis (see Chapter 56), help to determine what needs to be trained; others relate to the delivery of training (e.g., team simulation and exercise); while yet others deal with measurement, like the behavioral observation scales (BOS, see Chapter 54) or team error analysis. These tools are not isolated applications, but influence each other. The first step is to determine team training requirements using a variety of useful tools.

44.2.1.2 Step 2: Delineate Required KSAs

Research in the 1980s and 1990s emphasized the underlying processes that make teams work. These processes refer to the knowledge, skills, and attitudes (KSAs) of teams, i.e., team competencies (Salas and Cannon-Bowers, 2000). The first distinction to be made is that team competencies can be divided into two categories: teamwork competencies and taskwork competencies (Salas and Cannon-Bowers, 2001). Taskwork competencies refer to the knowledge, skills, and attitudes that the individual needs to accomplish his/her portion of the team task (e.g., operational skills). Although taskwork competencies are required, they alone are not sufficient to produce effective teams; teamwork competencies are also needed (Morgan et al., 1986). Essentially, team competencies refer to what team members need to know, how they need to behave, and what attitudes they need to hold (Salas, Smith-Jentsch, and Cannon-Bowers, 2001). Team knowledge competencies include accurate task models, team orientation, and shared task models. Team skill competencies consist of decision making, communication, and coordination. Team attitude competencies include team orientation, team cohesion, collective orientation, shared vision, and mutual trust. The competencies or KSAs provide the foundation for training goals and objectives. Essentially, in order to work in an interdependent, collaborative fashion, both taskwork and teamwork

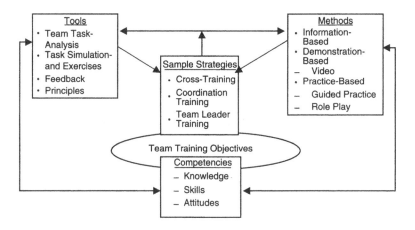

FIGURE 44.1 Structure of team training. (From Salas, E. and Cannon-Bowers, J.A. [1997], Methods, tools, and strategies for team training, in *Training for a Rapidly Changing Workplace: Applications of Psychological Research*, Quiñones, M.A. and Ehrenstein, A., Eds., American Psychological Association, Washington, D.C., pp. 249–279. With permission.)

skills must be encompassed in team training. These competencies are derived through the application of the tools mentioned in step 1.

44.2.1.3 Step 3: Select Appropriate Instructional Delivery Method

There are three basic methods for delivery of team training:

1. Information based
2. Demonstration based (e.g., video)
3. Practice based (e.g., guided practice)

Salas and Cannon-Bowers (1997) outline the components of these three methods. Information-based methods provide information (as opposed to practice or skill building) through applications such as lectures, slide presentations, or computer-based instruction. Information-based delivery serves as a passive method for conveying information about team goals, teammate roles, expectations, individual responsibilities, and task interdependencies, with its ultimate goal being to impart knowledge. Demonstration-based methods provide an opportunity for team members to observe the behaviors or actions that will be expected of them. In this method, the team member "is a passive observer of a situation, scenario, or exercise in which a number of learning points are embedded" (Salas and Cannon-Bowers, 1997). Demonstration-based methods (e.g., video) aim to develop shared mental models and expectations in team members. Practice-based methods are critical to team training. When done effectively, practice uses cueing, coaching, and feedback to foster understanding, organization, and assimilation of learning objectives. Although it is often misused, the practice method can be done properly by following guidelines developed in the literature (Cannon-Bowers et al., 1995; Salas et al., 1993; Swezey and Salas, 1992). Some methods commonly used in practice-based instruction include role playing, behavior modeling, computer-based simulations, and guided practice. The choice of methods (i.e., the media used to deliver instruction) is determined through the information derived from steps 1 and 2 as well as consideration of relevant organizational constraints and available resources.

44.2.1.4 Step 4: Design the Team Training Strategy

Tools, methods, and competencies combine to form the instructional strategies used for teams. Some of the strategies available for team training are shown in Table 44.1. The instructional method, tools, and content create the appropriate team training strategy. Furthermore, the task requirements, KSAs, and organizational factors (e.g., resources, policy, size) help determine which strategies should be applied. In most circumstances, a number of strategies may be appropriate, and any of the three methods (information, demonstration, and practice based) could be applied.

Training interventions also have some basic requirements outlined in the literature (Salas and Cannon-Bowers, 2001; Tannenbaum et al., 1992). Training interventions should have a theoretical basis and should be built around appropriate instructional strategies. Often, team training strategies combine the information, demonstration, and practice-based methods discussed above. Some strategies are more structured than others. For example, EBAT (see Chapter 47) requires the systematic identification of training opportunities to embed triggers, while team self-correction strategy is a naturally occurring phenomenon for effective teams where team members review events, correct errors, and plan for the future. A summary of some widely used team training strategies is shown in Table 44.1.

44.2.2 Team Training Principles

Combining the "structure" of team training with what has been learned over the last two decades, several principles have been delineated (Cannon-Bowers and Salas, 1998; Salas and Cannon-Bowers, 1997; Salas and Cannon-Bowers, 2001). Below is a representative sample:

- Team training should lead to teamwork skill development (e.g., leadership, adaptability, compensatory behavior).
- Both process and outcome measures are needed for teamwork diagnosis.

TABLE 44.1 Strategies for Team Training

Strategies	Definition	Benefit(s)	Source
Cross-training	Designed to develop commonly shared expectations about team functioning and a shared knowledge structure among team members; team members receive hands-on experience in performing other roles	Improves coordination and decreases process loss	Salas et al. (1997)
Event-based approach to training (EBAT)	Instructional strategy designed to structure training in complex, distributed environments. Provides guidelines and steps for training objectives, trigger events, measures of performance, scenario generation, exercise conduct and control, data collection, and feedback	When used appropriately, provides a meaningful framework with the opportunity to embed learning events into scenarios	Salas and Cannon-Bowers (2000)
Self-correction training	Instructional strategy that teaches team members how to evaluate and categorize their own and their team members' behavior to determine its effectiveness	Allows team members to provide constructive feedback within the team and correct deficiencies	Smith-Jentsch et al. (1998)
Stress exposure training	An integrated model of stress training that was influenced by a cognitive-behavioral approach to stress training, stress inoculation training (SIT); attempts to target coping strategies and stressors	Provides insight into links between stressors, the stress team members perceive they have, and their effect on performance	Driskell and Johnston (1998)
Team coordination training	Emphasizes underlying processes important to team effectiveness; used in crew resource management (CRM)	Improves team coordination, monitoring, and backup behavior	Entin and Serfaty (1999)
Scenario-based training	Similar to EBAT, scenarios contain trigger events that elicit targeted behaviors in contextually rich environments	Provides structure in complex environments	Osier et al. (1999)
Guided self-correction training	Emphasizes mutual performance monitoring, initiative, leadership, and communication; this strategy is driven by the team itself	Leads to better mental models of teamwork and more accurate situation assessment	Smith-Jentsch et al. (1998)
Simulation-based training and games	Widely used in business, education, and the military; range in cost, fidelity, and functionality; simulates real-world team environments and situations that can create instances of actual team content	Has the ability to mimic terrain, emergency situations, and dynamic environments	Cannon-Bowers and Salas (2000)
On-the-job training	Targets team members' procedurally based cognitive skills and allows team members to practice in actual job setting; the team interaction needed to accomplish actual job is practiced during training; this type of training is also typically self-guided	Training is provided in the same environment in which the team will be working	Ford et al. (1997)

- Simulations should allow team members to experience some alternative course of action.
- Response strategies, provided during training, should be linked to cues in the environment and from other team members.
- Team members should be provided sufficient opportunities to interact with novel environments in order to develop adaptive mechanisms.
- Training should be theoretically based.
- Team training should be more than "feel good" intervention.
- Guided practice is essential.
- Training should establish a mechanism to foster teamwork.
- Team training is embedded in an organizational system.
- Creation of an appropriate transfer climate for teamwork is essential.
- Team training should be ongoing.
- Both teamwork and taskwork competencies are needed for effective team functioning.

44.2.3 Team Training Misconceptions

Now that we know what team training consists of, it is important to clarify what team training is not. First of all, team training is not just a program that is applied. It entails several steps and considerations that must adapt to a variety of factors (e.g., organization, team members, individual characteristics, task characteristics). Furthermore, training evolves as it is implemented and is evolutionary in nature. Second, team training is not a "place," i.e., team members do not "go to" training. It is not some magical place that automatically arms team members with the KSAs they need. Third, team training is not a simulator. A simulator can be a tool used in training, but simulation alone cannot effectively train a team. Teams have additional training goals and content than a group of individuals, who will not need team competencies and KSAs. Lastly, team training, is not the same as team building (see Chapter 48). Team building, while effective in many circumstances, is not the same as team training, although it can be used in conjunction with team training.

44.3 Advantages

- Creates opportunities to practice and get feedback regarding teamwork KSAs.
- Provides a set of strategies to choose from (depending on requirements).
- Improves teamwork and related processes.
- Has a proven track record.

44.4 Disadvantages

- Not a simple structure to implement or design.
- Substantial effects on training found from individual differences.
- Consists of a number of processes that are unobservable and so must be assigned to observable behaviors or teachable skills.
- Notable gap between research and practice.
- Must be maintained by organizations.

44.5 Related Methods

Other methods of team instruction include team building, operational assignments, and work experience. Team building (e.g., a wilderness adventure that requires members of a team to depend on each other and promotes team development) consists of interventions that seek to improve the team's effectiveness (Tannenbaum et al., 1992). The main goal of team building is to improve team operations and processes by removing barriers, improving interpersonal relations, and clarifying roles. See Chapter 48 for more information on team building.

44.6 Approximate Training and Application Times

Training time varies depending on the strategy and method used.

44.7 Reliability and Validity

A series of studies evaluating team training strategies have offered a great deal of support for the validity of team training. Studies (Volpe et al., 1996; Blickensderfer et al., 1998) found a 12 to 25% improvement in team performance after cross-training interventions (Blickensderfer et al., 2000; Volpe et al., 1996). Regarding coordination training, a 12 to 15% team performance improvement was reported (Salas et al., 1999; Stout et al., 1997), while team leader training was found to lead to more participation, better briefings, and better teamwork behaviors following interventions (Tannenbaum et al., 1998).

Volpe et al. (1996) used four training conditions on a flight simulator and found that cross-training was a determinant of effective task coordination, communication, and performance.

Entin and Serfaty (1999) used naval officers in an adaptation-training experiment in which team members were trained to switch from explicit to implicit coordination and selection strategies. Researchers found that the adaptation training significantly improved performance.

Allen et al. (2001) examined how the fidelity of simulators affected the ability of teams to transfer training to real-world performance. They found that low functional fidelity resulted in longer problem solving and interresponse times.

Several studies developed and evaluated cockpit resource management (CRM) (Salas et al., 1999) in multiple aviation communities. CRM was found to be useful in identifying the needed KSAs, preparing the organization for the training, and applying sound instructional principles when designing training.

44.8 Tools Needed

The tools needed, as stated earlier, vary depending on a number of factors, including time, resources, content required, KSAs targeted, and organizational factors. Performance measurement, feedback, simulators, team task analysis techniques, and principles should be chosen *a priori* and used, as specified above, to structure methods, strategies, content, and competencies involved in team training.

References

Allen, R.W., Rosenthal, T.J., et al. (1999), Low Cost Virtual Environments for Simulating Vehicle Operation Tasks, TRB Paper 991136, Transportation Research Board, National Research Council, Washington, D.C.

Blickensderfer, E., Salas, E., and Cannon-Bowers, J. (2000), When the teams came marching home: U.S. military team research since World War II, in *Advances in Interdisciplinary Studies of Work Teams*, Beyerlein, M.M., Ed., Elsevier Science, New York, pp. 255–274.

Cannon-Bowers, J.A. and Salas, E. (1998), Team performance and training in complex environments: recent findings from applied research, *Curr. Directions Psychol. Sci.*, 7, 83–87.

Cannon-Bowers, J.A. and Salas, E. (2000), *Making Decisions under Stress: Implications for Individual and Team Training*, American Pschological Association, Washington, D.C.

Cannon-Bowers, J.A. and Salas, E. (2001), Team effectiveness and competencies, in *International Encyclopedia of Ergonomics and Human Feactors*, Karwoski, W., Ed., pp. 1384–1387.

Cannon-Bowers, J.A., Tannenbaum, S.I., Salas, E., and Volpe, C.E. (1995), Defining team competencies and establishing team training requirements, in *Team Effectiveness and Decision Making in Organizations*, Guzzo, R. and Salas, E., Eds., Jossey-Bass, San Francisco, pp. 333–380.

Driskell, J.E. and Johnston, J.H. (1998), Stress exposure training, in *Making Decisions under Stress: Implications for Individual and Team Training*, Cannon-Bowers, J.A. and Salas, E., Eds., American Psychological Association, Washington, D.C., pp. 191–217.

Entin, E.E. and Serfaty, D. (1999), Adaptive team coordination, *Hum. Factors*, 41, 312–325.

Ford, J. K., Kozlowski, S.W.J., Kraiger, K., Salas, E., and Teachout, M.S., Eds. (1997), *Improving Training Effectiveness in Work Organizations*, Lawrence Erlbaum Associates, Mahwah, NJ.

Guzzo, R.A. and Dickson, M.W. (1996), Teams in organization: recent research on performance and effectiveness, *Ann. Rev. Psychol.*, 47, 307–308.

Morgan, B.B., Glickman, A.S., Woodard, E.A., Blaines, A.S., and Salas, E. (1986), Measurement of Team Behaviors in a Navy Environment, NTSC technical report 86-014, Navy Training Systems Center, Orlando, FL.

NASA, Aviation Safety Program, System-Wide Accident Prevention, available on-line at http://avsp.larc.nasa.gov/program_swap.html; last visited on Dec. 2, 2002.

Oser, R.L., Gualtieri, J.W., Cannon-Bowers, J.A., and Salas, E. (1999), Training team problem-solving skills: an event-based approach, *Comput. Hum. Behav.*, 15, 441–462.

Salas, E. and Cannon-Bowers, J.A. (1997), Methods, tools, and strategies for team training, in *Training for a Rapidly Changing Workplace: Applications of Psychological Research*, Quiñones, M.A. and Ehrenstein, A., Eds., American Psychological Association, Washington, D.C., pp. 249–279.

Salas, E. and Cannon-Bowers, J.A. (2000), The anatomy of team training, in *Training and Retraining: A Handbook for Business, Industry, Government, and the Military*, Tobias, S. and Fletcher, J.D., Eds., Macmillan Reference, New York, pp. 312–335.

Salas, E. and Cannon-Bowers, J.A. (2001), The science of training: a decade of progress, *Annu. Rev. Psychol.*, 52, 471–499.

Salas, E., Cannon-Bowers, J.A., and Blickensderfer, E.L. (1993), Team performance and training research: emerging principles, *J. Wash. Acad. Sci.*, 83, 81–106.

Salas, E., Bowers, C., and Cannon-Bowers, J.A. (1995), Military team research: 10 years of progress, *Mil. Psychol.*, 7, 55–75.

Salas, E., Cannon-Bowers, J.A., and Johnston, J.H. (1997), How can you turn a team of experts into an expert team? Emerging training strategies, in *Naturalistic Decision Making*, Zsambok, C.E. and Klein, G., Eds., Lawrence Erlbaum Associates, Mahwah, NJ, pp. 359–370.

Salas, E., Fowlkes, J.E., Stout, R.J., Milanovich, D.M., and Prince, C. (1999), Does CRM training improve teamwork skills in the cockpit? Two evaluation studies, *Hum. Factors*, 41, 326–343.

Salas, E., Cannon-Bowers, J.A., and Smith-Jentsch, K. A. (2001), Principles and strategies for team training, in *International Encyclopedia of Ergonomics and Human Factors*, Karwoski, W., Ed., Taylor & Francis, London, pp. 1296–1298.

Smith-Jentsch, K.A., Cannon-Bowers, J.A., and Salas, E. (April 1998), The measurement of team performance, Master tutorial presented at the 13th annual meeting of the Society of Industrial and Organizational Psychology, Dallas, TX.

Stout, R.J., Cannon-Bowers, J.A., and Salas, E. (1997), Planning, shared mental models, and coordinated performance: an empirical link is established, *Hum. Factors*, 41(1), 61–71.

Swezey, R.W. and Salas, E. (1992), Guidelines for use in team-training development, in *Teams: Their Training and Performance*, Swezey, R.W. and Salas, E., Eds., Ablex, Norwood, NJ, pp. 219–245.

Tannenbaum, S.I., Beard, R.L., and Salas, E. (1992), Team building and its influence on team effectiveness: an examination of conceptual and empirical developments, in *Issues, Theory, and Research in Industrial/Organizational Psychology*, Kelley, K., Ed., Elsevier Science Publishers, Amsterdam.

Volpe, C.E., Cannon-Bowers, J.A., Salas, E., and Spector, P.E. (1996), The impact of cross-training on team functioning: an empirical investigation, *Hum. Factors*, 38, 87–100.

45

Distributed Simulation Training for Teams

Dee H. Andrews
U.S. Air Force Research Laboratory

45.1 Background and Applications

The complexity of today's work world has increased the need for teams and teamwork. Today's technology now allows large numbers of multidisciplinary workers to be placed in dispersed collaborative work teams to accomplish multifaceted tasks. Indeed, it is often the case that these complex work tasks can only be performed by teams whose members are in different locations.

Nowhere is this truer than military operations where multifaceted teams of thousands, and even hundreds of thousands, of operators and decision makers must interact in a time-constrained environment. While the military is a major user of distance technology (e.g., high-speed wide-area networks, capable routers, encryption of information), we are now seeing distance technologies applied to a host of applications broader than just the military. These applications include medicine, transportation, manufacturing, and warehousing.

This technology has not merely opened up new ways to conduct operations, but it has also provided a means by which operators and decision makers can be trained. By making use of distance technologies and the new methods for team training using those technologies, teams can be trained to requisite standards, even when they are dispersed. This training typically takes advantage of simulations, simulators, and computer-generated entities to create realistic synthetic work environments that can be used for training.

45.2 Example of Procedures and Equipment for Distributed Team Training

To illustrate the potential of distributed simulation for training, we will examine one application of this concept. The U.S. Air Force has embarked on a major distributed simulation-training program called distributed mission training (DMT). Originally, DMT was aimed at fighter pilot team training, but the

DMT concept has now been embraced by a variety of air force functions that need better team training. Examples include training of satellite operator teams, of decision makers, and of security police. In the fighter jet example, real-world training constraints (e.g., aircraft altitude restrictions, inability to frequently practice skills as a four-airplane team, security and safety restrictions on training ranges) are proving to be major impediments to training effectiveness for aircrew members around the world. The dangerous nature of fighter operations often prevents pilots from training as they intend to fight. DMT combines the following technologies to create a synthetic battlefield on which training can be done.

DMT consists of virtual (simulator) assets, constructive (computer generated) assets, and live (actual aircraft flying on training ranges) assets linked into the network. Specific components of DMT come from the following categories:

- Simulator cockpits of sufficient physical and functional fidelity to provide pilots with realistic interfaces that allow them to operate the simulator in much the same way that they operate their aircraft or weapons control console.
- Simulator visual systems that provide real-world visual cues. A variety of innovative approaches to solving this problem have been developed, including microlaser visual displays, PC-based high-end image generators, and photorealistic databases.
- Simulation threat systems that provide pilots with realistic threat actions and reactions. Threats could be enemy fighter aircraft or enemy surface-to-air missiles.
- Networking technologies that allow DMT entities at different locations to interact in a realistic fashion with almost no discernible latency. Latency refers to a delay between entities on the network. Obviously, a latency rate that is too high between fighter simulators could cause pilots to start behaving in an unrealistic manner.

Once the DMT network has been established, trainees can perform the same behaviors, with the same successes or failures, as they would in an actual combat environment. For fighter pilots, the core organization for combat is four aircraft teamed together, with one pilot designated as the lead pilot. This pilot is responsible for the overall performance of the four-pilot team. The lead pilot plans the mission; the lead pilot must get the mission tasking from leaders higher up; the lead must find out what is known about possible threats in the mission area; and the lead must coordinate with other friendly forces that will be in the mission area. The lead then briefs the other three pilots about the mission plan. They then get into their fighter simulators, which are networked to each other and to all the entities (simulators with live pilots and computer-generated friendly and enemy forces), and fly the mission.

During a 90-min session in the DMT network, the fighter pilots may get to fly five or six missions that can repeat the same scenario or consist of different scenarios. This compares to only one or two scenarios that they would be able to fly on live training ranges given the limited fuel supply. The pilots usually fly in a "four versus many" scenario, which means their four-fighter team fights a superior number of enemy threats. The missions can be as simple or as complex as the flight lead and the instructor pilot choose to make them. The instructor pilot sits at an exercise console during the training mission listening to the voice communication of the team and watching the virtual mission unfold on the console monitor. Typically, the instructor can see a "God's eye" plan view of the mission, and the instructor can see "out the cockpit window" scenes that the trainees are seeing as they fly.

Frequently, a fifth member of the mission team will be involved in the training mission. This member is the "weapons controller" who would operate a much larger radar system than the fighter pilots have. This controller has a better picture of the entire combat area than any of the pilots, and the controller provides information to the fighter pilots that they would not have without the controller's input. In training on live training ranges, it is seldom that the controller actually gets to go to the premission briefing and the postmission debriefing with the pilots. However, in DMT it is possible. This face-to-face contact is invaluable for training.

When the training mission is over, the instructor pilot, the four-ship team, and the controller go to a debriefing room, where they watch replays of the mission and discuss what went right and what went wrong. The replays include the radar picture of the mission, the "God's eye" plan view, and replays of the out-of-the-cockpit view that the pilots were watching during the mission. The instructor and trainee pilots can examine specific performance measurement data that were collected from the simulators while they flew the missions. These data are much more precise than the typical performance measurement data that the pilots get when they fly on the live training ranges. Because the modern flight simulator is based on digital computers, every control input can be measured precisely. In addition, tools are being developed to help the instructor exactly reconstruct the communications between the team members. This reconstruction is crucial because mission success or failure is usually dependent on the right thing being said at the right time. If the team flew over the network with other friendly forces in manned simulators, the team can debrief with them over the telephone or by video-teleconference. Finally, printouts of the mission and tapes of the communications can be given to the training teams to study after they leave the training center.

The graphic in Figure 45.1 shows how simulations and simulators for military aircraft can be networked to create realistic distributed team-training environments. However, one could substitute any number of nonmilitary entities (e.g., doctors, manufacturing employees, transportation operators) in place of the military objects to create a variety of distributed team-training environments. Figure 45.2 depicts a typical DMT network of virtual simulators, computer-generated forces, and live equipment assets.

45.3 Advantages

- Distributed simulation training creates job-relevant, realistic training and practice environments.
- Training can be done in these environments that can not be done in the real world due to safety, security, and cost factors.
- Performance measurement can be very precise.
- Feedback and remediation can be quickly and effectively accomplished.
- Resources (e.g., fuel, time, pollution) are saved.

FIGURE 45.1 Air force distributed mission training center. Four fighter cockpits inside wraparound 360-degree visual domes. An instructor/operator station outside of the four networked simulator domes.

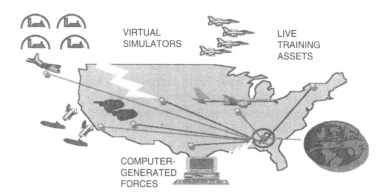

FIGURE 45.2 Typical DMT network.

45.4   Disadvantages

- May not be readily accepted by "gray beards/old-timers."
- Initial cost can be a disincentive if money cannot be saved by reducing the need for real-equipment training.
- Less than 100% fidelity of simulation could cause "negative training" if designers are not careful.
- Connectivity problems between simulation entities could cause a decrease in training effectiveness.
- Relatively immature state of capability to model human behavior means that computer-generated entities may lack the necessary fidelity.

45.5   Standards and Regulations

IEEE Std. 1278.2-1995, IEEE standard for distributed interactive simulation (DIS), provides tested protocols for linking simulations and simulators from various manufacturers and with various operating systems. Under the DIS standard, simulation entities send out protocol data units (PDUs) a number of times per second. These PDUs contain information such as the current and predicted (in the next few milliseconds) location and speed of each entity on the network. Each PDU goes to each simulation on the network, and then each simulation updates its database concerning the location and speed of all other entities on the network.

45.6   Approximate Training and Application Times

The time it takes to learn to use the distributed team-training tools depends on the experience of the trainee with the actual operational equipment and on the fidelity of the simulations. If the physical and functional fidelity of the simulators is high, an inexperienced operator on the actual equipment should be able to start training with the simulator immediately. However, high fidelity will probably cause a novice operator to have to spend considerable time learning the various subsystems of the simulation before s/he can start to train with the system in its entirety. Conversely, if the fidelity of the simulation is low, an experienced operator may need to spend considerable time learning the "sim-isms" of the simulator before meaningful training can take place. ("Sim-isms" refers to differences between the operational equipment and the simulator that need to be accommodated by the trainee, and their presence could lead to negative training.) A novice trainee may have less trouble learning the unique sim-isms because s/he does not yet have a good mental model of the how the operational equipment works. Because trainees typically do not use distributed simulation tools until they already are proficient with the operational equipment, novices typically are not exposed to the high-fidelity simulators associated with distributed simulation.

Another issue concerning training times involves the time it takes for an instructor to learn how to use the unique instructional tools associated with distributed simulation. Unique instructional characteristics and features, such as "freeze" (stopping the simulation during a training session for instruction), replaying the exercise, specific performance measurement, and other features may take some time for a new instructor to learn.

45.7 Reliability and Validity

Considerable work around the world has produced simulation systems and networks that are reliable. Latency rates (delay in the time an operator makes an input and the time the simulation system responds) are the key area of reliability concerns. Network tools have been developed to measure any reliability problems so that the operators of the networks can take appropriate action should delays become too noticeable.

Validity is a key issue with distributed simulation networks. The manner in which various entities on the network are represented is key to validity. The phrase "garbage in, garbage out" certainly applies here. If the simulations of equipment, human behavior, and the environment are not valid, as measured by comparison with their real-world counterparts, the ultimate training effectiveness could be less than desired.

45.8 Simulation Tools Needed

- Simulations/simulators of operational equipment (a visual representation of the real world, e.g., a truck driver's view of the highway, involves the use of image generators to create the image of the real world and visual displays on which the trainee will view the image)
- Instructor-operator stations
- Networking tools and network communication lines
- Database tools and real-world information (e.g., weather, terrain, communications, human behavior) that can be used to create realistic synthetic environments for team training
- Performance measurement tools for both the trainees and for the simulation network
- Network security tools, such as firewalls

References

Alluisi, E.A. (1991), The development of technology for collective training: SIMNET, a case history, *Hum. Factors*, 33, 343–362.

Andrews, D.H. and Bell, H.H. (2000), Simulation-based training, in *Training and Retraining: A Handbook for Business, Industry, Government and the Military*, Tobias, S. and Fletcher, J.D., Eds., Macmillan, New York.

Andrews, D.H. and Good, J.N. (2000), Overview of military aircrew training, in *Aircrew Training and Assessment*, O'Neil, H.F. and Andrews, D.H, Eds., Lawrence Erlbaum Associates, Mahwah, NJ.

Andrews, D.H., Waag, W.L., and Bell, H.H. (1992), Training technologies applied to team training: military examples, in *Teams: Their Training and Performance*, Swezey, R.W. and Salas, E., Eds., Ablex, Norwood, NJ.

Cannon-Bowers, J.A. and Bell, H.H. (1997), Training decision makers for complex environments: implications of the naturalistic decision making perspective, in *Naturalistic Decision Making*, Zsambok, C.E. and Klein, G.K., Eds., Lawrence Erlbaum Associates, Mahwah, NJ, pp. 99–110.

Crane, P. (1999), Implementing distributed mission training, *Commun. ACM*, 42, 91–94.

Moor, W.C. and Andrews, D.H. (2002), An empirical test of a method for comparison of alternative multiship aircraft simulation systems utilizing benefit-cost evaluation, in *Economic Evaluation of Advanced Technologies: Techniques and Case Studies*, Lavelle, J.P., Liggett, H.R., and Parsaei, H.R., Eds., Taylor & Francis, New York.

Morgan, B.B., Jr., Glickman, A.S., Woodard, E.A., Blaiwes, A.S., and Salas, E. (1986), Measurement of Team Behaviors in a Navy Environment, NTSC technical report TR-86-014, Naval Training Systems Center, Orlando, FL.

Orasanu, J. and Salas, E. (1993), Team decision making in complex environments, in *Decision Making in Action: Models and Methods*, Klein, G.A., Orasanu, J., Calderwood, R., and Zsambok, E.E., Eds., Ablex, Norwood, NJ, pp. 327–345.

Swezey, R.W. and Andrews, D.H., Eds. (2001), *Readings in Training and Simulation*, Human Factors and Ergonomics Society, Santa Monica, CA.

Weaver, J.L., Bowers, C.A., Salas, E., and Cannon-Bowers, J.A. (1995), Networked simulations: new paradigms for team performance research, *Behav. Res. Meth. Instruments Comput.*, 27, 12–24.

46

Synthetic Task Environments for Teams: CERTT's UAV-STE

Nancy J. Cooke
Arizona State University East

Steven M. Shope
U.S. Positioning Group, LLC

46.1 Background and Applications

Team tasks such as air traffic control, emergency response, and military command and control can be characterized as cognitively complex and embedded in a sociotechnical environment. Recent works have expressed opposition to studying behavior and cognition apart from the natural context in which it occurs (e.g., Hutchins, 1995; Zsambok, 1997), and these have prompted the search for a new research paradigm that preserves the task richness and complexity of work domain, yet provides more experimental control than typical field settings. STEs (synthetic task environments), or "research tasks constructed by systematic abstraction from a corresponding real-world task" (Martin et al., 1998), offer a solution. The objective of STEs is to be able to reproduce behavior and cognitive processes associated with these complex settings in the laboratory, where some experimental control and measurement capabilities can be preserved. An STE is a task environment in which a number of different task scenarios can be simulated. Compared with simulations, STEs tend to be more task-centric (i.e., faithful to behavioral and cognitive dimensions of task) and less equipment-centric.

A number of team STEs have recently been developed and funded, primarily by the U.S. military. In this chapter, we describe a specific STE for teams and use it to illustrate the methodology involved in developing STEs. The STE described here is based on the task of ground operations of a UAV (uninhabited air vehicle) by three interdependent individuals: the AVO (air vehicle operator), the PLO (payload operator), and the DEMPC (data exploitation mission planning and communications operator) or navigator and mission planner. These individuals work together to accomplish the goal of navigating the UAV to a position to take reconnaissance photos of designated targets. We use this STE at Arizona State University's CERTT (Cognitive Engineering Research on Team Tasks) Laboratory to study team cognition in this rich context.

46.2   Procedure

The procedure for designing the UAV-STE involved five steps, which we believe generally pertain to the development of STEs for other task domains:

1. Understanding the task in the field of practice
2. Understanding other constraints on the STE
3. Abstraction of task features
4. STE design
5. Validation of the STE

46.2.1   Step 1: Understanding the Task in the Field of Practice

The first step in designing a synthetic task is to acquire a full understanding of the task in the field of practice. There are a number of methods that can be used to achieve such an understanding, including interviews with subject-matter experts, cognitive task analysis, naturalistic observation, and perusing technical manuals and other documentation. More generally, the knowledge-elicitation techniques (Cooke, 1994) that are used to elicit and model the knowledge of domain experts can be helpful in this regard. It is likely that some research constraints will provide a filter for the information acquired under this step. For instance, because we were interested in developing an STE for the purpose of understanding team cognition, we attended most closely to aspects of the UAV task that were especially relevant to team performance or cognition.

For our project, information from the field of practice was gleaned largely through a cognitive task analysis done on UAV operations (Gugerty et al., 1999), information from actual operators, observation of an existing UAV-STE at Williams AFB in Mesa, AZ (Martin et al., 1998), and discussions with various investigators involved with these projects. Other information was obtained from various Internet sites and unpublished reports, especially training documentation for the UAV.

46.2.2   Step 2: Understanding Other Constraints on the STE

Exactly what features of the actual task are abstracted for the STE version of the task depends not only on understanding the task and work domain, but also on an awareness of the objectives of, and constraints on, the STE. As mentioned under the previous step, research objectives (e.g., understanding and measuring team cognition) provide one filter for task understanding and similarly serve as a filter for further selection of those features that will be represented in the STE. Another objective in our case was that the STE provide an experimenter-friendly research test bed. In other words, STE scenarios should be easy to modify; experimental manipulation and control should be facilitated; and cognitive and behavioral measurement should be supported. These kinds of objectives also serve as filters for the abstraction of task features.

Some constraints are more practical in nature. Some of ours included the expertise and scheduling constraints of the university participant population and various technological constraints in replicating task features in the CERTT laboratory. The CERTT facility was developed in parallel with the STE for the purpose of providing a state-of-the-art environment for STEs that are focused on the measurement of team cognition. In total, the lab contains four interconnected participant consoles and a large experimenter control center. Each participant console contains two Windows NT machines connected to other CERTT computers through a local area network (Figure 46.1 displays the six screens). Participant consoles contain video monitors, headsets, and an intercom. The experimenter control center contains a variety of monitoring and data recording equipment, including audio recording equipment, an intercom, headsets, video recording, and performance-monitoring software. Thus, this hardware configuration provided additional constraints on the STE design.

46.2.3 Step 3: Abstraction of Task Features

Once constraints imposed by research objectives and pragmatic considerations of the research setting are identified, features are abstracted from the actual task that are within the bounds of these constraints. To this end, we identified those aspects of the UAV ground-control task that we planned to emphasize (or maybe even exaggerate) in our STE. For instance, in the field of practice, background knowledge and information relevant to UAV operations are *distributed* across team members. In other words, some UAV knowledge is uniquely associated with individual team members. This feature of distributed knowledge and information across a team was one that seemed relevant to team cognition.

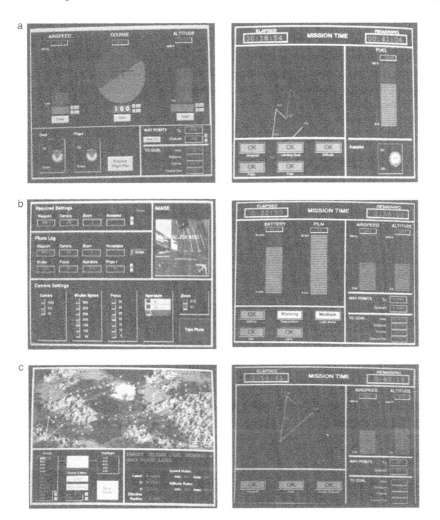

FIGURE 46.1 Two displays for each of the three participants in the CERTT UAV-STE. (a) Air vehicle operator screens; (b) Payload operator screens; (c) DEMPC screens.

Other team task features that we abstracted included knowledge and information-sharing requirements, extensive planning under multiple constraints, and dynamic replanning. In other cases, we deliberately chose to alter aspects of the original task due to constraints. For example, in the interest of minimizing training time, we altered the human–computer interface substantially. The original interface is highly complex and can require a lengthy acquisition period (over a year) to reach asymptotic levels of performance. As our objectives centered on performance of a team with members already versed in their individual interface operation, the replication of the interface was not only unnecessary (as functionality of the interface was preserved), but would be a hindrance to collecting data on teamwork and team cognition in our university setting.

46.2.4   Step 4: STE Design

The next step in the design process consists of transforming the ideas, constraints, and abstracted features generated in step 3 into a prototype STE. We chose to use cardboard and paper mock-ups, graphics software for screen designs, and extensive functional specifications. We started by determining the minimum set of functions that each team member and experimenter would be required to perform, and then began the task of designing the interface. Next, we developed a list of software data structures, dynamic variables, startup variables and files, and data stores needed to support the design. Functional specifications were drawn up in detail and provided information regarding interface displays and controls, measurement requirements, functional properties of subsystems, and information flow between startup files and other subsystems. Functional specifications were presented to the programmers, who reviewed them and provided feedback on technological constraints and software architecture that would impact the basic design. This process began a series of discussions between STE designers and programmers and concomitant feedback–redesign iterations.

46.2.5   Step 5: Validation of the STE

We view STE validity as a separate, but related, issue to the validity of the measures and experimental results developed in its context. However, demonstrations of the validity of the latter support the validity of the UAV-STE. The validity of an STE really concerns the issue of fidelity to the field of practice. Our view is that fidelity is multidimensional in that a task can be faithful to some aspects of the field, but not others. Thus, the UAV-STE should be faithful to the behavioral and cognitive aspects of the team task of operating a UAV.

Validity of this type can be assessed in various ways. First, to the extent that the cognitive task analysis and its translation into the STE are valid, we can infer that the UAV-STE that results is valid. So, any validation of the cognitive task analysis reflects on the STE validity. Second, face validity can be determined through expert assessment. Assessment of the face validity of the CERTT UAV-STE by one expert UAV operator and a host of other individuals knowledgeable in UAV operations has provided preliminary support for STE validity. Third, to the extent that the STE is faithful to a particular aspect of the actual task, then domain experts should excel on those aspects of the synthetic task. To this end, we are interested in benchmarking expert UAV operator performance on our UAV-STE to determine the extent to which the STE is faithful to team cognition.

46.3   Advantages

- STEs allow for experimental control.
- STEs facilitate measurement capabilities.
- STEs can be faithful to the dimensions of the task.
- STEs provide a rich test bed.
- Connecting STEs over the Internet enhances scope.

46.4　Disadvantages

- STEs are more complex than traditional lab settings, and so some of challenges of field research remain.
- STEs are effective research tools to the extent that they are valid representations of the task.
- STEs can be expensive to build and maintain, though less so than traditional high-fidelity simulators.
- STEs without the experimental infrastructure are ineffective research tools.

46.5　Related Methods

Because STEs abstract aspects of a task that are congruent with research objectives and other constraints, several very different STEs based on the same real task can emerge by virtue of these distinct filters. Such is the case with the UAV task, in which a variety of STEs have been developed that focus on various cognitive skills of individuals (e.g., Gugerty et al., 1999; Martin et al., 1998), and others, such as the one described in this paper, that focus on team cognition. Other related STEs exist that are based on other military command-and-control tasks such as Team Argus (Schoelles and Gray, 1998).

46.6　Standards and Regulations

46.6.1　Communications Protocols

The current CERTT Lab uses a combination of dynamic data exchange (DDE) for intralab communications and distributed interactive simulation (DIS) protocols for interconnecting the CERTT Lab to outside entities. Both are industry standard protocols. In next generation CERTT designs, we plan to use DIS almost exclusively. DIS seems ideally suited for interconnecting remote STE sites as well as intralab communications. HLA (high-level architecture) is also an acceptable protocol, but it is perhaps more stringent than can be presently justified for STE operations.

46.6.2　Software

We presently use custom user-defined objects in the task software that were developed in the Rapid™ Development Environment. This makes it difficult for outside parties to develop their own version of a participant console. One of our next-generation objectives is to use more ActiveX and other industry-standard objects to allow others to readily develop their own software modules for participating in the STE.

46.7　Approximate Training and Application Times

The hardware for the UAV-STEs was completed in one year, and the software was completed in another year. However, most of this time was spent in initial design. Given off-the-shelf software and hardware, the entire configuration could be set up in 90 days. Training experimenters to use the STE takes approximately 40 hours (most of it on-the-job training). Conducting experiments requires approximately 2 hours of participant training, followed by a series of 40-min missions per team in which teams typically reach asymptotic levels of performance after the fourth mission.

46.8　Reliability and Validity

Determining STE validity is an integral part of the procedure of designing an STE. Methods for determining validity were discussed in the earlier section. Thus far, validity of UAV-STE has been assumed based on its development using a cognitive task analysis of the actual task and several reports of sufficient face validity. In addition, research results established in this context have demonstrated predictive validity. Validity in terms of benchmarked expert performance remains to be evaluated.

46.9   Tools Needed

In the initial CERTT Lab design, we used a combination of custom-built and off-the-shelf hardware. We used custom hardware when cost and/or product capability did not meet our design objectives. Currently, our philosophy is to use off-the-shelf hardware almost exclusively, especially for the participant consoles. The driving force behind this line of thinking is to facilitate outside and remote research groups participating in CERTT Lab experiments and operations.

In the initial CERTT Lab design, we used Windows-based software that would run on any PC using Windows NT or higher operating systems. This included not only the STE task and participant applications, but also the experimenter control station software and the embedded measures. We plan to stay with Windows-based software, especially in view of the availability of image generation and virtual reality software components for Windows machines.

References

Cooke, N.J. (1994), Varieties of knowledge elicitation techniques, *Int. J. Hum.–Comput. Stud.*, 41, 801–849.

Gugerty, L., DeBoom, D., Walker, R., and Burns, J. (1999), Developing a simulated uninhabited aerial vehicle (UAV) task based on cognitive task analysis: task analysis results and preliminary simulator data, in *Proceedings of the Human Factors and Ergonomics Society 43rd Annual Meeting*, Human Factors and Ergonomics Society, Santa Monica, CA, pp. 86–90.

Hutchins, E. (1995), *Cognition in the Wild*, MIT Press, Cambridge, MA.

Martin, E., Lyon, D.R., and Schreiber, B.T. (1998), Designing synthetic tasks for human factors research: an application to uninhabited air vehicles, in *Proceedings of the Human Factors and Ergonomics Society 42nd Annual Meeting*, Human Factors and Ergonomics Society, Santa Monica, CA, pp. 123–127.

Schoelles, M.J. and Gray, W.D. (1998), Argus, a system for varying cognitive workload, in *Proceedings of the Human Factors and Ergonomics Society 42nd Annual Meeting*, Human Factors and Ergonomics Society, Santa Monica, CA, p. 1147.

Zsambok, C.E. (1997), Naturalistic decision making and related research issues, in *Naturalistic Decision Making*, Zsambok, C.E. and Klein, G., Eds., Lawrence Erlbaum Associates, Mahwah, NJ, pp. 3–16.

47

Event-Based Approach to Training (EBAT)

Jennifer E. Fowlkes
Chi Systems, Inc.

C. Shawn Burke
University of Central Florida

47.1 Background and Applications

The event-based approach to training (EBAT) is used to guide the design of scenario- or simulation-based training. It is usually incorporated into a training program encompassing information, demonstration, practice, and feedback, in which EBAT would be used to structure the practice and feedback portion. There are a number of variations of EBAT, but all seek to engineer training opportunities by systematically identifying and introducing events within training exercises that provide known opportunities to observe behaviors that have been targeted for training. Explicit links are maintained between training objectives, exercise design, performance assessment, and historical performance data, focusing training on the objectives and producing standardized training opportunities. While the methodology is generic and can be applied to train many types of skills in many types of contexts, the training products produced are highly context specific.

Examples of training approaches that incorporate EBAT include aircrew coordination training (ACT) (Salas et al., 1999), team dimensional training (TDT) (Smith-Jentsch et al., 1998), and the rapidly reconfigurable line-oriented flight training (RRLOE) (Jentsch et al., 1999).

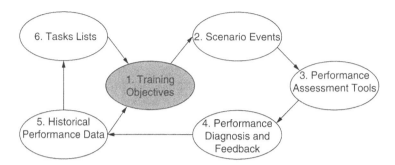

FIGURE 47.1 Steps for implementing EBAT. (Adapted from Zachary et al. [1997], Advanced embedded training concepts for chipboard systems, in *Proceedings of the 19th Interservice/Industry Training, Simulation and Education Conference*, CD-ROM, National Training Systems Association, Arlington, VA, pp. 670–679. With permission.)

47.2   Procedure

Implementing EBAT involves a six-step sequence shown in Figure 47.1 and discussed below. The sequence includes:

1. Identification of individual and team training objectives for a specific training event
2. Translation of training objectives into representative scenario events that provide opportunities for trainees to demonstrate competencies related to the training objectives
3. Development of performance criteria that are incorporated into event-based performance measures
4. Development of procedures that enable instructors (or systems) to observe and provide feedback on team processes and outcomes
5. Linking performance data to a historical performance database, allowing individual and team strengths and weaknesses to be diagnosed, and serving to focus future training events
6. Updating master task lists based on empirically identified training needs

47.2.1   Step 1: Identify Training Objectives

The identification of training objectives for a specific training event is the first step in the EBAT sequence. These will drive the development of the scenario events and ultimately the feedback that is provided to the trainees. Sources of training objectives include source documents (e.g., tasks lists, standard operating procedures), task analyses, and cognitive task analyses. In addition, as shown in Figure 47.1, historical data for individual and team strengths and weaknesses can be used to guide selection of training objectives for an event. Other considerations include the level of experience of the trainees, the input of training officers, and the requirements obtained from team leaders (e.g., supervisors, commanders in the military) (Stretton and Johnston, 1997).

47.2.2   Step 2: Develop Scenarios and Events

Once training objectives have been identified, scenarios and trigger events are developed to permit assessment of whether the competencies targeted for training have been achieved. Preplanned scenario events, deliberately linked to training objectives, provide known opportunities for participants to perform to those objectives, eliminating chance from driving what can be trained and assessed. Preplanned scenario events also serve to standardize training events across teams and individuals.

   Regarding the nature of events, in general:

- Events should present realistic situations.
- Events of appropriate difficulty should be presented at multiple points in the scenario.

- Events should provide multiple opportunities for exhibition of targeted objective or behavior.
- The introduction of events should be managed in some way. This can occur by scripting, in which the timing and nature of the events is precisely specified (e.g., by a master events list). It can also occur by designing scenarios so that events occur naturally, the result of the interactions among participants and simulated entities. In this case, the ability of the system or human to recognize events is more difficult.
- Events can include routine situations such as task phase. Events can also serve as prompts to observe behaviors that occur infrequently or that might not be observable (Fowlkes et al., 1998). For example, Johnston et al. (1995) use events to provide opportunities to observe the decision-making processes of shipboard teams.
- Sequential dependencies among events should be avoided; that is, responses to one event should have minimal impact on opportunities to respond to subsequent events in the scenario (Fowlkes et al., 1998).

47.2.3 Step 3: Develop Performance Measures that Capture Responses to Events

Performance measurement tools are developed around the scenario events to provide links between measurement objectives and diagnosis of performance. Johnston et al. (1995) described a variety of individual and team process measures designed to capture responses to events for navy shipboard teams. These included behavioral observation scales, assessment of latencies to events and errors, and ratings. Although the above measures were used in regard to navy shipboard teams, they can be adapted for use in a variety of settings. Additionally, Fowlkes et al. (1994) developed a checklist approach to capture responses to events (i.e., TARGETs, see Chapter 53). In general, indices of acceptable responses (e.g., behaviors, latencies) to events can be developed *a priori* and incorporated into measurement using subject-matter-expert input as well as task lists. This reduces the load on instructors during a training event, in that the judgment about what constitutes good or acceptable performance has already been accomplished either by the instructor or his/her peers.

47.2.4 Step 4: Develop Tools to Support Performance Diagnosis and Feedback

Instructors must observe performance during the event and diagnose performance strengths and weaknesses. This can be done manually, aided, or automated. The event-based observation aids this process, as it focuses the instructor's attention to events important for training. Tools enable instructors to track events, alert them to the occurrence of events, and facilitate data presentation and feedback after an event. The shipboard mobile aid to training and evaluation (ShipMATE) is an application, implemented on a handheld computer, that assists instructors in preparing for, implementing, and debriefing event-based training for shipboard teams (Zachary et al., 1997). Similarly, RRLOE performs similar functions for training commercial airline pilots (Jentsch et al., 1999). Zachary et al. (1997) described an advanced embedded training system in which automated event-based performance assessment would also occur.

47.2.5 Steps 5 and 6: Link Performance Data from Single Training Exercise to Historical Database and Task Lists

Ideally, in the implementation of EBAT, performance data from a single training event can be linked to a historical database that maintains past performance of teams and individuals. These types of data can be sorted by team, individual, training objective, time, and so forth. Observed deficiencies at the individual or team level can be used to focus future training opportunities. Finally, trends can be gleaned from historical databases to identify training shortfalls that can be translated into revised tasks and objectives.

47.3 Advantages

- Maintains links from measurement to training objectives, enhancing the relevance of training and feedback.
- Facilitates observation and diagnosis of complex performance by focusing measurement and feedback on specific events.
- Supports standardized training by controlling task content.
- Produces diagnostic performance scores.
- Provides a generalizable methodology.

47.4 Disadvantages

- Scenario development is labor intensive.
- Requires the support of domain experts for development.
- Difficult to develop scenarios and events that are of equal difficulty.
- Difficult to avoid sequential dependencies among events for large team-training exercises.
- With large exercises, instructor aids and automated tools are nearly essential.

47.5 Related Methods

Rapidly reconfigurable line-oriented evaluation (RRLOE) was developed to facilitate event-based training for the FAA. In this method, event sets, linked to training objectives, are used to automatically generate scenarios and associated tools such as scripts and performance standards (Bowers et al., 1997; Jentsch et al., 1999). The SALIANT (situational awareness linked instances adapted to novel tasks) tool is used to assess and train situational awareness (Dwyer and Salas, 2000; Muniz et al., 1998). Event-based approaches have been applied to facilitate test and evaluation (Fowlkes et al., 1999). Finally, event-based training is consistent with objective-based training in the military.

47.6 Standards and Regulations

Training objectives and performance standards can be obtained from published tasks lists. Examples include joint and mission essential-task lists, mission training plans, and unit standard operating procedures. Within an organizational context, these may include published task lists such as those contained in handbooks and prior training materials.

47.7 Approximate Training and Application Times

Development time depends on the resources available. Traditionally, scenario development has been a time-consuming process, typically requiring 8 hours at a minimum. For large exercises, scenario development can take weeks to accomplish (Stretton and Johnston, 1997). A number of tools have been developed to ease this process (e.g., RRLOE).

47.8 Reliability and Validity

The EBAT approach has not yet been formally evaluated. One reason for this is that attention to the design of scenario-based training is a recent phenomenon, coinciding with increased attention to the training of complex performances in realistic, context-specific settings. In addition, EBAT is part of a systematic approach to training that includes providing information and demonstration, in addition to providing practice and feedback. Data from programs incorporating EBAT report 6 to 20% improvements in performance during behavioral performance evaluations (e.g., Salas et al., 1999).

47.9 Tools Needed

EBAT can be implemented manually or supported with tools that include handheld instructor aids. It is also useful to incorporate a training management system to link performance data from a single training event to databases that can be used to drive future training events and initiatives.

References

Bowers, C., Jentsch, F., Baker, D., Prince, C., and Salas, E. (1997), Rapidly reconfigurable event-set based line operational evaluation scenarios, in *Proceedings of the Human Factors and Ergonomics Society 41st Annual Meeting*, Human Factors and Ergonomics Society, Santa Monica, CA, pp. 912–915.

Fowlkes, J.E., Dwyer, D.J., Milham, L.M., Burns, J.J., and Pierce, L.G. (1999), Team skills assessment: a test and evaluation component for emerging weapon systems, in *Proceedings of the 1999 Interservice/ Industry Training, Simulation and Education Conference*, CD-ROM, National Training Systems Association, Arlington, VA, pp. 994–1004.

Fowlkes, J., Dwyer, D.J., Oser, R.L., and Salas, E. (1998), Event-based approach to training (EBAT), *Int. J. Aviation Psychol.*, 8, 209–221.

Fowlkes, J.E., Lane, N.E. Salas, E., Franz, T., and Oser, R. (1994), Improving the measurement of team performance: the TARGETs methodology, *Mil. Psychol.*, 6, 47–61.

Jentsch, F., Abbott, D., and Bowers, C. (1999), Do three easy tasks make one difficult one? Studying the perceived difficulty of simulation scenarios, in *Proceedings of the Tenth International Symposium on Aviation Psychology*, Ohio State University, Columbus.

Johnston, J.H., Cannon-Bowers, J.A., and Jentsch, K.A. (1995), Event-based performance measurement system for shipboard teams, in *Proceedings of the First International Symposium on Command and Control Research and Technology*, The Center for Advanced Command and Technology, Washington, D.C., pp. 274–276.

Muniz, E., Stout, R., Bowers, C., and Salas, E. (1998), A methodology for measuring team situational awareness: situational linked indicators adapted to novel tasks (SALIANT), in *The First Annual Symposium/Business Meeting of the Human Factors and Medicine Panel on Collaborative Crew Performance in Complex Systems*, Edinburgh, U.K.

Salas, E., Fowlkes, J.E., Stout, R.J., Milanovich, D.M., and Prince, C. (1999), Does CRM training improve teamwork skills in the cockpit? Two evaluation studies, *Hum. Factors*, 41, 326–343.

Smith-Jentsch, K.A., Zeisig, R.L., Acton, B., and McPherson, J.A. (1998), Team dimensional training: a strategy for guided team self-correction, in *Making Decisions under Stress*, Cannon-Bowers, J.A. and Salas, E., Eds., American Psychological Association, Washington, D.C., pp. 271–297.

Stretton, M.L. and Johnston, J.H. (1997), Scenario-based training: an architecture for intelligent event selection, *Proceedings of the 19th Interservice/Industry Training Simulation and Education Conference*, CD-ROM, National Training Systems Association, Arlington, VA, pp. 108–117.

Zachary, W., Bilazarian, P., Burns, J., and Cannon-Bowers, J.A. (1997), Advanced embedded training concepts for chipboard systems, in *Proceedings of the 19th Interservice/Industry Training, Simulation and Education Conference*, CD-ROM, National Training Systems Association, Arlington, VA, pp. 670–679.

48

Team Building

Eduardo Salas
University of Central Florida

Heather A. Priest
University of Central Florida

Renée E. DeRouin
University of Central Florida

48.1 Background and Application

In recent years, managers have come to realize the importance of teams and of team attitudes, roles, and responsibilities to the productivity and efficiency of organizations. While team training has been applied to improve team competencies and processes, other methods can be applied to focus more on team affect and interpersonal relations. One supplement to team training is team building, an intervention designed to improve team functioning. Specifically, team building seeks to enhance team processes, improve individual and team characteristics, and alter organizational environments and structures (Tannenbaum et al., 1992). In this intervention, "intact work groups experientially learn, by examining their structures, purposes, norms, values, and interpersonal dynamics, to increase their skills for effective teamwork" (Liebowitz and De Meuse, 1982). Team building differs from other team training efforts in that it specifically seeks to clarify team member roles and responsibilities (Salas et al., 1999) in addition to focusing on the improvement of core team operations and processes.

Because team building focuses on achieving organizational productivity in addition to fostering team cohesion (Payne, 2001), team building has been widely adopted as an intervention for organizational development. In fact, Dyer (1977) suggests that team building can be used to help combat such various organizational problems as conflicts among organizational members, unclear roles and assignments, improperly carried out decisions, apathy of organizational members, lack of innovation in solving complex problems, dependence upon the manager for direction, customer complaints regarding quality of service, and cost increases that cannot otherwise be explained. Based on this, it appears that team building can be used to improve a diverse range of organizational and team member concerns.

Although some researchers are traditionally supportive of team building (e.g., Clark, 1994; Harrington-Mackin, 1994; Payne, 2001; Skopec, 1997), others are more cautious, noting a lack of empirical support for the technique (e.g., Buller and Bell, 1986; Liebowitz and De Meuse, 1982; Salas et al., 1999).

48.2   Procedure

The implementation of a team-building program varies, depending on the method used. However, each method generally involves six steps. In the first step of the intervention, team members provide information on their perception of team issues, relationships, and interactions. Teams are then given feedback regarding individual strengths and weaknesses. The second step of a team-building program focuses on the creation of the executive summary. In this step, team objectives are developed based on the issues of greatest importance to the team. The third step involves the design of the team mission and the generation of functional roles necessary to accomplish this mission. From the list of functional roles, each individual team member is then assigned specific roles and responsibilities. In addition, accountability agreements are created. After this is accomplished, the team-building development process moves to the fifth step, which involves the development of an action plan to guide the team-building intervention. The sixth and final step involves the evaluation of the team-building program for its overall effectiveness in either improving team processes or performance (Payne, 2001). (For further information on how to develop a successful team-building intervention, see Clark [1994], Harrington-Mackin [1994], and Skopec [1997].)

Team building often focuses on one or more of the following components: goal setting, interpersonal relations, role clarification, managerial grid, and problem solving. Although Beer (1976) outlined four of the team-building components (i.e., goal setting, interpersonal relations, role clarification, and managerial grid) in his description of various organizational development techniques, Buller and Bell (1986) developed the problem-solving component of team building to replace the managerial grid in a majority of the recent literature (e.g., Druckman and Bjork, 1994; Sundstrom et al., 1990; Tannenbaum et al., 1992).

We next briefly discuss each of the components:

- *Goal setting:* The goal-setting component is designed specifically to strengthen team member motivation to achieve team goals and objectives. By identifying specific outcome levels, teams can determine what future resources are needed. Individual characteristics (e.g., team member motivation) can also be altered by use of this intervention.

- *Interpersonal relations:* The interpersonal relations component of team building is based on the assumption that teams with fewer interpersonal conflicts function more effectively than teams with greater numbers of interpersonal conflicts. This component requires use of a facilitator to develop mutual trust and open communication between team members. In addition, as team members achieve higher levels of trust, cooperation, and cohesiveness, team characteristics can be changed as well.

- *Role clarification:* The role clarification component defines the team as comprising a set of overlapping roles. These overlapping roles are characterized as the behaviors that are expected of each individual team member. Role clarification can be used to improve team and individual characteristics (i.e., by reducing role ambiguity) and work structure by negotiating, defining, and adjusting team member roles.

- *Managerial grid:* The managerial grid component emphasizes a management context that promotes a joint concern for production levels and individuals. This approach typically involves a group of team members who compare and contrast the team's current situation to an ideal situation. Ways of moving the current situation toward the ideal situation are then discussed.

- *Problem solving:* Buller's (1986) problem-solving component subsumes aspects from all of the components described by Beer (1976). In this approach, team members practice setting goals, develop interpersonal relations, clarify team roles, and work to improve organizational characteristics through problem-solving tasks. In addition to team building, problem-solving approaches can also have the added benefit of enhancing team critical-thinking skills.

48.3 Advantages

- Appears to have a positive effect on member attitudes toward other team members.
- Increases team functioning when role clarification components are involved.
- Can be used as a supplement to team training (e.g., in conflict resolution, role clarification).

48.4 Disadvantages

- Research has typically been limited to white-collar/management teams.
- Suffers from methodological shortcomings (e.g., inadequate research design, small sample sizes, and inappropriate outcome measures).
- Is not theory-based.
- Inundated by pop psychology and popular culture (e.g., "how-to" books and consulting firms offering team-building exercises).
- Influence on team effectiveness appears to be modest.
- Effects are often short-lived.

48.5 Related Method

A related method of team building is team training, an example of which is crew resource management (CRM). CRM training is designed to provide skill-based instruction to high-impact work teams (although originally designed for airline cockpit crews [Helmreich et al., 1999]). According to Burke et al. (2001), CRM training generally results in producing positive reactions, increased learning, and improved team behaviors. The specific skills that are trained generally include motor skills, procedural skills, information skills, communication processes and decision-making skills, workload management and situational awareness skills, and team-building and maintenance skills (Driskell and Adams, 1992).

48.6 Standards and Regulations

A primary problem of team-building interventions has been the lack of uniformity in definitions of team building and inconsistency when implementing the approach. However, researchers have recommended several guidelines to aid in the standardization of team-building interventions (e.g., Druckman and Bjork, 1994; Dyer, 1977; Liebowitz and De Meuse, 1982). From these guidelines, we can generally conclude that:

- Team building does not stop after training is over. On the contrary, it should involve continuous efforts to improve team operations and processes and may call for refresher sessions when experienced team members leave and new members arrive.
- Team building should be compatible with existing organizational structures and systems (e.g., performance appraisal and selection systems). It is important that team building fit within these structures and systems so that team building can be supported after the intervention is completed.
- Top management support is necessary for team building to be successful. If top management does not support the team-building program, the intervention is likely to fail because its outcomes will not be rewarded in the workplace.
- Everyone affected by the team-building program should be involved in its development. In order for a team-building intervention to receive acceptance by team members and management, both of these groups need to be included in its design and implementation.

48.7　Approximate Training and Application Time

In order to administer team-building interventions, team leaders or managers should be trained as exercise facilitators, or consultants can be used to conduct the interventions. Training for facilitators of team-building interventions can be quite comprehensive and involved, because team-building exercises each generally require approximately 30 minutes to an hour to administer (Payne, 2001). In fact, many team-building workshops last for a day or more if the team-building interventions involve trips or wilderness adventures, both of which are becoming increasingly popular.

48.8　Reliability and Validity

Despite the fact that many organizations have been quick to adopt team-building interventions, the jury is still out on whether team building is an effective means of improving team performance. Team-building research has historically relied upon quasi-experimental studies. Although improvements to team-building research methods were developed in the 1980s, several problems exist with that research, including small sample sizes, internal validity threats, and uninterpretable findings (Tannenbaum et al., 1992). In addition, Druckman and Bjork (1994) found that team building is not supported by any strong empirical evidence for its use in organizational development. One of the biggest problems, however, is the reliance of team-building evaluations on affective, self-report measures designed to evaluate training satisfaction, training climate, and attitudes toward team building.

Although research suggests that team-building interventions improve affective responses to team building (e.g., Tannenbaum et al., 1992), little evidence exists as to whether team building significantly impacts team performance. In fact, Tannenbaum et al. reported that only one third of the studies they reviewed in their analysis of team building used objective measures of performance (e.g., organizational productivity, cost, and turnover). Therefore, until empirical research substantiates the effectiveness of team-building efforts, the reliability and validity of this intervention remains uncertain.

48.9　Tools Needed

Team-building tools vary, depending upon the specific features and technologies employed. For instance, tools can range from nothing at all (in the case of discussion-only team-building programs) to equipment and machinery actually used on the job (in the case of task-focused team-building programs). Because team norms and cooperation appear to be the most essential tools for implementing successful team-building interventions, many team-building programs consist of paper surveys designed to measure team member attitudes toward the task and toward other team members.

References

Beer, M. (1976), The technology of organization development, in *Handbook of Industrial and Organizational Psychology*, Dunnette, M.D., Ed., Rand McNally, Chicago, pp. 937–994.

Buller, P.F. and Bell, C.H., Jr. (1986), Effects of team building and goal setting on productivity: a field experiment, *Acad. Manage. J.*, 29, 305–328.

Burke, C.S., Wilson, K.A., Salas, E., and Bowers, C.A. (2001), Team training in the skies: does it really work? in *Enhancing Team Performance: Emerging Theory, Instructional Strategies, and Evidence*, Kozlowski, S.W.J. and DeShon, R., Chairs, symposium conducted as the 16th annual conference of the Society for Industrial and Organizational Psychology, San Diego, CA.

Clark, N. (1994), *Team Building: Practical Guide for Trainers*, McGraw-Hill, Berkshire, U.K.

Driskell, J.E. and Adams, R.J. (1992), *Crew Resource Management: An Introductory Handbook*, NTIS DOT/FAA/RD-92/26, Federal Aviation Administration, Washington, D.C.

Druckman, D. and Bjork, R.A. (1994), *Learning, Remembering, Believing: Enhancing Human Performance*, National Academy Press, Washington, D.C.

Dyer, W.G. (1977), *Team Building: Issues and Alternatives*, Addison-Wesley Publishing, Reading, MA.

Harrington-Mackin, D. (1994), *The Team Building Tool Kit*, AMACOM, New York.

Helmreich, R.L., Merritt, A.C., and Wilhelm, J.A. (1999), The evolution of crew resource management training in commercial aviation, *Int. J. Aviation Psychol.*, 9, 19–32.

Liebowitz, S.J. and De Meuse, K.P. (1982), The application of team building, *Hum. Relations*, 35, 1–18.

Payne, V. (2001), *The Team Building Workshop*, ANACOM, New York.

Salas, E., Mullen, B., Rozell, D., and Driskell, J.E. (1999), The effect of team building on performance: an integration, *Small Group Res.*, 30, 309–329.

Skopec, E.W. (1997), *The Practical Executive and Team Building*, 4th ed., NTC Contemporary, Lincolnwood, IL.

Sundstrom, E., De Meuse, K.P., and Futrell, D. (1990), Work teams: applications and effectiveness, *Am. Psychologist*, 45, 120–133.

Tannenbaum, S.I., Beard, R.L., and Salas, E. (1992), Team building and its influence on team effectiveness: an examination of conceptual and empirical developments, in *Issues, Theory, and Research in Industrial/Organizational Psychology: Advances in Psychology*, Elsevier Science, North-Holland, Amsterdam, pp. 117–153.

49

Measuring Team Knowledge

Nancy J. Cooke
Arizona State University East

49.1 Background and Applications

Advances in technology have increased the cognitive complexity of tasks, as well as the need for teamwork. Teams in cognitively rich domains (e.g., air traffic control, operating rooms, cockpits, command-and-control, disaster response) are required to detect and interpret cues, remember, reason, plan, solve problems, acquire knowledge, and make decisions as an integrated, coordinated unit. We refer to these collaborative cognitive processes as *team cognition*, and we propose that understanding team cognition is critical to understanding much team performance and intervening to prevent errors and improve effectiveness (Cooke et al., 2000; Cooke et al., in 2004).

Although numerous constructs have been proposed to capture aspects of team cognition (i.e., team mental models, team situation awareness, team decision making), measures are limited, albeit evolving. For example, measures of shared mental models have often focused on knowledge similarity (regardless of knowledge accuracy) in teams with heterogeneous knowledge backgrounds (e.g., Langan-Fox et al., 2000). Further, knowledge is typically elicited using a narrow range of the available methods at the individual level and aggregated across team members through averaging.

In this chapter we focus on a method for measuring team knowledge of the task (i.e., taskwork knowledge) that overcomes some of the shortcomings of traditional methods for measuring shared mental models. We use the term *team knowledge* to refer to the long-term knowledge of the task that experienced team members possess. Although, conceptually, this term overlaps considerably with the term *shared mental models*, it avoids ambiguities associated with the word *shared*. That is, *shared* can mean to have in common (e.g., shared beliefs, shared ancestors) or to apportion (e.g., share the dessert, time share).

Although shared mental models have been operationalized in terms of the former sense of shared, we propose that heterogeneous teams share knowledge in both senses.

49.2 Procedure

Our method for measuring team knowledge involves four steps. At each step along the way, there are multiple variations possible (largely dependent on domain and purpose of the measurement), although we present the series of techniques that we have used frequently and effectively. Some methodological possibilities are still undergoing development and evaluation (e.g., consensus rating task).

In describing our procedure we refer to the task of controlling a UAV (uninhabited air vehicle) by a three-person team. We have conducted research on a synthetic version of this task (Cooke et al., 2001; Cooke and Shope, 2002). The three team members who are involved include the AVO (air vehicle operator) who controls the airspeed, heading, and altitude of the UAV and monitors aircraft systems; the PLO (payload operator) who controls camera settings to take reconnaissance photos and also monitors camera equipment; and the DEMPC (data exploitation, mission planning, and communications operator) who is the chief navigator and mission planner. In our research we measure team knowledge relevant to this simulated task, but this is done in a session apart from actual task performance. The data collection procedures that occur in this session, as well as the associated preparatory steps, are subsumed under step 1. Steps 2 through 4 involve data analysis and interpretation.

49.2.1 Step 1: Knowledge Elicitation

49.2.1.1 Selecting the Elicitation Task

We have used several different methods, notably pairwise ratings of conceptual relatedness and multiple-choice tests of taskwork knowledge, to elicit knowledge from team members about the task. The goal is to elicit individual and, ultimately, team knowledge pertinent to the task at hand, but not specific to any one situation (i.e., situation models).

In our experience, multiple-choice-test accuracy (averaged across team members) and intrateam similarity of responses are not generally predictive of team performance. Instead, conceptual relatedness ratings have been predictive of team performance, though they show little evidence of change subsequent to training (e.g., Cooke et al., 2001). Converging evidence along these lines suggests that conceptual relatedness ratings are sensitive to team knowledge differences after training that are relatively stable and relevant to performance. The success of this elicitation method may be due to the fact that the ratings are indirect and accessible to team members with little background knowledge (i.e., even a novice may use the little knowledge they have to make a judgment of relatedness, but may not be able to explicitly justify the rating).

49.2.1.2 Selecting Concepts for Rating

In our rating task, judgments of relatedness are made on a five-point scale that ranges from slightly related to highly related. We also include a discrete option of unrelated. Concepts pairs are presented randomly to each individual (with item order in the pair counterbalanced). A rating is entered by clicking on the scale using a mouse. Pair presentation continues until all pairs have been presented (in one order only).

Concepts need to be identified that pertain to the task. For our example, we selected 11 task-related concepts (altitude, focus, zoom, effective radius, restricted operating zone (ROZ) entry, target, airspeed, shutter speed, fuel, mission time, and photos). Concept selection is the most critical and challenging step of the entire method. Although there are a number of ways to elicit concepts (Cooke, 1989), the problem is typically one of narrowing the field to 30 or fewer concepts in order to keep the number of pairwise ratings within the limits of participants' patience. The best approach to verifying that a proposed concept set is suitable is to hypothesize different knowledge structures based on those concepts that might result for the variables of interest (e.g., expert vs. novice; trained vs. untrained). If differences cannot be hypothesized, then the concept set needs reworking.

49.2.1.3 Construction of Referents

Referent structures are analogous to answer keys associated with multiple-choice tests. They depict an ideal or target knowledge structure. Thus, we can assess the accuracy of an individual's knowledge via the similarity between that individual's network structure and a referent representing overall task knowledge. However, for heterogeneous teams in which members specialize in a knowledge domain, this assessment can be misleading. Therefore, we also compare individual networks to referents associated with the knowledge pertaining to each role. Referent networks are constructed logically by a group of experimenters familiar with the task. Alternatively, such referents could be constructed empirically (e.g., an AVO referent could be constructed by averaging knowledge structures of the highest scoring AVOs). Our AVO, PLO, DEMPC, and overall referents are presented in Figure 49.1.

49.2.1.4 Consensus Ratings

We typically elicit ratings from individual team members and aggregate to characterize team knowledge. However, we have explored a procedure by which teams come to consensus on each rating as a group. First, the three team members enter individual ratings for all pairs. Then, for each pair, the three individual ratings are presented (on the computer screens) to each team member. The team members communicate over headsets and microphones to reach consensus on a team rating, which is entered by all three members. Other pairs are rated in the same way. This method is interesting in that it approaches knowledge elicitation at the holistic level and also provides a glimpse into team process behaviors (e.g., leadership or majority rule strategies can be identified).

49.2.2 Step 2: Pathfinder Analysis

The next step is to conduct a Pathfinder network analysis (Schvaneveldt, 1990) on each of the individual ratings (or the consensus ratings). We typically use KNOT (knowledge network organizing tool) software for this with default parameter settings (i.e., $r =$ infinite, $q = n - 1$). One result of this analysis is a graphical representation of a network, as in Figure 49.1, but KNOT also produces a list of links that can be used as input to calculate network similarities

49.2.3 Step 3: Deriving Accuracy and Similarity Metrics

Network similarities are derived (also using the KNOT tool) for the participant's network compared with each of the four referents. These similarities range from 0 to 1 and represent the proportion of shared links in the two networks. Three accuracy values are derived per individual team member. Overall accuracy is the similarity between the participant network and the overall referent. Role accuracy is the similarity between the participant network and the referent corresponding to that participant's role (e.g., DEMPC in Figure 49.1, or 0.33). Finally, interpositional accuracy is the average of the similarity between the participant network and each of the other two referents not associated with that person's role on the team (e.g., 0.41 in Figure 49.1). Intrateam similarity can be derived by averaging the three similarities derived for pairs of the three team members.

49.2.4 Step 4: Interpretation and Application

Overall role and interpositional accuracy metrics can be aggregated (e.g., averaged) across team members to generate team accuracy scores. The team accuracy and intrateam similarity scores reflect an assessment of team knowledge and should therefore correspond to other variables relevant to team knowledge or shared mental models. For instance, we have found, using these metrics, that team knowledge is predictive of team performance and that, in particular, the best teams have high overall and interpositional accuracy scores (Cooke et al., 2001). We also typically find that knowledge measured using this method is relatively

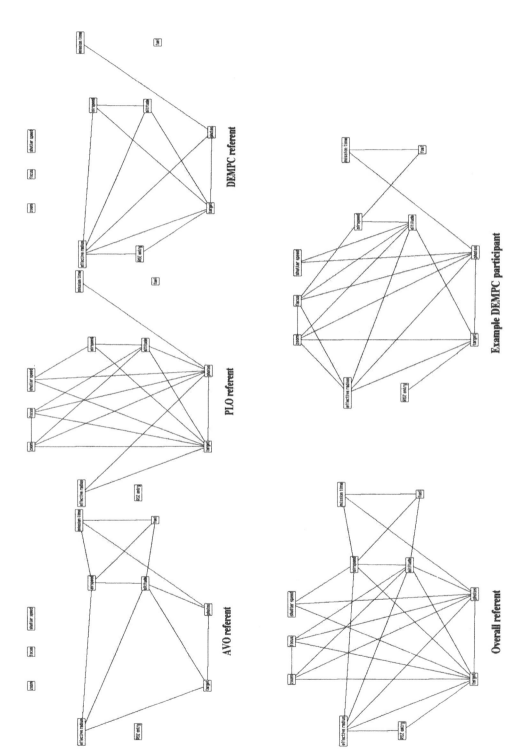

FIGURE 49.1 Overall and role referents used in the UAV team studies and sample data from a DEMPC participant in one of those studies. For this participant: DEMPC similarity to DEMPC referent = 0.33, average of DEMPC similarity to AVO referent and DEMPC similarity to PLO referent = 0.41, and DEMPC similarity to overall referent = 0.57.

stable after initial training (i.e., it does not change with mission experience). We interpret this as an indication that the method is capturing a kind of basic taskwork knowledge that a team member brings to the task, as opposed to a more fleeting knowledge associated with situation assessment. In general, this method provides information that can be diagnostic regarding the team knowledge associated with particularly effective or ineffective teams.

49.3 Advantages

- The method addresses heterogeneous knowledge in teams.
- There are possibilities for elicitation at the group level.
- Conceptual ratings are indirect and perhaps more sensitive as a result.
- Software is available to facilitate ratings and network analysis.
- Results have been predictive of team performance.

49.4 Disadvantages

- Elicitation is conducted apart from the task.
- Domain knowledge is required to identify concepts and generate referents.
- Ratings can get unwieldy with more than 30 concepts.
- Requires some experience to apply.
- Even long-term knowledge changes.

49.5 Related Methods

There are other approaches to eliciting the type of knowledge associated with mental models (e.g., interviews, think-aloud protocols) and other approaches to multivariate scaling (i.e., multidimensional scaling and cluster analysis) as well as direct methods of deriving concept maps from interview data (Cooke, 1994). This approach to measuring team knowledge is not restricted to concept ratings and Pathfinder. The uniqueness of this approach lies in the derivation of three team accuracy metrics appropriate for teams with heterogeneous knowledge backgrounds.

Thus, this approach can be applied to data elicited in a number of ways given that (1) there is some way to derive referents for accuracy assessment and (2) there is a way to quantitatively compare output of the method to referents and other output. Thus, multiple-choice tests, in which certain patterns of responses serve as referents and similarity is based on the number of matching responses, could similarly serve this purpose.

49.6 Standards and Regulations

There are no standards or regulations associated with this method.

49.7 Approximate Training and Application Times

Learning to use the tools to acquire ratings and generate Pathfinder networks and similarity values takes about 2 hours. Learning how to wisely select concepts and derive referents requires significant domain knowledge and more experience with the methodology (the exact time depending on the domain). Application time is about 1 hour to collect data and 1 hour to analyze data per team.

49.8 Reliability and Validity

Reliability and validity will vary depending on the elicitation method and concepts chosen. In our studies using concept ratings and Pathfinder, accuracy and similarity metrics are stable within teams, in that they show no statistically reliable change after training. Predictive validity has been established in a number of studies, with correlations between team knowledge and performance ranging from ($r = 0.54$ to 0.84, df = 9; Cooke et al., 2004). The consensus ratings have demonstrated somewhat less predictive validity ($r = 0.25$ to 0.55, df = 9), and this method is still in the early stages of development.

49.9　Tools Needed

There are tools available in the KNOT (knowledge network organizing tool) software to collect pairwise concept ratings, perform Pathfinder analysis, and generate network similarity values. Our consensus version of the rating task required headsets and an intercom systems and a ratings program adapted for that task. Other tools are available for other elicitation approaches (Cooke, 1994).

Acknowledgments

Thanks to Janie DeJoode, Rebecca Keith, and Harry Pedersen for their analytic and graphical assistance with Figure 49.1. This work was partially supported by AFOSR grant F49620-01-0261.

References

Cooke, N.J. (1994), Varieties of knowledge elicitation techniques, *Int. J. Hum.–Comput. Stud.*, 41, 801–849.

Cooke, N.M. (1989), The elicitation of domain-related ideas: stage one of the knowledge acquisition process, in *Expert Knowledge and Explanation*, Ellis, C., Ed., Ellis Horwood, Chichester, West Sussex, U.K., pp. 58–75.

Cooke, N.J., Kiekel, P.A., and Helm E. (2001), Measuring team knowledge during skill acquisition of a complex task, *Int. J. Cognitive Ergonomics*, 5 (special section: knowledge acquisition), 297–315.

Cooke, N.J., Salas, E., Cannon-Bowers, J.A., and Stout, R. (2000), Measuring team knowledge, *Hum. Factors*, 42, 151–173.

Cooke, N.J., Salas, E., Kiekel, P.A., and Bell, B. (2004), Advances in measuring team cognition, in *Team Cognition: Understanding the Factors That Drive Process and Performance*, Salas, E. and Fiore, S.M., Eds., American Psychological Association, Washington, D.C., pp. 83–106.

Cooke, N.J. and Shope, S.M. (2002), The CERTT-UAV task: a synthetic task environment to facilitate team research, in *Proceedings of the Advanced Simulation Technologies Conference: Military, Government, and Aerospace Simulation Symposium*, Society for Modeling and Simulation International, San Diego, CA, pp. 25–30.

Langan-Fox, J., Code, S., and Langfield-Smith, K. (2000), Team mental models: techniques, methods, and analytic approaches, *Hum. Factors*, 42, 242–271.

Schvaneveldt, R.W. (1990), *Pathfinder Associative Networks: Studies in Knowledge Organization*, Ablex, Norwood, NJ.

50

Team Communications Analysis

Florian Jentsch
University of Central Florida

Clint Bowers
University of Central Florida

50.1 Background and Application

Teams are distinguished from groups and other collectives largely due to a requirement for interdependence. That is, no one person on the team can perform the team's tasks alone. Rather, there is a need to exchange some type of resource among team members for the team to perform effectively. In many instances, the resource that must be shared is information. In other cases, a more tangible resource can be transferred among team members, such as a tool. In all cases, however, it is presumed that effective, efficient communications are required in order for teams to do their work. In fact, every existing theoretical model of team performance has included some aspect of communications as foundational to a team's performance. From their review of the literature in the area of intrateam communications, Kanki and Palmer (1993) concluded that research needed to expand our horizons with respect to "identifying communication links among teams and disentangling both the functions and variations with speech act patterns."

Since then, a number of studies have shown the utility of investigating communications when attempting to study team performance. Application areas have included aviation (e.g., Bowers et al., 1998b), military teams (Achille et al., 1995), and team training (Siegel and Federman, 1973), and have ranged as far as the analysis of communications in virtual teams (Carletta et al., 2000). Traditionally, frequency counts of speech and communications acts were used in team performance studies because frequency counts appeared to be intuitive and could be evaluated with relative statistical ease. Later, however, it was suggested that the analysis of communications patterns might offer important advantages for the study of team processes. A number of studies using either approach are discussed in the following sections.

50.1.1 Frequency of Communication Acts

Orasanu (1990) analyzed the communication of flight crews during a simulated mission with embedded aircraft problems. Crews had to diagnose the problems and alter their flight plans accordingly. Orasanu found that higher-performing crews made a significantly greater number of situational awareness statements than did poorer performing crews. A subsequent study by Orasanu and Fischer (1991) analyzed the relationship between communications and team performance across two separate aircraft types. Teams from each aircraft flew a simulated scenario that included poor weather, system failures, and an aborted landing. The results of this study indicated that the frequency of certain statements (e.g., those related to situational awareness) distinguished between good and poor teams in only one of the aircraft types, but not in the other.

A similar approach to that used by Orasanu (1990) and Orasanu and Fischer (1991) was employed by Mosier and Chidester (1991), who also attempted to identify a link between communication frequency counts and crew performance. They reasoned that the relationship between team performance and communications might be most apparent during emergency operations. They therefore investigated information transfer during two simulated emergency situations. Mosier and Chidester found that the number and type of communications were, in fact, related to crew performance. Similar findings were reported by Stout et al. (1994).

Frequency counts alone, however, can lead to surprising results, as shown by Thornton (1992), who analyzed team behaviors and performance in a study designed to assess the effects of automation in the cockpit. Thornton assessed frequencies of behavior along with subjective ratings made by trained raters. She found that, regardless of the level of automation, the frequency of certain communications, generally considered as helping team performance, was positively correlated with errors in flight. Thornton posited that poor crews might have employed a high number of these behaviors in order to correct previous mistakes, leading to the positive correlation between communications frequencies and errors. Observer ratings of the crew's performance, on the other hand, were negatively correlated with errors in flight, suggesting that observers based their subjective ratings at least in part on the number of errors a flight crew committed, in addition to the communications.

The data regarding the use of communications frequencies to assess team performance thus is equivocal. This has prompted the suggestion that frequency counts of communication alone are not adequate measures of teamwork (Kanki et al., 1991). Rather, it has been recommended that measurements that consider the nature and the timing of communication are more likely to contribute to our understanding of team processes.

50.1.2 Patterns of Communication

The analysis of sequential communication data has found widespread use in many areas of behavioral research (e.g., Allison and Liker, 1982; Bakeman, 1983; File and Todman, 1994; Hirokawa, 1980, 1982, 1983; Sackett, 1979, 1987; Salazar et al., 1994). With respect to teams, Kanki et al. (1991), for example, hypothesized that the homogeneity of a team's communication would facilitate the predictability of team process and would result in more effective performance. To test this hypothesis, teams were assigned to high- and low-performance groups based on the severity of operational errors that they committed during a simulation. Kanki et al. (1991) then analyzed the communication of the teams.

Rather than simply measuring the frequency of communication, the analyses focused on *a priori* communication patterns (Kanki et al., 1991). The patterns considered both the speaker and the content of the communication. The results indicated that communication patterns distinguished between high- and low-error crews. The poorer teams demonstrated little consistency in their communication patterns. Good crews, on the other hand, demonstrated very consistent speech in terms of the sequence of speakers and the content of communications. Similar results were obtained by Rhodenizer et al. (2000), who analyzed sequences of communications between emergency medical services units and their dispatchers.

The results from these studies suggest that reducing communication losses due to mismatches between expectations and actual communications can be achieved through the standardization of communication patterns in teams. In this context, what has been called "closed-loop" communication may be instrumental. In closed-loop communication, statements, commands, and actions of each individual in a team are acknowledged and responded to by other team members. In fact, Mosier and Chidester (1991) found preliminary evidence supporting this notion. Teams in their simulation collected information before and after decisions were made, indicating that problems were not "forgotten" after being solved. Later research by Bowers et al. (1998b) supported the utility of closed-loop communications for enhancing team performance when they studied communication sequences among pilot teams.

50.2 Procedure

Initially, communications are recorded and coded using a content categorization approach. A lag-sequential or Markov-chain analysis can then be used to identify the size of the pattern within which differences can be detected (Bowers et al., 1998a). A contingency table is then created that contains each chain of communications of the designated pattern size. *A priori* patterns can be evaluated using log-linear statistics. This is an effective method for identifying difference between known groups.

50.3 Advantages

- Allows a more thorough understanding of team processes.
- Allows an understanding of team responses to specific stimuli.
- Is useful in identifying patterns of interruptions or other negative team behaviors.

50.4 Disadvantages

- Communication coding is slow and tedious.
- Statistical analysis requires substantial computing power.
- Requires relatively large number of utterances (10,000+, depending on the number of factors and coding categories that must be analyzed).

50.5 Approximate Training and Application Times

The coding of communications requires substantial training to achieve adequate reliability. Training periods of 10 hours or more are not uncommon. Often, this training requires not only an explanation of each code, but several hours of practice, feedback, and consensus building. After coding, the statistical analyses can be completed in approximately 60 min.

50.6 Reliability and Validity

Interrater agreement of 0.8 to 0.9 (Pearson *r*, Spearman *rho*, or percent agreement, depending on the categorization scheme) is generally required to pursue this analysis approach. Achievement of this level of reliability is possible given appropriate training. Once archived, preliminary validity evidence is positive. For example, Bowers et al. (1998b) demonstrated that the occurrence of specific *a priori* communication sequences distinguished between effective and ineffective teams that completed a simulated aviation mission. Similar findings were reported by Jentsch et al. (1995).

50.7 Tools Needed

Video- and audio-recorder
Communications ratings sheets
Microsoft Excel or similar
SPSS advanced statistics log-linear analyses or similar

References

Achille, L.B., Schulze, K.G., and Schmidt-Nielsen, A. (1995), An analysis of communication and the use of military teams in navy team training, *Mil. Psychol.*, 7, 95–107.

Allison, P.D. and Liker, J.K. (1982), Analyzing sequential categorical data in dyadic interaction: a comment on Gottman, *Psychol. Bull.*, 91, 393–403.

Bakeman, R. (1983), Computing lag sequential statistics: the ELAG program, *Behavior Res. Methods Instrum.*, 15, 530–535.

Bowers, C.A., Jentsch, F.G., Braun, C.C., and Salas, E. (1998a), Studying communication patterns among aircrews: implications for team training, *Hum. Factors*, 40, 672–679.

Bowers, C., Jentsch, F., Salas, E., and Braun, C. (1998b), Analyzing communication sequences for team training needs assessment, *Hum. Factors*, 40, 672–679.

Carletta, J., Anderson, A.H., and McEwan, R. (2000), The effects of multimedia communication technology on non-collocated teams: a case study, *Ergonomics*, 43, 1237–1251.

File, P. and Todman, J. (1994), Identification of sequential dependencies in conversations: a Pascal program, *Behav. Res. Meth. Instruments Comput.*, 26, 65–69.

Hirokawa, R.Y. (1980), A comparative analysis of communication patterns within effective and ineffective decision-making groups, *Commun. Monogr.*, 47, 312–321.

Hirokawa, R.Y. (1982), Group communication and problem-solving effectiveness I: a critical review of inconsistent findings, *Commun. Q.*, 30, 134–141.

Hirokawa, R.Y. (1983), Group communication and problem-solving effectiveness. II. An exploratory investigation of procedural functions, *Western J. Speech Commun.*, 47, 59–74.

Jentsch, F., Bowers, C., Sellin-Wolters, S., and Salas, E. (1995), Crew coordination behaviors as predictors of problem detection and decision making, in *Proceedings of the 39th Meeting of the Human Factors and Ergonomics*, Human Factors and Ergonomics Society, Santa Monica, CA, pp. 1350–1354.

Kanki, B.G., Folk, V.G., and Irwin, C.M. (1991), Communication variations and aircrew performance, *Int. J. Aviation Psychol.*, 1, 149–162.

Kanki, B.G. and Palmer, M.T. (1993), Communication and crew resource management, in *Cockpit Resource Management*, Wiener, E.L., Kanki, B.G., and Helmreich, R.L., Eds., Academic Press, San Diego, CA, pp. 99–136.

Mosier, K.L. and Chidester, T.R. (1991), Situation assessment and situation awareness in a team setting, in *Proceedings of the 11th Congress of the International Ergonomics Association*, International Ergonomics Association, Paris.

Orasanu, J. (1990), Shared Mental Models and Crew Performance, Report CSLTR-46, Princeton University, Princeton, NJ.

Orasanu, J.M. and Fischer, U. (1991), Information transfer and shared mental models of decision making, in *Proceedings of the Sixth International Symposium on Aviation Psychology*, Jensen, R.S., Ed., Ohio State University, Columbus, pp. 272–277.

Rhodenizer, L.G., Jentsch, F., and Bowers, C. (2000), Emergency service dispatch: using sequential data analysis to study emergency services dispatchers' communications, *Ergonomics Design*, 8, 24–28.

Sackett, G.P. (1979), The lag sequential analysis of contingency and cyclicity in behavioral interaction research, in *Handbook of Infant Development*, Osofsky, J.D., Ed., Wiley, New York, pp. 623–649.

Sackett, G.P. (1987), Analysis of sequential social interaction data: some issues, recent developments, and a causal inference model, in *Handbook of Infant Development*, Osofsky, J.D., Ed., Wiley, New York, pp. 855–878.

Salazar, A.J., Hirokawa, R.Y., Propp, K.M., Julian, K.M., and Leatham, G.B. (1994), In search of true causes: examination of the effect of group potential and group interaction on decision performance, *Hum. Commun. Res.*, 20, 529–559.

Siegel, A.I. and Federman, P.J. (1973), Communications content training as an ingredient in effective team performance, *Ergonomics*, 16, 403–416.

Stout, R.J., Salas, E., and Carson, R. (1994), Individual task proficiency and team process behavior: what's important for team functioning? *Mil. Psychol.*, 6, 177–192.

Thornton, R.C. (1992), The Effects of Automation and Task Difficulty on Crew Coordination, Workload, and Performance, Ph.D. dissertation, Old Dominion University, Norfolk, VA.

51

Questionnaires for Distributed Assessment of Team Mutual Awareness

Jean MacMillan
Aptima, Inc.

Michael J. Paley
Aptima, Inc.

Eileen B. Entin
Aptima, Inc.

Elliot E. Entin
Aptima, Inc.

51.1 Background and Application

Teams, by definition, perform interdependent tasks that require team members to coordinate their decisions and actions in order to achieve their shared goals (Orasanu and Salas, 1993). To achieve the level of coordination that is required for successful interdependent performance, team members need a shared awareness of the situation and of the roles, tasks, and actions of the other team members. The existence of this "shared mental model" among team members has been suggested as an explanatory mechanism for effective teams and, as measured in various ways, has been shown to increase team performance (Cannon-Bowers et al., 1993; Stout et al., 1999).

Despite its importance, it is difficult to measure the extent to which team members are successful in developing and maintaining a shared, accurate awareness of the situation and of each other's roles. For distributed teams, with members who are not colocated, it is especially difficult both to develop and maintain this awareness and to measure it.

The notion of team mutual awareness — the extent to which team members are informed of other team members' behaviors — provides a measurable construct for assessing the presence of shared mental models. We propose a model, presented in Figure 51.1, that specifies three interrelated facets of team performance to provide a structure to assess team mutual awareness. Taskwork awareness refers to

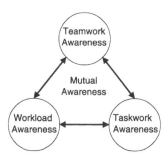

FIGURE 51.1 Team mutual awareness model.

awareness of what tasks other team members are completing and how important these tasks are. Workload awareness refers to awareness of the loading that the taskwork imposes on team members. Teamwork awareness, finally, refers to awareness of how well team members perform team-specific behaviors (i.e., coordination and backup).

We have developed a set of mutual-awareness questionnaires that are used to capture team-level data for each component of our model. The task mutual-awareness questionnaire asks team members to assess, for themselves and for each of the other team members, the most important task that each team member is performing at selected points in time. The team workload-awareness questionnaire asks for a subjective rating of (1) the workload that each respondent is experiencing and (2) ratings for the workload that he or she believes each of the other team members is experiencing. A third questionnaire, the teamwork-awareness questionnaire, asks the team members to rate the team on four dimensions of teamwork processes. A critical feature of all these questionnaires is that the results can be used to assess mutual awareness — the congruence of the team's perceptions about their workload and performance — not just the individual or team average levels of performance and workload.

Each of these questionnaires was originally developed and tested in a paper-and-pencil version for assessing the mutual awareness of colocated teams in order to identify factors that contribute to effective team performance (Entin et al., 2000). More recently, we have implemented Web-based versions of the workload and teamwork questionnaires for use in distributed team environments to simultaneously gather data from team members in multiple locations. Use of electronic collection methods greatly increases the utility of these measures because data collected with these questionnaires can be processed in near real time and used to provide immediate feedback for self-correction during team training. For example, the extent of agreement between each team member's rating of his or her own workload and the ratings provided by the other team members was recently used to provide the team with an indication of the team's mutual awareness as part of a system to train U.S. Army soldiers (Paley et al., 2002).

51.2   Procedure

51.2.1   Task Mutual Awareness

The task-awareness questionnaire requires each team member to retrospect at the end of a scenario trial about one or more salient events that occurred during the scenario. Each event acts as a common time marker so that all team members are focused at the same time within the scenario. For each event, team members report the tasks they were performing (or were just starting to perform) when the event occurred and then report the tasks each of the other team members was performing during the same event. As originally implemented, the team members report the tasks they and their teammates were performing using a free-response format. A subject-matter expert then classifies the team members' responses into task categories. An alternative to the free-response mode that has been used in subsequent administrations of this measure is to provide the team members with a list of task categories and have them check the category that represents the task they were performing. The former methodology has the advantage that it is not necessary to define the task categories ahead of time. The latter methodology, an example of which is shown in Figure 51.2, has the advantage of efficiency of response.

51.2.2 Team Workload Awareness

We have extended the task-load index (TLX) (Hart and Staveland, 1988), which provides a domain-independent assessment of individual team members' workload, to capture team as well as individual workload (Entin et al., 1998). We use a three-part questionnaire to assess individual and team workload. In the first part of the workload questionnaire, participants report their own workload in terms of five of the traditional items comprising the TLX: mental demand, temporal demand, performance, effort, and frustration. (We omit the sixth item, physical workload, because it is not applicable in most simulation-based situations.) In the second part of the questionnaire, each participant provides an estimate of the overall workload experienced by each of the other team members. In the third part of the

---

TASK AWARENESS QUESTIONNAIRE

Selection #	Category	Definition/Example
1	Hill	Attacking, taking, holding, etc.
2	Beaches	Attacking, taking, holding, etc.
3	Bridge	Blowing up correct bridge, etc.
4	Airport	Attacking, taking, holding, etc.
5	Seaport	Attacking, taking, holding, etc.
6	Ground Tasks	Includes mine clearing, clearing roads, attacking armor, etc.
7	Air Tasks	Includes AAW, CAS, engaging air threats, etc.
8	Sea Tasks	Includes mine clearing, gunfire support (NSFS), engaging PBs and Subs
9	Reconnaissance/ Identification	
10	Medevac	Calling for Medevac, launching Medevac helos
11	Don't Know	
12	Other (please specify)	Includes nonspecific moving, coordinating, attacking, deploying, launching, etc.

Each of the categories listed above describes an important task or group of tasks that team members perform during the scenario. Use these categories to identify what you were doing and what you think others were doing at specific times in the scenario.

1. Think back to when elements of the team had just completed taking the North Beach.
   a) What task were you performing (or had you just started to work on) at that time? Please select from the table above.

   Selection # (choose one) _____
   Comments _____

   b) What tasks do you believe each of the other positions in your team were performing (or had just started to perform) when you began working on the task you selected in (1a)?

   Select from the table above the category that best describes what each of the other positions were doing (omit yourself).

Other Positions	Selection #
Flag	
Green	
Blue	
Purple	
Red	
Orange	

2. Think back to when elements of the team were preparing to take the airfield.
   a) What task were you performing (or had you just started to work on) at that time? Please select from the table above.

   Selection # (choose one) _____
   Comments _____

   b) What tasks do you believe each of the other positions in your team were performing (or had just started to perform) when you began working on the task you selected in (2a)?

   Select from the table above the category that best describes what each of the other positions were doing (omit yourself).

Other Positions	Selection
Flag	
Green	
Blue	
Purple	
Red	
Orange	

FIGURE 51.2 Example of task mutual awareness questionnaire.

questionnaire, each participant responds to the five TLX items, but this time for the team as a whole (not just for themselves). The team workload questionnaire can be administered using a paper-and-pencil survey or as a Web-based electronic survey. An abbreviated electronic version, used by Paley et al. (2002), is shown in Figure 51.3.

51.2.3 Teamwork Awareness

Entin and Serfaty (1995) initially developed the teamwork assessment measure as an instrument used by subject-matter-expert observers to capture the quality of a team's teamwork processes on six dimensions of teamwork. Like the workload measure, the teamwork measure is independent of the task domain and mission objectives associated with any particular scenario. The teamwork measure has been used in a number of studies assessing team performance (Entin and Serfaty, 1999; Entin et al., 2000). In the traditional application of this measure, observers rate the team on each of the six dimensions using a behaviorally anchored seven-point scale. For each dimension, at one end of the scale are examples of behaviors indicating poor team process (for example, poor monitoring behavior), and at the other end are behaviors indicative of good team process on that dimension. During the scenario run, observers

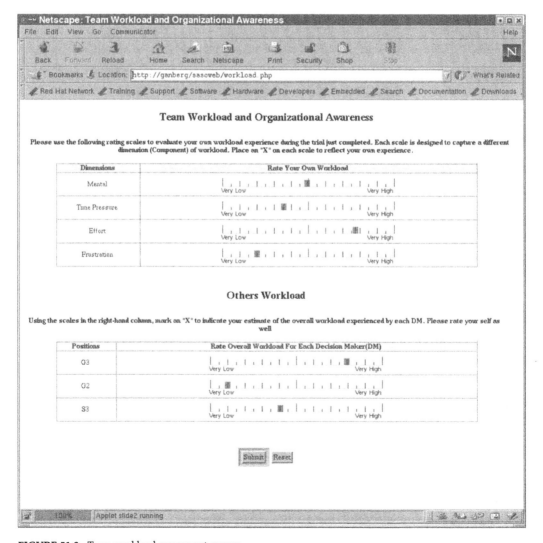

FIGURE 51.3   Team workload assessment survey.

take notes on team processes, and at the end of the scenario they complete the behavioral ratings based on their observations across the entire scenario.

We have recently instantiated this teamwork measure in a Web-based format and used it as a way of assessing *team members'* mutual assessment of their teamwork processes (Paley et al., 2002). That is, the team members, not external subject-matter experts, make subjective ratings of the team's performance. In this application, the team members rate the team on four teamwork behaviors: communication, backup, coordination and information management, and leadership/team orientation. Leadership score is related to team members' ability to agree on goals, tasks, and concepts involving the mission. Communication score is related to team members' ability to provide important information to others. The information-exchange score is related to team members' ability to pass critical information to the other members, thereby enabling them to accomplish their tasks. Backup-behavior score is related to team members' ability to be aware of each other's workload buildup and react to adjust division of task responsibilities to redistribute workload.

An example of the Web-based questionnaire is shown in Figure 51.4. Following the completion of a simulated training scenario, a Web browser automatically presents a survey (see Figure 51.4) on the players' computer screens. After all the players have completed this survey and the workload survey presented in Figure 51.1, results are tabulated and fed back to the players as part of an after-action review (AAR) to support team self-correction.

51.2.4 Scoring and Measure Development

51.2.4.1 Mutual Awareness of Tasks Performed

To develop the mutual task awareness measure, the task category representing what each team member said he or she was doing is compared with the task category representing what each of the other team

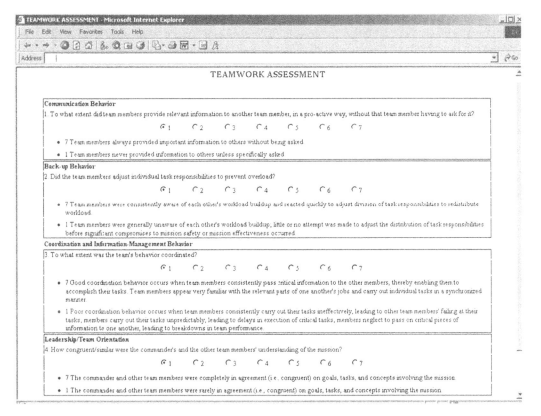

FIGURE 51.4 Teamwork assessment survey.

members said that team member was doing. The number of category matches is counted and a percentage agreement (congruence score) computed for each team.

To calculate mutual awareness based on workload scores, we calculate a congruence measure that reflects the difference between each team member's self-reported workload and the estimates of his or her workload made by each of the other team members. To compute this measure: (1) the self-reported workload for an individual is subtracted from each team member's estimate of the workload for that individual; (2) these difference scores are squared, summed, and averaged for the team; and (3) the square root of the average is taken.

51.2.4.2 Teamwork Assessment

For a team to achieve high levels of mission effectiveness, the members must perform as an effective team. The intent of the teamwork assessment survey is to provide feedback to team members to support self-correction in four key teamwork components: leadership, communications, information exchange, and backup behavior. This feedback can take two forms. The first is simply to provide a mean score of each rating across the team. This score, normalized on a 100-point scale, represents how well the team believes they are performing. The second feedback method is to calculate agreement scores within the team. This approach demonstrates to team members differences in perceived level of teamwork performance. In both feedback conditions, the objective is to stimulate conversation within the team as part of an AAR to foster self-correction.

51.3 Advantages

The mutual awareness questionnaires are quick and easy to complete, inexpensive to use, require little training of respondents, and can easily be collected at multiple points in time during a simulated mission or team training session without the need for trained observers. They provide information on a construct — the presence of mutual awareness and a shared mental model — that has been shown to be important for team performance but is difficult to measure. If implemented in the Web-based version, the questionnaire can quickly provide assessment data for self-correction in distributed team training.

51.4 Disadvantages

The data are subjective, based on self-ratings. The output measures the extent to which the team members have an internally consistent "picture" of what each person on the team is doing, how hard they are working, and how well the team is coordinating, but it does not indicate whether that picture matches reality. The questionnaires assess team process, not team performance, although task mutual awareness has been shown to be correlated with team performance outcomes (MacMillan et al., 2004). Also, the importance of mutual awareness depends on the design of the team structure. In particular, the extent to which team members are performing interdependent tasks affects the extent to which they need mutual awareness of the other team members (MacMillan et al., 2004).

51.5 Example Output

Feedback can be provided to team members at the conclusion of a simulation session. Figure 51.5 illustrates two types of output that can be provided from the teamwork awareness questionnaire: (1) teamwork assessment mean scores across team members on teamwork dimensions (based on self-assessment) and (2) teamwork awareness agreement scores for the team that show differences among team members in perceived levels of teamwork performance. The diagonal cells of the matrix represent a team member's rating of his or her own performance. The other cells in the same row represent how the other team members rated his or her performance.

Teamwork Assessment

Teamwork Score	75.0
Leadership Score	85.0
Communication Score	95.0
Information Exchange Score	70.0
Backup Behavior Score	50.0

Teamwork Score Interpretation

00–19%	Fair
20–39%	Moderately Good
40–59%	Good
60–79%	Very Good
80–100%	Excellent

Organizational Awareness

	G3	G2	S3	Discrepancy
G3	40.0	30.0	50.0	10.0
G2	60.0	50.0	20.0	29.15
S3	80.0	80.0	75.0	3.53
			Overall Discrepancy	14.23

Organizational Awareness Interpretation

00–19%	Excellent
20–39%	Very Good
40–59%	Good
60–79%	Moderately Good
80–100%	Fair

FIGURE 51.5 Example output for team mutual awareness: (1) teamwork assessment, (2) teamwork awareness.

51.6   Related Methods

The accuracy of mutual awareness affects how efficiently and effectively team members communicate to "push" and "pull" information, as measured by the team's anticipation ratio — the ratio of the number of communications transferring information to the number of communications requesting information. The anticipation-ratio measure has proved to be associated with effective team performance for a variety of different types of teams (Serfaty et al., 1998; Entin and Serfaty, 1999; Entin, 1999). More-accurate awareness of each others' roles and actions allows the team members to push information effectively, thus reducing the communication overhead for the team because only one message, rather than two, is needed for information transfer. This reduction in the time and resources required for communication is especially important when the team is experiencing periods of heavy task loading (Serfaty et al., 1998).

51.7   Standards and Regulations

Distributed assessment of the shared knowledge and team processes that lead to effective team performance is not yet covered by standards or regulations.

51.8   Approximate Training and Application Times

Several minutes of introductory training, which can be delivered through written materials, is needed to explain to respondents how they should complete the questionnaire. This training is needed only once in a data collection session, even if the questionnaire is administered several times during the session. Completion of the questionnaire requires less than 5 minutes for respondents, depending on the size of the team. The workload assessment can typically be completed in under a minute. The mutual task assessment can take several minutes if open-ended questions are used.

51.9   Reliability and Validity

The validity of the mutual-awareness measures is supported by their correlation to team performance (Entin and Entin, 2000; MacMillan et al., 2004) as well as their correlation to other process measures, such as the anticipation ratio, that have been shown to be associated with more effective levels of team performance (Entin, 1999; Entin and Serfaty, 1999; MacMillan et al., 2004). The measures also have considerable face validity, as they focus directly on observable aspects of team performance that the team members themselves see as important.

51.10   Tools Needed

The paper-and-pencil version requires no special tools. The Web-based version requires computers with Internet connections and Web browsers for respondents, and a Web server to deliver the questionnaire and collect, analyze, and provide the results.

References

Cannon-Bowers, J.A., Salas, E., and Converse, S. (1993), Shared mental models in expert team decision making, in *Current Issues in Individual and Group Decision Making*, Castellan, N.J., Jr., Ed., Lawrence Erlbaum Associates, Hillsdale, NJ, pp. 221–246.

Entin, E.B. and Entin, E.E. (2000), Assessing team situation awareness in simulated military missions, in *Proceedings of the International Ergonomics Association and 44th Annual Meeting of the Human Factors and Ergonomics Society*, San Diego, CA, pp. 73–76.

Entin, E.B., Entin, E.E., and Serfaty, D. (2000), Organizational Structure and Adaptation in the Joint Command and Control Domain, TR-915, Alphatech, Burlington, MA.

Entin, E.E. (1999), Optimized command and control architectures for improved process and performance, in *Proceedings of the 1999 Command and Control Research and Technology Symposium*, Newport, RI.

Entin, E.E., Serfaty, D., and Kerrigan, C. (1998), Choice and performance under three command and control architectures, in *Proceedings of the 1998 Command and Control Research and Technology Symposium*, Monterey, CA.

Entin, E.E. and Serfaty, D. (1999), Adaptive team coordination, *Hum. Factors*, 41, 312–325.

Entin, E.E. and Serfaty, D. (1995), Team Adaptation and Coordination Training: Emerging Issues in Distributed Training, TR-969, Alphatech, Burlington, MA.

Hart, S.G. and Staveland, L. (1988), Development of NASA-TLX (task load index): results of empirical and theoretical research, in *Human Mental Workload*, Hancock, P.A. and Mishkati, N., Eds., Elsevier, Amsterdam, pp. 139–183.

MacMillan, J., Entin, E.E., and Serfaty, D. (2004), Communication overhead: the hidden cost of team cognition, in *Team Cognition: Understanding the Factors That Drive Process and Performance*, Salas, E. and Fiore, S.M., Eds., American Psychological Association, Washington, D.C.

Orasanu, J. and Salas, E. (1993), Team decision making in complex environments, in *Decision Making in Action: Models and Methods*, Klein, G.A., Orasanu, J., Calderwood, R., and Zsambok, C.E., Eds., Ablex Publishing, Norwood, NJ.

Paley, M.J., Serfaty, D., Baker, K., Miller, P., Bailey, A., Ganberg, G., and Wan, L. (2002), Adaptive Performance in Warfighting and Peacekeeping Best Practices Report: Description of the DDD-SASO, Technical Report A003, Army Research Lab, Contract N61339-01-0049, Aptima, Inc., Woburn, MA.

Serfaty, D., Entin, E.E., and Johnston, J. (1998), Team adaptation and coordination training, in *Decision Making Under Stress: Implications for Training and Simulation*, Cannon-Bowers, J.A. and Salas, E., Eds., American Psychological Association Press, Washington, D.C.

Stout, R.J., Cannon-Bowers, J.A., Salas, E., and Milanovich, D.M. (1999), Planning, shared mental models, and coordinated performance: an empirical link is established, *Hum. Factors*, 41, 61–71.

52

Team Decision Requirement Exercise: Making Team Decision Requirements Explicit

David W. Klinger
Klein Associates Inc.

Bianka B. Hahn
Klein Associates Inc.

52.1 Background and Application

The decision requirements exercise (DRX) was developed by Klein Associates to help teams make explicit the critical decisions they made as they completed a task (Klein, 1992). This exercise method was developed as a tool to aid nuclear power-plant control-room crews in their debrief following a training session or actual event (Klinger and Klein, 1999). The foundation of the method is the decision requirement table, which has been used extensively to represent the decision requirements of individuals within a complex environment (Table 52.1). The method has been shown to be flexible and effective in a wide range of domains, and it serves a variety of application areas and research goals.

The outcome of the DRX is a calibrated understanding of critical decisions and information requirements across the team, which can hold implications for how each member shares information. Knowing the decision requirements of other team members enables individuals to share the right information with the right people at the right times. The team members maintain insights as to how their tasks contribute to the overall team's goals, and this prepares them to make better decisions (Zsambok et al., 1992). The DRX technique exposes difficulties related to particular decisions as a first step in remedying those portions of the team's process.

The DRX can be used for both recalibrating and training teams. Following the development of the technique within the nuclear power-plant domain, it was incorporated into a program of decision skills training for the U.S. Marine Corps (Klein et al., 1998). The marines currently use the DRX in the context

TABLE 52.1 Decision Requirements Table

What Decisions Were Made?	What Is Difficult about Making This Decision?	What Cues Did You Consider When You Made This Decision?	How Could You Do It Better Next Time?

of after-action reviews to highlight decisions and judgments, and for facilitating the implementation of tactics and procedures. As a training tool, the exercise facilitates a discussion about how to use various pieces of information from the environment (cues) and other factors to guide decision making. It also allows participants to hear how other individuals use those cues and factors to make decisions. In this way, it fosters shared knowledge and experience across a group (Miller et al., 1999).

52.2   Procedure and Advice

The DRX is best implemented immediately following a training simulation or actual incident in which the team was engaged. It can be conducted at any point within a debriefing session or after-action review. If no debriefing session is conducted, the DRX can be implemented as a stand-alone, postevent exercise.

52.2.1   Step 1: Introduce the Decision Requirements Exercise

The first step is to introduce the DRX to the group of participants. It is worthwhile to prime them for the type of questions they will be asked and the nature of the discussion they should anticipate. To do so, describe the purpose of the DRX and the objectives of the exercise. For example:

> I'd like to use a different debriefing approach that might not be familiar to you. It is called the decision requirements exercise, and the objective is to discuss the decisions and judgments you had to make during the simulation you just completed. We're going to look at each of the critical decisions you had to make. Then we're going to "unpack" those decisions so that everyone knows what everyone else was trying to accomplish, and so that you can figure out how to make these decisions more efficiently next time.

Make it clear that the overall objective is to improve the team's functioning. Give participants permission to adapt the exercise as it goes along by asking their own questions or by pointing out the questions that have the most and least impact.

52.2.2   Step 2: Draw a Decision Requirements Table (DRT) on the Whiteboard

Next, draw the table on the whiteboard, including a heading for each column (see Table 52.1). Prior to the DRX, determine which aspects of decision making are most relevant to the team. Choose column headings to match these decision-making aspects. The first column should always be "critical decision or judgment." Each of the other columns will provide additional depth pertaining to each decision entered into the table. The researcher can tailor the table to the needs of the team and can even change the column headings midcourse.

Suggested common column headings include:

- *Why difficult?* — the reasons why the decision is challenging (including barriers)
- *Common errors* — the errors inexperienced people tend to make when addressing the decision
- *Cues* — the pieces of information from the environment that are used to make the decision (e.g., readings of the radiation level at a certain time)

- *Factors* — the pieces of information known prior to the event that are used to make the decision (e.g., standard operating procedures)
- *Strategies* — how individuals say they make the decision (e.g., weigh certain cues and factors more heavily than others)
- *Information sources* — where the information used to make the decision comes from
- *Suggested changes* — how the team could better support the decision

Note that cues and factors often do not need to be separated into distinct columns. They both address information requirements, but in most cases it is inconsequential whether the information was known before or became evident during the event. Use your discretion to determine whether to combine them into one column or separate the two.

52.2.3 Step 3: Elicit Critical Decisions and Judgments

Ask the participants to state the critical decisions they had to make during the event. Make a list of those decisions on the whiteboard (in the first column of the table or outside the table, whichever you prefer). At this point, only ask questions to clarify the decision or judgment. Do not yet probe to fill out the remaining columns of the table.

It may be necessary to give participants an example of a critical decision. Be prepared to suggest a decision that you observed during the exercise that seemed critical to you. The purpose here is only to get the participants on the right track. Do not answer questions for them.

52.2.4 Step 4: Select the Most Critical Decisions to Expand Upon

Depending on time constraints, it may only be possible to address a subset of the decisions identified by the group. Ask the group to select the five (or whatever number seems appropriate) most critical decisions, or the five that they are most interested in pursuing. Then take the decisions, one at a time, and ask the participants questions to assist them in filling out the columns of the table for that decision.

The following are sample questions for eliciting items for each of the columns:

Why difficult?
- What is difficult about making this decision?
- What can get in the way when you make this decision?
- What might a less-experienced person have trouble with when making this decision?

Common errors
- What errors have you seen people make when addressing this decision?
- What mistakes do less-experienced people tend to make in this situation?
- What could have gone wrong (or did go wrong) when making this decision?

Cues and factors
(Cues are pieces of information that become available during the course of an event.)
- What cues did you consider when you made this decision?
- What were you thinking about when you made the decision?
- What information did you use to make the decision?
- What made you realize that this decision had to be made?

Strategies
- Is there a strategy you used when you made this decision?
- What are the different strategies that can be used for this kind of decision?
- How did you use various pieces of information when you made this decision?

Information sources
- Where did you get the information that helped you make this decision?
- Where did you look to get the information to help you here?

- What about sources, such as other team members, individuals outside the team, technologies and mechanical indicators, and even tools like maps or diagrams?

Suggested changes

- How could you do this better next time?
- What would need to be changed with the process or the roles of team members to make this decision easier next time?
- What will you pay attention to next time to help you with this decision?

Use the participants' responses to help them think about what gets in their way and how they could improve team functioning. As an example, the "why difficult?" and "common errors" columns provide valuable insights into the barriers that stand in the way of both experienced and inexperienced team members. The "cues and factors" category helps participants understand what information team members need to make decisions. Discussions around information requirements can give individuals insights into how they can assist other team members by giving them the information they need, in the right format, at the right time, without passing irrelevant information. The "suggested changes" column is a good segue into a discussion on improving the current process. Be sure to allow all the participants, not just the ones involved in the decision, to provide recommendations for making the decision easier.

52.3 Advantages

The DRX is a good technique to use if the objective of the researcher is one or more of the following:

1. *Calibrate a team's understanding of its overall objectives.* The DRX makes explicit the critical decisions that team leaders and other key members must make; why the decisions are challenging; and what cues, factors, and other pieces of information are necessary to make decisions. After team members see the requirements surrounding each decision, they are better able to introduce remedies to barriers and shift their own contributions to better support the leader's decision making.
2. *Calibrate understanding of roles and functions, and, subsequently, the requirements of each team member.* The DRX can provide a form of cross-training, in which individuals are made aware of the roles, functions, and decision requirements of the other team members. As a result, they will be better able to support the needs of subgroups within the team.
3. *Highlight barriers to information flow.* Some of the barriers to information flow may be the product of a lack of awareness of information requirements across the team. Other barriers may be the product of a suboptimal process, which can be remedied after the team as a whole can identify the obstacle.
4. *Act as a vehicle for sharing knowledge and expertise across team members.* As a training technique, the DRX can facilitate discussion of alternative interpretations of the same information. That can lead to discussion of how the team interprets events and makes decisions. Participants gain a better understanding of the decisions involved in an event from the perspective of each team member.

The DRX is best used in the context of a concrete event, training or actual. It can be a tool for understanding the decision making that occurred during an incident, and, as such, generally it is used as a supplement to a debriefing session or after-action review. The sooner the DRX is conducted following an event, the better it will work. Also, as a general rule the DRX will be more effective if you have observed the actual event or incident. When observation is not possible, as it may not be in many cases, we suggest that you prepare by becoming as familiar as possible with the incident and the roles that each team member or subgroup played within the incident.

52.4 Disadvantages

Although there is clearly a component within the DRX that relates to information requirements and information exchange across team members, this is not the best exercise to use to understand information flow within a team. Instead, use the DRX to expose areas in which information flow may be inefficient, reserving the analysis of the inefficiencies for a more appropriate technique.

The DRX is not an appropriate method to use if the team is unable to discuss a single incident or event that all team members have in common. It will be difficult for participants to fill in the decision requirements table if they cannot refer to a specific event.

Because it is used in the context of a single event, the findings of the DRX cannot necessarily be generalized to every event that team will encounter. But, the findings may have implications for decision making in similar events, especially if the team's goals remain constant across events (e.g., an emergency response organization will always have primary goals to restore safe conditions and pass information to relevant external agencies).

52.5 Required Time Frame

Plan for one to two hours to discuss the three or four most critical decisions identified by the team in the after-event or exercise debrief.

52.6 Tools and Participants

A whiteboard or comparable aid is necessary. One person serves as facilitator and recorder as the team creates a table of critical decisions and associated requirements. All key team members should attend the DRX.

References

Klein, G. (1992), Decision Making in Complex Military Environments, Contract N66001-90-C-6023 for the Naval Command Control and Ocean Surveillance Center, Klein Associates, Fairborn, OH.

Klein, G., Phillips, J.K., Klinger, D.W., and McCloskey, M.J. (1998), The Urban Warrior Experiment: Observations and Recommendation for ECOC Functioning, Contract SYN-S01-8291-54 for Synetics Corp., King George, VA, Klein Associates, Fairborn, OH.

Klinger, D.W. and Klein, G. (1999), Emergency response organizations: an accident waiting to happen, *Ergonomics Design*, 7, 20–25.

Miller, T.E., McDermott, P.L., Morphew, M.E., and Klinger, D.W. (1999), Decision-Centered Design: Cognitive Task Analysis, final annotated briefing prepared for Pacific Science and Engineering, Task Order 0048-0003, Klein Associates, Fairborn, OH.

Zsambok, C.E., Klein, G., Kyne, M.M., and Klinger, D.W. (1992), Advanced Team Decision Making: a Developmental Model, Contract MDA903-90-C-0117 for U.S. Army Research Institute for the Behavioral and Social Sciences, Klein Associates, Fairborn, OH.

53

Targeted Acceptable Responses to Generated Events or Tasks (TARGETs)

Jennifer E. Fowlkes
CHI Systems, Inc.

C. Shawn Burke
University of Central Florida

53.1 Background and Applications

The TARGETs (targeted acceptable responses to generated events or tasks) measurement approach uses a behavioral checklist (Figure 53.1) to record the occurrence or nonoccurrence of responses to trigger events embedded in job-relevant scenarios. The "trigger events" serve as opportunities for team members to perform behaviors or skills targeted for measurement. For each event, appropriate individual or team responses are identified *a priori* using domain experts and published performance standards. During the exercise, the instructor or evaluator uses a checklist that lists each event and the appropriate responses to the event.

- The use of trigger events ensures that measurement, and the resulting information obtained, is directly related to targeted measurement goals vs. being left to chance. Trigger events also serve to focus observation, so that not all aspects of performance have to be observed.
- To ensure that all intended events are presented, scenario scripts can be developed and used that detail when events will occur and the communications that should occur between the team and other entities included in the scenario. Scripts ensure that task conditions are maintained across teams observed.

EVENT: HEAVY ENEMY ARTILLERY	Score*
Pilots informed of situation	
Pilots instructed to orbit at safe location	
Aircraft time on station considered	
Informal ACA established/reestablished	
Alternate ACA considered	
ACA ¬ Protects fighters	
¬ Allows fighter maneuverability	
¬ Clearly defined (e.g., suitable landmarks identified)	
Time over target (TOT) revised	
Availability of assets to provide suppression of enemy air defenses established	
Pilots briefed ¬ ACA clearly communicated	
¬ TOT communicated	
CODING: 1 = Observed/performed satisfactorily; 0 = Omitted by team/failed to perform satisfactorily; N/A = No opportunity to perform/not required	

FIGURE 53.1 Segment of TARGETs checklist. (Adapted from Dwyer, D.J. and Salas, E. [2000], Principles of performance measurement for ensuring aircrew training effectiveness, in *Aircrew Training and Assessment*, O'Neil, H.F., Jr. and Andrews, D.H., Eds., Lawrence Erlbaum Associates, Mahwah, NJ. With permission.)

- Measurement is focused on behaviors that can be observed as either being present or absent. This functions to simplify the task for the data collector.
- Scoring can be performed in a number of ways. For example, overall performance scores can be obtained (proportion of TARGETs hit), or scores can be obtained for specific task segments or over specific groupings of TARGETs, which might be important, for example, when a factor structure is known or suspected.

Variations of the TARGETs method have been used to support evaluation of small teams such as aircrews (Fowlkes et al., 1994; Salas et al., 1999; Stout et al., 1997) and large distributed teams within military warfare exercises (e.g., Dwyer et al., 1997; Dwyer et al., 1999).

53.2   Procedure

The six steps involved in implementing the TARGETs methodology entail creating and maintaining linkages between measurement objectives, scenario design, and the development of the measurement tools (Table 53.1).

53.2.1   Step 1: Identify Measurement Objectives

Determination of measurement objectives is the first link in the assessment chain. The objectives will drive the development of the scenario events and, ultimately, the behavioral checklists that are developed. For training evaluation, training objectives have been used to drive the development of measurement objectives (Salas et al., 1999). Measurement objectives have also been derived in test and evaluation contexts in which the goal has been to assess the impact of system design on team performance (Fowlkes et al., 1999).

TABLE 53.1 Example of Links between Measurement Objectives, Scenario Events, and Acceptable Responses

Measurement Objective	Scenario Event	Acceptable Response
Assertiveness: states opinion on decisions/procedures	Flight leader describes procedure that violates standard operating procedure	Crew questions procedure
Assertiveness: asks questions when uncertain	Ship's air traffic control provides erroneous vector to first waypoint	Crew questions heading information

Note: The event was "heavy enemy artillery" during a close air-support mission scenario, causing the mission to be delayed while the plan was revised to provide a safe route or air coordination area (ACA) for the pilots.
Source: Adapted from Fowlkes et al., (1994), *Mil. Psychol.*, 6, 47–61.

53.2.2 Step 2: Develop Scenarios and Events

Once measurement objectives have been developed, scenarios and trigger events are developed to permit assessment of whether the targeted performance has been achieved. The development of scenarios is a crucial aspect of the methodology, as they fully determine what can be observed, recorded, measured, and analyzed. The types of events that are included control the validity of resulting measures. The number of events (and resulting observations) can be used to control reliability.

Scenario events are included in scenarios that have been systematically and intentionally designed *a priori* to permit assessment of whether training objectives have been met. Without preplanned scenario events that are deliberately linked to training objectives, the opportunity for participants to perform those objectives and the ability to observe that performance is left to chance. Events can include routine situations such as mission phase. Events also can include nonroutine prompts that are introduced to serve as cues for behaviors that occur infrequently (e.g., assertiveness) and to elicit behaviors that are not always observable (e.g., awareness of team member roles).

In individual or small-team settings, events can be presented at preplanned times that would be known to the data collectors, making it easy for them to assess responses to the events. Situations such as this, in which task content can be controlled, are ideal for measurement. TARGETs can also be implemented in exercises that cannot be fully scripted. In these cases, the scenarios must be crafted so that events important for measurement will occur naturally as a result of the interactions between participants and simulated entities, even though their timing cannot be known beforehand. Data collectors are instructed to sample some predetermined number of each type of event included in the scenario. In this case, the task for the data collector is more difficult, as he or she must remain vigilant to identify the occurrence of key events. Handheld data-collection devices have been used to alert data collectors to the occurrence of key events (Fowlkes et al., 1999) making this task easier.

53.2.3 Step 3: Develop Behavioral Checklists That Capture Acceptable Responses to Events

Performance measurement tools are developed around the scenario events to provide links between measurement objectives and diagnosis of performance. Specifically, behavioral checklists are developed that show each event and the acceptable responses to each event (Figure 53.1). These are scored as either present or absent. Acceptable responses are determined using the input of domain experts and through the application of performance standards and standard operating procedures. Performance scores represent the percent of acceptable subtasks performed (i.e., number of acceptable responses/number of valid opportunities). Scores can be aggregated, for example, by mission phase, by measurement objective, or by team, as the situation dictates.

53.2.4 Step 4: Develop Scenario Control Measures

Scenario control measures can be developed to ensure that desired events will occur. In scripted exercises, scripts specify when events should be introduced as well as communications that should occur between the team being studied and other teams, agencies, or personnel that are included in the scenario. When done well, scripts are transparent to participants.

53.2.5 Step 5: Pilot Test the Scenarios and Measures

If possible, the scenario and performance measures should be pilot tested prior to using them for measurement. Pilot testing can be used to ensure that events can be introduced as intended, responses to events can be observed, and that the scenario is realistic and at an appropriate level of difficulty.

53.2.6 Step 6: Perform Measurement

Depending on the application, performance can be scored during the scenario or after the performance period using videotapes or other recordings of performance. Both paper-and-pencil and computer-based formats have been used to support data collection using the TARGETs method. In addition, both domain experts and trained observers have been be used to score performance.

53.3 Advantages

- Maintains links from measurement to training objectives, enhancing the relevance of performance scores.
- Facilitates observation of complex performance by focusing measurement on specific events.
- Controls task content, creating comparable "tests" across teams.
- Produces diagnostic performance scores.
- Possesses good psychometric properties.

53.4 Disadvantages

- Labor-intensive effort required to develop the TARGETs checklists and scripts.
- Requires the support of domain experts for development.
- Checklists are scenario specific.
- Requires human observers for data collection vs. automated data collection.
- Difficult to develop scenarios and events that are equal in difficulty.

53.5 Related Methods

The SALIANT (situational awareness linked instances adapted to novel tasks) tool is based on the TARGETs methodology and is used to assess situational awareness (Dwyer and Salas, 2000; Muniz et al., 1998). TRACTs (tactically relevant assessment of combat teams) is another performance assessment methodology based on TARGETs that has been applied in synthetic warfare environments to assess the performance of distributed teams (Fowlkes et al., 1999). Johnston et al. (1995) describe event-based measures for shipboard teams. Also related to the TARGETS methodology is (a) an event-based knowledge-elicitation technique described by Fowlkes et al. (2000) and (b) training approaches such as team dimensional training (TDT) (Smith-Jentsch et al., 1998) and the event-based approach to training (EBAT, see Chapter 47) (Fowlkes et al., 1998), as they rely on scenario events to create training opportunities.

53.6 Standards and Regulations

Acceptable responses to events can be obtained from published tasks lists. Examples include mission-essential task lists, mission training plans, and unit standard operating procedures.

53.7   Approximate Training and Application Times

Domain experts can be trained to use the TARGETs method in 30 min to 1 hour (Fowlkes et al., 1999). Training typically encompasses a review of the checklists to be used, scoring procedure, guidelines for sampling events, and use of an automated data-collection tool, as applicable. Trained observers require several hours of training. In addition to the items required for domain experts, training entails defining each behavior to be observed followed by practice observing performance with the TARGETs checklist.

Implementing the TARGETs method requires at least 8 hours and may require considerably longer depending on the availability of domain experts.

53.8   Reliability and Validity

Interobserver reliabilities for the TARGETs method have been reported as $r = 0.94$ by Fowlkes et al. (1994) and as $r = 0.88$ for Study 1 and $r = 0.97$ for Study 2 by Salas et al. (1999). Fowlkes et al. (1994) reported that stability of performance scores across flight segments was $r = 0.81$. Finally, Stout et al. (1997) reported no difference in results when performance was scored by domain experts or trained observers.

53.9   Tools Needed

TARGETs can be implemented in a paper-and-pencil format. The methodology has also been implemented using handheld data-collection instruments for assessment of team performance in a distributed team (Fowlkes et al., 1999). Computer-based data collection facilitates the handling of multiple checklists by data collectors, alerting data collectors about upcoming events, automatic elicitation of checklists, and timely postexercise data analysis.

References

Dwyer, D.J., Fowlkes, J.E., Oser, R.L., Salas, E., and Lane, N.E. (1997), Team performance measurement in distributed environments: the TARGETs methodology, in *Team Performance Assessment and Measurement: Theory, Methods, and Applications*, Brannick, M.T., Salas, E., and Prince, C., Eds., Lawrence Erlbaum Associates, Hillsdale, NJ, pp. 137–153.

Dwyer, D.J., Oser, R.L., Salas, E., and Fowlkes, J.E. (1999), Team performance measurement in distributed environments: initial results and implications for training, *Mil. Psychol.*, 11, 189–215.

Dwyer, D.J. and Salas, E. (2000), Principles of performance measurement for ensuring aircrew training effectiveness, in *Aircrew Training and Assessment*, O'Neil, H.F., Jr. and Andrews, D.H., Eds., Lawrence Erlbaum Associates, Mahwah, NJ, pp. 223–244.

Fowlkes, J.E., Baker, D., Salas, E., Cannon-Bowers, J.A., and Stout, R.J. (2000), The utility of event-based knowledge elicitation, *Hum. Factors*, 42, 24–35.

Fowlkes, J.E., Dwyer, D.J., Milham, L.M., Burns, J.J., and Pierce, L.G. (1999), Team skills assessment: a test and evaluation component for emerging weapon systems, in *Proceedings of the 1999 Interservice/ Industry Training, Simulation and Education Conference*, (CD-ROM), National Training Systems Association, Arlington, VA, pp. 994–1004.

Fowlkes, J., Dwyer, D.J., Oser, R.L., and Salas, E. (1998), Event-based approach to training (EBAT), *Int. J. Aviation Psychol.*, 8, 209–221.

Fowlkes, J.E., Lane, N.E., Salas, E., Franz, T., and Oser, R. (1994), Improving the measurement of team performance: the TARGETs methodology, *Mil. Psychol.*, 6, 47–61.

Johnston, J.H., Cannon-Bowers, J.A., Smith-Jentsch, K.A. (1995), Event-based performance measurement for shipboard command teams, in *Proceedings of the First International Symposium on Command and Control Research and Technology*, The Center for Advanced Command and Technology, Washington, D.C., pp. 274–276.

Muniz, E., Stout, R., Bowers, C., and Salas, E. (1998), A Methodology for Measuring Team Situational Awareness: Situational Linked Indicators Adapted to Novel Tasks (SALIANT), paper presented at the First Annual Symposium/Business Meeting of the Human Factors and Medicine Panel on Collaborative Crew Performance in Complex Systems, Edinburgh, U.K.

Salas, E., Fowlkes, J.E., Stout, R.J., Milanovich, D.M., and Prince, C. (1999), Does CRM training improve teamwork skills in the cockpit? Two evaluation studies, *Hum. Factors*, 41, 326–343.

Smith-Jentsch, K.A., Zeisig, R.L., Acton, B., and McPherson, J.A. (1998), Team dimensional training: a strategy for guided team self-correction, in *Making Decisions under Stress*, Cannon-Bowers, J.A. and Salas, E., Eds., American Psychological Association, Washington, D.C., pp. 271–297.

Stout, R.J., Salas, E., and Fowlkes, J. (1997), Enhancing teamwork in complex environments through team training, *J. Group Psychotherapy, Psychodrama, and Sociometry*, 49, 163–186.

54

Behavioral Observation Scales (BOS)

J. Mathew Beaubien
American Institutes for Research

Gerald F. Goodwin
U.S. Army Research Institute

Dana M. Costar
American Institutes for Research

David P. Baker
American Institutes for Research

Kimberly A. Smith
University of Central Florida

54.1 Background and Applications

At the most basic level, team performance involves two primary classes of behavior: taskwork and teamwork. Taskwork refers to the task-specific behaviors that directly support a team's mission. For example, these may involve flying an airplane, extinguishing a fire, or performing surgery. Teamwork, on the other hand, refers to a broad class of interaction-based behaviors that allow the team members to effectively coordinate their individual tasks. Although several taxonomies of teamwork behaviors have been developed (e.g., Fleishman and Zaccaro, 1992; Prince et al., 1992; Smith-Jentsch et al., 1998), most include four primary dimensions: communication, information exchange, fostering a supportive team environment, and leadership. As might be expected, teamwork behaviors apply to all types of teams, regardless of their specific mission.

In this chapter, we describe a general class of measurement methods that are collectively referred to as behavioral observation scales (BOS). These methods are typified by their focus on measuring observable teamwork behaviors. When they differ, it usually involves their choice of measurement scale. For example, some measure teamwork behaviors using a checklist format, while others use frequency counts or rating scales. Regardless of how they are designed, behavioral observation scales are extremely useful for diagnosing team performance problems, especially in training contexts. Team performance feedback

— such as that provided by a trained rater using a BOS — is critical to successfully managing a team's long-term performance because it helps the team members to recognize those aspects of team performance that they had previously overlooked.

54.2   Procedure

There are ten major steps in developing a BOS.

54.2.1   Step 1: Conduct a "Critical Incident" Workshop

The critical incident technique (CIT) is a job analysis procedure that involves collecting and analyzing narrative reports of actual team performance. During the workshop, subject-matter experts (SMEs) — persons who have in-depth knowledge about the team task because of their specialized education, training, or experience — write narratives that describe examples of effective and ineffective team performance that they have personally witnessed. Each narrative includes a description of the conditions leading up to the event, the team's performance during the event, the event's outcome, and the reasons why the team's behavior was particularly effective or ineffective. The number of narratives required to describe a team task is proportional to the task's complexity (Anderson and Wilson, 1997).

54.2.2   Step 2: Summarize Each Narrative Using a Behavioral Statement

Critical incident narratives provide a wealth of information for understanding team performance. However, they are much too numerous and too detailed to be included on a rating scale. It is also quite likely that many of these narratives will be redundant. To reduce the amount of information to a more manageable level, the developer should summarize each narrative into a behavioral statement that describes the team's performance in general terms. For example, rather than describing a specific example of effective communication, the behavioral statement would describe the type of behavior (e.g., establishing a contingency plan) that was performed. To eliminate redundancies, statements that describe similar behaviors should be merged. Depending on the complexity of the team's task, this can be accomplished with or without the help of SMEs.

54.2.3   Step 3: Identify Teamwork Dimensions

Before the behavioral statements can be included on a rating scale, they need to be grouped into logical categories. The developer should assemble an independent panel of SMEs to perform this task. After reviewing the behavioral statements, SMEs should work together to develop a list of teamwork dimensions that can be used to organize the statements. Rather than developing a completely new list, we suggest that SMEs start by consulting previous taxonomies (Fleishman and Zaccaro, 1992; Prince et al., 1992; Smith-Jentsch et al., 1998), because they have already been validated. If necessary, SMEs can adapt the teamwork dimensions to their particular team task. When complete, each teamwork dimension should be accompanied by a definition and a list of task-specific example behaviors.

54.2.4   Step 4: Classify Incidents into Dimensions

The next step is to classify each behavioral statement into the most appropriate teamwork dimension. The developer should assemble a new panel of SMEs to perform this task. The SMEs should first review each dimension's definition and example behaviors. Next, the SMEs should independently classify each behavioral statement into one of the teamwork dimensions. Behavioral statements that cannot be reliably classified should be eliminated.

SMEs should also evaluate the effectiveness of each behavioral statement using a 5-point scale with anchors ranging from "highly ineffective" (1) to "highly effective" (5). This information can be used for a variety of purposes. For example, it can be used to identify the anchors on a behaviorally anchored rating scale (BARS) (Smith and Kendall, 1963). Alternatively, it can be used to eliminate behavioral statements that fail to differentiate between effective and ineffective teams (Anderson and Wilson, 1997).

54.2.5 Step 5: Choose a Scale Metric

Once the behavioral statements have been reliably classified into the teamwork dimensions, the next step is to choose a response scale. Latham and Wexley (1981) recommend using a 5-point rating scale with anchors ranging from "almost never" (1) to "almost always" (5). Dwyer and colleagues (1997) recommend using a checklist format. Smith and Kendall (1963) recommend using BARS. Each method has its unique strengths and limitations. The choice of which metric to use should be based on several considerations, such as minimizing rater workload or rater errors. When complete, each BOS should include a dimension name, a definition, a rating scale, and several behavioral statements.

54.2.6 Step 6: Pilot Test the BOS

Before the behavioral observation scales can be used to evaluate team performance, they need to be pilot tested. When possible, the pilot test should occur under conditions that simulate how the BOS will be used in the field. Emphasis should be placed on identifying problems that could interfere with data collection. For example, the pilot test may reveal additional team performance dimensions that were not reflected in the BOS. Alternatively, the pilot test may reveal that the BOS places an exceedingly high level of cognitive workload on the rater. The results of the pilot test should be used to revise the BOS prior to its full-scale use.

54.2.7 Step 7: Train the Raters

All rating scales require rater training. We recommend a combination of behavioral observation training (Thornton and Zorich, 1980) and frame-of-reference (FOR) training (Bernardin and Buckley, 1981). Behavioral observation training provides raters with information on how to accurately detect, perceive, recall, and recognize specific behavioral events. FOR training provides raters with a common set of standards for evaluating team performance. The amount of time devoted to rater training should be proportional to the complexity of the rating task. For example, if the raters will perform additional tasks while evaluating the team (e.g., manipulating the parameters of a flight simulator), their training should be more intense than if their sole job is to evaluate the team.

54.2.8 Step 8: Collect Performance Ratings

To make the rating process as efficient as possible, the order of the behavioral statements (within each BOS) should correspond to the natural work flow. For example, when evaluating pilot crews, behavioral statements that describe communication during takeoff should precede those that describe communication during landing.

54.2.9 Step 9: Analyze the Data

Once the performance ratings have been collected, the developer should assess the data's quality. For each behavioral statement, the variability and item–BOS correlation should be computed. These statistics will identify behavioral statements that do not differentiate between effective and ineffective teams.

54.2.10 Step 10: Compute BOS scores and total scores for each team

If any of the behavioral statements are negatively worded, the data should be recoded. BOS scores can then be calculated by summing across all the behavioral statements within a BOS. Each team's total score is calculated by summing their individual BOS scores.

54.3 Advantages

- Reliably and accurately measures observable teamwork behaviors.
- Capable of assessing teamwork behaviors, taskwork behaviors, or both.
- Provides diagnostic information for improving future team performance.
- Can be adapted to suit the requirements of specific training objectives.
- Flexible enough to accommodate multiple observations per performance episode.

54.4 Disadvantages

- Cognitive demands may limit each rater to rating only a few teamwork dimensions.
- Multiple raters may be required to evaluate each team.
- BOS development can be time consuming and costly.
- Rater training is required to obtain reliable and valid measures of team performance.
- Scales developed for one team task may not work well for other team tasks.

54.5 Example Behavioral Checklist

Title: Communication
Definition: Communication involves sending and receiving signals that describe team goals, team resources and constraints, and individual team member tasks. The purpose of communication is to clarify expectations, so that each team member understands what is expected of him or her. Communication is practiced by all team members.
Example behaviors (check all that apply):

- Team leader establishes a positive work environment by soliciting team members' input.
- Team leader listens nonevaluatively.
- Team leader identifies bottom-line safety conditions.
- Team leader establishes contingency plans (in case bottom line is exceeded).
- Team members verbally indicate their understanding of the bottom-line conditions.
- Team members verbally indicate their understanding of the contingency plans.
- Team members provide consistent verbal and nonverbal signals.
- Team members respond to queries in a timely manner.

54.6 Related Methods

Two of the most frequently used forms of behavioral observation scales are Likert-type rating scales and behavioral checklists. Likert-type rating scales include a list of teamwork behaviors that are required to perform the task. For each behavior, the team's performance is assessed using a 5-point scale with anchors ranging from "highly ineffective" (1) to "highly effective" (5) (Smith-Jentsch et al., 1998). Behavioral checklists also include a list of teamwork behaviors that are required to perform the task. However, behavioral checklists use a binary scale to indicate whether or not each teamwork behavior was performed (Dwyer et al., 1997). An example behavioral checklist appears in Section 54.5.

54.7 Standards and Regulations

No single set of standards covers the development and use of team performance information collected via behavioral observation scales. The most relevant standards to the development of BOS are the Standards for Educational and Psychological Testing (American Educational Research Association, 1999). BOS should be developed taking great care to ensure that the resultant scales can be reliably used; are valid measures of the desired performance; are not biased toward or against ethnic, racial, or gender subgroups in their content; and that the scales can be used fairly to assess performance. The standards also require that a qualified and credentialed individual supervise the development, revision, and use of psychological tests and measures, including BOS. Those intending on using BOS to assess team performance should carefully review the standards prior to undertaking their development. Additionally, if the information derived from BOS is to be used for employment-related decisions, such as performance management, selection into training, or compensation, several civil rights laws are applicable, including Title VII of the Civil Rights Act of 1964, the Civil Rights Act of 1991, and the Age Discrimination in Employment Act. Guidance for the development of tests and assessments for use in employment decision making can be found in the Uniform Guidelines on Employee Selection Procedures (1978).

54.8 Approximate Training and Application Time

Rater training should involve a combination of behavioral observation training and frame-of-reference training. Depending on the number of learning exercises, rater training may require up to 4 hours. Application time varies with the number of teamwork dimensions being assessed, but may require up to 3 hours per team.

54.9 Reliability and Validity

Previous research suggests that with appropriate rater training, behavioral observation scales can provide reliable and valid measures of team performance. For example, using a behavioral checklist, Dwyer et al. (1997) reported interrater reliabilities ranging between .50 and .90. Smith-Jentsch et al. (1998) reported similar results. Using a five-point Likert-type scale, Smith-Jentsch and colleagues reported inter-rater reliabilities ranging between .82 and .91. They also found evidence of convergent, discriminant, and predictive validity. Specifically, each teamwork dimension was significantly correlated with itself across multiple training events (all rs ranged between .32 and .67). Moreover, the average correlation between the four primary teamwork dimensions was .15, indicating that they were relatively distinct from one another. Finally, ratings on three of the four teamwork dimensions uniquely contributed to predicting overall team decision accuracy.

54.10 Tools Needed

Behavioral observation scales typically require only paper and pencil. However, they can be completed using handheld computers with custom-written software. This can facilitate the raters' job, for example, by automatically calculating overall team performance scores and digitally recording notes about the team's performance.

References

American Educational Research Association, American Psychological Association, and National Council on Measurement in Education (1999), *Standards for Educational and Psychological Testing*, American Educational Research Association, Washington, D.C.

Anderson, L. and Wilson, S. (1997), Critical incident technique, in *Applied Measurement Methods in Industrial Psychology*, Whetzel, D.L. and Wheaton, G.R., Eds., Davies-Black, Palo Alto, CA, pp. 89–112.

Bernardin, H.J. and Buckley, M.R. (1981), Strategies in rater training, *Acad. Manage. Rev.*, 6, 205–212.

Dwyer, D.J., Fowlkes, J.E., Oser, R.L., Salas, E., and Lane, N.E. (1997), Team performance measurement in distributed environments: the TARGETs methodology, in *Assessment and Management of Team Performance: Theory, Research, and Applications*, Brannick, M.T., Salas, E., and Prince, C., Eds., Lawrence Erlbaum Associates, Mahwah, NJ, pp. 137–154.

Fleishman, E.A. and Zaccaro, S.J. (1992), Toward a taxonomy of team performance functions, in *Teams: Their Training and Performance*, Swezey, R.W. and Salas, E., Eds., Ablex, Norwood, NJ, pp. 31–56.

Latham, G.P. and Wexley, K.N. (1981), *Increasing Productivity through Performance Appraisal*, Addison-Wesley, Reading, MA.

Prince, A., Brannick, M.T., Prince, C., and Salas, E. (1992), Team process measurement and the implications for training, in *Proceedings of the Human Factors Society 36th Annual Meeting*, Human Factors Society, Santa Monica, CA, pp. 1351–1355.

Smith, P.C. and Kendall, L.M. (1963), Retranslation of expectations: an approach to the construction of unambiguous anchors for rating scales, *J. Appl. Psychol.*, 47, 149–155.

Smith-Jentsch, K.A., Johnston, J.H., and Payne, S.C. (1998), Measuring team-related expertise in complex environments, in *Making Decisions under Stress: Implications for Individual and Team Training*, Cannon-Bowers, J.A. and Salas, E., Eds., American Psychological Association, Washington, D.C., pp. 61–87.

Thornton, G.C. and Zorich, S. (1980), Training to improve observer accuracy, *J. Appl. Psychol.*, 65, 351–354.

Uniform Guidelines on Employee Selection Procedures (1978), *Fed. Regist.*, 43, 38290–38315.

55

Team Situation Assessment Training for Adaptive Coordination

Laura Martin-Milham
University of Central Florida

Stephen M. Fiore
University of Central Florida

55.1 Background and Application

The question of concern is "How do teams respond to unexpected or unanticipated events?" This issue is best understood through areas of research in team cognition and team coordination (e.g., Salas and Fiore, 2004). A critical component of team cognition is that of team situation assessment (team SA$_s$), generally described as the process used (a) to assess risk and the time available for decisions and (b) to construct a mental picture of the operational environment. Fundamental to this process is the rapid apperception of cues and cue patterns in the environment (Salas et al., 2001). Environmental cues have been described as "situation observables" that get factored together with general knowledge (Noble, 1993), and they are said to be responsible for triggering rapid pattern-recognition processes that lead to initial decisions or problem solutions (Klein, 1997).

"Critical cues" form the basis with which the problem solver and/or decision maker matches the features in the environment with stored knowledge. From the team-coordination standpoint, a team can engage the appropriate behaviors only when an operational situation has been accurately assessed. When a team or team member recognizes that either the team has inaccurately assessed a situation, or when a new cue suggests that the initial assessment is incorrect, then the team must rapidly coordinate an adaptive response (i.e., change their behaviors). Thus, *team adaptability* describes how teams effectively deal with altered and/or unplanned events (Fiore et al., 2003). In particular, adaptability corresponds to the degree to which a team is able to modify behavior and/or plans in the presence of changing situational demands

(Prince and Salas, 1993; Zalesny et al., 1995). In this chapter, we describe a methodology for linking cue and cue pattern recognition to team behaviors in order to combine team SA_s processes with actual adaptive behaviors.

55.2 Procedure

We limit this approach to the cue detection process and the appropriate associated behavioral response, because this represents the critical first step in the processes that need to be engaged for adaptive team performance. Without the timely detection of an abnormality in the environment, team members may not have enough time to respond. To train these team SA_s components, we utilize synthetic task environments. Synthetic task environments (STEs) are designed to mimic complex multi-role team tasks, but they are scaled to be distributed using more-standard technology (Elliott et al., 2001). Through the use of STEs, we are able to systematically vary aspects of the task in order to train both cognitive processes and behavioral responses.

Our approach utilizes the distributed dynamic decision-making (DDD) platform, a synthetic task environment simulating military command and control (e.g., Fiore et al., 2002). Within the context of training teams for adaptability, synthetic task environments represent an effective means of exposing team members to the critical cues and cue patterns so that the appropriate behaviors can be elicited. For this effort, DDD was modified to simulate various elements of a bombing "strike" team to represent a fighter element, a command-and-control element, and a missile shooter. The trainee assumed the role of the command-and-control element, whose primary responsibility was to identify unknown aircraft, pass the information to other team members, and ensure that hostile aircraft do not enter forbidden zones. A confederate team member role-played the fighter aircraft and a missile shooter, whose roles were to drop bombs on a target, to target and kill enemy aircraft, and to destroy surface-to-air (SAM) sites.

We next describe the development of the critical subcomponents of team SA_s that were the foundation for our training methodology.

55.2.1 Configuring Training for Team SA_s

The following items are important for configuring training for team SA_s (see Figure 55.1):

1. *Situation events* are crafted as a foundation for team skills training.
2. *Cue categories* are determined and broken out into:
   - Task-oriented cues: provide information regarding the mission
   - Team-member cues: provide information regarding how other team members are doing (e.g., teammate is confused; teammate has high workload)
3. *Adaptive behaviors* are team behaviors that support a team member in cases where they are under high workload or are unaware of an unfolding situation.

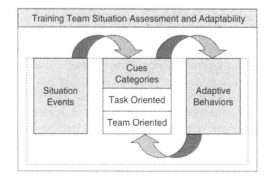

FIGURE 55.1   Framework for components of training team SA_s.

55.2.1.1 Determining Content of Cue Categories

Depending upon the domain in question, critical cues can be determined from the literature, culled from operational manuals, or determined from domain or cognitive task analyses (Neville et al., 2002). What is critical is that cue patterns important to successful team coordination are identified. These are cues and cue patterns that represent meaningful changes in the environment and that require coordination among various team members. These cue patterns are then structured around timelines, team members, and threats (see Table 55.1).

55.2.1.2 Embedding Adaptive Team Behaviors

In order for the critical cues to be instantiated within the team task, context is provided so that trainees can see a cue pattern unfolding (cf. Greeno, 1998). For example, a team member who is off the timeline may reach a checkpoint early. Being early to a checkpoint may mean that other team members may not have performed tasks ensuring the safety of the first team member. Further, this context is used to train two broad categories of knowledge (Marshall, 1995) that form the foundation for the implementation of critical team skills for adaptive coordination:

1. *Planning knowledge:* This is the process of using knowledge already gained through training to create expectations and plans of what will occur next. Based on an assessment of the situation, there are goals that guide specific execution behaviors. For example, if some team members have low situation awareness due to limitations of their equipment, then a planning goal of the trainee may be to provide environment information to those team members.
2. *Execution knowledge:* This includes carrying out the plan. Based upon the trainees' goals, there are specific ways to achieve "the plan." For example, if a trainee wants to provide information to a team member, s/he can use a team behavior (e.g., information exchange). This team behavior describes an observable indication of how the trainee is dealing with the situation and thus is a measurable indicator of knowledge.

To link these generic categories of knowledge to the rapid coordination of adaptive team behaviors, they are taught within the "team dimensional training" (see Table 55.2) framework (Smith-Jentsch et al., 1998). By linking these factors, the goal is to have trainees learn to use team skills adaptively. By this we mean that trainees are trained to:

- Recognize cue patterns
- Know what these cue patterns mean in that context
- Have goals to deal with these cue patterns
- Choose their team skills appropriately with respect to the cue patterns in that situation

TABLE 55.1 Types of Critical Cues To Be Trained for Team SA$_s$

Cue	Cue Definition
Timeline cues	Cues providing information for events in which the team is monitoring and utilizing the timeline as a way to coordinate the performance of the team
	Example: The team or a team member is engaging an enemy appropriately (e.g., within a certain distance of their location); the team reaches checkpoints on time
Threat cues	Cues providing information on threats, such as from where threats are coming and the priority of unknowns (whether enemy or friendly aircraft). Priority can indicate how much a threat the enemy could be
	Example: If an unknown aircraft is within two squares of a strike aircraft they are a high priority
Team member cues	Cues providing information about workload and the situation awareness of other team members
	Example: During the middle of an engagement, a team member is not responding/engaging a high-priority unknown; this suggests the team member is unaware of the threatening situation

TABLE 55.2 Team Dimensional Training Components

Team Dimension	Definition
Information exchange	The skill of passing information and gathering information from other team members in order to perform your task and meet mission goals
	Example: Providing big-picture updates about the threats, unknowns, and placement of other strike-team members
Supporting behaviors	The skill of providing or accepting assistance when one's self or a teammate are overtasked
	Example: Cross-checking information, such as comparing knowledge, and monitoring team members' behaviors to ensure they are acting appropriately
Initiative/leadership	The skill of performing an action or addressing a problem prior to being asked
	Example: Providing guidance and suggestions, such as giving a team member information without being asked
Communication	The skill of being able to get a message across clearly
	Example: Giving complete reports, not just if there is an enemy, but also where an enemy currently resides and/or is heading

Source: Smith-Jentsch et al., (1998), Team dimensional training: a strategy for guided team self-correction, in *Making Decisions under Stress: Implications for Individual and Team Training*, Cannon-Bowers, J.A. and Salas, E., Eds., American Psychological Association, Washington, D.C.

In complex operational environments, team members must be able to monitor not only their own task requirements, but they must also objectively and subjectively assess teammate actions. More specifically, trainees must:

- Learn to recognize when a team member has high workload and/or low situation awareness
- Know what that means (the team member may be busy, and if there are no threats, there is nothing that is of critical importance to that person)
- Know the goals for dealing with this (do not interrupt the team member unless there is time-critical environmental information)
- Determine the resulting choice of a team skill (halting information in the member's direction until he or she is ready to hear an update)

If the aforementioned factors are to be linked successfully, they must be integrated within the context of the situation events. Specifically, the team skills training is couched in situations occurring in the simulation and are designed such that the cues trigger a required team-related action that facilitates adaptive coordination. Thus, these team skills being trained are supportive, in that they require team members to step in and provide either backup, show initiative, or pass information when a team member needs that information.

55.2.2 Cue Assessment Training

For the training modules, snapshots and video were taken of various cue patterns as they played out in DDD. Cues critical to the DDD mission were used to illustrate stimulus patterns for the trainee (i.e., cues related to timeline and threats were presented to trainees). Timeline cues describe whether team members are or are not turning at a checkpoint and whether they arrive at a checkpoint on time. Threat cues describe various presentations of threats, such as from where threats are coming and how they need to be prioritized. Cue recognition training was provided to expose trainees to the task-relevant cues that were necessary to successfully identify the team relevant event. For example, a cue pattern might illustrate that a team member did not turn at a checkpoint.

55.2.3 Team Skill Training

For each cue pattern presented, there was an optimal set of team behaviors identified *a priori*. Based on cognitive task analyses of actual strike-team performance, these behaviors were suggested as those that would be most adaptive for a particular situation. Based on these behaviors, we developed training that presented: (a) cue patterns (cue pattern recognition training) and (b) behaviors in response to those cue patterns (team skills training). As mentioned, the goal was to develop training that would facilitate adaptive choice of behaviors in response to cue patterns. Toward this end, anchored examples were used to create a link between problem situations (i.e., the cue pattern) and solutions (i.e., adaptive behavior). The specific steps followed were:

- Build training to illustrate cue patterns and appropriate team behaviors to deal with the situations.
- Provide examples of how to show initiative as the simulation unfolds.
- Use confederates to provide practice and testing opportunities for trainees.
- Develop scenarios that provide opportunities for trainees to see specific cue patterns to which they must react.

55.2.4 Measures of Knowledge Acquisition and Application

55.2.4.1 Strategic Knowledge Acquisition

Trainees engage in team skill training in which they are presented with definitions and examples of team skills. Following team skills training, trainees are assessed on strategic knowledge. This knowledge involves the acquisition of a set of plans of action that can be applied in a variety of tasks within a domain (de Jong and Ferguson-Hessler, 1996; Stout et al., 1996). For this assessment, trainees were presented with videotaped cue patterns, asked to circle the cue patterns present, and then to write in the team skill they would use and specifically what they would say or do in that situation. Trainees are assessed on accurate perceptual identification, labeling of the "communication," and content of the communication. The process involved the following steps:

- Build performance measures to assess if training was effective, specifically to assess whether training resulted in successfully adaptive team performance
- Take events that were scripted to occur in the scenarios
- Chronologically list the events at the expected times throughout the mission
- For each of the events, list the possible behaviors, ranging from no response to specific supporting teamwork behavior(s) that indicated maximum awareness and responsiveness to the situation

55.2.4.2 Interactive Simulation Test (Adaptive Team Performance)

The final stage of testing is designed to assess the participants' adaptive behavior while playing an interactive DDD simulation scenario. Toward this end, trainees are engaged in brief scenarios with a confederate present. Trainees are instructed to identify unknowns, to minimize friendly losses, and to ensure that the team hits the target at a predetermined time. Through programming threat configurations and through confederate behavior, the cue patterns presented in training were illustrated throughout the scenarios. The confederates should be used to illustrate:

- Team members under varying conditions of workload (workload is illustrated by scripting whether or not the confederate was dealing with a number of events [e.g., flying through an area populated with a number of high-priority aircraft] or by being immersed in an engagement with an enemy aircraft)
- Team members with or without access to information
- Instantiation of cue patterns related to timeline, team members, and threats
- Level of awareness of situational events manipulated by the confederate reacting in one of several ways to high-priority aircraft (e.g., low awareness illustrated when the confederate would attack a friendly aircraft or ignore a high-priority enemy aircraft or SAM site)

55.2.4.3 Event-Based Adaptive Team Performance Measure

During the scenarios, trainees were assessed on their ability to use team skills adaptively. This was accomplished by using an event-based performance measure that listed the cue patterns as they would occur chronologically. Next to the cue patterns, several possible behaviors were listed, from the most adaptive choice to maladaptive choices (see Figure 55.2).

55.3 Results

In laboratory studies, participants trained using this method showed an increase in strategic knowledge. In addition, participants who learn components of team SA_s (i.e., cue and team skills training) make more-accurate calls during testing and also show superior adaptive team performance during the simulation trials.

55.4 Advantages

- Links empirically derived theories of learning and practice (e.g., situated learning, see Greeno, 1998) with actual training
- Links perception, cognition, and team performance within a single training method
- Use within a flexible synthetic task environment facilitates training evaluation

55.5 Disadvantages

- The method requires assistance by subject-matter experts in early stages of training development (e.g., for scenario development and assessment questions).
- Considerable up-front time is needed to develop the materials (e.g., identifying critical cue patterns; determining relevant coordinative behaviors).
- A confederate is needed for training.

Scenario 5

X = cue did not happen
S = striker was killed
F = Friendly was killed

Trainee #__________

Time	Coding	Situation	Subject's Label			Subject's Content		
0:00		HP friendly: 210	Identity	Monitor	Focus	Identifying aircraft (placement/ type)	Monitoring engagement	Focus: Watch out, Identity call if labeled focus
			Big Picture	Confirm	Direction	Route cut at checkpoint	Confirming destruction	Appropriate command
1:00		On time	Identity	Monitor	Focus	Identifying aircraft (placement/ type)	Monitoring engagement	Focus: Watch out, Identity call if labeled focus
			Big Picture	Confirm	Direction	Route cut at checkpoint	Confirming destruction	Appropriate command
1:00		Not turning	Identity	Monitor	Focus	Identifying aircraft (placement/ type)	Monitoring engagement	Focus: Watch out, Identity call if labeled focus
			Big Picture	Confirm	Direction	Route cut at checkpoint	Confirming destruction	Appropriate command

FIGURE 55.2 Sample of portion of event-based performance measure.

55.6 Related Methods

A variety of training methods address the subcomponents described here. Adaptive team coordination has been investigated in the context of learning how to modify coordination strategies, depending on situational demands (Entin and Serfaty, 1999). The skills associated with team dimensional training have been trained in a number of varying situations, ranging from, for example, shipboard systems training (e.g., Smith-Jentsch et al., 1997) to coordinating effective containment during mock prison riots. Others have examined methodologies for linking perceptual knowledge to more-complex decision making, but only at the level of the individual decision maker (e.g., Kirlik et al., 1996). As such, these approaches have all addressed the components we have presented. Nonetheless, our present methodology incorporates these into an integrated program to link the core cognitive processes with associated team behaviors that are fundamental to team SA_s and adaptive coordination.

55.7 Standards and Regulations

There are no standards or regulations for this methodology.

55.8 Approximate Training and Application Time

The time for developing the materials for the cue-recognition portions of the training will vary depending upon the complexity of the task. Because standard methodologies for the TDT implementation already exist, there is little time required other than the generation of examples relevant to the given operational environment.

55.9 Reliability and Validity

Experimentally based reliability and validity of this methodology has been demonstrated in laboratory studies of team training. Specifically, laboratory studies show significant correlations between measures and across trials for this methodology.

55.10 Tools Needed

To develop the training materials, there are no specific tools needed other than standard office-based software.

References

de Jong, T. and Ferguson-Hessler, M.G.M. (1996), Types and qualities of knowledge, *Educ. Psychol.*, 31, 105–113.

Elliott, L.R., Dalrymple, M., Regian, J.W., and Schiflett, S.G. (2001), Scaling scenarios for synthetic task environments: issues related to fidelity and validity, in *Proceedings of the 45th Annual Meeting of the Human Factors and Ergonomics Society*, Human Factors and Ergonomics Society, Santa Monica, CA, pp. 377–381.

Entin, E.E. and Serfaty, D. (1999), Adaptive team coordination, *Hum. Factors*, 41, 312–325.

Fiore, S.M., Cuevas, H.M., Scielzo, S., and Salas, E. (2002), Training individuals for distributed teams: problem solving assessment for distributed mission research, *Comput. Hum. Behav.*, 18, 125–140.

Fiore, S.M., Salas, E., Cuevas, H.M., and Bowers, C.A. (2003), Distributed coordination space: toward a theory of distributed team process and performance, *Theor. Issues Ergonomic* Sci., 4(3–4), 340–363.

Greeno, J.G. (1998), The situativity of knowing, learning, and research, *Am. Psychologist*, 53, 5–26.

Kirlik, A., Walker, N., Fisk, A.D., and Nagel, K. (1996), Supporting perception in the service of dynamic decision making, *Hum. Factors*, 38, 288–299.

Klein, G. (1997), The recognition-primed decision (RPD) model: looking back, looking forward, in *Naturalistic Decision Making*, Zsambok, C.E. and Klein, G., Eds., Lawrence Erlbaum Associates, Mahwah, NJ, pp. 285–292.

Marshall, S.P. (1995), *Schemas in Problem Solving*, Cambridge University Press, New York.

Neville, K., Fowlkes, J., Milham, L., Bergondy, M., and Glucroft, B. (2002), Team coordination expertise in complex distributed teams: a preliminary cognitive task analysis of the navy carrier air wing strike team, in *Proceedings of International Symposium of Aviation Psychology*, (CD-ROM), Columbus, OH.

Noble, D. (1993), A model to support development of situation assessment aids, in *Decision Making in Action*, Klein, G., Orasanu, J., Calderwood, R., and Zsambok, C., Eds., Ablex, Norwood, NJ, pp. 287–305.

Prince, C. and Salas, E. (1993), Training and research for teamwork in the military aircrew, in *Cockpit Resource Management*, Wiener, E.L., Kanki, B.G., and Helmreich, R.L., Eds., Academic Press, San Diego, CA, pp. 337–366.

Salas, E. and Fiore, S.M., Eds. (2004), *Team Cognition: Understanding the Factors That Drive Process and Performance*, American Psychological Association, Washington, D.C.

Salas, E., Cannon-Bowers, J.A., Fiore, S.M., and Stout, R.J. (2001), Cue-recognition training to enhance team situation awareness, in *New Trends in Collaborative Activities: Understanding System Dynamics in Complex Environments*, McNeese, M., Salas, E., and Endsley, M., Eds., Human Factors and Ergonomics Society, Santa Monica, CA, pp. 169–190.

Smith-Jentsch, K.A., Johnston, J.H., Cannon-Bowers, J.A., and Salas, E. (1997), Team dimensional training: a methodology for enhanced shipboard training, in *Proceedings of the 19th Annual Interservice/ Industry Training System and Education Conference*, (CD-ROM), Orlando, FL.

Smith-Jentsch, K.A., Zeisig, R.L., Acton, B., and McPherson, J.A. (1998), Team dimensional training: a strategy for guided team self-correction, in *Making Decisions under Stress: Implications for Individual and Team Training*, Cannon-Bowers, J.A. and Salas, E., Eds., American Psychological Association, Washington, D.C., pp. 271–297.

Stout, R.J., Cannon-Bowers, J.A., and Salas, E. (1996), The role of shared mental models in developing team situational awareness: implications for training, *Training Res. J.*, 2, 85–116.

Zalesny, M.D, Salas, E., and Prince, C. (1995), Conceptual and measurement issues in coordination: implications for team behavior and performance, in *Research in Personnel Human Resources Management*, Vol. 13, Ferris, G.R., Ed., JAI Press, Greenwich, CT, pp. 81–116.

56

Team Task Analysis

C. Shawn Burke
University of Central Florida

56.1 Background and Applications

The last several decades have witnessed an increased use of teams as a key organizational strategy. Accompanying this increase is a focus on training and evaluating the effectiveness of organizational teams. The foundation of both training and evaluation relies on a determination of the competencies (i.e., knowledge, skills, and attitudes) that need to be trained and/or evaluated. Perhaps one of the most common and theoretically grounded methods for determining the requisite competencies is the use of a task analysis. Goldstein (1993) defines a task analysis as a tool that is used to "determine the instructional objectives that will be related to performance of particular activities or job operations." (p. 54)

While there are numerous methods for analyzing team tasks, most do not typically capture the requisite cues, conditions, and standards that provide the basis for team tasks (Swezey et al., 1998). Most methods of analyzing team tasks focus on identifying only one of the two behavioral tracks needed when completing team tasks (i.e., the taskwork track). However, Glickman et al. (1987) have shown that both taskwork (i.e., task-oriented skills) and teamwork skills (i.e., behavioral, attitudinal, and cognitive responses needed to coordinate with fellow team members) are needed to complete team tasks effectively. Thus, procedures for analyzing team tasks need to capture both tracks of skills. In light of this need, several researchers have been working over the past decade to develop and refine a procedure known as team task analysis (Bowers et al., 1993; Bowers et al., 1994; Levine and Baker, 1990, 1991; McNeese and Rentsch, 2001). This procedure not only allows researchers and practitioners to identify the operational skills needed within team tasks, but also the teamwork skills needed for smooth coordination among team members. However, the procedure is still not widely used among organizations (except for the military and aviation communities) because of a lack of prescriptive guidance and the effort required to complete team task analyses.

56.2   Procedure

As stated earlier, despite the large interest and guidance pertaining to task-analysis procedures for individuals, there is much less guidance for team task analysis, and the little that does exist is not well integrated. This chapter attempts to integrate the information that currently exists and present it in a condensed format. A complete description of all the variations are beyond the scope of this chapter, so the reader is referred to key citations within the text for more detailed information.

56.2.1   Step 1: Conduct a Requirements Analysis

The first step in conducting a team task analysis is a requirements analysis. This is where the target job is defined by creating a narrative that describes the duties and conditions under which the job is to be performed. Defining the target job is an important first step because job and/or position titles are often misleading in terms of the actual composition of the job. After the target job is defined, the next step is to identify which of the various knowledge-gathering methodologies will be used within the team task analysis (see Table 56.1). Selection of the appropriate methodologies is based on the characteristics identified in the target job and the purpose of the team task analysis (training, selection, design, evaluation). Once the appropriate methods have been selected, a protocol for conducting the team task analysis is prepared. Finally, the participants who will serve as subject-matter experts (SMEs) during the actual analysis are identified. Subject-matter experts are often job incumbents and supervisors, depending on the stage of the team task analysis. The exact number of subject-matter experts used is often a function of organizational constraints and resources (time, money, availability). The requirements analysis may be the easiest step to accomplish within a team task analysis, but it is also one of the most crucial, for it lays the foundation for the entire effort.

TABLE 56.1   Key Methods Used to Gather Information during Task Analysis

Methods	Description	Pros/Cons
Observation	Observation of job incumbents performing the job or simulation of the job	Minimizes interruption of work; *but* takes a highly skilled observer
Questionnaire(s)	Surveys given to either job incumbents or supervisors. These can take a variety of forms, most often seen as forced choice, priority ranking, or rating scales. Surveys given early in the process often include instructions for the addition of missing tasks	Relatively inexpensive, can reach a large number of people, easy to summarize data reported; *but* little room for free expression, takes time and expertise to develop, and suffers low return rates
Source documents	Gathering of information from journals, handbooks, manuals, annual reports, and essential task lists	Relatively simple and easy to conduct, serves as a good starting point; *but* often reflects the past and what should be done as opposed to what *is* done
Interviews	Conducted with subject-matter experts (job incumbents or supervisors) and can be structured, semistructured, or structured. Can be done one-on-one or in a group discussion format.	Provides much information, allows participants to express themselves and clarify misunderstandings; *but* time consuming and results can be difficult to quantify
Work samples	Similar to observation, but in written form.	Carry some of the advantages of observation; *but* take time away from actual work

Source: Adapted from Goldstein, I.L. (1993), *Training in Organizations: Needs Assessment, Development, and Evaluation*, 3rd ed., Brooks/Cole Publishing, Pacific Grove, CA.

56.2.2   Step 2: Identify the Tasks That Compose the Target Job

Now begins the actual team task analysis. This step involves identifying the full spectrum of tasks that are performed on the target job. As such, this process begins in the same manner as that of a job analysis conducted for individual tasks. (For a thorough review, see Goldstein [1993] and Goldstein and Ford [2002].) Typically, source documents and interviews with SMEs are used to identify an initial set of tasks that compose the target job. At this point, the SMEs that are used are often job incumbents, since the focus is on identifying the full range of tasks that are actually done on the job. The identification of tasks that compose the target job is often an iterative process that is conducted with multiple sets of SMEs in an effort to ascertain the full range of tasks. Once tasks are identified, task statements are written for each. These statements should have the following characteristics: (a) be direct and avoid long sentences, (b) begin with a verb that describes the type of work to be accomplished, and (c) describe what the worker does, how it is done, to whom it is done, and why it is done (Goldstein and Ford, 2002).

56.2.3   Step 3: Identify Teamwork Taxonomy

Once the initial tasks have been identified, the next step is to determine which of those tasks relate to the taskwork track and which relate to the teamwork track. To do this, one must first behaviorally define what is meant by teamwork. Although there is no one established methodology for identifying such behaviors (Bowers et al., 1993), there are several teamwork taxonomies available in the literature (Cannon-Bowers et al., 1995; Fleishman and Zaccaro, 1992; Smith-Jentsch et al., 1998; Stevens and Campion, 1994). These taxonomies can be used to provide the foundation for step 4, coordination analysis.

56.2.4   Step 4: Conduct a Coordination Analysis

After a teamwork taxonomy has been identified, it can be used to conduct a coordination analysis. The purpose of this analysis is to determine the extent to which each of the previously identified tasks places a requirement on the team to coordinate its activities (i.e., use the teamwork track of behaviors) so that it can accomplish the task. In other words, it determines the extent that team members must interact with and rely on one another to accomplish a given task.

Although many methods can be used to conduct a coordination demand analysis, perhaps the most common is the use of surveys. One example of the use of surveys to gather such information is illustrated by the work of Bowers et al. (1993). These researchers used an adaptation of the Cannon-Bowers et al. (1995) teamwork dimensions as a framework to ask SMEs the extent to which each of the previously identified tasks (gathered in step 2) required each of seven teamwork behaviors (representing aspects of coordination), as well as an overall assessment of coordination (see Figure 56.1). Once this information

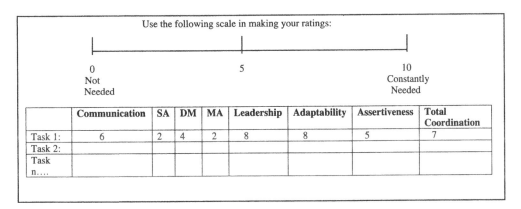

FIGURE 56.1 Prototypic survey used within coordination demand analysis. (Adapted from Bowers et al., 1993.)

is collected, it can then be subjected to techniques such as cluster analysis to identify task clusters based on coordination demand. Typically, subject-matter experts during this phase of the analysis are job incumbents and supervisors, depending on organizational resources.

56.2.5   Step 5: Determine Relevant Taskwork and Teamwork Tasks

After conducting the coordination demand analysis, the practitioner has a list of the job tasks that specify whether the tasks require taskwork and/or teamwork. The next step is to determine which of the tasks are the most relevant. The exact indices that are used to determine relevance will vary along two dimensions. First, it has been suggested that the indices used for the taskwork tasks do not directly transfer to teamwork tasks (Bowers et al., 1994; Dyer, 1984). For example, team tasks typically are not difficult to learn, whereas taskwork tasks may be difficult to learn (Franz et al., 1990). Second, the purpose of the team task analysis (selection, design, training, evaluation) may impact the indices that are most appropriate to use.

To determine the relevancy of the initially identified tasks, a questionnaire is often developed using a Likert-type response format. The left-hand side of this questionnaire shows the tasks that make up the job. Each task is then rated according to the chosen indices. For the rating of taskwork tasks, the relevant indices typically include importance of training, task frequency, task difficulty, difficulty of learning, and importance to the job. Conversely, no one has yet developed a set of standard indices for rating the relevance of teamwork tasks. Typically, some subset of the indices used for taskwork tasks are used. However, Bowers et al. (1994) found that many of these taskwork indices have low reliability and validity when applied to team tasks. These researchers calculated five indices of task importance: (1) multiplying criticality of error and task difficulty, then adding relative time spent (Levine, 1983); (2) summing task criticality and difficulty of learning (Sanchez and Levine, 1989); (3) task criticality and importance of training; (4) task frequency; and (5) overall task importance. Overall task importance was best predicted by a composite measure that included task criticality augmented with ratings of the importance of training. Depending on the purpose of the team task analysis, additional analyses can be performed (skill decay analysis, practice analysis) (Swezey et al., 1998). See Figure 51.2 for an example of a rating scale to determine task relevance.

56.2.6   Step 6: Translation of Tasks into KSAs

Once the relevant tasks have been identified and formed into task clusters, the next step is to translate the tasks into knowledge, skills, abilities, and attitudes (KSAs). (For definitions, see Goldstein and Ford,

---

TASK CRITICALITY

Task criticality refers to "the degree to which a failure in the task causes negative consequences" (Bower et al., 1994, p. 208).

Please rate the task criticality of each task using the scale below:

1=Not critical   (Improper task performance results in no harm or consequence)

2=Slightly critical   (Improper task performance results in slightly serious
                            consequences or damage)

3=Critical   (Improper task performance results in moderate damage or consequences)

4=Very critical   (Improper task performance results in serious consequences or damage)

5=Essential   (Improper task performance results in tremendous damage
                            and very serious consequences)

FIGURE 56.2   Prototypic scale used to determine task relevance.

2002.) In making this translation, interviews with subject-matter experts (often supervisors) are commonly used. Questions that can be asked during this process (Goldstein and Ford, 2002) are:

- What are the characteristics of a good employee? A poor employee?
- What does a person need to know to complete the tasks in the cluster?
- Identify both good and bad critical incidents related to this task cluster. Why did you choose these?
- What are the attitudes needed to perform the tasks in this cluster?

In addition, in terms of the teamwork tasks, the taxonomies of Stevens and Campion (1994) and Cannon-Bowers et al. (1995) can be used as a guide, since these works delineate teamwork into its component knowledge, skills, and attitudes. Of these two works, Cannon-Bowers et al. is more detailed in its breakdown.

56.2.7 Step 7: Link KSAs to Team Tasks

The final step is to link the identified KSAs to individual tasks. This is often done through the use of surveys given to subject-matter experts. For each identified task, the SME would be asked whether the KSA is essential, helpful, or not relevant for the particular task.

56.3 Advantages

- Aids in ensuring that the intervention is targeting relevant tasks and KSAs.
- Provides a systematic view of the tasks that compose the target job.
- Provides various methodologies that can be used to reach the end state.
- Provides the foundation for improvement of team task performance.

56.4 Disadvantages

- Knowledge-elicitation methods are labor intensive.
- Requires the support of domain experts.
- Expensive to conduct a full team task analysis (i.e., taskwork, teamwork).
- No established methodology for the teamwork portion.
- Requires skill in developing and/or application of knowledge-elicitation methods.

56.5 Example Output

Please give an estimate on the degree to which each of the coordination dimensions in Table 56.2 is needed by the team for optimum performance for each given task. Estimates should be given using a rating scale of 0 to 10, where 0 indicates the dimension is not needed and 10 indicates that the dimension is needed constantly for maximum performance.

56.6 Related Methods

Methods that are related to team task analysis include such variations as collaborative task analysis, which holistically elicits and assesses interpersonal and cognitive task information (McNeese and Rentsch, 2001). TTRAM (task and training requirements analysis methodology) (Swezey et al., 1998) and MAP (multiphase analysis of performance) (Levine and Baker, 1990, 1991) are also procedures that incorporate team task analysis. TTRAM is a broader methodology that incorporates team task analysis to identify potential application areas for the use of networked simulation. MAP is a methodology for analyzing multioperator tasks and can incorporate team task analysis to inform training requirements. Finally,

TABLE 56.2 Definition of Coordination Dimensions

Coordination Dimension	Definition
Communication	Includes sending, receiving, and acknowledging information among team members
Situational awareness (SA)	Refers to identifying the source and nature of problems, maintaining an accurate perception of one's location relative to the external environment, and detecting situations that require action
Decision making (DM)	Includes identifying possible solutions to problems, evaluating the consequences of each alternative, selecting the best alternative, and gathering information needed prior to arriving at a decision
Mission analysis (MA)	Includes monitoring, allocating, and coordinating the personnel and material resources of the team; prioritizing tasks; setting goals and developing plans to accomplish the goals; creating contingency plans
Leadership	Refers to directing activities of others, monitoring and assessing the performance of team members, motivating members, and communicating mission requirements
Adaptability	Refers to the ability to alter one's course of action as necessary, maintain constructive behavior under pressure, and adapt to internal or external changes
Assertiveness	Refers to the willingness to make decisions, demonstrating initiative, and maintaining one's position until convinced otherwise by facts
Total coordination	Refers to the overall need for interaction and coordination among team members

Source: Adapted from Bowers, C.A., Morgan, B.B., Jr., Salas, E., and Prince, C. (1993).

cognitive task analysis is a methodology that is arguably related to team task analysis, in that it captures the knowledge portion (Klein, 2000).

56.7   Standards and Regulations

Team task analysis is a new methodology, and there are not yet any published standards or regulations pertaining to this method.

56.8   Approximate Training and Application Time

Team task analysis is a new procedure, so there are few estimates on approximate training time. In terms of application time, conducting a team task analysis is very labor intensive because two task analyses are being completed: one identifying tasks requiring taskwork skills and one identifying tasks requiring teamwork skills. However, there are several times throughout when these procedures overlap. Generally, each time data are collected from a subject-matter expert, the application time is approximately 45 min to 1 hour. Bowers et al. (1993) reported that with a subset of 38 core tasks to be rated, SMEs required approximately 40 min. Similarly, Bowers et al. (1994) reported a time of approximately 45 min when rating between 42 and 56 items on four different scales. It is difficult to provide a total application time because the number of available SMEs tends to vary depending on organizational constraints.

56.9   Reliability and Validity

Team task analysis is a new procedure, so there is little data about the method's reliability or validity. Furthermore, much of this methodology's reliability and validity is contingent on the ability of the person conducting the analysis and the willingness of the SMEs to contribute. Bowers et al. (1993) found that pilots' self-reports appeared to be a valid method for obtaining coordination demand data, in that the pilots employed a reasonable range when responding to task items, and the overall pattern of results was consistent with expert opinions and earlier research. Results also indicated that different tasks within a target job required different levels and types of teamwork (i.e., coordination). Bowers et al. (1994) reported low reliability and validity when standard indices for individual task analysis are applied to the teamwork portion of a team task analysis. Reported validity estimates ranged from 0.08 to 0.38, with task frequency demonstrating significantly lower levels of reliability and a composite index (task criticality, importance to train) yielding the highest. Using the TTRAM methodology, Swezey et al. (1998) reported that it appeared to discriminate effectively among tasks that required high levels of internal and external teamwork. Finally, SMEs have rated the MAP system as user friendly, efficient, and useful (Levine and Baker, 1990). Obviously more work needs to be conducted in this area.

56.10 Tools Needed

The tools needed to implement a team task analysis depend primarily on the methodologies that are chosen to elicit knowledge from the SMEs. Common tools include paper and pencil, tape recorder, and often a statistical package to help analyze the information gathered.

References

Bowers, C.A., Morgan, B.B., Jr., Salas, E., and Prince, C. (1993), Assessment of coordination demand for aircrew coordination training, *Mil. Psychol.*, 5, 95–112.

Bowers, C.A., Baker, D.P., and Salas, E. (1994), Measuring the importance of teamwork: the reliability and validity of job/task analysis indices for team-training design, *Mil. Psychol.*, 6, 205–214.

Cannon-Bowers, J.A., Tannenbaum, S.I., Salas, E., and Volpe, C.E. (1995), Defining competencies and establishing team training requirements, in *Team Effectiveness and Decision Making in Organizations*, Guzzo, R. and Salas, E., Eds., Jossey-Bass, San Francisco, pp. 333–380.

Dyer, J. (1984), Team research and training: a state-of-the-art review, in *Human Factors Review: 1984*, Muckler, F.A., Ed., Human Factors Society, Santa Monica, CA, pp. 285–323.

Fleishman, E.A. and Zaccaro, S.J. (1992), Toward a taxonomy of team performance functions, in *Teams: Their Training and Performance*, Swezey, R.W. and Salas, E., Eds., Ablex, Norwood, NJ, pp. 31–56.

Franz, T.M., Prince, C., Cannon-Bowers, J.A., and Salas, E. (1990), The identification of aircrew coordination skills, in *Proceedings of the 12th Annual Department of Defense Symposium*, U.S. Air Force Academy, Colorado Springs, CO, pp. 97–101.

Glickman, A.S., Zimmer, S., Montero, R.C., Guerette, P.J., Campbell, W.S., Morgan, B.B., Jr., and Salas, E. (1987), The Evolution of Teamwork Skills: An Empirical Assessment with Implications for Training, tech report 87-016, Naval Training Systems Center, Orlando, FL.

Goldstein, I.L. (1993), *Training in Organizations: Needs Assessment, Development, and Evaluation*, 3rd ed., Brooks/Cole Publishing, Pacific Grove, CA.

Goldstein, I.L. and Ford, K.J. (2002), *Training in Organizations: Needs Assessment, Development, and Evaluation*, 4th ed., Brooks/Cole Publishing, Pacific Grove, CA.

Klein, G. (2000), Cognitive task analysis of teams, in *Cognitive Task Analysis*, Schraagen, J.M. and Chipman, S.F., Eds., Lawrence Erlbaum Associates, Mahwah, NJ, pp. 417–429.

Levine, E.L. (1983), *Everything You Always Wanted To Know about Job Analysis*, Mariner, Tampa, FL.

Levine, E.L. and Baker, C.V. (1990), Team Task Analysis for Training Design: A Procedural Guide to the Multiphase Analysis of Performance (MAP) System and a Tryout of the Methodology, contract DAAL03-86-D-001, Scientific Services Program, funded by Army Research Office.

Levine, E.L. and Baker, C.V. (1991), Team task analysis: a procedural guide and test of the methodology, in *Methods and Tools for Understanding Teamwork: Research with Practical Implications?* Salas, E., Chair, Proceedings of Sixth Annual Conference for the Society of Industrial and Organizational Psychology, St. Louis, MO.

McNeese, M.D. and Rentsch, J.R. (2001), Identifying the social and cognitive requirements of teamwork using collaborative task analysis, in *New Trends in Cooperative Activities: Understanding System Dynamics in Complex Environments*, McNeese, M., Salas, E., and Endsley, M., Eds., Human Factors and Ergonomics Society, San Diego, CA, pp. 96–113.

Sanchez, J.I. and Levine, E.L. (1989), Determining important tasks within jobs: a policy-capturing approach, *J. Appl. Psychol.*, 74, 336–342.

Smith-Jentsch, K.A., Zeisig, R.L., Acton, B., and McPherson, J.A. (1998), Team dimensional training: a strategy for guided team self-correction, in *Making Decisions under Stress: Implications for Individual and Team Training*, Cannon-Bowers, J.A. and Salas, E., Eds., American Psychological Association, Washington, D.C., pp. 271–297.

Stevens, M.J. and Campion, M.A. (1994), The knowledge, skill, and ability requirements for teamwork: implications for human resource management, *J. Manage.*, 20, 503–530.

Swezey, R.W., Owens, J.M., Bergondy, M.L., and Salas, E. (1998), Task and training requirements analysis methodology (TTRAM): an analytic methodology for identifying potential training uses of simulator networks in teamwork intensive task environments, *Ergonomics*, 41, 1678–1697.

57

Team Workload

Clint A. Bowers
University of Central Florida

Florian Jentsch
University of Central Florida

57.1 Background and Application

The proliferation of teams in the workplace can be attributed to the increasing complexity of many work tasks. The requirements of many modern work systems simply exceed the resources possessed by individuals. Therefore, multiple operators are often used to accomplish what individuals cannot. When these workers share a common goal and work interdependently toward that goal, they are considered to be a team.

Workload for teams, as for individuals, can be considered to be an index of the available resources possessed by the team relative to the demands placed upon it. In both cases, it is presumed that performance deteriorates when demands exceed the available resources.

One must exercise some caution, however, in extending the construct of individual workload to teams. As pointed out by Bowers et al. (1997), individuals in teams are confronted with something akin to a dual-task situation. That is, they must not only perform their individual task-related duties, but also attend to the team-coordination demands, such as communication among the team, coordination of resources, back-up of one another, and so forth. Thus, the overall workload capacity of a team is almost certainly not the sum of the task-related resources of the individual members. Rather, each member added to a team brings in not only additional resources, but also represents a process cost. Thus, the goal for a measure of team workload is to estimate the capacity of a team, bearing in mind the workload associated with the team processes required to perform the overall team task.

57.2 Procedure

It should be noted at the outset that the measurement of team workload has received very little research attention. This is unfortunate, given the complexity of the issue. There are a variety of issues that must be evaluated before one can express confidence in an estimate of team workload.

That said, it appears that the NASA (National Aeronautics and Space Administration) task load index (TLX) is the best of the existing workload measures for teams. However, the optimal delivery of the measure is somewhat different for teams than it is for individuals. Optimal results were obtained when researchers asked individual team members to rate their own workloads using the traditional form of the measure as well as a subjective evaluation of the overall team's workload using an adapted form of the measure.

The approach was validated against a team performance decrement in a medium- vs. high-workload manipulation (Urban et al., 1995). Using this approach, the lowest workload value obtained among the individuals was the best predictor of the team's overall performance. That is, the higher the workload level of the least-loaded team member, the lower was the performance of the overall team.

Two additional significant predictors of team performance were also identified, although they were not quite as strong predictors as the workload of the least-loaded team member: (1) the highest level of workload found among the individuals and (2) the averaged team subjective workload. Again, higher levels of workload among the highest-loaded team member and higher levels of team subjective workload were associated with lower team performance. It is important to note, however, that these relationships were not validated on the other side of the workload–performance curve, i.e., there was no comparison between low and medium levels of workload in this study.

57.3 Advantages

- Provides an estimate of overall team capacity
- Leverages existing measurement approaches
- Inexpensive

57.4 Disadvantages

- Does not provide separate estimates for teamwork vs. taskwork
- Cumbersome
- Perceived as redundant by subjects

57.5 Related Methods

One common approach to measuring workload involves the use of secondary-task paradigms. Using this approach, workload is estimated by observing the decrement in performance of a secondary task added to the primary team task as workload levels change. This approach has been used only rarely with teams, however. This might be due to the inherent complexity of team tasks, which can preclude the use of secondary-task approaches. The few studies that have attempted this method have been disappointing. For example, Bowers et al. (1992) found that this method was sensitive to the workload imposed by only one of three secondary tasks added to a team resource-allocation task.

57.6 Approximate Training and Application Times

The NASA TLX is easy to administer and use (see Chapter 39, Mental Workload). The individual and team measures take about 10 min each. Scoring of the TLX is not intuitive, but the scoring template is easy to follow.

57.7 Reliability and Validity

The reliability of the TLX is acceptable for individuals, but it has not been evaluated in team settings. Subjective estimates of workload have typically demonstrated only moderate reliability, and such estimates are not likely to be much better in the team application. In fact, it is likely that they are worse.

The validity of these approaches has most often been evaluated by their ability to predict performance decrements. Subjective measures of team workload have been effective in this regard in settings such as air traffic control (Bailey and Thompson, 2000) and process control (Sebok, 2000). However, neither the

TLX nor other subjective measures predicted team performance in a simulated flight task (Thornton et al., 1992).

References

Bailey, L.L. and Thompson, R.C. (2000), The effects of performance feedback on air traffic control team coordination: a simulation study, DOT-FAA-AM-00-25, FAA Office of Aviation Medicine Reports, Washington, D.C.

Bowers, C.A., Urban, J.M., and Morgan, B.B., Jr. (1992), The Study of Crew Coordination and Performance in Hierarchical Team Decision Making, Team Performance Laboratory Technical Report 92-1, University of Central Florida, Orlando, FL.

Bowers, C.A., Braun, C.C., and Morgan, B.B., Jr. (1997), Team workload: its meaning and measurement, in *Team Performance Assessment and Measurement: Theory, Research, and Applications*, Brannick, M.T., Salas, E., and Prince, C., Eds., Lawrence Erlbaum Associates, New York.

Sebok, A. (2000), Team performance in process control: influences of interface design and staffing levels, *Ergonomics*, 43, 1210–1236.

Thornton, C., Braun, C.C., Bowers, C.A., and Morgan, B.B., Jr. (1992), Automation effects in the cockpit: a low-fidelity investigation, in *Proceedings of the 36th Annual Meeting of the Human Factors Society*, Human Factors Society, Santa Monica, CA, pp. 30–34.

Urban, J.M., Bowers, C.A., Monday, S.D., and Morgan, B.B., Jr. (1995), Workload, team structure, and communication in effective team performance, *Mil. Psychol.*, 7, 123–135.

58

Social Network Analysis

James E. Driskell
Florida Maxima Corporation

Brian Mullen
Syracuse University

58.1   Background and Application

A social network is a set of entities or actors (such as members of a medical staff, a classroom, a terrorist cell, or a set of organizations) who have some type of relationship with one another. Social network analysis is a method for analyzing relationships between social entities. Although the roots of social network analysis can be traced to early work in sociometry (Moreno, 1934) and graph theory (Harary et al., 1965), it has emerged over the past century as an extensive method of analysis used in the social and behavioral sciences, political science, economics, health research, communications, and engineering (Scott, 1991; Wasserman and Faust, 1994).

Social network analysis is a set of procedures — mathematical and graphical techniques — that use indices of relatedness among entities to represent social structures in a compact and systematic manner. There are several general goals of network analysis. The first goal is to represent relationships of interest visually as a network or graph, and to display information in a way that allows the user to see relationships among actors embedded in the overall network. A second goal is to examine basic properties of relationships in a network, such as density, centrality, and prestige. A third goal is to test hypotheses regarding the structure of connections among actors. Social network analysts can examine the effects that relationships have on constraining or enhancing individual behavior or network efficiency. A major advantage of the network approach is that it focuses on the relationships among actors embedded in their social context.

58.2 Procedure

Social network analysis can be carried out in four steps: (1) defining the network, (2) measuring relations, (3) representing relationships, and (4) analyzing relationships.

58.2.1 Step 1: Defining the Network

The first step in conducting a social network analysis is to define the network or population to be studied. Membership rules, which determine whether an actor is to be included in a network, are defined by whether or not the actor falls within some boundary. Some networks, such as students in a classroom, may have well-defined, preexisting boundaries. Other networks, such as users of computer bulletin boards, are not as easily circumscribed. Typically, the researcher is interested in studying all members of a given network, representing a complete or whole network analysis. However, there are strategies for sampling from a larger population. For example, in an ego-centered network sample, a set of respondents is asked to provide a list of people that they are linked with through the relationship in question, and this list can define the network to be studied.

58.2.2 Step 2: Measuring Relations

The researcher must determine what relationships are to be measured for the network that has been defined. Actors can be defined by any number of relationships. Broad types of relationships that researchers can investigate include affective relations (e.g., respect, liking), roles (e.g., parent, supervisor), interactions (e.g., communication), transfer of resources (e.g., work materials), and environmental relations (e.g., proximity). The researcher selects what relationships to study based on relevance to the research question posed and the theoretical interests of the researcher.

Data can be collected on a relationship of interest by any number of means, including surveys, questionnaires, interviews, direct observation, or archival records. Data can be collected at nominal, ordinal, or interval levels of measurement. The most common method of measuring a relationship is through a binary measure, coded as 0 if the relationship is absent and 1 if the relationship is present. More advanced levels of measurement allow the researcher to measure strength or intensity of relationships (such as counts of the number of interpersonal contacts) and allow the use of a more sophisticated array of statistical approaches. Although binary data represent a loss of information relative to ordinal or interval measures, many common algorithms in network analysis are designed for simple binary data.

58.2.3 Step 3: Representing Relationships

Social networks can be represented as matrices or graphs. A network matrix is a square array of measurements, as illustrated in Table 58.1, which presents two simple five-person networks that represent the presence or absence of communication links between actors. The data for these simple networks are arranged in a 5 × 5 matrix, with the rows and columns representing the nodes within the network and with the cells indicating the presence (1) or absence (0) of a link between the nodes.

Although Table 58.1 presents all of the information required for analysis of these simple networks, this same information can be represented graphically, as shown in Figure 58.1. In Figure 58.1, the nodes

TABLE 58.1 Matrices for Two Social Networks

	Network A						Network B				
	A	B	C	D	E		A	B	C	D	E
A	—	0	1	0	0	A	—	1	1	0	1
B	0	—	1	0	0	B	1	—-	1	0	0
C	1	1	—-	1	1	C	1	1	—-	1	1
D	0	0	1	—-	0	D	0	0	1	—	1
E	0	0	1	0	—	E	1	0	1	1	—

represent actors and the lines represent a relationship between pairs of actors. The simple graphs shown in Figure 58.1 allow the researcher to more readily visualize the relationships among actors and the overall structure of the network. More elaborate graphs can be used to represent other dimensions. For example, the relations in Figure 58.1 are undirected, meaning that the relationship of person A to B is the same as the relationship of person B to A. Relationships such as "is related to" or "is in contact with" are necessarily symmetric. In contrast, directed relationships are nonsymmetric. For example, person A may be the supervisor of B but B is not the supervisor of A. Note that when directed relationships are studied, data would be recorded in the matrix with the row nodes "taking action to" the column nodes. Thus, in the network A matrix shown in Table 58.1, if person A supervises person B, the row entry for person A would reflect this relationship with B, and the row entry for person B would reflect the lack of a supervisory relationship with A. Graphs can incorporate arrowed lines to represent directed relationships. Valued relationships (that vary in intensity or strength) can also be represented. If the relationships studied are valued rather than binary, the lines in the graph can have attached values, and the cells in the matrix will contain values other than 0 and 1.

58.2.4   Step 4: Analyzing Relationships

Communication networks as depicted in Figure 58.1 can serve as an effective model of teams exchanging information over telecommunication systems, of computer-mediated communication systems, or of hierarchically structured organizations. Networks can differ in terms of basic structural properties. For example, early research (Bavelas, 1950; Shaw, 1955) indicated that individuals in a more central network position were more likely to emerge as the group leader, to engage in the group task, and to perform more efficiently. Researchers have identified three distinct components of network centrality: degree (the number of other positions in the network in direct contact with a given position), betweenness (the frequency with which a position falls between pairs of other positions in the network), and closeness (the extent to which a position is close to all other positions). Degree, betweenness, and closeness measure different facets of the same general construct "centrality," and they are typically highly interrelated. However, these components of centrality are distinct. Closeness is the component of centrality considered in the original social psychological studies of communication network centrality (Bavelas, 1950; Shaw, 1955). However, Mullen et al. (1991) found betweenness to be the most powerful independent predictor of group performance.

Degree, betweenness, and closeness can be used to characterize the centrality of a particular position within a given network (positional centrality) or the centrality of a communication network as a whole (network centrality). For illustrative purposes, we have derived these three indices of centrality for the networks shown in Figure 58.1. These indices, in standardized form, range from 0.00 (low centrality) to 1.00 (high centrality). The value of standardizing these indices of centrality is that a number representing centrality will be comparable from one setting to another, regardless of the number of positions in the network.

Even with these simple graphs, it is easy to discern some basic properties of these networks. For example, position C occupies the most central position in both networks, and thus we would expect that C would be more likely to emerge as a leader and to participate more in group activities. Position C has the maximum number of channels open to it (degree); it falls between the maximum number of other positions in the network (betweenness); and it is located the shortest possible distance from all of the other positions in the network (closeness). In network B, positions B and D are the least central according to all three indices of centrality, and persons B and D are likely to be less involved in group activity. These peripheral positions have fewer channels open to them (degree); they fall between fewer pairs of positions in the network (betweenness); and they are located at longer distances from all of the other positions in the network (closeness).

In terms of network centrality, we find that network A exhibits a higher centrality than network B according to all three indices, and thus network A would be expected to exhibit the most efficient task performance. We can also easily calculate network density by dividing the sum of all ties in the network by the number of possible ties. For network A, the sum of all ties is 8 (the sum of the cell entries in Table

58.1) and the number of possible ties is 20, and thus the density is 0.40. For network B, the sum of all ties is 14 and the number of possible ties is 20, and thus the density is 0.70. Therefore, network B is denser that network A, and we would expect higher rates of participation by all members in network B than in network A.

It should be recognized that these examples are extremely simplified and straightforward. Actual team communication networks are invariably larger, less symmetrical, more complex, and more heterogeneous by an order of magnitude compared with the simple networks used for the sake of illustration here.

58.3   Advantages

- Allows analysis of actors embedded in a social context.
- Applicable to a wide range of settings in which individual, team, or organizational interactions and relationships are salient, from counseling (Koehly and Shivy, 1998) to computer-mediated communications (Garton et al., 1997).
- Supports multiple levels of analysis (e.g., individuals embedded in teams, teams embedded in organizations, etc.).

58.4   Disadvantages

- Data collection can be labor intensive, both for complete networks where the researcher collects data on all members of a defined network and for ego-centered networks.
- It is often difficult to identify the boundaries of the network under study.
- Network studies are subject to similar concerns regarding methodology of data collection as are other types of studies. These include concerns with self-reported data, where respondents may be asked to provide complex and detailed information that they may have difficulty recalling.
- Nonindependence of data violate assumptions of standard statistical tests, currently addressed by the advent of permutation-based tests.

58.5   Example Output

Network A	Position	Degree	Betweenness	Closeness
Wheel	A	.25	.00	.57
	B	.25	.00	.57
	C	1.00	1.00	1.00
	D	.25	.00	.57
	E	.25	.00	.57
	Network A	1.00	1.00	1.00
Network B				
Double-barred circle	A	.75	.08	.80
	B	.50	.00	.67
	C	1.00	.33	1.00
	D	.50	.00	.67
	E	.75	.08	.80
	Network B	.50	.29	.62

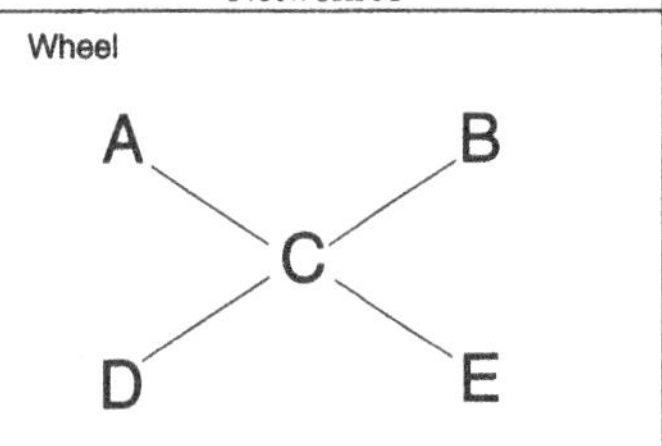

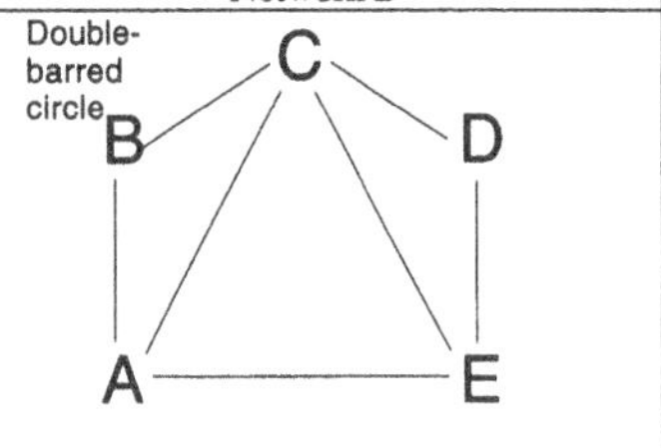

FIGURE 58.1 Graphs of two social networks.

58.6 Related Methods

A number of alternative approaches exist to operationalizing other structural aspects of groups. Other approaches include Mullen's (1983, 1991) approach to operationalizing the effects of relative group size, Latane's (1981) social impact theory, and Blau and Schwartz's (1997) macrostructural sociological consideration of intergroup contact.

58.7 Standards and Regulations

There are no applicable standards and regulations for social network analysis.

58.8 Approximate Training and Application Times

There are organizations (International Network for Social Network Analysis) and journals (e.g., *Social Networks*) devoted to network analysis that can provide informal training for the neophyte. There are also comprehensive texts on the subject (e.g., Scott, 1991; Wasserman and Faust, 1994). The major software programs such as UCINET offer user manuals and online help.

58.9 Reliability and Validity

The data employed in network analysis include a variety of subjective and objective measures. These data are subject to the same psychometric concerns, including reliability and validity, as other behavioral science data.

58.10 Tools Needed

Software packages for network analysis include UCINET (Borgatti et al., 2002) and STRUCTURE (Burt, 1991).

References

Bavelas, A. (1950), Communication patterns in task oriented groups, *J. Acoustical Soc. Am.*, 22, 272–283.

Blau, P.M. and Schwartz, J.E. (1997), *Crosscutting Social Circles: Testing a Macrostructural Theory of Intergroup Relations*, Transactions, New Brunswik, NJ.

Borgatti, S.P., Everett, M.G., and Freeman, L.C. (2002), *UCINET for Windows*, Analytic Technologies, Harvard, MA.

Burt, R.S. (1991), *STRUCTURE*, Columbia University, Center for the Social Sciences, New York.

Garton, L., Haythornthwaite, C., and Wellman, B. (1997), Studying online social networks, *J. Comput. Mediated-Commun.*, 3; available on-line at http://www.ascusc.org/jcmc/vol3/issue1/garton.html.

Harary, F., Norman, R.Z., and Cartwright, D. (1965), *Structural Models: An Introduction to the Theory of Directed Graphs*, Wiley, New York.

Koehly, L.M. and Shivy, V.A. (1998), Social network analysis: a new methodology for counseling research, *J. Counseling Psychol.*, 45, 3–17.

Latane, B. (1981), The psychology of social impact, *Am. Psychologist*, 36, 343–356.

Moreno, J.L. (1934), *Who Shall Survive?* Beacon Press, Beacon, NY.

Mullen, B. (1983), Operationalizing the effect of the group on the individual: a self-attention perspective, *J. Exp. Soc. Psychol.*, 19, 295–322.

Mullen, B. (1991), Group composition, salience, and cognitive representation: the phenomenology of being in a group, *J. Exp. Soc. Psychol.*, 27, 297–323.

Mullen, B., Johnson, C., and Salas, E. (1991), Effects of communication network structure: components of positional centrality, *Soc. Networks*, 13, 1–17.

Scott, J. (1991), *Social Network Analysis: A Handbook*, Sage, London.

Shaw, M.E. (1955), A comparison of two types of leadership in various communication nets, *J. Abnormal Soc. Psychol.*, 50, 127–134.

Wasserman, S. and Faust, K. (1994), *Social Network Analysis: Methods and Applications*, Cambridge University Press, New York.

Environmental
Methods

Alan Hedge
Cornell University

59

Environmental Methods

This section of the handbook deals with environmental conditions. Information on each topic is presented in a different format than in the other handbook sections because each topic area encompasses multiple methods, and no single method is favored or complete. Thus each chapter reviews the main methods used and the main methodological considerations for selecting and using methods in that topic area. Consequently, each of the topic contributions presents the rationale behind the main methods that are available, and gives advice on issues to consider in the choice and use of various methods, rather than trying to describe each method in detail or selecting just a single method. Each chapter also provides guidance on where to look for additional methodological information.

In the first half of the 20th century, studies of the impact of modifying the physical environmental conditions at work helped to establish the groundwork for the emergence of the discipline of ergonomics. From the infancy of the discipline, ergonomists realized that our ability to perform work is inextricably linked to the prevailing environmental conditions in the workplace. The human body has adaptive physiological mechanisms that allow us to tolerate a range of physical environmental conditions, but often at a cost to the body. When conditions exceed the capabilities of the body's adaptive mechanisms, then performance and health deteriorate, and in extreme situations conditions could even prove fatal. This section does not deal with the full gamut of environmental agents, excluding obvious risks such as ionizing radiation. Rather, this section deals with (a) the ergonomics considerations for those environmental factors most often designed into a workplace and (b) the impact that these can have on human performance and well-being. This section also deals with those conditions that can arise in normal terrestrial workplaces; it does not address unique issues found in extreme environments, such as deep sea diving or in space.

The goal of the ergonomic design of the workplace environment is to create prevailing ambient conditions that are comfortable, acceptable, and do not compromise work performance or worker health. In some situations, this design involves modifying the physical characteristics of the workplace, e.g., providing sufficient light, while in other situations it involves modifying worker behavior to regulate exposures, e.g., minimizing heat stress. As such, the ergonomic design of environments can be thought of as the science of moderation, because it attempts to create sustained exposures that fall within the regulatory range of the body's physiological processes (Figure 59.1). However, ergonomists seldom have complete responsibility for creating work environment conditions. Modern work environments are constructed to meet various health and safety standards for appropriate thermal, visual, aural, and other conditions. The design changes that the ergonomist can make have to harmonize with the requirements of these various standards. To be successful, the ergonomist also needs to work with other professionals who specialize in creating specific aspects of the physical workplace, such as ventilation engineers, illuminating engineers, acoustical engineers, health and safety professionals, and architects and designers. In such interactions, the ergonomist can play two important roles: one involves setting the environmental conditions limits for acceptable human performance; the second involves bringing a systems perspective

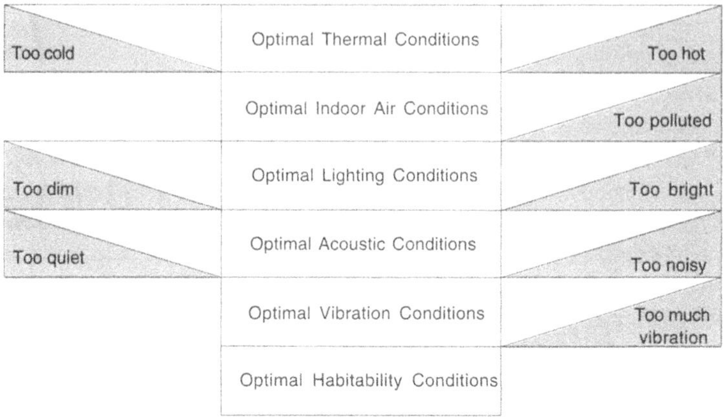

FIGURE 59.1 Ergonomic design of the work environment conditions.

to the design process to assess the interplay between the different designed environmental conditions and the nature of the work tasks, the work technology, and the characteristics of the workers.

Assessing the physical environment can be a complex task. Decisions have to be made about what variables to measure, where and when to take measures, what instruments to use and how to use them, and how to interpret and combine objective measures of conditions along with subjective reports of conditions. It is beyond the scope of this section to present details on measurement instruments and techniques. Rather, the methods described here give the ergonomists guidance on how to conduct environmental assessments, what to look for, what pitfalls to avoid.

The first chapter in this section (Chapter 60) considers the thermal environment. We are homeothermic organisms, and our thermoregulatory systems, both physiological and behavioral, allow us to actively adapt to quite a wide range of climate conditions. To augment our physiological capabilities, we wear clothing as artificial skin to regulate thermal insulation: when we feel cold we can don clothing, and when we feel hot we can doff clothes. We also use artificial heat sources, such as fires, and cool sources, such as air conditioning. Measurement of the thermal environment is, however, more complicated than simply quantifying air temperature. Several measures and methods are required to adequately characterize thermal conditions in terms of their impact on the heat balance of the human body. But simply quantifying conditions does not provide guidance on what is safe and acceptable in the workplace. The various cold-stress and heat-stress indices that have been developed over the past 50 years are reviewed in Chapters 61 and 62, respectively, and those that best allow prediction of the limits of human endurance in cold or hot work conditions are presented.

Maintaining environmental conditions within these thermal extremes is critical to the ability to successfully perform work. Thankfully, now that most work occurs indoors, most of the modern workforce seldom, if ever, gets to experience extremes of work climate conditions. Also, with the dramatic changes in the nature of the workplace over the past 50 years, moving from a heavy industrial work base to a light industrial and knowledge work base, the number of workers exposed to extremes of thermal conditions in the workplace has substantially diminished. The focus of much recent work on thermal conditions has been to try to determine conditions for optimal comfort and performance. Chapter 63 discusses the concept of thermal comfort and the methods for gauging this.

Throughout the past century, the nature of the workplace and our associated behavioral patterns have changed. Nowadays we spend most of our time in artificial environments. We live in houses; we learn in schools; we work in offices and factories; we commute to and from work in cars, buses, or trains; we fly in planes; we shop in malls, etc. Diary studies of U.S. residents show that the typical person is indoors for over 90% of the day. The past 25 years have seen an upsurge of interest in studying the impact of indoor air quality on human health and performance. The modern specter of chemical and biological terrorist weapons has also raised awareness of and interest in understanding human requirements for healthful air. Our perception of thermal conditions also affects our judgments of indoor air quality. Moreover, the thermal characteristics of the air impact the effects of many air pollutants, especially volatile organic compounds (Chapter 64), and the growth of indoor microorganisms, such as bacteria and fungi (Chapter 65), that can have irritant, allergenic, toxic, and even fatal effects. Thermal conditions also impact our sense of smell, which is one of the body's first lines of defense against potentially hazardous air quality. There are standardized test methods for many hundreds of indoor air gases, vapors, or inorganic particulates, and rather than attempting to select a handful of these, attention has been focused on how to conduct an indoor air quality investigation of a workplace and what kinds of instruments and sampling methods can be used. Similarly, there are many thousands of hazardous organisms that can adversely affect us indoors, and the topic of biological contaminants gives an overview of the main methodological issues as well as useful guidance for what may constitute an acceptable environment.

Olfactometry methods are presented in Chapter 66. Our sense of smell will either encourage or discourage certain behaviors. We can use deodorants to decrease the chance of emitting unpleasant bioeffluents; we can use fragrances to create more pleasing odors; and we can use unpleasant odors as both a warning, such as adding an odorant to natural gas so that gas leaks can be smelled, or even to repel people and to deter certain behaviors. Indeed, in the future, the use of unpleasant odors may prove to be a successful nonlethal crowd-control strategy.

Although smell is the most primitive of the our senses, human beings are predominantly visual organisms, and it is estimated that as much as 80% of the brain's capacity to process information is devoted to creating our personal visual world. Assuming that we have normally corrected vision, the ability to see depends on the light available in the environment. Since Edison's introduction of the electric light bulb, artificial lighting has permeated every aspect of our environment. Even though we have been designing buildings with electric lighting systems for over a century, complaints about unsatisfactory indoor lighting are quite common in today's workplaces. Indeed, studies show that among modern computer workers, reports of eyestrain top the list of complaints. Eyestrain is a complaint that is associated with glare caused by a mismatch between the indoor lighting systems and the characteristics of the visual task and of the display technology. Consequently, the good ergonomic design of the luminous environment needs to consider a variety of factors, well beyond simply quantifying the amount of light, that can impact our capacity to work successfully under artificial lighting. The topics included here start with a review of the various kinds of lighting installations that might be encountered by the ergonomist seeking to set the proper context for the use of subsequent methods (Chapter 67). Then, the various measures used to quantify lighting are presented (Chapter 68), and finally the way in which lighting evaluations should be conducted is described (Chapter 69).

Our ability to hear sounds is critical to many human behaviors. We use sound in the form of speech to communicate; we listen to sounds, such as music, to affect our emotional state; and we use sounds, such as a car horn or a siren, as ways of warning others of impending danger. Seldom is our environment devoid of sound. However, not all sounds are desirable, and a sizable part of the workforce is exposed to noise, or unwanted sound, on a daily basis. When this noise is very loud, we can suffer hearing impairment and even permanent hearing loss. When this noise is unpredictable, it can have a serious adverse impact on our work performance and well-being. The concept of noise is more than simply the quantification of the intensity of a sound; it involves measuring psychological and behavioral variables as well as the acoustic characteristics of the sound exposure. For example, what may be heard as highly desirable loud rock music by one person might be regarded as deafeningly loud noise by another person. If the sound is loud enough, both may experience hearing loss as a consequence of exposure, but unwittingly the first person might seek out such loud noise exposure, while the second person will shun this. The study of acoustics is a discipline in its own right, and rather than focusing on methods for measuring all aspects of sound, the chapters presented here (Chapters 70, 71, 72) focus on both the physical and subjective measurement of noise and the consequent annoying and disruptive effects that this can have on human performance and well-being.

Sound is the vibration of atoms, and sound and vibration are associated. Vibration can impact a body segment, such as the hand and arm when using a vibrating power tool, or the whole body, such as the vibration that occurs when driving over a rough road. Exposure to vibration disrupts human performance (try typing on a laptop on a bus traveling over a bumpy road). It can also accelerate the onset of injuries, such as vibration-induced white finger, which can produce permanent manual impairment and loss of fingers. The methods and standards for assessing vibration exposure are described in Chapter 73.

The final topic considered is the concept of habitability and its measurement, which is covered in Chapter 74. While we can characterize the various components of the work environment and assess each separately, the body is exposed to all physical factors simultaneously. Thus we need methods that address desirable conditions across different environmental settings. Nowhere is the concept of habitability more important than in trying to evaluate and predict optimal conditions as human beings leave the relative safety of terrestrial earth and venture into inhospitable places, such as deep under the oceans or out in space. As we plan for the possible colonization of extraterrestrial settings, such as the moon and the planet Mars, we need to have a clear and consistent way of evaluating the habitability of the artificial places that we create. The idea of a habitability index offers great promise to improve the quality of the ergonomic design of environments everywhere.

60

Thermal Conditions Measurement

George Havenith
Loughborough University

60.1 Background and Application

Thermal comfort and strain can be assessed using a variety of objective and subjective methods.

60.1.1 Objective Heat-Exchange Analysis

Assessment via heat exchange requires measurement of the climatic parameters affecting the human heat balance (Figure 60.1). For comfort or low thermal strain, this balance should be close to equilibrium (heat production equals heat loss), resulting in a relatively stable body temperature.

Heat production is determined by metabolic activity. For a body at rest, this is the amount of energy needed for the body's basic functions. Metabolic activity increases when the body is working. When active muscles burn nutrients for mechanical activity, some energy is liberated outside the body as external work, but most energy is released internally as heat. The ratio between this external work and the total energy consumed is the efficiency with which the body performs the work. Efficiency is close to zero for most tasks. In the cold, shivering can produce additional heat, increasing metabolic rate and heat production up to fourfold.

Several pathways are available for heat loss, expressed in $W \cdot m^{-2}$, from the body between skin and environment (Figure 60.2). For each pathway, the amount of transferred heat depends on the driving force (e.g., temperature or vapor pressure gradient), the body surface area involved, and the resistance to that heat flow (e.g., insulation via clothing). Of the pathways shown in Figure 60.2 (unless working in water, in special gas mixtures [prolonged deep-sea dives], in supine positions, or handling cold products), conductivity is not a relevant factor. Convection heat loss occurs when air cooler than the skin flows along the skin and carries body heat away. When there is a difference between the body's surface temperature and the temperature of the surfaces in the environment, heat is exchanged by radiation. Finally, heat can be lost by evaporation of moisture (sweat) on the skin. Evaporative and

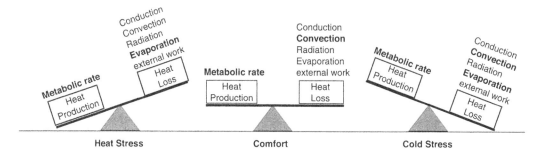

FIGURE 60.1 The heat balance: the sum of all heat losses and gains of the body; left: in heat stress, heat loss is lower than heat production, causing heat to be stored in the body (increase in body temperature); middle: in comfort, loss and gain are balanced; right: in cold stress, more heat is lost than produced, leading to body cooling. The size of the characters of the individual thermal heat exchange factors represents their importance in the three situations.

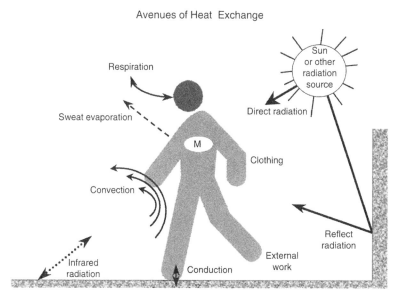

FIGURE 60.2 Schematic representation of the pathways for heat loss from the body. M = metabolic heat production. (Reproduced with permission from Havenith, G. [1999], *Ann. Occup. Hyg.*, 43, 289–296.)

convective heat losses also occur in the lungs during respiration, as inspired air is usually cooler and dryer than the lung's internal surface. This heat loss can be up to 10% of the total heat production.

The body gains heat when air temperature, radiant temperature, or vapor pressure around the body is higher than the skin value, and this can result in heat stress.

60.1.1.1 Measurement of Relevant Factors in Heat Exchange

External parameters that need to be assessed to determine heat or cold stress levels or comfort conditions include:

- Temperature
- Air humidity
- Air speed

60.1.1.1.1 Temperature

As temperature increases, body heat loss by convection, conduction, and radiation decreases. The overall effect of temperature can be assessed by measuring three relevant properties:

1. Air temperature (t_a)
2. Mean radiant temperature $(\bar{t}_r)$
3. Surface temperature (t_s)

60.1.1.1.1.1 Air Temperature (t_a) — Air temperature (t_a) can be measured by a conventional alcohol-filled thermometer or by an electronic thermometer. Smaller sensors react faster to variations in the climate. Use a shielded sensor (Figure 60.3) when radiation is present (sun or other heat source). Typically, a (polished) shield is used in combination with a device to suck air (blowing could add heat from the fan motor) over the sensor to ensure a true air temperature measurement (see Assman psychrometer, Figure 60.4). Sensor range/accuracy: to measure comfort: 10 to 40°C/±0.5°C; to measure stress: –40 to +120°C/ outside comfort range ±1°C; desirable accuracy ±0.2°C.

60.1.1.1.1.2 Mean Radiant Temperature $(\bar{t}_r)$ — Mean radiant temperature $(\bar{t}_r)$ is the mean temperature of all walls and objects in the space (including sky outdoors). When mean radiant temperature exceeds skin temperature (e.g., in steel mills, or in work in the sun), heat transfers from the environment to the skin. Mean radiant temperature $(\bar{t}_r)$ is most commonly measured indirectly using a matte black globe (typically 15-cm diameter) with a temperature sensor in its center (Figure 60.5). Given the size of the globe, it needs a long period to equilibrate (>20 min), and is therefore not sensitive to fast fluctuations in radiation. The globe temperature can be used to calculate $\bar{t}_r$.

$$\text{If} \quad v_a > 0.15 \text{ (air movement)}: \bar{t}_r = \left[\left(t_g + 273 \right)^4 + \frac{1.1 \cdot 10^8 v_a^{0.6}}{\varepsilon_g \cdot D^{0.4}} \left(t_g - t_a \right) \right]^{0.25} - 273$$

where $\bar{t}_r$ = mean radiant temperature (°C)

$\quad t_g$ = globe temperature (°C) (60.1)

$\quad t_a$ = ambient temperature (°C, shielded)

$\quad v_a$ = air speed (ms^{-1})

$\quad D$ = globe diameter (standard = 0.15 m)

$\quad \varepsilon_g$ = emission coefficient (matte black paint = 0.95)

If other equipment is used (Figure 60.5), refer to ISO 7726 for calculation of $\bar{t}_r$.

Sensor $\bar{t}_r$ range/accuracy: comfort: 10 to 40°C ± 2°C; stress: –40 to +50°C ± 5°C; above +50°C: linear increase from 5 to 13°C at 150°C; desirable accuracy for comfort: ±0.2°C; desirable accuracy for stress: ±0.5°C.

For measurement of plane radiant temperature, typically used in comfort assessment, the accuracy required in the range of 0 to 50°C is ±0.5°C.

60.1.1.1.1.3 Surface Temperature (t_s) — Surface temperature (t_s) is measured with special sensors (Figure 60.3) that ensure a good contact with the surface while insulating the sensor from the environment, or with a noncontact infrared sensor (Figure 60.5). With a contact sensor, the conduction between surface and sensor must be much higher than that from the sensor to the environment, and sometimes conductive paste helps. However, surfaces with very low conductivity (e.g., wood, styrofoam) may yield false values, and in such cases a noncontact infrared sensor is better. Accuracy is dependent on the emissivity/ reflectivity of the surface required to calculate the actual t_s. For most matte surfaces, this is rather constant,

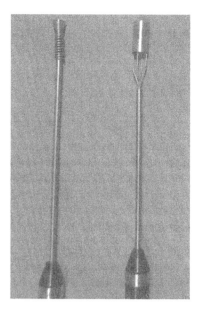

FIGURE 60.3 Example of shielded (radiation) ambient temperature sensor (right) and contact temperature sensor (left). This shielded sensor cannot be used with very high radiation levels because the single shield is in close proximity to the actual sensor, and without ventilation of the space in the shield, the shield would heat up and affect the sensor through radiation.

but shiny surfaces can act as a mirror, and the sensor may measure reflected radiation from other objects as well.

Sensor t_s range/accuracy: comfort: 0 to 50°C, ±1°C; stress: –40 to +120°C, between –10 and +50°C: ±1°C, below –10°C and above +50°C: linear increase from 1 to 3.5°C and 4.5°C, respectively, to range limit; desirable accuracy comfort: ±0.5°C; desirable accuracy for stress: ±0.5°C

60.1.1.1.2 Air Humidity

The amount of moisture present in the environmental air (the moisture concentration in $g \cdot kg^{-1}$, $g \cdot m^{-3}$, or vapor pressure in pascals [Pa]) determines whether moisture (sweat) in vapor form flows from the skin to the environment or vice versa. Often air humidity is expressed as relative humidity, i.e., the actual amount of moisture in the air compared with the maximum amount possible at that temperature:

$$RH = 100\frac{p_a}{p_{as}} \quad (\%)$$

where p_a = ambient vapor pressure   (60.2)

p_{as} = saturated vapor pressure at ambient temperature

Relative humidity can be measured with hair hygrometers, but these have a very limited accuracy and react slowly. Electronic sensors (Figure 60.4; capacitance sensors with relative-humidity-sensitive dielectricum; lithium chloride hygrometers) are available, but when exposed to extreme climates, these tend to show a slow drift (over several days) and need regular recalibration. Very accurate, but costly, are dew-point sensors that cool a smooth surface and detect the temperature at which condensation occurs, i.e., the dew point. The moisture concentration of the environment is then equal to the saturation vapor pressure at this dew point.

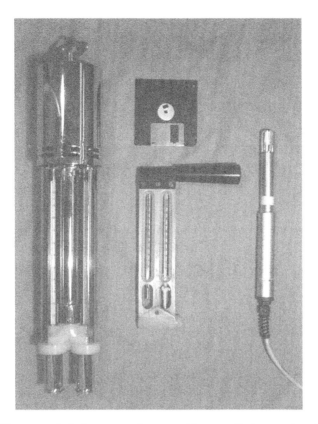

FIGURE 60.4 Humidity sensors. Left: Assman psychrometer. The top cylinder contains a fan that draws air over the two thermometers. These can be seen in the left and right columns. The sensor heads are at the bottom, shielded against radiation by a double (ventilated) shield. One thermometer is covered with a muslin wick that is moistened and will cool the sensor (aspirated wet-bulb temperature) in relation to relative humidity. The other sensor measures ambient (dry bulb) temperature (shielded for radiation). Middle: sling or whirling psychrometer; this works on the same principle of the Assman psychrometer, though air is drawn over the sensors by whirling it around in the air. There is no shielding against radiation, and this should be considered if the sensor is used in radiant environments. Right: an electronic humidity sensor that can be read with an electronic display unit.

The psychrometer (Figure 60.4) is both accurate and affordable. It combines an ambient-temperature sensor (dry bulb) and a temperature sensor with a wet cotton wick around it (wet bulb). Evaporation from the wet wick cools the thermometer and lowers the wet-bulb temperature relative to the dry-bulb air temperature. The difference is used to calculate the relative humidity. Whirling the sensor in the air (a whirling hygrometer), or having a fan suck air over it as in an Assman psychrometer, helps the instrument attain stability. The Assman instrument also shields the sensors from thermal radiation (Figure 60.4).

When no forced air movement is applied to the wick and the sensor is freely exposed to the environment, it is sensitive to both air movement and radiation, and it is called a natural wet-bulb thermometer. This is part of a WBGT (wet bulb globe temperature) meter.

Relative humidity can be converted to vapor pressure as:

$$p_a = \frac{RH}{100} \cdot p_{as} = \frac{RH}{100} \cdot e^{\left[23.5613 - \frac{4030.183}{t_a + 235}\right]} \quad (Pa) \qquad (60.3)$$

If the dew point needs to be converted into vapor pressure, this is simply done using Equation 60.3 with RH = 100% and the dew-point temperature instead of t_a.

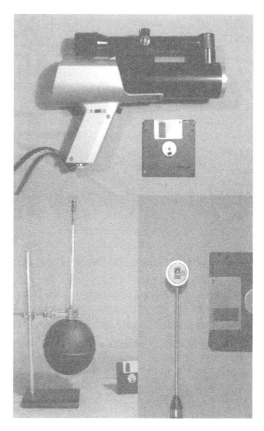

FIGURE 60.5 Three types of radiation measurements: bottom left: black globe with mercury thermometer in center; bottom right: unidirectional sensor, using the temperature measurement of a reflective and an absorbing surface to calculate radiation levels; top: infrared radiation meter, a spot-meter with a small measuring angle, hence the visor to aim precisely at the correct spot.

For the calculation of vapor pressure from the whirling hygrometer or psychrometer, i.e., from t_a or dry-bulb temperature (t_{db}) and aspirated wet-bulb temperature (t_{wb}), the following equation is used:

$$p_a = p_{as}(at\ wet\ bulb\ temperature) - 66.7 \cdot \left(t_a - t_{wb}\right) = e^{\left[23.5613 - \frac{4030.183}{t_{wb}+235}\right]} - 66.7 \cdot \left(t - t\right) \quad (60.4)$$

Other conversions can be found in ISO 7726 (1998).

The moisture concentration or vapor pressure, not the relative humidity, is the determining factor for skin evaporation. Saturated air that has a relative humidity of 100% can contain different amounts of moisture (Equation 60.3), depending on its temperature. The higher the temperature, the higher is the moisture content at equal relative humidities. When the air temperature is lower than the skin temperature, sweat will always be able to evaporate from the skin, even at 100% relative humidity, as the skin's vapor pressure will be higher than that of the air.

Absolute humidity p_a range/accuracy: comfort: 500 to 3000 Pa ± 150 Pa; stress: 500 to 6000 Pa ± 150 Pa; accuracy to be guaranteed for a range of $\left|\bar{t}_r - t_a\right|$ ≤10°C for comfort, and ≤20°C for stress.

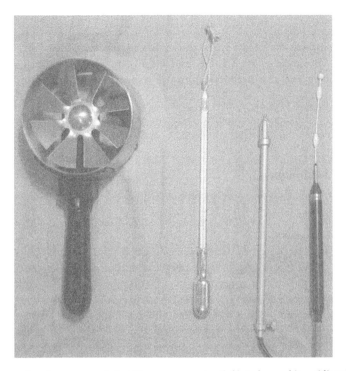

FIGURE 60.6 Sensors for air movement/wind: Vane anemometer (left), to be used in unidirectional air movement. Kata-thermometer (second from left), to be freely hanging in air after preheating. Cooling speed is related to air speed. Unidirectional hot wire anemometer (second from right): the actual wire protrudes at the top. It is extremely thin and is not really visible in this view. The shaft is used to protect the sensor when not in use. Spherical sensor (right), designed to be sensitive to multidirectional wind and with a fast response time.

60.1.1.1.3 Air Speed

The magnitude of air movement (v_a) and its direction and turbulence affects both convective and evaporative heat losses, and heat exchange increases with increasing wind speed. In a cool environment, the body cools faster in the presence of wind, whereas in an extremely hot, humid environment, it will heat up faster. In a very hot but dry environment, it will promote dry heat transfer toward the body, but will also increase evaporative heat loss from the body.

Air speed can be measured using a vane or cup anemometer (Figure 60.6) if the air movement is coming from a fixed direction and fluctuates only slowly. A hot-wire anemometer is used if the wind is not unidirectional or fluctuates rapidly (Figure 60.6). In some designs, the heated wire is replaced by other heated shapes (e.g., sphere; Figure 60.6). Air velocity can also be measured with a Kata-thermometer, which has a very large fluid reservoir (Figure 60.6). It is first heated and then hung in the relevant location and allowed to cool to air temperature. The cooling speed measured by the drop of the fluid level in the stem over time is a measure of the air movement.

Sensor v_a range/accuracy: comfort: 0.05 to 1 msec^{-1} ± (0.05 + 0.05 v_a) msec^{-1}; stress: 0.2 to 20 msec^{-1} ± (0.1 + 0.05 v_a) msec^{-1}; response time (90% of final value reached in this time): comfort ≤0.5 sec, desirable ≤0.2 sec (for measurement of turbulence intensity).

60.1.1.1.4 Measuring Locations

60.1.1.1.4.1 Spatial Considerations — The exact location of thermal measurements should represent the actual workers' locations in the space, i.e., their workstations. If this is not practical, conditions should be measured over a grid of locations in the workplace. The higher the thermal variations in space, the denser this grid will need to be. A grid width with 5-m intervals should be sufficient for most situations. Apart from this horizontal distribution, it is also relevant to repeat measurements at different heights for each location: typically at the height of the worker's head, trunk, and legs/feet (for standing work: 1.7, 1.1, and 0.1 m; for sitting work: 1.1, 0.6, and 0.1 m). In very homogeneous environments, a measurement at abdomen level (1.1 m for standing, 0.6 m for sitting) is sufficient.

60.1.1.1.4.2 Temporal Considerations — Measurement at one point in time can be misleading. It may be important to investigate the climate profile for daily and seasonal patterns. Leaving equipment to log climate conditions for at least a day is the minimum. Using questionnaires to identify the problem times can save a lot of effort.

60.1.1.2 Personal Parameters

Some thermal comfort and stress assessment methods require information on clothing insulation and metabolic heat production.

60.1.1.2.1 Clothing Insulation

Clothing resists heat and moisture transfer between skin and environment. This can protect against extreme heat and cold, but it can also hamper heat loss during physical effort. Clothing insulation (measured in units of m^2 °C W^{-1} or in units of clo [1 clo = 0.155 m^2 °C W^{-1}]) is expressed either as total insulation (I_T, includes surface air layer) or so-called intrinsic insulation (I_{cl}, clothing with enclosed air layers only). Values for clothing insulation (I_T for clothing including surface air layer, or I_{cl} for clothing only) and vapor resistance (R_e) can be measured (Havenith, 1999; Havenith et al., 2002), but usually these are estimated from lists of insulations for numerous clothing ensembles. Extensive lists of clothing insulations as well as different methods for its estimation can be found in ISO 9920 (1995). Some examples are presented in Table 60.1.

Hardly any listings of clothing vapor resistance are available, but once insulation is known, vapor resistance can be estimated as:

$$R_e = \frac{I_t}{0.0165 \cdot i_m}$$

where R_e = clothing vapor resistance ($m^2 Pa°CW^{-1}$)

I_t = clothing heat resistance ($m^2°CW^{-1}$)   (60.5)

i_m = clothing permeability index (n.d.)

0.0165 = Lewis constant (0.0165 °C.Pa^{-1})

The value for i_m can be estimated from Table 60.2.

Clothing insulation and vapor resistance are affected by many factors, including movement of the wearer, wind, wetting, etc. For a more detailed description, see Havenith et al. (1999, 2002), Holmér et al. (1999), Nilsson et al. (2000), and Havenith and Nilsson (2004). Table values for clothing insulation for most normal and warm-weather clothing (1.9 clo > I_T > 1.2 clo) can be corrected:

$$I_{T,dynamic} = \text{correction factor} \cdot I_{T,static}$$

$$\text{correction factor} = e^{(-0.281 \times (v_{ar} - 0.15) + 0.044 \times (v_{ar} - 0.15)^2 - 0.492w + 0.176w^2)}$$

where $I_{T,static}$ = clothing insulation obtained from tables

w = walking speed; maximally 1.2 ms^{-1} (60.6)

if other movement, derive from metabolic rate as

w = 0.0052 × (Metabolic rate [Wm^{-2}] - 58); limited to $w \le 0.7$ ms^{-1}

v_{ar} = relative air speed; minimally 0.15; maximally 3 ms^{-1}

TABLE 60.1

Sample Values of Clothing Insulation Values of Various Work Wear (left) and Daily Wear (right) Ensembles

Work Clothing	(clo)	I_{cl} (m² °C W⁻¹)	I_T (m² °C W⁻¹)	Daily Wear Clothing	(clo)	I_{cl} (m² °C W⁻¹)	I_T (m² °C W⁻¹)
Underpants, boiler suit, socks, shoes	0.7	0.11	0.196	Panties, T-shirt, shorts, light socks, sandals	0.3	0.05	0.145
Underpants, shirt, trousers, socks, shoes	0.75	0.115	0.200	Panties, petticoat, stockings, light dress with sleeves, sandals	0.45	0.07	0.162
Underpants, shirt, boiler suit, socks, shoes	0.8	0.125	0.209	Underpants, shirt with short sleeves, light trousers, light socks, shoes	0.5	0.08	0.171
Underpants, shirt, trousers, jacket, socks, shoes	0.85	0.135	0.218	Panties, stockings, shirt with short sleeves, skirt, sandals	0.55	0.085	0.175
Underpants, shirt, trousers, smock, socks, shoes	0.9	0.14	0.222	Underpants, shirt, light-weight trousers, socks, shoes	0.6	0.095	0.183
Underwear with short sleeves and legs, shirt, trousers, jacket, socks, shoes	1	0.155	0.235	Panties, petticoat, stockings, dress, shoes	0.7	0.105	0.192
Underwear with short legs and sleeves, shirt, trousers, boiler suit, socks, shoes	1.1	0.17	0.248	Underwear, shirt, trousers, socks, shoes	0.7	0.11	0.196
Underwear with long legs and sleeves, thermo jacket, socks, shoes	1.2	0.185	0.262	Underwear, track suit (sweater and trousers), long socks, runners	0.75	0.115	0.200
Underwear with short sleeves and legs, shirt, trousers, jacket, thermo jacket, socks, shoes	1.25	0.19	0.266	Panties, petticoat, shirt, skirt, thick kneesocks, shoes	0.8	0.12	0.205
Underwear with short sleeves and legs, boiler suit, thermo jacket and trousers, socks, shoes	1.4	0.22	0.293	Panties, shirt, skirt, round-neck sweater, thick knee socks, shoes	0.9	0.14	0.222
Underwear with short sleeves and legs, shirt, trousers, jacket, thermo jacket and trousers, socks, shoes	1.55	0.225	0.297	Underpants, singlet with short sleeves, shirt, trousers, V-neck sweater, socks, shoes	0.95	0.145	0.226
Underwear with short sleeves and legs, shirt, trousers, jacket, heavy quilted outer jacket and overalls, socks, shoes	1.85	0.285	0.352	Panties, shirt, trousers, jacket, socks, shoes	1	0.155	0.235
Underwear with short sleeves and legs, shirt, trousers, jacket, heavy quilted outer jacket and overalls, socks, shoes, cap, gloves	2	0.31	0.375	Panties, stockings, shirt, skirt, vest, jacket	1	0.155	0.235
Underwear with long sleeves and legs, thermo jacket and trousers, outer thermo jacket and trousers, socks, shoes	2.2	0.34	0.403	Panties, stockings, blouse, long skirt, jacket, shoes	1.1	0.17	0.248
Underwear with long sleeves and legs, thermo jacket and trousers, parka with heavy quilting, overalls with heavy quilting, socks, shoes, cap, gloves	2.55	0.395	0.454	Underwear, singlet with short sleeves, shirt, trousers, jacket, socks, shoes	1.1	0.17	0.248
				Underwear, singlet with short sleeves, shirt trousers, vest, jacket, socks, shoes	1.15	0.18	0.257
				Underwear with long sleeves and legs, shirt, trousers, V-neck sweater, jacket, socks, shoes	1.3	0.2	0.275
				Underwear with short sleeves and legs, shirt, trousers, vest, jacket, coat, socks, shoes	1.5	0.23	0.302

Note: I_{cl} = intrinsic clothing insulation (without adjacent surface air layer); I_T = total insulation (clothing + surface air layer); clo = unit of insulation defined relative to the value of an American business suit (1 clo = 0.155 m² °C W⁻¹).

TABLE 60.2 Example Data for Estimating the Static Clothing Permeability Index (i_m) Using Description of the Clothing Type

Clothing Description	Estimated i_m Static	
	In the Cold (<15°C)	In the Heat (>30°C)
Nude	0.5	0.5
Normal, permeable clothing, regardless of number of layers	0.38	0.38
As 1, with tightly woven jacket	0.34	0.34
As 1, with coated jacket OR trousers	0.31	0.31
As 1, with two-piece semipermeable overgarment	0.17	0.15
As 1, with one-piece semipermeable overgarment	0.14	0.13
As 1, with two-piece impermeable overgarment	0.12	0.07
As 1, with one-piece impermeable overgarment	0.1	0.06
As 1, with one-piece impermeable overgarment, covered head except face, gloves, openings sealed (e.g., immersion suit)	0.06	0.02
Completely encapsulating suit, all openings sealed, no skin exposed	0.05	0.0

Note: Difference between cold and heat values represents the effect of condensation at the inner clothing surface; intermediate values can be interpolated as $i_m = [\alpha\, i_{m15} + (1 - \alpha)\, i_{m30}]$, with $\alpha = (30 -$ air temperature$)/15$.

Source: Clothing descriptions are from Havenith, G., Holmér, I., Den Hartog, E.A., and Parsons, K.C. (1999), *Ann. Occup. Hyg.*, 43, 339–346.

For specialized, insulating cold-weather clothing, which typically has low air permeability and where high wind speeds occur more frequently, the equation for the correction factor to be used is:

$$\text{correction factor} = 0.54 \cdot e^{(-0.15v_a\ -0.22w)} \cdot p^{0.075} - 0.06 \cdot \ln(p) + 0.5$$

where:

v_{ar} = relative air speed; from 0.4 m / s to 18 ms^{-1}

w = walking speed (ms^{-1}); from 0 to 1.2 ms^{-1} $\qquad\qquad$ (60.7)

p = air permeability of outer fabric (lm^{-2}s^{-1}) ; from 1 to 1000 lm^{-2}s^{-1}

low: (e.g., coating or laminate) 1; medium 50; high (open weave) 1000 lm^{-2}s^{-1}

Here it is assumed that head and hands are covered with hood/hat and gloves; i.e., the body is totally covered.

For correction of vapor resistance, the relations are slightly more complex. For most applications, however, a reduction in vapor resistance equal to 1.3 times that in heat resistance (Equation 60.6 or Equation 60.7) is a good approximation. For more details, see Havenith et al. (1999, 2002).

60.1.1.2.2 Metabolic Rate

For most ergonomics applications, the efficiency of external work (energy released outside the body) performed is close to zero. Only on cycle ergometers, when one continuously walks upward on stairs, does the efficiency climb to significant values (maximum around 0.23). Hence, almost all metabolic energy is released as heat in the body. It can be measured using indirect calorimetry (measuring oxygen uptake; ISO 8996, 1990). Metabolic rates for a large number of activities can be estimated using tables describing activities, professions, postures, etc. An overview is given in Table 60.3, with an example for a quite coarse classification in Table 60.4. Data in both tables are taken from the extensive listings in ISO 8996 (1990) and from Spitzer et al. (1982).

TABLE 60.3 Six Methods for Estimating Metabolic Heat Production

Level	Method	Accuracy	Inspection of the Workplace
I	A: classification according to kind of activity	Rough information where the risk of error is very great	Not necessary
	B: classification according to occupation		Information on technical equipment, work organization
II	A: use of tables of group assessment	High error risk: accuracy ±15%	Time study necessary
	B: use of estimation tables for specific activities		
	C: use of heart rate under defined conditions		Not necessary
III	Measurement	Risk of errors within the limits of the accuracy of the measurement and of the time study: accuracy ±15%	Time study necessary

Source: ISO 8996/EN28996 (1990), Ergonomics: Determination of Metabolic Heat Production, International Standardization Organization, Geneva; Spitzer et al., 1982.

TABLE 60.4 Classification of Metabolic Rate Based on General Work Description

Class	Value To Be Used for Calculation of Mean Metabolic Rate (W/m²)	(W)	Examples
0 Resting	65	115	Resting
1 Low metabolic rate	100	180	Sitting at ease: light manual work (writing, typing, drawing, sewing, bookkeeping); hand and arm work (small bench tools, inspection, assembly or sorting of light materials); arm and leg work (driving vehicle in normal conditions, operating foot switch or pedal) Standing: drilling (small parts); milling machine (small parts); coil winding; small-armature winding; machining with low power tools; casual walking (speed up to 3.5 km/h)
2 Moderate metabolic rate	165	295	Sustained hand and arm work (hammering in nails, filing); arm and leg work (off-road operation of lorries, tractors, or construction equipment); arm and trunk work (work with pneumatic hammer, tractor assembly, plastering, intermittent handling of moderately heavy material, weeding, hoeing, picking fruits or vegetables, pushing or pulling lightweight carts or wheelbarrows, walking at a speed of 3.5 km/h to 5.5 km/h, forging)
3 High metabolic rate	230	415	Intense arm and trunk work; carrying heavy material; shoveling; sledgehammer work; sawing; planing or chiseling hard wood; hand mowing; digging; walking at a speed of 5.5 km/h to 7 km/h Pushing or pulling heavily loaded hand carts or wheelbarrows; chipping castings; concrete block laying
4 Very high metabolic rate	290	520	Very intense activity at fast to maximum pace; working with an axe; intense shoveling or digging; climbing stairs, ramp, or ladder; walking quickly with small steps; running; walking at a speed greater than 7 km/h

Source: ISO 8996/EN28996 (1990), Ergonomics: Determination of Metabolic Heat Production, International Standardization Organization, Geneva; Spitzer et al., 1982.

60.1.2 Subjective Methods

For thermal environments, subjective assessment methods have been developed suggesting scales to be used in questionnaires (ISO 10551, 1995). The subjective assessment is split in several categories for which the related questions are:

Category	Question
Perceptual:	How are you feeling (at this precise moment)?
Evaluation:	Do you find this … ?
Preference:	Would you prefer to be … ?
Acceptability:	Do you find this acceptable?
Tolerance:	Is it tolerable?

The relevant scales to be used are shown in Table 60.5 through Table 60.8.

Note that many factors other than workplace climate (e.g., stress, problems with management, general working conditions) affect dissatisfaction with the thermal environment. Wherever possible, objective data should complement subjective assessments.

TABLE 60.5 Scale of Subjective Descriptions of Personal Thermal State in Response to the Question "How are you feeling now?"

Poles	Degrees	English	French	Spanish
Hot	(+4)	very hot	extrêmement chaud	calor excesivo
	+3	hot	très chaud	mucho calor
	+2	warm	chaud	calor
	+ 1	slightly warm	légèrement chaud	algo de calor
Indifference	0	neutral	ni chaud ni froid	ni calor ni frio
	−1	slightly cool	légèrement froid	algo de frio
	−2	cool	froid	frio
	−3	cold	très froid	mucho frio
Cold	(−4)	very cold	extrêmement froid	frio excesivo
Common introductory term(s)		I'm feeling/I'm …	J'ai….	Tengo…..

Note: The central tendency of the perceptual judgments obtained by applying one of the above-mentioned scales yields an observed mean vote that can be compared with the predicted mean vote (PMV) index determined according to ISO 7730.

TABLE 60.6 Scale of Subjective Evaluative Judgments of Personal Thermal State in Response to the Question "Do you find this … ?"

Pole	Degrees	Wording of Degrees
Comfort	0	comfortable
	1	slightly uncomfortable
	2	uncomfortable
	3	very uncomfortable
Discomfort	4	extremely uncomfortable
Common introductory terms		I find it ….

Note: By summing up the judgments that express discomfort, one obtains an observed percentage of dissatisfied people, which can be compared with the predicted percentage of dissatisfied (PPD) index determined according to ISO 7730.

TABLE 60.7 Scale of Subjective Thermal Preference in Response to the Instruction "Please state how you would prefer to be now"

Poles	Degrees	Wording of degrees for 7-degree scale	Equivalent for 3-degree scale
Warmer	+3	much warmer	warmer
	+2	warmer	
	+1	a little warmer	
	0	neither warmer nor cooler	
	−1	slightly cooler	cooler
	−2	cooler	
Cooler	−3	much cooler	
Common introductory terms		I would prefer to be	

TABLE 60.8 Personal Acceptability Statement Form

Categories	(a) Explicit Wording of the Degrees After the question: "How do you judge this environment (local climate) on a personal level?"	(b) Wording of the Degrees after Initial Statement 1 or 2 After the common instructions: "Taking into account only your personal preference …"	
	"On a personal level, this environment is for me …"	**Initial statement 1:** "Would you accept this environment (local climate) rather than reject it?"	**Initial statement 2:** "Would you reject this environment (local climate) rather than accept it?"
0	… acceptable rather than unacceptable"	Yes	No
1	… unacceptable rather than acceptable"	No	Yes

Note that instead of using a two-category statement form (yes/no), personal acceptability can also be expressed on a continuous scale, such as the following four-category statement: Clearly acceptable, Just acceptable, Just unacceptable, and Clearly unacceptable.

Note: The preliminary instructions would be as follows (after the question, "How do you judge this environment (local climate) on a personal level?" or after the instruction, "Taking into account only your personal preference …"): "Please mark the appropriate place on the scale to express your acceptance of the environment (local climate). Do not mark the middle of the scale, but express either acceptance or unacceptance."

60.2 Standard and Regulations

The most relevant standards for this area are developed in the committees ISO TC159/WG1 and CEN TC122/WG11.

60.2.1 Basic Standards

ISO 7726: 1998 (EN27726), Thermal environments: instruments and methods for measuring physical quantities

ISO 8996: 1990 (EN28996), Ergonomics: determination of metabolic heat production

ISO 9920 (EN9920): 1995, Ergonomics of the thermal environment: estimation of the thermal insulation and evaporative resistance of a clothing ensemble (under revision)

ISO 11399: 1995 (EN11399), Ergonomics of the thermal environment: principles and application of international standards

ISO 13731 (prEN13202) Ergonomics of the thermal environment: vocabulary and symbols

ISO 12894: 1993 (EN12894), Ergonomics of the thermal environment: medical supervision of individuals exposed to hot or cold environments

60.2.2   Thermal Comfort, Stress and Strain

ISO 7730: 1994 (EN7730), Moderate thermal environments: determination of the PMV and PPD indices and specification of the conditions for thermal comfort

ISO 7243: 1995 (EN7243) Hot environments: estimation of the heat stress on working man, based on the WBGT-index (wet bulb globe temperature)

ISO 7933: 1989 (EN12515), Hot environments: analytical determination and interpretation of thermal stress using calculation of required sweat rate

ISO 9886: 1992 (EN9886), Evaluation of thermal strain by physiological measurements

ISO 10551: 1995 (EN10551), Ergonomics of the thermal environment: assessment of the influence of the thermal environment using subjective judgement scales

ISO TR 11079 (technical report): 1993 (ENV11079), Evaluation of cold environments: determination of required clothing insulation, IREQ

60.2.3   Risk of Burns and Cold Injury

ISO/NP 13732 Part 1, Ergonomics of the thermal environment: methods for the assessment of human responses to contact with surfaces, Part 1: hot surfaces

ISO CD 13732 Part 2, Ergonomics of the thermal environment: methods for the assessment of human responses to contact with surfaces, Part 2: moderate surfaces

ISO/NP 13732 Part 3, Ergonomics of the thermal environment: methods for the assessment of human responses to contact with surfaces, Part 3: cold surfaces

60.2.4   Special Applications

ISO NP 14405, Ergonomics of the thermal environment: evaluation of the thermal environment in vehicles

ISO NP 14415, Ergonomics of the thermal environment: application of international standards to the disabled, the aged and handicapped persons

ISO NP 15743, Ergonomics of the thermal environment: working practices for cold indoor environments.

60.3   Approximate Training and Application Time

Training in the use of instruments for thermal assessment should take around 2 to 4 hours, depending on the user's background, and should include recognizing problems with the measurement instruments. A staged approach to thermal assessment can be used: first assessments are done by lay persons. If any thermal problems are indicated, more highly trained people become involved, and these again could call upon experts in critical cases (Malchaire, 2000).

The application time is mainly determined by the number of locations; the type, amount, and response times of the equipment (e.g., black globe: 20 to 30 min), and by the expected variations in the climate. The more variable the climate, the longer the logging should be continued to catch average and worst-case data. For seasonal effects, repeated measurements will be required.

60.4   Reliability and Validity

Thermometers (mercury/alcohol thermometer, thermometer in black globe) will require calibration against a standard, but then they are very stable and reliable. Humidity measurements based on wet-bulb temperature are sensitive to pollution of the wick (replace regularly; use distilled water) and problems with the air ventilation rate. The sensor itself is a thermometer and is thus reliable.

Electronic sensors (temperature, humidity, wind speed) need regular calibration of both sensor and electronic circuits. Electronic temperature sensors are usually quite stable, but electronic humidity measurements need regular checks and calibrations, especially when used in extremes.

Estimations of metabolic rate and of clothing insulation can show large errors (±10%). This should be considered when evaluating the outcome of the analyses.

60.5 Tools Needed

A basic tool set should allow for the determination of the four climatic parameters t_a, $\bar{t}_r$, p_a (or RH), and v_a. Hence it should consist of:

- Ambient-temperature sensor: a low-cost, regular thermometer or electronic sensor, preferably shielded for radiation
- Radiant-temperature sensor: a low-cost, black globe or a medium/high-cost radiometer
- Humidity sensor: a medium-cost whirling hygrometer, Assman psychrometer, electronic RH sensor; or a high-cost dew-point sensor
- Wind-speed/air-movement sensor: a medium-cost Kata-thermometer, hot-wire anemometer, vane anemometer (unidirectional); or a high-cost heated sphere (multidirectional)

60.6 Steps in Method

Steps that need to be taken for the objective assessment are:

- Identify and select appropriate equipment to measure t_a, $\bar{t}_r$, p_a (or RH), and v_a considering range, accuracy, and response time of apparatus and sensors.
- Calibrate equipment.
- Survey expected climate fluctuations in time (seasons, weather) and space within work space by interviews with workers.
- Locate relevant workstations as measurement locations or define locations using a grid of the total work space.
- Define locations and measurement timing over day, season, or weather conditions.
- Measure and register climatic parameters at three heights (one height in highly uniform environments) at all locations.
- Survey the work load (metabolic rate) and clothing (insulation) worn at the various workplaces.
- Correct clothing insulation for movement and wind.
- Take all data and use heat-stress, cold-stress, or comfort evaluation methods, as described in following chapters.

References

Havenith, G. (1999), Heat balance when wearing protective clothing, *Ann. Occup. Hyg.*, 43, 289–296.

Havenith, G. and Nilsson, H. (in press), Correction of clothing insulation for movement and wind effects, a meta-analysis, *J. Appl. Physiol.*

Havenith, G., Holmér, I., Den Hartog, E.A., and Parsons, K.C. (1999), Clothing evaporative heat resistance-proposal for improved representation in standards and models, *Ann. Occup. Hyg.*, 43, 339–346.

Havenith, G., Holmér, I., and Parsons, K. (2002), Personal factors in thermal comfort assessment: clothing properties and metabolic heat production, *Energy Buildings*, 34, 581–591.

Holmér, I., Nilsson, H., Havenith, G., and Parsons, K.C. (1999), Clothing convective heat exchange: proposal for improved prediction in standards and models, *Ann. Occup. Hyg.*, 43, 329–337.

ISO 7726/EN27726 (1998), Thermal Environments: Instruments and Methods for Measuring Physical Quantities, International Standardization Organization, Geneva.

ISO 8996/EN28996 (1990), Ergonomics: Determination of Metabolic Heat Production, International Standardization Organization, Geneva.

ISO 9920/EN9920 (1995), Ergonomics of the Thermal Environment: Estimation of the Thermal Insulation and Evaporative Resistance of a Clothing Ensemble, International Standardization Organization, Geneva.

ISO 10551/EN10551 (1995), Ergonomics of the Thermal Environment: Assessment of the Influence of the Thermal Environment Using Subjective Judgement Scales, International Standardization Organization, Geneva.

Malchaire, J. (2000), Strategy for the management of the thermal working conditions, in *Proceedings of the XIVth Triennial Congress of the International Ergonomics Association and 44th Annual Meeting of the Human Factors and Ergonomics Society*, (CD-ROM), Human Factors and Ergonomics Society, Santa Monica, CA.

Nilsson, H., Anttonen, H., and Holmér, I. (2000), New algorithms for prediction of wind effects on cold protective clothing, in *Ergonomics of Protective Clothing, Proceedings of Nokobetef 6*, Stockholm, pp. 17–20.

Spitzer, H., Hettinger, T., and Kaminsky G. (1982), Tafeln für den Energieumsatz bei körperlicher Arbeit (Tables for the energy usage in physical work), Beuth Verlag, Berlin (in German).

61

Cold Stress Indices

Hannu Rintamäki
*Oulu Regional Institute of
Occupational Health*

61.1 Background

Cold stress is developed when some, or several, of the basic avenues of heat loss (radiation, evaporation, conduction, and convection) exceed the heat production of human tissues. The available methods of assessing cold stress analyze one or several forms of the above-mentioned avenues of heat loss, or they quantify the signs of cold stress in human body. There are several indices and standards to quantify cold stress. Assessment of cold exposure and cold stress are reviewed by Holmér (1993, 2001) and Parsons (2003). An ISO standard under development (ISO CD 15743, unpublished) aims to give a practical three-level approach to analyze cold stress starting from simple checklists and continuing, if necessary, to simple measurements and, finally, to extensive measurements.

61.2 Methods to Analyze Cold Strain

61.2.1 Wind-Chill Indices

There are several wind-chill indices that quantify the convective heat loss caused by different combinations of low ambient temperature and air movements. The original wind-chill index (WCI) was developed by Siple and Passel (1945) based on their empirical experiments. WCI gives the amount of heat loss in a given combination of ambient temperature and air movements. It also gives a temperature equivalent in calm conditions. Wind-chill indices developed since then (e.g., Steadman, 1984) are typically based on heat-balance equations. To serve better the users of weather reports, a new wind-chill index was developed in 2000–2001. The new index (W) is based on heat transfer theory, and it is calculated as follows:

$$W = 13.12 + (0.6215 \times T_{air}) - (11.37 \times V^{0.16}) + (0.3965 \times T_{air} \times V^{0.16})$$

where
W = the wind-chill index (°C)
T_{air} = air temperature (°C)
V = wind speed at 10 m (km/h)

To calculate W in Fahrenheit:

$$W = 35.74 + (0.6215 \times T_{air}) - (35.75 \times V^{0.16}) + (0.4275 \times T_{air} \, V^{0.16})$$

where
W = the wind-chill index (°F)
T_{air} = air temperature (°F)
V = wind speed at 10 m (mph)

The new wind-chill index uses the wind speed at the average height of the human face (1.5 m). Therefore, the wind speed at 10 m (standard anemometer height) is multiplied by a factor of two thirds. Wind-chill index is not a temperature, but it expresses human sensation (equivalent temperature). The calm wind threshold is 4.8 km/h (1.3 m/sec), which is based on the speed at which people walk. Considering the risk of frostbite, W can be interpreted as follows:

- Risk of frostbite in prolonged exposure (30 minutes) at –28°C
- Frostbite possible in 10 minutes at –40°C (shorter time if sustained wind greater than 50 km/h)
- Frostbite possible in 5 minutes at –48°C (shorter time if sustained wind greater than 50 km/h)
- Frostbite possible in 2 minutes or less at –55°C

The active development of wind-chill indices is ongoing, and the effect of solar radiation may be added to the index (Environment Canada's Wind Chill Program: http://www.msc.ec.gc.ca/education/windchill/charts_tables_e.cfm).

61.2.2 Required Clothing Insulation

Required clothing insulation — insulation required (IREQ) (see Holmér, 1984; ISO TR 11079, 1993) — is an assessment method based on heat-balance equations. The IREQ equation aims to quantify all forms of heat loss while also accounting for heat production. It calculates the required thermal insulation of clothing in a given thermal environment and activity level (Figure 61.1). The IREQ can be calculated for

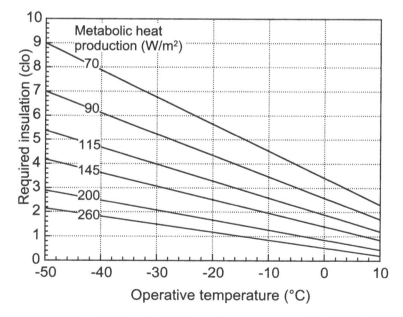

FIGURE 61.1   Neutral IREQ as a function of ambient operative temperature at six levels of metabolic heat production. (Modified from ISO TR 11079 [1993], Ergonomics of the Thermal Environment: Cold Environments, Determination of Required Clothing Insulation [IREQ], International Organization for Standardization, Geneva.)

two levels of thermal strain: low strain (neutral IREQ) and high strain (minimal IREQ). At the same time, IREQ gives guidelines for cold protection and serves as a measure of cold stress. The computer program for solving the IREQ equation is listed in ISO TR 11079 and the program is available at the Web site: http://www.eat.lth.se/Research/Thermal/Thermal_HP/Klimatfiler/IREQ2002alpha.

If the thermal insulation of the clothing is not sufficient for given conditions, the IREQ equation can be used to calculate the duration-limited exposure time (DLE time), which expresses the allowable exposures time before reaching the threshold of low or high thermal strain.

It is important to note that the IREQ equation gives the required thermal insulation for the whole body, but the local cold strain — e.g., in hands, feet, and respiratory system — should always be evaluated separately. This is especially true if there is high local cold stress, such as contact with cold items or high ventilation due to heavy exercise.

61.2.3 Physiological Measurement

Physiological measurements of cold stress are based on measurements of superficial skin and body core temperatures and sometimes also heat loss. The measured skin and deep-body temperatures are usually interpreted in five levels (comfort, discomfort, performance degradation, adverse health effects, and tolerance). The interpretation is based on the classification developed by Goldman and expanded by Lotens (1988). The general principles, procedures, and interpretation of physiological methods of cold stress assessment is presented in ISO 9886 (1992). The methods cited in ISO 9886 (1992) include measurement of skin temperature, body core temperature, heart rate, and body mass loss.

61.2.4 Thermal Sensation

Recording thermal sensations is a simple but practical method to analyze cold stress. ISO 10551 (1995) gives the guidelines for recording thermal sensations and thermal preferences.

Thermal sensations are recorded after the question "How are you feeling now?" The sensations are classified as follows: hot (code +3), warm (+2), slightly warm (+1), indifference (0), slightly cool (−1), cool (−2), and cold (−3). If necessary, very hot (+4) and very cold (−4) can also be used.

The degree of thermal discomfort is recorded after the question "Do you find this…?" The classification is as follows: comfortable (0), slightly uncomfortable (1), uncomfortable (2), very uncomfortable (3), and extremely uncomfortable (4).

Scales for thermal preference and personal acceptability statement are also presented in ISO 10551 (1995).

61.2.5 Cold Strain Index

A recently developed cold strain index (CSI), based on core (T_{core}) and mean skin temperatures (T_{sk}), is capable of indicating cold strain in real time by analyzing existing databases. This index rates cold strain on a universal scale of 0 to 10 using the following equation:

$$\text{CSI} = 6.67 \, (T_{\text{core t}} - T_{\text{core 0}})/(35 - T_{\text{core 0}})^{-1} + 3.33 \, (T_{\text{sk t}}) - T_{\text{sk 0}}/(20 - T_{\text{sk 0}})^{-1}$$

where $T_{\text{core 0}}$ and $T_{\text{sk 0}}$ are initial measurements, and $T_{\text{core t}}$ and $T_{\text{sk t}}$ are simultaneous measurements taken at any time t; when $T_{\text{core t}} > T_{\text{core 0}}$, then $T_{\text{core t}} - T_{\text{core 0}}$ is assumed as 0. CSI has the potential to be widely accepted and used universally, but it requires further development (Moran et al., 1999; Castellani et al., 2001).

61.2.6 Conductive Heat Loss

A recently developed ISO standard (ISO NP 13732, unpublished) quantifies the cold stress caused by conductive heat loss. It also gives time limits for contacting different materials in different temperatures.

61.2.7 Other Methods

- Vocabulary and symbols (ISO DIS 13731, 2001)
- Principles and applications of relevant international standards (ISO 11399, 1995)
- Instruments and methods for measuring physical quantities (ISO 7726, 1998)
- Determination of metabolic heat production (ISO 8996, 1990)
- Medical supervision of individuals exposed to extreme hot or cold environments (ISO DIS 12894, 2001)

References

Castellani, J.W., Young, A.J., O'Brien, C., Stulz, D.A., Sawka, M.N., and Pandolf, K.B. (2001), Cold strain index applied to exercising men in cold-wet conditions, *Am. J. Physiol. Regul. Integrated Comput. Physiol.*, 281, R1764–1768.

Holmér, I. (1984), Required clothing insulation (IREQ) as an analytical index of cold stress, *ASHRAE Trans.*, 90, 1116–1128.

Holmér, I. (1993), Work in the cold: review of methods for assessment of cold exposure, *Int. Arch. Occup. Environ. Health*, 65, 147–155.

Holmér, I. (2001), Assessment of cold exposure, *Int. J. Circumpolar Health*, 60, 413–421.

ISO 7726 (1998), Thermal Environments: Instruments and Methods for Measuring Physical Quantities, International Organization for Standardization, Geneva.

ISO 8996 (1990), Ergonomics: Determination of Metabolic Heat Production, International Organization for Standardization, Geneva.

ISO 9886 (1992), Evaluation of Thermal Strain by Physiological Measurements, International Organization for Standardization, Geneva.

ISO 10551 (1995), Ergonomics of the Thermal Environment: Assessment of the Influence of the Thermal Environment Using Subjective Judgement Scales, International Organization for Standardization, Geneva.

ISO 11399 (1995), Ergonomics of the Thermal Environment: Principles and Applications of Relevant International Standards, International Organization for Standardization, Geneva.

ISO CD 15743 (unpublished), Ergonomics of the Thermal Environment: Working Practices in Cold, Strategy for Risk Assessment and Management, International Organization for Standardization, Geneva.

ISO DIS 12894 (2001), Ergonomics of the Thermal Environment: Medical Supervision of Individuals Exposed to Extreme Hot or Cold Environments, International Organization for Standardization, Geneva.

ISO DIS 13731 (2001), Ergonomics of the Thermal Environment: Vocabulary and Symbols, International Organization for Standardization, Geneva.

ISO NP 13732 (unpublished), Ergonomics of the Thermal Environment: Methods for Assessment of Human Responses to Contact with Surfaces, Part 3, Cold Surfaces, International Organization for Standardization, Geneva.

ISO TR 11079 (1993), Ergonomics of the Thermal Environment: Cold Environments, Determination of Required Clothing Insulation (IREQ), International Organization for Standardization, Geneva.

Lotens, W.A. (1988), Comparison of thermal predictive models for clothed humans, *ASHRAE Trans.*, 94, 1321–1340.

Moran, D.S., Castellani, J.W., O'Brien, C., Young, A.J., and Pandolf, K.B. (1999), Evaluating physiological strain during cold exposure using a new cold strain index, *Am. J. Physiol.*, 277, R556–564.

Parsons, K.C. (2003), *Human Thermal Environments: The Effects of Hot, Moderate and Cold Environments on Human Health, Comfort and Performance*, Taylor & Francis, London.

Siple, P.A. and Passel, C.F. (1945), Measurements of dry atmospheric cooling in subfreezing temperatures, *Proc. Am. Philos. Soc.*, 89, 177–199.

Steadman, R.G. (1984), A Universal Scale of apparent temperature, *J. Clin. Appl. Meteorol.*, 23, 1674–1687.

62

Heat Stress Indices

Alan Hedge
Cornell University

62.1 Introduction

Heat stress occurs when the body absorbs or produces more heat than can be dissipated through thermoregulatory processes, and illness and death can result from the increases in core temperature (Parsons, 1993). Outdoor conditions can present heat stress risks for people in hot climates, e.g., in desert or tropical locations. Heat stress can occur in unique situations, such as firefighting. Indoors, heat stress conditions occur in many workplaces, such as iron and steel foundries, glassmaking plants, bakeries, commercial kitchens, laundries, power plants, etc. Behavioral factors can amplify the risks, e.g., wearing impermeable clothing such as a protective suit. Individual susceptibility to heat stress varies as a function of several physiological risk factors. Table 62.1 summarizes all of these factors. Heat stress creates a progression of heat disorders with increasingly severe symptoms that can culminate in death (see Table 62.2).

62.2 Guidelines for Investigating Heat Stress

The following procedural guidelines are useful for evaluating employee heat stress risks. These guidelines assume that "workers should not be permitted to work when their deep body temperature exceeds 38°C" (ACGIH, 1992).

1. Review occupational injury logs/records for indications of heat stress problems.
2. Conduct employee/employer interviews to ascertain the nature of employee complaints, what and where the potential heat sources are, and what action has been taken to prevent heat stress problems.
3. Conduct a walk-around inspection of heat sources; perform temperature measurements; calculate relative heat load per employee; determine the need for engineering controls.
4. Determine the workload category of each job performed in hot conditions (see Table 62.3). Calculate/estimate average metabolic rates for the tasks (formula 1) and sum these to determine workload category. Overall job categories are light work (up to 200 kcal/h), medium work (200 to 350 kcal/h), and heavy work (350 to 500 kcal/h).

TABLE 62.1 Risk Factors for Heat Stress

Environmental Risks	Physiological Risks	Behavioral Risks
High air temperature	Age	Strenuous activity
High humidity	Weight	Dehydration
High radiant temperature	Physical fitness	Alcohol consumption
Direct contact with hot objects	Metabolism	Drug use
Still air	Degree of acclimatization	Type of clothing (semipermeable, impermeable)
	Hypertension	
	Prior heat injury	

TABLE 62.2 Heat Disorders and Symptoms

Disorder	Processes	Symptoms
Heat fatigue	Lack of acclimatization to hot conditions	Impaired performance of skilled sensorimotor, mental, or vigilance tasks
Heat rashes	Sweat glands plugged by salt deposits from evaporated sweat; unevaporated sweat accumulates in glands	Small, red, blisterlike papules on skin areas where the clothing is restrictive; produce a prickling sensation ("prickly heat"); symptoms reversed when in cooler conditions
Heat collapse	Loss of consciousness because blood pools in the extremities, causing brain anoxia	Rapid and unpredictable fainting
Heat cramps	Excessive sweating leads to electrolyte imbalance caused by salt loss	Painful muscle spasms (abdomen, arms, legs) during or soon after a period of work
Heat exhaustion	Dehydration, poor heat acclimatization, poor physical fitness	Headache, nausea, vertigo, weakness, thirst, and giddiness
Heat stroke	Thermoregulation fails and core temperature rises to critical levels	Confusion; irrational behavior; loss of consciousness; convulsions; a lack of sweating (usually); hot, dry skin; abnormally high body temperature ($\geq 41°C$); death

TABLE 62.3 Workload Categories

Work	Example
Light hand work	Writing, knitting
Heavy hand work	Typing
Heavy work with one arm	Hammering in nails (shoemaker, upholsterer)
Light work with two arms	Filing metal, planing wood, raking the garden
Moderate work with the body	Cleaning a floor, beating a carpet
Heavy work with the body	Railroad track laying, digging, barking trees

Formula 1

$$\text{Average } M = [(M_i)(t_i)]/(t_i)$$

where
M = metabolic rate in kilocalories (see Table 62.4)
t = time in min
i = each task

   5. Calculate a heat stress index.
   6. Choose engineering controls.

TABLE 62.4 Assessment of Workload

Body Position and Movement		Workload (kcal/min) [a]
Sitting		0.3
Standing		0.6
Walking		2.0–3.0
Walking uphill		add 0.8 for every meter rise

Type of Work		Average (kcal/min)	Range (kcal/min)
Hand work	light	0.4	0.2–1.2
	heavy	0.9	
Work: one arm	light	1	0.7–2.5
	heavy	1.7	
Work: both arms	light	1.5	1.0–3.5
	heavy	2.5	
Work: whole body	light	3.5	2.5–15.0
	moderate	5	
	heavy	7	
	very heavy	9	

[a] For a "standard" worker of 70 kg body weight and 1.8 m² body surface.
Source: ACGIH (1992), 1992–1993 Threshold Limit Values for Chemical Substances and Physical Agents and Biological Exposure Indices, American Conference of Governmental Industrial Hygienists, Cincinnati, OH

62.3 Heat Stress Indices: Body Measurement

The most direct measure of heat stress is to record the core temperature of the body. The most accurate instrument is a rectal thermometer, but often this is impractical in work situations. Ear or skin temperature measurements do not provide accurate and reliable estimates of core temperature.

62.4 Heat Stress Indices: Environmental Measurements

Environmental heat measurements should be made at, or as close as possible to, the specific work area where the worker is exposed. When a worker is not continuously exposed in a single hot area but moves between two or more areas having different levels of environmental heat, or when the environmental heat varies substantially at a single hot area, environmental heat exposures should be measured for each area and for each level of environmental heat to which employees are exposed.

62.4.1 Wet-Bulb Globe Temperature Index

Wet-bulb globe temperature (WBGT) forms the basis of ISO 7243 (1989) and the ACGIH (1992) guidelines. This is the simplest method to assess workplace heat stress, and it is preferred method used in the U.S.

For indoor and outdoor conditions with no solar load, WBGT is calculated as:

$$WBGT = 0.7 \, NWB + 0.3 \, GT$$

For outdoor work with a solar load, WBGT is calculated as

$$WBGT = 0.7 \, NWB + 0.2 \, GT + 0.1 \, DB$$

where
WBGT = wet-bulb globe temperature index
NWB = natural wet-bulb temperature

DB = dry-bulb temperature
GT = globe temperature (mean radiant temperature)

For long (several hours) or continuous heat exposures, the WBGT should be averaged over a 60-min period. For intermittent exposures, it should be averaged over a 120-min period. WBGT averages are calculated using formula 2.
Formula 2

$$\text{Average WBGT} = [(\text{WBGT}_i)(t_i)]/(t_i)$$

where
WBGT = wet-bulb globe temperature index
t = time in min
i = each task

Permissible hourly exposure limits are shown in Table 62.5. These limits assume that workers will be acclimatized and physically fit, fully clothed in light clothing, and have adequate water and salt intake. If employees are wearing heavy clothing or clothing that impedes sweat evaporation, the limits must be adjusted for the additional clothing insulation using the criteria in Table 62.6.

Ideally, the WBGT should be measured at the head, abdomen, and ankle levels, and the weighted average WBGT calculated as follows:

$$\text{WBGT} = [\text{WBGT}_{head} + \text{WBGT}_{ankles} + (2 \times \text{WBGT}_{abdomen})]/4$$

The WBGT reference values from ISO 7243 (1989) are shown in Table 62.7.

62.4.2 Heat Stress Instruments

Portable area instruments and personal heat stress monitors can be used (Figure 62.1). These instruments calculate the WBGT index and often allow this information to be logged. The instruments allow a quick determination of the duration a person can safely work or remain in a particular hot environment.

TABLE 62.5 Permissible Heat Exposure Threshold Limit Values

	Work Load [a]		
Work/rest regimen	Light	Moderate	Heavy
Continuous work	30.0°C	26.7°C	25.0°C
75% work, 25% rest, each hour	30.6°C	28.0°C	25.9°C
50% work, 50% rest, each hour	31.4°C	29.4°C	27.9°C
25% work, 75% rest, each hour	32.2°C	31.1°C	30.0°C

[a] Values are in °C WBGT.

TABLE 62.6 WBGT Correction Factors (°C)

Clothing Type	Clo [a] Value	WBGT Correction
Summer lightweight working clothing	0.6	0
Cotton coveralls	1	−2
Winter work clothing	1.4	−4
Water barrier, permeable	1.2	−6
Fully encapsulating suit, gloves, boots, and hood	1.2	−10

[a] clo: insulation value of clothing. One clo = 5.55 kcal/m²/h (radiation and convection) for each 1°C temperature difference between the skin and the adjusted dry-bulb temperature.

TABLE 62.7 WBGT Reference Values

Metabolic Rate (W·m⁻²)	Acclimatized (°C)		Not acclimatized (°C)	
Resting M < 65	33		32	
65 < M < 130	30		29	
130 < M < 200	28		26	
200 < M < 260	25	26 [a]	22	23 [a]
M > 260	23	25 [a]	18	20 [a]

[a] Values with sensible air movement. Other values refer to no sensible air movement.
Source: ISO 7243 (1989), Hot Environments: Estimation of the Heat Stress on Working Man, Based on the WBGT-Index (Wet Bulb Globe Temperature), International Organization for Standardization, Geneva.

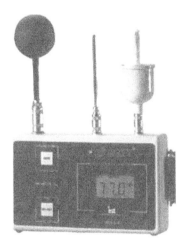

FIGURE 62.1 Heat stress monitors.

Separate instruments can be used to determine WBGT:

1. *Black globe thermometer:* a 15-cm-diameter hollow copper sphere painted matte black on the outside and with a central thermometer or sensor with a range of –5°C to +100°C ± 0.5°C. Allow at least 25 min for the globe thermometer to equilibrate before taking a reading.
2. *Natural (static) wet-bulb thermometer* with a range of –5°C to +50°C ± 0.5°C. Keep the whole wick of the natural wet-bulb thermometer wet with deionized or distilled water for at least 30 min before the reading the temperature. Ensure the wick is clean.
3. *Dry-bulb thermometer* with a range of –5°C to +50°C ± 0.5°C. Shield the dry-bulb thermometer from radiant sources while maintaining normal airflow around the bulb.

These instruments should be suspended on a stand at around chest height in representative work or rest areas.

62.5 Other Heat Stress Indices

62.5.1 Heat Stress Index (HSI)

The heat stress index (HSI), formulated by Belding and Hatch (1955), compares the sweat evaporation required to maintain thermoneutrality (E_{req}) with the maximum evaporation achievable in that setting (E_{max}) as follows:

$$HSI = (E_{req}/E_{max}) \times 100\%$$

TABLE 62.8 Heat Stress Index (HSI) Values and Symptoms

HSI	Effect of 8-h Exposure	Work Impact
–20	Mild cold strain	Recovery from heat exposure
0	No thermal strain	None
10–30	Mild to moderate heat strain	Little impact on physical work; possible impairment of skilled work.
40–60	Severe heat strain	Health threat for unfit workers; acclimatization necessary
570–90	Very severe heat strain	Workers selected by medical examination; adequate water and salt intake necessary
100	Maximum daily thermal strain	Tolerable only by fit, acclimatized young workers
>100	Limited exposure time tolerable	Body core temperature rises

Source: Adapted from Parsons, K.C. (1993), *Human Thermal Environments*, Taylor & Francis, Bristol, PA.

The HSI and the required skin wetness value (W_{req}) are equivalent. Table 62.8 shows the associations between the HSI values and symptoms. The HSI considers the effects of all environmental factors and also work rate, but it is difficult to use to determine an individual worker's heat stress.

The allowable exposure time (AET) in minutes can be calculated from:

$$AET = 2440/(E_{req} - E_{max}) \text{ mins}$$

62.5.2 Index of Thermal Stress (ITS)

The index of thermal stress (ITS) improves on the HSI because it recognizes that not all sweat evaporates and that some is lost as droplets (Givoni, 1976). McIntyre (1980) shows that the ITS can be calculated as follows:

$$ITS = [H - (C+R) - R_s]/0.37_{sc}$$

where
H = metabolic heat from work
C = convective heat loss per unit area
R = radiative heat loss per unit area
R_s = solar load
$_{sc}$ = efficiency of sweating

62.5.3 Required Sweat Rate (SW$_{req}$)

This index forms the basis of the international standard ISO 7933 (1989). It is a development of both the HSI and ITS, and it calculates the sweating required for heat balance (Vogt et al., 1981):

$$E_{req} = M - W - C_{res} - E_{res} - C - R$$

and

$$S_{req} = E_{req}/r_{req}$$

where
E_{req} = required sweat evaporation for thermal equilibrium
S_{req} = required sweat rate for thermal equilibrium
M = metabolic rate ($W \cdot m^{-2}$); M is estimated at 70 $W \cdot m^{-2}$
W = mechanical power ($W \cdot m^{-2}$); W is estimated, and when task details are unavailable W = 0
C_{res} = convective respiratory heat exchange
E_{res} = evaporative respiratory heat loss

C = convective heat exchange
R = radiative heat exchange
r_{req} = evaporate efficiency at required sweat rate

ISO 8996 (1990) describes methods for determining both M and W. The index has been evaluated (Mairiaux and Malchaire, 1988).

62.6 Standards

62.6.1 ISO 7243

ISO 7243 (1989) is entitled "Hot environments: estimation of the heat stress on working man, based on the WBGT-index (wet bulb globe temperature)." This standard uses the WBGT heat stress index to evaluate hot environments. In the U.S., this is equivalent to the ACGIH 1992 standard

62.6.2 ISO 7933

ISO 7933 (1989) is entitled "Hot environments: analytical determination and interpretation of thermal stress using calculation of required sweat rate." This uses the required sweat rate to estimate heat stress in hot environments.

62.7 Strategies to Reduce Heat Stress

• *Engineering controls:* increased ventilation, air cooling, fans, shielding and/or insulation from heat source.
• *Administrative and work controls:* reduce metabolic effort of work, e.g., by using power assists and tools that require less effort; take frequent rest breaks in a cool rest area. Automate tasks. Implement a heat stress training program (NIOSH, 1986). Provide ample supplies of cool liquids (10 to 15°C) close to the work area and encourage workers to frequently drink small amounts, e.g., one cup every 20 min. Provide loosely worn heat reflective clothing or clothing with an auxiliary cooling system (e.g., ice vest).
• *Worker controls:* acclimatize the worker to the hot environment. This reduces cardiovascular demands and allows more efficient sweating. NIOSH (1986) recommends an exposure regimen (Table 62.9) that depends on the worker's previous experience with hot environments. Personal monitoring (e.g., heart rate, recovery heart rate, oral temperature, extent of body water loss, personal heat stress monitor) of workers in extraordinary conditions, e.g., wearing semipermeable or impermeable clothing in temperatures >21°C, working at extreme metabolic loads (>500 kcal/h).

TABLE 62.9 Recommended Regimen (percent of 8-h workday) for Acclimatizing Workers to Hot Environments

Days	Worker with Previous Hot-Environment Experience	New Worker with No Previous Heat Experience
1	50%	20%
2	60%	40%
3	80%	60%
4	100%	80%
5		100%

Source: Adapted from NIOSH (1986), Working in Hot Environments (revised criteria 1986), publication 86-113, DHHS, National Institute for Occupational Safety and Health.

References

ACGIH (1992), 1992–1993 Threshold Limit Values for Chemical Substances and Physical Agents and Biological Exposure Indices, American Conference of Governmental Industrial Hygienists, Cincinnati, OH.

Belding, H.S. and Hatch, T.F. (1955), Index for evaluating heat stress in terms of resulting physiological strain, *Heat, Piping Air Cond.*, 27, 129–136.

Givoni, B. (1976), *Man, Climate and Architecture*, 2nd ed., Applied Science, London.

ISO 7243 (1989), Hot Environments: Estimation of the Heat Stress on Working Man, Based on the WBGT-Index (Wet Bulb Globe Temperature), International Organization for Standardization, Geneva.

ISO 7933 (1989), Hot Environments: Analytical Determination and Interpretation of Thermal Stress Using Calculation of Required Sweat Rate, International Organization for Standardization, Geneva.

ISO 8996 (1990), Ergonomics: Determination of Metabolic Heat Production, International Organization for Standardization, Geneva.

Mairiaux, P. and Malchaire, J. (1988), Comparison and validation of heat stress indices in experimental studies, in *Heat Stress Indices Seminar Proceedings*, Commission of the European Communities, Luxembourg, pp. 81–110.

McIntyre, D. (1980), *Indoor Climate*, Applied Science, London.

NIOSH (1986), Working in Hot Environments (revised criteria 1986), publication 86-113, DHHS, National Institute for Occupational Safety and Health; available on-line at http://www.cdc.gov/niosh/86-113.html, reprinted with minor revisions in 1992 (http://www.cdc.gov/niosh/hotenvt.html).

OSHA Technical Manual (http://www.osha.gov/dts/osta/otm/otm_iii/otm_iii_4.html#3).

Parsons, K.C. (1993), *Human Thermal Environments*, Taylor & Francis, Bristol, PA.

Vogt, J.J., Candas, V., Libert, J.P., and Daull, F. (1981), Required sweat rate as an index of thermal strain in industry, in *Bioengineering, Thermal Physiology and Comfort*, Cena, K. and Clark, J.A., Eds., Elsevier, Amsterdam, pp. 99–110.

63

Thermal Comfort Indices

Jørn Toftum
Technical University of Denmark

63.1 Background and Application

Numerous indices for the assessment and design of thermal comfort conditions have been developed during the past 50 to 60 years. One of the most widely used indices in moderate thermal environments, the PMV index (predicted mean vote), predicts the mean value of the overall thermal sensation of a large group of persons as a function of activity (metabolic rate), clothing insulation, and the four environmental parameters: air temperature, mean radiant temperature, air velocity, and air humidity (Fanger, 1970). The PPD index (predicted percentage dissatisfied) is derived from the PMV index and predicts the percentage of thermally dissatisfied persons among a large group of people.

The thermal sensation of humans is mainly related to the thermal balance of the body as a whole. Thermal balance exists when the internal heat production in the body is equal to the loss of heat to the environment. The studies underlying the PMV model showed that thermal sensation could be described as a function of the thermal load on the effector mechanisms of the human thermoregulatory system (vasodilation, vasoconstriction, sweating, shivering). In the model, the thermoregulatory response has been related statistically to thermal-sensation votes collected from more than 1300 subjects.

Thermal dissatisfaction can be caused by a too warm or too cool overall thermal sensation. But even for a person who is thermally neutral for the body as a whole, thermal dissatisfaction may be the result of unwanted cooling or heating of local body parts. Separate indices exist for the assessment of the different types of local thermal discomfort.

The PMV and PPD indices can be used to assess overall thermal comfort in a wide range of buildings and vehicles with differing HVAC (heating, ventilation, and air-conditioning) systems as well as for different combinations of activity, clothing habits, and environmental parameters. The indices are used widely for the evaluation and design of indoor thermal environments. In standards and guidelines, the indices are used to specify comfort criteria.

63.2 Procedure

The PMV can be determined when the metabolic rate and the clothing insulation are estimated and the air temperature, mean radiant temperature, relative air velocity, and air humidity are measured or estimated. The PMV integrates the effects of the two personal parameters and the four environmental parameters on the thermal balance, and it predicts the mean thermal sensation on a seven-point thermal-sensation scale, as seen in Figure 63.1.

63.2.1 Metabolic Rate

Metabolic rate varies over a wide range, depending on the activity, the person, and the conditions under which the activity is performed. It can be very roughly assessed from knowledge of the occupation or from analysis of a task or activity. A more precise method, with an accuracy of around ±20% (ISO/WD 8996-1999), involves observation of the activity and the use of tabulated values of metabolic rates for specific activities, as shown in Table 63.1.

The data in Table 63.1 are based on measurement of metabolic rates (oxygen consumption) performed on human subjects continuously occupied with a specific activity. A detailed description of the evaluation and measurement of metabolic rate as well as a comprehensive collection of metabolic rates for typical activities can be found in ISO/WD 8996-1999.

It may be necessary to determine a time-weighted average metabolic rate for individuals with activities that vary over a period of one hour or less. New research results show that steady-state models for the prediction of thermal sensation, such as the PMV model, seem to be applicable after approximately 15 min of constant activity (Goto et al., 2002). Metabolic rates that differ considerably between individuals in a space should not be averaged; the thermal comfort conditions for subjects at each activity level also should be evaluated separately. For example, people attending a lecture have a considerably lower metabolic rate than the lecturer, and thus they may have a completely different thermal sensation.

63.2.2 Thermal Insulation of Clothing

Clothing insulation varies between occupants in a space due to differences in clothing preferences, company dress code, season, etc. Clothing insulation can be measured with a heated thermal manikin or with human subjects, but in practice, thermal comfort estimates based on tables may be sufficiently

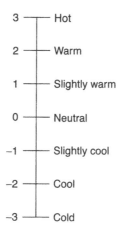

FIGURE 63.1 Seven-point thermal-sensation scale.

TABLE 63.1 Metabolic Rate of Typical Activities

Activity	Metabolic Rates	
	W/m²	met
Reclining	46	0.8
Seated, relaxed	58	1.0
Sedentary activity (office, dwelling, school, laboratory)	70	1.2
Standing, light activity (shopping, laboratory, light industry)	93	1.6
Standing, medium activity (shop assistant, domestic work)	116	2.0
Walking on the level:		
2 km/h	110	1.9
5 km/h	200	3.4

Note: The metabolic rate, typically designated M, is the rate of energy production of the body by metabolism, which varies with activity. Metabolic rate can be quantified by the met unit, where 1 met is defined as the metabolic rate of a sedentary person (seated, quiet); 1 met = 58.2 W/m². The unit W/m² refers to the area of the nude body. The most commonly used measure of nude-body surface area is described by: $A_D = 0.202 \, m^{0.425} \cdot h^{0.725}$, where A_D = Dubois surface area (m²), m = body mass (kg), and h = height (m).

Source: ISO 7730-1994 (1994), Moderate Thermal Environments: Determination of the PMV and PPD Indices and Specification of the Conditions for Thermal Comfort, International Organization for Standardization, Geneva.

accurate. Table 63.2 lists the insulation provided by clothing ensembles composed of some typical combinations of garments (ISO 7730-1994). If no matching clothing ensemble can be found in Table 63.2, tabulated insulation values of a wide variety of individual garments are provided in ISO 9920-1990. Summation of these partial insulation values for individual garments can be used as an estimate of the insulation of the entire clothing, I_{cl}.

The data in Table 63.2 are for standing persons. For sedentary persons, a typical chair may contribute an additional insulation of 0.1 to 0.3 clo; a typical computer chair adds around 0.15 clo (McCullough et al., 1994).

63.2.3 Thermal Environment Parameters

Measurement of the thermal parameters of the environment should be made in the occupied zones of the building at locations where the occupants are expected to spend their time, i.e., at their workstations or in seating areas. For the determination of PMV, the thermal parameters should be measured at the

TABLE 63.2 Thermal Insulation for Typical Combinations of Garments

Clothing Ensemble	Insulation of Entire Clothing (I_{cl})	
	clo	m²·K /W
Panties, T-shirt, shorts, light socks, sandals	0.30	0.050
Panties, petticoat, stockings, light dress with sleeves, sandals	0.45	0.070
Underpants, shirt with short sleeves, light trousers, light socks, shoes	0.50	0.080
Panties, stockings, shirt with short sleeves, skirt, sandals	0.55	0.085
Underwear, shirt, trousers, socks, shoes	0.70	0.110
Panties, petticoat, stockings, dress, shoes	0.70	0.105
Panties, shirt, skirt, sweater, thick kneesocks, shoes	0.90	0.140
Panties, shirt, trousers, jacket, socks, shoes	1.00	0.155

Note: 1 clo unit is the insulation required to keep a sedentary person comfortable at 21°C; 1 clo = 0.155 m²·K /W

Source: ISO 9920-1990 (1990), Ergonomics of the Thermal Environment: Estimation of the Thermal Insulation and Evaporative Resistance of a Clothing Ensemble, International Organization for Standardization, Geneva.

center of gravity, which is 0.6 m for sedentary occupants and 1.1 m for standing activity. The PMV can also be measured directly by an integrating sensor.

During sedentary activity, the mean air velocity near the person determines the convective heat loss. If the person moves, it is the velocity of air relative to the body's movement. On average, it can be assumed that the relative air velocity is a function of the metabolic rate (Fanger, 1970):

$$v_{ar} = v + 0.005(M - 58)$$

The mean radiant temperature is a complex variable defined as the uniform surface temperature of an imaginary black enclosure in which an occupant would exchange the same amount of radiant heat as in the actual nonuniform space. When a room has no strong radiant sources and if the occupant is not located close to cold windows or other cold or warm surfaces, the mean radiant temperature can often be approximated by the air temperature.

Olesen (1995), ISO 7726-1998, and ASHRAE 55-1992R provide detailed descriptions of the requirements for the measuring instrumentation and for thermal comfort measurement procedures.

With the personal parameters of metabolic rate and clothing insulation and the thermal environment parameters as input variables, the PMV can be expressed by the equation:

$$PMV = (0.303 \cdot e^{-0.036 \cdot M} + 0.028)\{(M - W) - 3.05 \cdot 10^{-3} \cdot [5733 - 6.99 \cdot (M - W) - p_a]$$
$$-0.42 \cdot [(M - W) - 58.15] - 1.7 \cdot 10^{-5} \cdot M \cdot (5867 - p_a) - 0.0014 \cdot M \cdot (34 - t_a)$$
$$-3.96 \cdot 10^{-8} \cdot f_{cl} \cdot [(t_{cl} + 273)^4 - (\bar{t}_r + 273)^4] - f_{cl} \cdot h_c \cdot (t_{cl} - t_a)\}$$

where

$$t_{cl} = 35.7 - 0.028 \cdot (M - W) - I_{cl} \cdot \{3.96 \cdot 10^{-8} \cdot f_{cl} \cdot [(t_{cl} + 273)^4 - (\bar{t}_r + 273)^4] + f_{cl} \cdot h_c \cdot (t_{cl} - t_a)\}$$

$$h_c = 2.38 \cdot (t_{cl} - t_a)^{0.25} \text{ for } 2.38 \cdot (t_{cl} - t_a)^{0.25} > 12.1 \cdot \sqrt{v_{ar}}$$

$$12.1 \cdot \sqrt{v_{ar}} \text{ for } 2.38 \cdot (t_{cl} - t_a)^{0.25} < 12.1 \cdot \sqrt{v_{ar}}$$

$$f_{cl} = 1.00 + 1.290 \cdot I_{cl} \text{ for } I_{cl} \leq 0.078 \text{ m}^2 \cdot °C/W$$
$$1.05 + 0.645 \cdot I_{cl} \text{ for } I_{cl} > 0.078 \text{ m}^2 \cdot °C/W$$

and where
PMV is the predicted mean vote
M is the metabolic rate, W/m^2
W is the external work (zero for most indoor activities), W/m^2
I_{cl} is the thermal resistance of the clothing, (m$^2 \cdot$°C)/W
f_{cl} is the ratio of the clothed surface area to the nude surface area
t_a is the air temperature, °C
$\bar{t}_r$ is the mean radiant temperature, °C
v_{ar} is the air velocity relative to the human body, m/sec
p_a is the partial water vapor pressure, Pa
h_c is the convective heat transfer coefficient, W/(m$^2 \cdot$°C)
t_{cl} is the surface temperature of the clothing, °C

The mathematical expression derived for the calculation of PMV is complicated and not very suitable for use in practice. For that purpose, ISO 7730 and ASHRAE 55-1992R include a computer code for calculation of PMV as well as tabulated values of PMV covering a wide range of combinations of the six input variables. The American Society of Heating, Refrigerating and Air-Conditioning Engineers (ASHRAE) has published a PC tool for easy calculation of several thermal comfort indices, including the PMV (Fountain and Huizenga, 1997). In addition, a thermal comfort index calculator is available at http://atmos.es.mq.edu.au/~rdedear/pmv/ (de Dear, 1999).

It is recommended to use the index only for PMV values in the range −2 to +2, metabolic rates from 0.8 met to 4 met, clothing insulation from 0 clo to 2 clo, air temperatures from 10 to 30°C, mean radiant temperatures from 10 to 40°C, and relative air velocities from 0 to 1 m/sec.

In non-air-conditioned buildings in warm climates, the occupants may sense warmth as being less severe than PMV predicts. For such buildings, an extension of the PMV model that includes an expectancy factor to account for this phenomenon has been introduced (Fanger and Toftum, 2002).

Occupants of buildings are not alike, and therefore the individual thermal-sensation votes of the occupants of a given environment will be scattered around the mean. The PPD index predicts the number of people likely to feel uncomfortably warm or cool, i.e., those voting hot (+3), warm (+2), cool (−2), or cold (−3) on the seven-point thermal-sensation scale (Figure 63.1). When the PMV value is known, the PPD index can be calculated from the equation:

$$PPD = 100 - 95 \cdot e^{-(0.03353PMV^4 + 0.2179PMV^2)}$$

Typically, a 10% dissatisfaction criterion for whole-body thermal comfort is used for the determination of acceptable thermal conditions (ISO 7730-1994; ASHRAE 55-1992R). This corresponds to a PMV in the range −0.5 to +0.5. Note that the minimum attainable PPD is 5%, even when the result is a neutral thermal sensation (PMV = 0). Because of inter-individual differences, it is not possible to satisfy everyone.

63.2.4 Local Thermal Discomfort

Thermal neutrality for the body as a whole is a necessary, but not sufficient, condition for thermal comfort. Local thermal discomfort due to draft, vertical temperature gradient, radiant asymmetry, or warm or cold floors may cause occupants to find the thermal conditions unacceptable. The most common cause of complaint is draft, which is defined as an unwanted, local cooling caused by air movement. Criteria to assess local thermal discomfort are given in ISO 7730-1994 and ASHRAE 55-1992R.

63.3 Advantages

The PMV index:

- Is a flexible tool that predicts overall thermal sensation and discomfort under many different conditions indoors and outdoors
- Is very well known and has been used extensively by professionals for many years
- Is incorporated in several standards and guidelines

63.4 Disadvantages

The PMV index:

- Requires expensive measuring instrumentation or qualified assessment of the thermal environment
- Has clothing and metabolic rate as input, which, in practice, may be difficult to assess in buildings
- Performs best near thermal neutrality and at low to moderate activity levels
- Builds on a complex equation

63.5 Example Output

In a factory hall, the employees perform work in a standing position along an assembly line. Two of the walls, facing the back and front of the employees, are equipped with windows that extend the whole width of the hall. The location and size of the windows are shown in Figure 63.2 and Figure 63.3.

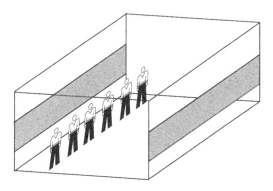

FIGURE 63.2 View of the factory hall.

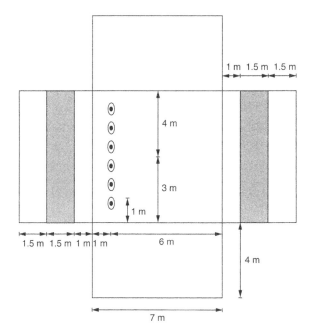

FIGURE 63.3 Plan of the factory.

Due to the processing of perishable goods, hall temperatures are rather low. No instrumentation for the measurement of mean radiant temperature was available, and therefore the temperature of the surfaces was measured at a number of points. The surface temperatures of the walls were rather uniform and were measured to be 17°C. The temperature of the inner surface of the windows was also uniform (disregarding edge effects) and was measured at 5°C.

There was no direct sunlight through the windows. With direct sunlight, the calculations would be more complex (see Fanger [1970] for calculation of mean radiant temperature with strong radiant sources).

The air temperature was 17°C, and the air velocity 0.25 m/sec. An evaluation of the comfort conditions for the employees is provided in the following discussion.

If all the surfaces of an enclosure cannot be considered isothermal, they should be divided into smaller surfaces, each of which is essentially isothermal. In the current situation, the inner surface temperature of the windows and the walls is different. If the temperature differences between the surfaces are small,

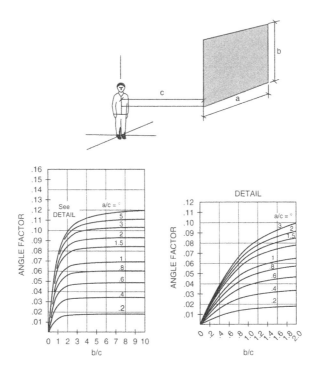

FIGURE 63.4 Angle factors between a standing person and a vertical rectangle in front of or behind the person. (From Fanger, P.O. [1970], *Thermal Comfort*, Danish Technical Press, Copenhagen. With permission.)

the mean radiant temperature can be approximated by the mean value of the surrounding temperatures weighted according to the magnitude of the respective angle factors between the person and each surface.

For the employees working near the center of the room, there is a distance of approximately 3 m to one side wall and approximately 4 m to the other side wall. From their back there is 1 m to the window behind them and 6 m to the window they face. It is assumed that the distance between the employees is sufficiently large to ignore radiation exchange between employees.

The angle factors between the employees near the middle of the hall and the window surfaces can be determined from Figure 63.4 (Fanger, 1970). According to Figure 63.4, each window must be divided into two adjacent surfaces, to the left and right of the person's centerline, to be considered separately.

F_{P-X} is the angle factor from the person (P) to the surface in question (X).

Person to front, left: $b/c = 1.5/1 = 1.5$ $a/c = 3/1 = 3$ $F_{P\text{-front, left}} = 0.085$
Person to front, right: $b/c = 1.5/1 = 1.5$ $a/c = 4/1 = 4$ $F_{P\text{-front, right}} = 0.085$
Person to back, left: $b/c = 1.5/6 = 0.25$ $a/c = 3/6 = 0.5$ $F_{P\text{-back, left}} = 0.015$
Person to back, right: $b/c = 1.5/6 = 0.25$ $a/c = 4/6 = 0.67$ $F_{P\text{-back, right}} = 0.02$

The sum of all angle factors between a person and the surrounding surfaces in an enclosure is 1, i.e., $\sum F_P = 1$. Therefore, the angle factor from the center employees to the window surfaces, $F_{P\text{-win}}$, is:

$$F_{P\text{-win}} = 0.085 + 0.085 + 0.015 + 0.02 = 0.20$$

and to the remaining isothermal surfaces (walls), $F_{P\text{-walls}}$, is:

$$F_{P\text{-walls}} = 1 - 0.2 = 0.8$$

TABLE 63.3 Values for A Corresponding to Different Relative Velocities of Air, v_{ar}

v_{ar}	<0.2 m/sec	0.2 m/sec to 0.6 m/sec	0.6 m/sec to 1 m/sec
A	0.5	0.6	0.7

The mean radiant temperature thus yields:

$$t_{mrt} = F_{P\text{-win}} \cdot t_{win} + F_{P\text{-walls}} \cdot t_{walls} = 0.2 \cdot 5 + 0.8 \cdot 17 = 14.6°C$$

The operative temperature, t_o, is defined as the uniform temperature of an imaginary black enclosure in which an occupant would exchange the same amount of heat by radiation and convection as in the actual nonuniform environment. When the relative air velocity is small (<0.2 m/sec) and the difference between air and mean radiant temperature is less than 4°C, the operative temperature can be approximated by the mean of the air and mean radiant temperatures. At higher relative air velocities, the following expression can be used to calculate operative temperature:

$$t_o = A \cdot t_a + (1 - A) \cdot t_{mrt}$$

where A is given in Table 63.3.

Here, the relative air velocity is:

$$v_{ar} = 0.25 + \{0.005[(1.6 \times 58) - 58]\} = 0.42 \text{ m/sec}$$

and the operative temperature is:

$$t_o = 0.6 \cdot 17 + 0.4 \cdot 14.6 = 16°C$$

For the employees standing nearest to the side walls, similar calculations yield an angle factor to the windows, $F_{P\text{-win}} = 0.17$, a $t_{mrt} = 15°C$, and a $t_o = 16.2°C$. Since there is only a 0.2°C difference in operative temperature between the two locations, and since the employees' location along the assembly line may not be fixed, the operative temperature is assumed to be 16°C for all employees.

From Figure 63.5, which is based on the PMV equation, it appears that if the employees should be thermally neutral, they should wear a clothing insulation of around 1.4 clo. This clothing insulation corresponds to a working clothing ensemble composed of briefs, thermal undershirt, thermal underpants, insulated coveralls, calf-length socks, and shoes, providing an insulation of 1.36 clo (ISO 9920-1990). Figure 63.5 also shows that the operative temperature should be in the range 12 to 20°C (16°C ± 4°C) with 1.4 clo in order to limit the PPD (predicted percentage of dissatisfaction) to 10%, corresponding to a PMV in the range −0.5 to +0.5.

Air humidity mostly influences human heat transfer when evaporation of sweat is necessary to maintain thermal balance, for example at high temperatures or during exercise at high activity levels. Under the prevailing conditions, air humidity has only a limited influence on the occupants' thermal balance. Additional indices should be used to assess local thermal discomfort due to draught, radiant asymmetry, vertical air temperature difference, and floor temperature.

63.6   Related Methods

Other methods for the assessment of moderate thermal environments include the new effective temperature (ET) and the standard effective temperature (SET) (Nishi and Gagge, 1977). These indices are based on a simple model of the human body. Today, advanced models are available that allow for the transient prediction of very detailed thermoregulatory parameters and, for some of the models, subjective responses to a wide range of environmental conditions (e.g., Fiala, 1998).

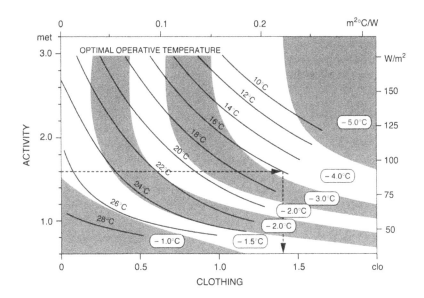

FIGURE 63.5 Optimal operative temperature as a function of clothing and metabolic rate.

For non-air-conditioned buildings, an adaptive model has been proposed that determines the neutral temperature indoors based on the monthly average temperature outdoors (de Dear and Brager, 1998). Additional indices have been developed for the assessment of thermal strain in hot environments (wet-bulb globe temperature [WGBT], required sweat rate [RSW]) and in cold environments (required clothing insulation [IREQ]).

63.7 Standards and Regulations

The PMV model has been included in numerous standards and guidelines as a basis for specifying acceptable thermal environmental conditions. Such standards include ISO 7730-1994, ASHRAE 55-1992R, DS 474-1995, and CEN CR 1752-1998. The standards address methods for the design and assessment of thermal environmental conditions within occupied spaces and are not written in mandatory language. A discussion of thermal comfort standards is provided by Olesen and Parsons (2002).

The purpose of ISO 7730-1994 is to present a method for predicting the thermal sensation and the degree of discomfort of people exposed to moderate thermal environments, and to specify environmental conditions for comfort. The standard can be used in the design of new environments or in assessing existing ones.

ASHRAE 55-1992R specifies conditions in which a given fraction of the occupants will find the environment thermally acceptable. The standard is intended for use in design, commissioning, operation, and testing of buildings and other occupied spaces and their HVAC systems. Because it is not possible to prescribe the metabolic rate of occupants, and because of variations in occupant clothing levels, operating setpoints for buildings cannot be practically mandated by the standard.

DS 474-1995 prescribes that the code is applicable to rooms where the wish to achieve an acceptable thermal indoor climate during occupancy is decisive for the choice of temperature conditions, air movement, clothing, etc.

A more recent approach is to include more aspects of the indoor environment in standards, seeking to integrate indoor environment issues. CEN CR 1752-1998, for example, covers the thermal, air quality, and acoustic conditions.

63.8   Approximate Training and Application Time

Based on experience with university students, it takes approximately 8 to 10 hours for training in the measurement of thermal parameters, assessment of clothing insulation and metabolic rate, and determination and evaluation of the PMV index.

63.9   Reliability and Validity

The PMV index has been validated as a reliable predictor of thermal sensation in numerous comprehensive field studies in different climate regions during both summer and winter as well as in a wide range of climate-chamber studies. However, in warm climates in buildings without air-conditioning, field studies have shown that the index predicts a warmer thermal sensation than the occupants actually feel (de Dear and Brager, 1998). An extension of the PMV model for such buildings has now been proposed (Fanger and Toftum, 2002).

Inaccuracies in the methods for measuring thermal parameters and for estimating clothing insulation and metabolic rates may influence reliability. Individual differences between users of the index may to some extent affect the determination and interpretation of the index.

63.10   Tools Needed

- Instrumentation for the measurement of the thermal parameters. If only some parameters can be measured, the others must be estimated.
- Tables with values of clothing insulation for typical ensembles/garments and metabolic rates of different activities (e.g., ISO 7730-1994, ISO 9920-1990, ISO/WD 8996-1999).
- Tabulated values of the PMV index for different combinations of the input parameters (ISO 7730-1994), or a PC with thermal-comfort software installed or connected to the Internet.

References

ASHRAE 55-1992R (1992), Thermal Environmental Conditions for Human Occupancy, American Society of Heating, Refrigerating and Air-Conditioning Engineers, Atlanta, GA.

CEN CR 1752-1998 (1988), Ventilation for Buildings: Design Criteria for the Indoor Environment, European Committee for Standardization, Brussels, Belgium.

de Dear, R. (1999), WWW Thermal Comfort Index Calculator; available on-line at http://atmos.es. mq.edu.au/~rdedear/pmv/.

de Dear, R.J. and Brager G.S. (1998), Developing an adaptive model of thermal comfort and preference, *ASHRAE Trans.*, 104, 145–167.

DS 474-1995 (1995), Code for Indoor Thermal Climate, Danish Standards, Copenhagen.

Fanger, P.O. (1970), *Thermal Comfort*, Danish Technical Press, Copenhagen.

Fanger, P.O. and Toftum, J. (2002), Extension of the PMV model to non-air-conditioned buildings in warm climates, *Energy Build.*, 34, 533–536.

Fiala, D. (1998), Dynamic Simulation of the Human Heat Transfer and Thermal Comfort, Ph.D. thesis, De Montfort University, Leicester, U.K.

Fountain, M.E. and Huizenga, C. (1997), A thermal sensation prediction tool for use by the profession, *ASHRAE Trans.*, 103, 130–136.

Goto, T., Toftum, J., de Dear, R., and Fanger, P.O. (2002), Thermal sensation and comfort with transient metabolic rates, in *Proceedings of Indoor Air 2002*, 9th International Conference on Indoor Air Quality and Climate, Monterey, CA.

ISO 7726-1998 (1998), Ergonomics of the Thermal Environment: Instruments for Measuring Physical Quantities, International Organization for Standardization, Geneva.

ISO 7730-1994 (1994), Moderate Thermal Environments: Determination of the PMV and PPD Indices and Specification of the Conditions for Thermal Comfort, International Organization for Standardization, Geneva.

ISO/WD 8996-1999 (1999), Ergonomics: Determination of Metabolic Heat Production, International Organization for Standardization, Geneva.

ISO 9920-1990 (1990), Ergonomics of the Thermal Environment: Estimation of the Thermal Insulation and Evaporative Resistance of a Clothing Ensemble, International Organization for Standardization, Geneva.

McCullough, E.A., Olesen, B.W., and Hong, S. (1994), Thermal insulation provided by chairs, *ASHRAE Trans.*, 100, 795–802.

Nishi, Y. and Gagge, A.P. (1977), Effective temperature scale useful for hypo- and hyperbaric environments, *Aviation, Space Med.*, 48(2), 97–107.

Olesen, B.W. (1995), Measurements of the physical parameters of the thermal environment, *Ergonomics*, 38, 138–153.

Olesen, B.W. and Parsons, K.C. (2002), Introduction to thermal comfort standards and to the proposed new version of EN ISO 7730, *Energy Build.*, 34, 537–548.

64

Indoor Air Quality: Chemical Exposures

Alan Hedge
Cornell University

64.1 Introduction

The quality of indoor air can have a dramatic effect on the comfort, performance, and health of workers (O'Reilly et al., 1998). Workers may report a variety of symptoms, such as headache, nasal congestion, lethargy, etc. If these symptoms are accompanied by clinical signs of illness, such as a fever, and if these symptoms are not resolved upon leaving the workplace, the worker may be suffering from a building-related illness. If the symptoms are not accompanied by clinical signs and if they do resolve within hours of leaving the building the worker may be experiencing sick building syndrome. Building-related illnesses usually are associated with exposure to an allergenic or pathogenic organism in the workplace, such as Legionnaire's disease (see Chapter 65, which addresses biological contaminants). Sick building syndrome complaints may be associated with inadequate indoor air quality (IAQ) and are symptoms of irritation. There is a diverse range of indoor air pollutants, and testing and monitoring all of these is prohibitive. Consequently, the initial stages of an indoor air quality investigation involve determining the nature of the worker problems, assessing the potential workplace indoor air hazards, and then testing for specific contaminants. Once the investigator has formulated hypotheses for possible sources of the worker complaints, then appropriate testing and monitoring methods can be implemented.

64.2 Guidelines for Indoor Air Investigations

A general strategy for approaching any indoor air investigation has been developed by the U.S. Environmental Protection Agency (EPA, 1991), and this is shown in Figure 64.1. These guidelines should be followed to define the scope of the problem both spatially, temporally, and in terms of possible indoor air pollutants responsible for the human effects. An investigation usually begins in response to worker complaints about air quality. The investigator takes an initial walk-through of the workplace to look for obvious signs of poor ventilation or of pollutant sources. At this time, interviews with the workers who are experiencing problems often can help to define the possible pollutants. After following the flowchart

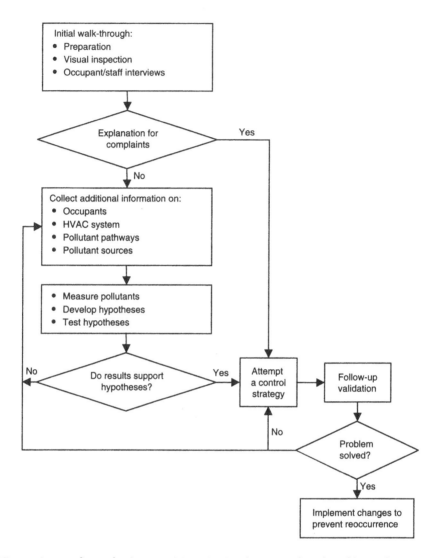

FIGURE 64.1 Strategy for conducting an IAQ investigation. (From EPA [1991], Building Indoor Air Quality: A Guide for Building Owners and Facility Managers, document reference no. 402-F-91-102, Dec., U.S. Government Printing Office, Washington, D.C.)

and formulating a hypothesis about the suspected indoor air pollutant or pollutants, based on the worker and workplace information that has been gathered, a sampling plan must be developed. This plan specifies the nature of the sampling and the location and timing of sampling. If the source of the pollutant produces a continuous emission, then timing is less critical, but if this is an intermittent source, e.g., the pollutant is only emitted when a specific machine is running, then appropriate timing of the sampling is critical.

64.3 Indoor Air Pollutants

The range of indoor air pollutants commonly investigated in workplaces includes:

- *Combustion gases:* Any combustion source in the workplace can emit various toxic gases. If the workplace has a combustion source, such as a gas fire, a wood fire, a furnace, or a gasoline engine, or if combustion is part of the work process, then the air should be tested for the combustion gases listed in Table 64.1.

TABLE 64.1 Indoor Air Pollutants that Most Commonly Are Tested for in Indoor Air Quality Investigations

Pollutants	TWA	CLG /STEL	Health Effects
		Combustion Gases	
Carbon monoxide	35 ppm	2,000 ppm	Headache, drowsiness, confusion, death
Carbon dioxide	5,000 ppm	30,000 ppm	Headache, drowsiness, confusion, asphyxia
Nitric oxide	25 ppm	5 ppm	Mucus membrane irritant
Nitrogen dioxide	—	5 ppm	Mucus membrane irritant
Ozone	0.1 ppm	—	Mucus membrane irritant
Sulfur dioxide	2 ppm	5 ppm	Mucus membrane irritant
		Volatile Organic Compounds (VOCs)	
Benzene	0.1 ppm	1 ppm (15 min)	Mucus membrane irritant, carcinogen
Formaldehyde	0.016 ppm	0.1 ppm	Mucus membrane irritant, possible carcinogen
Styrene	25 ppm	15 ppm	Mucus membrane irritant, carcinogen
Toluene	100 ppm	150 ppm	Mucus membrane irritant, carcinogen
Total VOCs (screening)	5 ppm	—	Mucus membrane irritant, neurotoxic symptoms, carcinogen
		Airborne Fibers	
Asbestos	0.1 fiber/cm³/400 l [a]	—	Carcinogen
Fibrous glass dust	15 mg/m³	—	Mucus membrane irritant, skin irritant

[a] Fibers >5 mm long.
Source: Mostly NIOSH values are shown; http://www.skcinc.com/NIOSH1/NIOSH.HTM.

- *Volatile organic compounds:* There is a large number of organic compounds that can be present in a workplace. Volatile organic compounds (VOCs) may be used as part of the work process, e.g., the solvents used in dry-cleaning and the adhesives used in furniture manufacturing, or they may be integral to the work process, e.g., a petrochemical refinery. Testing for each of a wide range of VOCs can be costly, and consequently, unless the presence of a specific VOC is suspected, the indoor air can be screened for the total VOCs. The VOCs normally tested for are listed in Table 64.1.
- *Airborne fibers:* Mineral-based fibers, such as fiberglass or asbestos, can become airborne from degradation or disruption of building materials, such as thermal insulation. These fibers also can become airborne if they are part of the work process, e.g., a glass fiber production factory. The two types of fibers most often tested for are listed in Table 64.1.

Table 64.1 also lists the occupational exposure limits for these air pollutants. The U.S. EPA provides additional good guidance on undertaking IAQ investigations. There are good-practice standard methods that should be followed for monitoring each type of pollutant of interest (Bisesi and Kohn, 1995). Each country or regulatory body may have its own standard methods for measuring different air pollutants. For example, in the U.S., different air sampling and analysis methods are published by the Occupational Health and Safety Administration (OSHA), by the National Institute of Occupational Safety and Health (NIOSH), by the American Society for Testing and Materials (ASTM), and by the U.S. EPA. In the U.K., the Health and Safety Executive (HSE) publishes its own guideline to methods. Local methodology guidelines should always be consulted for an investigation.

64.4 Exposure Measures

The U.S. Occupational Health and Safety Administration (OSHA) regulates workplace safety in the U.S., including indoor air quality concerns, and it uses various regulatory measures of worker exposure to pollutants:

- *Short-term exposure limit (STEL):* This is the pollutant concentration that workers can be exposed to continuously for a short period of time without suffering adverse effects. STEL exposures are very brief, usually 15 min. When a pollutant has both acute and chronic effects, the STEL supplements other exposure limits. Workers can be exposed to a maximum of four STEL periods per 8-h shift, with at least 60 min between exposure periods.

- *Permissible exposure limit (PEL):* This is the maximum pollutant concentration that a worker can be exposed to without adverse consequences. PELs can be defined as either:

  Ceiling values (VLG): At no time should this pollutant exposure limit be exceeded.

  Time-weighted average (TWA): The TWA is usually calculated for an 8-hour work shift. The TWA combines the measured pollutant concentration (C) with the sample time (T) in hours as follows:
  For a single sample:

  8-h TWA = $(C \times T)/8$

  For multiple samples of the same pollutant:

  8-h TWA = $(C_i \times T_i)/8$

- *Threshold Limit Values (TLVs):* TLVs are available for more than 700 pollutants and are published by the American Conference of Governmental Industrial Hygienists (ACGIH, 2003). These guidelines indicate the worker exposure that can occur without an unreasonable risk of disease or injury. TLVs are used to assist decision making regarding safe workplace exposure levels.

64.5 Pollutant Monitoring

Indoor air pollutants can be monitored in several different ways:

- *Instantaneous monitoring:* also termed "real-time" monitoring. Pollutant levels are measured for short periods, usually <10 min. This testing requires the availability of real-time monitoring equipment, which is not necessarily available for every potential pollutant. Unless the pollutant is continuously released at a constant rate, results from instantaneous monitoring cannot be used to quantify a worker's exposure. However, if the timing and location of collecting these "spot" samples are varied, the results can be used to quickly screen a workplace for a contaminant.
- *Integrated monitoring:* Also termed "continuous" monitoring. Pollutant levels are measured for longer periods, usually >15 min to several hours. This testing requires the availability of continuous-monitoring equipment. Results from integrated monitoring can be used to quantify a worker's exposure to a pollutant in several ways. Integrated monitoring is best used to assess exposure in an area once the pollutant of interest and the area of concern in the workplace are known.
- *Personal monitoring:* This involves pollutant sampling by equipment on the worker. Both instantaneous and integrated monitoring assess the pollutant concentration in an area, but results can be misleading if monitoring is not conducted in the "breathing zone," the area within a 30-cm radius of the worker's nose and mouth. Also, area measures alone can prove difficult if the worker's job requires movement between different areas in a workplace. Personal monitoring can be undertaken using either passive devices, such as monitoring badges, or active devices, such as wearing portable sample pumps.
- *Surface sampling:* When the contaminant is a fiber or particle, there can be exposure from direct contact as well as from inhalation. Sampling the dust that accumulates on work surfaces provides an estimate of possible exposures.
- *Grab sampling:* A sample of the air from the workplace can be collected in a sealed, inert bag for subsequent analysis.

64.6 IAQ Measurement Instruments

A variety of different instruments can be used to measure the levels of most air pollutants. The instrument chosen should be capable of indicating the levels of the air pollutant in appropriate units, such as parts

FIGURE 64.2 Handheld pump and gas detector tube.

FIGURE 64.3 Direct-reading formaldehyde meter.

per million (ppm), mg·m^{-3}, or particles·m^{-3}. Generally, these instruments fall into one of the following categories:

- *Gas detector tubes:* More than 350 gases, vapors, and aerosols can be directly measured using a handheld pump and samples tubes that change color depending on the pollutant concentration (Figure 64.2). These are most useful for instantaneous monitoring.
- *Electronic detectors:* The concentrations of many gases and vapors and the counts of particulates can be directly measured with electronic devices. Specific instruments normally are available for each pollutant (Figure 64.3 and Figure 64.4). Typically, the instruments incorporate a pump, and a known volume of air is sampled. The concentration of the pollutant can then be read directly. Recent instruments often combine the ability to detect several pollutants, log the data, and compute statistics (Figure 64.5). These are most useful for instantaneous monitoring.
- *Sorbent tubes:* Many different contaminants can be measured using sorbent tubes, but the results are not immediately available. Usually, these tubes are attached to a low-volume sample (Figure 64.6) that is programmed to run at a specific flow rate for a set time period. The sorbent tubes are then analyzed using laboratory methods. These are most useful for integrated exposure monitoring.
- *Filters:* Particulates and fibers can be sampled by drawing air through a suitable filter and then either weighing or counting the materials that accumulates on the filter. Filters usually are placed in a sealed cassette that can be uncapped for sampling (Figure 64.7). When the particulates of interest are fine, such as environmental tobacco smoke or welding fumes, this sampling is under-

FIGURE 64.4 Direct-reading particle counter.

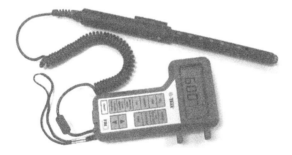

FIGURE 64.5 Multichannel IAQ monitor and data logger.

taken using a low-volume sample pump. When sampling for heavier particulates or fibers, such as asbestos, a high-volume sample pump is used (Figure 64.8).

- *Passive samplers:* Various kinds of passive samplers can be used to measure either area concentrations of a pollutant or personal exposures. Sampling badges are often worn for personal monitoring (Figure 64.9). These badges respond to pollutant concentrations by changing color, or they may require laboratory analysis.

64.7 Occupant Surveys

Occupant surveys of workers' indoor-air-quality complaints and health symptoms are often conducted in workplaces either before or along with indoor-air-pollutant testing. Many worker complaints relate to unsatisfactory thermal conditions. A simple, self-report questionnaire has been developed that can be used, and this survey provides normative data for comparison purposes (Hedge and Erickson, 1998). The questionnaire gathers data on worker perceptions of ambient environmental conditions and on symptoms of sick building syndrome (Figure 64.10). The 14 core items are derived from analyses of larger survey data sets (Hedge et al., 1995, 1996).

FIGURE 64.6 Low-volume programmable sample pump.

FIGURE 64.7 Filter cassette for collecting dust samples.

The normative data survey data showing the percentages of men and women reporting indoor air quality and sick building syndrome problems for each workplace quartile are shown in Table 64.2 and Table 64.3.

64.8 Standards

There are no international indoor-air-quality standards. However, the International Standards Organization has published two IAQ testing standards:

- ISO 16000-3:2001: Indoor air, Part 3: Determination of formaldehyde and other carbonyl compounds — Active sampling method

FIGURE 64.8 High-volume sample pump.

FIGURE 64.9 Personal monitor, passive badge sampler.

- ISO 16017-1:2000: Indoor, ambient and workplace air: Sampling and analysis of volatile organic compounds by sorbent tube/thermal desorption/capillary gas chromatography, Part 1: Pumped sampling

In the U.S., there is a voluntary standard for acceptable indoor air quality:

- ASHRAE Standard 62-2001: Ventilation for acceptable indoor air quality (ANSI Approved)

In the U.K., there are several indoor air standards:

CORNELL OFFICE ENVIRONMENT SURVEY

(Short form)

Building ID# ☐☐   Floor # ☐☐   Department/Area # ☐☐   Case ID# ☐☐☐

Please answer the following questions about environment comfort conditions and health symptoms that you may have experienced in the office during the past month (4 weeks).

1. What is your gender?   Woman ☐   Man ☐

2. Please indicate whether you have experienced each of the following environment conditions in the office during the past month:

Condition experienced at least once per week during the past month (4 weeks).

	YES	NO
a. Air temperature too cold	☐	☐
b. Air temperature too warm	☐	☐
c. Too little air movement	☐	☐
d. Air too dry	☐	☐
e. Unpleasant odor in air	☐	☐
f. Air too stale	☐	☐
g. Air too dusty	☐	☐

3. Please indicate whether you have experienced any of the following symptoms on at least a weekly basis during the past month (4 weeks), and whether this symptom got better when you were away from the office (e.g. evenings, weekends):

Symptom experienced at least once per week during the past month (4 weeks) and symptom got better when away from work.

	YES	NO
a. Irritated, sore eyes	☐	☐
b. Sore, irritated throat	☐	☐
c. Hoarseness	☐	☐
d. Stuffy, congested nose	☐	☐
e. Excessive mental fatigue	☐	☐
f. Headache across forehead	☐	☐
g. Unusual tiredness, lethargy	☐	☐

FIGURE 64.10   Cornell IAQ short-form survey. (From Hedge, A. and Erickson, W.A. [1998], *Int. J. Facilities Manage.*, 1, 1–8. With permission.)

TABLE 64.2   Quartiles for the Percentages of Men and Women Reporting Weekly Perceived Indoor Air Quality Conditions in Office Buildings

Perceived Indoor Air Quality	Men (percentiles)			Women (percentiles)		
	25th	50th	75th	25th	50th	75th
Temperature too cold	12.2	19.4	26.9	42.4	49.7	54.7
Temperature too warm	22.1	28.7	49.2	27.4	39.9	50.9
Too little air movement	23.0	30.3	39.8	41.6	54.5	61.8
Air too dry	15.3	20.7	29.0	35.9	49.2	59.7
Unpleasant odor in air	4.0	6.4	8.3	9.5	13.5	18.7
"Stale" air	16.6	21.3	27.0	31.2	44.9	52.1
Dusty air	8.9	11.9	14.0	22.9	31.0	34.5

Source: Hedge, A. and Erickson, W.A. (1998), *Int. J. Facilities Manage.*, 1, 1–8.

TABLE 64.3 Quartiles for the Percentages of Men and Women Reporting
Weekly Work-Related SBS Symptoms in Office Buildings

Work-Related SBS Symptoms	Men (percentiles)			Women (percentiles)		
	25th	50th	75th	25th	50th	75th
Irritated, sore eyes	12.9	15.3	20.8	24.4	30.5	34.7
Sore, irritated throat	2.7	5.1	8.3	9.1	13.0	16.3
Hoarseness	2.5	3.5	5.3	6.6	8.7	11.6
Stuffy, congested nose	10.2	11.1	15.3	18.9	26.9	32.4
Excessive mental fatigue	13.3	16.7	23.7	21.9	27.1	30.3
Headache across forehead	7.1	8.3	14.5	18.9	24.8	29.6
Unusual tiredness, lethargy	9.4	12.7	14.8	14.6	20.0	27.2

Source: Hedge, A. and Erickson, W.A. (1998), *Int. J. Facilities Manage.*, 1, 1–8.

- PD CR 1752:1999: Ventilation for buildings: Design criteria for the indoor environment
- BS ISO 16000-3:2001: Indoor air: Determination of formaldehyde and other carbonyl compounds, Active sampling method
- BS EN ISO 16017-1:2001: Indoor, ambient and workplace air: Sampling and analysis of volatile organic compounds by sorbent tube/thermal desorption/capillary gas chromatography, Pumped sampling

Where there is no indoor air standard, the investigator should use any outdoor air standard as a guide for what is acceptable for a prolonged exposure.

References

ACGIH (2003), 2002–2003 Threshold Limit Values for Chemical Substances and Physical Agents and Biological Exposure Indices, American Conference of Governmental Industrial Hygienists, Cincinnati, OH.

ASHRAE Standard 62-2001, Ventilation for Acceptable Indoor Air Quality (ANSI Approved), American Society of Heating and Refrigerating Engineers, Atlanta, GA.

Bisesi, M.S. and Kohn, J.P. (1995), *Industrial Hygiene Evaluation Methods*, Lewis Publishers, Boca Raton, FL.

BS ISO 16000-3:2001, Indoor Air: Determination of Formaldehyde and Other Carbonyl Compounds, Active Sampling Method, British Standards Institute, London.

BS EN ISO 16017-1:2001, Indoor, Ambient and Workplace Air: Sampling and Analysis of Volatile Organic Compounds by Sorbent Tube/Thermal Desorption/Capillary Gas Chromatography, Pumped Sampling, British Standards Institute, London.

EPA (1991), Building Indoor Air Quality: A Guide for Building Owners and Facility Managers, document reference no. 402-F-91-102, Dec., U.S. Government Printing Office, Washington, D.C.

Hedge, A., Erickson, W.A., and Rubin, G. (1995), Psychosocial correlates of sick building syndrome, *Indoor Air*, 5, 10–21.

Hedge, A., Erickson, W.A., and Rubin, G. (1996), Predicting sick building syndrome at the individual and aggregate levels, *Environ. Int.*, 22, 3–19.

Hedge, A. and Erickson, W.A. (1998), Indoor environment and sick building syndrome complaints in air conditioned offices: benchmarks for facility performance? *Int. J. Facilities Manage.*, 1, 1–8.

ISO 16000-3:2001 (2001), Indoor Air, Part 3: Determination of Formaldehyde and Other Carbonyl Compounds — Active Sampling Method, International Organization for Standardization, Geneva.

ISO 16017-1:2000 (2000), Indoor, Ambient and Workplace Air — Sampling and Analysis of Volatile Organic Compounds by Sorbent Tube/Thermal Desorption/Capillary Gas Chromatography, Part 1: Pumped Sampling, International Organization for Standardization, Geneva.

O'Reilly, J.T., Hagan, P., Gots, R., and Hedge, A. (1998), *Keeping Buildings Healthy: How To Monitor and Prevent Indoor Environmental Problems*, John Wiley & Sons, New York.
PD CR 1752:1999 (1999), Ventilation for Buildings: Design Criteria for the Indoor Environment, British Standards Institute, London.

65

Indoor Air Quality: Biological/Particulate-Phase Contaminant Exposure Assessment Methods

Thad Godish
Ball State University

65.1 Nature of Building-Related Exposure/Health Concerns

There is increasing evidence that exposures to particulate-phase substances, most notably those of a biological origin, are responsible for a variety of illness symptoms associated with indoor environments. Biological contaminants include airborne mold spores/hyphal fragments, fungal glucans, mycotoxins, lower and higher bacteria (actinomycetes), bacterial endotoxins, antigens associated with the fecal wastes and body parts of dust mites and cockroaches, antigens associated with pet dander, and components of surface dust described as macromolecular organic dust (MOD). Exposures to airborne mold spores and fragments, dust mite and cockroach antigens, and pet dander can cause chronic allergic rhinitis and, in severe cases, asthma (Weissmann and Schuyler, 1991). Symptoms of chronic allergic rhinitis or common allergy are, in most cases, indistinguishable from many of the common respiratory and general symptoms reported in buildings associated with sick building syndrome (Godish, 1995).

Mold infestation of building materials associated with water leaks and intrusion through the building fabric are common. As such, mold exposures and concomitant mold-related illness symptoms are prevalent (Dales et al., 1991; Flannigan and Miller, 1994; Levetin, 1995). Building-related illness symptoms have also been reported to be associated with exposures to fungal glucans and components of the cell wall of fungi found in building surface dusts (Rylander et al., 1992). More problematic are potential exposures to toxigenic fungi and their mycotoxins. Particular public health concern has focused on *Stachybotrys chartarum*, a fungal species that produces potent mycotoxins and commonly infests repeatedly wetted materials made from processed wood fibers (e.g., gypsum-board facing, ceiling tiles, etc.) (Sorenson et al., 1987; Miller, 1992; Etzel, 1995; Sorenson, 1999).

Illness symptoms associated with building environments have in some cases been associated with airborne bacteria. These include communicable diseases such as tuberculosis and meningitis (Nardell, 2001). Such exposure concerns are primarily associated with high-density or high-risk environments, such as homeless shelters, prisons, army barracks, nursing homes, and hospitals.

Problem building concerns associated with exposures to bacteria include outbreaks of Legionnaire's Disease associated with contaminated cooling towers, spas, and hot water heaters (Imperato, 1981). They also include outbreaks of hypersensitivity pneumonitis (Fink, 1983; Cookingham and Solomon, 1995), an inflammatory lung ailment associated with exposures to multiple-order-of-magnitude concentrations of bacteria, particularly thermophilic actinomycetes, and in a number of cases small-diameter fungal spores of the genera *Penicillium* and *Aspergillus*.

Gram-negative bacteria produce endotoxins on lyses of their cell walls. Exposure to endotoxins appears to cause an inflammatory lung disease in office workers described as humidifier fever (Rylander et al., 1980). Endotoxins found in building surface dust have been reported to cause inflammatory responses in building environments at much lower concentrations than those required to produce humidifier fever (Rylander, 1997). Such exposures may be responsible for a portion of the illness spectrum of sick building syndrome (SBS).

A limited number of studies have shown significant relationships between building-related illness symptoms and airborne particles and surface dusts (Hedge et al., 1993; Raw et al., 1993; Gyntelberg et al., 1994). Because particulate-phase substances are heterogeneous, it is quite probable that either organic (such as MOD) and/or inorganic components of settled/resuspended dust are responsible for inducing illness symptoms upon exposure in indoor environments. The relationship between airborne-particle and surface-dust exposures and building-related illness symptoms have been reviewed by Godish (1995).

65.2 Contaminant Assessment Methodologies

A variety of methodologies have been developed and utilized for conducting assessments of particulate-phase contaminant levels in both airborne and surface samples. Of particular note are those that are applied to biological contaminants.

65.2.1 Sampling Analysis Methods for Airborne Biological Contaminants

A number of techniques are commonly utilized to collect and subsequently analyze samples of airborne mold/mold fragments, bacteria, fungal glucans, mycotoxins, endotoxins, and allergens associated with mites, cockroaches, and pets. In many, if not most, indoor-air-quality investigations, assessments are limited to collecting and analyzing samples of airborne mold/mold fragments and bacteria. These include collection by impaction on culture media, impaction on greased microscope slides or coverslips, filtration on membrane filter surfaces, and impingement in liquid samples.

Mold sampling/analysis methods can be described as culturable-viable or total mold spore/structures methods. The former quantifies all airborne mold or bacteria that are alive and grow on the culture media employed. After a period of maturation, colonies are identified to genus or species and counted. Concentrations are expressed as colony-forming units per cubic meter (CFU/m^3) for total airborne culturable-viable mold/bacteria and/or for individual genera or types (Dillon et al., 1996; Crooks, 1995; Flannigan, 1997).

A number of sampling devices are used for culturable-viable sampling. These include slit or slit-to-agar impactors, single and multistage sieve impactors, impingers, and filters. Most research studies and problem-building investigations report culturable-viable airborne mold and bacteria concentrations using the Andersen N-6 single-stage sampler. In this device, air is drawn through a sievelike plate with 400 holes at a rate of 28.3 l/min, with recommended sampling durations of 1 to 2 min. Spores/bacteria are collected on a culture medium by impaction on a nutrient-rich culture plate below. Most commonly, the culture medium used for general-purpose airborne mold sampling is malt extract agar (MEA), with dicloran glycerol agar (DG-18) being increasingly used to support the growth of xerophilic fungi such as *Aspergillus*. Special media such as cornmeal agar or cellulose Chapex agar are used for the toxigenic fungus *Stachybotrys chartarum*. Trypticase soy agar (TSA) is commonly used to collect and enumerate culturable-viable bacteria concentrations.

Though widely used and in many cases exclusively used by building investigators, culturable-viable techniques have significant limitations. Since most airborne mold and bacteria are no longer viable or alive, such sampling underestimates exposures/potential exposures, in most cases by an order of magnitude or more. This is due to the fact that allergenicity is independent of viability.

Total airborne mold levels can be quantified by using sampling devices that impact airborne particles on a greased microscope slide or coverslip or collect them on a filter surface (Flannigan, 1997). The former are the most widely used total mold sampling devices in building investigations. Their use has increased dramatically in the U.S., as they are the sampling method of choice in conducting clearance sampling after mold remediation.

Three samplers are commonly used (Burkard™, Allergenco™, and Air-O-Cell™). All three have a similar intake orifice that impacts particles on defined deposition area on a greased microscope slide in an automated instrument or adhesive-coated coverslip in a specially designed sampling cassette. Mold spores/fungal structures are counted on a fraction of the deposition area at 1000× magnification (Burkard, Allergenco) or at 400 to 600× magnification (Air-O-Cell). Concentrations are expressed as S/m^3. Total culturable-viable and total mold spore concentrations in residential and mechanically ventilated buildings are summarized in Table 65.1 (Godish, T., 2000).

Because of their relatively high cutoff diameters (circa 2.4 μm), these samplers cannot be used for determining total airborne bacteria levels. Total airborne bacteria concentrations can be determined by sampling with a filter cassette and subsequent counting using epifluorescence microscopy (Palmgren et al., 1986). This method is particularly appropriate for heavily contaminated workplace environments, but it is less sensitive to low-concentration office, institutional, and residential environments.

Airborne levels of fungal glucans, mycotoxins, endotoxins, and a variety of allergens can be determined from samples collected on membrane or fiber filters loaded in 25- and 37-mm cassettes. Sampling rates and durations reflect minimum sampling volumes necessary to provide detectable contaminant levels. Sampling rates of 1.5 to 3 l/min for a 4- to 8-h sampling period can be used to determine endotoxin levels in nonindustrial building environments. These sampling rates/volumes may in many cases be sufficient for fungal glucans, mycotoxins, and allergens that remain suspended (such as cat allergen).

After collection, samples are analyzed by methods specifically developed for each contaminant. Concentrations are typically expressed as $\mu g/m^3$. Major analysis methods are described under the following sections.

TABLE 65.1 Culturable-Viable and Total Mold Spore Concentration Ranges Observed in Buildings

		Concentration Ranges	
Building Types	Conditions	Culturable-Viable Mold Spores (CFU/m³)	Total Mold Spores (S/m³)
Residential	new structure	<300	1000–3000
	not mold infested (avg.)	>300–<1000	>3000–<6500
	moderately infested	>1000–<3000	>6500–<20,000
	heavily infested	>3000	>20,000
Mechanically ventilated nonresidential	not mold infested (avg.)	<300	1000–3000
	moderate, localized infestations	>1000–<3000	>3000–<6500
	heavily infested	>3000	>20,000

Source: Godish, T. (2000), *Indoor Environmental Quality*, CRC Press, Boca Raton, FL.

65.2.2   Surface Sampling/Analysis

For many environmental assessments, potential exposures are inferred from qualitative and quantitative evaluations made from surface samples. This reflects the fact that particulate-phase substances settle out rapidly, and in a number of cases, exposures occur episodically as a result of surface dust resuspension. In the case of mold, the organism is often found actively growing on building surfaces.

Surface sampling can be conducted to collect dust samples for subsequent analyses and quantification of culturable-viable mold, culturable-viable bacteria, endotoxins, fungal glucans, mycotoxins, and allergens such as those of house dust mites, cockroaches, pet dander, and specific fungal species. This includes the collection of dust samples using vacuuming techniques, wipe methods, and, in the case of mold, the application and removal of transparent sticky tape to an infested surface.

Building-dust samples collected for mold and bacteria identification/quantification can be washed and plated out on one or more culture media (Macher, 2001). Wipe samples are commonly collected by rubbing a prepackaged swab over a relatively small surface area (e.g., 2.5 cm²) and then applied it to the surface of a culture medium. Concentrations in both cases are reported as CFUs/unit surface area.

The transparent-sticky-tape method is widely used in problem building investigations to identify major mold genera present on infested surfaces. Identifications are conducted by means of light microscopic analysis at magnifications of 100 to 1000×. The technique is especially useful in evaluating whether mold types of special concern (such as *S. chartarum* and *Aspergillus* species) are present and what specific remediation practices are to be employed.

Vacuum sampling is increasingly being used to assess surface dust concentrations of endotoxins, fungal glucans, and mycotoxins. Such sampling and subsequent analyses provide an indication of the degree of contamination present and the potential exposure upon resuspension of settled dusts (Bischof et al., 2002). Endotoxin and fungal glucan concentrations are determined using the Limulus amoebocyte lysate (LAL) bioassay. Several variations of the LAL assay are used for endotoxins. These include the gel-clot, turbidimetric, and kinetic chromogenic methods. Because of its relative accuracy, sensitivity, and reproducibility, the kinetic chromogenic method is the analytical method commonly used by commercial laboratories (Milton et al., 1996). Endotoxin concentrations can be expressed as EU/unit surface area or as ng/unit surface area. One nanogram is equal to 10–15 EUs (endotoxin units).

The LAL method is not as specific for fungal glucans as might be desired. Consequently, concentrations might better be determined using an immunochemical method. Mycotoxins can be identified and quantified using a variety of analytical procedures. Concentrations are expressed as µg/mg per unit surface area.

Surface sampling for common allergens is the exposure assessment method of choice of medical scientists and those conducting building assessments. Sampling is conducted on surfaces using a vacuum cleaner with special filter attachments. Vacuuming is conducted over a standard surface area (1 m²) or for a standard duration (5 min). After screening and weighing, samples are analyzed for specific allergens using monoclonal or polyclonal antibody techniques, with the former being the most commonly used. Concentrations are usually expressed as µg allergen/g dust collected; for antigens associated with cockroaches, concentrations are expressed as biological units (BU)/g dust. Monoclonal antibody assays are commercially available for two dust mite antigens (Der p1 and Der f1), cat antigens (fel d1), dog antigens (can f1), German cockroach (Bla gII) (Chapman, 1995; Squillace, 1995; Luczynska, 1995), and for three fungal species: *Alternaria alternata* (Wijnands et al., 2000b) and *Aspergillus fumigatus* (Wijnands et al., 2000a). Relative concentrations of dust mite, cat, and dog allergens in house dust are summarized in Table 65.2 (European Communities, 1993).

Most allergen exposure assessments are conducted in houses. Several recent studies indicate that potentially significant exposure risks to allergens associated with pets can also occur in school buildings as a result of passive transport (Munir, 1993; Godish, D., 2000).

65.2.3   Bulk Sampling

Bulk sampling techniques are occasionally used to identify dominant mold types in infested building materials. Infested materials are broken up, washed, and placed on culture plates. Colonies that subse-

TABLE 65.2 Relative Categorization of Mite, Cat, and Dog Allergens in House Dust (μg/g)

| Category | House-Dust Mite | | Cat | Dog |
	Der p 1	Der f 1	Fel d 1	Can f 1
Very low	<0.5	<0.5	<0.1	<0.3
Low	0.5–5	0.5–5	0.1–1.0	0.3–10.0
Intermediate	>5–15	>5–15	>1–10	>10–100
High	>15–20	>15–20	>10–100	>100–1000
Very high	>20	>20	>100	>1000

Source: European Communities (1993), Biological Particles in Indoor Environments, 12 EUR 14988EN, European Communities, Luxembourg.

quently grow are identified. They can also be analyzed using QPCR (quantitative polymerase chain reaction) methods.

65.2.4 Limitations of Biological Contaminant Exposure Assessment Methodologies

Sampling of airborne and surface contaminants are routinely conducted in problem building investigations. It is assumed that the results of such sampling can be meaningfully interpreted relative to the exposures that may be occurring and that these exposures are reflective of the health risks involved. This is particularly the case for airborne contaminants. For the most part, clear dose–response relationships between airborne contaminant concentrations and reported respiratory health effects have not been demonstrated. This in part reflects the nature of exposures and difficulties in their assessment, and it is partly due to the absence of standardized protocols.

Biological contaminants in indoor air vary considerably over time. Airborne mold concentrations, for example, can vary by an order of magnitude or more over the course of a single day (Flannigan, 1997). Such variation and the limited time period over which samples are taken make it difficult to demonstrate dose–response relations. Such relationships may be further confounded by differential sensitivity to mold species present and by responses to other contaminants such as endotoxins, fungal glucans, and a variety of allergens. Most health studies of biological contaminants have been based on small-study populations using statistical designs that are not sufficiently powerful to detect dose–response relationships that may exist.

In the absence of such relationships, potential exposure health risks are often inferred from a combination of factors including the presence of visible contamination (e.g., a mold-infested surface), relatively high airborne or surface concentrations (as compared with the range of concentrations in reported studies), and statistical relationships between surface dust concentrations and disease risk. It is notable that exposure guideline values (when they are available) are based on surface dust concentrations (e.g., endotoxins, allergens) (Rylander, 1997; Platts-Mills and deWeck, 1989). Such concentrations are at best an indirect indicator of exposure potential. They do, however, have the advantage of having been shown to be statistically related to respiratory disease.

65.2.5 Airborne Particles

A number of instrumental sampling methods have been used to measure airborne particle concentrations in buildings, particularly in the inhalable (≤10 μm) and respirable (≤2.5 μm) size ranges. These include gravimetric methods wherein air is drawn through a fibrous filter medium with subsequent gravimetric analysis. Concentrations are expressed as μg/m³. Sampling rates, durations, and volumes employed depend on the sampling device/system used (Cohen and Herring, 1995).

Respirable particles are commonly measured using real-time or quasi-real-time instruments that report particle concentrations on a weight basis ($\mu g/m^3$) or by particle counts. Several physical principles are used to determine particle concentrations. These include piezoelectric resonance devices and those that use optical procedures. In the former, nonrespirable particles are removed by an impactor or cyclone. Respirable particles are then electrostatically deposited on a quartz crystal sensor. The difference in oscillating frequency between sensing and reference cells is determined/displayed as $\mu g/m^3$. A number of sampling devices also use optical procedures. Forward light scattering by particles in the size range of 0.1 to 10 μm is used to determine concentrations in some instruments. Light scattering is used to count ultrafine particles (<0.1 μm diameter) after they have been made to grow by condensation of water on their surfaces.

Concentrations of total surface dust can be determined from vacuum samples, tape samples, or optical techniques over a standard surface area (Holopainen et al., 2002). Vacuum samples can be used to identify/ quantify one or more dust components (Hedge et al., 1993; Gyntelberg et al., 1994). Such concentrations are used as indirect measures of potential human exposures.

Despite the fact that a number of indoor-air-quality studies have shown statistical relationships between indoor particle levels and respiratory symptoms, particle levels in indoor environments are rarely measured in investigations of problem buildings. This may reflect the fact that building investigators do not have sufficiently detailed knowledge of the potential adverse health effects of airborne particle exposures in the indoor environment.

References

Bischof, W., Koch, A., Gehring, U., Fahlbusch, B., Wichmann, H.E., and Heinrich, J. (2002), Predictors of high endotoxin concentrations in the settled dust of German homes, *Indoor Air*, 12, 2–9.

Chapman, M.D. (1995), Analytical methods: immunoassay bioaerosols, in *Bioaerosols*, Burge, H.A., Ed., Lewis Publishers, Boca Raton, FL.

Cohen, B.S. and Herring, S.V., Eds. (1995), *Air Sampling Instruments for Evaluation of Atmospheric Contaminants*, 8th ed., American Conference of Governmental Industrial Hygienists, Cincinnati, OH.

Cookingham, C.E. and Solomon, W.R. (1995), Bioaerosol-induced hypersensitivity diseases, in *Bioaerosols*, Burge H.A., Ed., Lewis Publishers, Boca Raton, FL, 205–234.

Crooks, B. (1995), Inertial samplers: biological perspectives, in *Bioaerosols Handbook*, Cox, S. and Wathes, C.M., Eds., CRC Lewis Publishers, Boca Raton, FL, 247–264.

Dales, R.E., Zwanenburg, H., Burnett, R., and Franklin, C.A. (1991), Respiratory health effects of home dampness and molds among children, *Am. J. Epidemiol.*, 134, 196–293.

Dillon, H.K., Heinsohn, P.A., and Miller, J.D. (1996), *Field Guide for the Determination of Biological Contaminants in Biological Samples*, American Industrial Hygiene Association, Fairfax, VA.

Etzel, R.A. (1995), Acute pulmonary hemorrhage/hemesiderosis among infants: Cleveland, January 1993 – November 1994, *Morbidity Mortality Wkly. Rep.*, 43, 881–833.

European Communities (1993), Biological Particles in Indoor Environments, 12 EUR 14988EN, European Communities, Luxembourg.

Fink, J. (1983), Hypersensitivity pneumonitis, in *Allergy: Principles and Practice*, 2nd ed., Middleton, E. et al., Eds., C.V. Mosby, St. Louis, pp. 1085–1099.

Flannigan, B. and Miller, J.D. (1994), Health implications of fungi in indoor environments: an overview, in *Health Implications of Fungi in Indoor Environments*, Samson, R.A., Flannigan, B., Flannigan, M.E., Verhoeff, A.P., Adan, O.C.G., and Hoekstra, E.S., Eds., Elsevier, Amsterdam, pp. 1–28.

Flannigan, B. (1997), Air sampling for fungi in indoor environments, *J. Aerosol Sci.*, 28, 381–392.

Godish, D. (2000), Allergens in elementary schools: their prevalence, quantity, and association with schools' socioeconomic characteristics and asthma episodes, in *Engineering Solutions to Indoor Air Quality Problems*, Symp. Proc., July 2000, pp. 121–129.

Godish, T. (1995), Sick buildings, in *Definition, Diagnosis and Mitigation*, Lewis Publishers, Boca Raton, FL.

Godish, T. (2000), *Indoor Environmental Quality*, CRC Press, Boca Raton, FL.

Gyntelberg, F., Sudicani, P., Nielsen, J.W., Skov, P., Bisgaard, H., and Gravesen, S. (1994), Dust and sick building syndrome, *Indoor Air*, 4, 223–230.

Hedge, A., Erickson, W.A., and Rubin, G. (1993), Effects of man-made mineral fibers in settled dust on sick building syndrome in air-conditioned offices, *Indoor Air*, 1, 291–296.

Holopainen, R., Asikainen, V., Pasanen, P., and Seppanen, O. (2002), The field comparison of the surface dust level in ventilation ducts, *Indoor Air*, 12, 47–54.

Imperato, P.J. (1981), Legionellosis and the indoor environment, *Bull. NY Acad. Med.*, 57, 922–935.

Levetin, E. (1995), Fungi, in *Bioaerosols*, Burge, H.A., Ed., Lewis Publishers, Boca Raton, FL, pp. 87–120.

Luczynska, C.M. (1995), Mammalian aerollergens, in *Bioaerosols*, Burge, H.A., Ed., Lewis Publishers, Boca Raton, FL, pp. 149–161.

Macher, J.M. (2001), Review of methods to collect settled dust and isolate culturable microorganisms, *Indoor Air*, 11, 99–110.

Miller, J.D. (1992), Fungi as contaminants in indoor air, *Atmos. Environ.*, 26A, 2163–2172.

Milton, D.K., Johnson, D.K., and Park, J.H. (1996), Environmental endotoxin measurement: interference and sources of variation in the Limulus assay of house dust, *Am. Ind. Hyg. Assn. J.*, 58, 861–867.

Munir, A.K.M., Einarsson, R., Schou, C., and Dreborg, S.K.G. (1993), Allergens in school dust. I. The amount of the major cat (Fd d1) and dog (can f1) allergens in dust from Swedish schools is high enough to probably cause perennial symptoms in most children with asthma in most children sensitized to cat and dog, *J. Allergy Clin. Immunol.*, 177–181.

Nardell, E. (2001), Tuberculosis, in *Indoor Air Quality Handbook*, Spengler, J.D., Gamet, J.M., and McCarthy, J.F., Eds., McGraw-Hill, New York, pp. 47.1–47.12.

Palmgren, U., Strom, G., Blomquist, G., and Malmberg, P. (1986), Collection of airborne microorganisms on nucleopore filters, estimation and analysis: CAMNEA method, *J. Appl. Bacteriol.*, 61, 401–406.

Platts-Mills, T.A.E. and deWeck, A.L. (1989), Dust mite allergens and asthma: a worldwide problem, *J. Allergy Clin. Immunol.*, 83, 416–427.

Raw, G.J., Roys, M.S., and Whitehead, C. (1993), Sick building syndrome: cleanliness is next to healthiness, *Indoor Air*, 3, 237–245.

Rylander, R., Haglind, P., Lundholm, M., Mattsby, I., and Stenqvest, K. (1980), Humidifier fever and endotoxin exposure, *Clin. Allergy*, 8, 511–516.

Rylander, P., Persson, K., Git, H., Yaasa, K., and Tanaka, S. (1992), Airborne P $(1{\rightarrow}3)$ d-glucan may be related to symptoms in sick buildings, *Indoor Environ.*, 1, 263–267.

Rylander, R. (1997), Endotoxins in the indoor environment: a criteria document, *Int. J. Occup. Environ. Health*, 3, S1–S48.

Sorenson, W.S., Frazer, D.G., Jarves, B.B., Simpson, J., and Robinson, V.A. (1987), Trichotcene mycotoxins in aerosolized conidia of Stachybotrys atra, *Appl. Environ. Microbiol.*, 53, 1370–1375.

Sorenson, W.S. (1999), Fungal spores: hazardous to health? *Environ. Health Perspect.*, 107 (Suppl. 3), 469–471.

Squillace, S.P. (1995), Allergens of arthropods and birds, in *Bioaerosols*, Burge, H.A., Ed., Lewis Publishers, Boca Raton, FL, pp. 133–148.

Weissman, D.N. and Schuyler, M.R. (1991), Biological agents and allergic diseases, in *Indoor Air Pollution: A Health Perspective*, Gamet, J.M. and Spengler, J.D., Eds., Johns Hopkins University Press, Baltimore, MD, pp. 285–305.

Wijnands, L.M., Deisz, W.D.C., and van Leusden, F.M. (2000a), Marker antigens to assess exposure to molds and their allergens. I. *Aspergillus fumigatus*, *Allergy*, 55, 850–855.

Wijnands, L.M., Deisz, W.D.C., and van Leusden, F.M. (2000b), Marker antigens to assess exposure to molds and their allergens. II. *Alternaria alternata*, *Allergy*, 55, 856–864.

66

Olfactometry: The Human Nose as Detection Instrument

Pamela Dalton
Monell Chemical Senses Center

Monique Smeets
Utrecht University

66.1 Background and Applications

Despite common perceptions that the human olfactory system is relatively insensitive in comparison with that of other species, humans can detect many odorants in extremely low concentrations. For example, at concentrations where many odors can just be detected or even recognized, there are few health consequences from inhalation exposure, yet simple awareness of an odor can produce annoyance or concern about the potential consequences of exposure and thereby impact mood and well-being (Shusterman et al., 1991). For this reason, the establishment of methods for evaluating the detectability, intensity, and hedonic characteristics of environmental odorants is a critical step in their regulation and management.

A set of methods, collectively known as olfactometry, can be used to evaluate the perceptual characteristics of odors in both indoor and outdoor environments when such emissions form the basis for annoyance, complaints, and discomfort. These include evaluations of the impact of odors from human bodies, tobacco smoke, volatile building materials, and molds, all of which are major determinants of the quality and acceptability of indoor air in residences or workplaces (Cone and Shusterman, 1991). Olfactometry can also be used to evaluate the air quality near industrial facilities such as refineries, pulp mills, or wastewater treatment plants. More recent applications of olfactometry have included the evaluation of the sensitivity and response of chemically exposed workers in order to evaluate the safety and comfort associated with their exposure or to evaluate the degree to which such exposure has altered their olfactory capability (Gagnon et al., 1994; Wysocki et al., 1997).

TABLE 66.1 Overview of Relevant Odor Characteristics and Assessment Methods

Levels of Exposure	Odor Characteristics	Index	Method
Perithreshold	Detectability	Odor detection threshold	Static olfactometry Dynamic olfactometry
Suprathreshold	Intensity Hedonics Quality	Intensity rating Acceptability and annoyance ratings Descriptive labels	Scaling methods Qualitative descriptive analysis

66.1.1 Odor Characteristics

The process of odor perception begins with the binding of volatile chemical molecules to specialized receptors in the nasal cavity. The process ends when information from the many different receptors is organized into patterns that can be recognized by the brain as distinct odors, with different characteristics.

No one method is optimal for measuring odor impact in all situations. Techniques for measuring odor characteristics include those that evaluate concentrations at perithreshold and suprathreshold levels. (Table 66.1, first column). Beyond this, the choice of method will depend largely on the equipment and technical resources available and the nature of the odor to be evaluated.

At perithreshold level, the odorant is at a concentration where its presence can first be differentiated from ambient air (Schiffman et al., 2001), but little other information about the odorant is available. This level may vary widely from one person to another, and it is dependent on the individual's sensitivity to that particular compound or mixture. At higher concentrations (suprathreshold), the detectability of the odor stimulus is assumed, and interest is focused on characteristics such as perceived intensity, quality (categorization of odorant sensations into categories such as "pungent," "putrid," or "floral"), and hedonics (on dimensions such as "pleasant–unpleasant" or "acceptable–unacceptable"). Methods for evaluating the basic odor characteristics are listed in Table 66.1. In our continued discussion of these characteristics, the emphasis will be mostly on detectability, intensity, and hedonics, and less on odorant quality. Most odor evaluations will seek first to determine the concentration at which the odorant becomes detectable and then to determine the subsequent intensity and hedonic response.

66.1.2 Irritation Potential

In addition to odor sensation, many odorants are also capable of stimulating sensations of burning, tingling, stinging, warmth, or cooling in the eyes, nose, and throat. The potential for a chemical to produce these irritant sensations is largely a matter of concentration. For example, acetic acid at low concentrations will elicit a smell of vinegar; however, at higher concentrations the smell will be accompanied by a pungent sensation in the eyes and nose. The difference between the concentration at which a chemical is detectable by odor and the concentration at which it begins to elicit irritation can be considered a measure of its irritant potency (Cometto-Muñiz and Cain, 1995). Because irritant potency varies considerably across different odorant chemicals as a function of carbon chain length, lipophilicity, and perhaps other unidentified factors, it can impact the hedonic acceptability of odorants at increased concentrations.

66.2 General Prerequisites

With a modality as sensitive as the olfactory modality, the following issues should receive careful attention during testing:

- *Research environment:* The environment in which the tests are conducted should be odor-free, maintained at constant temperature, and equipped with a ventilation system for the expedient turnover of air.

- *Selection of research subjects or panelists:* Sensitivity to different odorants can vary widely among individuals and can be impacted by many factors. Subjects should be screened for such factors, which include: gender; age (general olfactory ability); the presence of respiratory disease, such as asthma, seasonal allergies, or active colds; medication use; smoking; occupational history; and chemical sensitivities. Panelists are often screened for their sensitivity to n-butanol, eliminating those panelists with extreme performance (either sensitive or insensitive) from the final sample. However, this practice may create a false sense of precision. Because sensitivity to odors is specific rather than general, the response to n-butanol is not necessarily a good predictor of sensitivity for other compounds (Köster, 1986). An adjunct or alternative tool, such as the seven-odor screening test, which evaluates panelists' ability to detect and identify odors drawn from seven different categories (such as putrid, pungent, floral, camphoraceous) may therefore present a more complete picture of general olfactory ability (NASA, 1991).
- *Methods for stimulus delivery and measurement:* There is a variety of methods of stimulus delivery and measurement, each having advantages and disadvantages. These are discussed in the following sections.

66.3 Procedures

66.3.1 Olfactometry: Methods of Delivery

There are three general methods that are used to deliver odorants to the nasal cavity for detection and evaluation: static olfactometry, dynamic olfactometry, and chamber exposure. In this chapter, we will focus on static and dynamic olfactometry methods only, as they are the most widely used.

66.3.1.1 Static Olfactometry

In static olfactometry, the odorant to be smelled is presented as a fixed concentration of the liquid chemical in a closed container, and the stimulus is the odorized air or headspace over the liquid (Prah et al., 1995). A series of plastic squeeze or glass bottles is typically used, in which each bottle of the series contains a different concentration of the odorant dissolved in a diluent according to a predetermined dilution sequence (e.g., binary, tertiary, or logarithmic series, in which each bottle contains 1/2, 1/3, or 1/10, respectively, of the concentration of the previous bottle). The choice of diluent depends on the chemical solute, but it is typically odorless mineral oil or polyethylene glycol (Sigma) or, less frequently, water. Although many investigators using this technique report the concentration of the odorant in solution, it is important to recognize that the actual stimulus of interest — the amount of odorant that is available to the subject's nose — is the concentration of the odorant in the vapor phase (i.e., headspace), not the amount of odorant in the solution. Vapor-phase concentration in the headspace of the bottles cannot be simply predicted from the amount of odorant in the solution because it is dependent on many factors. It is most reliably assessed using gas chromatography (Dalton, 2002) to express the concentration in parts per million (ppm) or parts per billion (ppb) in air.

66.3.1.2 Dynamic Olfactometry

Dynamic olfactometry involves delivering a continuous, well-regulated gas flow that contains odorized air mixed in varying proportions with a carrier gas, typically odorless air or nitrogen (Prah et al., 1995). Although dynamic olfactometry can be used to determine a response to a single, known chemical, it has perhaps most usefully been applied in the context of gaseous mixtures sampled from the environment (e.g., agricultural emissions), the composition of which is not (fully) known and which is difficult to reproduce in the laboratory. The method characteristically contains the following elements: a sample of odorous air is collected in a large container made of nonadsorbent material (e.g., glass, stainless steel, or Teflon), from which it is extracted and directed to the olfactometer. The sample is then increasingly diluted with odorless air in sequential steps (e.g., 1/2, 1/4, 1/8) and presented to the panel via sniffing ports for evaluation at each concentration (see Figure 66.1).

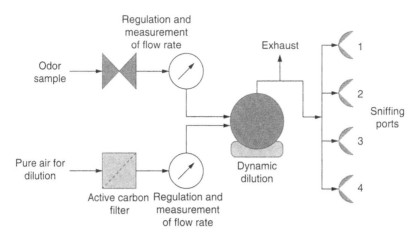

FIGURE 66.1 Schematic representation of an air dilution olfactometer.

66.4   Advantages and Disadvantages of Olfactometry

66.4.1   Advantages

- Static olfactometry can be a technically simple, rapid tool of measurement that can be used when rapid or field assessment of olfactory sensitivity is the goal.
- The bottle series, which is used in static olfactometry, is prepared in the laboratory from stock solutions of known chemicals. Thus this approach is most suitably employed for odor evaluation when the chemical of interest (preferably, a single compound) is known. This situation may apply to occupational exposures by industrial workers to known and single-compound chemicals, such as acetone (Dalton et al., 1997), or for studies on limit setting of known chemicals (Smeets et al., 2002).
- Dynamic olfactometry is best used when the odor of interest is coming from a mixture of unknown composition, such as is often the case in environmental exposure to the odors from agricultural or industrial emissions. The ability to dilute the sample to determine a sensory threshold is the main advantage of dynamic olfactometry.

66.4.2   Disadvantages

- In static olfactometry, equilibration time is needed after trials to replenish the quantity of odorant that was lost during earlier trials.
- In both static and dynamic olfactometry, the vapor-phase concentration tends to be unstable over time, whether the sample was prepared in a bottle or gathered in the field. Both samples are subject to degradation and adsorption and can be used only for limited periods of time.

66.5   How To Measure Odor Sensitivity

66.5.1   Concept of the Odor Detection Threshold

Regardless of how odorant stimuli are presented, the most widely used method for evaluation of sensitivity is the odor detection threshold (ODT), which is the minimum amount of stimulus concentration necessary to discriminate an odor from a blank sample of ambient air (Dalton, 2002). Although the concept of ODT is strongly suggestive of a fixed threshold, it is widely recognized that an individual's threshold for any odorant is but an estimate of sensitivity obtained at one point in time, and it is inherently variable. Thus it is important that any obtained threshold for an individual or a group should be considered as representing a range of sensitivity.

In ODT assessment, odor stimuli are usually presented in a sequence going from weakest to strongest (ascending series) rather than from strongest to weakest (descending series). This is done because exposure to high concentrations of the odor early on may cause olfactory adaptation in the panelists. Using either bottles (static) or sniffing ports (dynamic), the participant is presented with two kinds of samples in succession — the odorant sample and one or more samples of clean air — and is instructed to identify in which sample they smelled an odor. This "forced choice" procedure is generally preferred over the yes–no procedure, in which the subject is simply asked whether or not a bottle containing an odorant contains the odor or not.

A fixed number of trials are presented at each concentration (typically three to five) until the subject is able to correctly identify the odor stimulus on all trials. Then the procedure may be repeated for reliability. Often, judgments of confidence are integrated into calculation of the ODT, such that the cutoff for reaching the ODT is at a predetermined level of confidence.

66.5.2 Calculation of the Odor Threshold

The literature contains studies that have calculated odor thresholds in one of two different ways. From a psychophysical tradition, the ODT is an absolute threshold, i.e., a concentration of the odorant that an individual can detect on 50% of the trials (McNicol, 1972). Thus, the concentration that lies between the step at which the participant could not identify the odorant sample and the step where the participant could always identify the odorant sample is considered to be the odor threshold for that participant. The individual thresholds can be averaged across the group of participants who are tested.

An alternative method of estimating odor thresholds for a population is to determine the concentration step at which 50% of the panelists can reliably detect the odor (Frechen, 1994; Miedema and Ham, 1988). Whereas the former definition is consistent with the original psychophysical concept of absolute threshold, the latter follows the exact probit method (Finney, 1971). Clearly, different values will result from either formula. Thus, since these thresholds cannot be used interchangeably, one should always pay attention to how the ODT was determined in a particular instance.

A hypothetical example depicting the difference in the two methods of threshold calculation is presented in Table 66.2. The ODT at which 50% of the panel can smell odor mixture X is between dilution

TABLE 66.2 A Hypothetical Example of a Response Sheet Obtained from Eight Panelists Using Dynamic Dilution Olfactometry with Odorant Mixture X

Dilution step	S1	S2	S3	S4	S5	S6	S7	S8	Mean ODT
15	0	0	0	0	0	0	0	0	
14	0	0	0	0	0	0	0	0	
13	1	0	0	0	0	0	0	0	
12	1	0	0	1	0	0	0	0	50% panel
11	1	1	0	1	0	1	0	0	10.5
10	1	1	0	1	1	1	1	0	
9	1	1	1	1	1	1	1	1	
8	1	1	1	1	1	1	1	1	
7	1	1	1	1	1	1	1	1	
6	1	1	1	1	1	1	1	1	
5	1	1	1	1	1	1	1	1	
...	...	...	...	...	...	...	...	...	
...	...	...	...	...	...	...	...	...	
...	...	...	...	...	...	...	...	...	Individual ODTs
Individual ODT	13.5	11.5	9.5	12.5	10.5	11.5	10.5	9.5	11.125

Note: Every judgment of no odor perceivable has been marked as "0," and every judgment of odorant as clearly perceivable as "1."

step 10 and 11, which would amount to 10.5. However, the average of all eight individual ODTs is 11.125, which is a lower concentration.

In applications of dynamic olfactometry, mostly within the environmental context where exposure from agricultural or industrial odors are assessed, the number of dilutions necessary to reach the ODT may be expressed as the odorant's concentration, c_{od}, and expressed as odor unit per cubic meter, or ou/m^3 (Frechen, 1994; Miedema et al., 1988).

Measures of odor thresholds as determined by olfactometry are rarely used in isolation, but they are often used in combination with dispersion models in an effort to predict atmospheric dilution of odorous air at different geographical locations to prevent nuisance. (For an overview of olfactometric procedures and dispersion models, see Nielsen et al. [1986] and the Belgian Filtration Society's conference proceedings [SBF, 1984].) More importantly, odor units have been used to regulate odor emissions from facilities that may emit malodors. Many states in the U.S. require the fence-line odor to be undetectable at a dilution ranging from 4 to 24, depending on the zoning of the surrounding region (ASTM, 2002).

66.5.3 Suprathreshold Measures

For the measurement of odor impact at concentrations above threshold, odorants can be presented using either static or dynamic methods, but they are typically evaluated using scalar ratings of intensity or hedonics. (See Engen [1971] for a review of scaling methods.) The following methods are used:

- *Category scales:* Category scales usually employ seven or more labeled categories (e.g., from "not at all" to "extremely") from which the subject can select to express the intensity of the odor or the extent to which the odor is experienced as annoying (e.g., see Köster et al., 1986; Steinheider et al., 1998). The number of response alternatives is sometimes considered as too limited.
- *Line scales (visual analog scales):* Here, the subject places a mark anywhere along a line of 100 mm length to indicate, e.g., odor intensity. The measured distance in mm to the mark is used as the numerical measure of intensity.
- *Ratio scales (e.g., magnitude estimation/production):* Ratio scales require subjects to assign numbers relative to the magnitude of sensations elicited by numerous stimuli (Berglund, 1977; Cain, 1969).
- *Category-ratio scales:* Category-ratio scales (e.g., the labeled magnitude scale; see Green et al. [1996]) serve as alternatives to magnitude estimation in the scaling of olfactory stimuli. They yield ratio data while allowing subjects to use natural language descriptors, rather than numbers, to scale sensation.

66.5.4 Caveats in Using Values Derived from Olfactometry

If large changes in concentration are required for the subject to detect a difference in perceived intensity, the shape of the psychophysical function for that odorant, relating change in sensory experience to stimulus concentration, is shallow. In contrast, if small changes suffice to yield a difference in perception, the function is steep. This is significant because the shape of the function will also determine the relationship between the odor unit of a given concentration (i.e., the number of dilutions to threshold) and the perceived intensity of that concentration. Figure 66.2 illustrates this using the compounds acetic acid and PEA. An odor concentration of 100 ou/m^3, for example, may be related to radically different perceived odor intensities, depending on the compound studied.

For this reason, odor concentrations expressed in units of ou/m^3 are primarily useful when comparing the effects of odor reduction measures or in studies where dispersion models are used to find the distance to the source at which threshold is reached (Köster, 1986). A similar warning can be made for comparing measures of annoyance to odor concentration. Evaluations of environmental odor impact have found little evidence of a relation between odor concentration and degree of annoyance (Deane and Sanders, 1978; Köster et al., 1986). Rather, studies have shown a stronger correlation between odorant source and reported discomfort, even for compounds with comparable odor intensities (Winneke and Kastka, 1987). Thus, annoyance and intensity should be assessed independently of odor concentration or ODT.

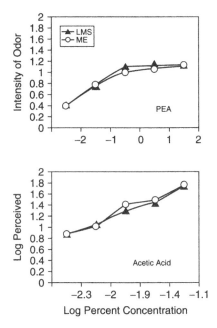

FIGURE 66.2 Psychophysical functions relating perceived intensity to chemical concentration for two chemicals that vary in their irritancy potential.

66.6   Summary

The term *olfactometry* refers to a collection of methods that can be used to evaluate the perceptual characteristics of odors in both indoor and outdoor environments when such emissions form the basis for annoyance, complaints, and discomfort. We distinguish between three relevant odor characteristics: detectability (the level at which the odor can be differentiated from ambient air), intensity (or perceived odor strength), and hedonics (across dimensions of pleasantness or annoyance). Different methods of stimulus delivery are used, typically referred to as static or dynamic olfactometry.

The odor detection threshold (ODT) is used as an index of detectability. For ODT measurement, a commonly used method involves the participation of panelists who indicate which sample (among one odor sample and one or more blanks) contains the odor when odor concentration is varied from below threshold to slightly above threshold. The measurement of odor intensity and hedonics typically involves the use of scaling methods, in which humans rate the intensity, pleasantness, or annoyance of suprathreshold (above threshold) concentrations of odorants on a numbered or labeled category scale along a continuum.

As odor issues continue to increase in their ability to generate attention and concern in residential, industrial, and agricultural settings, standardized methods for evaluating the impact of odorous emissions will be necessary to guide their regulation and management.

References

ASTM International (2002), Standard of Practice E679-91, Determination of Odor and Taste Threshold by a Forced-Choice Ascending Concentration Series Method of Limits, American Society for Testing and Materials, Philadelphia.

Berglund, B. (1977), Quantitative approaches in environmental studies, *Int. J. Psychol.*, 12, 111–123.

Cain, W.S. (1969), Odor intensity: differences in the exponent of the psychophysical function, *Percept. Psychophysics*, 6, 349–354.

Cometto-Muñiz, J.E. and Cain, W.S. (1995), Relative sensitivity of the ocular trigeminal, nasal trigeminal and olfactory systems to airborne chemicals, *Chem. Senses*, 20, 191–198.

Cone, E.J. and Shusterman, D. (1991), Health effects of indoor odorants, *Environ. Health Perspect.*, 95, 53–59.

Dalton, P. (2002), Olfaction, in *Stevens' Handbook of Experimental Psychology*, Vol. 1, *Sensation and Perception*, 3rd ed., Yantis, S., Ed., Wiley, New York, pp. 691–756.

Dalton, P., Wysocki, C.J., Brody, M.J., and Lawley, H.J. (1997), Perceived odor, irritation and health symptoms following short-term exposure to acetone, *Am. J. Ind. Med.*, 31, 558–569.

Deane, M. and Sanders, G. (1978), Annoyance and health reactions to odor from refineries and other industries in Carson, California, *Environ. Res.*, 15, 119–132.

Engen, T. (1971), Psychophysics, in *Woodworth and Schlosberg's Experimental Psychology*, 3rd ed., Kling, J.W. and Riggs, L.A., Eds., Holt, Rinehart and Winston, New York, pp. 11–86.

Finney, D.J. (1971), *Probit Analysis*, 3rd ed., Cambridge University Press, Cambridge, U.K.

Frechen, F.B. (1994), Odour emissions of wastewater treatment plants: recent German experiences, *Water Sci. Technol.*, 30, 35–46.

Gagnon, P., Mergler, D., and Lapare, S. (1994), Olfactory adaptation, threshold shift and recovery at low levels of exposure to methyl isobutyl ketone (MIBK), *Neurotoxicology*, 15, 637–642.

Green, B.G., Dalton, P., Cowart, B.J., Shaffer, G., Rankin, K.R., and Higgins, J. (1996), Evaluating the "Labeled Magnitude Scale" for measuring sensations of taste and smell, *Chem. Senses*, 21, 323–334.

Köster, E.P. (1986), Limitations imposed on olfactometric measurements by the human factor, in *Odour Prevention and Control of Organic Sludge and Livestock Farming*, Nielsen, V.C., Voorburg, J.H., and L'Hermite, P., Eds., Elsevier Applied Science, London, pp. 86–93.

Köster, E.P., Punter, P.H., Maiwald, K.D., Blaauwbroek, J., and Schaefer, J. (1986), Direct scaling of odour annoyance by population panels, *VDI Berichte*, 561, 299–312.

McNicol, D. (1972), *A Primer of Signal Detection Theory*, George Allen and Unwin, London.

Miedema, H.M.E. and Ham, J.M. (1988), Odor annoyance in residential areas, *Atmos. Environ.*, 22, 2501–2507.

NASA (1991), Flammability, Odor, Off Gassing, and Compatibility Requirements and Test Procedures for Materials in Environments that Support Combustion, Handbook NHB 8060.1C: Office of Safety and Mission Quality, National Aeronautics and Space Administration, Washington, D.C.

Nielsen, V.C., Voorburg, J.H., and L'Hermite, P. (1986), *Odour Prevention and Control of Organic Sludge and Livestock Farming*, Elsevier Applied Science, London.

Prah, J.D., Sears, S.B., and Walker, J.C. (1995), Modern approaches to air dilution olfactometry, in *Handbook of Olfaction and Gustation*, Doty, R.L., Ed., Marcel Dekker, New York, pp. 227–255.

SBF (Belgian Filtration Society) (1984), Characterization and Control of Odoriferous Pollutants in Process Industries, paper presented at SBF international symposium, Louvain-la-Neuve, Belgium, April.

Schiffman, S.S., Walker, J.M., Dalton, P., Lorig, T.S., Raymer, J.H., Shusterman, D., and Williams, C.M. (2001), Potential health effects of odor from animal operations, wastewater treatment, and recycling of byproducts, *J. Agromed.*, 7, 7–81.

Shusterman, D., Lipscomb, J., Neutra, R., and Satin, K. (1991), Symptom prevalence and odor-worry interaction near hazardous waste sites, *Environ. Health Perspect.*, 94, 25–30.

Smeets, M.A., Maute, C.M., and Dalton, P. (2002), Acute sensory irritation from exposure to isopropanol in workers and controls: objective versus subjective effects, *Ann. Occup. Hyg.*, 46, 359–373.

Steinheider, B., Both, R., and Winneke, G. (1998), Field studies on environmental odors inducing annoyance as well as gastric and general health-related symptoms, *J. Psychophysiol. Suppl.*, 12, 64–79.

Winneke, G. and Kastka, J. (1987), Comparison of odour-annoyance data from different industrial sources: problems and implications, in *Environmental Annoyance: Characterization, Measurement, and Control*, Koelega, H.S., Ed., Elsevier Science, Amsterdam, pp. 95–104.

Wysocki, C.J., Dalton, P., Brody, M.J., and Lawley, H.J. (1997), Acetone odor and irritation thresholds obtained from acetone-exposed factory workers and from control (occupationally non-exposed) subjects, *Am. Ind.. Hyg. Assn. J.*, 58, 704–712.

67

The Context and Foundation of Lighting Practice

Mark S. Rea
Rensselaer Polytechnic Institute

Peter R. Boyce
Rensselaer Polytechnic Institute

67.1   A Brief History of Lighting Recommendations

It is sometimes assumed that lighting recommendations are determined simply by the inherent capabilities of the visual system. This is untrue. The characteristics of the visual system are undeniably important, but the available lighting technology, the cost of light, and the social context of the times all affect lighting recommendations. This influence of nonvisual factors on lighting recommendations is evident from a simple review of the history of lighting practice.

In the first quarter of the last century, electric lighting began to be widely used. The typical illuminance recommended by "experts" was between about 30 and 100 lx. Indeed, it was argued by some that illuminance higher than this was harmful to the retina. Figure 67.1 shows "good" electric lighting in a post office in 1906. It is interesting to observe the somewhat shielded incandescent lamp fixtures suspended at different heights over each desk on flexible cords. One can infer from this old photograph that these post office workers sometimes placed these fairly dim incandescent light sources close to the task to achieve much higher illuminance than the recommended levels of 30 to 100 lx. Perhaps the "experts" did not notice.

Figure 67.2 shows "good" lighting in a drafting office in 1947. By this time, recommended illuminance had increased dramatically to between 300 and 500 lx, an increase made possible by the availability of low-cost fluorescent lighting systems and a plentiful supply of electricity. Importantly too, fluorescent lighting complemented the new, sleek, modern architecture by providing high illuminance throughout the office space from fixed positions in the ceiling. No longer was it necessary to place the dim light sources close to the task to ensure getting adequate light. At that time, "experts" began to argue for light levels that "reached for the sun."

By 1972, recommended illuminance for office work in the U.S. had increased to between 500 and 2000 lx. Fluorescent lighting systems and electric energy were very inexpensive, and the research showed that "more light gave better sight," so why not "reach for the sun" in interior lighting? Figure 67.3 shows typical lighting for four 1972 offices. Soon after the energy crisis of 1973, recommended illuminances

FIGURE 67.1 Money order room, New York post office in 1906. (From Woodwell, J.E. [1906], Data on indoor illumination, *Illuminating Eng.*, 1, 645–666. With permission.)

FIGURE 67.2 Drafting under an artificial skylight in 1947. (From IES [1947], *IES Lighting Handbook*, 1st ed., Illuminating Engineering Society, New York. With permission.)

became a central topic of discussion among the "experts," and by 1981 they had dropped to levels similar to those recommended in the late 1940s. In addition, a revolutionary office technology was beginning to emerge. Offices started to move from an age of paper, seen by reflected light, to an age of computer displays that were self-luminous. Early computer terminals, with dark backgrounds and shiny screens, produced poor-quality displays. Reflections of lighting systems — designed to provide high, general illuminance for paper-based tasks — interfered with, rather than enhanced, visibility of alphanumeric

FIGURE 67.3 Four office environments from 1972. (From Kaufman, J.E. and Christensen, J.F., Eds. [1972], *IES Lighting Handbook*, 5th ed., Illuminating Engineering Society, New York. With permission.)

text displayed on the screen. To reduce this problem, both lighting fixtures and screen technology were redesigned to reduce reflected glare.

Current lighting practice demonstrates a continued emphasis on computer tasks and energy efficiency. Figure 67.4 shows a modern, open-plan office with relatively low general lighting (200 to 300 lx) complemented by under-cabinet task lighting that produces 500 lx on the desk. There also has been a growing emphasis on "mood" lighting in commercial spaces because, in today's expanding service environment, face-to-face negotiations often define productivity. The number of independent lighting designers has grown over the past two decades to serve this expanding market. It is not at all uncommon for these new "experts," in a desire to create a pleasing atmosphere, to recommend lower lighting levels than those that were recommended in 1946.

This very brief review illustrates how lighting practice reflects contemporary technology, lighting costs, work organization, and social concerns. Even "expert" opinions reflect the context in which they are made. Yet, it is safe to say that the visual system has not changed at all in the past 100 years. How then does the modern "expert" fairly and competently evaluate lighting conditions?

67.2. Visibility

The answer is to concentrate on the essentials of visibility before considering the social and the technological contexts of the visual environment. The factors that affect good visibility are the same today as they were in 1900. These factors are visual size, luminance contrast, color differences, retinal image quality, and retinal illumination.

The visual size of a target describes how big it is. In general, the larger a target is, defined in terms of the solid angle (in microsteradians, $\mu\omega$) it subtends at the eye, the easier it is to detect. A lowercase letter on this page is between approximately 4 and 5 $\mu\omega$ at a 16-in. (40 cm) viewing distance. Print half this

FIGURE 67.4 A modular workstation with lighting that reflects concerns for energy consumption and computer terminal visibility. (From Rea, M.S., Ed. [2000], *IESNA Lighting Handbook*, 9th ed., Illuminating Engineering Society of North America, New York. With permission.)

size is very difficult to read. Lighting can induce a change in the visual size of the target. People will often lean closer to a task, increasing its visual size, if the illumination on the task is too low, although this may eventually lead to discomfort (Rea et al., 1985). Lighting can also change the visual size of a three-dimensional target directly by casting a shadow, thus increasing the apparent size of the object and its shadow.

The luminance contrast of a target quantifies its luminance relative to its background. The higher the luminance contrast, the easier and faster it is to process the target. A letter on this page has a contrast of approximately 0.7. Contrasts lower than about 0.3 become difficult to see. Lighting can reduce the luminance contrast of a target by producing direct glare into the eyes or reflected glare from the task materials.

Luminance contrast is specified without regard for the apparent colors of the target, its background, or the light emitted from the light source. It is possible to have a target with zero luminance contrast that can still be detected because it differs in color from its background. Lighting can, to some degree, alter both the luminance contrast and the apparent color differences between the target and its background by using light sources of different spectral power distributions. The visibility of achromatic (black and white) targets, like the letters printed on this page, will be largely unaffected by the spectral power distribution, particularly if the color rendering index (CRI) of the source is greater than 70 (e.g., daylight, fluorescent, incandescent lamps). Indeed, only rarely will the color of a light source affect the visibility of targets in the work environment.

The visual system is an image processing system and works best when it is presented with a clear, sharp, focused image. The sharpness of the retinal image is determined by the stimulus itself, by the extent to which the medium through which it is transmitted scatters light, and by the ability of the visual system to focus the image on the retina. Lighting can do little to alter any of these factors; however, vivid colors will sometimes appear to be out of focus or appear at different distances due to chromatic aberrations of the eye. It is beyond the scope of this chapter to discuss causes and remedies to this situation, but, in general, less-saturated colors will be less problematic.

The amount of light entering the eye will largely determine the state of adaptation of the visual system and, therefore, its ability to see detail and color as well as the speed and accuracy with which it can

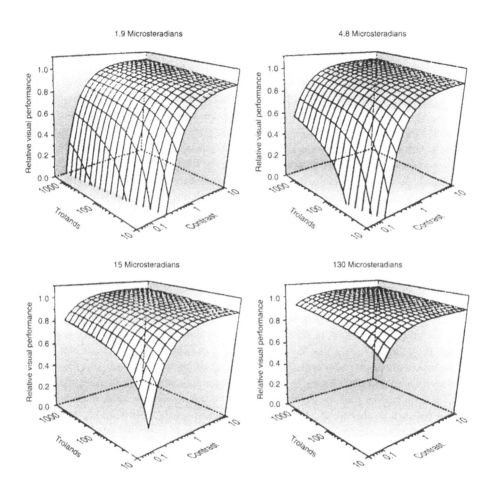

FIGURE 67.5 Relative visual performance plotted as a function of target contrast and retinal illumination for different target sizes. (From Rea, M.S., Ed. [2000], *IESNA Lighting Handbook*, 9th ed., Illuminating Engineering Society of North America, New York. With permission.)

process visual information. Retinal illumination is determined by the luminance in the visual field, modified by the pupil size. In general, the higher the background luminance for a target, the higher is the level of visibility that can be expected. Assuming that this page is being read under the commonly recommended office illumination of 500 lx, the luminance of the white paper is about 125 cd/m². Reading becomes difficult if the background luminance drops below about 50 cd/m². Of course luminance can also reach a level high enough to cause discomfort. It is uncommon to find visual tasks so bright as to cause discomfort, but windows and bare light sources are common sources of discomfort in the workplace. As a rough guide, bare fluorescent lamps can be expected to cause discomfort, and they produce luminances of about 10,000 cd/m². So any luminous object brighter than 10,000 cd/m² can be expected to cause discomfort. Pupil size has very little effect on visibility or discomfort.

From this discussion of significant factors, it is evident that the visibility of a target is not determined by the lighting system alone. The visual abilities of the worker and the inherent visual characteristics of the task are also important. Therefore, the first stage in evaluating a lighting installation is to determine whether the task characteristics are likely to be problematic. The above text gives some guidelines on the levels of light-source color, background luminance, luminance contrast, and visual size below which performance difficulties might be experienced. The rationale behind these guidelines is evident in Figure 67.5. This figure shows the quantitative relationship between background luminance (retinal illumination, in Trolands) and relative visual performance for tasks of different luminance contrast and visual size (Rea and Ouellette, 1991). Although there is a broad plateau of good visual performance, there can be a very rapid deterioration in one's ability to perform visual tasks if the task characteristics approach those described above.

67.3 Examples of Tasks That Require Special Consideration

No single set of lighting recommendations (e.g., Rea, 2000; CIBSE, 2002) can possibly ensure satisfactory levels of visual performance in every work environment (Boyce, 1996). Recommendations represent a consensus based on scientific knowledge, practical experience, assumptions about the visual environment, and, of course, the social and technological contexts in which they were made. In the following subsections, we seek to give some examples of special tasks and areas that would not be adequately addressed by simply complying with recommended illumination levels.

67.3.1 Drafting

Drafting involves seeing and generating fine detail, often of low contrast. For paper-based drafting, the important aspects of the lighting are the amount and distribution of illumination on the task. Usually, a task light, adjustable in position, is used to provide enough light together with a light distribution that eliminates reflected glare from the drafting paper and minimizes shadows cast by the worker's hands and by drafting instruments. Of course, this lighting solution is not suitable for computer-assisted drafting (CAD) work. The problem here is that the high-resolution monitors usually operate with a dark background. This makes the display very sensitive to reflections of light in the room, so much so that it is a common experience to find CAD rooms lit only by the light emitted from the screens. This is a rational solution to the problem, because any light produced in the room will only serve to reduce the contrast of the display, but such an absence of lighting is not helpful where some work on paper is also required. To provide some flexibility in the lighting of CAD rooms, two approaches are possible: either the use of well-shielded task lights at each workstation, or the use of indirect lighting in the whole room. In either case, it is desirable to avoid incident light on the screens and to provide a dimming control so that the workers can adjust the lighting as required (Rea, 1991).

67.3.2 Transilluminated Displays

Another task that requires careful consideration is the viewing of transilluminated displays, such as x-rays in doctors' offices and images in a graphic display business. The problem here is that light incident on the front of the display from the general room lighting reduces the contrast of the displays, making detail more difficult to see. Again, the solution is to provide a means of dimming the general room lighting when necessary.

67.3.3 Conference Rooms

Conference rooms are locations where the perception of facial expression and body language is almost as important as being able to hear what is being said and to see what is being presented. Lighting recommendations intended for general offices do not consider these subtleties. The important factor for revealing facial expression is the direction of the flow of light (Cuttle et al., 1967). If the flow of light is predominantly downward, as occurs when the lighting is provided by recessed downlights, shadows are cast under the eyes, nose and chin, producing an unattractive facial appearance. To eliminate this problem, the illumination on the plane of the face should be of the same order as the illumination on the top of the head. This can be achieved, without creating glare, by using direct/indirect lighting, high-reflectance (80%) room surfaces, and medium reflectance (30 to 50%) work surfaces.

67.3.4 Training Rooms

The training room is a variation on the conference room found in most office buildings. The tasks to be performed here can vary from conventional reading to viewing projected material or television. The other requirement is that the students can see the instructor clearly and that the instructor can see the students clearly. The key characteristic for the lighting of such spaces is flexibility, so that the amount and distribution of light can be easily matched to the task requirements. Often it is desirable to have two separate lighting installations in such spaces, one for tasks that use materials seen by reflection, and another for tasks that use self-luminous materials. Typically, the former will concentrate the lighting onto the working surfaces, while the latter will primarily light the walls. The visual environment of the training room will benefit if both installations can be dimmed.

67.3.5 Video Conferencing

A special facility that is found in increasing numbers is a video-conferencing room. Ideally, video-conferencing enables people in two or more different locations to hold a conference in which both the participants and the materials they are using can be seen by all as if they were present in the same space. What distinguishes a video-conferencing facility from a conventional conferencing room is the limitations of the video technology. Typically, the cameras used have a more limited dynamic range than does the human visual system. This means that any shadows produced by the lighting are exaggerated, and any glare is enhanced when seen through the video system. The solution to this problem is to use indirect lighting and to have high-reflectance matte surfaces in the room, their combined effect being to produce very diffuse lighting, without glare (Carter and Boyce, 2003).

67.3.6 Lobbies

The lobby is the public face of an office building. As such, the lighting has not only to enable the receptionist to see what needs to be seen, but it also has to help to send a "message" about the nature of the business. There is no unique answer to what the lighting should be for lobbies; it will depend on what sort of "message" is desired. Often lighting is used to create luminous points of interest, incorporating color and/or texture, to communicate status or prosperity.

67.4 Checklist

Below is a list of questions to be asked when assessing the lighting requirements for any given space in an office. The answers to these questions can be used to determine whether the conventional recommendations for office lighting are appropriate or whether other factors need to be considered.

1. Are the visual tasks to be done typical of those in an office?
2. Are the tasks located on a desk or workstation?
3. Is it easy to see the detail necessary to do the tasks?
4. Does video-conferencing take place?
5. Is the appearance of people very important?
6. Is the appearance of the space very important?
7. Are the tasks seen by reflected light, or are they self-luminous?
8. Is the display (paper or monitor) glossy?
9. Do the screens have a bright or dark background?
10. Are the surfaces of the space light or dark?

If the answers to questions 1 to 3 are positive and the answers to questions 4 to 6 are negative, the conventional lighting recommendations are usually appropriate.

If the answers to any of questions 1 to 3 are negative or the answers to any of the questions 4 to 6 are positive, the lighting needs special consideration. Generic design advice on lighting suitable for different situations is given in Chapter 10 of the *IESNA Lighting Handbook* (Rea, 2000), while advice for specific applications is contained in Chapter 68 of this handbook. Selection among the alternatives offered there can be based on the answers to questions 7 to 10.

References

Boyce, P.R. (1996), Illuminance selection based on visual performance, and other fairy stories, *J. Illuminating Eng. Soc.*, 25, 41–49.

Carter, C.B. and Boyce, P.R. (2003), Lighting standards for video-conferencing in distance-learning environments, *J. Illuminating Eng. Soc.*, 32, 37–51.

Chartered Institution of Building Services Engineers (2002), *Code for Lighting*, Butterworth Heinemann, Oxford, U.K.

Cuttle, C., Valentine, W.B., Lynes, J.A., and Burt, W. (1968), Beyond the working plane, in *Proceedings of the CIE, 16th Session*, Commission Internationale de l'Éclairage, Paris.

Illuminating Engineering Society (1947), *IES Lighting Handbook*, 1st ed., Illuminating Engineering Society, New York.

Kaufman, J.E. and Christensen, J.F., Eds. (1972), *IES Lighting Handbook*, 5th ed., Illuminating Engineering Society, New York.

Rea, M.S., Ed. (2000), *IESNA Lighting Handbook, Reference & Application,* 9th ed., Illuminating Engineering Society of North America, New York.

Rea, M.S. (1991), Solving the problem of VDT reflections, *Progressive Architecture*, 72, 35–40.

Rea, M.S. and Ouellette, M.J. (1991), Relative visual performance: a basis for application, *Lighting Res. Technol.*, 23, 135–144.

Rea, M.S., Ouellette, M.J., and Kennedy, M.E. (1985), Lighting and task parameters affecting positive, performance and subjective ratings, *J. Illuminating Eng. Soc.*, 15, 231–238.

Woodwell, J.E. (1906), Data on indoor illumination, *Illuminating Eng.*, 1, 645–666.

68

Photometric Characterization of the Luminous Environment

Mark S. Rea
Rensselaer Polytechnic Institute

68.1 Introduction

Photometry is the international system of light measurement, and every organization or technical society concerned with the visual environment employs photometric quantities in their recommendations. Two of the most important of these bodies in North America are the Commission Internationale de l'Éclairage (CIE) and the Illuminating Engineering Society of North America (IESNA).

Photometric quantities are characteristics of the physical world that can be measured by calibrated instruments. Photometry is fundamentally rooted in human vision, and without human vision, photometric quantities would have no meaning. The circularity between physical measurement of the luminous environment and human visual perception has been a great source of confusion to the users of photometry. Users often implicitly assume that photometric quantities are synonymous with visual perception and are surprised when careful photometric measurements do not correlate well with subjective impressions of light. Discrepancies are common, for example, between brightness perception and photometric measurements. This chapter differentiates and explains different photometric quantities and describes their strengths and limitations in characterizing visual environments.

68.2   Photometric Quantities

68.2.1   Luminous Flux

The total amount of light, or luminous flux, emitted by a lamp (lightbulb) is measured in lumens. Manufacturers must print the "rated lumens" produced by a given lamp type on the package. The rated lumen value is the average amount of light produced by a large sample of lamps of that type when operated at their rated voltage. In general, a light source with higher rated lumens will make a space look brighter than one with lower rated lumens, but, as discussed later, this is not always the case. Figure 68.1 gives typical rated lumens for different commercially available light sources. Lumen values are always obtained in a photometric laboratory and are never measured in the field.

68.2.2   Luminous Intensity

Two sources emitting the same number of lumens can have very different intensity distributions, measured in candelas (lumens per solid angle). Some light sources emit light equally in all directions and have the same luminous intensity in all directions. Other light sources, such as an automobile headlamp, emit most of their light in a particular direction. These latter types of sources will have a higher luminous intensity in one direction than at other angles. Figure 68.2 shows luminous intensity diagrams for several types of luminaires (light fixtures) used in commercial environments. Like luminous flux, luminous intensity values are always obtained in the laboratory.

68.2.3   Illumination

Illumination, measured in lux or footcandles, is the density of luminous flux falling on a surface area, and is the most common photometric quantity found in lighting recommendations. For a diffusely reflecting object (paper, painted walls, carpet, wood, brick, etc.), the higher the illumination falling on a surface, the brighter the object will appear. Objects that are hard to see, such as small print on paper, will generally be associated with high recommended illumination values. Figure 68.3 shows examples from the current IESNA recommendations for different types of visual tasks. Unlike values of luminous flux and luminous intensity, illumination values are almost always obtained in the field.

68.2.4   Luminance and Luminance Contrast

Of all the photometric quantities, luminance — measured in nits (candelas per square meter) or footlamberts (1/candelas per square foot) — is best correlated with brightness perception. The higher the illuminance falling on a surface, the greater is the luminance of that surface. For a given illuminance, a material with a greater reflectance (e.g., white paper) will have a higher luminance than a material with

	Lumens (lm)	Luminance (cd/m2)
Sun (at noon)	3×10^{28}	1.6×10^{9}
Full moon (at zenith)	7.5×10^{16}	2.5×10^{3}
Candle flame	12	1.0×10^{4}
60-W frosted incandescent lamp	1060	1.2×10^{5}
T-8 (1-inch diameter) fluorescent lamp	2850	1.1×10^{4}
T-12 ($1\,^{1}/_{2}$-inch diameter) fluorescent lamp	2475	8.2×10^{3}

FIGURE 68.1 Light source lumens and luminances. (From Rea, M.S., Ed. [2000], *IESNA Lighting Handbook: Reference and Application*, 9th ed., Illuminating Engineering Society of North America, New York. With permission.)

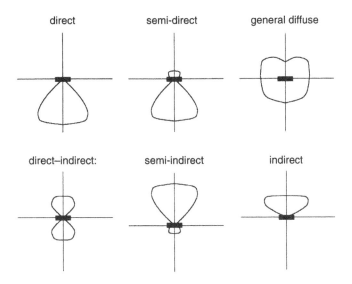

FIGURE 68.2 Luminous intensity distributions for different types of light fixtures. Every black rectangle represents a light fixture centered along vertical and horizontal axes. The solid lines represent the luminous intensity distribution emitted from the different types of fixtures. All but the "direct" fixture would be mounted below the ceiling plane. (Image courtesy of the Lighting Research Center, Rensselaer Polytechnic Institute.)

Application	Horizontal Illuminance (lx)	Vertical Illuminance (lx)
Health Care Facilities		
Operating table	3,000–10,000	500
Scrub room	1,000	300
Waiting areas, general	100	30
Waiting areas, reading	300	50
Offices		
Filing	500	100
Lobby, lounge, reception area	100	30
Mail sorting	500	30

FIGURE 68.3 Example illuminance recommendations. (From Rea, M.S., Ed. [2000], *IESNA Lighting Handbook: Reference and Application*, 9th ed., Illuminating Engineering Society of North America, New York.)

a low reflectance (e.g., gray paper). Equation 68.1 defines the relationship between illuminance and luminance for a matte (diffusely reflecting) surface.

$$L = E\rho/\pi \qquad (68.1)$$

where
L = luminance in nits (cd/m²)
E = illuminance in lux
ρ = reflectance of the object in the direction of measurement or view

Many surfaces exhibit some gloss or sheen, and because luminance is dependent upon the geometric relationship between the light source, the surface, and the viewing angle, it is often important to stipulate the location of the luminance measurement. Figure 68.1 lists the luminances of various light sources.

Luminance is also used in quantifying luminance contrast. The luminance contrast of a target against its background is particularly important for human vision and can be defined in at least two ways. Luminance contrast of periodic stimuli, such as gratings used to determine contrast sensitivity, is usually defined as:

$$C = (L_{max} - L_{min})/(L_{max} + L_{min}) \qquad (68.2)$$

where
C = luminance contrast
L_{max} = maximum luminance of the grating
L_{min} = minimum luminance of the grating

The luminance contrast of small objects against a large background, such as alphanumeric text on white paper, is usually defined as:

$$C = (L_b - L_t)/L_b \qquad (68.3)$$

where
C = luminance contrast
L_b = background luminance
L_t = target luminance

Like illuminance measurements, luminance measurements are usually made in the field. Although some bodies making lighting recommendations promulgate luminance measurements, they are much less common than illuminance recommendations, in part because the instrumentation for making accurate luminance measurements is relatively expensive, and in part because this quantity is often directionally sensitive. When luminance values are recommended, they are usually offered in terms of luminance ratios for different parts of the visual field, assuming matte surfaces. Figure 68.4 provides

Angle from Vertical (degrees)	Preferred Luminance (cd/m²)	Maximum Luminance (cd/m²)
55	850	—
65	350	850
75	175	350
85+	175	175

FIGURE 68.4A Luminance recommendations for luminaires in offices with computers. (From Rea, M.S., Ed. [2000], *IESNA Lighting Handbook: Reference and Application*, 9th ed., Illuminating Engineering Society of North America, New York.)

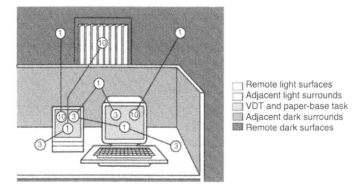

FIGURE 68.4B Luminance ratios for computer work areas. When the circled values are connected by a line, the values represent the maximum recommended luminance ratio between the surfaces identified by the circles. For example, the luminance of the computer screen should not exceed ten times the luminance of the far wall. (From Rea, M.S., Ed. [2000], *IESNA Lighting Handbook: Reference and Application*, 9th ed., Illuminating Engineering Society of North America, New York.)

IESNA-recommended luminances and luminance ratios for office lighting with computer terminals. Luminance contrast values are rarely if ever recommended, even though contrast is arguably the most important physical stimulus for vision. The primary reason for this omission is probably because contrast is very hard to measure accurately (see Section 68.4.3).

68.3   Inherent Spectral Responses of All Photometric Quantities

All photometric quantities weight radiant power in the electromagnetic spectrum according to idealized spectral sensitivities of different populations of photoreceptors in the human retina. Figure 68.5 shows the two spectral weighting functions, the photopic and scotopic luminous efficiency functions, officially sanctioned by the CIE for use in photometry. The photopic function (V_λ), representing the spectral sensitivity of cones in the human fovea, is shown on the right, and the scotopic luminous efficiency function (V'_λ), representing the spectral sensitivity of rods in the human retina, is shown on the left. The photopic luminous efficiency function is used almost exclusively in lighting recommendations and in commercially available photometric instruments.

The photopic luminous efficiency function has proved to be very useful since it was adopted in 1924 because most visual environments illuminated by natural or electric light sources are illuminated to levels where the cones in our fovea function quite well. However, the ubiquity of the photopic luminous efficiency function invites discrepancies between photometry and human perception. The scotopic luminous efficiency function was adopted in 1951 as a partial response to this problem, but the scotopic luminous efficiency function is never used in photometric instruments because human visual environments are only very rarely illuminated to light levels so low that only rods are active. It must be stressed, however, that accurate photometric measurements based either upon the photopic or the scotopic functions can be performed at any light level because, again, photometry is a physical characterization of the visual environment. Obviously, however, the utility of photometric measurements for understanding human perception will vary considerably, depending upon the light level in the visual environment.

The relevance of photometry for analyzing human perception is, unfortunately, much more complicated than simply keeping track of light levels. Some of the more important discrepancies will be discussed here.

The photopic luminous efficiency function does not adequately capture the spectral sensitivity of the human cones except under certain conditions. Cones in the peripheral retina, for example, have a different spectral sensitivity to light than the cones in the fovea; peripheral cones are more sensitive to short wavelength (blue) light. More significantly, however, perceptions of brightness can be radically different than would be inferred from photometric measurements.

As previously noted, luminance is the photometric quantity most closely associated with perceived brightness. Luminance is not synonymous with brightness, however. In general, for "white" light sources

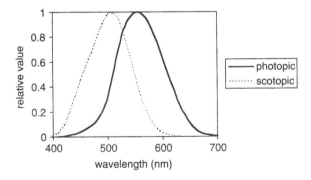

FIGURE 68.5   Photopic and scotopic luminous efficiency functions.

used in commercial applications, the higher the correlated color temperature (CCT), the brighter the light emitted by the lamps will appear for the same luminance. Yellow-white light sources have a low CCT (2500 to 3500K), whereas blue-white light sources will have a high CCT (4000 to 6500K). Confusingly, lamps with a low CCT are referred to as having a "warm" appearance, and those with a high CCT are referred to as being "cool" in appearance. This confusion probably stems from the observations that open flames, appearing yellow, feel "warm" to the touch, whereas daylight, appearing blue, feels "cool" through a window. In any event, Figure 68.6 shows the relationship between perceived brightness and luminance for lamps of different CCTs (Alman, 1977).

It is very important to point out that perceived brightness does not correlate well with visual performance. Luminance, and thus luminance contrast, defined in terms of the photopic luminous efficiency function, seems to be a robust rectifying variable for describing stimuli important to visual performance. Visual tasks performed in most commercial environments rely heavily upon processing by the retinal foveae. The photopic luminous efficiency function that underlies luminance measurements characterizes the spectral sensitivity of the fovea and, thus, visual performance is largely independent of the CCT of the light source illuminating the visual task, as long as the light levels provided on the task are equivalent. In other words, the CCT of the light source can have a dramatic effect on perceived brightness in a space, but it will have little or no effect on visual performance as long as the luminance and contrast of the visual task is constant.

Photophobia is an aversion to bright light and is probably related to perceived brightness. Continuously viewing a light source with a luminance higher than 10,000 cd/m² can cause damage to the retina (Sliney and Wolbarsht, 1980). Presumably, the discomfort experienced by people looking at sources this bright, and brighter, is an overt response to protect the retina from damage. "Cool" light sources generally have more energy that can cause damage to organic materials than "warm" ones. "Cool" light sources are also generally seen as more glaring than warm light sources for the same luminance, as might be expected from Figure 68.6.

68.4 Measurement Guidelines

68.4.1 General Comments

Photometric measurements are indispensable for characterizing the visual environment. Without physical measurements, there can be no sound basis for diagnosis and improvement of this environment. Photometric measurements have inherent limitations, however, and these should be considered when drawing inferences from the values obtained from photometric instruments. Naturally, every instrument will be an imperfect embodiment of photometric ideals such as spectral and spatial sensitivity. These limitations,

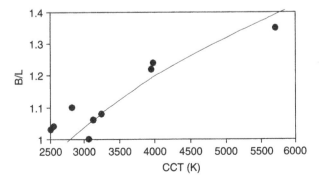

FIGURE 68.6 Brightness/luminance ratios of light sources as a function of correlated color temperature (CCT). The reference light source was 2856K. (From Alman, D.H. (1977), *J. Illuminating Eng. Soc.*, 7, 55–62.)

while important, are generally second-order phenomena when compared with the proper utilization of the instrument during measurement. The following section provides some guidelines for measuring illuminance and luminance. More significantly, however, is the fact that measurements of illuminance and luminance are poor surrogates for representing different degrees of visibility or comfort. Users of photometric data must acknowledge that the richness of the human visual system can never be reduced to a handful of photometric measurements. Photometry does, however, form the foundation for any meaningful discussion of the stimuli for human visual perception.

68.4.2 Illuminance

Illuminance provides a crude representation of light level in the visual environment. In general, as light level increases, adaptation level increases and humans can see objects of finer detail and can process visual information faster and more accurately. For this reason, high illuminance is recommended for difficult (e.g., threading a needle) or critical (e.g., brain surgery) visual tasks. Since most visual tasks experienced in commercial environments are relatively large and are produced in high contrast (Dillon et al., 1987), illuminance measurements are a simple, practical, if somewhat inaccurate, way of characterizing adaptation level.

One of the reasons that illuminance is the most common quantity used in lighting recommendations is that instruments that measure illuminance are relatively inexpensive. Figure 68.7 shows several illuminance meters. Properly designed and manufactured illuminance meters will have a spatial sensitivity to light that follows a cosine distribution. This means that light reaching the device from angles nearly parallel to the surface will have little effect on the illuminance reading, whereas those incident perpendicular to the surface will have the greatest effect. Figure 68.8 shows the cosine spatial sensitivity of a well-designed illuminance meter.

Illuminance measurements are usually specified for the assumed location of the visual task, often a horizontal desk surface. One measurement is usually not enough to properly characterize the illuminance on the task area. The IESNA, for example, is quite explicit as to the number and location of measurements that need to be taken to represent the light levels in a space (Rea, 2000). In practice, however, only a few "representative" values are usually obtained. The paucity of measurements is usually sufficient to give a general indication of light levels because, again, illuminance is inherently a crude indicator of adaptation level.

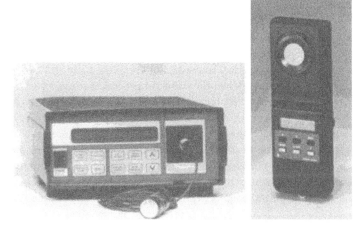

FIGURE 68.7 Illuminance meters. (From Rea, M.S., Ed. [2000], *IESNA Lighting Handbook: Reference and Application*, 9th ed., Illuminating Engineering Society of North America, New York.)

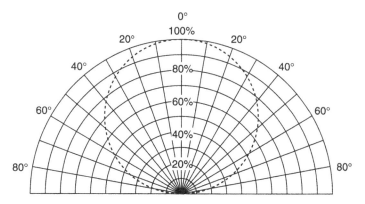

FIGURE 68.8 Cosine response of a high-quality illuminance meter. This figure shows that light incident on the illuminance meter at, for example, 50° will be approximately 65% as effective as the same light intensity at 0°. (Image courtesy of the Lighting Research Center, Rensselaer Polytechnic Institute.)

Illuminance measurements are generally obtained with a handheld instrument. When measuring illuminance, it is important to recognize its cosine spatial sensitivity (Figure 68.8). A common mistake made in measuring illuminance is holding the instrument too close to the body or, worse, placing part of one's body between a source of illumination and the instrument. Ideally, the instrument should be placed on a tripod and illuminance values read from a remote display. Again, illuminance is only a crude surrogate for adaptation level, so deviations of 20 to 50% from recommended illuminance (e.g., Figure 68.3) are probably not important for most applications, assuming that precautions have been made to minimize gross errors in measurement of the type described above.

An important consideration in measuring illuminance is the orientation of the instrument. Usually the illuminance meter will be placed on a horizontal surface at the height assumed or specified in the recommendations (e.g., on the work surface). However, it should be realized that the illuminance on vertical surfaces in the field of view is also important and occasionally recommended (Figure 68.3). In general, recommendations of vertical illuminance are lower than those recommended for horizontal illuminance because electric light sources in the ceiling direct most of their light in a downward direction. Vertical illuminance is specified to minimize shadows on faces, shelves, and important vertical tasks.

68.4.3   Luminance and Luminance Contrast

Luminance is a better indicator of adaptation level than illuminance because it includes both the amount of light falling on the task and the proportion of light reflected from the task. Adaptation level is not the only variable important for predicting visual response, however. Target size and luminance contrast are two of the most important stimulus variables for vision. In general, it is important to provide a visual environment that avoids small, low-contrast targets seen against a low-luminance background. This can be accomplished with magnifiers, properly designed and maintained computer displays (Rea, 1991) and photocopiers, as well as lighting fixtures that provide recommended illuminance levels without producing direct or reflected glare. illuminance measurements are simply too crude to assess direct and reflected glare, so luminance measurements must be made to assess visibility losses. An important consideration when making luminance measurements is to record the position from which the measurements are being made, because of the geometric relationship between the light source, the task, and the observer, as well as the directional reflection characteristics of many surfaces in architectural spaces.

There are no commercially available instruments that measure brightness, but there are many that measure luminance. Instruments that measure luminance are more expensive than those that measure illuminance because they require image-forming optics. Figure 68.9 shows a handheld luminance meter. The luminance of an object is obtained after the instrument operator focuses it through the viewfinder.

FIGURE 68.9 Luminance meter. (From Rea, M.S., Ed. [2000], *IESNA Lighting Handbook: Reference and Application*, 9th ed., Illuminating Engineering Society of North America, New York.)

Proper focus is extremely important for accurate luminance measurements. However, image formation is imperfect in all luminance meters, so stray light (defocus) and scattered light within the instrument (internal reflections) will always influence the photometric reading. Usually the size of the light-sensitive area in the viewfinder of the luminance meter is, nominally, one degree of visual arc (1°). Several instruments have smaller light-sensitive areas (e.g., 1/3°), and some can be fitted with close-up lenses that provide the operator with an opportunity to measure the luminance of very small objects, such as alphanumeric text. Supplemental lenses will add further to stray light and scattered light, however.

Scattered light and stray light can be a serious problem for luminance measurements. If all objects within the instrument's field of view are about the same, luminance measurements can be quite accurate. Scatter and defocus are much more serious problems when it is important to measure the contrast of small, dark objects on a bright background, such as alphanumeric text on white paper. Scattered light and stray light from the white background can increase the measured luminance of the dark text. Although the human eye also suffers from scattered light and several optical aberrations that increase stray light, there are a number of anatomical and neurological processes in the retina that significantly reduce their impact on image contrast. No analogous processes are available in a luminance meter, so it is common to underestimate the perceived contrast of small targets with luminance contrast measurements. Placing black cloth around the recording area can significantly reduce scattered light and thus increase the measured contrast of the target. Still, the measured contrast may not necessarily reflect perceived contrast. Indeed, the relationship between measured contrast and perceived contrast is always uncertain.

68.5 Final Comment

Modern lighting equipment is much more sophisticated than ever before. This sophistication has made a marked improvement in visual environments by reducing glare and flicker and increasing color perception. Just as importantly, computer screen technology and print quality have improved significantly. As a result, good visual environments are becoming more and more common. Of course, modern technology does not always ensure good applications. It is still possible to find reflected glare and shadows in many visual environments. Moreover, there are many visual tasks that are inherently difficult to see and require special care and planning of the luminous environment. In every case, however, from the routine to the unusual, photometry has been needed to design, specify, and implement satisfactory luminous environments. Indeed, photometry is the only meaningful language that can be used to describe the liabilities and the assets of luminous environments. Photometry, then, despite its inherent uncertainties, is the foundation for achieving excellent visual environments.

Acknowledgment

Technical support was provided by my colleague John D. Bullough with the Lighting Research Center.

References

Alman, D.H. (1977), Errors of the standard photometric system when measuring the brightness of general illuminance light sources, *J. Illuminating Eng. Soc.*, 7, 55–62.

Dillon, R., Pasini, I., and Rea, M.S. (1987), Survey of visual contrast in office forms, in *Proceedings of the CIE*, Commission Internationale de l'Éclairage, Vienna, Austria.

Rea, M.S., Ed. (2000), *IESNA Lighting Handbook: Reference and Application*, 9th ed., Illuminating Engineering Society of North America, New York.

Rea, M.S. (1991), Solving the problem of VDT reflections, *Progressive Architecture*, 72, 35–40.

Sliney, D. and Wolbarsht, M. (1980), *Safety with Lasers and Other Optical Sources*, Plenum Press, New York.

69

Evaluating Office Lighting

Peter R. Boyce
Rensselaer Polytechnic Institute

69.1 Introduction

The method presented here allows the user to make rapid on-site office lighting evaluations and to compare the results against normative criteria. The method also is inexpensive and easy to use. It can be used to identify poor-quality lighting, for troubleshooting, and as an element in a performance-contracting system.

69.2 Some Philosophy

The method presented here is comprehensive and, in its most complete form, it allows the measurement of occupants' opinions about the lighting, photometric quantities, power metrics, and the state of maintenance (Boyce and Eklund, 1995). It is designed to be flexible, in that any or all of these aspects can be evaluated. Here, attention is limited to the measurements of occupants' opinions and photometric quantities. No attempt has been made to combine various measures into a single number despite the obvious attractions of a single-number index of lighting quality (Bean and Bell, 1992). Single-number indices are simple to use, but they inevitably involve a loss of information, which impedes understanding.

The purpose of office lighting is to enable people to do their work quickly, accurately, and in comfort. Peoples' opinions of office lighting, although labile, are of prime importance. To avoid biased conclusions, the method uses a converging-operations approach to interpret the information collected by comparing subjective lighting assessments with the appropriate photometric quantities. For sound conclusions, peoples' opinions and the photometric measurements should be congruent.

TABLE 69.1 Normative Data for Each Statement in the Office Lighting Survey

Statement	Average Percentage Agreeing
Overall, the lighting is comfortable	69
The lighting is uncomfortably bright for the tasks I perform	16
The lighting is uncomfortably dim for the tasks I perform	14
The lighting is poorly distributed here	25
The lighting causes deep shadows	15
Reflections from the light fixtures hinder my work	19
The light fixtures are too bright	14
My skin is an unnatural tone under the lighting	9
The lights flicker throughout the day	4
How does the lighting compare to similar workplaces in other buildings?	
Worse	19
About the same	60
Better	22
Reading 8 point print and larger	99
Reading 6 point print and larger	94
Reading 4 point print and larger	76

Note: Results based on responses from 1259 office workers working in 13 different offices.
Source: Eklund, N. and Boyce, P. (1996), The development of a reliable, valid and simple office lighting survey, *J. Illuminating Eng. Soc.*, 25, 25–40.

69.3   The Office Lighting Evaluation Toolkit

69.3.1   Tools

The tools needed to carry a simple evaluation of an office lighting installation are:

- A copy of the office lighting survey
- An illuminance meter
- A piece of white office paper laminated between two sheets of transparent plastic
- A small plane mirror

69.3.2   The Office Lighting Survey

The office lighting survey consists of a postcard-sized questionnaire designed to measure occupants' visual comfort and reading performances. On one side are a series of statements with which the occupants express their agreement or disagreement (Table 69.1). Tests of the office lighting survey in six office buildings have demonstrated that the survey is reliable and valid (Eklund and Boyce, 1996). The normative criteria against which the results of the office lighting survey in a particular building are compared were obtained from the responses of 1259 office workers occupying 13 different office buildings in the northeastern U.S., covering a range of lighting installations of various ages (Table 69.1).

The other side of the office lighting survey has a reading test that uses sentences printed in 10-, 8-, 6-, and 4-point print. All people have to do is to indicate the smallest print they can read. This test is based on the finding that printing of twice the size of the threshold visual acuity will be easily read (Bailey et al., 1993). For good-quality lighting, 95% of occupants should be able to read 6-point print or larger. This percentage is based on the results of the reading test carried out by 252 occupants of five buildings.

Although the office lighting survey is usually presented in a single-postcard format, with tear-out wedges for indicating the response,[1] there is no reason why it should not be presented in other formats,

---

[1] Available from Lighting Research Center, 21 Union Street, Troy, NY 12019.

provided that the same wording of the statements is used and the answers are given as a simple agree-or-disagree response.

69.3.3 The Illuminance Meter

The illuminance meter to be used should be cosine and color corrected, and have sufficient sensitivity for the illuminances likely to be found in offices (300 to 700 lx).

69.3.4 Laminated Paper and Mirror

The transparent plastic sheet covering the white paper makes the surface specular and hence capable of producing veiling reflections. If veiling reflections are seen on the paper, then the source of the reflections can be revealed by placing the mirror on the veiling reflections.

69.4 Method

Before starting:

- Define the office area to be evaluated.
- Define the people whose opinions are of interest.
- Define the operating state of the lighting installation during the evaluation by asking such questions as:

  Is daylight present?

  If daylight is present, what lighting condition is of interest: lighting during daytime or after dark or both?

  Is the lighting installation well maintained? If not, as shown by the presence of dirty luminaires and failed lamps, should the luminaires be cleaned and the failed lamps replaced before starting the evaluation?

  If there is an automatic switching or dimming system, what state should it be in during the evaluation?

  If the occupants have some control over their lighting — by positioning, switching, or dimming of task lights — what should be done about it?

The answers to these questions will determine whether more than one evaluation is required in the same space and whether the evaluation will answer the questions that it is supposed to answer. Having determined the lighting conditions to be evaluated, proceed with the following steps.

69.4.1 Step 1

Administer the office lighting survey. The method of distributing and collecting this should be decided on the basis of the number of participants, convenience, the need to ensure confidentiality, and the acceptability of the method to the participants and the organization employing them.

69.4.2 Step 2

Before starting any photometric measurements, be sure the lights have been on for at least 20 min. Given that most office lighting is designed to produce a uniform illuminance over the whole office, there is usually no need to make photometric measurements at every workstation. If there are fewer than ten workstations in the area being evaluated, take photometric measurements at each workstation. If there are more than ten workstations, take photometric measurements at a representative sample of workstations, the minimum number being ten. Sit at each workstation and identify the main working area. Measure the illuminance on the work surface. This can be done using a grid of points at 30 cm (1 ft) separation, arranged so as to cover the working area. Calculate the average illuminance and then the ratio of the minimum to the average illuminance for each workstation. This ratio quantifies the illuminance uniformity for the working area of each workstation. Calculate the overall mean illuminance for all the workstations from the average illuminance for each workstation. The ratio of the lowest average illuminance for any workstation to the overall mean illuminance for all the workstations quantifies the illuminance uniformity between workstations.

69.4.3 Step 3

While sitting at the workstation, place the laminated paper on the desk where the occupant would typically place reading materials. If a conspicuous, bright image of a luminaire or a window is apparent in the paper, then veiling reflections are likely to be present when glossy materials are used. If veiling reflections are present, place the mirror on the location of the reflection, and look into the mirror. Whatever is seen in the mirror will be the source of that veiling reflection. If the workstation has a computer, turn the computer system on, sit at the workstation as if you were using the computer, and look for a conspicuous, bright image of a luminaire or a window on the screen. If one is apparent, place the mirror on the image and look into it. Whatever is seen in the mirror will be the source of that image.

69.4.4 Step 4

Identify the light sources used in the lighting installation. To do this, it may be necessary to open the luminaire. The value of this information is that it can be used to identify the color-rendering properties of the light source. Fluorescent lamps, which are by far the most common light source used in offices, usually have a manufacturer's name and code, e.g., F32T8/735, printed on one end. From this code and the appropriate manufacturer's catalogue, the color-rendering index (CRI) of the light source can be identified. As for the probability of discomfort glare occurring, this can be crudely identified by looking at the type of lamp and luminaire. Figure 69.1 provides a simple flow chart to determine if discomfort glare is likely for different lamp/luminaire types.

69.5   Criteria

Table 69.2 lists the metrics that can be derived from the measurements made, the criteria that they should meet for the lighting installation to be considered satisfactory, and the origins of those criteria.

69.6   Interpretation

If a lighting installation passes all the criteria listed in Table 69.2, then it can be considered successful. However, if one or more of the metrics do not pass the relevant criterion, the lighting may be considered unsuccessful and remedial action may be necessary. What form any remedial action should take depends on which criterion is not met. The sections below provide some guidance on how to interpret and check a " fail" on every criterion.

69.6.1 Occupant Comfort

If the installation fails on the percentage of occupants who find the lighting comfortable, the installation is worse than average. For suggestions about what needs to be changed to improve the lighting, examine the answers given to the other statements on the office lighting survey (Table 69.1). These answers will indicate what aspects of lighting need to be improved.

69.6.2 Visual Performance

If the installation fails on the percentage of occupants who can read 6-point type, examine the results for the "percentage of occupants who find the lighting comfortable." If this metric passes, no further

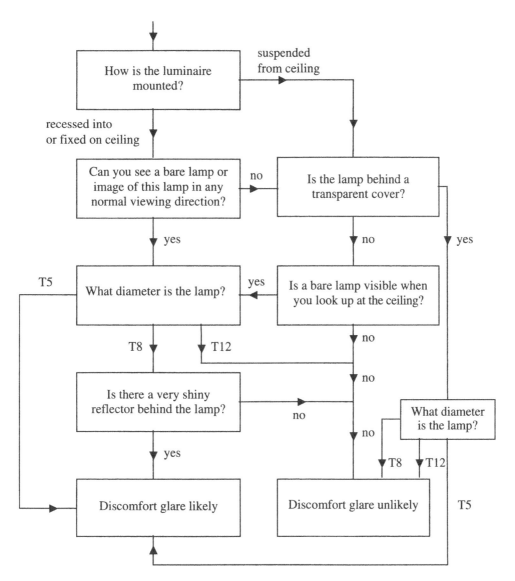

FIGURE 69.1 A flow chart for identifying fluorescent lamp/luminaire combinations likely to produce discomfort glare. The answers are given by an observer sitting at a workstation. For identification purposes, the diameters of the three most common fluorescent lamp types are T12 = 38 mm (1.5 in.); T8 = 25 mm (1.0 in.); T5 = 16 mm (0.63 in.).

action is required. If this metric also fails, examine the answers given to the other questions on the office lighting survey (Table 69.1). If the lighting is described as "uncomfortably dim" by a significant number of occupants, consider how to increase the average illuminance on the workstations.

69.6.3 Overall Mean Illuminance

If the installation fails on the overall mean illuminance at the workstations, examine the results for the "percentage of users who find the lighting comfortable." If this metric passes, no action is needed. If this metric also fails, examine the answers given to the other questions in the office lighting survey. If the lighting is considered "uncomfortably dim" by a significant number of occupants, consider how to increase the average illuminance on the workstations.

TABLE 69.2 Metrics, Methods of Measurement, the Criteria Used, and the Source of Each Criterion

Metric	Measurement Method	Criterion to Pass	Criterion Source
Percent finding the lighting comfortable	Office lighting survey questionnaire	>70%	normative data (Table 69.1)
Percent reading 6-point print and larger	Office lighting survey reading test	>95%	normative data (Table 69.1)
Overall mean illuminance	Mean of the average illuminance measured on the sample of workstations	variable according to situation	Lighting Handbook (IESNA, 2000)
Illuminance uniformity between workstations	Minimum average/overall mean illuminance for the sample of workstations	>0.8	CIBSE Code for Interior Lighting (CIBSE, 1994)
Illuminance uniformity across workstations	Number of workstations with a minimum/average illuminance ratio <0.7	0	(Boyce and Slater, 1990)
Number of workstations with veiling reflections	Observation of conspicuous veiling reflections in specular material	0	—
Number of VDT monitors with veiling reflections	Observation of conspicuous reflections in monitor	0	—
Color rendering	Number of lamp types with a color-rendering index <70	0	CIBSE Code for Interior Lighting (CIBSE, 1994)
Discomfort glare	Number of inappropriate lamp/luminaire types	0	Lighting Handbook (IESNA, 2000)

69.6.4 Illuminance Uniformity and Veiling Reflections

If the installation fails on the illuminance uniformity across a workstation or between workstations, the number of workstations with veiling reflections, or the number of computer screens with veiling reflections, examine the result for "percentage of occupants who find the lighting comfortable." If this metric passes, no further action is required. If this metric also fails, examine the answers given to the other statements on the office lighting survey. If the lighting is considered "poorly distributed" or as causing "deep shadows" by a significant number of occupants, consider how to improve the light distribution on and around the workstations.

69.6.5 Color-Rendering Index and Discomfort Glare

If the installation fails by having lamp types with a color-rendering index <70 or by having lamp/luminaire types likely to produce discomfort glare, examine the answers given to other statements in the office lighting survey. If a significant number of occupants indicate that the lighting makes their skin look an unnatural tone, consider installing lamps with higher CRI values. If a significant number of occupants describe the luminaires as "too bright," discomfort glare is probably occurring. Consider how to reduce the luminance of the luminaires.

69.7 Caveats

Like all "one-size-fits-all" methods of evaluation, this method will inevitably fail to detect a problem in some situations. It has been developed to cope with the most common types of office spaces, where people work on paper- or screen-based tasks, at desks or in cubicles, with electric lighting that uses fluorescent lamps and that is designed to produce the same illuminance over the whole office. It is not appropriate for other spaces found in offices where the working requirements are very different, e.g., conference rooms and CAD (computer-aided design) rooms.

Note that following this method does not excuse the user from thought or from making observations. This latter is particularly valuable, as the presence of any ad hoc modifications to the lighting, such as lamps removed or paper taped over windows, is a clear indication that there is something wrong with the lighting.

References

Bailey, I., Clear, R., and Berman, S. (1993), Size as a determinant of reading speed, *J. Illuminating Eng. Soc.*, 22, 102–117.

Bean, A. and Bell, R. (1992), The CSP Index: a practical measure of office lighting quality as perceived by the office worker, *Lighting Res. Technol.*, 24, 215–226.

Boyce, P.R. and Eklund, N.H. (1995), Evaluating lighting quality, in *Proceedings Right Light Three Conference*, Northern Power, Newcastle-upon-Tyne, U.K.

Boyce, P. and Slater, A.I. (1990), Illuminance uniformity on desks, *Lighting Res. Technol.*, 22, 165–174.

CIBSE (1994), *CIBSE Code for Interior Lighting, 1994*, Chartered Institution of Building Services Engineers, London, U.K.

Eklund, N. and Boyce, P. (1996), The development of a reliable, valid and simple office lighting survey, *J. Illuminating Eng. Soc.*, 25, 25–40.

IESNA (2000), *Lighting Handbook*, 9th ed., Illuminating Engineering Society of North America, New York.

70

Rapid Sound-Quality Assessment of Background Noise

Rendell R. Torres
Rensselaer Polytechnic Institute

70.1 Background and Applications

Many methods for measuring noise typically focus on sound *level*, as level is a general indicator of how disturbing a noise may be. One of the most common indices is the A-weighted decibel level, dB(A), which uses a frequency-dependent weighting function to account for the ear's frequency-dependent sensitivity to sound (Harris, 1998). However, there are few simple methods (i.e., not requiring special equipment or signal processors) that indicate sound *quality*, in particular, the spectral characteristics (or the relative levels of low, mid, and high frequencies) of a given noise signal. As discussed in the next section, a set of noise spectra with the same dB(A) level may each be perceived very differently with respect to annoyance or acceptability.

The measurement method presented in this chapter describes both sound level and sound quality by indicating the relative balance of low-frequency noise to high-frequency noise within a given space (Blazier, 2002). Low-frequency noise can be a significant problem, even though human hearing is less sensitive to low-frequency sounds, as shown by equal-loudness-level contours (Harris, 1998). Because the loudness contours crowd closer together at lower frequencies, however, we see that fluctuations in level at lower frequencies (e.g., 63 Hz and below) are perceptually larger than similar fluctuations at higher frequencies (e.g., near 1 kHz) and are thus more sensitively perceived. Moreover, the corresponding annoyance or disturbance by these fluctuations is also greater.

The proposed method is fast and simple enough to be used by laypersons, i.e., workers without significant acoustical expertise. It can be applied to any situation that requires an approximate but indicative estimate of noise level and sound quality. It assumes relatively continuous (as opposed to transient) noise signals and only requires a sound-level meter with spectral filters corresponding to standard A- and C-weighting. For additional discussion, refer to Blazier (2000, 2002) and the ASHRAE Handbook (1999, 2001).

70.2   Procedure

There are three steps in the (C − A) technique:

1. Measure the A-weighted dB(A) level.
2. Measure the C-weighted dB(C) level.
3. Calculate the difference: dB(C) − dB(A).

If the difference is

- 10 dB or less, the noise will likely be perceived as hiss-like noise
- ≈15 dB, the noise will be perceived as neutral
- 20 dB or greater, the noise will likely be perceived as rumble noise

70.2.1   Step 1

Set the sound-level meter to A-weighting and record the value. The A-weighted decibel level, dB(A), indicates the overall sound level in a room by using a weighting filter to represent the frequency-dependent sensitivity of human hearing. Moreover, it reflects the fact that our ears perceive sounds more sensitively at mid frequencies (e.g., around 1 kHz) than at lower frequencies (e.g., below 125 Hz). The acceptable A-weighted level depends on the usage of a space, but as a rule of thumb, it should not exceed about 45 dB(A) (Blazier, 2002). More critical spaces such as performance halls or classrooms usually require more stringent noise control (and thus more complex descriptors than single-value dB(A) levels). To illustrate the range of noise levels tolerated in a variety of spaces, Table 70.1 lists some typical (but not necessarily standardized) preferred values (Egan, 1988).

70.2.2   Step 2

Set the sound-level meter to C-weighting and record the value. The C-weighted level is a "flatter" standardized weighting filter; moreover, the weighting factors have less variation over frequency.

70.2.3   Step 3

Compute the difference dB(C) − dB(A). The difference between the dB(C) and dB(A) values indicates the sound quality (or tonal character) of the noise and is referred to here by the condensed notation (C − A), expressed in units of dB. As shown in Table 70.2, because C-weighting has a "flatter" frequency

TABLE 70.1   Examples of Typical Spaces and Corresponding Ranges of Allowable dB(A) Values

Type of Space (examples)	dB(A) (approx.)
Concert hall, recording studio	<30
Residences, sleeping quarters, hospitals, bedrooms, classrooms	34–42
Private offices, small conference rooms	38–42
Larger offices	42–47
Lobbies, common work spaces (fair listening conditions)	47–52
Kitchens, laundry (moderately fair listening conditions)	52–61

Note: These are only guidelines; the dB(A) value is often a necessary but insufficient indicator of whether the background noise is tolerable.

Source: Egan, M.D. (1988), *Architectural Acoustics*, McGraw-Hill, New York, p. 233.

TABLE 70.2 Differences in Weighting Factors between A-Weighting and C-Weighting

Weighting Factors	16 Hz	32 Hz	63 Hz	125 Hz	250 Hz	500 Hz	1 kHz	2 kHz	4 kHz	8 kHz	16 kHz
dB(C)	−8	−3	−1	0	0	0	0	0	−1	−3	−8
dB(A)	−56	−39	−26	−16	−9	−3	0	1	1	−1	−7
Difference	48	36	25	16	9	3	0	−1	−2	−2	−1

Note: The most significant differences occur at lower frequencies (e.g., up to about 250 Hz). Thus, large positive values of (C − A) indicate the presence of low-frequency noise.

spectrum than A-weighting (which is more severe at lower frequencies), the difference between the two values reflects the relative proportion of low-frequency energy in the noise (up to about 250 Hz).

As an example, Figure 70.1 depicts three different noise spectra with the same dB(A) value. They are perceptually distinct because of the differing spectral distribution of sound energy. The curve with square boxes represents a "neutral" sounding spectrum. It has a frequency slope of about −5 dB/octave and has been found to be a perceptually acceptable background noise spectrum (Blazier, 2000). When the slope of the frequency spectrum varies significantly from this, one has either a perceptually dominant low-frequency range (causing "rumble"-sounding noise) or a perceptually dominant high-frequency range (causing "hiss"-sounding noise).

This has led to guidelines for correlating (C − A) values to approximate assessments of the sound quality of a given noise (Blazier, 2002). If the (C − A) value is approximately 15 dB, the noise will be perceived as neutral. If the (C − A) value is 10 dB or less, the noise will likely be perceived as hiss-like noise, whereas if the (C − A) value is 20 dB or greater, the noise will be perceived as rumble noise. The ASHRAE Fundamentals Handbook (2001) comments that a (C − A) value greater than 30 dB indicates excessive low-frequency content.

Blazier (2002) has suggested that this method could be used for screening rooms on the basis of quality for speech communication. As an additional example of the importance of detecting low-frequency noise, it has also been shown that low-frequency noise can have measurable, and possibly subliminal, effects on the ability of people to cope with demanding cognitive tasks (Persson-Waye et al., 1997).

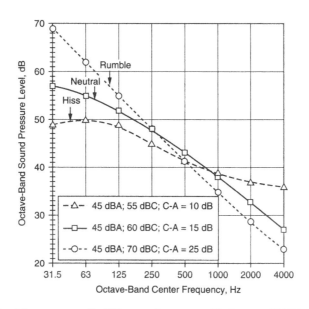

FIGURE 70.1 Example of three perceptually different noise spectra with the same dB(A) level. Even at lower dB(A) levels, a given spectrum may be perceived as annoying due to an excessively imbalanced frequency spectrum, indicated by its (C − A) value.

In addition, if one wishes to roughly estimate in which low-frequency bands the noise dominates, one can compare the (C – A) value to the (Lin – A) value, where Lin refers to the linear (unweighted) filter setting on a sound-level meter. Large differences between (C – A) and (Lin – A) would indicate low-frequency content in the 16- and 32-Hz-octave bands, where C-weighting diverges significantly from 0 dB. In fact, using the linear (unweighted) setting instead of C-weighting would theoretically be preferred except that, in practice, the lower cutoff frequency is not standardized. For example, the lower cutoff frequency may vary on sound-level meters from 2 to 10 Hz. To make a meaningful comparison of (C – A) to (Lin – A), it is recommended that the lower cutoff frequency for the linear (unweighted) setting be below approximately 10 Hz.

70.3   Example Output Table

The data in Table 70.3 illustrate how this method can be used. The dBA value is measured first to indicate overall level. The (C – A) value indicates the sound quality (e.g., whether neutral, rumble-noise, or hiss-noise). Using the steps described above, one can judge whether the background noise is acceptable or is in some way annoying.

70.4   Advantages

- Indicates sound quality in addition to overall sound level.
- Simplicity of the procedure makes it a de facto standard because of its utility.
- Does not require complex equipment apart from a basic sound-level meter.
- Measurements can be done rapidly (in less than 5 min).
- Avoids inconsistencies in nonstandardized linear filters, which can have minimum cutoff frequencies ranging from 2 to 10 Hz.

70.5   Disadvantages

- Not as informative as a more involved measurement and calculation of loudness level, as defined by ISO Standard 532, Acoustics: Method for Calculating Loudness Level.
- Does not give detailed shape of noise spectrum, e.g., for diagnosis of sound source.
- Does not indicate the amplitude of noise fluctuation.

70.6   Related Methods, Standards, and Regulations

This measurement method describes both sound level and sound quality. Methods that are more comprehensive may utilize noise criteria (NC) curves or RC curves (ASHRAE, 2001). It is also possible to compute loudness level over critical bands, as described in ISO Standard 532 (Acoustics: Methods for Calculating Loudness Level). However, these methods require somewhat more sophisticated frequency analyzers and additional calculation. They are not as rapid as the (C – A) method, but they do offer additional ways of investigating or diagnosing a problem when one is found. A standard on classroom

TABLE 70.3 Example Data and Analysis for Three Different Background Noise Situations

Type of Space	dB(C) (dB)	dB(A) (dB)	dB(C) – dB(A) (dB)	Sound Quality Assessment
Conference room	75	67	8	hissy, loud level
Classroom	45	20	25	rumble, soft level
Private office	50	35	15	neutral, acceptable level

acoustics is currently being drafted by the American National Standards Institute (ANSI) and the Acoustical Society of America (ASA).

70.7 Approximate Training and Application Times

The approximate training time is related to the complexity of the sound-level meter. Many sound-level meters utilize a power switch to display the measured level and an additional switch or button to choose the weighting filter. Training time should take less than 30 min. Application time should take less than 5 min per measurement.

70.8 Reliability and Validity

The method's reliability is based on the fact that the A- and C-weighting filters are standardized; therefore, the frequency ranges that give rise to large values of (C – A) are relatively well understood. However, the validity of the method would benefit from more extensive investigation of a wide variety of frequency spectra.

70.9 Tools Needed

One of the strengths of this method is that it only requires a sound-level meter with the common A- and C-weighting filters. (If a linear setting is included with a cutoff frequency below 10 Hz, this can also be used to estimate the approximate frequency bands of the low-frequency energy.)

Acknowledgment

The author is grateful to W. Blazier and C. Ebbing for helpful discussions and assistance. The described method is based on previous work and publications by W. Blazier.

References

ASHRAE (1999), *ASHRAE Handbook: HVAC Applications*, American Society of Heating, Refrigerating and Air-Conditioning Engineers, Atlanta, GA, chap. 46.

ASHRAE (2001), *ASHRAE Handbook: Fundamentals*, American Society of Heating, Refrigerating and Air-Conditioning Engineers, Atlanta, GA, chap. 7.

Blazier, W. (2000), Room noise criteria, in *Noise Manual*, 5th ed., AIHA Press, Fairfax, VA, chap. 13.

Blazier, W. (2002), Room noise criteria (update), Council for Accreditation on Occupational Hearing Conservation (CAOHC), Vol. 14.

Egan, M.D. (1988), *Architectural Acoustics*, McGraw-Hill, New York, p. 233.

Harris, C. (1998), *Handbook of Acoustical Measurements and Noise Control*, 3rd ed., Acoustical Society of America, Woodbury, NY.

Persson-Waye, K. et al. (1997), Effects on performance and work quality due to low-frequency ventilation noise, *J. Sound Vib.*, 205, 467–474.

71

Noise Reaction Indices and Assessment

R.F. Soames Job
University of Sydney

71.1 Background and Application

Noise is a major environmental pollution that produces negative emotional reactions and loss of quality of life (Berglund and Lindvall, 1995; Job, 1988), cognitive deficits (Haines et al., 2001; Hygge et al., 1996), and probably physical and mental health effects (Berglund and Lindvall, 1995; Job, 1996). Noise experienced in and around the home is an especially potent environmental stressor because of people's attitudes toward their homes and recreational time, disturbance of sleep, and lack of control over the noise (Hatfield et al., in press). Research in the area has two essential purposes: (1) to determine (and thus regulate for) reasonable levels of noise exposure, balancing the needs of noisemakers (transport, construction, industry) against the negative effects on residents, and (2) to understand human reaction to noise and its underlying causal factors, including many psychological factors (see Fields, 1992; Job, 1988). Both purposes demand measurement of the extent (and possibly the form) of reaction to noise.

Reaction to noise is a subjective matter that currently is measurable only by self-report. In many studies, noise reaction in residents, also called community reaction, is assessed only as annoyance (see Fields, 1992; Job, 1988). However, reaction to noise involves much more than annoyance. People may react with frustration, depression, apathy, anxiety, disturbance, distress, fear, or many other emotions (see Job, 1993). Thus, valid measurement of reaction must include assessment of more than just annoyance. Rather than attempting to measure each of the large variety of negative reactions that may be elicited by noise, a general, nonspecific assessment is recommended (such as asking people how "dissatisfied" they are with the noise, or how "bothered, disturbed, or annoyed" they are with the noise). The use of more than one question to assess reaction also demonstrably increases the reliability of the measurement of noise reaction (Job, 1988; Hatfield et al., in press).

71.2   Procedure

The measurement of reaction requires that residents (or employees, if the method is to be adapted to occupational settings) in the areas of interest be asked to rate their reactions to the noise. Bias of the outcome is minimized if residences are randomly selected and a resident is randomly selected within each randomly selected residence (e.g., by employing the last-birthday technique). This approach also minimizes the sample loss, especially the refusal rate. For examples of procedures that address these issues, see Hatfield et al. (in press).

A personal approach by an experienced interviewer, possibly with a financial incentive for the respondents to participate, is likely to produce a small refusal rate, and having the questions asked by an interviewer (rather than through a self-completed questionnaire) reduces sample loss due to literacy problems while also ensuring that certain critical questions are answered before other information is revealed in subsequent questions. In socioacoustic surveys, the beginning of the interview offers an important opportunity to ask about reaction to the relevant noise before the respondent knows that this is the focus of the survey. To achieve this, the respondent is informed that the survey is about conditions in the area. In particular, the interview may begin with questions on what is most liked about the neighborhood and what the respondent would most want to change. (The issue of most relevance here is whether the respondents identify noise as one of the features they would most want to change.) Asking one of the questions (e.g., question 2 below) in relation to many noise sources possibly heard in the area (dogs, lawnmowers, cars, neighbors' televisions) — with the noise source of critical interest placed into the list — also allows assessment of reaction prior to the focus of the rest of the survey on this noise being apparent to the respondent. This alleviates concern that respondents may report more reaction than really occurs when they discover that the particular noise is the focus of the interview. The concern arises from the view that respondents may be motivated to overreport their reactions in the belief that this will cause something to be done about the noise. However, there is little evidence that this is a valid concern.

During the interview, four explicitly worded questions are recommended (with the first two based directly on the work of the International Commission of the Biological Effects of Noise Community Response). (See Fields et al., 1997, 1998, 2001, for reports of the relevant research, basis for these wordings, and other suggested standardizations.) These questions are ideally interspersed with other questions and not presented together.

1. Thinking of the (last 12 months or so), when you are here at home, how much does the noise from (source) bother, disturb, or annoy you: extremely, very, moderately, slightly, or not at all?
2. Next is a zero-to-ten opinion scale for how much (source) noise bothers, disturbs, or annoys you when you are here at home. If you are not at all annoyed, choose zero; if you are extremely annoyed, choose ten; if you are somewhere in between, choose a number between zero and ten. Thinking about (the last 12 months or so), what number from zero to ten best shows how much you are bothered, disturbed, or annoyed by noise from (source)?
3. Thinking of the (last 12 months or so), when you are here at home, how dissatisfied are you with the noise from (source): extremely, very, moderately, slightly, or not at all?
4. Again, using the zero-to-ten opinion scale, rate how dissatisfied you are with the noise from (source) when you are here at home. If you are not at all dissatisfied, choose zero; if you are extremely dissatisfied, choose ten; if you are somewhere in between, choose a number between zero and ten. Thinking about (the last 12 months or so), what number from zero to ten best shows how dissatisfied you are with the noise from (source)?

Answer cards are recommended to help respondents answer these questions, as shown in Figure 71.1.

While four questions are offered above, the number to be employed in a given survey may depend on the length of the questionnaire and the exact purposes of the study. In some jurisdictions, noise regulations are couched specifically in terms of annoyance as the only relevant reaction. If the study is purely

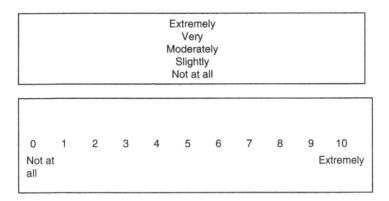

FIGURE 71.1 Recommended answer cards for questions 1 and 3 and for questions 2 and 4.

for noise regulation purposes in such a jurisdiction, then the terms *bothers* and *disturbs* should be deleted from the first two questions, and questions 3 and 4 are not appropriate.

A number of options exist for analysis of responses, depending on the purposes of the research. A check on internal consistency can be applied, which previous research suggests would reveal high internal consistency across questions 1 and 2 as a scale and across questions 3 and 4 as a scale, pointing to combining these pairs of questions to produce reliable indices of reaction to noise. The latter index may reflect more of the range of possible negative reactions by virtue of not specifying any particular emotions or outcomes but allowing that any negative reaction can contribute to dissatisfaction.

Although not a useful method of analysis from many perspectives, for regulatory purposes, it may be necessary to dichotomize the data into those who are seriously affected or highly annoyed vs. those who are moderately affected/annoyed or who are not so affected/annoyed. This can be achieved via conventional cut points, with scores of 8, 9, and 10, and "very" or "extremely" ranking in the highest category of effect. Noise can be considered to be excessive if 10% or more of the population is seriously affected or highly annoyed.

71.3 Advantages

The recommended method:

- Minimizes bias due to sampling problems and due to people falsifying their responses
- Is available and has been checked in a number of languages, such as English, Dutch (Flemish), French, German, Hungarian, Japanese, Norwegian, Spanish, and Turkish (see Fields et al., 2001)
- Can be employed for noise regulatory purposes
- Is more valid than questions that narrow reaction to annoyance only
- Affords greater reliability than does a single question

71.4 Disadvantages

The recommended method:

- Is dependent on obtaining an unbiased sample
- Is more expensive than employing a convenience sample or a single-question measure
- May be compromised by poorly trained or unmotivated interviewers
- Dichotomizes the data for regulatory purposes, thus losing information and statistical power
- May be overly prescriptive for research purposes that require the separation of various specific components of reaction or other factors

71.5   Related Methods

Many alternative question wordings are suggested in the literature. The most common alternative is to employ question wordings that focus only on annoyance, although this approach does not validly reflect the numerous other negative reactions that may be elicited by noise.

The only apparent alternative to asking questions is to record data on complaints, which may be available in cases such as airports. However, complaints are quite weakly related to noise exposure (Avery, 1982) and are influenced by many variables other than the extent of negative reaction (e.g., education, the person's belief that s/he is capable of complaining, the belief that a complaint will have an effect, knowledge of how and to whom to complain). Thus, complaints are not a valid measure of noise reaction.

71.6   Standards and Regulations

In most jurisdictions there are no set standards for surveys of reaction to noise. Rather, a general principle is applied such that the survey should be unbiased and of good quality.

Noise exposure standards are often set on the principle that a noise is excessive if 10% or more of the exposed population is seriously affected (or highly annoyed in cases where the regulatory focus is purely on annoyance). This principle, combined with dose–reaction curves (see Fidell et al., 1991), allows the setting of noise exposure limits in decibels.

However, in many instances the nature of the noise (impulsivity, tonality) or the extent of complaint from residents may force consideration of whether a noise is acceptable on the basis of the "10% seriously affected" rule, even when set decibel levels exist. In such instances, a survey is required to resolve the issue.

71.7   Approximate Training and Application Time

Interviewers with extensive prior experience will require little training explicitly for this type of fully structured questionnaire interview. A general briefing on the purposes of the survey, consideration of how to handle questions from respondents, and an overview of a brief questionnaire may require 2 to 3 hours.

A typical questionnaire containing many questions on the neighborhood and the particular noise of interest may take 20 to 30 min to complete with each respondent.

71.8   Reliability and Validity

Reliability, in terms of both internal consistency and test–retest stability, has been assessed for a variety of noise reaction questions. As a guide to internal consistency for questions on annoyance, the mean interitem correlation is 0.58, compared with 0.81 for general reaction questions (couched in terms of dissatisfaction and extent of effect of the noise), suggesting that the general scales are more internally consistent (for review, see Job et al., 2001). These mean interitem correlations result in high internal consistencies when several items are employed. Job et al. (2001) also found that the general reaction questions were significantly more stable than the specific annoyance questions when directly compared within the one survey (mean test–retest correlations of 0.82 and 0.66, respectively).

Construct validity, assessed by comparison with other measures of reaction, such as reported disturbances of activity by the noise, is good for both types of question (see Job et al., 2001).

71.9   Tools Needed

The required tools are the questionnaire and response cards described in this chapter.

References

Avery, G.C. (1982), Comparison of telephone complaints and survey measures of noise annoyance, *J. Sound Vibration*, 82, 215–225.

Berglund, B. and Lindvall (1995), Community noise, *Arch. Cent. Sensory Res.*, 2, 1–95.

Fidell, S., Barber, D.S., and Schultz, T.J. (1991), Updating a dosage-effect relationship for the prevalence of annoyance due to general transportation noise, *J. Acoustical Soc. Am.*, 89, 221–233.

Fields, J.M. (1992), Effect of Personal and Situational Variables on Noise Annoyance: With Special Reference to En Route Noise, Contractor Report CR-189670, NASA, Hampton, VA.

Fields, J.M., de Jong, R.G., Brown, A.L., Flindell, I.H., Gjestland, T., Job, R.F.S., et al. (1997), Guidelines for reporting core information from community noise reaction surveys, *J. Sound Vibration*, 206, 685–695.

Fields, J.M., de Jong, R.G., Flindell, I.H., Gjestland, T., Job, R.F.S., Kurra, S., et al. (1998), Recommendations for shared annoyance questions in noise annoyance surveys, in *Noise Effects: Proceedings of the 7th International Congress on Noise as a Public Health Problem*, Carter, N.L. and Job, R.F.S., Eds., Noise Effects, Sydney, Australia, pp. 481–486.

Fields, J.M., DeJong, R.G., Gjestland, T., Flindell, I.H., Job, R.F.S., Kurra, S., Lercher, P., Vallet, M., Yano, T., Guski, R., Flescher-Suhr, U., and Schomer, R. (2001), Standardized general-purpose noise reaction questions for community noise surveys: research and a recommendation, *J. Sound Vibration*, 242, 641–679.

Haines, M.M., Stansfeld, S.A., Job, R.F.S., Berglund, B., and Head, J. (2001), Chronic aircraft noise exposure, stress responses mental health and cognitive performance in school children, *Psychol. Med.*, 31, 265–277.

Hatfield, J., Job, R.F.S., Hede, A.J., Peploe, P., Carter, N.L., Taylor, R., and Morrell, S. (in press), The role of learned helplessness in human response to noise, *Int. J. Behav. Med.*

Hygge, S., Evans, G.W., and Bullinger, M. (1996), The Munich Airport noise study: cognitive effects on children from before and after the change over of airports, in *Proceedings of Inter-Noise '96*, Vol. 5, Institute of Acoustics, Liverpool, U.K., pp. 2189–2192.

Job, R.F.S. (1988), Community response to noise: a review of factors influencing the relationship between noise exposure and reaction, *J. Acoustical Soc. Am.*, 83, 991–1001.

Job, R.F.S. (1993), The role of psychological factors in community reaction to noise, in *Noise and Man: Noise as a Public Health Problem*, Vallet, M., Ed., INRETS, Lyon, France, pp. 48–79.

Job, R.F.S. (1996), The influence of subjective reactions to noise on health effects of the noise, *Environ. Int.*, 22, 93–104.

Job, R.F.S., Hatfield, J., Peploe, P., Carter, N.L., Taylor, R., and Morrell, S. (2001), General scales of community reaction to noise (dissatisfaction and perceived affectedness) are more reliable than scales of annoyance, *J. Acoustical Soc. Am.*, 110, 939–946.

Masden, K., Fields, J.M., Yano, T., Hatfield, J., and Job, R.F.S. (2000), Effect of social survey question wording on measured noise reactions and hypothetical reactions to environmental factors: a comparison of Japanese and English speakers, in *Proceedings of Westprac VII*, Kumamoto, Japan, Oct. 2000, pp. 951–954.

72

Noise and Human Behavior

Gary W. Evans
Cornell University

Lorraine E. Maxwell
Cornell University

72.1 Introduction

Noise affects a wide array of human behaviors that have implications for health and well-being. This chapter briefly describes these impacts, focusing on methods of assessment. Noise impacts can be divided into auditory and nonauditory effects. Auditory impacts refer to impacts of noise that directly interfere with hearing and are caused primarily by the energy level (i.e., decibels) of the stimulus. Because of widespread, detailed information on audiometry and the need for training, persons desiring to assess hearing damage from noise are advised to retain a certified audiologist. Thus we will not cover audiology herein but touch on two other auditory impacts of noise: sleep disturbance and speech interference. Nonauditory impacts are outcomes of noise exposure that are independent of auditory effects. They occur at sound intensity levels below those necessary to produce hearing damage and appear to have nothing to do with auditory interference with listening, such as masking of sounds.

72.2 Auditory Impacts of Noise

72.2.1 Speech Interference

Difficulty in speech communication can be attributed to two causes: normally audible speech is masked by concurrently present noise, or there is a loss of hearing sensitivity in the listener. The first cause is the subject of this section. Understanding speech is primarily a function of the signal-to-noise ratio, which includes the speech level of the speaker, the distance between the speaker and the listener, and the concurrent noise level (Lazarus, 1990). The most common source of noise that interferes with speech perception is speech itself, i.e., other people talking (Kryter, 1994).

There are two methods to test for the effects of noise on speech perception: speech intelligibility methods (SI) and articulation index methods (AI). The speech intelligibility methods assess the number of correctly identified words, phrases, or sentences under noise conditions. The most commonly used speech intelligibility method is a simplified version of the articulation index, the speech interference level (SIL) (Beranek, 1947). With this method, the distance between two persons that is required for reliable speech communication using unexpected word material under a steady noise level background is measured. The SIL is calculated by taking the arithmetic average of the octave-band sound-pressure levels of the ambient noise with central frequencies of 500, 1000, 2000, and 4000 Hz (ASA, 1977; Cunniff, 1977). See Lazarus (1990) for a detailed explanation of the calculation of the SIL. The SIL method may not be valid in real, as opposed to laboratory, communication situations. Such real-life conditions include individual variability in speech level, poor articulation of the speaker, various types of speech items (sentences, text, commands), visual interactions, high-level fluctuations, reverberation, and being able to understand a second language (Lazarus, 1990). The SIL method, however, does provide a reasonably accurate way of measuring speech communication under various noise conditions. However, unlike the AI, it should not be used with noise spectra that have intense low- or high-frequency components (Kryter, 1994).

The articulation method (AI) assesses the number of correctly identified individual phonemes, i.e., consonants and vowels in mono- or polysyllabic real or artificial words (Kryter, 1994). The AI was developed as an index based on the physical measurements of the spectra of speech, thresholds of audibility, and competing noise sources (French and Steinberg, 1947). It provides accurate predictions of speech intelligibility of most European languages, including English (Kryter, 1994). The AI describes the possibility that certain speech signals will be equally understood by anyone with specific senso-rineural ability and under certain ambient conditions. The AI assumes that speech intelligibility is proportional to the average weighted differences between the masking level, in decibels, of noise and the long-term level, in decibels plus 12, of the speech signal in each of a number of relatively narrow frequency bands. AI predicts speech intelligibility (Egen and Wiener, 1946; French and Steinberg, 1947). Note that AI intelligibility tests are influenced by the proficiency and training of the talker and listening group and by the difficulty of the speech material. Therefore, test scores may not be identical with a specific communication system when different groups of listeners and talkers are involved, even though the AI of the system remains constant. The most stringent tests of speech intelligibility are articulation methods that use artificial words, since language or context can not be used as an aid in the perception of correctly identified words (Kryter, 1994).

72.2.2 Sleep Interference

The effects of sleep deprivation are annoying, unpleasant, and potentially harmful to one's health. Typical effects of sleep deprivation from noise exposure are sleepiness, irritability, disruption of circadian rhythms, and impaired performance for complex, difficult, or monotonous tasks that require vigilance (Tilley and Brown, 1993). The four stages of sleep are drowsiness, rapid eye movement or REM sleep, and two stages of deep sleep. Partial sleep deprivation occurs when individuals either do not get a full night's sleep or sleep is interrupted. One of the causes of sleep interference when one is awakened is noise. Research indicates that noise can also shift sleep from one stage to another (Cunniff, 1977).

There are several methods of measuring sleep interference, such as behavioral awakening, where the subject presses a switch when awakened, and physiological measures of electrical brain activity using an electroencephalogram (EEG) to determine the stage of sleep of the subject. Other measures include heart rate and vasoconstriction in the finger, and subjective measurement of the quality of sleep when noises are present (Kryter, 1994). Sleep interference research in the home uses subjective measures of quality of sleep (surveys, rating scales) as well as behavioral awakening and changes in EEG activity or other physiological changes. The most popular subjective measure of sleepiness is the Stanford sleepiness scale (SSS). The SSS is a seven-point Likert self-rating scale that measures psychological or subjective sleepiness (Dement, 1993). The scale works best when sleepiness ratings are averaged over an hour-long period rather than 15-min periods. It is most accurate in documenting subjective sleepiness of normal people and less so with those who have clinical sleep disorders (Hoddes, 1993).

72.3 Nonauditory Impacts

72.3.1 Physiological

Noise reliably elevates multiple physiological indices of stress. Acute physiological impacts are exacerbated among those who are noise sensitive or engaged in difficult, demanding cognitive tasks (Evans, 2001). Chronic noise exposure particularly over longer periods of time elevates physiological stress. The most common methods of assessing physiological stress impacts of noise are cardiovascular and neuroendocrine parameters. Blood pressure and heart rate should be assessed with automated monitors that are significantly more accurate than the use of a stethoscope and manual cuff. Purchase of a unit designed for medical institutional use or clinical research is recommended rather than relatively inexpensive home-monitoring devices as sold in pharmacies. In order to obtain an accurate assessment of resting cardiovascular parameters, the subject should be seated with the properly sized cuff placed approximately at heart height. At least seven readings should be taken, and the average of the final six readings should be employed as the estimate (Kamarck et al., 1992; Krantz and Falconer, 1995).

Both the catecholamines (epinephrine, norepinephrine) and corticosteroids (cortisol) are sensitive to noise (Evans, 2001). Acute reactions can be assessed in plasma, but the half-life of the catecholamines is very short (1 to 3 min). Longer term catecholamine reactions can be assessed in urine, whereas cortisol can be assessed in urine or saliva. Because catecholamines are unstable, steps must be taken to minimize oxidation. In plasma, preparation must occur under refrigerated conditions; urine samples should be collected with a preservative, then acidified. In both cases, the samples should be deep frozen at ultralow temperatures (−70°C or colder). Although cortisol is rather stable, another challenge is presented because of the marked diurnal rhythm of this substance. Thus very careful matching of time-of-day assessments is required. Urinary assessments require the collection of total volume over a set duration (typically 12-h or 24-h periods). Incomplete voiding is a problem, but this can be largely dealt with by incorporating an index of kidney function, creatinine. Saliva collection is extremely easy, consisting of chewing on a small cylinder of cotton for approximately a minute. Eating and brushing of teeth are prohibited for 30 min prior to the sample collection. For an assessment over a relatively short time period, salivary cortisol peaks approximately 30 min past the event. For more chronic assessments, several salivary collections beginning 15 min after awakening, and then at three more 15-min intervals, are recommended. For more details on cortisol measurement procedures, see http://www.macses.ucsf.edu (MacArthur Foundation, 2001).

Catecholamines are typically assayed by high-performance liquid chromatography (HPLC) or with radioenzymatic assay. Cortisol is typically assayed with radioimmune assays and less commonly with HPLC. Since considerable training and infrastructure is needed to conduct these bioassays, use of a medical school clinical laboratory is recommended. For additional details on neuroendocrine assays, see Baum and Grunberg (1995) and Grunberg and Singer (1990).

72.3.2 Performance

A subset of cognitive tasks that are sensitive to noise include ones that demand processing of multiple sources of information simultaneously. A typical example of such a task is a dual task wherein one must monitor and respond to one signal (e.g., manually tracking a moving stimulus) while also responding to a second signal (e.g., reaction time to visual signals). In this type of task, noise reliably interferes with the secondary task but has little impact on the primary task. Task primacy is defined by instructions or incentive payoffs ("Your main job is to manually track the moving stimulus, but do your best to also respond quickly to the lights").

Under noise, subjects process information faster in working memory but at a cost in capacity. For example, in a running memory procedure in which subjects are required to recall in sequence letters that they just saw, recent items are recalled better under noise, but greater errors occur farther back in the list. Visual search tasks in which target stimuli held in memory must be found in arrays (e.g., cross out any of the following letters e, t, c, o, p, w in the text) are also sensitive to noise, especially when the memory load is high. Long-term memory, particularly for more complex information such as meaning, appears to be impaired by noise. There is also evidence that incidental memory can be impaired by noise. For example, if subjects are shown slides and asked to recall the words shown or the contents of the slides, little or no effect of noise during the presentation will be seen. However if subjects are asked for incidental information (e.g., what color was the ink, what corner of the slide was the word shown in), deficits emerge. See Evans and Hygge (in press) and Smith and Jones (1992) for reviews of noise and performance.

Chronic exposure to loud ambient noise (e.g., transportation noise sources such as airports) reliably interferes with reading acquisition in children in the primary grades (Evans and Hygge, in press). The number of standardized reading tests is too large to review here, but the characteristics of reading that should be assessed include comprehension and phonological processing. There is some suggestion in the literature that comprehension of more difficult material will be especially sensitive to noise. Thus in addition to overall measures of reading comprehension, researchers should examine comprehension at varying levels of difficulty. Two common phonological processing tasks are nonword repetition and phoneme recognition. Nonword repetition consists of presentations of phonetically representative nonwords from the child's language (e.g., "nigong" in English). Nonwords vary in length of sounds, and the child is asked to repeat exactly what s/he hears. Phoneme recognition consists of listening to words and then choosing from among other words that contain the same designated phoneme (e.g., for the target word *bat*, choose the word with same initial phoneme from the following list: boat, clown, deer). Two good starting points are the Woodcock/Johnson reading battery (Woodcock et al., 2001) and the Comprehensive Test of Phonological Processing (Wagner et al., 1999). Both batteries are comprehensive, appropriate across a range of ages, and have been normed on large, diverse samples within the U.S. Furthermore, both test batteries have extensive psychometric information available, indicating excellent validity and reliability.

72.3.3 Motivation

Both acute and chronic exposure to noise produces decrements in task motivation. Experimental work indicates that the uncontrollability of noise exposure is a critical element in this effect. For example, Glass and Singer (1972) showed that when adults were exposed to uncontrollable noise for about 30 min, immediately following the noise exposure period, they persisted for a shorter period of time on challenging geometric puzzles in comparison with individuals who had previously been in quiet for the same 30-min period. Interestingly, if participants were informed that they could shut the noise off by pushing a button (i.e., perceived control over the noise), their task persistence closely resembled the no-noise comparison group.

Task motivation has been assessed primarily through the use of geometric line-tracing puzzles. The participant's task is to trace over a complex line drawing without lifting his or her pencil or going back over any line twice. Typically, the initial puzzle is either unsolvable or extremely difficult and is followed by a more readily solvable puzzle. The pattern is repeated for a total of two unsolvable and two solvable puzzles. Each of these four puzzles is presented in piles of multiple copies of the same puzzle. The primary dependent measure is the number of puzzles attempted until the person gives up on the two unsolvable puzzles. The solvable puzzles are usually not examined and are employed more for ethical reasons to ensure that the participant has some success. Details on task procedure can be found in Glass and Singer (1972). Cohen (1980) also provides a review covering use of this task with several other stressors. This task has also been adapted for use by elementary school children exposed to chronic noise (Bullinger et al., 1999).

References

ASA (1977), American National Standard for Rating Noise with Respect to Speech Interference, American Institute of Physics for Acoustical Society of America, Paramus, NJ.

Baum, A. and Grunberg, N.E. (1995), Measurement of stress hormones, in *Measuring Stress*, Cohen, S., Kessler, R.C., and Gordon, L., Eds., Lawrence Erlbaum Associates, Mahwah, NJ, pp. 175–192.

Beranek, L.L. (1947), The design of speech communication systems, *Inst. Radio Eng.*, 35, 880–890.

Bullinger, M., Hygge, S., Evans, G.W., Meis, M., and von Mackensen, S. (1999), The psychological cost of aircraft noise for children, *Zentralbl. Hyg. Umweltmed.*, 202, 127–138.

Cohen, S. (1980), Aftereffects of stress on human performance and social behavior: a review of research and theory, *Psychol. Bull.*, 88, 82–108.

Cunniff, P.F. (1977), *Environmental Noise Pollution*, John Wiley & Sons, New York.

Dement, W.C. (1993), Sleepiness, in *Encyclopedia of Sleep and Dreaming*, Carskadon, M.A., Ed., Macmillan, New York, pp. 554–559.

Egen, J.P. and Wiener, F.M. (1946), On the intelligibility of bands of speech in noise, *J. Acoustical Soc. Am.*, 18, 435–441.

Evans, G.W. (2001), Environmental stress and health, in *Handbook of Health Psychology*, Baum, A., Revenson, T., and Singer, J.E., Eds., Lawrence Erlbaum Associates, Mahwah, NJ, pp. 365–385.

Evans, G.W. and Hygge, S. (in press), Noise and performance in children and adults, in *Noise and Its Effects*, Luxon, L. and Prasher, D., Eds., Whurr, London.

French, N.R. and Steinberg, J.C. (1947), Factors governing the intelligibility of speech sounds, *J. Acoustical Soc. Am.*, 19, 901–919.

Glass, D.C. and Singer, J.E. (1972), *Urban Stress*, Academic Press, New York.

Grunberg, N.E. and Singer, J.E. (1990), Biochemical measurement, in *Principles of Psychophysiology*, Cacioppo, J.T. and Tassinary, L.G., Eds., Cambridge University Press, New York, pp. 149–176.

Hoddes, E. (1993), Stanford sleepiness scale, in *Encyclopedia of Sleep and Dreaming*, Carskadon, M.A., Ed., Macmillan, New York, p. 595.

Kamarck, T., Jennings, R., Debski, T., Glicksman-Weis, E., Johnson, P., Eddy, M., and Manuck, S. (1992), Reliable measures of behaviorally evoked cardiovascular reactivity from a PC-based test battery, *Psychophysiology*, 29, 17–28.

Krantz, D.S. and Falconer, J. (1995), Measurement of cardiovascular responses, in *Measuring Stress*, Cohen, S., Kessler, R.C., and Gordon, L., Eds., Oxford University Press, New York, pp. 193–212.

Kryter, K.D. (1994), *The Handbook of Hearing and the Effects of Noise*, Academic Press, New York.

Lazarus, H. (1990), New methods for describing and assessing direct speech communication under disturbing conditions, *Environ. Int.*, 16, 373–392.

MacArthur Foundation (2001), Network on Socioeconomic Status and Health, available on-line at http://www.macses.ucsf.edu.

Smith, A.P. and Jones, D.M. (1992), Noise and performance, in *Handbook of Human Performance*, Vol. 1: The Physical Environment, Jones, D.M. and Smith, A.P., Eds., Academic Press, London, pp. 1–28.

Tilley, A. and Brown, S. (1993), Sleep deprivation, in *Handbook of Human Performance*, Vol. 3, Smith, A.P. and Jones, D.M., Eds., Academic Press, New York, pp. 237–259.

Wagner, R.K., Torgesen, J., and Rashotte, C. (1999), *Comprehensive Test of Phonological Processing (CTOPP)*, Pro-Ed, Austin, TX.

Woodcock, R.W., McGrew, K., and Mather, N. (2001), *Woodcock Johnson III: Tests of Achievement*, Riverside Publishing, Itasca, IL.

73

Occupational Vibration: A Concise Perspective

Jack F. Wasserman
University of Tennessee

Donald E. Wasserman
University of Tennessee

David Wilder
University of Tennessee

73.1 Introduction

In the U.S. alone, there are some 8 million to 10 million workers who daily are exposed to occupational vibration (Wasserman et al., 1974). These vibration exposures are usually categorized into two groups (Wasserman, 1987). Whole body vibration (WBV), or head-to-toe exposure, affects such employees as truck, bus, heavy equipment, farm vehicle, forklift, and overhead-crane operators. The second group is hand-arm vibration (HAV), or localized vibration exposure, mainly, but not exclusively, affecting employees who use all manner of vibrating pneumatic, electrical, hydraulic, and gasoline-powered hand tools. Rarely does one speak of "crossover exposures" between WBV and HAV except in the case of certain hand-tool usage such as road rippers or jackhammer-type tools, where the worker can choose either to grasp the tool with the hands, extending the tool away from the torso (HAV exposure), or to let the tool rest against the torso (WBV exposure) in an attempt to damp the vibration (Shields and Chase, 1988; Wasserman, 1989). Sometimes both can occur simultaneously, such as in motorcycling or mountain bike use. The medical effects of HAV and WBV are distinctly very different, as are their vibration exposure patterns and physical characteristics such as acceleration levels, vibration frequencies, and pathways into the human body. Thus it is common practice to discuss HAV and WBV separately because, although they share a common physics, they do not share a common physiology, nor do they share the same safety and health effects.

73.2   Health and Safety Effects of Occupational Vibration Exposure

73.2.1   Hand-Arm Vibration

Hand-arm vibration exposure has been causally linked to a generally irreversible condition of the fingers and hands called hand-arm vibration syndrome (HAVS) (Pelmear and Wasserman, 1998). HAVS was discovered in the U.S. and characterized more than eight decades ago by the famous pioneering occupational physician, Alice Hamilton, during an early investigation of hand-arm maladies stemming from the daily use of vibrating pneumatic hand tools by workers in the quarrying, cutting, and carving of oolitic limestone in Indiana (Hamilton, 1918). Hamilton termed this condition "Raynaud's phenomenon of occupational origin"; later the name would be changed to "vibration white finger" or "dead hand" disease; and finally this condition became known as hand-arm vibration syndrome.

Major symptoms of HAVS (Pelmear and Wasserman, 1998) are initially characterized by tingling and/or numbness in the fingers, similar to but not the same as carpal tunnel syndrome. As vibration exposure continues, the appearance of a single "white" or blanched fingertip occurs, usually, but not always, in the presence of cold. This seemingly innocuous attack of "white finger" marks the beginning of the dreadful and later irreversible finger blanching process. Often these attacks are mistaken as frostbite by workers. Initial finger blanching attacks last about 5 to 15 min and are widely spaced apart. As vibration exposure continues, especially in cold conditions, these attacks increase rapidly in number, intensity, duration, and finger pain. In the later stages of HAVS, attacks can and do occur in all seasons as well as on and off the job. HAVS interferes in both the patient's work and nonwork lives, in the latter case while doing tasks like mowing the lawn or when touching cold objects such as a vehicle steering wheel early in the morning or when cold water strikes the fingers, etc. Cold helps trigger HAVS attacks; and the simultaneous combination of vibration, cold, and nicotine from smoking is particularly deadly, since all three tend to act as vasoconstrictors and thus help "close down" blood vessels. In extreme conditions, the loss of blood supply to the fingers can lead to gangrene, which may require finger amputation. Thus HAVS can quickly become a serious occupational disease.

For the most part, HAVS is irreversible. Workers are advised to cease vibration exposure and seek medical attention immediately if they notice HAVS signs and symptoms. Prevalence of HAVS in U.S. tool users has been reported as high as 50% (NIOSH, 1983; Wasserman et al., 1982). Medical treatment is generally palliative and can include the use of blood pressure control medications known as calcium channel blockers (Pelmear and Wasserman, 1998).

73.2.2   Whole-Body Vibration

Whole-body vibration exposure is quite different from HAV exposure and enters the human body through different pathways, such as via the spine while an operator drives a vehicle, for example. There can be potential acute safety effects as well as chronic health effects resulting from WBV exposure. WBV exposure has been causally linked, but not limited to, severe low-back pain (Wilder et al., 2002) (lumbar spine) and degeneration, moisture loss, buckling, and herniation of the lumbar discs. Generally, WBV chronic exposure take some time before low-back problems develop. Sometimes workers report that they have kidney pain from WBV exposure; simulated animal studies have shown this mostly to be referred pain from the lumbar spine (Badger et al., 1974). Finally, poor vehicle seating, awkward postures, and manual cargo handling in addition to WBV exposure tend to exacerbate low-back pain misery. Recent studies have indicated that female workers who become pregnant and are exposed to WBV can possibly have added risk factors such as miscarriages and other gynecological disorders (Abrams, 1990; Peters et al., 1996).

The WBV safety issue concerns vehicle operators who are subjected to whole-body resonant conditions while driving vehicles and the possibility of their losing control of their vehicles due to the mechanical decoupling action between the steering wheel and the driver's hands as he or she attempts to hold and safely control the steering wheel (Wasserman, 1987).

Finally, there have been recent attempts to use WBV as part of the so-called return-to-work regimes called "work hardening." However, as of this writing, neither we nor our colleagues know of any scientific/ therapeutic value of using WBV exposure, and thus we strongly advise against its use in rehabilitation medicine (Wasserman et al., 1997).

73.3 Vibration Measurements

73.3.1 Basics

This section is too brief to provide the details of occupational vibration measurements in any great depth. Interested readers are advised to investigate in-depth presentations available in the literature (Griffin, 1990; Wasserman, 1987). Here are just a few basics:

Vibration is a description of motion. As such, vibration is called a vector quantity, which simply means motion is described by both a magnitude and intensity (i.e., acceleration or velocity or displacement) and a direction in which the motion moves. Vibration at any given point is defined by six vectors: three mutually perpendicular "linear" motions that move in a line (i.e., front-to-back, up-down, side-to-side) and three rotational vectors (i.e., pitch, yaw, roll). For occupational vibration, rotational motions are not measured; only the three (triaxial) linear axes are simultaneously measured. For HAV, vibration is measured from tool handles, where the worker grasps the tool, while for WBV, it is measured from the top of seat cushion where a vehicle driver sits. The measure of vibration intensity is usually acceleration or, more precisely, a form of average acceleration known as rms, or root-mean-squared, acceleration. This is measured separately and simultaneously for each of the three mutually perpendicular axes. There are uniform coordinate systems and measurement methods described in standards for both HAV and WBV, and if these standards are followed, it is usually possible for HAV or WBV measurements to be directly compared.

Finally, there is the concept of resonance, or natural frequency, wherein the human body as well as other physical structures respond by acting as a sort of a vibration "tuner," rejecting certain impinging vibration frequencies and responding or "tuning" to other vibration frequencies by actually amplifying and exacerbating these impinging vibration frequencies. For example, human WBV resonance occurs in the vertical (up-down) direction from 4 to 8 Hz. What is the concern? Simply this: if a vehicle, for example, contains spectral components of 4- to 8-Hz frequencies and these vibrations reach the operator's spine via the driver's seat, then the spine will most likely involuntarily respond by actually amplifying and exacerbating the effects of the WBV exposure. In other words, our body has the ability to select, accept, and amplify certain vibration frequencies over others, and in doing so it can worsen the effects of the vibration. The various vibration standards attempt to define and compensate for these unwanted and potentially troublesome human resonant frequencies. Resonance may seem unusual to many, but it affects virtually all physical structures, such as bridges. For example, this is the reason why soldiers never march across a bridge, lest it absorb the vibration excitation from marching and then internally amplify the vibration, causing the bridge to sway and eventually collapse.

The frequency ranges of measurement for hand-arm vibration and whole-body vibration are significantly different, which affects the selection of accelerometer transducers and their method of placement.

73.3.2 Whole-Body Measurements

Because transducers cannot be placed effectively and directly on the body, the standard approach is to use a hard-rubber seat pad/disc that contains the triaxial accelerometers. The pad is attached to the seat of the operator, commonly using duct tape (Wasserman, 1987). Because the frequency range of the measurements extends to below 1 Hz, measurements initially were made using devices such as piezore-sistive accelerometers or capacitive accelerometers, which could measure steady state (i.e., 0 Hz). However, piezoelectric accelerometers are now available that will meet the low-frequency requirements. During measurements, the seat must be occupied so that the seat pad reflects the accelerations actually being transmitted to the body. The data collection needs to be a minimum of 2 to 5 min so that the processed values obtained have statistical significance. If the peaks are large (relative to the average values that normally occur), the standards will underestimate the actual hazard, thus indicating that these averages may not be adequate predictors of potential harm.

73.3.3 Hand-Arm Vibration Measurements

To obtain meaningful data, triaxial accelerometers must be mounted in close proximity to the hands during functional use of the tool (Wasserman, 1987). The transducers need to be mounted as rigidly as possible to the tool to insure the full transmission of tool vibration to the transducer. The range of accelerations produced by the tool needs to be initially measured to ensure that the transducer measure-ment range is appropriate. The time of measurement needs to exceed 90 sec of continuous vibration to provide statistical significance. All three axes of measurement must be independently captured and stored simultaneously.

73.3.4 Data Processing

The data can be processed in a number of ways to provide meaningful information. Figure 73.1 and Figure 73.2 are examples of third-octave curves that are commonly used. Data for each axis are separately examined and then combined for the calculation of a single value that provides a quick determination of potential problems. Special analyzers are available that can produce these numbers for comparison with requirements such as the EU (European Union) directives.

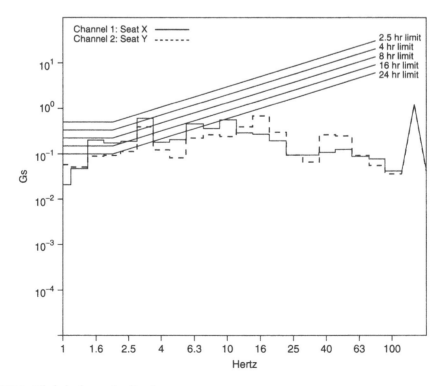

FIGURE 73.1 Whole-body x and y directions.

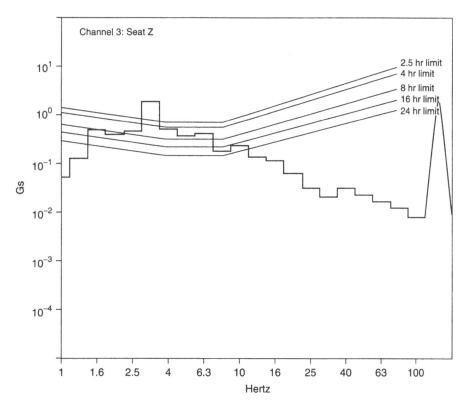

FIGURE 73.2 Whole-body *z* direction.

73.4 Advantages

- The demand for these measurements is going to significantly increase because of the impending implementation of the EU vibration directives into law.
- Changes in technology are enhancing the ability to measure vibration with a higher degree of confidence.
- Adoption of world standards will increase the need for measurements of occupational vibration. New information on the effects of multiple shocks will lead to the development of new requirements and thus more measurements.
- The manufacturers (and major users) of both vehicles and tools will need individuals to provide guidance in this area.

73.5 Disadvantages

- The ability to accurately perform these measurements requires specialized training because of the complexity of the process.
- The measurement and processing instrumentation used to obtain this information are improving, but they are still somewhat complex.
- The cost of the measurements is significant because of both the equipment and the time required to set up and perform the measurements on site.
- Although much of the world is considering the adoption of the EU directives as standard values, this process will take time, and some of the current standards use methods that are more complex than a single comparison of values.

73.6   Related Methods

73.6.1   Occupational Hand-Arm and Whole-Body Vibration Control

Controlling workplace vibration is usually a multifaceted effort that is undertaken when vibration measurements indicate that some appropriate vibration standard has been exceeded. Whole-body vibration control (Wasserman et al., 1982; Wasserman, 1987) in vehicles such as trucks, buses, and heavy equipment usually centers around the use of so-called air-ride seats that are designed primarily for maximum vertical vibration control in attenuating the particularly hazardous 4- to 8-Hz resonance frequencies. Some manufacturers also offer seats that provide vertical vibration control as well as front-to-back and side-to-side control. Seats alone are not necessarily a panacea and should be supplemented, where possible, in vehicles with suspended cabs, properly inflated tires, and properly functioning shock absorbers. In plant situations where vibrating machinery is used, air-ride seats are a possibility, as well as mechanically isolating the vibrating equipment from the floor. Where possible, remote operation also should be considered using inexpensive closed-circuit TV. Try to keep workers away from WBV exposure.

Hand-arm vibration control (Pelmear and Wasserman, 1998) is primarily concerned with replacing conventional vibrating hand tools with so-called reduced-vibration or antivibration (A/V) tools. A word of caution: power tools advertised as "ergonomically designed" are not necessarily vibration reduced! An ergonomically designed power tool is usually a product where the tool handle's characteristics allow the tool to be used with the hand-wrist maintained in the so-called neutral position, thus minimizing the tendency toward carpal tunnel syndrome and *not* HAVS. To minimize the vibration generated by the tool, it should be internally mechanically damped and/or isolated. Thus in our view, the proper tool to seek is one that is both antivibration and ergonomically designed. The use of so-called tool wraps around the handles of conventional tools is not recommended and should only be considered as a last-resort measure and only for the shortest time possible. In our view, the problems with wraps are basically twofold: they tend to increase tool handle diameter, thus creating the possibility of introducing other cumulative trauma disorders into the hand; and wraps do not necessarily attenuate enough (lower frequency) vibration to bring the tool(s) into compliance with the HAV standards. Next, to help protect workers from HAVS, we recommend using only full-finger-protected antivibration (A/V) gloves that meet or exceed ISO 10819 (antivibration glove standard). A/V gloves will reduce vibration, but they must also: (a) keep the fingers and hands warm and dry, (b) help prevent cuts and lacerations, and (c) be properly fitted. Remember: using finger-exposed A/V gloves is not recommended, since HAVS usually begins at the finger tips and then moves downward toward the palm.

Finally, good work practices should be used, which includes letting the tool do the work by gripping it as lightly as possible, consistent with safe work practices. Workers should not use the tool more than necessary; refrain from smoking; keep the fingers and hands warm and dry to avoid HAVS attacks; keep (cold) tool exhaust away from hands; and consider taking vibration-free breaks, about 10 min per hour. If signs and symptoms of HAVS appear, workers should seek medical help immediately.

73.6.2   Occupational Hand-Arm and Whole-Body Standards

There are numerous occupational HAV and WBV standards used throughout the world (Griffin, 1990; Pelmear and Wasserman, 1998; Wasserman, 1987). The U.S. government has not issued either regulatory HAV or WBV standards as of this writing. The major-consensus occupational HAV and WBV standards that are used in the U.S. as of this writing are as follows:

- HAV standards: American Conference of Governmental Industrial Hygienists (ACGIH), standard for hand-arm vibration (1984–2003); American National Standards Institute (ANSI), S3.34 (1986–2002); National Institute for Occupational Safety and Health (NIOSH), 89-106, criteria for a recommended standard for hand-arm vibration (1989)
- WBV standards: ACGIH standard for whole-body vibration (1994–2003); International Standards Organization (ISO) 2631 (1997); ANSI S3.18 (2002)

Note that for each of the HAV and WBV standards, there are prescribed uniform/universal methods for collecting and computer-analyzing so-called triaxial linear vibration acceleration data. However, these standards are not uniform in their presentation or interpretation of said data and of what constitutes excessive values of acceleration and vibration frequencies. Thus the reader is cautioned not to attempt any vibration measurements before obtaining the actual standard and reading/understanding it thoroughly.

73.7 EU Directives

The EU directives contain specific values that require action by the employer. The initial values require medical monitoring, while the final values require job modification.

73.8 Approximate Training and Application Time

The minimum training involves a 2- to 4-day training course. It is also advisable to take training offered by the manufacturers of the equipment that will be used.

For example, a typical evaluation of a set of tools will require a day to set up and measure when the agent provides tool information and videos of tool use. The processing of the information can be done either on site or off site. The advantage of recording the data and processing off site is the ability to take the time to validate the information. However, on-site evaluation of data quality is essential for a successful completion of the task in a timely manner.

References

Abrams, R. (1990), Sound and vibration in pregnancy, in *Seminars in Perinatology*, Part 11, W.B. Saunders, Philadelphia, pp. 273–334.

Badger, D., Sturges, D., Slarve, R., and Wasserman, D. (1974), Serum and Urine Changes in Macaca Mulatta following Prolonged Exposure to Whole-Body Vibration, Paper 1310-1, AGARD Conference on Vibration, Oslo, Norway.

Griffin, M.J. (1990), *Handbook of Human Vibration*, Academic Press, London.

Hamilton, A. (1918), A Study of Spastic Anemia in the Hands of Stonecutters: The Effect of the Air Hammer on the Hands of Stonecutters, Report 236, No. 19, U.S. Department of Labor, Washington, D.C.

NIOSH (1983), Vibration Syndrome, Current Intelligence Bulletin 38: NIOSH Pub. 83-110, National Institute for Occupational Safety and Health, Washington, D.C.

Pelmear P. and Wasserman, D.E. (1998), *Hand-Arm Vibration: A Comprehensive Guide for Occupational Health Professionals*, 2nd ed., OEM Medical Publishers, Beverly Farms, MA.

Peters, A., Abrams, R., Gearhardt, K., and Wasserman, D. (1996), Acceleration of the fetal head induced by vibration of the maternal abdominal wall in sheep, *Am. J. Obstet. Gynecol.*, 174, 552–556.

Shields, P. and Chase, K. (1988), Primary torsion of the omentum in a jackhammer operator: another vibration injury, *J. Occup. Med.*, 30, 892–894.

Wasserman, D., Badger, D., Doyle, T., Margolies, L. (1974), Industrial vibration: an overview, *J. Am. Soc. Saf. Eng.*, 19, 38–43.

Wasserman, D. (1987), *Human Aspects of Occupational Vibration*, Elsevier, Amsterdam.

Wasserman, D. (1989), Jackhammer usage and the omentum, *J. Occup. Med.*, 31, 563.

Wasserman, D., Taylor, W., Behrens, V., et al. (1982), Vibration White Finger Disease in U.S. Workers Using Chipping and Grinding Hand Tools, Vol. 1, Epidemiology, NIOSH Pub. 82-101, National Institute for Occupational Safety and Health, Washington, D.C.

Wasserman, D., Wilder, D., Pope, M., Magnusson, M., Alekslev, A., and Wasserman, J. (1997), Whole-body vibration and occupational work hardening, *J. Occup. Environ. Med.*, 39, 403–407.

Wilder, D., Wasserman, D., and Wasserman, J. (2002), Occupational vibration exposure, in *Physical and Biological Hazards of the Workplace*, 2nd ed., Wald, P. and Stave, G., Eds., Wiley, New York.

74

Habitability Measurement in Space Vehicles and Earth Analogs

Brian Peacock
National Space Biomedical Research Institute

Jennifer Blume
National Space Biomedical Research Institute

Susan Vallance
Johnson Engineering

74.1 The Context

Historically, people have built progressively more elaborate habitats for protection and to facilitate their purposeful activities. Nowadays, people live in tents, submarines, houses, mansions, prison cells, hospitals, space vehicles, and sometimes on the street. The physical, operational, and social environments and activities vary greatly, and people have a remarkable capability to adapt, within certain limits. Features of these environments, activities, and their corresponding habitats are qualitatively very different. People weight the importance of different features according to their particular purposes and their state of adaptation. As humans begin to explore outer space, we face the greater challenges of different gravities and radiation and lack of oxygen, water, and food. Spacecraft have to carry all their life-support systems with them. Ironically, in outer space, interior space restriction of habitats is a major challenge because of the gravitational, temporal, and financial challenges of transporting materials into space. For the foreseeable future, spacefarers will be cooped up in a can or space suit for long periods of time. They will also face limited resupply opportunities and communication with earth.

74.2   The Challenge

The challenge is to produce an appropriate, comprehensive, and reliable measure/index of habitability that addresses several important issues such as the following:

- Too much light, noise, and heat are not conducive to sleep.
- Overcrowding and a surfeit of information interfere with physical and intellectual activity.
- Communication is hampered by distance and intervening technology.
- Staying still and moving about are hampered by lack of anchors or masses with which to react.
- Radiation is an invisible threat.
- Crewmembers need to act as a team but also need privacy.

The habitability index is one of a family of indices that adopts a common currency to facilitate habitat evaluation and design. The common currency, like the "gold standard" in international finance, allows valid amalgamations of and comparisons between different features of the habitat. The index can also be used to delegate many basic human factors decisions to the engineers and mission designers and free up the limited human factors resources to deal with more complex situations, which greatly simplifies the requirements management process.

74.3   Habitability Measurement

Various authors (Harrison, 2001; Stuster, 1996) have contributed extensive qualitative descriptions of space environments and earth analogs, with particular emphasis regarding the physical and social environments and their changing effects on behavior and performance as the length of space missions increases. Celantano et al. (1963) used a quantitative approach in their development of a "habitability index for space stations and planetary bases." They started by articulating the challenges of complexity and then developed an index number that was a "weighted average of relatives." First they established, with reference to the literature, minimum, maximum, and optimum values for each habitability factor, assuming continuous exposure. The scale of each variable was anchored at zero (either a maximum or minimum), with the optimal level being 100.

Their example assumed a maximum pCO_2 of 20-mm Hg and an optimal level less than 5 mm. Given a system design level of 8 mm, then the relative value is $[(20 - 8)/(20 - 5)] \times 100 = 80\%$.

The next step involved averaging the scores within the following subsystems:

- Environmental control
- Nutrition and personal hygiene
- Gravitation
- Living space
- Crew work–rest cycles
- Fitness

Finally, the overall index was calculated as the weighted average of the subsystem scores, with the sum of the weights being set to ten. The Celantano paper (Celantano et al., 1963) went into considerable detail in the establishment of the separate variables, their measurement, and their optimum values.

The anchoring and scaling process serves to standardize the scores prior to weighting; however, the mechanism for establishing the weights was not discussed in detail. This standardization and weighting approach is somewhat similar to the one used by Peacock and Stewart (1985) in the context of grading students and Chen et al. (1989) in the context of development of a physical work stress index. In the former context, weightings were established either by individual students (with some restrictions) or by the instructor; in the latter context, weightings were established by consensus of experts.

Fraser (1968) described "intangibles of habitability during long duration space missions." In this paper, he deliberately ignored those variables for which standards were well established, such as environmental control systems, and concentrated on:

- Volume
- Configuration
- Privacy
- Personal hygiene
- Illumination, color, and décor
- Leisure and recreation

The results of these efforts were a compilation of qualitative and quantitative statements relating to the many features of these intangibles.

In 2001 and 2002, the American Bureau of Shipping published guides for "Passenger Comfort on Ships," "Crew Habitability on Ships," and "Crew Habitability on Offshore Installations" (ABS, 2001a, 2001b, 2002). Their approach was first to use a traditional "standards" approach by the presentation of tables and short paragraphs describing the ideal ranges of groups of variables such as accommodations, vibration, noise, indoor climate, and lighting together with extensive references. In the appendices, they reduced the complexity to a long sequence of qualitative/quantitative statements that could be rated on an acceptable/not acceptable binary scale. In the "Crew Habitability on Ships" guide, they added another cutoff value to the scales to differentiate between acceptable and ideal. Although these guides are very well researched and comprehensive, they do not attempt to address the challenges of interactions, exposure duration, variable weighting, and variability in human tolerance. Furthermore, there is no attempt to reduce the complexity to a single index.

Peacock (2002) described another precedent for index development, in the context of manufacturing operations. After the identification of the variables of interest, a process of consensus was used to establish high and low cutoff points and their relationship to a four-point "outcome prediction" scale (0,1,2,3). A variable value below the low cutoff point scored "0" and was considered to be ideal in terms of convenience and innocuous in terms of health and safety, whatever the context and duration of exposure. A variable value above the high cutoff point scored "3" and was considered to be unacceptable in its own right with regard to productivity and safety, irrespective of the conditions. The "region of uncertainty" or "region of interaction" between the high and low points separated the grades of "1" and "2." A score of "2" meant that interaction with other variables was highly likely, and a grade of "1" indicated a possibility of interactions. A rationalization for dealing with individual variables independently, even though they were known to interact, was that the resolution of a problem by the engineering community would necessarily deal with individual variables separately.

74.4 Amalgamations and Indices

Various attempts have been made to amalgamate separate ratings of habitability variables into a single score, which raises some conceptual and logical issues. In the previous example, summing ratings, such as a "2" for noise, plus a "2" for weight, plus a "2" for reach, plus a "2" for duration (making a total of "8"), make no sense unless they are measured on a true common-currency scale. Multiplying some ratings could be justified mathematically when an interaction is likely (e.g., lift moment or decibel hours). The amalgamation process that was adopted was simple counting of the number of "0s," "1s," "2s," and "3s." A superimposed management policy on a worksheet containing about twenty questions was that any single "3" or more than five "2s" was unacceptable. Given this prioritization to address the worst jobs, a process of continuous improvement was instituted to remove all stressors based on the elimination of "2s," where feasible. Coincidentally, the Toyota corporation developed a similar device that resulted in a numerical index with a continuous-improvement mandate for annual reduction in the overall scores. Again in the physical ergonomics arena, the NIOSH lift equation uses a multiplicative discounting process to calculate a "lift index" (Waters et al., 1993). The parameters of this index were developed by expert opinion with regard to the biomechanical, physiological, psycho-physical, and epidemiological literature.

74.5 Interactions

The scaling approach can be extended to deal with logically appropriate two-way interactions by presenting the checklist as a series of 4 × 4 matrices (Peacock and Orr, 2000). A similar logic permits assessment of the temporal exposure variables of repetition, frequency, and exposure duration. A further extension of the interaction assessment allows complex combinations of variables to be mapped onto the basic scale. Multiple variables can be amalgamated in a single chart, as in predetermined time charts for industrial operations or in nomograms used to assess the thermal environment. The principal advantage of the common-currency scale was that a whole, very large organization quickly came to understand the meaning of "0s," "1s," "2s," and "3s."

74.6 Scaling

A methodological shortcoming of this short, discrete (0,1,2,3) scale is that many assessments end up with "2s." The outcome scale does not discriminate sufficiently, even though the engineering measurements do allow greater resolution. Consequently, the following seven-point (ten-point) scale, based on the Cooper–Harper rating with links to the familiar green, yellow, red color-coding, is preferable (Table 74.1).

74.7 Outcomes

The purpose of measurement and a calculated index is that it should be predictive in terms of some outcome(s), such as effective activities (quality), productivity (efficiency), safety, health, behavior, or preference. The operations research community has developed many mathematical approaches to the optimization of outcomes, such as mathematical programming. Conversely, the individual is adept at choosing which outcome to emphasize under which conditions. Unfortunately, the optimal mathematical programming approaches are not always available in a timely manner, and the individual is not always optimal in his or her judgments due to human quirks such as the "halo effect" (in which individuals' emphasis on one of a set of criteria dominates the other criteria) and primacy/recency effects in operational memory. But in cases of habitability assessment, it is the human subjective response that dominates.

74.8 Activity and Adaptation

Human factors assessments are usually context and activity dependent. They are also dependent on the level of adaptation of the individual(s) concerned. For example, the optimal temperature for sedentary physical activity is higher than that for strenuous exercise. Similarly, the adaptation of the eyes to different light levels creates different requirements for highly precise visual work and general activity. Humans adapt either permanently or temporarily to heat, light, noise, spatial context, gravity changes, and equipment availability. This human variability, coupled with fundamental differences among individuals, can be addressed by an activity-dependent approach to habitability description or an approach that depends on the adaptation level of the individuals concerned. Activities that comprise a day on a space vehicle have been identified as shown in Table 74.2.

74.9 The Big Picture and Common Currency

It is useful to view the multiple aspects of the habitation environment on common scales as a collective picture. A profile of the habitat can be developed by simply counting the number of variables that are associated with a particular value or color on the outcome scale. Where appropriate, it is also possible to calculate a weighted value or index of all the variables for the purposes of comparing different habitats or the same habitat with hypothetical or real modifications. Thus the display can be a useful tool in the assessment of where the best opportunities lie for intervention and how these interventions will affect

TABLE 74.1 Common-Currency Scaling

Score	Code	Implication	Action
0	—	irrelevant	—
1	green	ideal	accept
2	green	acceptable	accept
3	yellow	tolerable	investigate
4	yellow	marginal	investigate
5	yellow	undesirable	investigate
6	red	unacceptable	reject
7	red	intolerable	reject
8–10	—	unthinkable	reject

TABLE 74.2 Activity Basis for Space Habitat and Operations Evaluation

Sustaining	Working	Contingencies
Sleeping	Assembly	Emergencies
Eating	Maintenance	Routine caution and warning response
Exercising	Logistics	Malfunctions
Health checks	Science (payload)	Other interruptions
Hygiene	Training	
Recreation	Housekeeping	
	Planning	
	Monitoring	
	Communications	

the overall situation. The level of granularity of the assessment can be modified according to the activity of interest. A very high level of granularity is shown in the example in Table 74.3. The choice of variables and cutoff points are for illustration only. In practice, a consensus approach to the establishment of context-dependent decision thresholds is warranted.

Note that the choice of variables and cutoff points is dependent on the situation of interest. The example in Table 74.3 also demonstrates a mechanism for viewing environmental, equipment, and operational variables simultaneously. A more elaborate interactive, color-enhanced version of this display is readily developed using Excel™

74.10 Exposure

Temporal exposure generally consists of two elements: duration and frequency. A duration measure can range from microseconds to years, and on occasion it may be appropriate to address both short- and long-term durations. Similar considerations can also be given to the frequency variables. For example, it may be of interest to consider very-short-duration high-intensity (impulse) noise and long-duration exposure to moderate-intensity noise. These two interactions may have different outcomes — the former causing acute damage to hearing (temporary or permanent), and the latter causing gradual hearing damage and short-term annoyance — before adaptation sets in. Similarly, an intrusion into a translation corridor may be a minor inconvenience, but if the translations are carried out very frequently, then this temporal interaction may rise to the level of acute annoyance.

Exposure to physical stressors over long time periods inevitably causes adaptation in the people concerned. This adaptation may be useful or harmful. For example, exercise may cause short-term fatigue, but when it is repeated, it can result in a long-term increase in strength, agility, and endurance. On the other hand, inappropriate physical demands can result in chronic fatigue, with associated decrements in performance.

TABLE 74.3 Illustrative Interface for Habitability Index

Name	Date	Place	Context	Activity	Importance	1	2	3	4	5	6	7
Spatial	height	width	depth	intrusions	orientations							
	1234567	1234567	1234567	1234567	1234567							
	70 30	70 30	70 30	0 10	F B							
Environmental	thermal	acoustic	visual	gravity	pressure							
	1234567	1234567	1234567	1234567	1234567							
	70 95	50 80	1000 10	1 0	14.3 4							
Equipment	access	displays no.	controls no.	labels	instructions							
	1234567	1234567	1234567	1234567	1234567							
	70 5	1 10	1 10	5 20	5 20							
Operations	mission	duration	frequency	team size	training							
	1234567	1234567	1234567	1234567	12345677							
	10 30	1 10	1 50	1 10	high none							
Activity	complexity	time limit	failure cost	failure risk	other							
	1234567	1234567	1234567	1234567	1234567							
	low high	hr sec	10^1 10^6	10^1 10^6	low high							

Temporal interactions in the context of index development can be addressed in a number of ways. First, it may be appropriate to multiply (divide) the physical variable by the temporal variable or use the concepts familiar in engineering measurement and calculus, such as velocity, impulse, or cumulative strain. A second way is to use simple matrices in which both the temporal (duration or repetition) and physical variables are converted to common-currency scales and amalgamated by expert consensus.

A final, more sophisticated approach is to plot the outcome scale against the time variable and use the slopes of the associated regression equations to reflect the effect of the physical variable of interest. This approach can also be simplified into the common-currency format by portraying the relationships between time, physical context, and common-currency outcome as a simple nomogram (Table 74.4).

74.11 The Policy Overlay and Decision Making in Design

Ultimately, the decision regarding cutoff points is not a technical decision. Rather it is the responsibility of managers or lawmakers to take the evidence presented by the analysts and impose cutoffs based on costs, risks, and benefits. These decisions must also acknowledge variability and level of protection within the population of interest. The usual requirements statement should be that all design values should be below the low cutoff and that in no case should any be above the high cutoff. The intermediate region is the region of uncertainty, where design tradeoffs are made. Presentation of the array of variables with their associated outcome ratings allows the decision makers to make decisions about the trees with due regard to the full picture of the forest (Table 74.3).

TABLE 74.4 Illustrative Table for Assessing Time, Contextual, and Outcome Variables

Outcome

7						
6						
5						
4						
3						
2						
1						
	0.1	1	10	100	1,000	10,000

Mission Duration (days)

Note: The interaction between the contextual and temporal variable can be portrayed by the slope of the regression line, including nonlinearities.

74.12 Verification and Validation of Designs

A major hurdle in the design of complex systems, such as a space vehicle or a closed-environment earth analog, is verification of requirements. Where the consensus process of cutoff development involves all concerned, including crewmembers, engineers, and managers, the resulting quantitative assignment provides an unequivocal basis for verification. Validation of designs can only be made in full context, either in an analog environment or the real environment. Because of the ever-present temporal elements, validation must also take place in the operational context. For example, one may judge a car seat to be comfortable in the showroom, but this assessment may not be "validated" on a long journey. Similarly, space vehicle habitability must first be assessed in earth analog contexts with extended stays. Validation should also employ valid testing methods with an appropriate subject sample involved in representative activities. For example, passageways may conform to spatial specifications. However, the formal testing should involve a spectrum of subjects involved in appropriate use and foreseeable misuse (unplanned activities.) In the space context, a foreseeable use could be the normal movement of a large crewmember through a hatch. An unplanned use could be emergency egress of a fully suited crewmember carrying a colleague. In both instances, performance requirements associated with the validation exercise would involve quantitative performance statements of allowable times and interferences.

74.13 Conclusions

The reduction of complexity is an important aid to personal and managerial decisions. The gold standard is somewhat useful in international finance, as are the various stock market indices. The task of assessing human factors involves similar levels of complexity, and specialists in this field are often called upon to make tradeoffs with no common basis for comparison. The evaluation of habitats is a particular challenge for human factors in space, and the concepts illustrated above go some way toward rationalizing the human factors decision-making process. The remaining feasible challenge is to populate the assessment tools with cutoff values based on a consensus of informed inputs from engineering, human factors, and operations personnel as well as the eventual inhabitants of these habitats.

References

ABS (2001a), Crew Habitability on Ships, American Bureau of Shipping, Houston, TX.
ABS (2001b), Passenger Comfort on Ships, American Bureau of Shipping, Houston, TX.
ABS (2002), Crew Habitability in Offshore Installations, American Bureau of Shipping, Houston, TX.

Celantano, J.T., Amorelli, D., and Freeman, G.G. (1963), Establishing a Habitability Index for Space Stations and Planetary Bases, paper presented at the AIAA/ASMA Manned Space Laboratory Conference, Los Angeles.

Chen, J.G., Peacock, J.B., and Schlegel, R.E. (1989), An observational technique for physical work stress analysis, *Int. J. Industrial Ergonomics*.

Harrison, A.A. (2001), *Spacefaring*, University of California Press, Berkeley.

Fraser, T.M. (1968), The Intangibles of Habitability during Long Duration Missions, NASA, Washington, D.C.

Peacock, J.B. (2002), Measurement in manufacturing ergonomics, in *Handbook of Human Factors Testing and Evaluation*, Charlton, S.G. and O'Brien, T.G., Eds., Lawrence Erlbaum Associates, Mahwah, NJ.

Peacock, J.B. and Stewart, B. (1985), Structured Grading Software, paper presented at Human Factors Society Annual Conference, Baltimore, MD.

Peacock, J.B. and Orr, G. (2000), A Checklist on Industrial Ergonomics Checklists, paper presented at Applied Ergonomics Conference, Orlando, FL.

Stuster, J. (1996), *Bold Endeavors*, Naval Institute Press, Annapolis, MD.

Waters, T.R., Putz-Anderson, V., Garg, A., and Fine, L.J. (1993), Revised NIOSH equation for the design and evaluation of manual lifting tasks, *Ergonomics*, 36, 749–776.

Macroergonomic
Methods

75
Macroergonomic Methods

Hal W. Hendrick
Hendrick and Associates

The macroergonomics domain deals with the overall design of work systems. Since the early days of the discipline, organizational design and management factors have sometimes been considered in ergonomic analysis and design, but it was not until the beginning of the 1980s that the area began to receive formal recognition as a distinct subdiscipline of ergonomics.

A major impetus to formal consideration of organizational design and management factors in ergonomics — and its application to the design of the overall work system — was the work of the U.S. Human Factors Society (HFS) Select Committee on the Future of Human Factors, 1980–2000. This committee was organized by former HFS president and distinguished ergonomist, Arnold Small, to look at trends in all aspects of life and project how they were likely to impact on the discipline of human factors or ergonomics over the next 20 years. At the 1980 annual meeting of the Human Factors Society (now known as the Human Factors and Ergonomics Society), the committee presented its findings, including the following anticipated developments in the areas of management and technology:

1. Breakthroughs in technology that would fundamentally change the nature of work, particularly microelectronics, automation, and development of the desktop computer
2. The "graying" of the workforce — with a related increase in education, experience, and maturity — and the need for organizations to adapt to the expectations and needs of this more experienced and mature workforce
3. Fundamental differences between the post–World War II baby boomers and their older colleagues in industrialized nations regarding their expectations about the nature of the workplace, with the baby boomers expecting to participate in decision making about their work, to have meaningful jobs, and to have satisfying social relationships at work (Yenkelovich, 1979)
4. The inability of purely microergonomic interventions to fully achieve expected reductions in lost-time accidents and injuries and increases in productivity
5. Increasing workplace and product liability litigation based on ergonomic safety design deficiencies.

In addition, it was clear that increasing world competition was going to require increasingly efficient work-system structures and processes in order for companies to remain competitive. It is interesting to note that all of these predictions from 1980 have come to pass and are continuing.

Based on these findings, it was clear that if ergonomics was to realize its potential and be responsive to the needs of industry, the discipline was going to have to formally integrate organizational design and management factors into its research and practice. Since that report in 1980, the subdiscipline of macroergonomics has come into being and has enjoyed rapid growth.

By 1996, the development of new methods (and the adaptation of old methods) for macroergonomic analysis, design, and evaluation of work systems had reached the point where the U.S. Human Factors and Ergonomics Society's Organizational Design and Management (ODAM) Technical Group formally changed its name to the Macroergonomics Technical Group. Today, the macroergonomics concept, approach, and methodology for the analysis, design, and evaluation of work systems is widely accepted internationally and has sometimes achieved truly remarkable results. Reductions of 60 to 90% or more in work-related musculoskeletal disorders, accidents, injuries, and scrap rates have been typical, leading to similarly impressive productivity improvements. For some documented case examples, see Hendrick (1996) and Hendrick and Kleiner (2001, 2002).

This section of the handbook presents 16 of the most commonly used and effective macroergonomic methods. Each of the 15 chapters devoted to these methods (Chapters 76 through 90) was written by a leading ergonomics authority specializing in the use of that method. Indeed, eight of the methods were actually developed and validated by the authors.

The first six of the methods described in this section (Chapters 76 through 81) are adaptations of well-known organizational and behavioral research methods for macroergonomic application.

The Macroergonomic Organizational Questionnaire Survey (MOQS) is an adaptation of the organizational questionnaire survey method by Pascale Carayon and Peter Hoonakker of the University of Wisconsin, who describe it in Chapter 76. These surveys can be very useful for quickly and inexpensively identifying symptoms of work-system design problems and locating where these problems may be occurring within the work system. Sometimes a problem can be identified in some work-system unit, and a MOQS can be developed and used to determine how widespread the problem is throughout the organization. In addition, when used in its survey feedback form as an integral part of a participatory ergonomics process (see Chapter 81), the MOQS results can provide both managers and employers with data to help them identify work-system design problems and/or to suggest what needs to be done to correct or improve the work system's functioning.

Semistructured interview surveys (Chapter 77) can be a particularly effective way of identifying and gaining insight into problems with a current work system's design. They also can help reveal specific kinds of macroergonomic intervention that might be effective in either redesigning the work system or implementing the intervention. The interview method also can be useful in identifying mismatches between the macroergonomic design of the overall work system and the microergonomic design of individual jobs and/or related human–machine and human–environment interfaces.

When applied within the context of macroergonomics, the focus group (Chapter 78) brings people from a particular work system together to be interviewed about specific aspects of the work system or its sociotechnical environment. Sometimes, a work-system change can be simulated, and the group members then are brought together in a focus group to gain their collective perceptions and opinions about specific aspects of the change.

In macroergonomic applications of the laboratory approach (Chapter 79), a simulated work-system environment often is created to enable the ergonomist to systematically manipulate either work-system or sociotechnical variables of interest (e.g., work-system complexity, formalization, or centralization) and then systematically observe and record the impact on various performance indices of interest.

Field studies and field experiments are covered in Chapter 80. By systematically observing work-system structures and processes in actual work settings (i.e., the "field") and measuring relevant performance variables, it is possible to tease out work-system design deficiencies and identify an appropriate macroergonomic strategy for intervening and correcting them. Field experiments are an efficient way of trying out macroergonomic interventions within a given work unit to determine their potential effectiveness for the total work system. Often, the results will suggest ways of fine-tuning the intervention before implementing the changes within the entire organization.

Participatory ergonomics (PE) (Chapter 81) is an adaptation of participatory management that was developed for both micro- and macroergonomic interventions. When the technique is applied to evaluate the overall work system, the employees work with an ergonomics professional who serves as both the facilitator and resource person for the group. A major advantage of this approach is that the employees are in the best position to know the problem symptoms and to identify the macroergonomic intervention approach that will be most acceptable to them. Equally important, having participated in the process, the employees are more likely to support the work-system changes, even if their own preferred approach is not adopted. Finally, the participatory approach has proved to be particularly effective in establishing an ergonomics and/or safety culture that sustains the performance and safety improvements resulting from the macroergonomic intervention. For an excellent case example, see Imada (2002). Participatory ergonomics is perhaps the most widely used method in macroergonomic interventions. It often is used in combination with the other methods described in this section.

Two methods described in this section are microergonomic methods used in product design that also are applicable to macroergonomic design of work systems: the cognitive walk-through method (CWM) (Chapter 82) and kansei engineering (Chapter 83).

The cognitive walk-through method (CWM) is a usability inspection method that rests on the assumption that evaluators are capable of taking the perspective of the user and can apply this user perspective to a task scenario to identify design problems. As applied to macroergonomics, evaluators can assess the usability of conceptual designs of work systems to identify the degree to which a new work system is harmonized or the extent to which workflow is integrated.

Kansei engineering, a method for translating consumers' affective responses to new products into ergonomic design specifications, was developed by Mitsuo Nagamachi (the author of Chapter 83) while he was at Hiroshima University. As applied in macroergonomics, kansei engineering can be used to translate worker's affective responses to proposed changes to a work system into macro- and microergonomic design specifications.

The remaining seven methods covered in this section (Chapters 84 through 90) have been developed specifically for macroergonomic applications. Three of these deal with manufacturing work systems. The first two — high integration of technology, organization, and people (HITOP) and TOP Modeler — were both developed by Ann Majchrzak of the University of Southern California and her colleagues. Majchrzak describes them in Chapters 84 and 85, respectively. The third of these — computer-integrated manufacturing, organization, and people (CIMOP) system design — was developed by Waldemar Karwowski and J. Kantola of the University of Louisville, who describe CIMOP in Chapter 86.

HITOP is a step-by-step manual procedure for implementing technological change. The procedure is designed to enable managers to be more aware of the organizational and human implications of their technology plans, thus improving their ability to integrate technology within its organizational and human context. TOP Modeler is a decision-support system designed to help manufacturing organizations identify the organizational changes required when new process technologies are being considered. CIMOP is a knowledge-base system for evaluating computer-integrated manufacturing, organization, and people system design. The intended users of CIMOP are companies designing, redesigning, or implementing a CIM system.

The final four chapters (Chapters 87 through 90) describe additional new methods designed for macroergonomic applications: anthropotechnology, systems analysis tool (SAT), macroergonomic analysis of structure (MAS), and macroergonomic analysis and design (MEAD).

Anthropotechnology deals specifically with analysis and design modification of systems for effective technology transfer from one culture to another. It was developed by the distinguished French ergonomist, Alain Wisner. Like Wisner and others, Philippe Geslin of the Institut National de la Recherche Agronomique (INRA), who describes the method in Chapter 87, has been highly successful in applying anthropotechnology to systems transferred from industrially developed to industrially developing countries.

Systems analysis tool (SAT) is a method developed by Michelle Robertson of the Liberty Mutual Research Center for conducting systematic trade-off evaluations of work-system intervention alternatives. Robertson describes the method in Chapter 88. It is an adaptation, elaboration, and extension of the basic steps of the scientific method. SAT has proved useful in enabling both ergonomists and managerial decision makers to determine the most appropriate strategy for making work-system changes.

Macroergonomic analysis of structure (MAS) is described in Chapter 89. MAS was developed by the editor of this section (Hendrick) for the purpose of assessing the structure of work systems in terms of their compatibility with their unique sociotechnical characteristics. These include the key aspects of the work system's technology, personnel subsystem, and the external environment to which the organization must respond to survive and be successful. MAS integrates empirically developed models that evaluate the key characteristics of each of these three work-system elements in terms of their implications for work-system design. By "plugging in" the values for each key variable for a given work system, a given model suggests what amount of organizational complexity, formalization, and centralization is optimal. Comparison of the MAS results with the actual work-system structure thus can identify deficiencies and suggest how to correct them for more optimal work-system performance.

Although macroergonomic analysis and design (MEAD) addresses work-system structure, the main value of MEAD is its ten-step process for evaluating work-system processes. MEAD, like the other methods developed specifically for macroergonomic application, is based on sociotechnical systems theory and the supporting body of research literature. See Hendrick and Kleiner (2001, 2002) for a comprehensive description of sociotechnical systems theory. MEAD is described in Chapter 90 by its developer, Brian Kleiner, of Virginia Technological Institute and State University.

References

Hendrick, H.W. (1996), *Good Ergonomics Is Good Economics*, Human Factors and Ergonomics Society, Santa Monica, CA.

Hendrick, H.W. and Kleiner, B.M. (2001), *Macroergonomics: An Introduction to Work System Design*, Human Factors and Ergonomics Society, Santa Monica, CA.

Hendrick, H.W. and Kleiner, B.M., Eds. (2002), *Macroergonomics: Theory, Methods and Applications*, Lawrence Erlbaum Associates, Mahwah, NJ.

Imada, A.S. (2002), A macroergonomic approach to reducing work-related injuries, in *Macroergonomics: Theory, Methods and Applications*, Hendrick, H.W. and Kleiner, B.M., Eds., Lawrence Erlbaum Associates., Mahwah, NJ, pp. 151–172.

Yenkelovich, D. (1979), *Work, Values and the New Breed*, Van Nostrand Reinhold, New York.

76

Macroergonomic Organizational Questionnaire Survey (MOQS)

Pascale Carayon
University of Wisconsin

Peter Hoonakker
University of Wisconsin

76.1 Background and Application

Questionnaire surveys can be used to collect information on a range of ergonomics variables (Salvendy and Carayon, 1997). Macroergonomic organizational questionnaire surveys collect information on various aspects of the work system (Carayon and Smith, 2000), including tasks, organizational conditions, environmental issues, tools, technologies, and individual characteristics. In addition, macroergonomic organizational questionnaire surveys collect information on various outcomes, such as quality of working life (e.g., job satisfaction), physical and psychological stress, physical and mental health, performance, and attitudes (e.g., intention to leave the job).

Macroergonomic organizational questionnaire surveys can be a useful tool at several stages, such as the diagnosis stage, benchmarking an organization on key characteristics of interest, evaluating the impact of a change on key characteristics, and monitoring worker opinions during the implementation of a change.

Carayon and Hoonakker (2001) have emphasized the following issues for macroergonomic organizational surveys: objectivity/subjectivity, reliability and validity, development of the questionnaire, and conducting the survey. At the questionnaire development stage, it is important to clearly define the concepts to be measured and to explore the range of questions that can be used to measure those concepts. In particular, attention should be paid to the measures' degree of objectivity/subjectivity, i.e., the degree to which cognitive and emotional processing influences answers to the questions (Carayon and Hoonakker, 2001).

76.2   Procedure

Particular attention should be paid to the development of questionnaires. The methods used to develop, implement, and use questionnaires are very important for the quality and usefulness of the data collected. Before developing a questionnaire survey, it is important to clearly specify the objective of the questionnaire: What will the questionnaire be used for?

Carayon and Hoonakker (2001) have defined five stages for developing a questionnaire survey:

1. Conceptualization
2. Operationalization
3. Sources of questionnaire
4. Constructing the questionnaire
5. Pretesting of the questionnaire

Each step is described in detail by Carayon and Hoonakker (2001). Table 76.1 describes the major questions to ask for each step and lists background information or materials that can be useful to complete successfully each step.

The actual conduct of a macroergonomic organizational questionnaire survey involves the following steps (Carayon and Hoonakker, 2001; Church and Waclawski, 2001):

1. Pooling resources
2. Communicating objectives
3. Administering

TABLE 76.1   Development of a Macroergonomic Organizational Questionnaire Survey

Steps	Questions To Ask	Background
Step 1: Conceptualization	What concepts will be measured by the macroergonomic organizational questionnaire survey?	Babbie (1990)
	• What elements of the work system will be evaluated: tasks, organizational conditions, physical environment, tools and technologies, individual characteristics? • What outcomes will be evaluated: quality of working life, physical and psychological stress, physical and mental health, performance and attitudes?	
	What are the objectives of the survey and how do the concepts to be measured fit with the objectives?	
Step 2: Operationalization	What are the dimensions of each concept? Check for overlap Review previous work	Babbie (1990)
Step 3: Sources of questionnaire	What questionnaire surveys are available that can be used?	Cook et al. (1981)
	• Office worker survey — University of Wisconsin-Madison • NIOSH job stress questionnaire • Karasek's job strain questionnaire	
Step 4: Constructing the questionnaire	What forms of questions will be used? What are the rating scales? How should be questionnaire be organized (i.e., order of questions, instructions, layout)?	Converse and Presser (1986) Dillman (2000)
Step 5: Pretesting of the questionnaire	Who will participate in the pretest? What are the objectives of the pretest (e.g., check for clarity of questions, test the questionnaire format, assess the duration of the questionnaire)?	

4. Analyzing and interpreting
5. Delivering results
6. Transferring and action planning

Each step is described in detail by Carayon and Hoonakker (2001) and by Church and Waclawski (2001). Table 76.2 describes the major questions to ask for each step.

The process for developing and conducting a macroergonomic organizational questionnaire survey can be summarized by the following questions:

- *What?* What are the objectives? What measures will be used? How reliable and valid are the measures?
- *How?* What is the process used to develop the questionnaire? What is the process used to conduct the survey? How is feedback provided to the organization and its employees?
- *Who?* Who is to participate in the survey? Who is involved in developing and conducting the survey? Who should be informed at what step?
- *When?* When will the survey start and end? What is the timeline for the entire process?
- *Where?* Where will the survey be conducted? Where will the data be stored?

76.3 Advantages

Surveys allow the researcher to obtain large amounts of data from large numbers of people at relatively low cost and relatively fast (Sinclair, 1995). In addition, macroergonomic organizational questionnaire surveys provide structured data that can be easily scored, analyzed, and compared.

76.4 Disadvantages

A major issue in the development of a macroergonomic organizational questionnaire survey is to identify the objectives of the survey and, therefore, define the concepts to be measured in the survey. It is

TABLE 76.2 Conducting a Macroergonomic Organizational Questionnaire Survey

Steps	Questions To Ask
Step 1: Pooling resources	Are the organization's management and key stakeholders committed to the successful completion of the questionnaire survey?
	Who are the key stakeholders?
	How should be they be involved and informed?
Step 2: Communicating objectives	What will be communicated?
	How will the communication occur? How often? To whom?
	How will the survey be presented to the respondents? What are the expected benefits and costs of the survey for the respondents?
Step 3: Administering	What is the best time to conduct the survey?
	What procedures will be used to administer the survey?
	What methods will be used to collect the data (e.g., paper vs. electronic survey)?
Step 4: Analyzing and interpreting	What software will be used to enter the data, to analyze the data, and to present the data analysis?
	What statistical methods will be used? How do the statistical methods help achieve the survey objectives?
Step 5: Delivering results	To whom will the results be presented? When? In what order?
	How will the report be structured? Who will read the report?
	How will confidentiality of respondents be assured?
	How will the integrity of the results be maintained?
Step 6: Transferring and action planning	How is commitment to actions secured?
	What follow-up activities are necessary to ensure that the data are used for planning and implementing actions?
	Is any follow-up survey scheduled? If yes, how will this be done? When?

sometimes difficult to know how to ask the right questions and ascertain the right response categories. This is why pretesting a questionnaire is important. Two major weaknesses of questionnaire surveys are the limited space to ask questions and the limited time to answer questions.

76.5 Examples

A couple of examples are shown to demonstrate the potential uses of the macroergonomic organizational questionnaire survey.

76.5.1 Evaluating Employee Opinion Regarding a Technology Implementation

Figure 76.1 shows results of an employee questionnaire survey performed before the implementation of an information technology (Carayon and Smith, 2001). The results of the survey showed many negative opinions regarding the technology implementation: lack of information received about the technology, lack of inputs regarding the design and implementation of the technology, lack of understanding of the technology impact on one's job, and negative attitudes toward the technology. This information provides diagnostic data that management can use to improve the technology implementation. For instance, based on these data, management might consider various means for communicating with employees regarding the design and implementation of the technology.

76.5.2 Impact of a Work Organization Intervention

A series of work organization interventions aimed at improving quality of working life were implemented for three groups of computer users and office workers in an organization (Carayon et al., 2000). Group 1 received the intervention right after round 1 (R1), whereas Groups 2 and 3 received the intervention after round 2 (R2). Data was collected at three times, separated by about 6 to 8 months (R1, R2, and R3). Several elements of the work system and outcomes were evaluated in this study. Figure 76.2 shows data for one element of the work system, i.e., pace control. The results in Figure 76.2 show that group 1 employees perceived an increase in pace control after the work organization intervention, whereas pace

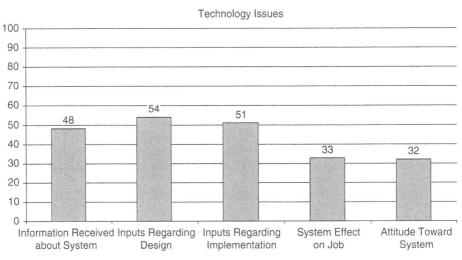

FIGURE 76.1 Employee opinions regarding a technology implementation (measured before the implementation). (From Carayon, P. and Smith, P.D. [2001], Evaluating the human and organizational aspects of information technology implementation in a small clinic, in *Systems, Social and Internationalization Design Aspects of Human–Computer Interaction*, Smith, M.J. and Salvendy, G., Eds., Lawrence Erlbaum, Mahwah, NJ, pp. 903–907. With permission.)

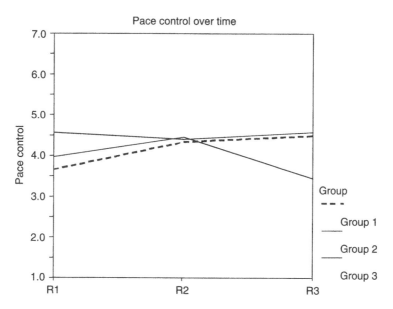

FIGURE 76.2 Employee perceptions of pace control over time. (From Carayon, P. et al. [2000], Intervention research for reducing musculoskeletal injuries, in *Proceedings of IEA 2000/HFES 2000 Congress*, Vol. 2, Human Factors and Ergonomics Society, Santa Monica, CA, pp. 169–171. With permission.)

control did not increase for group 2 employees. Group 3 employees actually reported decreased pace control. When presenting the results to management and employees, the increase in pace control for group 1 was confirmed to be a result of the work organization intervention. The decrease of pace control in group 3 was explained by major organizational changes occurring in this group at the time of the R3 data collection.

76.6 Reliability and Validity

The reliability and validity of macroergonomic organizational questionnaire surveys have been studied extensively and found to be satisfactory in many studies. Specific data on reliability and validity can be found for various forms of MOQS (e.g., Cook et al., 1981).

In the study used as the second example above (Section 76.5.2), the various measures of psychosocial work factors were found to have reliability Cronbach-alpha scores ranging from 0.45 to 0.94. A Cronbach-alpha score of 0.70 is accepted as an appropriate minimal level (Nunnaly, 1978). In the Carayon et al. (2000) study, 20% of the psychosocial work factors measures had Cronbach-alpha scores below 0.70. The main reason for the low reliability scores for these measures was the low number of questionnaire items addressing the measures (Carmines and Zeller, 1990). Cronbach-alpha scores are dependent on the number of questionnaire items that address a measure, and many of these measures with low-reliability scores consisted of only two items. In this study, many of the measures of psychosocial work factors were found to change over time because of the work organization interventions (see Figure 76.2 for data on the psychosocial work factor of pace control). This result is an indication of the validity of these measures of psychosocial work factors. A longitudinal study of psychosocial work factors, ergonomic factors, and health by Hoonakker et al. (1998) showed the predictive validity of these variables on absenteeism among a large sample of Dutch construction workers. For instance, a positive value for experienced health was related to short absenteeism, and higher levels of turbulence on the job were related to more frequent absences.

Because much effort, expertise, and time are necessary to ensure the quality of questionnaire surveys, it is recommended that established questionnaires be used as much as possible (see, for example, Cook et al. [1981] for various questionnaires relevant to macroergonomics). The following questionnaire surveys have been shown to be reliable and valid, and they have been used extensively in various macroergonomic studies:

- NIOSH job stress questionnaire
- University of Wisconsin office worker survey
- Job-content questionnaire by Karasek

76.7   Tools Needed

Macroergonomic organizational questionnaire surveys can be conducted using different methods. With the increasing use of the Internet and other electronic communication technologies, questionnaire surveys are increasingly being conducted via electronic methods. There are several means of distributing surveys electronically:

- Survey embedded in an e-mail
- Survey attached as a document
- Survey attached as a program (exe. file)
- Web-based surveys

The advantages of electronic surveys as compared with mail surveys include the following:

- Easy access to (worldwide) samples
- Speed and response time
- Lower costs (time and money, only the cost of constructing and posting the survey)
- Reducing errors (e.g., data entry)
- High response quality
- More completed questionnaires
- Higher probability of answers to open-ended questions
- Greater flexibility (respondents can choose how and when to respond)
- Ease of administration
- Ease of formatting (color, sound, images)
- Skip patterns, pop-up boxes, etc.

The disadvantages of electronic surveys as compared with mail surveys include the following:

- Coverage error/bias (limited and biased population, class, race, age, income, and gender)
- Sampling error
- Measurement error (PC [il]literacy, multiple copies)
- Nonresponse error
- Anonymity
- Nondeliverability (estimated at 20% [Bachman et al., 1996, 1999] and at 19% and 24% [Weible and Wallace, 1998])
- Security

Table 76.3 lists studies that have used mail surveys and various forms of electronic surveys. The results of these studies show that (1) mail surveys tend to lead to higher response rates than e-mail and/or Web surveys and (2) response rates to e-mail have gone down over time. However, Web-based surveys are getting better results (Cobanoglu et al., 2001). Therefore, when deciding whether to conduct a mail survey or an electronic survey, attention should be paid to the various factors that contribute to response rates: salience of the issue addressed by the survey, prenotification, personalized communication, length of survey, sponsorship and management commitment, and follow-up (Dillman, 2000).

TABLE 76.3 Studies Using Mail Surveys and Various Forms of Electronic Surveys

Author(s)	Sample	Survey Topic	Method	Sample Size	Response Rate	Response Time (days)	Response Quality
Kiesler and Sproull (1986)	Employees of a Fortune 500 company	Corporate communication	Mail E-mail[a]	115 115	67% 75%	10.8 9.6	E-mail had fewer mistakes and a higher completion rate
Parker (1992)	Employees of AT&T	Internal communication	Mail E-mail[a]	70 70	38% 68%	NA NA	NA NA
Schuldt and Totten (1994)	Marketing and MIS professors	Shareware copying	Mail E-mail	200 218	56.5% 19.3%	NA NA	NA
Mehta and Sivadas (1995)	Usenet users	Internet communication	Mail E-mail	309 182	56.5% 54.3%	NA NA	Both groups had similar number of item omissions, but e-mail respondents wrote more
Tse et al. (1995)	University population	Business ethics	Mail E-mail	200 200	27% 6%	9.8 8.1	No significant difference in number of item omissions
Bachman et al. (1996)	Business school deans	TQM	Mail E-mail	224 224	65.6% 52.5%	11.2 4.7	E-mail respondents were more willing to answer open-ended questions
Weible and Wallace (1998)	MIS professors	Internet use	Mail Fax E-mail Web form	200 200 200 200	35.7% 20.9% 29.8% 32.7%	12.9 8.8 6.1 7.4	NA
Schaefer and Dillman (1998)	University faculty	Unknown	Mail E-mail	226 226	57.5% 58.0%	14.4 9.2	E-mail surveys had fewer item omissions and longer answers to open-ended questions
Bachman et al. (1999)	Business school deans and division chairpersons	TQM	Mail E-mail	250 250	66.0% 19.1%	18.3 4.3	No differences in response patterns; e-mail respondents were more willing to answer open-ended questions
Sheehan and McMillan (1999)	Creators of health-related Web sites	Values of site creators, site purpose, and funding	E-mail (individual)[b]	834	47%	5.0	More salient issues and prenotification did increase response rates; reminder increased response rates, varying from 23 to 48%
	Faculty, staff, and students	Attitudes toward on-line privacy	E-mail (batch)	580	47%	4.6	
	Individuals with personal e-mail accounts	Attitudes and behaviors associated with on-line privacy	E-mail (merge)	3,724	24%	3.6	

TABLE 76.3 Studies Using Mail Surveys and Various Forms of Electronic Surveys (continued)

Author(s)	Sample	Survey Topic	Method	Sample Size	Response Rate	Response Time (days)	Response Quality
Dommeyer and Moriarty (2000)	CSUN students	Attitudes toward binge drinking	E-mail (embedded)	150	37%	4.3	No significant difference in number of item omissions
			E-mail (attached)	150	8%	5.7	
Couper et al. (2000?)	Employees in several government statistical agencies	Organizational climate	Agency A: mail	2,699	68%	NA	E-mail response rate was much lower, mostly due to technical problems (different e-mail software); no significant difference in number of item omissions
			Agency A: e-mail	2,699	37%	NA	
			Agency B: mail	790	76%	NA	
			Agency B: e-mail	396	63%	NA	
			Agency C: mail	266	74%	NA	
			Agency C: e-mail	265	60%	NA	
			Agency D: mail	216	75%	NA	
			Agency D: e-mail	221	53%	NA	
			Agency E: mail	216	76%	NA	
			Agency E: e-mail	215	55%	NA	
Cobanoglu et al. (2001)	Hospitality professors	Hospitality, education	Mail	100	26%	16.5	80.7% (completed surveys)
			Fax	100	17%	4.0	76.4% (completed surveys)
			E-mail (Web-based survey [WBS])	100	44%	6.0	81.4% (completed surveys)

[a] Mail = postal mail; e-mail (unless defined otherwise) = questionnaire embedded in an e-mail.

[b] Respondents were given the option to return a paper copy, and 3% made use of this option.

Source: For a complete list of references of these studies, please contact the authors.

References

Babbie, E. (1990), *Survey Research Methods*, Wadsworth, Belmont, CA.

Bachmann, D., Elfrink, J., and Vazzana, G. (1996), Tracking the progress of e-mail vs. snail-mail, *Market. Res.*, 8, 30–36.

Bachmann, D., Elfrink, J., and Vazzana, G. (1999), E-mail versus snail-mail face off in rematch, *Market. Res.*, 11, 10–15.

Baruch, Y. (1999), Response rate in academic studies—a comparative analysis, *Hum. Relations*, 52(4), 421–438.

Bosnjak, M. and Tuten, T.L. (2001), Classifying response behaviors in Web-based surveys, *J. Comput. Mediated Commun.*, 6(3).

Carayon, P., Haims, M.C., Hoonakker, P.L.T., and Swanson, N.G. (2000), Intervention research for reducing musculoskeletal injuries, in *Proceedings of IEA 2000/HFES 2000 Congress*, Vol. 2, Human Factors and Ergonomics Society, Santa Monica, CA, pp. 169–171.

Carayon, P. and Smith, M.J. (2000), Work organization and ergonomics, *Appl. Ergonomics*, 31, 649–662.

Carayon, P. and Smith, P.D. (2001), Evaluating the human and organizational aspects of information technology implementation in a small clinic, in *Systems, Social and Internationalization Design Aspects of Human-Computer Interaction*, Smith, M.J. and Salvendy, G., Eds., Lawrence Erlbaum Associates, Mahwah, NJ, pp. 903–907.

Carayon, P. and Hoonakker, P. (2001), Survey design, in *International Encyclopedia of Ergonomics and Human Factors*, Karwowski, W., Ed., Taylor & Francis, London, pp. 1899–1902.

Carmines, E.G. and Zeller, R.A. (1990), *Reliability and Validity Assessment*, Sage, Beverly Hills, CA.

Church, A.H. and Waclawski, J. (2001), *Designing and Using Organizational Surveys*, Jossey-Bass, San Francisco.

Cobanoglu, C., Warde, B., and Moreo, P.J. (2001), A comparison of mail, fax and web-based survey methods, *Market Res. Soc.*, 43, 441–52.

Commerce, U.S.D.o. The Digital Revolution. The Emerging Digital Economy.

Commerce, U.S.D.o. Building Out the Internet. The Emerging Digital Economy.

Converse, J.M. and Presser, S. (1986), *Survey Questions: Handcrafting the Standardized Questionnaire*, Sage, Beverly Hills, CA.

Cook, J.D., Hepworth, S.J., Wall, T.D., and Warr, P.B. (1981), *The Experience of Work*, Academic Press, London.

Cook, C., Heath, F., et al. (2000), A meta-analysis of response rates in web- or internet-based surveys, *Educ. Psychol. Meas.*, 60(6), 821–836.

Couper, M.P., Blair, J., et al. (1999), A comparison of mail and e-mail for a survey of employees in federal statistical agencies, J. Off. Stat., (15), 39–56.

Dillman, D.A. (1978), Mail and Telephone Surveys: The Total Design Method., John Wiley & Sons, New York.

Dillman, D.A. (1991), The design and adminstration of e-mail surveys, Annu. Rev. Sociol., 17, 225–249.

Dillman, D.A. (2000), *Mail and Internet Surveys: The Tailored Design Method*, John Wiley & Sons, New York.

Dillman, D.A., Tortora, J.C., et al. (1998), Influence of plain versus fancy design on response rates for Web surveys, Annual Meeting of the American Statistical Association, Dallas, TX.

Dillman, D.A., Tortora, J.C., et al. (1998), Principles for Constructing Web Surveys, Washington State University Social and Economics Sciences Research Center, Pullman.

Dommeyer, C.J. and Moriarty, E. (2000), Comparing two forms of an e-mail survey: embedded vs. attached, Mark. Res. Soc., 42(1), 39–50.

Fox, R.J., Crask, M.R., et al. (1988), Mail survey response rate: a meta-analysis of selected techniques for inducing responses, Public Opinion Q., 52, 467–491.

Heberlein, T.A. and Baumgartner, R. (1978), Factors affecting response rates to mailed questionnaires: a quantitative analysis of the published literature, Am. Sociol. Rev., 43, 447–462.

Hoonakker, P.L.T., van Dierendonck, D., van der Molen, H.F., and van Ginkel, A. (1998), The relation between experienced workload, working conditions, health and absenteeism in construction industry, in *Human Factors in Organizational Design and Management VI*, Vink, P., Ed., Elsevier Science, Amsterdam.

Kiesler, S. and Sproull, L.S. (1986), Response effects in the electronic survey, Public Opinion Q., 50, 402–413.

Kittleson, M.J. (1995), An assesment of the response rate via the postal service and e-mail, Health Values, 18(2), 27–29.

Kittleson, M.J. (1997), Determining effective follow-up of e-mail surveys, Am. J. Health Behav., 21(3), 193–196.

Martin, C.L. (1995), The impact of topic interest on mail survey response behavior, J.Mark. Res. Soc., 36(4), 327–337.

McAbe, S.E., Boyd, C.J., et al. (2002), Mode effects for collecting alcohol and other drugs data: web and U.S. mail, J. Stud. Alcohol, 63(6), 755–761.

Mehta, R. and Sivadas, E. (1995), Comparing response rates and response content in mail versus electronic mail surveys, J. Mark. Res. Soc., 37(4), 429–439.

Nunnaly, J.C. (1978), *Psychometric Theory*, McGraw-Hill, New York.

Parker, L. (1992), Collecting data the e-mail way, Training & Dev., July, 52–54.

Raine, H. (2001), The Changing Online Population: It's More and More Like the General Population, Pew Internet and American Life Project report.

Roberson, M.T. and Sundstrom, E. (1990), questionnaire design, return rates, and response favorableness in a employee attitude questionnaire, J. Appl. Psychol., 75(3), 354–357.

Salvendy, G. and Carayon, P. (1997), Data-collection and evaluation of outcome measures, in *Handbook of Human Factors and Ergonomics*, Salvendy, G., Ed., Wiley, New York, pp. 1451–1470.

Schaefer, D.R. and Dillman, D.A. (1998), Develeopment of standard e-mail methodology: results of an experiment, Public Opinion Q., 62: 378-97.

Schuldt, B.A. and Totten, J.W. (1994), Electronic mail versus mail survey response rates, Mark. Res., 6(1), 36–39.

Selwyn, N. and Robson, K., Using E-Mail as a Research Tool, http://www.soc.surrey.ac.uk/sru/SRU21.html.

Sheehan, K.M. and McMillan, S.J. (1999), Response variation in e-mail surveys: an exploration, J. Advertising Res., July/August, 45–54.

Sinclair, M.A. (1995), Subjective assessment, in *Evaluation of Human Work: A Practical Ergonomics Methodology*, Wilson, J.R. and Corlett, E.N., Eds., Taylor & Francis, London, pp. 68–100.

Solomon, D.J. (2001), Conducting Web-Based Surveys. Practical Assessment, Research & Evaluation. 7.

Steele, T.J., Schwendig, W.L., et al. (1992), duplicate responses to multiple survey mailings: a problem? J. Advertising Res., 32(2), 26–34.

Surveys, N.I. (2002), http://www.nua.ie/surveys/how_many_online.

Tse, A.C.B., Tse, K.C., et al. (1995), Comparing two methods of sending out questionnaires: e-mail versus mail, J. Mark. Res. Soc., 37, 441–446.

Weible, R. and Wallace, J. (1998), The impact of the Internet on data collection, *Mark. Res.*, 10, 19–27.

Yammarino, F.J., Skinner, S., et al. (1991), Understanding mail survey response behavior, Public Opinion Q., 55(4), 613–639.

Yun, G.W. and Trumbo, C.W. (2000), Comparative response to a survey executed by post, e-mail, & web form, J. Comput. Mediated Commun., 6(1).

77

Interview Method

Leah Newman
The Pennsylvania State University

77.1 Background and Application

The interviewing process has been used as a tool for gathering information for decades. The roots of the interview process can be traced back to the Egyptians during the gathering of census data (Fontana and Frey, 2000). More formally, the process was extensively used in clinical diagnoses and counseling, and it became a widespread technique during World War I, when interviews were conducted to perform psychological tests (Fontana and Frey, 2000).

The interviewing process allows for rich, highly informative, detailed exchanges to take place between researchers and participants. Through the interview, researchers can delve deeply into the world of their subject(s) because they believe the information to be gathered is important. The interview is a window to an understanding of the behaviors of those being interviewed (Seidman, 1998). In macroergonomics applications, the interview method is particularly useful in identifying work-system problems or problem symptoms.

77.2 Procedure

Once a researcher has decided upon using interviews as a method for collecting data, there are a few things that must be done to ensure a quality outcome. Conducting an interview is very subjective in nature, and much of the process depends upon the individual administering the session. However, there are a few details one must understand when preparing for an interview.

An interviewer must first locate and enlist the cooperation of participants (also called gaining entry), and this might also involve locating someone who is willing to act as an informant or guide for the researcher. This individual will provide information and direction to the researcher as s/he tries to understand the experiences of the participants (Fontana and Frey, 2000). The researcher must also be able to gain the trust of, and establish rapport with, those being interviewed. This will help to ensure that the responses are candid and honest, especially if the participants are being asked to divulge sensitive information (Fontana and Frey, 2000). Due to time and financial constraints, it is not feasible to interview large numbers of individuals within an organization. Therefore, an approach often used is that of a stratified random sample of individuals (Wengraf, 2001). This stratified sample allows one to interview a small number of individuals representative of the larger group. When conducting the actual interview, the researcher has to take into account the attitude of the participants while also concealing his or her personal feelings and biases. The research objectives must be clearly defined to focus the attention of the participants so that the discussion is concentrated toward the goals of the study. Also, a detailed outline or guide of the topics to be covered during the meeting must be developed (Smith, 1995).

Next, one must determine the interview content. A literature review should be conducted to gain an understanding of previous relevant research and to prepare a script for the interview. In general, the session should begin with questions that gradually draw the participants into the discussion, continue with transition questions and key questions that are more focused on the research, and end with questions that tie the session together and bring closure. The interviewer should prepare for a session that will last anywhere from 90 min to 2 h (Smith, 1995).

Once the interview has ended, the researcher must analyze and report the results. Field notes should be taken immediately following each session. Ideally, someone should be available during the meeting to take notes on the discussions, as well as to track the flow of the discussion to get a better idea of participant interaction. The researcher is then charged with creating codes and constructs from the data retrieved from the conversations. This involves several iterations of reading through transcripts to gain an understanding of the topics being discussed from the viewpoint of the participants. Each new reading is likely to generate new thoughts and ideas, hence new codes and constructs (Smith, 1995).

From this analysis, a master list of themes will be generated. These themes should be those keywords that best represent the thoughts presented by the participant. Many of these themes will originate from the interview protocol; others will be created as the transcripts are read (Smith, 1995). This master list also may include other codes that help to further explain an overarching or master theme (Smith, 1995). As themes are emerging from the transcripts, it is important to keep track of where these themes were found within the transcripts.

77.2.1 Interviewer Role

Researchers have increasingly become more concerned with the role of the interviewer with regard to this data collection process. In particular, they are concerned with what appears to be the controlling nature of the interviewer (Fontana and Frey, 2000). These concerns have required researchers to focus more attention on the voice of the participant; the relationship between the interviewer and the participant; as well as the importance of gender, race, socioeconomic background, and age in the interviewing process (Fontana and Frey, 2000). Many of these concerns can be addressed in the approach one takes in conducting the interview.

When conducting interviews, it is important that the interviewer remain neutral, never providing an opinion regarding a participant's responses. The interviewer must work in a manner that encourages participation without providing guidance or evaluation of responses (Fontana and Frey, 2000). An interviewer must be flexible, objective, persuasive, a good listener, open-minded, and able to think on one's feet in order to clarify confusion and allay concerns.

77.2.2 Interview Formats

Interviews can be developed in a structured, semistructured, or unstructured format (Fontana and Frey, 2000). Structured interviews require questions that are very specific in nature with a closed or limited number of available responses. The same questions are asked of all of the participants, in the same sequence, with little flexibility allowed for variation or elaboration of responses (Fontana and Frey, 2000).

Semistructured interviews require a predetermined set of open-ended questions. Participants are able to elaborate upon their responses rather than being forced to choose between a predetermined set of answers.

Unstructured interviews require very little guidance on the part of the researcher. It is an open and flexible process that allows for the collection of very rich, voluminous data (Fontana and Frey, 2000). The semistructured interview format is the approach that is most commonly used (Fontana and Frey, 2000).

77.2.3 Guidelines

Specific guidelines on how to conduct an interview would be difficult to provide because technique depends upon the research being conducted, the individual conducting the interviews, the interviewees who are participating in the research, as well as the questions being asked. Interviews can be very subjective, depending a great deal upon the aforementioned characteristics. However, there are a few basic, almost commonsense, guidelines that can be utilized:

- Talk less/listen more
- Establish a safe and positive environment for participants where responses are free from criticism
- Encourage participation from everyone
- Follow hunches and follow up on participants' responses
- Be spontaneous
- Make participants feel that their input will make a difference in the research
- Do not drag the meeting on for too long
- Provide incentives (e.g., food, money)

77.2.4 Macroergonomic Applications

Interviews often are used within a macroergonomic framework in an effort to identify problems associated with the design of a specific work system (Hendrick and Kleiner, 2001). The interview process allows the researcher to ask questions of the employees to gain a better understanding of how the system works and what design changes can be implemented to improve the individual job and/or the work system. These changes will, ideally, improve levels of comfort, safety, and efficiency for employees. Interviews also can be used as a means of gathering rich pools of information about an organization that can later be incorporated into a questionnaire survey (see Chapter 76, Macroergonomic Organizational Questionnaire Survey) and distributed to larger numbers of individuals across larger geographical settings (Hendrick and Kleiner, 2001).

77.3 Advantages

- A useful technique for collecting rich and highly informative data (e.g., problems or symptoms of problems in work systems and their possible causes, problems with the design of products, and insights into how to correct perceived deficiencies)
- Opportunity for an effective interviewer to develop a rapport with the interviewee, thus obtaining a more precise description of the interviewee's experience and more accurate results
- A powerful technique for gaining a thorough understanding of the experiences of others (Seidman, 1998)

77.4 Disadvantages

- The process can be very expensive and time-consuming.
- Interviews may be subject to biases; specifically, the interviewer's beliefs may affect the way in which questions are asked and subsequently influence participant responses (Madriz, 2000).
- Interviews are open-ended with regard to coding and interpretation.

- Interviews often are difficult to summarize (Seidman, 1998).
- The interviewee may be prone to tell the interviewer what s/he thinks the interviewer wants to hear rather than his or her true beliefs, perceptions, or feelings.

77.5 Example Output

The initial output from an interview will be a transcript of conversations where each participant is given some type of identifier followed by his or her comments. The subsequent output gleaned from these transcripts can take many forms, depending on the preference of the researcher. Table 77.1 is an example of how the data can be presented. Table 77.1 shows statements made by participants and the resultant codes assigned by the researcher (Newman, 1997).

77.6 Related Methods

Interviews are used along with other methods in an effort to access the opinions, viewpoints, attitudes, and experiences of others. Related methods include focus groups, participant observation, and organizational questionnaire surveys. These methods often are used along with interviews to collect additional data to enable a better understanding of the research topic. For example, interviews can be used to supplement and/or corroborate information collected from surveys and/or participant observations (Madriz, 2000). As indicated by Seidman (1998), one can have what he terms an "observational understanding" of a particular job or situation. However, this observational understanding might not necessarily coincide with how a particular individual views his/her job or situation. By conducting interviews, the researcher is now in a position to gather more information, thereby garnering more understanding and appreciation of the individual's experience.

77.7 Standards and Regulations

There are no specific standards and regulations regarding interviews. In general, interviewees should be told the specific purpose of the interview, how the data will be used, the extent to which responses will be kept anonymous, and be provided feedback about the results of the study.

77.8 Approximate Training and Application Time

Most interviews run anywhere from 1 to 2 h. The time frame depends a great deal upon those being interviewed and what is being accomplished. Because the purpose of interviewing is to present an opportunity for individuals to share and elaborate on their particular experiences, it is necessary to provide a time frame that will accomplish this goal. Typically, a meeting under 90 min would be too short a period of time to accomplish the goal, and a meeting that lasts longer than 2 h might be too long for someone to sit at one time (Seidman, 1998). Because interviews are conducted with each interviewee individually, the process of interviewing a number of people can be quite time-consuming.

TABLE 77.1 Participant Statements and Codes Assigned by Researcher

Statement	Code	Subcode
Racist or race-based comments by instructors	Faculty educational climate	Attitude
Bad faculty	Faculty educational climate	Academics
Difficulty in finding time to study at the computer labs; computer labs overcrowded	Institutional	Resources
Not getting into a department	Institutional	Academics
No academic support from engineering system	Institutional	Alienation

77.9 Reliability and Validity

Validity in qualitative research refers to description and explanation, and whether or not a given description matches a given explanation. Is the explanation trustworthy? Merriam (1988) suggests six basic strategies to ensure truth-value:

1. Triangulation, or using multiple sources of data to confirm emerging themes
2. Member checks, or taking the data and interpretations back to the team members and seeing if the results are trustworthy; or having an outsider read the field notes and transcripts (Janesick, 1994)
3. Long-term or repeated observations of same phenomenon
4. Peer examinations, or sharing emerging themes with colleagues
5. Participatory modes of research involving participants in all phases of the research
6. Examination of researcher bias, or clarification of the researcher's assumptions, worldview, and theoretical orientation

The history surrounding qualitative research suggests that the value of the research lies in its uniqueness. Therefore, reliability in the conventional sense of replicability denies the very characteristic that makes it most valuable (Janesick, 1994). Generalizability should be left to those "who wish to apply the findings in their own situation" (Merriam, 1988). Still, there are those who believe that because the findings gleaned from qualitative methods represent the ideals, attitudes, and thoughts of a specific group at a specific point in time, the results cannot be generalized to an entire population.

77.10 Tools Needed

Very few tools are needed to conduct interviews. All that is required are writing utensils, paper, and a reliable tape recorder.

References

Fontana, A. and Frey, J.H. (2000), The interview from structured questions to negotiated text, in *Handbook of Qualitative Research Methods*, Denzin, N. and Lincoln, Y., Eds., Sage, Thousand Oaks, CA, pp. 645–672.

Hendrick, H.W. and Kleiner, B. (2001), *Macroergonomics: An Introduction to Work System Design*, Human Factors and Ergonomics Society, Santa Monica, CA.

Janesick, V.J. (1994), The dance of qualitative research design: metaphor, methodology, and meaning, in *Handbook of Qualitative Research Methods*, Denzin, N. and Lincoln, Y., Eds., Sage, Thousand Oaks, CA, pp. 209–219.

Madriz, E. (2000), Focus groups in feminist research, in *Handbook of Qualitative Research Methods*, Denzin, N. and Lincoln, Y., Eds., Sage, Thousand Oaks, CA, pp. 645–672.

Merriam, S.B. (1988), *Case Study Research in Education: A Qualitative Approach*, Jossey-Bass, San Francisco.

Newman, L. (1997), Quality Improvement/Assessment of Educational System for Students of Color in the University of Wisconsin College of Engineering, University of Wisconsin, Madison.

Seidman, I. (1998), *Interviewing as Qualitative Research: A Guide for Researchers in Education and the Social Sciences*, Teachers College Press, New York.

Smith, J.A. (1995), Semi-structured interviewing and qualitative analysis, in *Rethinking Methods in Psychology*, Smith, J.A., Harré, R., and Van Langenhove, L., Eds., Sage, London, pp. 9–26.

Wengraf, T. (2001), *Qualitative Research Interviewing*, Sage, Thousands Oaks, CA.

78

Focus Groups

Leah Newman
Pennsylvania State University

78.1 Background and Application

Social scientists initially used focus groups during the 1920s, most often to assist in the development of questionnaires (Morgan, 1996). Robert Merton and Paul Lazarsfeld pioneered the use of the focus group method in social science research in the 1940s (Madriz, 2000). Between World War II and the 1970s, focus groups were used primarily in market research to ascertain the wants and needs of the public. They continue to be used for this purpose today. Lastly, from the 1980s until recently, focus groups have been used to collect data in a variety of social areas (Morgan, 1996). Recently, focus groups have gained greater popularity among qualitative researchers as a method of choice (Madriz, 2000). Researchers view the focus group process as a window into human interaction (Madriz, 2000).

Focus groups allow one to interview small groups of individuals simultaneously, allowing those being interviewed to feel more comfortable speaking openly on what often are sensitive subjects. Individuals typically indicate feeling less intimidated during focus group interviews. They have also indicated that focus groups are more rewarding than individual interviews (Madriz, 2000) because there is a certain level of camaraderie felt among the subjects sharing during a focus group interview. Focus groups provide participants a safe environment in which to share information with individuals from similar ethnic, gender, and socioeconomic backgrounds (Madriz, 2000).

78.2 Procedure

When conducting a focus group, it is imperative to select an experienced and effective moderator as well as a suitable site. The moderator must have excellent communications skills, must be able to lead and control conversations without being overbearing, and must possess a strong working knowledge of the research being conducted. The moderator is considered the most important part of a successful focus group. The moderator should make sure all participants are sharing their thoughts and experiences and not let anyone dominate the conversation (Smith, 1995). The moderator should ideally be someone who can remain completely objective regarding the research, and someone who is willing to provide an unbiased interpretation of the conversation (Smith, 1995). Regarding the focus group site, it is important to create an environment that is relaxed and comfortable.

For focus groups, it is also important to determine the ideal size of the group(s), as well as the number of sessions. Typical focus groups contain anywhere from 6 to 12 participants. Keep in mind that these numbers are not definitive. It may be difficult to encourage discussions within small groups, while a group that is too large may be difficult to manage. The size of the group will depend, again, on the purpose of the research (Smith, 1995). Ideally, the researchers should conduct as many focus groups as are necessary to present trustworthy results. Three to five sessions with different groups is often the target, which helps to provide very rich data as well as comparisons between groups (Smith, 1995). The analysis of the information garnered from focus groups is similar to that described for the interview method, covered in Chapter 77.

78.2.1 Facilitator Role

Researchers have increasingly become more concerned with the role of the facilitator/moderator with regard to the data collection process. In particular, they are concerned with what appears to be the controlling nature of focus group facilitators (Fontana and Frey, 2000). These concerns have required researchers to focus more attention on the voice of the participants; the relationship between the facilitator and the participants; as well as the importance of gender, race, socioeconomic background, and age in the interviewing process (Fontana and Frey, 2000). Many of these concerns can be addressed in the approach one takes in facilitating the focus group.

When facilitating focus groups, it is important that the moderator remain neutral, never providing an opinion regarding a participant's responses. The moderator must work in a manner that encourages participation without providing guidance or evaluation of responses (Fontana and Frey, 2000). Moderators or facilitators must be flexible, objective, persuasive, good listeners, open-minded, and able to think on their feet in order to clarify confusion and probe responses. The facilitator also must be able to discourage the development of a dominant individual or coalition of individuals; ensure that all participants are able to contribute to the discussion; and encourage obstinate individuals to participate without turning them off to the process (Seidman, 1998).

78.2.2 Forms of Focus Groups

Focus groups can take various forms depending upon the purpose, including brainstorming sessions with very little structure or direction, semistructured sessions with some preestablished open-ended questions, or more structured meetings as when using the nominal group technique (NGT), Delphi method, or focus groups for marketing research (Fontana and Frey, 2000). The NGT involves individuals making written responses to problems, with some time allowed for discussion. In some cases, respondents are in different places while participating, which explains the name, in that a group is formed in name only (Fontana and Frey, 2000). The Delphi method is a group communication process used to draw forth information regarding a specific task or problem. The method usually involves anonymity of responses, with the information usually exchanged via e-mail, regular mail, or fax (Fontana and Frey, 2000).

78.2.3 Guidelines

Specific guidelines on how to conduct a focus group would be difficult to provide because technique depends upon the research being conducted, the individual facilitating the focus groups, the group members who are participating in the research, as well as the questions being asked. However, as with the interview process (see Chapter 77), there are a few basic, almost commonsense, guidelines that can be utilized.

- Talk less/listen more
- Establish a safe and positive environment for participants where responses are free from criticism
- Encourage participation from everyone
- Follow hunches and follow up on participants' responses
- Be spontaneous
- Make participants feel that their input will make a difference in the research
- Do not drag the meeting on for too long
- Provide incentives (e.g., food, money)

78.2.4 Macroergonomic Applications

Focus groups can be an important part of gaining valuable information regarding work systems within organizations. Groups of individuals can be brought together in an effort to better understand how specific systems work or interact (Hendrick and Kleiner, 2001). From this understanding, interventions can be developed to improve working conditions, thereby improving the function of the organization as a whole. Oftentimes, subsets of individuals from a particular area of an organization are brought together to provide insight as to how best to conduct research to improve that area. This group of individuals can assist in the development of surveys that will be distributed to a larger group. Focus groups also can help in the design and implementation of specific interventions or any changes that are proposed to improve working conditions for the larger group.

78.3 Advantages

In addition to the advantages mentioned for the interview method (see Chapter 77), the advantages of using focus groups are:

- The researcher can observe the process of interaction between and among the participants.
- Comments made by one individual may spark thoughts from other participants.
- Interaction among the participants limits the level and amount of contribution made by the researcher or facilitator. This enables the researcher to collect more information related specifically to the attitudes, ideas, and concerns of the participants (Madriz, 2000).
- Data from a number of persons can be obtained more economically and efficiently than with the interview method.

78.4 Disadvantages

- Focus groups are often carried out in a neutral space instead of the place where the social interaction being observed usually takes place, which thus limits the amount of "behavioral information" that is collected (Fontana and Frey, 2000).
- The presence of the interviewer or facilitator can affect participant behavior, thereby altering the genuineness of the data (Madriz, 2000).
- The group culture that is established may impede individual responses and "groupthink" may emerge (Fontana and Frey, 2000).
- One individual may dominate the session (Fontana and Frey, 2000).

78.5 Example Output

The initial output from an interview will be a transcript of conversations, with each participant given some type of identifier followed by his or her comments. The subsequent output gleaned from these transcripts can take many forms, depending on the preference of the researcher. Table 78.1 is an example

TABLE 78.1 Participant Statements and Codes Assigned by Researcher

Statement	Code	Subcode
Not knowing how to study when you get here	Personal facts	Precollege
Lack of computer resources at home	Personal facts	Resources
Lack of confidence	Personal facts	Self worth
Lack of computer training	Personal facts	Skills

of how the data can be presented. Table 78.1 shows statements made by participants and the resultant codes assigned by the researcher (Newman, 1997).

78.6 Related Methods

The two primary methods used when conducting qualitative research are participant observations and individual interviews (Madriz, 2000). Focus groups are used along with other methods in an effort to access the opinions, viewpoints, attitudes, and experiences of others. These processes are used along with other methodologies to the extent to which it maximizes the possibility of collecting additional data that allows for a better understanding of the research topic.

Focus groups are also often used in an effort to develop better questionnaire surveys (see Chapter 76, Macroergonomic Organizational Questionnaire Survey). Frequently, the responses available on surveys do not adequately represent the position of the respondent. By facilitating a focus group, the researcher would enable the participant to elaborate on his/her responses regarding specific questions of interest, thereby providing a more accurate depiction of those experiences (Fontana and Frey, 2000). Focus groups also are used during the developmental stage of survey design. In addition, a researcher can enlist a focus group to test the wording as well as the measurement scales of a survey (Fontana and Frey, 2000).

78.7 Standards and Regulations

There are no specific standards and regulations regarding focus groups.

78.8 Approximate Training and Application Time

No specialized training beyond general knowledge and experience as a facilitator are required. Persons without training or experience in facilitating group discussions would benefit from a training course or practicum with an experienced facilitator.

Most focus group meetings run anywhere from 1 to 2 h. The time frame depends a great deal upon those being interviewed and what is being accomplished. Because the purpose of a focus group is to present an opportunity for individuals to share and elaborate on their particular experiences, it is necessary to provide a time frame that will accomplish this goal. Typically, a meeting under 90 min would be too short a period of time to accomplish the goal, and a meeting that lasts longer than 2 h might be too long for someone to sit at one time (Seidman, 1998).

78.9 Reliability and Validity

Any type of formal research being conducted requires the researcher to consider the validity and reliability of the data that are collected and the measurement tools being used. Reliability and validity play a similar role with regard to the focus group method and the interview method. Researchers often become lost in the qualitative process or methodology, and as a result, the issues of validity, reliability, and generalizability are not addressed or evident (Janesick, 1994). When well-planned focus groups are carried out by a skilled facilitator, the results can be consistent and their application successful. For further information regarding the issues of reliability, validity, and generalizability, see Chapter 77, Interview Method.

78.10 Tools Needed

Very few tools are needed to facilitate focus groups. All that is required are writing utensils, paper, and a reliable tape recorder. Some individuals may wish to use a video recorder in an effort to obtain a more accurate depiction of who is responding to a question. A video recorder would also provide a better portrayal of the participant interaction and nonverbal responses. However, depending upon the venue, one may not have access to video cameras and high-quality recording devices. Researchers often find themselves in positions where they have to rely on their ability to quickly take notes, mental as well as written (Fontana and Frey, 2000).

References

Fontana, A. and Frey, J.H. (2000), The interview from structured questions to negotiated text, in *Handbook of Qualitative Research Methods*, Denzin, N. and Lincoln, Y., Eds., Sage, Thousand Oaks, CA, pp. 645–672.

Hendrick, H.W. and Kleiner, B. (2001), *Macroergonomics: An Introduction to Work System Design*, Human Factors and Ergonomics Society, Santa Monica, CA.

Janesick, V.J. (1994), The dance of qualitative research design: metaphor, methodology, and meaning, in *Handbook of Qualitative Research Methods*, Denzin, N. and Lincoln, Y., Eds., Sage, Thousand Oaks, CA, pp. 209–219.

Madriz, E. (2000), Focus groups in feminist research, in *Handbook of Qualitative Research Methods*, Denzin, N. and Lincoln, Y., Eds., Sage, Thousand Oaks, CA, pp. 645–672.

Morgan, D.L. (1996), *Focus Groups as Qualitative Research*, Sage, Newbury Park, CA.

Newman, L. (1997), Quality Improvement/Assessment of Educational System for Students of Color in the University of Wisconsin College of Engineering, University of Wisconsin, Madison.

Seidman, I. (1998), *Interviewing as Qualitative Research: A Guide for Researchers in Education and the Social Sciences*, Teachers College Press, New York.

Smith, J.A. (1995), Semi-structured interviewing and qualitative analysis, in *Rethinking Methods in Psychology*, Smith, J.A., Harré, R., and Van Langenhove, L., Eds., Sage, London, pp. 9–26.

79

Laboratory Experiment

Brian M. Kleiner
Virginia Polytechnic Institute
and State University

79.1 Background and Applications

Laboratory experiments typically include: a research hypothesis that predicts causal effects of one or more variables on others; at least two levels of one or more independent variables; objective assignment of subjects to conditions; systematic procedures for empirically testing hypothesized causal relationships; and specific controls to reduce threats to internal validity (Graziano and Raulin, 2000). In work-system design, experimentation and quasi-experimentation are needed to build an understanding of what works and why (Hendrick and Kleiner, 2001). Although much research on groups and teams is not empirical, it is entirely plausible to empirically investigate factors from the personnel, technological, and environmental subsystems of a work system as well as their interactions. For example, actual work systems can be simulated in the laboratory, and specific independent work-system variables can be manipulated to study their effect on outcome (dependent) variables of interest. Typically, it is desirable to return to the field following laboratory investigation for field validation, usually in the form of quasi-experiments, using organizational members and natural work groups as subjects (see Chapter 80, Field Study and Field Experiment).

79.2 Procedure

79.2.1 Review Literature

Typically, the first step in the laboratory research process is to understand the literature within the domain of interest. This is usually preceded by an initial idea or question. Specifically, it is useful to discover what is known in the domain and what knowledge remains to be learned. This leads to a problem statement. For macroergonomics or work-system design, the relevant bodies of knowledge are diverse. Domains include industrial and organizational psychology, systems engineering, ergonomics and human factors, and management and business. Research that involves groups and teams can be found in these domains as well as education, computer science, and other disciplines.

79.2.2 Pose Research Questions and Hypotheses

From the literature, specific testable research questions are derived. Hypotheses are specific predictions of the effects of independent variables on dependent variables and are based on the research questions. Theories from the literature often guide decisions about the research. In group and team laboratory experimentation, there is typically a manipulation and associated prediction of the effect(s) upon the group or team. In order to optimize technological and organizational design variables of groups or teams, an understanding of the underlying team process dimensions that translate into successful team outcomes must first be achieved. To understand and objectively measure group or team performance, we must operationalize and, when possible, quantify these underlying processes.

79.2.3 Test Research Hypotheses

Hypotheses involve predictions about the effect of the independent variable(s) on the dependent variable(s). The null hypothesis states that there is no effect beyond that caused by chance. The alternative or causal hypothesis states that the independent variable causes a significant change in the dependent variable(s). It also is possible that confounding-variable hypotheses will be confirmed, i.e., that observed differences are due to the effects of confounding factors. An experimental design establishes which factors or variables will be manipulated. Measured variables or dependent variables are also identified. Most laboratory experiments then manipulate independent variables to assess their effect(s) upon dependent variables. Subjects are selected from a population. Sometimes, specific selection and/or sampling criteria guide the process.

79.2.4 Using Groups and Teams as Subjects

In macroergonomic laboratory studies, groups or teams of subjects are often employed. A nominal group is a group in name only. In such studies, subjects are typically recruited as individuals and then combined into groups for experimentation. Alternatively, teams can be used. Teams can be recruited as units based on their previous experience operating as teams. It also is possible to train groups to become cohesive teams, but this can be time consuming and costly. In general, one of the difficulties in group and team research is to control for team experience and characteristics. Intra- and interteam variability can thus surface as a potential confound.

This type of research requires the development of a valid set of measures (Prince et al., 1992), and these measures must include sociotechnical, process, and outcome dimensions in order to reveal the relationship among team characteristics, processes, and outcomes, and to provide a framework for further research on virtual-team design. With virtual teams, the traditional physical infrastructure needed, including geographical collocation and means for physical interactions, are replaced by information and communication technology.

Modern organizational theory has as one of its basic assumptions the need to group people and units together for coordination and supervision purposes. Lucas and Baroudi (1994) note that current directions in information technology are increasingly invalidating this assumption. Research in this area is scarce, and the results already point to conflicting conclusions.

When using groups or teams as subjects, it is important to consider different levels of performance. Most typically, it is desirable to measure performance at the individual, group, and possibly organizational levels of performance. Within-and-between analysis (WABA) is one method of statistical analysis that can address these different levels of performance. Thus, in macroergonomic studies we manipulate independent variables and measure their effect(s) on individual, group, and organizational performance indices.

79.3 Advantages

- Ability to answer questions about causality
- Ability to exercise control over variables
- Systematic process
- Use of groups or teams is realistic

79.4 Disadvantages

- Generalizablity to the real world often questioned (i.e., external validity)
- Sometimes difficult to control extraneous and confounding variables
- Sometimes slow and tedious process
- Difficult to control for group or team variability

79.5 Related Methods

Experiments and quasi-experiments also can be performed in the field (e.g., in organizations). Another popular method is survey research. These methods are covered in Chapters 80 and 76, respectively.

79.6 Standards and Regulations

Institutional review boards (IRB) provide oversight for laboratory studies in universities, research institutes, hospitals, and schools. Military research organizations have similar bodies for oversight. IRBs typically comprise peer researchers and stakeholders from the community. By law, federally sponsored research involving humans as subjects must be approved by IRBs. The federal policies and procedures for laboratory experimentation include such areas as beneficence, selection, and risk. In short, the benefits must outweigh the costs, selection must be fair and equitable, and risk must be minimized.

79.7 Approximate Training and Application Times

An educational course in experimental design is helpful. Ideally, the researcher also has a background in probability and statistics. Oversight for laboratory experimentation should be provided by a Ph.D. level researcher, typically called the "principal investigator."

79.8 Reliability and Validity

A properly designed and performed laboratory experiment will yield high reliability and validity. Reliability has to do with the consistency of measurement. While absolute validity cannot be guaranteed, the laboratory experimenter will anticipate and reduce potential threats to validity. This typically involves the appropriateness and soundness of the methodology and its inherent procedures. The term *construct validity* refers to the extent to which the results support underlying theory and constructs. Given the artificial environment of the laboratory, external validity or the extent to which results are generalizable is of particular concern. Statistical and internal validity is confirmed through statistical analysis. Using ANOVA or other appropriate tests, the researcher confirms that results are more than the product of chance.

79.9 Tools Needed

There are many useful statistical software packages available, including SAS, SPSS, BMDP, JMP, etc. Three orthogonal dimensions of group support can be derived from previous taxonomies and the results of empirical studies. These are the level of support for the communication process, the level of support for the decision-making process, and the level of support for the sense of presence or virtuality of the group (Cano et al., 1998). For group and team research, there are several off-the-shelf software packages to assist with computer-supported collaborative work. An example of a communication-support package is Group Systems. An example of a decision-support package is Group Expert Choice.

References

Cano, A.R., Meredith, J.W., and Kleiner, B.M. (1998), Distributed and collocated group communication vs. decision systems support, in *Proceedings of the Human Factors and Ergonomics Society 42nd Annual Meeting*, Human Factors and Ergonomics Society, Santa Monica, CA.

Graziano, A. and Raulin, M. (2000), *Research Methods: A Process of Inquiry*, Allyn & Bacon, Boston.

Hendrick, H.W. and Kleiner, B.M. (2001), *Macroergonomics: An Introduction to Work System Design*, Human Factors and Ergonomics Society, Santa Monica, CA.

Lucas, H.C. and Baroudi, J. (1994), The role of information technology in organizational design, *J. Manage. Inf. Syst.*, 10, 9–23.

Prince, A., Brannick, M.T., Prince, C., and Salas, E. (1992), Team process measurement and implications for training, in *Proceedings of the Human Factors Society 36th Annual Meeting*, Human Factors Society, Santa Monica, CA, pp. 1351–1355.

80

Field Study and Field Experiment

Hal W. Hendrick
University of Southern California

80.1 Background and Application

The field study method and its variation, the field experiment, along with the laboratory and question-naire survey methods, are the oldest and most classic means of studying organizations. The field study method is used to gather information on organizational or work-system functioning through systematic direct observation. This information is most often used to identify possible causal relationships between work-system variables and to identify problems with organizational functioning.

The field experiment differs from a pure field study in that one or more variables in the field or organizational setting is manipulated so that its effect on the dependent variable(s) of interest — usually some aspect of work-system performance — can be observed directly. Often, the variable that is manip-ulated or changed is some aspect of the work system's structure or processes. That change may have been suggested as the result of a field study, organizational questionnaire survey, or laboratory simulation. Often, the field experiment is done for part of the organization's work system and, if the change proves effective, it can be implemented throughout the organization.

80.2 Field Study

80.2.1 Procedure

The field study method also is variously referred to as systematic or naturalistic observation and as real-life research. These terms, taken together, provide a good description of this approach. It involves going out into the field to systematically observe events as they occur naturally in real life. The primary concern of the researcher in using this method is *realism*. By systematically observing events as they occur naturally in the work setting, the researcher eventually can tease out causal relationships or, at least, identify correlates among variables in the work system that suggest causation. This information can be highly useful in understanding how aspects of the work system's structure, processes, or management and supervision affect various organizational effectiveness dimensions, such as employee health, safety, or job satisfaction, and system productivity or quality of production. In a pure field study, no attempt is made to control potentially causal variables in the field (i.e., work system) situation (Vercruyssen and Hendrick, 1990).

When the researcher is interested in using the field study method to determine causal relationships, s/he must control variables that can influence the dependent variable of interest, but the researcher does so as the control occurs naturally. For example, let us assume that variables A, B, C, and D exist in a given work system, and we wish to determine which one primarily is responsible for changes in some dependent variable, such as employee performance. The researcher would have to wait for situations to occur where different values of A, B, and C, were present, but D was constant. Similarly, the researcher would have to observe situations where the only variables *changing* were ABD, ACD, and BCD. By this procedure, the researcher eventually would be able to determine which variable was the primary causal one. A classic example of this kind of application was an investigation to determine factors that distinguish high-producing from low-producing work units, conducted by the Survey Research Center of the University of Michigan (Kahn and Katz, 1963). The Survey Research Center conducted field study investigations of a wide variety of work units, including in an insurance company, automotive manufacturer, tractor company, electric utility, and maintenance-of-way section gangs on a railroad. Through this extensive investigation over a period of years, four factors were isolated that appeared to distinguish high-producing from low-producing work units. Three of these were found to be characteristics of the work supervisor. The supervisors of high-producing units (a) tended to clearly distinguish their job as supervisor from that of their subordinates, (b) were employee oriented rather than just production oriented (i.e., they treated their subordinates as persons rather than simply as units of production), and (c) did not oversupervise (i.e., they told their people *what* to do, not step-by-step *how* to do it). The fourth factor was a general feeling among the high-producing workers that their unit was better than others in the organization.

A common use of the field study approach in macroergonomics is to examine existing performance records for a given work system and the specific conditions or nature of the work system under which those performances occurred. Through this approach, it is often possible (a) to identify problem areas with the work system's design that are amenable to macroergonomics intervention and (b) to gain insight as to the nature of the macroergonomics intervention that is needed.

80.2.2   Advantages

- The field study method's primary advantage is realism. By observing things as they occur naturally, the researcher avoids the sterility and artificiality of the laboratory. However, a caution is in order here: it is important for the researcher to recognize that his or her very presence changes the situation and thus can affect what happens. The researcher has to take great care to be as unobtrusive as possible.
- Because of the realism of this method, when the researcher is able to establish cause–effect relationships, we can have high confidence in the practical usefulness of the results.

80.2.3   Disadvantages

- The researcher has to wait for things to occur naturally. Consequently, the observation process may take a long time and incur considerable expense before any cause–effect relationships can be established.

- The researcher may have to observe things occurring naturally many times under various conditions before extraneous variables can be eliminated as causal factors and the true causal variables teased out and their interactions identified.

80.3 Field Experiment

Perhaps the most widely used of the classical methods in macroergonomics interventions is the field experiment. The field experiment differs from the field study in that, instead of passively observing events as they occur naturally, the ergonomist acts as a change agent. Selected variables are deliberately and systematically manipulated, and the effect on the outcome or performance variables of interest is observed. Often, the variable(s) to be manipulated has been identified through a field study or other classical methods, such as a laboratory simulation, interviews, or organizational questionnaire survey (Vercruyssen and Hendrick, 1990).

In one such study by the author, a field study of several operational units in a technical training organization had suggested that the division supervisors were maintaining too much decision authority, rather than delegating more of the authority to the branch chiefs who, through training and experience, had become quite professionalized. A related factor was that the branch chiefs were being restricted by formalized rules and procedures that were inhibiting innovation and the branch chiefs' ability to respond quickly to unanticipated events. As a result, the branch chiefs felt their skills were being wasted, and they were frustrated, lacking in motivation, and performing poorly. In one division, a field experiment was conducted where many of the formal rules and procedures were eliminated and much of the tactical (day-to-day) decision authority was delegated to the branch chiefs. The performance improvement in both the behavior of the branch chiefs and their respective units was dramatic. In addition, the change freed up the division chief to concentrate on strategic planning and decision making, thus making better use of his experience and skills. As a result of this field experiment, the entire work system was modified to incorporate these same changes in all divisions and branches.

80.3.1 Macroergonomics Applications

As described in the example above, one effective way of using a field experiment is as a follow-on to an initial field study. An initial field study can suggest possible dependent variables that can then be manipulated in a field experiment. For example, a review of performance records and an analysis of the work system's organizational structure (see Chapter 89, Macroergonomic Analysis of Structure) may indicate where a structural change is needed. The change then could be made and the impact on the work system's performance observed. Sometimes this work-system change will be done in a particular part of the organization to test it. If the change proves effective, it then will be implemented on a larger scale. Because of the potential problems of employee acceptance and support of the change, the field experiment often is combined with a participatory ergonomics approach to effect the intervention. As noted in Chapter 81, Participatory Ergonomics, employees are far more likely to accept and support changes where they are actively involved in the change process.

80.3.2 Advantages

To a large extent, the field experiment combines the advantages of the field study and the laboratory experiment while overcoming their major disadvantages.

- Like the laboratory study, the researcher deliberately manipulates the dependent variable(s) of interest, thus overcoming the field study's problem of having to wait for things to occur naturally.
- The field experiment gains the field study's advantage of realism.
- In comparison with the field study, the field experiment is more efficient in terms of time and related costs.

80.3.3 Disadvantages

- By "artificially" causing a desired change to occur, the researcher or change agent may be introducing extraneous variables that influence the effects of the change. In macroergonomics field experiments, one such problem can be the employees' perceptions of the purpose of the change. How employees perceive the intervention can alter how they respond to it and their related motivation.
- How changes are implemented can sometimes determine the success or failure of the intervention.
- Sometimes, the cost of using workers in field experiments is seen by organizations as too prohibitive. For this reason, in presenting such a proposal to management, it is important to include a sound cost-benefit analysis.

80.4 Related Methods

- Laboratory experiment
- Organizational questionnaire survey
- Interview and focus group methods
- Participatory ergonomics
- Macroergonomic analysis of structure (MAS)
- Macroergonomic analysis and design (MEAD)

80.5 Standards and Regulations

There are no direct standards or regulations governing field studies or field experiments. In some cases, both the field study and field experiment methods can be used in helping employers provide a safer work environment in compliance with the U.S. OSHA General Duty clause and various state health and safety regulations. These methods should be equally applicable to helping meet health, safety, and ergonomics regulations in other countries.

80.6 Approximate Training and Application Time

Training requirements can vary, depending on the sophistication of the application and general educational background of the individual. For a professionally educated ergonomist without specific training and experience conducting field studies or field experiments, approximately 20 contact hours of training and practice should suffice. Application time can vary greatly from a few days to several years, depending on the scope of the project.

80.7 Reliability and Validity

Both the field study and field experiment have demonstrated good reliability and validity in literally thousands of organizational studies. Where they have not shown good reliability or validity has been in those situations where the presence of the researcher altered the situation and thereby affected the outcome. With the field experiment, *how* the change is introduced and implemented can affect the employees' behavior, such as creating a Hawthorne effect or, if the change is resented, inciting deliberately poor performance.

80.8 Tools Needed

While some field studies and field experiments can be conducted with nothing more than a simple means of manually recording observations of behavior and/or performance outcomes, such as a paper pad and pencil, there are numerous off-the-shelf devices that can aid in recording both behavior and performance outcomes. Video cameras and audio recording devices often are used to record performance behavior. For performance outcomes, a given situation might lend itself to the use of automatic means of recording productivity and/or errors. For example, if the task involves the use of computers, then the number of words per unit of time or the number of errors per given number of keystrokes could be compiled automatically. Production often can be measured in terms of units produced per given period of time using automatic counting devices. Where applicable, physiological measures, such as EEG, EKG, EMG, etc., can automatically be recorded by the equipment.

References

Kahn, R.L. and Katz, D. (1963), Leadership practices in relation to productivity and morale, in *People and Productivity*, Suttermeister, R.A., Ed., McGraw-Hill, New York.

Vercruyssen, M. and Hendrick, H.W. (1990), *Behavioral Research and Analysis: An Introduction to Statistics within the Context of Experimental Design*, 3rd ed., Ergosyst, Lawrence, KS.

81

Participatory Ergonomics (PE)

Ogden Brown, Jr.
University of Denver

81.1 Background and Application

Participation and participatory practices are the principal methodologies in the design and analysis of work systems (Brown, 2002), and they are also important methodologies for product design (see Hendrick, 1996). There are several closely related concepts and terms that appear throughout the literature in the fields of ergonomics, psychology, and management. Such terms as participation, employee involvement, participatory ergonomics, participative management, and other participatory approaches are often used interchangeably. For example, Cotton (1993, p.3) defines the term "employee involvement" as "a participative process to use the entire capacity of workers, designed to encourage employee commitment to organizational success." He points out that it is not a true unitary scientific concept, but rather a useful catchall term for a variety of approaches, all of which employ participation. Thus, participatory ergonomics (PE) can be considered as an approach to employee involvement that is concerned with ergonomics design and analysis. From this point of view, PE is an approach or scheme that belongs in the catchall category noted above and fits the definition given by Cotton. One can thus infer that participatory ergonomics is the involvement approach unique to the field of ergonomics (Brown, 1994, 2002).

Participatory ergonomics as a concept has been defined in differing but complementary ways. Wilson and Haines (1997) point out that PE can be regarded as a philosophy, an approach or strategy, a program, or a set of techniques and tools. They have defined it as the involvement of people in planning and controlling a significant amount of their own work activities, with sufficient knowledge and power to influence both processes and outcomes to achieve desirable goals. Imada (1991) defines it as a macroergonomic approach to the implementation of technology in organizations that requires end users to be highly involved in developing and implementing the technology. And as Hendrick and Kleiner (2001) observe, when participation or involvement involves ergonomic design and analysis, worker involvement can be said to constitute participatory ergonomics.

In sum, participatory ergonomics is a complex concept involving many different dimensions (Haines and Wilson, 1998). PE can be found in virtually all other ergonomic methods to some degree, and its applications in ergonomic design and analysis are endless.

81.2   Procedure

A careful examination of various approaches to participation reveals at least three different approaches to participatory arrangements. Each is designed to encourage worker participation, even though each results in a very different type of involvement (Brown, 1994, 2000, 2002). Organizations that are interested in implementing some form of worker participation and involvement should be aware of the differences between these approaches and select the approach that offers the best *fit* with the organization (Brown, 2002). The concept of *fit* is very important in organizational design. Sociotechnical systems theory holds that no one part of an organization should be changed without an awareness and consideration of its role in the entire system. Major elements that need to "fit" are the people, the information processes, the technology, the reward system, and the organizational structure (Lawler, 1992). Structure is a critical variable in determining how involvement-oriented the organization should and can be. Some organizational designs make it virtually impossible to create an involvement-oriented organization; others almost demand that the organization be involvement-oriented (Brown, 2002; Lawler, 1992).

The three major approaches to participation (Brown, 1994, 2000, 2002) are:

1. Parallel suggestion involvement (consultative participation)
2. Job involvement (substantive participation)
3. High involvement

These approaches basically differ in the degree to which they propose that four key features of an organization should be moved to the lowest possible level. These features are:

1. Information about worker knowledge
2. The reward system
3. Organizational performance
4. The power to act and reach decisions that influence organizational practices and policies

These key features are also a useful way to consider *fit* among different parts of the organization with respect to how the parts affect them (Lawler, 1992). When these key features are moved downward in an organization, worker participation is being practiced. The parallel suggestion approach does the least to move them downward; the high involvement approach does the most. There is no more basic change in an organization than to move power, knowledge, rewards, and information to lower levels. This is the very essence of participation and involvement. It serves to alter the basic nature of the work itself and directly impacts the job of every worker through empowerment and enablement. It also directly impacts the effectiveness of the entire organization (Brown, 1994, 2000, 2002).

81.2.1   Parallel Suggestion Involvement

Parallel suggestion involvement programs ask workers to solve problems and produce ideas that will influence routine operation of the organization. Such programs are parallel structures to normal activities, since they place people in a separate new situation or structure that operates differently from the traditional organization. The most widely used approach is the quality circle or worker problem-solving group. They have no formal authority, direct rewards are not given, and they do not have the power to implement their own decisions

Another suggestion involvement approach is that of quality of work life programs. They often employ a parallel structure at multiple levels in the organization and can serve to bring two adversarial groups together. These programs serve to change the relationship between worker and organization. They may afford workers the opportunity to influence things they would not normally influence. However, it is expensive and difficult to maintain momentum due to resistance by middle levels of management and a lack of expertise on the part of workers to solve more complex problems (Brown, 2000, 2002; Lawler, 1992).

81.2.2 Job Involvement

Job involvement approaches focus on designing work in ways that better motivate job performance. One such strategy, job enrichment, focuses on the creation of individual tasks that provide feedback to people, require them to employ a variety of skills, increase their influence over how work is performed, and give them a complete piece of work to perform. By involving the worker in enriching a job, the organization stands not only to achieve job satisfaction and increased motivation, but also to realize better job and workstation design (or redesign), improved workspace arrangement, and perhaps job modification or change (Brown, 1994, 2000).

An increasingly popular job involvement approach is that of the self-directed work team. Self-directed work teams feature a formal system of worker involvement, direct worker participation, and a high degree of control (Cotton, 1993). The team approach differs from individual job enrichment in that the work group is the primary unit of involvement. The job involvement approach has significant implications for how an organization is structured and managed. Involvement in this case is not a special activity, as in the parallel-suggestion approach; it is the way the organization conducts its business (Brown, 1996, 2002).

81.2.3 High Involvement

The high involvement or commitment approach builds upon what has been learned from the suggestion and job involvement approaches. It structures an organization so that those at the lowest levels will have a sense of involvement not just in how well they do their jobs or how effectively their team performs, but in terms of total performance of the entire organization. High involvement goes much further than the other approaches toward moving information, knowledge, rewards, and power to the lowest organizational level. It can create an organization in which workers care about organizational performance because they know about it, are able to influence it, are rewarded for doing so, and possess the knowledge and skills to contribute to it (Brown, 2000, 2002).

High involvement ergonomics requires consistent and continuing change in virtually every part of the organization. Workers must be involved in decisions about their jobs and work activities. There are major implications here for job and workstation design/redesign and even for job modification. Workers also should be empowered to play a meaningful role in organizational level decisions concerned with strategy, structure, and other such important decisions. This would seem to argue for a structure with few hierarchical levels and wide spans of control. Expertise must be gained by all members of the organization in problem analysis, decision making, group process, and self-management. This in turn requires expanded training programs for both technical training as well as for interpersonal and team skills. Clearly, creating a high involvement organization is a complicated and complex task (Brown, 1996, 2000, 2002).

81.2.4 Role of the Ergonomist

The role of the ergonomist in the PE design and analysis process is varied and complex, but a role common to all participatory approaches is that of change agent. In some approaches, the ergonomist is primarily a trainer. Several authors have emphasized the need to provide training as part of a PE initiative. Members of ergonomics committees or problem-solving groups should be given training in teamwork and interpersonal skills to permit them to perform effectively as a group, and management may need instruction in how to relate to workers who now may be making decisions (Wilson and Haines, 1997). The ergonomist also may function as a resource person, familiar with the technology in use and available with the skills and knowledge to help solve a problem or contribute to analysis and design of an ergonomic intervention. Finally, the ergonomist can be a facilitator of change. The facilitator may be the "owner" of a project or the person who wants the intervention and can be someone from inside or outside the organization. Whatever the role of the ergonomist, and whatever the focus and level of the PE process, participation will by definition involve team-based activities. This can vary from a design team or a workplace ergonomics group through to a participative structure with project teams and steering committees (Wilson and Haines, 1997).

81.2.5 Implementation Issues

The following are among the many issues that should be addressed if participatory practices are to be implemented (Brown, 2000):

- *Planning:* To implement PE successfully and to realize the rewards possible, a definite plan should be devised. The plan should fit the type of involvement approach being considered and the specific circumstances of the work-system intervention or product development.
- *Impact on structure:* It is critical to examine this important aspect because most organizational systems are based upon belief and value systems that differ markedly from participatory structures.
- *Resourcing:* In resourcing the participative process, there are costs associated with PE that must be accounted for but not taken as additional business expenses. These costs must be budgeted, tracked, and managed.
- *Impact on decision making:* This can be a threatening issue to managers and decision makers in an organization. To many managers, being in control of the decision making process is evidence of competence and their most basic prerogative.
- *Impact on organizational roles:* Implementation of PE will cause certain roles to be changed. Roles change and evolve over the course of change interventions, but the evolution (like the change process itself) should be done in a planned way.
- *Top management commitment:* For work-system intervention programs, top management commitment is perhaps the most important implementation issue of all. Top management must assume an active role, because the support of powerful advocacy is vital in any organizational change program. There must be support to overcome resistance, to gain training funds, to make decisions on new reward systems and other structural and rules changes, and to enable the organization to realize its envisioned potential.

81.3 Advantages

There is no one best approach to the use of participation and worker involvement. Each participatory approach offers certain advantages, some unique to a particular approach and some common across PE approaches.

- The use of PE techniques in ergonomic analysis and design interventions and their subsequent implementation tend to result in greater feelings of solution "ownership" among those involved and affected, in increased job satisfaction, and in a greater commitment to the changes being implemented.
- Workers are the "experts" at what they do. They have the necessary skills and knowledge and know their jobs and work environments better than anyone else. They are in the best position to identify and analyze problems, and thus are able to provide and/or evaluate ergonomic solutions that not only will improve a given situation but also will be acceptable to those involved.
- Employment of a PE approach likely will lead to more appropriate and more acceptable ergonomic solutions than ergonomic design and development interventions that do not involve worker participation.

- Involvement in an ergonomic development and implementation process can lead to faster and more thorough learning of a new system or procedure that, in turn, may result in reduced training costs and improved job performance.
- The process of participation may have a systemic effect beyond its original focus and dimensions, impacting other parts of the organization either through the content or process of participation strategies.

81.4 Disadvantages

Each participatory approach also includes certain disadvantages, some unique to a particular approach and some common to all.

- Participation of any type and at any level (micro or macro) can be difficult to promote and to gain support from both labor and management.
- Organizational structure may limit the degree of worker involvement possible, or even make it impossible to create an involvement-oriented organization.
- For work-system PE intervention programs, top management commitment is needed, but it may be difficult to obtain. In the case of high involvement programs, obtaining top management commitment is absolutely essential. An organizational philosophy must be adopted that advocates active participation and involvement.
- Ergonomic analysis and design interventions/projects that are planned and developed participatively may be more costly because of greater effort and more time expended.

81.5 Example Output

No specific example of output is appropriate here. The implementation of participatory ergonomics is (and must be) unique to each particular situation and organization.

81.6 Related Methods

In addition to participatory ergonomics, another increasingly popular approach is that of total quality management (TQM). One of the more important principles of TQM concerns worker involvement or empowerment. It is common for TQM programs to state that involvement is an important part of any successful TQM program (Lawler, 1994). This leads to an interesting and speculative line of inquiry. One possibility is that participation and involvement is an activity that supports a TQM program. Another possibility is that TQM practices are best used to support PE programs. It would appear that there are important implications here for the way an organization is structured and managed (Brown, 1997).

81.7 Standards and Regulations

No specific standards or regulations apply.

81.8 Training and Application Time

In addition to a professional background in ergonomics, the ergonomist (trainer/resource person/facilitator) also should have whatever technical expertise and competence may be called for in a specific ergonomic analysis and design intervention/project as well as a firm knowledge of the technology involved in that intervention. Further, the ergonomist should have experience in the use of participative techniques in change analysis, development, and intervention.

Participation often may be a learning experience for the ergonomist that will improve current and future analyses and designs or implementations. Serving as the resource person/facilitator, the ergonomist not only is learning, but also is providing additional training throughout the PE intervention by fostering understanding and expertise in ergonomics among those involved in the project.

Application and implementation times vary widely and depend upon many variables, such as focus and level of the intervention, size of the intervention, type of involvement approach to be used, design and structural factors, and a myriad of other considerations.

81.9   Reliability and Validity

The fundamental problem in research on participation is that there is no single form of participation. Many studies use indices of participation encompassing a wide range of participatory approaches, and this masks or confounds the effects of any one form of participation. Further, there are many diverse forms of participation, each with its own issues and results (Brown, 2002; Cotton, 1993). This helps to explain why there is a paucity of reliability and validity studies dealing with participation and participatory ergonomics.

Outcome measures, reduction of risk factors, process measures, and cost-benefit analyses are the usual methods for evaluating participatory programs (Haines and Wilson, 1998). There are many studies in the literature reporting on the successful use of participation and participatory interventions, many being of a testimonial nature. Some, however, have been supported by good documentation. For example, Hendrick (1996) has described 24 well-documented, highly successful ergonomics case studies, of which 20 involved some form of participatory ergonomics. Unfortunately, failed interventions are very rarely (if at all) reported in the literature.

In an effort to provide clarity and organization to participatory ergonomics, a framework was developed by Haines and Wilson (1998) that defined dimensions across which to describe PE initiatives. Haines et al. (2002) then developed an adapted set of dimensions as a conceptual framework for understanding PE — the participatory ergonomics framework (PEF). The PEF was then validated within a structured framework. Results indicated that the PEF provides a successful approach for capturing the diversity of a range of PE projects. The authors concluded that the PEF has an applied role for setting up and supporting PE initiatives and programs and that it contributes to a better understanding of what is involved in participative processes for ergonomics change (Haines et al., 2002).

81.10   Tools Required

The tools required for a PE analysis and design project will depend upon the participants' expertise. The ergonomist can intervene by supporting or guiding the participatory process; by being involved with a variety of workers as a member of a multifunction group; or by personally doing much of the analysis, perhaps through the use of a research team. From a macroergonomic perspective, tools also are needed to "sell" participation to stakeholders and/or to maximize training, team working, and interpersonal skills (Wilson and Haines, 1997). Both Imada (1991) and Wilson and Haines (1997) describe a number of useful tools and techniques.

References

Brown, O., Jr. (1994), High involvement ergonomics: a new approach to participation, in *Proceedings of the Human Factors and Ergonomics Society 38th Annual Meeting*, Human Factors and Ergonomics Society, Santa Monica, CA, pp. 764–768.

Brown, O., Jr. (1996), Participatory ergonomics: from participation research to high involvement ergonomics, in *Human Factors in Organizational Design and Management V*, Brown, O., Jr. and Hendrick, H.W., Eds., North-Holland, Amsterdam, pp. 187–192.

Brown, O., Jr. (1997), High involvement ergonomics and total quality management: a comparison and evaluation, in *Proceedings of the Human Factors and Ergonomics Society 41st Annual Meeting*, Human Factors and Ergonomics Society, Santa Monica, CA, pp. 729–733.

Brown, O., Jr. (2000), Participatory approaches to work systems and organizational design, in *Proceedings of the XIVth Triennial Congress of the International Ergonomics Association and 44th Annual Meeting of the Human Factors and Ergonomics Society*, Vol. 2, Human Factors and Ergonomics Society, Santa Monica, CA, pp. 535–538.

Brown, O., Jr. (2002), Macroergonomic methods: participation, in *Macroergonomics: Theory, Methods, and Applications*, Hendrick, H.W. and Kleiner, B.M., Eds., Lawrence Erlbaum Associates, Mahwah, NJ, pp. 25–44.

Cotton, J.L. (1993), *Employee Involvement*, Sage, Newbury Park, CA.

Haines, H.M. and Wilson, J.R. (1998), *Development of a Framework for Participatory Ergonomics*, Health and Safety Executive, Sudbury, Suffolk, U.K.

Haines, H.M., Wilson, J.R., Vink, P., and Koningsveld, E. (2002), Validating a framework for participatory ergonomics, *Ergonomics*, 45(4), 309–327.

Hendrick, H.W. (1996), *Good Ergonomics Is Good Economics*, Human Factors and Ergonomics Society, Santa Monica, CA.

Hendrick, H.W. and Kleiner, B.M. (2001), *Macroergonomics: An Introduction to Work System Design*, Human Factors and Ergonomics Society, Santa Monica, CA.

Imada, A.S. (1991), The rationale and tools of participatory ergonomics, in *Participatory Ergonomics*, Noro, K. and Imada, A.S., Eds., Taylor & Francis, London, pp. 30–49.

Lawler, E.E. (1992), *The Ultimate Advantage: Creating the High-Involvement Organization*, Jossey-Bass, San Francisco.

Lawler, E.E. (1994), Total quality management and employee involvement: are they compatible? *Acad. Manage. Exec.*, 8, 68–76.

Wilson, J.R. and Haines, H.M. (1997), Participatory ergonomics, in *Handbook of Human Factors and Ergonomics*, 2nd ed., Salvendy, G., Ed., Wiley, New York, pp. 490–513.

82

Cognitive Walk-Through Method (CWM)

Tonya L. Smith-Jackson
*Virginia Polytechnic Institute
and State University*

82.1 Background and Application

The cognitive walk-through method (CWM) is a usability inspection method (also known as discount usability methods) that rests on the assumptions that evaluators are capable of taking the perspective of the user and can apply this user perspective to a task scenario to identify design problems. It is analytical and is similar to task analysis. Other usability inspection methods are heuristic evaluation, claims analysis, and design reviews. Researchers have used CWMs to identify usability problems related to product or system learnability. The user is viewed as approaching the task in an exploratory manner. According to Polson et al. (1992), users interact with a technology by exploring within the context of use, rather than using a systematic, highly structured approach. The best evaluators are capable of reasoning hypothetically and are able to explore a product from the perspective of a user. Likewise, a good evaluator for a CWM is one who can identify with users' preexisting schemas and mental models that will influence exploration and product use.

CWM is neither informal nor easy to implement. The method is structured and requires careful planning and design. In previous practice, CWM relied heavily on evaluators' knowledge of cognitive psychology. However, the recommended approach to be discussed here does not require extensive knowledge of cognition as long as the tasks are specified in detail (Sears, 1997; Sears and Hess, 1999).

82.2   Procedure

Typically, a group of evaluators who are usability specialists and/or designers will walk through several scenarios and identify the cognitive steps that are required for the user to successfully complete a scenario. The method requires the designers to use imagery and dialogue to literally "walk in a user's shoes." Polson et al. (1992) viewed the cognitive walk-through as analogous to the design reviews conducted in product development. Rather than using informal paper-and-pencil sketches to determine functional flow in software code, the CWM requires the evaluator to imagine what the user will have to think about, know, learn, react to, or comprehend to operate the system successfully. Although reasoning from another perspective is the norm in a CWM, these processes are conducted in a structured assessment environment. A decision has to be made of whether to conduct a CWM jointly as a group or to conduct a CWM separately and then share the results. Because of problems with reliability of the CWM (discussed later) — and due to the benefits that group dynamics can contribute when identifying and solving problems — group walk-throughs are recommended.

There are six steps to the design of a CWM (Lewis and Wharton, 1997; Polson et al., 1992; Sears, 1997; Sears and Hess, 1999; Wharton et al., 1994):

1. Develop a full understanding of target users' prior knowledge
2. Identify tasks that represent what users will do in the real world
3. Create detailed task-based scenarios
4. Walk through the correct action sequences necessary to complete the selected tasks
5. Identify and discuss the task-relevant cognitive processes the user must undergo to successfully complete the action sequences
6. Identify the learning or adaptive responses that are most likely to occur when the user is exploring the product

The first four steps are common to a number of other usability design practices. Steps 5 and 6 require more attention and elaboration.

82.2.1   Step 5: Identifying Relevant Cognitive Processes

When action sequences are identified, each sequence must be matched to the type of processing required to perform the sequences. Group-display media (e.g., whiteboards) are needed so that the entire group of evaluators is focused on each action sequence. For a hypothetical handheld device interface, the action sequences for a text-messaging task are shown in Table 82.1. The matching theoretical behaviors relevant to exploratory learning proposed by Polson et al. (1992) also are included.

CWM evaluators should discuss the cognitive processing required to complete each action sequence. For example, in step 4 of a CWM, if an action sequence begins with the device turned off, both experienced and novice users will likely look for the "Applications" menu to be displayed right after the power-up sequence has been completed (in standby mode) (see Figure 82.1). However, it is important to note that a user may have exited the system from another high-level menu such as "Internet," and the software may be designed such that it returns to the most recently activated function when the user powered off and does not go through a power-up sequence when turned on again. Experienced users may search for the applications menu after backing out of the task display that reflected the most recent activity, and they may not be surprised by the recent task-related display after turning the device on. Novice users may be confused by the display and may be unable to get back to the applications menu because of a lack of previous exposure to this kind of software flow.

Possible usability problems such as the one just described should be given priority when design solutions are explored. The evaluators should discuss all possible scenarios that lead to an action sequence. Each action sequence should be discussed in terms of the general cognitive tasks the user will have to undergo to complete the sequence. Issues to consider include user feedback to understand what to do

TABLE 82.1 Exploratory Behavioral Tendencies and Example for a PDA (Personal Digital Assistant)

Exploratory Behavior	Description	PDA Example
Goal setting	User determines goal to be accomplished	Develop and send a text message
Interface exploration	User searches for actions to support goal (icons, menus)	Find "Applications" menu
Action selection	User determines action to move toward goal; selects action	• Select "Applications" menu (level 1), press \<OK\> • Select "Text Message" from menu (level 2 or submenu), press \<OK\> • Select "Edit" • Select mode — keypad or stylus (activated upon selection)
Performance and evaluation	User performance action and evaluates system feedback	Enter message, press \<DONE\> Review displayed message, press \<SEND\> or \<EDIT\>

Source: Polson, P.G. et al. (1992), *Int. J. Man-Machine Stud.*, 36, 741–773.

next, recovery aids to exit out of a sequence, or interface affordances to help the user recognize or identify what action needs to be conducted at any point.

It is useful to have a scribe who can use a simple schematic to illustrate the walk-through output. The action-sequence–cognitive-processing schematic can take any form that is useful to the evaluators. For example, the plan–goal graph is one structure that provides a useful mapping between actions and cognitions relevant to the task. In completing step A, a schematic such as the one shown in Figure 82.1 could be used to illustrate actions and cognitions (circles indicate cognitive requirements).

Note that the schematic uses a deductive approach, moving from a general activity to the specific cognitions that must be completed to complete the action. Each of the cognitive activities can also be detailed further to include additional cognitive processes. For example, to identify and then execute the navigation requirement to get to the correct menu icon, a user must know where the navigation keys are located and must have an understanding of how the keys support movement through the menu system. But, suppose one of the supported design features of the proposed product is a four-way navigation key. Novice users may not have a concept of lateral/horizontal movement within a menu structure, so navigation errors could occur.

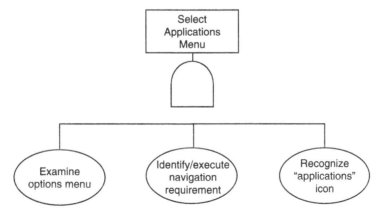

FIGURE 82.1 Excerpt of a schematic for Step A of an action sequence for a hypothetical handheld device. (From DuBois, D. et al. [1997], *Training Res. J.*, 3, 103–142. With permission.)

82.2.2 Step 6: Identifying Likely Learning or Adaptive Responses

When action sequences and their cognitions have been mapped out, even though problems may have been discovered previously, the subsequent schematic usually reveals only the best-case scenarios. Step 6 focuses on what could go wrong when a user explores the product, encounters unexpected problems and confusion, or gets stuck in an operational quagmire (e.g., lost in the menu structure). It is assumed in CWM that users will fall back on previously learned procedures or generalize assumptions from prior experience to the operational setting. For example, if the targeted users of a PDA design are believed to be frequent users of desktop computers, then assumptions about the desktop can be applied to the PDA, especially when the user gets lost or confused during exploration. The evaluators should try to predict changes in usage that may occur with experience, such as using shortcut keys, and the consequences of these adaptive responses.

These predictions can be proposed as stories describing (a) how the user will respond when conducting different activities and (b) the types of problems or the recovery procedures they are likely to use. For experienced users, it is possible that actions that were successful with previous models could cause major problems if applied to the proposed model under development. Step 6 can only be successful if the previous steps have been completed, and in particular, evaluators must have a good understanding of users' declarative knowledge, relevant experience, contexts of use, and assumptions or rule-based thinking to predict the likely procedural output.

82.2.3 Supporting Information

Every evaluator conducting a CWM should have usability information available. This could include archival reports and research documents or documents containing relevant usability guidelines and principles. The level of detail in a CWM needs to be predetermined by the evaluators. CWMs can be extensive and detailed or broad and general, depending on the needs of the design team.

CWMs designed to predetermine general usability problems can be supported by any number of documents that provide a description of the concept or product (e.g., customer requirements survey or preliminary function list). Trade journals that describe features and functions that reflect the product concept can be useful. Interface mock-ups and storyboards are useful as well.

More-specific analyses require more-detailed information. In combination with items available for a more-general CWM, a user requirements specification can be very helpful in supporting greater specificity. Requirements specifications are based upon focus groups, interviews, or other user survey methods that can be conducted early before conceptual design. Requirements specifications can also be produced using information from marketing documents and marketing surveys. Requirements specifications provide detailed information on the expectations, assumptions, prior experiences, and needs of potential users. This information will help evaluators develop a more accurate model of the user and, in turn, conduct a higher quality CWM. Kwahk et al. (2001) developed preferred menu items, menu formats, and navigation aids by using focus groups and Web-based assessment tools. These preferences were included in a user requirements specification and were used to create preliminary design concepts that could be used in design reviews and walk-throughs.

82.2.4 Macroergonomic Applications

Although the CWM method is most often applied to product design, the procedure can be applied to evaluate and improve the design of work systems. There is some historical precedent. Researchers of diffusion within organizations conduct inspections on the group level to identify conflicts or organizational usability problems. Although conducted on a larger scale, this practice differs only slightly from what would be done in a cognitive walk-through. Evaluators can assess the usability of conceptual designs of work systems to identify the degree to which a new work system is harmonized or the extent to which work flow is integrated. The evaluators would need to have an understanding of the assumptions, previous knowledge, and experience of the personnel. A detailed CWM could identify problems that could occur based on workers' overreliance on, or lack of experience with, technologies coupled with their work activities. The CWM can be useful in isolating problems and conflicts before changes are implemented within the organization, thereby serving as an important part of change management. A detailed CWM also could identify key variances where performance might be undermined by incompatibilities introduced by a new work system.

82.3 Advantages and Disadvantages

Table 82.2 is a list of advantages and disadvantages of CWM.

82.4 Example Output

Besides a plan–goal graph or some other schematic representing user actions and possible problem areas (Figure 82.1), the common output from a cognitive walk-through is a usability problem list. All problems identified by evaluators are presented in a list that is categorized on the basis of either interface issues (e.g., navigation problems) or product functions (e.g., text messaging or phone book). Table 82.3 shows an example of a usability problem list resulting from a cognitive walk-through. If the CWM is done

TABLE 82.2 Advantages and Disadvantages of CWM

Issue	Advantage	Disadvantage
Involves expert evaluators	Expertise supports usability problem identification	Problems differ from actual user reports
Problem identification	Significant and realistic problems can be identified	Cannot be used alone; must be combined with other methods
Tools	Preliminary documents can prove useful	During early development stages (i.e., conceptual design), these documents can be difficult to acquire
Cost	Cost and most resource demands relatively lower	Time demands can be high, depending upon specificity
Effectiveness	Effective at capturing usability problems	Low consistency across evaluators and when compared with usability testing

TABLE 82.3 Sample Problem List Generated from a Cognitive Walk-Through

Function	Problem Description	User Assumption
On/Off	At the power-up sequence, there is a long delay (5 sec) before the standby display is activated	User may assume product is not functioning and may press the <POWER> button again, which will then turn the device OFF
On/Off	At power-off, a short press leads to a selectable options list requiring the user to select "SHUT DOWN"	Experienced users may not have a problem. Inexperienced users may not understand the term "SHUT DOWN" and may continue to press other buttons to turn off the phone. A common reaction at that point is to remove the battery pack. Repeating this action could eventually damage the phone.
Soft keys	Activators for the three soft keys do not physically map to the display key	Novices may not understand soft-key assignment and will not know the functions of the three hardware keys on the keypad. During exploration, other keys will be pressed, including navigation arrows.
Receiving a text message	There is no information available after reading a message that allows the user to delete the message	After the message has been read, the user will not know which button to press to delete the message. It is possible that a user will press the <END> key assuming that since <END> terminates voice calls, it will also delete text messages. But, <END> returns the user to standby without eliminating a text message.

separately rather than as a group, it is always a good idea to tag the usability problems with the name of the evaluator(s) who identified the problem in case further clarification is needed.

82.5 Related Methods

The CWM is an analytical usability inspection method. As mentioned in the first paragraph of this chapter, the other usability inspection methods are heuristic evaluation, claims analysis, and design reviews. Concise descriptions of these methods can be found in Nielsen (1989). Heuristic evaluation uses existing principles or guidelines to analyze the usability of a product. Claims analysis makes use of customer reports or claims cited against products to identify problems with usability and identify solutions. Design reviews are informal sessions conducted by the design team and often involve exploration of different design alternatives and evaluation of existing design decisions to assess usability.

82.6 Standards and Regulations

No standards or regulations apply.

82.7 Approximate Training and Application Time

Training for the CWM varies based upon the experience of the potential evaluators. The process and output require little training, but the critical and abstract thinking demands of the CWM are the most challenging aspect and require some degree of training. Evaluators with strong usability backgrounds could learn the method in a matter of hours. Evaluators with little experience in usability and cognition could require at least one week of intense training to understand usability, hypothetical reasoning relevant to consumer behavior, and the role of cognition in system use and exploration.

If done as a group, the amount of time for a comprehensive CWM could take only a few hours, depending on product complexity and the level of detail in the conceptual design. When done separately, evaluators can take weeks to finish a CWM, since they often conduct the walk-through in several separate and short sessions.

82.8 Reliability and Validity

There are few studies on the reliability and validity of CWMs. Most of the studies have produced mixed results and have been challenged in terms of validity (Desurvire, 1994; Jeffries et al., 1991; Nielsen and Phillips, 1993). A study by Jacobsen and John (2000) tested the reliability of the CWM by examining the problems identified by two different evaluators who were trained to use the method. Evaluators did not interact during the evaluation process. The number of problems reported differed dramatically (42 vs. 9), and only 3 problems were identical. Validity was tested by comparing the problems identified by the evaluators to problems later identified through usability testing with potential users. Not surprisingly, very few of the problems predicted by the evaluators were found in the usability tests.

Although some report that reliability and validity were low, this is not necessarily a counterargument for the effectiveness of CWMs. In terms of reliability, CWMs differ between evaluators who conduct evaluations separately. Although there was little overlap, the types of problems reported by both evaluators were indeed usability problems. The same case can be made for the low validity. CWMs are a supplemental method, and duplicate other methods only to a limited degree. The CWM is designed to add to and expand, rather than duplicate, other methods, such as usability tests. Achieving an acceptable match between the problems identified in the CWM and those discovered during usability testing is almost impossible, since the CWM examines a system that is in a very preliminary state (such as in concept only), while usability testing occurs later, when at least a system mock-up is available.

82.9   Tools Needed

Usability inspection methods are attractive because they demand few resources. The level of CWM specificity determines the required tools. All evaluators should have relevant usability guidelines available in documents or in a database. In addition, current usability research in journals or periodicals should be reviewed, since they contain new guidelines that may be relevant to the product concept. Finally, illustration or flowcharting software is very helpful when conducting detailed cognitive walk-throughs. These tools help the group to understand the reasoning of other members of the group.

References

Desurvire, H.W. (1994), Faster, cheaper!! Are usability inspection methods as effective as empirical testing? in *Usability Inspection Methods*, Nielsen, J. and Mack, R.L., Eds., Wiley, New York, pp. 173–199.

DuBois, D., Shalin, V.L., Levi, K., and Borman, W.C. (1997), A cognitively oriented approach to task analysis, *Training Res. J.*, 3, 103–142.

Jacobsen, N.E. and John, B.E. (2000), Two Case Studies in Using Cognitive Walkthrough for Interface Evaluation, Tech. Rep. CMU-HCII-00-100, HCI Institute, Carnegie Mellon and Nokia Mobile Phones, Copenhagen.

Jeffries, R., Miller, J.R., Wharton, C., and Uyeda, K.M. (1991), User interface evaluation in the real world: a comparison of four techniques, in *Human Factors in Computing Systems CHI '91*, Robertson, S.P., Olson, G.M., and Olson, J.S., Eds., ACM, New York.

Kwahk, J., Smith-Jackson, T.L., and Williges, R.C. (2001), From user-centered design to senior-centered design: development of a health information portal for seniors, in *Proceedings of the Human Factors and Ergonomics Society 45th Annual Meeting*, Human Factors and Ergonomics Society, Santa Monica, CA, pp. 580–584.

Lewis, C. and Wharton, C. (1997), Cognitive walkthroughs, in *Handbook of Human–Computer Interaction*, 2nd ed., Helander, M., Landauer, T.K., and Prabhu, P., Eds., Elsevier, Amsterdam.

Nielsen, J. (1989), Usability engineering at a discount, in *Designing and Using Human–Computer Interfaces and Knowledge Based Systems*, Salvendy, G. and Smith, M.J., Eds., Elsevier, Amsterdam, pp. 394–401.

Nielsen, J. and Phillips, V.L. (1993), Estimating the relative usability of two interfaces: heuristic, formal, and empirical methods compared, in *Proceedings of INTERCHI 93*, ACM, New York, pp. 214–221.

Polson, P.G., Lewis, C., Rieman, J., and Wharton, C. (1992), Cognitive walkthroughs: a method for theory-based evaluation of user interfaces, *Int. J. Man-Machine Stud.*, 36, 741–773.

Reason, J.R. (1990), *Human Error*, Cambridge University Press, Cambridge, U.K.

Sears, A. (1997), Heuristic walkthroughs: finding the problems without the noise, *Int. J. Hum.–Comput. Inter.*, 9, 213–234.

Sears, A. and Hess, D.J. (1999), Cognitive walkthroughs: understanding the effect of task-description detail on evaluator performance, *Int. J. Hum.–Comput. Interact.*, 11, 185–200.

Wharton, C., Rieman, J., Lewis, C., and Polson, P. (1994), The cognitive walkthrough method: a practitioner's guide, in *Usability Inspection Methods*, Nielsen, J. and Mack, R.L., Eds., John Wiley & Sons, New York, pp. 105–141.

83

Kansei Engineering

Mitsuo Nagamachi
Hiroshima International University

83.1 Background and Application

Kansei engineering is defined as translating consumers' affective responses to new products into ergonomic design specifications (Nagamachi, 1988, 2002). Similarly, kansei engineering can be used to translate workers' affective responses to proposed changes to a work system into macro- and microergonomic design specifications. The term *kansei* is Japanese and implies feelings and emotions that the customer has in his or her mind. When buying something, or "buying into" a new or modified work design, customers (including employees in a work system) have some kind of preconceived image. This is the kansei. Kansei engineering implements the human's kansei in the design fields to produce a product matching the customers' feelings in an effort to maximize customer satisfaction. The customers' psychological responses (the kansei) are more general qualitative characteristics. To transfer kansei to the design field, the qualitative data must be quantified. In other words, the qualitative psychological phenomena should be changed to quantified characteristics. To design a new product, relational rules are needed between the kansei and design specifications. This procedure requires kansei engineering technology to link kansei to design specifications.

Manufacturers typically produce products (and work systems) according to their own philosophy. This is called "product-in philosophy," which means that companies manufacture the product based on their own ideas, regardless of the customers' demands and feelings. The opposite philosophy is "market-in philosophy," whereby the manufacturers focus on the customer and produce products that match the customers' desires and needs. Kansei engineering helps to realize development of a new product by implementing the customers' desires and feelings (kansei). This is why kansei engineering is called "a customer-oriented product development technology" in terms of ergonomics. The products developed from the viewpoint of kansei engineering invariably lead to satisfaction, because customers are willing to buy goods matching their needs and feelings. Therefore, the purpose of kansei engineering is to enhance quality of life through customer satisfaction.

83.2   Procedure

83.2.1   Kansei Survey of Customers

Kansei engineering starts with a survey of customers. The customer kansei is unveiled in the styles of dialogue, attitude and behavior, or psychophysiological phenomena. When a customer describes a product, s/he uses words with positive or negative attitude. In the former, s/he chooses preferable words; in the latter case, s/he complains. The positive or negative dialogue about the existing products is recorded, and a revised product or design of a new product results from these cues. Note that the "product" can be a consumer product, or it can be a work system, where the employees are the "customers."

If the kansei is largely dependent on psychophysiological fields, medical measurements are collected to obtain the data needed. In this case, EEG, EMG, ECG, HR, GSR, or RMR measurements are utilized to find which ergonomic factors fit the customers' feelings and demands. In this stage, it is important to select the best method leading to the most useful outcome (designing). For example, if kansei relates to muscle load, EMG measurements from appropriate body parts are taken. If it regards operational function, the customers' movements are recorded. An intelligent air-conditioner that provides the most comfortable room temperature needs to approximate human skin temperature through nontactile means. Therefore, kansei engineering uses both psychological and psychophysiological measurements to determine the ergonomic factors needed for the design specifications. If the kansei relates to the exterior design, the customers' dialogue is recorded as they view the exterior design of a product.

83.2.2   Analysis of Collected Data

In the psychological measurement, the existing products belonging in the same domain are collected and evaluated using a semantic differentials scale (Osgood et al., 1957), and these data are analyzed by a multivariate statistical analysis. Cluster analysis groups the products with similar factors. Factor analysis and multidimensional scaling reveal the meaningful data structure with several factors, and these are used to position each product by mapping on the factor axis. Since most kansei data are qualitative, Quantification Theory I, II, III, and IV, developed by Hayashi (1974), are useful as data analysis tools for treating these data (Nagamachi, 1988).

83.2.3   Relational Rules

Kansei-design relational rules are needed for making an appropriate design based on the customer's kansei. In using statistics for the data, the relational rules between the kansei and the design elements are obtained. For example, when implementing the kansei of "easy handling," "strong," and "attractive" for a new product, a good product fit to this kansei is constructed from the relational rule base obtained by data analysis.

Quantification Theory Type I is an excellent method for obtaining the kansei-design relational rules. Similar kansei-design relational charts can be revealed by genetic algorithm (Tsuchiya et al., 1996), which is able to construct a decision tree leading to the design, and by rough set theory, which is able to group interrelational factors with reference to the new product design.

83.2.4   Kansei Engineering System

Kansei engineering system (KES) implies a computerized system that has databases of word processing, knowledge rules, image construction, and an inference engine. If a customer inputs her or his kansei into the system, it displays a candidate design as the calculated outcome through this expert system. The KES is useful to designers for suggesting the new trends of a coming product or for the customer to select a product that best fits her or his feelings (Nagamachi, 1995).

A hybrid kansei engineering system (HKES) is a more sophisticated version of KES that has double kansei systems: the forward kansei engineering system from the kansei to design and the backward kansei system from the candidate design to the kansei assessment. If a designer uses the hybrid kansei engineering system, s/he first inputs some kansei words into the system and it produces a candidate design. After looking at the candidate design, the designer catches an idea and inputs it into the system. The system can also be used to assess the kansei scores through the backward kansei system. This system is a very powerful and efficient support for the designer's activity (Matsubara and Nagamachi, 1997).

A third category of KES is the virtual kansei engineering system (VKES). VKES combines KES and virtual-reality technology. KES first translates the kansei into the design, and then the virtual reality system provides the customers with the virtual experience. This system supports the customer's decision making very efficiently (Nomura et al., 1998).

83.3 Advantages

- Takes into account the customer's kansei
- Constructs a new product based on the customer's kansei
- Leads to greater customer satisfaction
- Can suggest the future trend of a new product domain
- Enhances the design sense of the designer group

83.4 Disadvantages

- Can be difficult to capture the customer's kansei, as the customer's behavior and psychophysiological settings can vary
- Requires kansei engineers to have a sophisticated knowledge and understanding of how to use statistical methodology
- Requires kansei engineers to be able to read design sense from the numbers calculated by statistical analysis
- There are no good deterministic statistical tools for treating the nonlinear features of the kansei

83.5 Case Examples

The products and organizations described below are typical cases developed using kansei engineering technology, and they illustrate the usual style of application.

83.5.1 Shampoo and Treatment Developed from a Market Survey

Success was achieved in producing a new shampoo and treatment for women. This activity began with a survey of women regarding their hair complaints and hair conditions. These qualitative data were analyzed by Quantification Theory III, and the relations among hair conditions, hair complaints, and customer desires with regard to their hair were obtained. The analyzed kansei progressed to a chemical material research stage, where these materials were examined using monitors. Finally, the special ingredients needed for the final product were determined. In related research, a variety of shampoo containers were evaluated on the semantic differential (SD) scale, and a container and color were chosen based on Quantification Theory I analysis. This demonstrated how kansei used customers' complaint surveys to develop a new product (Nagamachi, 2000a).

83.5.2 Kansei Community Ergonomics

The plan was to support a small Japanese village to address the "digital divide" issue. The name of the village is Kimita Village, where about 2000 people live. The percentage of people over 65 years of age is 36%, which makes it the oldest village in Hiroshima Prefecture. To support the village and enhance motivation, computers were introduced to help prevent "digital divide." First, the Kimita Elderly Society

(KES), the village office staff, the board members of KES, and a university researcher discussed ways to support the village. The elderly did not like being isolated from culture and technology. Therefore, information technology was introduced into their lives. One hundred computers were donated to this small village by Nippon Telephone and Telegraph Company (NTT). These computers were distributed to the elderly, who were motivated to learn how to use the computers. University students taught people how to manage the computer, which took almost 1 year to complete. Because of this voluntary activity, the elderly learned to use computers; even an 81-year-old person sent e-mail to his grandchild.

The elderly people reported the kinds of information that should be distributed to their computers. Those items most desired were the agricultural news, current news about the residence, hospital information, and health care news. These topics correspond to kansei, similar to a new product development. At present, the elderly in the small village are enjoying their computers and find them useful in their daily lives. Kimita Village is now called "Cyber Village." This is a good example of community ergonomics applying kansei ergonomics (Nagamachi, 2000b).

83.6   Approximate Training and Application Time

It takes approximately 5 hours to train a professional ergonomist with a background in multivariate analysis. Training includes the concept of kansei engineering, kansei engineering methodology, how to apply multivariate analysis to the kansei data, and how to analyze and apply the statistical results to design.

Application time depends on the product design domain. For example, it can vary from 3 to 4 months in the cosmetics field and from to 1 to 2 years in the automotive industry.

83.7   Reliability and Validity

The reliability and validity of the results depends on the skill and experience level of the kansei engineers. For excellent kansei engineers, the probability of success can be consistent and high.

83.8   Tools Needed

A good statistical software package is required, especially a good multivariate analysis program.

References

Hayashi, C. (1974), *Method of Quantification*, Toyokeizai, Tokyo.

Matsubara, Y. and Nagamachi, M. (1997), Hybrid kansei engineering system and design support, *Int. J. Ind. Ergonomics*, 19, 81–92.

Nagamachi, M. (1988), Kansei engineering: a new consumer-oriented technology for product development, in *The Occupational Ergonomics Handbook*, Karwowski, W. and Marras, W.S., Eds., CRC Press, New York, pp. 1835–1848.

Nagamachi, M. (1995), Kansei engineering: an ergonomic technology for product development, *Int. J. Ind. Ergonomics*, 15, 3–11.

Nagamachi, M. (2000a), Application of kansei engineering and concurrent engineering to a new cosmetic product, in *Proceedings of Ergonomics and Safety for Global Business Quality and Productivity*, pp. 101–104.

Nagamachi, M. (2000b), Kansei of the elderly and community ergonomics, in *Proceedings of the 14th Triennial Congress of the International Ergonomics Association and 44th Annual Meeting of the Human Factors and Ergonomics Society*, Vol. 6, Human Factors and Ergonomics Society, Santa Monica, CA, pp. 368–371.

Nagamachi, M. (2002), Kansei engineering in consumer product design, *Ergonomics Design*, 10, 5–9.

Nomura, J., Imamura, K., Enomoto, N., and Nagamachi, M. (1998), Virtual space decision support system using kansei engineering, in *Cyberworlds*, Kunii, J. and Luciani, A., Eds., Springer, New York, pp. 273–288.

Osgood, C.E., Suci, G.I., and Tannenbaum, P.H. (1957), *The Measurement of Meaning*, University of Illinois Press, Urbana.

Tsuchiya, T., Maeda, T., Matsubara, Y., and Nagamachi, M. (1996), A fuzzy rule induction method using genetic algorithm, *Int. J. Ind. Ergonomics*, 18, 135–145.

84

HITOP Analysis™

Ann Majchrzak
University of Southern California

M.M. Fleischer
University of Southern California

D. Roitman
University of Southern California

J. Mokray
University of Southern California

84.1 Background and Application

HITOP is a step-by-step manual procedure for industry practitioners who must implement technological change. The basic premises of the procedure are that technological change will be more successfully implemented when managers and designers are aware of the organizational and human implications of their technology plans, and that this knowledge is used to design systems that integrate technology within its organizational and human context. The procedure is called HITOP, which is an acronym for high integration of technology, organization, and people. HITOP was developed in direct response to an industry-identified need for simple tools that identify likely organizational and human implications of technology plans. These industry experts reiterated the accumulating evidence that technology changes are generally likely not to succeed in U.S. industry today and that a failure to effectively plan for the organizational and human implications of the technology is a primary cause of the problem. With HITOP, companies preparing for technology changes are provided with the forms, checklists, terms, and questions for assessing the human and organizational changes needed to support the technology changes. An early version of HITOP was created by Majchrzak (1988).

84.2 Procedure

The manual (Majchrzak et al., 1991) may be obtained from the first author. Ideally, a team of individuals responsible for the redesign of the facility or work area will work through the procedure. At Hewlett-Packard, a new process/product introduction team was formed, and the team conducted the HITOP analysis. The team consisted of the development engineer, process engineer, maintenance technician, supervisor, and the operator of the new process technology being introduced simultaneously with the new product. The steps in a HITOP analysis are:

1. Organizational readiness assessment: assess factors that inhibit the organization's ability to be open to possible change resulting from the HITOP analysis and how to make changes to these factors, if needed.

2. Technology assessment: analyze critical technical features (CTFs) likely to have the greatest impact on organizational design. CTFs include mechanical integration with other systems, information integration with other systems, reliability, input/process flexibility, self-diagnostic and correction capability, and fault tolerance.

3. Job requirements and design assessment: analyze essential task requirements (ETRs) for operational, support, and supervisory roles. ETRs include task interdependence, manual work, decision authority, problem solving, proactive opportunity seeking, and setting complex goals. Job classifications are organized as solo or teams.

4. Skill requirements and how requirements will be met: analyze skill needs, including perceptual, conceptual, manual dexterity, problem-solving, human relations, and technical knowledge. Training and selection options for meeting skill requirements are discussed.

5. Reward systems: consider pay-for-performance, reward, recognition systems, and career paths.

6. Organization design: assess reporting lines, formality of procedures, unit grouping, cross-unit coordination, and organizational culture.

7. Implementation plan.

84.3 Advantages

The following advantages are based on the results obtained in the Hewlett-Packard case study of the use of HITOP:

- Faster time to market, since products and processes are developed jointly by engineering and production and because production ramp-ups are more aggressive
- Completion of manufacturing training and documentation before release of the new product to manufacturing because the manufacturing specs are known in advance
- Realistic expectations of technology, since reliability, integration, flexibility, and capabilities are defined early
- Early delivery of training and documentation designed for support jobs
- Enhanced quality of technology, design, and layout through the concurrent design of organization and processes
- Process improvement before release of the new product to manufacturing

84.4 Disadvantages

HITOP does not contain a knowledge base of best practices. Instead, it has a series of forms to be completed by the design team. Thus, if the design team fills in the wrong information into the forms, there is no way to correct their error. For this reason, TOP-Modeler was developed. TOP-Modeler is described in Chapter 85.

84.5 Example Output

Table 84.1 shows a worksheet for job requirements.

84.6 Approximate Training

This includes reading the HITOP reference manual (Majchrzak, 1991).

84.7 Reliability and Validity

HITOP is a procedure and not a knowledge base, so predictive validity is inappropriate. The previously cited Hewlett-Packard case study indicates the usefulness of the procedure.

TABLE 84.1 Worksheet for Job Requirements

Critical Technical Features (CTF)	Essential Task Requirements				
	Task Interdependence	Information Exchange	Manual Work	Decision Authority	Complexity of Prob-Solv
Information integration Low	Low	Low	Low	Low	Low
Mechanical integration Low	Low	Low	Low	Low	Low
Reliability High	Low	Low	Low	Low	Low
Flexibility in programming and setup Medium	High with programming	Medium with maintenance	Low	High with programming	Medium
Flexibility difficulty Medium	High with programming	Medium with maintenance	Low	High	Medium
Feedback-automated monitoring and diagnosis Low	Low	Low	Low	High	Low
Feedback-automated correction Low	Low	Low	Low	High	Low
Fault tolerance Low	Low	Low	High in emergencies	High in emergencies	Low
Cross-CTF consistency?	Yes	Yes	Yes	Yes	Yes
Cross-CTF total	Low except high with programming	Low except med with maintenance	Low except in emergencies	High	Low

84.8 Tools Needed

No tools are needed other than the reference manual (Majchrzak, 1991).

References

Majchrzak, A. (1988), *The Human Side of Factory Automation*, Jossey-Bass, San Francisco.
Majchrzak, A., Fleischer, M., Roitman, D., and Mokray, J. (1991), *Reference Manual for Performing the HITOP Analysis*, Industrial Technology Institute, Ann Arbor, MI.

85

TOP-Modeler©

Ann Majchrzak
University of Southern California

85.1 Background and Application

TOP-Modeler was developed under contract from the National Center for Manufacturing Sciences, in conjunction with Hughes, General Motors, EDS, DEC, and Hewlett-Packard. TOP-Modeler is a decision-support system for manufacturing organizations to help them identify the organizational changes required when new process technologies are being considered.

85.2 Procedure

TOP-Modeler must first be downloaded for free from www-rcf.usc.edu/~majchrza and installed on any local personal computer. Ideally, a team of individuals responsible for the redesign of the facility or work area are seated around a table with TOP-Modeler on the screen. Ideally, this team consists of engineers responsible for the design of the new technology, factory workers who must operate the new technology, a human resources staff person (for changes to job descriptions), as well as other staff who may be affected by the change (e.g., quality control, material handling, supervisors, etc.). The team then responds to TOP-Modeler's requested inputs by clicking the screen to specify which business objectives the team has for the redesign, and the process (also called production) variances that the team either desires or is willing to live with. Example variances include mass customization (which may be desired), and equipment unreliability (which may be undesired but which the team is willing to live with). In addition, the team clicks on inputs to describe their current as-is design (in terms of norms, assignment of job responsibilities, workforce skills, customer-involvement strategies, etc.). TOP-Modeler contains an extensive knowledge base of the best-practice organizational design elements for different business objectives and under different types of process variances. Therefore, TOP-Modeler is able to take the team's inputs and compute a to-be/as-is gap analysis, indicating alternative ways to prioritize the gaps to determine which ones to fix first in the team's redesign plans. The team can rapidly change any of their inputs (such as inputting not their actual as-is state but rather their intended planned to-be state) to see if gaps remain.

85.3 Advantages

- Identifies gaps in organizational changes as new technologies are being designed.
- Prioritizes gaps to determine which ones to fix first.
- Identifies lack of agreement among team members about the as-is state when joint ventures are being considered.
- Ensures that certain issues such as job descriptions, customer involvement strategies, and employee empowerment strategies are not ignored during the design of new technologies.
- Encourages challenging the as-is state: does it have to be like this for the future?
- Provides fast analysis that takes only minutes with no training in how to use the system.

85.4 Disadvantages

- Does not provide the silver-bullet quick fix to a poorly designed factory or redesign plan; presents only a set of prioritized gaps.
- Does not provide the catalyst for change; only analyzes what changes are needed. The catalyst must come from inside the organization.
- Garbage in/garbage out. If the as-is state of the organization is incorrectly described, outputs will be meaningless.
- Provides only one of many inputs into a complex decision-making process.
- Does not describe precisely how to make the needed changes. For example, it suggests that additional training in specific skills are needed, but it does not specify a plan for achieving the training.

85.5 Approximate Training

None is required.

85.6 Reliability and Validity

Predictive validity of the knowledge base in TOP-Modeler was assessed by comparing the predictions made by TOP-Modeler about how well a manufacturing organization would perform given its as-is organizational and technology design against actual data collected from the manufacturing organization about its performance. Data on the as-is design of 86 U.S.-based manufacturing firms were collected, along with the speed of throughput time (discounted for rework rates). TOP-Modeler accurately predicted the firms' throughput time 95% of the time, based on the as-is design of each firm. In addition, observations were made of 23 users located in five different organizations redesigning their manufacturing operations. Results of the observations indicated that 18 of the 23 users reported that they obtained learning in one or more of the following areas as a result of their use of TOP-Modeler: identifying a lack of consensus about an issue among the management team, identifying that an organizational design feature they thought was present in the as-is organization was absent, and prioritizing gaps that needed to be resolved. In addition, eight users were able to track back decisions made about their organizational redesigns to these learnings.

85.7 Tools Needed

TOP-Modeler software is the only tool needed. It can be downloaded for free from www-rcf.usc.edu/ ~majchrza and installed on any local personal computer.

References

Majchrzak, A. (1997), What to do when you don't have it all: toward a theory of sociotechnical dependencies, *Hum. Relations*, 50, 535–565.

Majchrzak, A. and Finley, L. (1995), A practical theory and tool for specifying sociotechnical requirements to achieve organizational effectiveness, in *The Symbiosis of Work and Technology*, Benders, J., de Haan, J., and Bennett, D., Eds., Taylor & Francis, London, pp. 95–116.

Majchrzak, A. and Gasser, L. (2000), TOP-MODELER: supporting complex strategic and operational decision making, *Inf. Knowledge Syst. Manage.*, 2, 95–110.

Majchrzak, A. and Meshkati, N. (2000), Aligning technological and organizational change when implementing new technology, *The Handbook of Industrial Engineering*, 3rd ed., Salvendy, G., Ed., Wiley, New York.

Markus, M.L., Majchrzak, A., and Gasser, L. (2002), A design theory for systems that support emergent knowledge processes, *MIS Q.*, September.

86

The CIMOP System©

Waldemar Karwowski
University of Louisville

Jussi Kantola
University of Louisville

86.1 Background and Application

The CIMOP (computer-integrated manufacturing, organization and people) system is a knowledge-base evaluation system. It was developed for evaluating computer-integrated manufacturing, organization, and people system design. The intended users of CIMOP are companies with an existing CIM system; companies designing, redesigning, or implementing a new CIM system; and consultants, researchers, or any other parties engaged with CIM system design and evaluation (Kantola and Karwowski, 1998; Karwowski et al., 2002).

86.2 Procedure

In CIMOP, the critical design aspects are quantified and are called the design factors (DFs). These factors relate a CIM system design to practice and allow for quantitative:

- Evaluation of a CIM system design
- Evaluation of an existing CIM system
- Comparison between CIM system designs
- "What if"-type analyses of the effect of alternative system design improvements

CIMOP utilizes 75 design factors representing critical design aspects, including:

- Organization subsystem (12 DFs)
- Technology subsystem (17 DFs)
- Information systems (IS) subsystem (22 DFs)
- People subsystem (24 DFs)

Five evaluation modes are used to determine the overall system design quality:

1. CIM system design
2. Organization subsystem
3. Technology subsystem
4. IS subsystem
5. People subsystem

The DFs, DF groups, CIM subsystems, and overall CIM design quality are defined as fuzzy linguistic variables. Three linguistic states — low, medium, and high — are used to describe the state of DFs as well as the state of inputs (intermediate results) to the next levels of the evaluation process. Five linguistic states — low, LTM, medium, LTH, and high — are used to describe the design quality of DF groups (intermediate result), the design quality of CIM subsystems, and the design quality of the CIM system. The reasoning scheme of the CIMOP method utilizes a hierarchical fuzzy logic control (FLC) consisting of four modules:

1. Fuzzy rule base
2. Fuzzy engine
3. Fuzzification module
4. Defuzzification module

The evaluation screen allows the quantitative evaluation of a CIM system design. This is done by evaluating the state of DFs of the system design. DFs are arranged as a hierarchy at the left side of the screen. The tree can be expanded or collapsed by clicking the nodes. The rightmost branches are the DFs. Predefined levels of the DFs appear in the text box on the right side of the tree. After all DFs have been evaluated, the results can be viewed. Results are presented graphically on the results screen. The CIMOP method has two alternative output scales for the results: numerical and verbal. The evaluation in CIMOP is best conducted by a team comprising the people in the organization who are most knowledgeable about the CIM system design. The team should have representatives (at least) from top management/factory level, human resources, IS (information system) management, and manufacturing cells/shop floor level. The evaluation team then compares the state of each DF against the predefined levels. After the state of each DF in the project has been estimated, graphical results representing the design quality can be examined. Such results indicate the design quality of the CIM system, its subsystems, and DF groups.

The CIMOP method operates on a PC in the Windows™ operating system. The method has an online help including the following help topics: the CIMOP method, DFs, evaluation, comparison, analysis, results, and output scale. Results can be printed. Printouts include general information, evaluation inputs, and graphical results. Contact Waldemar Karwowski at karwowski@louisville.edu to obtain the CIMOP program.

86.3 Advantages

- Ability to evaluate CIM subsystems individually (from their integration point of view) in addition to the whole system
- Ability to select specific DFs for inclusion in the evaluation criteria
- Ability to evaluate CIM structure prior to its implementation to avoid possible design failures
- Useful as an aid in determining whether a particular CIM project should be implemented or improved
- Ability to determine the uncertainty of imprecisely defined, qualitative, or subjective DFs

86.4 Disadvantages

- Helps to pinpoint the problem areas in the design, but does not give the solutions to the problems.
- Does not provide any quick-fix prescription for the improvement of a system (or subsystem) that is evaluated as "needs improvement" or "unacceptable."
- Only compares the state of each DF to a predefined level.

86.5 Approximate Training

No special training is required.

86.6 Reliability and Validity

The fuzzy-logic controller (FLC) of CIMOP was validated against TIL Shell 3.0.0 (Togai Infralogic, Inc.), which is an FLC building and testing tool. FLCs of both CIMOP and TIL Shell produced identical results from 30 sets of the same random inputs.

86.7 Tools Needed

No special tools are required.

References

Kantola, J. and Karwowski, W. (1998), A fuzzy-logic based tool for the evaluation of computer-integrated manufacturing, organization and people system design, in *Manufacturing Agility and Hybrid Automation II*, Karwowski, W. and Goonetilleke, R., Eds., IEA Press, HKUST, Hong Kong, pp. 43–46.

Karwowski, W., Kantola, J., Rodrick, D., and Salvendy, G. (2002), Macroergonomics aspects of manufacturing, in *Macroergonomcis: Theory, Methods, and Applications*, Hendrick, H.W. and Kleiner, B.M., Eds., Lawrence Erlbaum Associates, Mahwah, NJ, pp. 223–258.

87

Anthropotechnology

Philippe Geslin
*Institut National de la Recherche
Agronomique (INRA)*
Université de Neuchâtel

87.1 Background and Application

In 1979, Alain Wisner, one of the founders of French ergonomics, referring to the quasi-failure of technology transfers to industrially developing countries, declared there was an urgent need to develop "a true anthropotechnology, i.e., an adaptation of technology to the people, which just as ergonomics associates knowledge from the human sciences to improve the design of technical systems ... and since the scale considered is different, so the sources needed must also be different."

Wisner reactivated the debate on technology transfers. He broke away from conventional partitioned, single-discipline approaches and initiated a new approach that considers the multiple relationships between the microscopic features of human activities and the macro factors describing societal functioning. Analyses that focus on one of these levels will prove relevant only if they also consider the mechanisms occurring at the complementary level (Wisner, 1997).

From now on, Wisner declared, "another scientific domain must be included in a range of ergonomic studies: that of anthropology." Technological systems are increasingly being transferred to other countries, other regions, and other firms. This means that differences in social and industrial fabrics must be taken into account in addition to various fields of anthropology: physical anthropology (body dimensions, physical strength), cultural anthropology (value systems), and cognitive anthropology (linguistics, cognitive models, cognitive load in work situations). However, the link between anthropology and ergonomic analysis of work was not incorporated in anthropotechnological interventions until the early 1990s (Geslin, 1999).

Encouraged by the results and acknowledgment of this work, we developed the anthropotechnological approach in other situations, chiefly in the agricultural sphere, that were applied in France, the Philippines, Brazil, and Africa. What follows is therefore based on our experience in the rural sphere. The frameworks for combining ergonomic analyses of work and the anthropotechnological approach are progressively being formalized (Geslin, 1999–2002). These frameworks contribute to producing technology that takes into account the sociocultural environments and demands of the populations concerned (i.e., both the producers and receivers of the technology).

87.2   Procedure

87.2.1   Analyzing and Reformulating Requirements

This stage is designed to highlight the stakes underpinning the demand and expectations of the local partners and populations. The intervener will judge at this stage whether the demand formulated is such as to allow his/her participation. If an intervention is deemed possible, then s/he will define objectives and the type of approach to be implemented in alignment with the requirements set down by the contract (time, costs, and national and international legislation). The specific demand to be met may emerge through progressive collaboration between the intervener and the local partners. It then is coconstructed up to its formulation in terms of action. Three situations have arisen in our interventions:

1. The technology has not been transferred yet. In designing the future system, one must take into account the specific features of the future receiving end. In this case, the anthropotechnologist will support the design process.
2. The technology has been transferred, but corrupted forms of functioning appear in the transfer process. Here, the anthropotechnologist's approach consists in altering an existing state.
3. The technology has not been transferred yet, but the technological choice has been made and is operational in the supplying country. As in the second situation, the approach will consist in altering the system to fit the receiving country.

Given these three possible situations, some changes may occur in the dynamic of the stages described hereafter.

87.2.2   Comprehensive Observation of the System on which the Anthropotechnologist Intervenes

The observation stages are always carried out in real-life situations. The anthropotechnologist endeavors to grasp the technical system as a whole so as to better focus his/her future action and produce a correct diagnosis of the situation. This involves observing and analyzing the existing situation, taking care to set the observed facts into their context and giving prominent place to the actors' points of view. The intervener consults available documents and interviews the different partners. S/he seeks to understand the way the system operates, the strategies used by the different actors, and tries to highlight "situations of characteristic action." These serve as markers to be used in structuring the following stages as well as exchanges among the actors.

Reference to the historical background is crucial when seeking to alter a system. Knowing what the system was in the past and the changes that have occurred under different influences (political, development schemes, economic, social, environmental, etc.) is a precondition for intervening in a given situation. Comprehensive observation of the system may coincide with the demand analysis stage. It familiarizes the intervener with local conditions.

In this long-term learning process, attention should be paid to opportunities to permanently combine local dimensions (practices, technical know-how, social organization, etc.) with wider dimensions (economic, political, regional, national, networks of actors, etc.). In some cases a several-months-long ethnographic approach to a given situation may profitably supply a detailed view of the situation locally. This is, for instance, the case for interventions in the agricultural environment, where the full crop year often needs to be considered. This raises the issue of the competences needed in such situations.

87.2.3 Setting Up a College of "Experts"

One constraint in the anthropotechnological approach stems from the need to gain in-depth knowledge of the sociocultural background in an intervention time span that is often all too limited. An ad hoc collective of permanent experts must systematically be established for the program duration according to the criteria that are to be considered. This collective should consist of scientists or of local or international specialists with recognized competence in the domains to be considered. Combining both approaches will help to produce a local diagnosis of the system being investigated and to propose potential courses of action to deal with the reformulated demand.

87.2.4 Creating a Monitoring Group

The monitoring group is different from the college of experts and is responsible for steering the intervention and its dynamic (Garrigou, 1999). It is present locally, which is not necessarily the case with the "experts." This is the group that decides about the organizing of the working group to be set up at a later date (when the markers for the design/alteration of the system will need to be developed) based on "situations of characteristic action."

87.2.5 Formulating the Diagnosis

One outcome of comprehensive observation is a diagnosis on the causes of system malfunctioning and on the "situations of characteristic action" to be investigated in greater depth. Whatever the situation, the diagnosis must be elaborated in collaboration with the local actors. Obtaining *a posteriori* acceptance by the actors of the standpoint that is produced goes counter to the inclusive anthropotechnological approach.

87.2.6 Setting Up a Working Group

A working group includes a specific set of actors plus the anthropotechnologist. Its role is technical. In establishing it, the monitoring group must be aware of the position held by each member inside his/her firm as well as, *sensu lato*, in society. In some countries, it may be necessary to include in these groups actors who are not necessarily directly involved in the project but whose standing in society (age, family ties, social status, etc.) makes them obligatory parties in any decision making. With the anthropotechnologist and certain experts, the working group contributes to elaborating markers for the design or alteration of the technical system. Throughout the intervention period, each meeting of the group deals with a specific topic as systematic observations progress. The anthropotechnologist's role is to create "space for discussion " (Dejours, 1996) and give the partners involved as much of a free hand as possible, since they subsequently will have to live on with the group's choices.

87.2.7 Systematic Observation of "Situations of Characteristic Action"

The anthropotechnologist's study concentrates during this stage on one or several "situations of characteristic action" (i.e., situations that seem to best reflect the difficulties or questions of the different actors involved in the diagnosis). S/he collects the operators' views on their own activity. S/he observes their practices with reference to a given context. "Auto confrontation" (self-report) sessions (Pinsky and

Theureau, 1987) help enrich the intervener's observations. Filming a "situation of characteristic action," then showing the filmed situation to the concerned actor and asking for his/her comments, allows much finer insight into the actor's activity than what could have been obtained from observations and interviews alone. In any design or transformation approach, systematic observations must be carried out in firms or sites in which similar equipment is being operated. These are called "reference situations." The interveners and their local partners will visit these similar sites. Visits to such sites allow the local partners to get a better understanding of the problems raised. They also enrich the partners' practices and will often lead to exchanges via the organization of "visits" to their respective sites.

87.2.8 Validation of Results

Each significant result should be validated by the actors in the working and monitoring groups, as each has its own validation criteria. For example, what matters for the ergonomist is the operators' health, while the anthropologist regards the stability of social structures within the project framework as paramount, and the actors are concerned with improving their work conditions, etc. The range of validation criteria is not easy to deal with. Their confrontation during the different group meetings needs to be managed if a consensus among the actors is to be reached. The results of the systematic observations are either validated or challenged. Adjustments are possible. The initial diagnosis is reassessed. Complementary systematic observations may be needed to complete or specify some of the information.

87.2.9 Creating Scenarios in Experimental Situations

Simulation of the "space of possible forms of future activities" (Daniellou, 1999) is performed by writing up scenarios, which are then analyzed and discussed within the working group. Their purpose is to illustrate different operating conditions of the chosen system over time. They also may bring out new problems for which it may be necessary to carry out new systematic observations locally or in reference situations. These scenarios also enable the prediction, inasmuch as possible, of probable impacts of the chosen solutions on the actors, the firm, or certain sections of society in which they move (systems of representation, gender distribution of tasks, religion, alliances, etc.). In some countries subject to major political or economic instability, the scope of these scenarios will be unavoidably limited. Anthropotechnology does not allow real-size experimenting of future means of work (Daniellou, 1999) in relation to sociocultural features in large industrial or agricultural systems.

87.2.10 Presentation of the Sociotechnical Markers

At this stage in the process, these markers, which result from the different observations, form proposals that the project members will rely on to change the situation. They apply to the technological process to be transferred or altered and the social conditions for its use. They serve to write up the list of specifications. They are produced in collaboration with the actors. These markers consist of material changes in the means of work, which also involve actions in the wider area of social and economic relationships — both within the considered society and between that society and its close or, in some cases, more distant environment.

87.2.11 Monitoring the Transformation or Design Operations

Both the working and the monitoring groups and the anthropotechnologist must follow the design and transformation stages, taking care that the markers listed in the specifications are observed. Alterations are still possible at this stage.

87.2.12   Monitoring the New System at the Startup and Stabilization of Production and Action Stages

The markers to be proposed must take into account the high mobility of social features in some situations. Their accompaniment must therefore be planned at the elaboration stage and not after they have been presented. This is why we propose an "anthropotechnological monitoring" process. Dysfunction may arise at these stages. The anthropotechnologist and his/her partners seek to remedy them by reassessing/ adjusting the proposed markers.

87.2.13   Anthropotechnological Monitoring

This stage is designed to enable the anthropotechnologist to respond to possible changes in the social setting in unstable situations, particularly in the agricultural environment or within minorities subjected to political or economic pressure. It involves sensitizing the local actors to the anthropotechnological approach and training national or international experts to perform readjustments in the system in response to events that could not be anticipated in the initial intervention. These training procedures are based on the principle of "training through practice" (hands-on training), which relies on a specific methodology.

87.3   Advantages

- In the anthropotechnologist's approach, attention is focused on work, i.e., on the activities of people who produce something. The approach traces back to the causes based on knowledge of their effects in the work situation. The work-analysis results make it possible to construct a tree of causes. This tree enables detection of critical anomalies that can be dealt with in a simple way without, however, overlooking sources situated higher upstream (Wisner, 1997). The phenomena observed are placed back in a context that extends beyond the workplace to include the sociological, cultural, political, and economic aspects of a given region and country.
- The anthropotechnological method increases the likelihood that technology transferred from the designer's country will be suitable for and successfully applied within the receiving country/culture.

87.4   Disadvantages

- The method is time-consuming, as it requires looking beyond the action framework to apprehend contextual dimensions that demand multidisciplinary work within the project.
- This approach does not produce prescriptions as outputs. A prescriptive stance is out of the question when dealing with social phenomena, as societies proceed from continuous and uncertain creation and not from an order that has been set once and for all (Balandier, 2002).
- The need to implement multidisciplinary expert knowledge and points of view that demand long-term monitoring may increase the cost and duration of the project.

87.5   Example Output

In past years, dissertations in anthropotechnology have been produced at the Centre National des Arts et Métiers (CNAM) (Wisner, 1997). The research work involved did not, however, correspond to intervention demands by local partners. The first intervention research approaches in this field (Geslin, 1999; Nourroudine, 2001) dealt with the rural world. Nowadays, institutional recognition of anthropotechnology in a range of national and international bodies enables us to extend our action from rural spheres to small businesses (including small enterprises in the rural environment or farm holdings) and issues regarding the environment in different countries.

87.6   Related Methods

The anthropotechnological approach partly derives from methods developed in applied anthropology (Ervin, 2000) and from ergonomic analysis of work as regards the intervention procedure. Approaches developed by ethnomethodologists are brought in to understand the cognitive processes at work in the activities being analyzed.

87.7   Standards and Regulations

Particular attention should be given to current legislation in the country of intervention, since technological policies in many countries are included in national or international legislation on technology transfers. When intervening in populations defined as minority populations, the specific legislation concerning these populations in international bodies must be respected.

87.8   Approximate Training and Application Time

Anthropotechnology is not taught as a specific field at the moment. In Switzerland, at the Institut d'ethnologic de l'Université de Neuchâtel, it is to be included in academic curricula and is of interset to students in anthropology (M.A. and Ph.D levels). In France, it is to be included in academic curricula in 2004/2005 at the Ecoles des Hautes Etudes en Sciences Sociales (EHESS) de Paris, and will be of interest to students in anthropology and ergonomics (M.A. and Ph.D. levels). Meanwhile, in addition to having a professional background in ergonomics, a course in cultural anthropology or social psychology should be helpful.

Application time is typically 2 to 3 years.

87.9   Reliability and Validity

This is an innovative approach in the domain of technology transfers. Ongoing programs chiefly concern the agricultural sphere. The SICOMOR (Sociétés, Changements Techniques et Connaissances dans les Mondes Ruraux) research unit, headed by Philippe Geslin, in the French Institut National de la Recherche Agronomique, devotes part of its work to developing this approach in partnership with several research and development institutions within and outside France. Past programs have delivered conclusive results that have induced our institutional partners and the populations concerned to formulate new demands in other fields.

87.10   Tools

Analyses are based on video recording and video restitution of activities and tape recording of verbal communications.

References

Balandier, G. (2002), Toute société procède d'une création incertaine et continuelle, non d'un ordre fixé une fois pour toute, in *La Compagnie des Contemporains*, Droit, R.P, Ed., Odile Jacob, Paris, pp. 201–206.

Daniellou, F. (1999), Le statut de la pratique et des connaissances dans l'intervention ergonomique de conception, in *Collection Thèses et Mémoires*, Université Victor Segalen, Bordeaux, France.

Dejours, C. (1996), Epistémologie concrète et ergonomie, in *L'Ergonomie en quête de ses principes*, Daniellou, F., Ed., Octarès Éditions, Tolouse, France.

Ervin, A.M. (2000), *Applied Anthropology*, Allyn and Bacon, Boston.

Garrigoou, A. (1999), Les apports des confrontations d'orientations socio-cognitives au sein de processus de conception participatifs: le rôle de l'ergonomie, in *Collection Thèses et Mémoires*, Université Victor Segalen, Bordeaux, France.

Geslin, P. (1999), *L'apprentissage des mondes, une anthropologie appliquée aux transferts de technologies*, Maison des Sciences de l'Homme, Paris.

Geslin, P. (2001), Anthropology, ergonomics and technology transfers: some methodological perspectives in light of a Guinean project, *Pract. Anthropol.*, 23, 23–27.

Geslin, P. (2002), Les formes sociales d'appropriations des objets techniques ou le paradigme anthropo-technologique, Ethnographiques.org [en ligne] no. 1 (April 2002); http://www.ethnographiques.org/documents/articles/arGeslin.html.

Geslin, P. and Salembier, P. (2000), Anthropology and ergonomics in designing innovations: theoretical and methodological foundations of a transdisciplinary research, in *Transdisciplinarity: Joint Problem-Solving among Science, Technology and Society*, Häberli, R., Scholz, R., et al., Eds., Swiss Federal Institute of Technology, Zurich, pp. 488–491.

Hutchins, E. (1981), Reasoning in Trobriand's discourse, in *Language, Culture and Cognition*, Casson, R.W., Ed., Macmillan, New York, pp. 481–489.

Nourroudine, A. (2001), *Techniques et cultures*, Octarès Éditions, Toulouse, France.

Pavard, B. (1997), Introduction, in *Anthropotechnologie: vers un monde industriel pluricentrique*, Wisner, A., Ed., Octarès Éditions, Toulouse, France, pp. 1–3.

Pinsky, L. and Theureau, J. (1987), *Conception des situations de travail et étude du cours d'action, 2° partie de l'étude du cours d'action*, Rapport n° 88, CNAM, Laboratoire d'Ergonomie, Paris.

Roqueplo, Ph. (1997), *Entre savoir et décision, l'expertise scientifique*, INRA Editions, Collection Sciences en Questions, Paris.

Wisner, A. (1979), Vers une anthropotechnologie, CNAM (Laboratoire d'ergnomie), Paris.

Wisner, A. (1994), Using Imported Technology as a Foreign Cultural Artefact, paper presented during the panel on cultural ergonomics at the 4th International Symposium on Human Factors in Organizational Design and Management, Stockholm.

Wisner, A. (1997), *Anthropotechnologie: vers un monde industriel pluricentrique*, Octarès Éditions, Toulouse, France.

88

Systems Analysis Tool (SAT)

Michelle M. Robertson
Liberty Mutual Research Institute for Safety

88.1 Background and Application

One methodology used for macroergonomic assessment of work-system processes is the systems analysis tool (SAT). Traditionally, approaches to work-system design have focused on technology-centered design and adopted a "leftover" approach to function and task allocation (Hendrick and Kleiner, 2001). Based on systems engineering (Hall, 1969) and a system analysis framework for policy decision making (Mosard, 1982), the SAT integrates these methodologies with that of a macroergonomic approach to identify problems and probable causal factors as related to work environments. SAT also can provide a process for developing strategic and systematic solutions to problems arising in a work environment. Development of potential solution alternatives and evaluation of the costs and benefits of these potential solutions are accomplished by applying the SAT.

Critical to the application of SAT is the development of an evaluation scorecard table (EST) and a decision criteria table (DCT). The EST first serves as an evaluation tool for considering the trade-offs of the micro- and macroergonomic intervention alternatives. For each alternative, the DCT provides an assessment of each of the intervention alternatives according to the probability of a future condition in the organizational environment. The available level of funding (or budget) could serve as a future condition along with the expected probability of that condition. The value of SAT is that it provides a systematic evaluation process for examining the trade-offs and selecting an intervention alternative. After the optimal solution alternative is selected, specific measures of success are defined, and feedback on the results of the selected intervention are obtained.

The SAT can be applied in a variety of work environments, including industrial and office environments.

88.2   Procedure

In applying the SAT, the level of analysis is targeted at the business-unit or departmental level, where the business mission and objectives are identified as well as individual and group goals in support of that mission. The SAT analytic steps are:

1. Define the problem: create a problem factor tree (PFT)
2. Develop an objectives/activities tree (OAT)
3. Model alternatives: the input–output flow diagram (IOFD)
4. Evaluate alternatives: evaluation scorecard table (EST)
5. Select an alternative: decision criteria table (DCT)
6. Plan for implementation: scheduling and management of project flow
7. Evaluation, feedback, and modification process

88.2.1   Defining the Problem: The Problem Factor Tree

From system performance data previously collected or recently observed, the problem factor tree (PFT) is constructed. The PFT identifies the problems, subproblems, and potential causal factors, along with their interactions, in a logical, hierarchical structure (Hall, 1969; Mosard, 1982). To develop a PFT, work issues and problems are precisely stated and linked together through an iterative process (Mosard, 1982). The lower-level causal factors in the tree contribute to the major problem. Feedback loops also can be incorporated.

The problem factor tree integrates both microergonomic and macroergonomic aspects of work systems and subsystems, including organizational and job design issues. This initial step of the systems analysis develops a better understanding of the technical, social, and work environment subsystems.

88.2.2   Setting the Objectives and Developing Alternatives: The Objectives/Activities Tree

With the problem defined, objectives and evaluation criteria are developed for use in selecting the best alternative (intervention) to address the causal factors. The objectives/activities tree (OAT) is a hierarchical, graphical depiction of objectives and solution alternatives developed to address the problems identified in the PFT (Mosard, 1982). The tree is created by identifying a major goal, objective, and subobjectives. Four potential solution alternatives along with a specified set of supporting activities are then identified to address the objective. Interactions between objectives and alternatives involved in the process should also be analyzed (Mosard, 1982).

Intervention alternatives are identified that could represent typical micro- and macroergonomic approaches. Additionally, hybrid alternatives incorporating one or more of the defined alternatives can be created to incorporate the best features of any of the initially identified alternatives. For each alternative, several potential activities (action steps) are then proposed. Solution alternatives and related activities are based on case studies, field research, and longitudinal studies and represent typical approaches companies implement to achieve the major defined objective (listed at the top of the OAT). One of these solution alternatives will be selected in step 5 and become the subobjective in the objectives tree.

After the objectives and alternatives are selected, a preliminary criteria table is developed. This table is used to define the criteria on which the final selection decision will be made. Decision criteria typically include risks, costs, expected benefits, and a measure of effectiveness based on both a short-term and long-term perspective (Mosard, 1983). These decision criteria should be considered as alternatives and are modeled in step 3. They also serve as the basis for the final decision criteria table, which is developed in step 4.

88.2.3   Modeling the Alternatives: The Input–Output Flow Diagram

In this step, each solution alternative and its associated activities are modeled to estimate the gross resource requirements and to assess the alternative's potential effectiveness. Typically, a descriptive or predictive model is used, and the modeling techniques include flowcharts, simulations, etc. One modeling approach that captures the essential resources and creates a flow diagram of where these resources are needed is the input–output flow diagram (IOFD). The IOFD is created by identifying the inputs of the required resources that one can reasonably expect for the completion of a specified solution and set of activities. Resources may include people, finances, and information. Outputs are the results or products of the activities. Some outputs may become sources of inputs to other subsystems, permitting a more complete representation of the entire system. The IOFD has two phases: a redesign phase and an operation phase. Inputs for the redesign phase include contributions from two general areas: human resources and financial resources. Outputs from the redesign phase become the inputs for the operation phase.

88.2.4   Evaluating the Alternatives: The Evaluation Scorecard Table (EST) — Economic Advantage Analysis

Each of the modeled alternatives and its associated set of activities is evaluated according to several major decision criteria. These criteria include: project cost, risk of failure, potential effectiveness, and benefits for all appropriate future conditions. These criteria may include both short- and long-term perspectives. The criteria are assigned equal weights, which then are summed across all of the criteria to provide an overall rating for each alternative. Alternately, weights can be assigned to each criterion as determined by its importance.

88.2.5   Selecting an Alternative: The Decision Criteria Table (DCT)

Step 5 involves creating decision tables. Each decision table includes an assessment of the alternatives according to the probability of a future condition in the organizational environment. A decision criteria table (DCT) can be created by using the available level of funding (or budget) as the future condition along with the expected probability of that condition. Each alternative then is assessed against the level of the future condition (in this case high, medium, or low available funds) and the probability of that level. For each future condition level, the analyst ranks each alternative indicating a first, second, third choice, etc. When available funding is high, a more comprehensive and aggressive alternative is preferred; while, when funding is low, less ambitious approaches may be more realistic. (See Table 88.1.) The solution alternative that emerges as the winner at this point becomes the subobjective in the objectives/activities tree and proceeds to implementation.

88.2.6   Implementation

A schedule and sequence of tasks, responsibilities, and requirements is developed for implementation. This schedule may include a contingency plan with scheduled decision points and corresponding responsibilities. There are several available scheduling techniques that can be used as well as various software programs for creating new databases containing effectiveness measures, costs, and other pertinent metrics collected during the systems analysis.

88.2.7   Evaluation, Modification, and Feedback

Several activities occur in this step in order to define, establish, and develop the evaluation processes, all of which provide feedback to the appropriate strategic decision maker regarding the results of the selected intervention alternative(s). Establishing an evaluation measurement process could include a five-step/

TABLE 88.1 Decision Criteria Table: Selecting an Alternative Based on Anticipated Future Funding Conditions

Alternatives	Ratings[a] Based on Future Funding Conditions (probability of funding at each level)		
	High Level of Funding (0.50)	Moderate Level of Funding (0.65)	Low Level of Funding (0.80)
Alternative A: Redesign job/job content	3	2	1
Alternative B: Ergonomically redesign workstation and environment	4	1	2
Alternative C: Redesign job/job content, ergonomically redesign workstation and environment, train managers, and distribute office ergonomics manual	1	4	4
Alternative D: Ergonomically redesign workstation and environment, train managers, and distribute office ergonomics manual	2	3	3

[a] The ratings score indicates the selection preference rankings based on the probability-assigned criteria and the overall rating scores for each of the alternatives from the preliminary criteria table previously developed. Each alternative is subjectively rated on a 0 to 10 scale where a 0 rating indicates a low preference and a 10 rating indicates a high preference.

level process used by organizations to assess the effectiveness of workplace and training interventions (Gordon, 1994):

1. Baseline assessment prior to intervention
2. User's reaction to the intervention
3. User's degree of learning
4. User's performance (individual as well as unit level)
5. Organizational results (business goals and productivity)

A multidisciplinary team consisting of various subject-matter experts ensures a diversity of viewpoints and critical inputs when developing evaluation measures for each of the five levels. At each evaluation level, specific measures are defined. These measures may already exist in the organization, or they may be developed. Using the evaluation process, the outcome measures that are most likely to reflect the impact of a selected alternative can be identified. This critical information then can be used to provide feedback to the program team, an individual management decision maker, or the organization as a whole.

Feedback regarding program effectiveness begins the cycle of continuous improvement. Using information gathered from the evaluation and feedback process, the analyst can assess and then implement potential modifications and changes to the program.

88.3 Advantages

- Provides a deep understanding of the casual factors, on a micro- and macroergonomic level, that contribute to an identified work environment problem.
- Helpful in developing a range of intervention alternatives to solve the work-system problem focusing at both the micro- and macroergonomic level.
- Enables trade-off analyses of each alternative solution, at the micro- and macroergonomic level, based on cost benefit data and future conditions.
- Provides a robust analytical method that can be applied to a variety of work environments and identified work-system problems.

- Provides decision makers with a systematic viewpoint of the work-system problem and solutions by presenting either a graphical diagram, flowchart, or matrix table for each SAT step. This allows decision makers to understand not only the complexity of the problem, but the level of impact each alternative solution will have on solving the work-system problem.

88.4 Disadvantages

- Difficulty in obtaining a multidisciplinary viewpoint to create the problem factor tree and formulate the range of alternative solutions to the defined work-system problem.
- Difficulty in finding reliable and valid trade-off (costs/benefit) and effectiveness data for each identified alternative solution to construct the evaluation criteria scorecard.
- Time required to apply the SAT in a thorough and systematic manner and to create the graphical illustrations for presentation to the decision maker.

88.5 Example Outputs

Two examples of outputs of the SAT are provided in Figure 88.1 and Table 88.1. The graphical output for the SAT step 1 is the PFT (Figure 88.1), where this particular PFT is focused on examining an office work system and the hypothesized problem of human performance and well-being. For a complete example of the SAT as applied to an office work system, see Robertson et al. (2002).

For step 5 of the SAT, an example of a decision criteria table (DCT) is given in Table 88.1. This table presents each alternative intervention and is evaluated according to the probability of a future condition within the organizational environment, such as level of funding.

88.6 Related Methods

There are several related systems-driven methodologies, including macroergonomic analysis of structure (MAS) and macroergonomic analysis and design (MEAD), described in Chapters 89 and 90, respectively. For the SAT steps of creating a problem factor tree and objective activities tree, there are similar methods such as conducting a systems-safety fault-tree analysis, cause–effect diagrams, fishbone diagram, hierarchical-task analysis, and multiattribute utility analysis. Stochastic decision modeling and Monte Carlo simulations are other related modeling and decision analytical tools.

88.7 Standards and Regulations

There are no specific standards and regulations for SAT. However, the model can be used in support of compliance with the draft standard, ANSI Z365: Management of Work-Related Musculoskeletal Disorders (WMSDs), as well as other workplace ergonomics standards and regulations (e.g., California and Washington OSHA regulations in the U.S.).

88.8 Approximate Training and Application Time

A professional safety, human factors, and/or industrial engineer, ergonomist, or risk manager would possibly have the background to implement SAT. In order to create a transdisciplinary perspective, other team members with specialized expertise (e.g., operations-modeling experts and production specialists) are necessary to assist the analyst. Depending on the level of expertise and knowledge of the analyst, resource availability, and the complexity of the defined problem and alternative solutions, the application time could range from 40 to more than 300 hours.

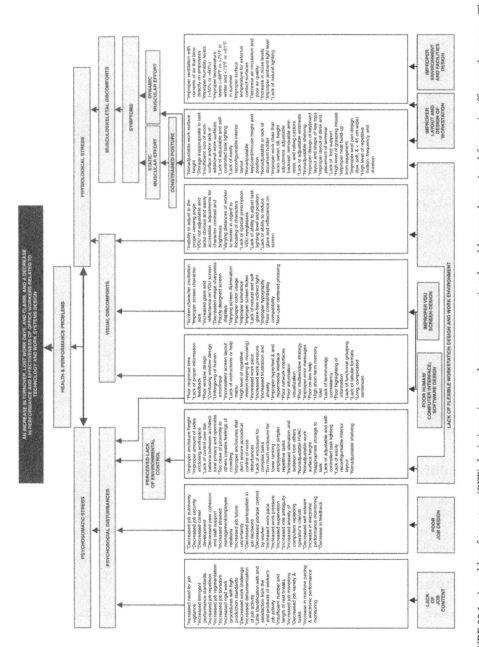

FIGURE 88.1 A problem factor tree (PFT) depicting micro- and macroergonomic subproblems and associated factors in office work systems. The major problem is defined at the top, with the subproblems and associated factors shown below indicating the hierarchical, logical structure of the encompassing problem elements.

88.9   Reliability and Validity

The reliability and validity of the outcome when implementing the SAT is dependent on several factors: the data used in applying SAT (economic costs, benefits, and effectiveness data), reliability of the decision maker and analyst in judging the evaluation process, and the consistency and stability of the outcome given the same conditions and decision maker. SAT was successfully applied in creating alternative solutions for designing a new office work environment and training program for a large customer service company. The model identified the potential ROI for each alternative and hybrid alternatives, which drove the decision makers to choose the higher-cost work environment change and training program (including a workplace change program). The alternative chosen ultimately produced the expected ROI in performance, customer satisfaction, and efficient business processes.

88.10   Tools Needed

Scheduling and modeling tools, graphical production tools, literature searches, and access to databases are required. Development of the problem factor tree and the objective/activities tree can be accomplished using various desktop publishers and organizational software. Operations-modeling computer tools are beneficial in creating the models of each alternative solution.

References

Gordon, S. (1994), *Systematic Training Program Design: Maximizing and Minimizing Liability*, Prentice-Hall, Englewood Cliffs, NJ.

Hall, A.D. (1969), A three dimensional morphology of systems engineering, *IEEE Trans. Syst. Sci. Cybernetics*, 5, 156–160.

Hendrick, H.W. and Kleiner, B.M. (2001), *Macroergonomics: An Introduction to Work System Design*, Human Factors and Ergonomics Society, Santa Monica, CA.

Mosard, G. (1982), A generalized framework and methodology for systems analysis, *IEEE Trans. Eng. Manage.*, 29, 81–87.

Mosard, G. (1983), Problem definition: tasks and techniques, *J. Syst. Manage.*, 34, 16–21.

Robertson, M.M., Kleiner, B.M., and O'Neill, M.J. (2002), Macroergonomic methods: assessing work system processes, in *Macroergonomics: Theory, Methods, and Applications*, Hendrick, H.W. and Kleiner, B., Eds., Lawrence Erlbaum Associates, Mahwah, NJ, pp. 67–96.

89

Macroergonomic Analysis of Structure (MAS)

Hal W. Hendrick
Hendrick and Associates

89.1 Background and Application

The macroergonomic analysis of structure (MAS) method combines empirically developed analytical models of the effect of three major sociotechnical system elements — the technological subsystem, personnel subsystem, and relevant external environment — on the fourth major element, the structure of the organization's work system. Through the analysis of the key characteristics of these three sociotechnical system elements, the basic design of the work system's structure for effective functioning can be determined. The MAS analysis results then can be compared with the existing structure of a given organization's work system to identify discrepancies for correction. The model's results not only aid in identifying discrepancies, but also provide guidance on what to do to correct them for more optimal work-system functioning. This, in turn, sets the stage for the related analysis and refinement of the work system's processes.

89.2 Procedure

89.2.1 Structural Dimensions of Work Systems

The structure of a work system often is conceptualized as having three core dimensions (Stevenson, 1993; Bedeian and Zammuto, 1991; Robbins, 1983):

1. Complexity

2. Formalization
3. Centralization

89.2.1.1 Complexity

Complexity refers to the degree of differentiation and integration that exist within a work system's structure. Differentiation refers to the degree to which the work system is segmented into parts. Integration refers to the number and types of mechanisms that are used to integrate the segmented parts for the purposes of communication, coordination, and control.

89.2.1.1.1 *Differentiation*

Three common types of differentiation are employed in structuring work systems: vertical, horizontal, and spatial. Increasing any one of the three increases a work system's complexity.

89.2.1.1.1.1 Vertical Differentiation — Vertical differentiation refers to the number of hierarchical levels separating the chief executive position from the jobs directly involved with the system's output.

89.2.1.1.1.2 Horizontal Differentiation — Horizontal differentiation refers to the degree of departmentalization and specialization within a given work system. There are two commonly used ways to determine whether or not a work group should be divided into one or more departments. These are the degree of commonality of:

1. Goals
2. Time orientation

The more that subgroups differ either in goals or time orientation, the greater is the likelihood that they should be structured as separate departments.

89.2.1.1.1.3 Spatial dispersion — Spatial dispersion refers to the degree that an organization's activities are performed in more than one location. Three different measures commonly are used to quantify the degree of spatial dispersion (Hall et al., 1967):

1. The number of geographic locations comprising the total work system
2. The average distance of the separated locations from the organization's headquarters
3. The proportion of employees in these separated units in relation to the number in the headquarters

In general, increasing any of these three measures increases complexity.

89.2.1.1.2 *Integration*

Integration refers to the number of mechanisms designed into a work system for ensuring communication, coordination, and control among the differentiated elements. In general, as a work system increases in differentiation, the need for integrating mechanisms also increases. This happens because greater differentiation increases the number of units, levels, and departments that must communicate with one another, coordinate their separate activities, and be controlled for efficient operation. The most common types of integrating mechanisms that can be designed into a work system are formal rules and procedures, committees, task teams, liaison positions, and system integration offices. Vertical differentiation, in itself, is a primary integrating mechanism (i.e., a manager at one level typically serves to coordinate and control the activities of several lower-level groups). It also should be noted that computerized information and decision-support systems can be designed to serve as integrating mechanisms. Having too few integrating mechanisms will result in inadequate coordination and control among the differentiated elements; too many will stifle the work system's functioning and increase costs.

89.2.1.2 Formalization

From a macroergonomics perspective, formalization can be defined as the degree to which jobs within the work system are standardized. Highly formalized work systems allow for little employee discretion over what is to be done, how it is to be accomplished, or when it is to be done (Robbins, 1983). Highly formalized designs are characterized by explicit job descriptions, extensive rules, and clearly defined standardized procedures covering work processes. As a general rule, the simpler and more repetitive the jobs to be designed into the work system, the higher should be the level of formalization. However, consideration should be given to *not* making the work system so highly formalized that jobs lack any intrinsic motivation, fail to effectively utilize employee skills, or degrade human dignity. In contrast, the more nonroutine or unpredictable the work tasks and related decision making, the less amenable the work system is to high formalization.

89.2.1.3 Centralization

Centralization has to do with where formal decision making occurs within the work system. In highly centralized work systems, formal decision making is concentrated in a relatively few individuals, group, or level, usually high in the organization; and lower-level supervisors and employees have only minimal input into the decisions affecting their jobs (Robbins, 1983). In highly decentralized work systems, decisions are delegated downward to the lowest level having the necessary expertise. Highly decentralized work systems thus require lower-level employees to have a relatively higher level of education and training or professionalism. It should be noted that there are two general classes of organizational decision making, *tactical* or day-to-day operational decisions and *strategic* or long-range planning decisions, and that where they are made in an organization can differ greatly.

89.2.2 Analysis of the Sociotechnical System Elements

Effective design of a work system's structure involves consideration of three major sociotechnical system elements that interact and affect optimal work-system functioning. These are (DeGreene, 1973):

1. Technological subsystem
2. Personnel subsystem
3. Relevant external environment, or that portion of the external environment that permeates the organization and upon which it depends for its survival and success

Each of these elements has been studied in relation to its effects on the three organizational design dimensions described above — complexity, formalization, and centralization — and empirical models have emerged. These models can be used as macroergonomics tools in assessing and developing or modifying the design of a given work system. The models included in the MAS method have proven particularly useful to the author.

89.2.2.1 Technological Subsystem Analysis

Perrow (1967) developed a generalizable model of the technology–work-system structure relationship that uses a knowledge-based classification scheme. Perrow begins by defining technology as the action one performs upon an object in order to change the object. This action requires some form of technical knowledge. Using this approach, Perrow identified two underlying dimensions of knowledge-based technology: task variability and task analyzability. Task variability is the number of exceptions encountered in one's work, and task analyzability concerns the type of search procedures that are available for responding to task exceptions. These search procedures can range from "well defined" to "ill defined." At the "well-defined" end of the continuum, problems are solved using rational–logical, quantitative, and analytical reasoning. At the "ill-defined" end there are no readily available formal search procedures, and one must rely on experience, judgment, and intuition to solve problems. Dichotomizing these two dimensions yields a matrix having four cells, as shown in Table 89.1. Each of the four cells represents a different knowledge-based technology.

The four knowledge-based technologies in this schema are:

1. *Routine* technologies, which have well-defined problems with few exceptions. Mass-production units typify this category, as do some types of service organizations, where the nature of the servicing is largely repetitive. Routine technologies lend themselves to standardized coordination and control procedures and, thus, are associated with high formalization and centralization.

TABLE 89.1 Perrow's Four-Category Classification Scheme for Technologies

		Task Variability	
		Routine with few exceptions	High variety with many exceptions
Problem Analyzability	Well defined and analyzable	Routine	Engineering
	Ill-defined and unanalyzable	Craft	Nonroutine

2. *Nonroutine* technologies, which have many exceptions and problems that are difficult to analyze. Combat aerospace operations is an example. Critical to these technologies is flexibility. Therefore, they need to be highly decentralized and have low formalization.
3. *Engineering* technologies, which have many exceptions but can be handled using well-defined rational–logical processes. Consequently, they lend themselves to moderate centralization but need the flexibility that is achievable through low formalization.
4. *Craft* technologies, which involve fairly routine tasks, but problem solving relies heavily on the experience, judgment, and intuition of the individual craftsperson. Thus, those with the particular expertise must make decisions. This requires decentralization and low formalization.

Perrow's model has been supported by empirical research in both the private and public sectors (e.g., Van deVen and Delbecq, 1979; Magnusen, 1970; Hage and Aiken, 1969). I have found this model to be particularly useful for analyzing an organization's technology to determine its implications for the work system's structure

89.2.2.2 Personnel Subsystem Analysis

There are at least three major characteristics of the personnel subsystem that are sensitive to the design of an organization's work-system structure. These are the degree of professionalism, the cultural factors, and the psychosocial aspects of the work force.

89.2.2.2.1 Degree of Professionalism

Robbins (1983) notes that formalization can take place either on the job or through the process of professionalization. On the job, formalization is external to the employee. Indeed, the term *formalization* applies to a situation where rules, procedures, and human–system interfaces are designed into the work system to limit employee discretion. As a result, the skill requirements of such jobs tend to be low. Professionalism, on the other hand, creates internal formalization of behavior through a socialization process that is an integral part of the education and training process. From a macroergonomics design standpoint, there is a trade-off between formalization of the work system and professionalization of the jobs in the work system. Where the work system is designed to allow for low formalization and, thus, considerable employee discretion, jobs should be designed to require persons with a relatively high level of professionalism. Most often, it is the need to have employees that can deal with unique, nonroutine, or unanticipated situations that creates the need for low formalization and more highly professionalized jobs.

89.2.2.2.2. Cultural Factors

It is important to consider the values, perceptions, mores, and attitudes of the culture(s) from which the workforce is drawn. For example, if autonomy and individualism are greatly valued, then providing for autonomy and individual recognition for accomplishment will be important in the design of the work system and the jobs that it comprises.

89.2.2.2.3 Psychosocial Factors

I have found a very useful assessment model to be that of *cognitive complexity* (Harvey et al., 1961; Harvey, 1963). Because a full explanation of this model is beyond the scope of this description, the reader is referred to Hendrick (1996, 1997) or Hendrick and Kleiner (2001). In general, as persons are (a) encouraged through child rearing and early school education to explore and think for themselves and (b) exposed to a diversity of experiences through education, media, travel, etc., they develop progressively more conceptual categories for storing their experiential information, more "shades of gray" within categories, and more rules and rule combinations for integrating information from the various conceptual categories. In short, they become more cognitively complex. By comparison, cognitively concrete persons tend to have a high need for structure and order, a low tolerance for ambiguity, closed belief system, are fairly authoritarian, and rely on rules regulations and traditions in their decision making. They tend to see the world as relatively static and unchanging, and they become upset or stressed when things do change — including organizational structures and processes. In contrast, more cognitively complex persons have a low need for structure and order, are tolerant of ambiguity, open in their belief system, not highly authoritarian, very people oriented, and are flexible. They see the world as highly dynamic and expect things to change, including organizational structures and processes. In light of the above, it is not surprising that concrete-functioning persons tend to prefer work systems that are relatively slow to change, have a clear unambiguous structure, and are fairly formalized. In contrast, cognitively complex persons are comfortable in organizations with low formalization and centralization, and where organizational structures may be ambiguous and subject to frequent change.

89.2.2.3 Relevant External Environment

Critical to the survival of organizations is their ability to adapt to their external environment. As open systems, organizations require monitoring and feedback mechanisms to follow and sense changes in their relevant external environments and a capacity to make timely, responsive adjustments. *Relevant task environments* refer to those aspects of the external world that can positively or negatively influence the organization's effectiveness.

Neghandi (1977) has identified five types of external environments that impact on organizational functioning.

1. *Socioeconomic:* particularly the degree of stability, nature of the competition, and availability of materials and qualified workers
2. *Educational:* the availability of facilities and programs, and the educational level and aspirations of workers
3. *Political:* governmental attitudes toward (a) business (friendliness vs. hostility), (b) control of prices, and (c) "pampering" of industrial workers
4. *Cultural:* social status and caste system, values and attitudes toward work, management, etc., and the nature of trade unions and union–management relationships
5. *Legal:* the degree of legal controls, restrictions, and compliance requirements

These categories are based on the results of field studies of 92 industrial firms in five different countries.

89.2.2.4 Environmental Uncertainty

Of particular importance to work-system design is the fact that all specific task environments vary along two highly critical dimensions: change and complexity (Duncan, 1972). The degree of change refers to the extent to which a given specific task environment is dynamic or remains stable and predictable over time. The degree of complexity is defined operationally by the number of components that constitute an organization's specific task environment (i.e., does the company interact with few or many government agencies, customers, suppliers, competitors, etc.?). Environmental change and complexity, in combination, determine the environmental uncertainty of an organization. Table 89.2 illustrates this relationship for four different levels of uncertainty.

TABLE 89.2 Environmental Uncertainty as a Function of Complexity and Change

		Degree of Change	
		Stable	Dynamic
Degree of Complexity	Simple	Low uncertainty	Moderately high uncertainty
	Complex	Moderately low uncertainty	High uncertainty

Based on studies of 20 English and Scottish industrial companies, Burns and Stalker (1961) found that the type of work-system structure that worked best in a relatively stable and simple organizational environment differed from that required for a more dynamic and complex environment. For stable, simple environments, mechanistic structures — characterized by relatively high to moderately high vertical and horizontal differentiation, formalization, and centralization — worked best. Mechanistic structures typically have routine tasks, programmed behaviors, and can respond to change only slowly. A strong emphasis is placed on stability and control. Conversely, for dynamic, complex environments, organic structures, characterized by flexibility and quick adaptability, worked best. Organic work systems stress lateral rather than vertical communication, influence based on knowledge and expertise rather than position and authority, information exchange rather than directives from above, conflict resolution by interaction rather than by superiors, and loosely, rather than tightly defined job descriptions and responsibilities. Organic work systems thus require low vertical differentiation and formalization, decentralized tactical decision making, and a relatively high level of professionalism. These findings have been supported by the research of many others.

89.2.3 Integrating the Results of the Separate Assessments

The separate analyses of the key characteristics of a given organization's technological subsystem, personnel subsystem, and specific task environment each should have provided guidance about the structural design for the work system. Frequently, these results will show a natural convergence. At times, however, the outcome of the analysis of one sociotechnical system element may conflict with the outcomes of the other two. When this occurs, the ergonomist is faced with the issue of how to reconcile the differences. Based both on the suggestions from the literature and my personal experience in evaluating over 200 organizational units, the outcomes from the analyses can be integrated by weighting them approximately as follows:

If the technological subsystem analysis is assigned a weight of "1," give the personnel subsystem analysis a weight of "2," and the relevant external environment analysis a weight of "3." For example, let us assume that the technological subsystem falls into Perrow's "routine" category, the personnel subsystem's jobs call for a high level of professionalism, and the external environment has moderately low environmental uncertainty. Weighting these three as suggested above would indicate that a moderately formalized and somewhat centralized work system would work best. Accordingly, the results would indicate that most jobs should be redesigned to require a somewhat lower level of professionalism and that attendant hardware and software interfaces should be designed/redesigned to be compatible.

It should be noted that the specific functional units of an organization might differ in the characteristics of their technology, personnel, and relevant task environments, particularly within larger organizations. Therefore, the separate functional units may, themselves, need to be analyzed as though they were separate organizations, and the resultant work systems designed accordingly.

89.3 Advantages

- Enables the ergonomist or organizational design specialist to take into account the effect of an organization's unique sociotechnical characteristics in determining an optimal work-system design.
- By comparing the MAS results with the current work-system structure, dysfunctional discrepancies often can be identified.
- Where discrepancies are noted, the MAS results also provide guidance for correcting them.

89.4 Disadvantages

- Use of the MAS requires training and experience in conducting organizational assessments.

- Determining the amount of a key sociotechnical variable present or absent is not a simple quantitative process; it requires subjective judgment, based on education and experience.

89.5   Example Output

The output generally takes two forms: a series of tables and a narrative report utilizing the data in the tables. The tables are as follows:

1. A first table depicting the MAS results for each of the three key sociotechnical variables for the organization (see Table 89.3).
2. A second table depicting the work-system structure indicated by the MAS. The structural dimensions for each of the three sociotechnical elements are rated using the information in the first table, and a weighted average of the ratings is obtained (see Table 89.4).
3. A third table provides a comparison of the MAS results from the second table with the data for the work system as it currently exists, noting any discrepancies between the two (see Table 89.5).

TABLE 89.3   MAS Results for Each Key Sociotechnical Variable for Organization X

Sociotechnical Variable	Rating Score [a]
Technological Subsystem	
Task variability	5
Task analyzability	3
Personnel Subsystem	
Level of professionalism	5
Cultural factors	4
Psychosocial factors	4
External Environment	
Environmental complexity	4
Environmental uncertainty	5

[a] Rating scale: 1 = low, 3 = intermediate, 5 = high.

TABLE 89.4   Work-System Structure Indicated by the MAS for Organization X

Structural Dimensions	Technological Subsystem [a]	Personnel Subsystem [a]	External Environment [a]	Weighted Rating [b]
Vertical differentiation	3	2	2	2.2
Horizontal differentiation	4	4	4	4.0
Integrating mechanisms	4	3	4	3.7
Formalization	2	1	1	1.2
Centralization: Tactical	2	1	1	1.2
Centralization: Strategic	4	3	4	3.7

[a] Weights: Technological Subsystem =1; Personnel subsystem = 2; External Environment = 3.
[b] Rating scale: 1 = low, 3 = intermediate, 5 = high.

TABLE 89.5   Comparison of the MAS Results with the Current Work System

Structural dimensions	MAS	Current	Difference
Vertical differentiation	2.3	4.0	+1.7
Horizontal differentiation	4.0	4.0	—
Integrating mechanisms	3.7	2.5	−1.2
Formalization	1.2	3.0	+1.8
Centralization: Tactical	1.2	3.5	+2.3
Centralization: Strategic	3.7	4.0	+0.3

The narrative report of the analysis, citing the tables, provides a narrative description of the procedure, findings, and recommendations.

The results in Table 89.5 indicates that the organization has too many hierarchical levels, not enough integrating mechanisms for its complexity, too much formalization for such a highly professionalized work force, and needs to decentralize more of its tactical decision making. Correction of these discrepancies should result in a marked improvement in the organization's functioning.

89.6   Related Methods

The MEAD ten-step method for assessing and designing work-system processes, discussed in Chapter 90, is the most closely related method. However, the other methods for assessing organizational design — HITOP, TOP-Modeler, CIMOP, and SAT, described in Chapters 84, 85, 86, and 88, respectively — also are related and, where applicable, can be used in conjunction with MAS.

89.7   Standards and Regulations

There are no standards or regulations that specifically apply to MAS.

89.8   Training and Application Time

For ergonomists with professional education in organizational psychology, approximately 20 contact hours of hands-on training should be sufficient. For those without education and training in organizational psychology, a graduate level course in organizational psychology or macroergonomics and a practicum in conducting organizational assessments with MAS are advisable.

The time to apply the method to a given organization depends largely on the size of the organization. Several weeks of data gathering from organizational records, interviews with key management personnel, and direct observation usually will be enough for a small organization. Large complex organizational units could require much longer. Once the data are gathered, inputting it into the various models and deriving the output can be done within several days for a small organization and within several weeks for a large, complex one.

89.9   Reliability and Validity

Each of the models used in the MAS method has been extensively validated. To date, the integration of these models into the MAS method has been validated in only a few cases. I personally have validated the method in two applications: a high-technology aerospace company and a university college. (See Hendrick and Kleiner [2001] for a summary of the university college study.)

89.10   Tools Required

The primary tools needed are methods for recording data as it is gathered. It can be very helpful to develop forms for each type of data required (e.g., for each sociotechnical dimension, listing the key variables and providing space for recording the data collected for each). For large, complex organizations, it might be advisable to supplement interviews, records, and direct observation by developing and administering a questionnaire survey to obtain some of the needed data (see Chapter 76, Macroergonomic Organizational Questionnaire Survey).

References

Bedeian, A.G. and Zammuto, R.F. (1991), *Organizations: Theory and Design*, Dryden Press, Chicago.

Burns, T. and Stalker, G.M. (1961), *The Management of Innovation*, Tavistock, London.

DeGreene, K. (1973), *Sociotechnical Systems*, Prentice-Hall, Englewood Cliffs, NJ.

Duncan, R.B. (1972), Characteristics of organizational environments and perceived environmental uncertainty, *Administrative Sci. Q.*, 17, 313–327.

Hage, J. and Aiken, M. (1969), Routine technology, social structure, and organizational goals, *Admin. Sci. Q.*, September, 366–377.

Hall, R.H., Haas, J.E., and Johnson, N.J. (1967), Organizational size, complexity and formalization, Am. Sociol. Rev., December, 905–912.

Harvey, O.J. (1963), System structure, flexibility and creativity, in *Experience, Structure and Adaptability*, Harvey, O.J., Ed., Springer, New York.

Harvey, O.J., Hunt, D.E., and Schroder, H.M. (1961), *Conceptual Systems and Personality Organization*, Wiley, New York.

Hendrick, H.W. (1996), Cognitive complexity, conceptual systems, and behavior, *J. Wash. Acad. Sci.*, 84, 53–67.

Hendrick, H.W. (1997), Organizational design and macroergonomics, in *Handbook of Human Factors and Ergonomics*, 2nd ed., Salvendy, G., Ed., Wiley, New York, pp. 594–636.

Hendrick, H.W. and Kleiner, B.M. (2001), *Macroergonomics: An Introduction to Work System Design*, Human Factors and Ergonomics Society, Santa Monica, CA.

Magnusen, K., 1970, Technology and Organizational Differentiation: A Field Study of Manufacturing Corporations, Ph.D. dissertation, University of Wisconsin, Madison.

Negandhi, A.R. (1977), A model for analysing organization in cross-cultural settings: a conceptual scheme and some research findings, in *Modern Organization Theory*, Negandhi, A.R., England, G.W., and Wilpert, B., Eds., University Press, Kent State, OH.

Perrow, C. (1967), A framework for the comparative analysis of organizations, *Am. Sociol. Rev.*, 32, 194–208.

Robbins, S.R. (1983), *Organization Theory: The Structure and Design of Organizations*, Prentice-Hall, Englewood Cliffs, NJ.

Stevenson, W.B. (1993), Organizational design, in *Handbook of Organizational Behavior*, Golembiewski, T., Ed., Marcel Dekker, New York, pp. 141–168.

Van de Van, A.H. and Delbecq, A.L. (1979), A task contingent model of work-unit structure, *Adm. Sci. Q.*, 24, 183–197.

90

Macroergonomic Analysis and Design (MEAD)

Brian M. Kleiner
Virginia Polytechnic Institute
and State University

90.1 Background and Application

The macroergonomic analysis and design (MEAD) approach was developed based on the contributions of Emery and Trist (1978), Taylor and Felton (1993), and Clegg et al. (1989) and experience with large-scale change in academia, industry, and government (Kleiner, 1996). It is more fully described in Hendrick and Kleiner (2001, 2002). The approach integrates sociotechnical systems (STS) theory and ergonomics.

90.2 Procedure

There are ten steps in the MEAD approach, as follows:

90.2.1 Scanning the Environmental and Organizational Subsystems

The first phase of sociotechnical analysis of work-system process is to scan the system. Then, the environment and organizational subsystems are scanned. Achieving a valid organization/environment fit and joint optimization are essential.

Formal company statements about mission (i.e., purpose), vision, and principles are identified and evaluated with respect to their components in an effort to assess variances between what is professed and what is practiced. With regard to performance objectives, it is instructive to see whether and to what extent the organization places emphasis on targeted criteria.

System scanning involves defining the workplace in systems terms, a process that includes defining boundaries. The organization's mission is detailed in systems terms (i.e., inputs, outputs, processes, suppliers, customers, internal controls, and feedback mechanisms). The system scan also establishes initial boundaries of the work system. As described by Emery and Trist (1978), throughput, territorial, social, and time boundaries should be considered.

In the environmental scan, the organization's subenvironments and the major stakeholders within these subenvironments are identified. Their expectations for the organization are identified and evaluated. Entities outside the boundaries identified during the system scan are part of the external environment. Conflicts are viewed as opportunities for process or interface improvement. Variances are evaluated to determine design constraints and opportunities for change. The work system itself can be redesigned to align itself with external expectations, or conversely, the work system can attempt to change the expectations of the environment to be consistent with its internal plans and desires. According to STS theory, the response in part will be a function of whether the environment is viewed by the organization as a source of provocation or inspiration (Pasmore, 1988). Much of the time, the gaps between work-system and environmental expectations are gaps of perception, and communication interfaces need to be developed between subenvironment personnel and the organization. Design focuses on the interfaces among the organizational system and relevant subenvironments to improve communication and/or decision making. These interfaces are referred to as organization–environment or work system–environment interfaces. As can be seen — consistent with an STS approach — variances are the focal point. Several variances are noted between the current state and the future state. Optimal levels of complexity (both differentiation and integration), centralization, and formalization can be hypothesized at this point.

90.2.2 Defining Production System Type and Setting Performance Expectations

The work system's production type can help determine optimal levels of complexity, centralization, and formalization as well. The system scan performed in the previous phase should help in this regard, and the analyst should consult available production models. In this context, key performance criteria related to the organization's purpose and technical processes are itemized. This entails a determination of success factors for products and services, but it can also include performance measures at other points in the organization's system, especially if decision making is important to work-process improvement. As described in Kleiner's (1997) framework adapted from Sink and Tuttle (1989), specific standardized performance criteria guide the selection of specific measures that relate to different parts of the work process. Measures can be subjective, as in the case of self-reports, or measures can be objective.

Sink and Tuttle (1989) proposed that organizational performance can be measured or assessed using seven performance criteria or clusters of measures. These are efficiency, effectiveness, productivity, quality, quality of work life, innovation, and profitability or budgetability. Within a given performance criterion, specific measures can be derived. Data sources for each measure can be subjective, as in the case of self-reports, or can be based on objective data. Kleiner (1997) added a criterion related to flexibility that related to each of these checkpoints, as well, due to the increasing need to manage and measure flexibility in systems. According to Sink and Tuttle (1989), quality of work life (QWL) includes safety as a criterion. However, it is proposed that the need for a healthy and safe working environment should be differentiated from QWL, which is the affective perception of the total work environment. The efficiency criterion focuses on input or resource utilization. Effectiveness focuses on whether objectives are realized. Productivity is defined as outputs/inputs. Innovation refers to creative changes to process or product that result in performance gains. Profitability is a standard business management criterion. For not-for-profit organizations, Sink and Tuttle (1989) introduced "budgetability" or expenditures relative to budget to replace the profitability criterion. Quality checkpoints 2 and 4 correspond to traditional measures of quality control, traditionally assured through inspection of inputs and outputs. Quality checkpoints 1, 3, and 5 are quality criteria popularized by Deming and the TQM (total quality management) movement. In essence, a TQM approach to quality moves resources from checkpoints 2 and 4 exclusively and shares resources at the other system checkpoints. Checkpoint 1 emphasizes the quality of suppliers, which has been operationalized within the quality movement in the form of supplier certification programs and processes. Checkpoint 3, in-process control, pertains to the use of statistical quality control charts to monitor and control processes. Checkpoint 5 refers to customer satisfaction, operationalized as the customer getting what is wanted and needed. Checkpoint 6 corresponds to a TQM approach to managing the remaining checkpoints.

The organizational design hypotheses generated in the previous phase should be supported or modified until the personnel subsystem can be thoroughly analyzed as well. In terms of function allocation, requirements specifications can be generated, including microergonomic requirements, at this juncture. Also included are system design preferences for complexity, centralization, and formalization.

90.2.3 Defining Unit Operations and Work Process

Unit operations are groupings of conversion steps that together form a complete piece of work and are bounded from other steps by territorial, technological, or temporal boundaries. Unit operations often can be identified by their own distinctive subproduct and typically employ 3 to 15 workers. They also can be identified by natural breaks in the process (Hendrick and Kleiner, 2002). For each unit operation or department, the purpose/objectives, inputs, transformations, and outputs are defined. If the technology is complex, additional departmentalization may be necessary. If collocation is not possible or desirable, spatial differentiation and the use of computerized integrating mechanisms may be needed. If the task exceeds the allotted schedule, then work groups or shifts may be needed. Ideally, resources for task performance should be contained within the unit, but interdependencies with other units may complicate matters. In these cases, job rotation, cross training, or relocation may be required.

The current work flow of the transformation process (i.e., conversion of inputs to outputs) should be flowcharted, including material flows, workstations, and physical as well as informal or imagined boundaries. In linear systems, such as most production systems, the output of one step is the input of the next. In nonlinear systems, such as many service or knowledge work environments, steps may occur in parallel or may be recursive. Unit operations are identified. Also identified at this stage are the functions and subfunctions (i.e., tasks) of the system (Clegg et al., 1989). The purpose of this step is to assess improvement opportunities and coordination problems posed by technical design or the facility. Identifying the work flow before proceeding with detailed task analysis can be helpful. Once the current flow is charted, the macroergonomist or analyst can proceed with a task analysis for the work-process functions and tasks.

90.2.4 Identifying Variances

A variance is an unexpected or unwanted deviation from standard operating conditions, specifications, or norms. STS distinguishes between input and throughput variances. For the ergonomist, identifying variances at the process level as well as the task level can add important contextual information for job and task redesign to improve safety and quality performance. By flowcharting the current process and the detailed task analysis, which corresponds to the flowchart, the macroergonomist or analyst can identify variances.

90.2.5 Creating the Variance Matrix

Key variances are those variances that significantly impact performance criteria and/or may interact with other variances, thereby having a compound effect. The purpose of this step is to display the interrelationships among variances in the transformation work process to determine which ones affect which others. The variances should be listed in the order in which they occur on the *y*-axis and the horizontal *x*-axis. The unit operations (groupings) can be indicated, and each column represents a single variance. The ergonomist can inspect each column to see if this variance causes other variances. Each cell thus represents the relationship between two variances. A blank cell implies that two variances are unrelated. The analyst or team also can estimate the severity of variances by using a Likert-type rating scale (e.g., 5- or 7-point scale). Severity would be determined on the basis of whether a variance, or combination of variances, significantly affects performance. This should help identify key variances.

A variance is considered "key" if it significantly impacts quantity of production, quality of production, operating costs (utilities, raw material, overtime, etc.), social costs (dissatisfaction, safety, etc.), or if it has several relationships with other variances (matrix). Typically, 10 to 20% of the variances are significant determinants of the quality, quantity, or cost of product.

90.2.6 Creating the Key-Variance Control Table and Role Network

The purpose of this step is to discover how existing variances currently are controlled and whether personnel responsible for variance control require support. The key-variance control table includes: the unit operation in which variance is controlled or corrected; who is responsible; what control activities are currently undertaken; what interfaces, tools, or technologies are needed to support control; and what communication, information, special skills, or knowledge are needed to support control.

A "job" is defined by the formal job description that is a contract or agreement between the individual and the organization. This is not the same as a work role, which comprises the actual behaviors of a person occupying a position or job in relation to other people. These role behaviors result from actions and expectations of a number of people in a role set. A role set is a group of people who send expectations and reinforcement to the role occupant. Role analysis addresses who interacts with whom, about what, and how effective these relationships are. This relates to technical production and is important because it determines the level of work-system flexibility (Hendrick and Kleiner, 2002).

In a role network, the role responsible for controlling key variances is identified. Although there may be multiple roles to satisfy this criterion, there is often a single role without which the system could not function. With the focal role identified within a circle, other roles can be identified and placed on the diagram in relation to the focal role. Based upon the frequency and importance of a given relationship or interaction, line length can be varied, where a shorter line represents more or closer interactions. Finally, arrows can be added to indicate the nature of the communication in the interaction. A one-way arrow indicates one-way communication and a two-way arrow suggest two-way interaction. Two one-way arrows in opposite directions indicate asynchronous (different time) communication patterns. To show the content of the interactions between the focal role and other roles and to illustrate the presence or absence of a set of functional relationships for functional requirements, the following labels are used to indicate: the **Goal** of controlling variances; **Adaptation** to short-term fluctuations; **Integration** of activities to manage internal conflicts and promote smooth interactions among people and tasks; and **Long-term development** of knowledge, skills, and motivation in workers. Also the presence or absence of particular relationships is identified as **Vertical** hierarchy; **Equal** or peer; **Cross-boundary**; **Outside**; or **Nonsocial**.

The relationships in the role network are now evaluated. Internal and external customers of roles can be interviewed or surveyed for their perceptions of role effectiveness. Also, the organizational design hypotheses can be tested against the detailed analysis of variance and variance control. The role analysis and variance control table may suggest, for example, a need to increase or decrease formalization or centralization. If procedures are recommended to help control variances, this increase in formalization must be evaluated against the more general organizational design preferences suggested by the environmental and production system analyses.

90.2.7 Performing Function Allocation and Joint Design

After specifying system objectives, requirements, and functions, it now is time to systematically allocate functions and tasks to human and machine or computer. It is helpful to review the environmental scan data to check for any subenvironment constraints before making any mandatory allocations (Clegg et al., 1989). Preliminary allocations can be made to the human(s), machine(s), both, or neither. In the latter case, a return to developing requirements may be appropriate using four categories of criteria: technical feasibility, health and safety, operational requirements (i.e., physical, informational, performance), and function characteristics (i.e., criticality, unpredictability, psychological).

Technical changes are made to prevent or control key variances. Human-centered design of the following may be needed to support operators as they attempt to prevent or control key variances: interfaces, information systems to provide feedback, job aids, process control tools, more-flexible technology, redesigned workstation or handling system, or integrating mechanisms.

After considering human-centered system changes in the previous step, it is time to address knowledge and/or skill requirements of key variances and any selection issues that may be apparent. In the variance control table, we identified who controls variances and the tasks performed to control these variances. At this stage, we suggest personnel system changes to prevent or control key variances. This may entail specific skill or knowledge sets that can be acquired through technical training, formal courses, workshops, or distance learning.

At this juncture, organizational design hypotheses have been generated and iteratively adjusted as new analyses are performed. Now it is time to take the specifications for organizational design levels of complexity, centralization, and formalization and produce specific structures. Depending upon the level of work-system process analysis, this may require design/redesign at the organizational level, the group/team level, or at both levels.

90.2.8 Understanding Roles and Responsibilities Perceptions

It is important to identify how workers perceive their roles documented in the variance control table. Through interviews, role occupants can participate in an analysis of their perceptions of their roles. Using the previously constructed table, expected roles, perceived roles, and any variances can be identified. Variances can be managed through training and selection as well as technological support. Essentially, two role networks are operating: the one needed and the one perceived. Any variation between the two can be reduced through participatory ergonomics, training, communication, interface design, or tool design.

90.2.9 Designing/Redesigning Support Subsystems and Interfaces

Now that the work process has been analyzed and jointly designed, other internal organizational support subsystems may require redesign. The goal is to determine (a) the extent to which a given subsystem impacts the sociotechnical production system, (b) the nature of the variance, (c) the extent to which the variance is controlled, and (d) the extent to which tasks should be taken into account in the redesign of operating roles in the supporting subsystem units.

According to Clegg's et al. (1989) method of function allocation, individual and cumulative allocations made on a provisional basis earlier can be further evaluated with respect to requirements specifications, including resources available at the time of implementation (including human and financial) and the total outcome. In addition to an audit of function allocation, interfaces among subsystems should be checked and redesigned at this juncture.

Especially at the team and individual levels of work, the internal physical environment should be ergonomically redesigned if necessary to promote human well-being, safety, and/or effectiveness. Evaluating the technical and personnel variance analyses, we can assess whether there are physical environmental changes that will promote improvement.

90.2.10 Implementing, Iterating, and Improving

At this point, it is time to execute or implement the work-process changes prescribed, design interfaces, and allocate functions. Proposals with recommendations for change may be required for presentation within the formal organizational structure. These proposals should be consistent with the macroergonomic principles and should include, for example, both technical and social objectives; will likely include participatory ergonomics; and should predict multidimensional performance improvement. Based on the proposal feedback, modifications to the proposal may be necessitated, which will require a return to the previous step that represents a challenged assumption or design.

90.3   Advantages

- Integrates organizational analysis with ergonomics analysis.
- Systematic, comprehensive methodology that reflects the principles of macroergonomics.
- Methodology has an underlying theoretical framework.
- Unlike microergonomics, MEAD addresses the larger system's environmental and organizational issues.

90.4   Disadvantages

- Because of its comprehensive nature, it takes time to execute.
- Still could benefit from additional reliability and validity testing in a diverse array of domains.

90.5   Related Methods

Macroergonomics analysis of structure (MAS), covered in Chapter 89, is a related method. There are other large-scale organizational change methodologies and ergonomic intervention methods available. The uniqueness of MEAD lies in its integration of these levels of analysis and design.

90.6   Standards and Regulations

There are no standards or regulations pertaining to MEAD.

90.7   Approximate Training and Application Times

Ideally, a course or workshop in macroergonomics will precede the application of MEAD. Since macroergonomics is a subfield within ergonomics, ergonomists are well suited to implement MEAD.

90.8   Reliability and Validity

MEAD's underlying theoretical framework, sociotechnical systems theory, has a strong history of consistent and accurate results. In addition, MEAD has been reviewed as applied to large-scale change in academia, industry, and government (Kleiner, 1996).

The MEAD process is iterative. For continuous improvement (i.e., STS principle of "incompletion"), evaluations may suggest a return to an earlier step in the process for renewed partial or full redesign. Once the proposal for change is accepted and implementation begins, regular reviews of progress are required. To complement the weekly formative evaluations performed by the implementation team, semiannual formative evaluations should be performed by an objective outside party. This evaluation should be presented to the implementation team, and a constructive dialogue about expectations and progress to date should be conducted.

90.9 Tools Needed

While various aspects of MEAD can and have been automated through computerization, the entire method can be applied manually. Additionally, the analyst can perform the process qualitatively or can perform statistical analyses on such data as variances.

References

Clegg, C., Ravden, S., Corbertt, M., and Johnson, S. (1989), Allocating functions in computer integrated manufacturing: a review and new method, *Behav. Inf. Technol.*, 8, 175–190.

Emery, F.E. and Trist, E.L. (1978), Analytical model for sociotechnical systems, in *Sociotechnical Systems: A Sourcebook*, Pasmore, W.A. and Sherwood, J.J., Eds., University Associates, La Jolla, CA, pp. 120–133.

Hendrick, H.W. and Kleiner, B.M. (2001), *Macroergonomics: An Introduction to Work System Design*, Human Factors and Ergonomics Society, Santa Monica, CA.

Hendrick, H.W. and Kleiner, B.M., Eds. (2002), *Macroergonomics: Theory, Methods and Applications*, Lawrence Erlbaum Associates, Mahwah, NJ.

Kleiner, B.M. (1996), Macroergonomics lessons learned from large-scale change efforts in industry, government, and academia, in *Human Factors in Organizational Design and Management, V*, Brown, O., Jr. and Hendrick, H.W., Eds., North-Holland, Amsterdam, pp. 483–488.

Kleiner, B.M. (1997), An integrative framework for measuring and evaluating information management performance, *Int. J. Comput. Ind. Eng.*, 32, 545–555.

Pasmore, W.A. (1988), *Designing Effective Organizations: The Sociotechnical Systems Perspective*, Wiley, New York.

Sink, D.S. and Tuttle, T.C. (1989), *Planning and Measurement in Your Organization of the Future*, Industrial Engineering and Management Press, Norcross, GA.

Taylor, J.C. and Felton, D.F. (1993), *Performance by Design*, Prentice-Hall, Englewood Cliffs, NJ.

Index